Lecture Notes in Computer Science 521

Edited by G. Goos and J. Hartmanis

B. Bouchon-Meunier R. R. Yager
L. A. Zadeh (Eds.)

Uncertainty in Knowledge Bases

3rd International Conference on Information Processing and Management of Uncertainty in Knowledge-Based Systems, IPMU '90
Paris, France, July 2-6, 1990
Proceedings

Springer-Verlag
Berlin Heidelberg New York
London Paris Tokyo
Hong Kong Barcelona
Budapest

Series Editors

Gerhard Goos
GMD Forschungsstelle
Universität Karlsruhe
Vincenz-Priessnitz-Straße 1
W-7500 Karlsruhe, FRG

Juris Hartmanis
Department of Computer Science
Cornell University
Upson Hall
Ithaca, NY 14853, USA

Volume Editors

Bernadette Bouchon-Meunier
LAFORIA, Université Paris VI
Boite 169, 4 place Jussieu
75252 Paris Cedex 05, France

Ronald R. Yager
Machine Intelligence Institute, Iona College
New Rochelle, N.Y. 10801, USA

Lotfi A. Zadeh
Computer Science Division, University of California
Berkeley, CA 94720, USA

CR Subject Classification (1991): I.2.3-6, I.2.1, H.4, H.1.1, G.3

ISBN 3-540-54346-5 Springer-Verlag Berlin Heidelberg New York
ISBN 0-387-54346-5 Springer-Verlag New York Berlin Heidelberg

Printed in Germany

Typesetting: Camera ready by author
Printing and binding: Druckhaus Beltz, Hemsbach/Bergstr.
2145/3140-543210 - Printed on acid-free paper

Foreword

The management and processing of uncertain information has shown itself to be a crucial issue in the development of intelligent systems, beginning with its appearance in such seminal systems as Mycin and Prospector. The papers in this volume reflect the current range of interests of researchers in the field. Currently, the major approaches to uncertainty include fuzzy set theory, probabilistic methods, mathematical theory of evidence, non-standard logics such as default reasoning, and possibility theory.

The initial part of this volume is devoted to papers dealing with the foundations of these approaches. Recent attempts have been made to develop systems combining multiple approaches. Work in this direction is also reflected in this book. A significant portion of the book looks at the management of uncertainty in a number of the paradigmatic domains of intelligent systems such as expert systems, decision-making, data bases, image processing and reasoning networks.

The papers in this volume are extended versions of presentations at the third International Conference on Information Processing and Management of Uncertainty in knowledge-based systems (IPMU Conference). Two other companion volumes exist, based on the first two IPMU Conferences, entitled "Uncertainty in Knowledge-Based Systems" (LNCS 286) and "Uncertainty and Intelligent Systems" (LNCS 313). We would like to express our thanks to all the authors in this volume, as well as all the participants at IPMU '90. Special thanks are given to Ecole Nationale Supérieure de Techniques Avancées, who hosted the Conference, and to all the other sponsors of the Conference.

June 1991

B. Bouchon-Meunier
R. R. Yager
L. A. Zadeh

Contents

1. MATHEMATICAL THEORY OF EVIDENCE

ON SPOHN'S THEORY OF EPISTEMIC BELIEFS

PRAKASH P. SHENOY
School of Business, University of Kansas, Lawrence, KS 66045-2003, USA
Tel: (913)-864-7551, Fax: (913)-864-5328, Bitnet: PSHENOY@UKANVM

ABSTRACT

This paper is about Spohn's theory of epistemic beliefs. The main ingredients of Spohn's theory are (i) a functional representation of an epistemic state called a disbelief function, and (ii) a rule for revising this function in light of new information. The main contribution of this paper is as follows. First, we provide a new axiomatic definition of an epistemic state and study some of its properties. Second, we state a rule for combining disbelief functions that is mathematically equivalent to Spohn's belief revision rule. Whereas Spohn's rule is defined in terms of the initial epistemic state and some features of the final epistemic state, the rule of combination is defined in terms of the initial epistemic state and the incremental epistemic state representing the information gained. Third, we state a rule of subtraction that allows one to recover the addendum epistemic state from the initial and final epistemic states. Fourth, we study some properties of our rule of combination. One distinct advantage of our rule of combination is that besides belief revision, it can also be used to describe an initial epistemic state for many variables when this information is provided in the form of several independent epistemic states each involving a small number of variables. Another advantage of our reformulation is that we are able to demonstrate that Spohn's theory of epistemic beliefs shares the essential abstract features of probability theory and the Dempster-Shafer theory of belief functions. One implication of this is that we have a ready-made algorithm for propagating disbelief functions using only local computation.

KEY WORDS: Spohn's theory, consistent epistemic state, content of an epistemic state, disbelief function, Spohnian belief function, Spohn's rules for belief revision, A,α-conditionalization, λ-conditionalization, rule of combination for disbelief functions, rule of subtraction for disbelief functions, axioms for local computation of marginals.

1. INTRODUCTION

This paper is about Spohn's theory of epistemic beliefs (Spohn 1988, 1990). Spohn's theory is an elegant, simple and powerful calculus designed to represent and reason with plain human beliefs.

The main ingredients of Spohn's theory are (i) a functional representation of an epistemic state called a natural (or ordinal) conditional function, and (ii) a rule for revising this function in light of new information. Since the values of a natural conditional function represent degrees of disbeliefs, we call such a function a disbelief function. Like a probability distribution function, a disbelief function for a variable is completely specified by its values for the singleton subsets of configurations of the variable. Spohn (1990, pp. 152-154) has interpreted the values of a disbelief function as infinitesimal probabilities (see also (Pearl 1989)). Smets (private communication) and Dubois and Prade (1990) have pointed out that a disbelief function can be interpreted as the negative of the logarithm of a possibility function as studied by, e.g., Zadeh (1978) and Dubois and Prade (1988).

The main contribution of this paper is as follows. First, we provide an axiomatic definition of a consistent epistemic state. Some of the axioms we propose are different from the ones proposed by Spohn. Our axioms are a little easier to understand, but we show that the two sets of axioms are mathematically equivalent. These axioms are also found in (Gardenfors, 1988).

Second, we state a rule of combination for disbelief functions that is mathematically equivalent to Spohn's belief revision rule. Whereas Spohn's rule is defined in terms of the initial epistemic state and some features of the final epistemic state, the rule of combination is defined in terms of the initial epistemic state and the incremental epistemic state representing the information gained. The rule of combination for disbelief functions is pointwise addition.

Third, we state a rule of subtraction that allows one to always recover the addendum epistemic state from the initial and final epistemic states. This rule is useful in cases where the information gained is expressed in terms of the final epistemic state as it allows us to recover the addendum epistemic state that is combined with the initial to obtain the final. The subtraction rule is also useful for non-monotonic reasoning when it becomes necessary to retract the conclusion of an earlier inference without influencing conclusions drawn using other means. The rule of subtraction for disbelief functions is pointwise subtraction.

Fourth, we study some properties of our rule of combination. Spohn's belief revision rule is formulated to revise an epistemic state in light of new information. On the other hand, the rule of combination described in this paper is more flexible. It can also be used to describe an initial epistemic state for many variables when this information is provided in the form of several independent epistemic states each involving a small number of variables. The initial epistemic state for all variables is then obtained by combining these independent epistemic states using the rule of combination. This is a distinct advantage of our reformulation of Spohn's belief revision rule.

Another advantage of our reformulation is that we are able to demonstrate that Spohn's theory of epistemic beliefs shares the essential abstract features of probability theory and the Dempster-Shafer theory of belief functions as described in Shenoy and Shafer (1988, 1990). These features are (i) a functional representation of knowledge (or beliefs), (ii) a rule of marginalization, and (iii) a rule of combination. In Shenoy and Shafer (1988, 1990), we also describe three axioms for the marginalization and combination rules that enable one to use local computation in the calculation of the marginals of a joint function without explicitly having to compute the joint function. In this paper we show that the rules of marginalization and combination for disbelief functions satisfy the required three axioms. One implication of this is that we have a ready-made algorithm for propagating disbelief functions using only local computation as described in detail in Shenoy and Shafer (1990).

2. CONSISTENT EPISTEMIC STATES

In this section, we will define axiomatically a consistent epistemic state. Some of our axioms are different from those used by Spohn but the two sets of axioms are equivalent. Next, we describe a characterization of a consistent epistemic state due to Spohn.

Consider a variable X. Let $\mathcal{W}_X$ denote a finite set of all possible values of X such that exactly one of the values is true. We shall call $\mathcal{W}_X$ a *frame* for X. Assume further that $\mathcal{W}_X$ is defined such that the propositions regarding X that are of interest are precisely those of the form 'The true value of X is in A' where A is a subset of $\mathcal{W}_X$. Thus the propositions regarding X that are of interest are in a one-to-one correspondence with the subsets of $\mathcal{W}_X$ (Shafer 1976, p. 36).

The correspondence between subsets and propositions is useful since it translates the logical notions of conjunction, disjunction, implication and negation into the set-theoretic notions of intersection, union, inclusion and complementation (Shafer 1976, pp. 36-37). Thus, if A and B are two subsets of $\mathcal{W}_X$, and A' and B' are the corresponding propositions, then $A \cap B$ corresponds to the conjunction of A' and B', $A \cup B$ corresponds to the disjunction of A' and B', $A \subseteq B$ if and only if A' implies B', and A is the set-theoretic complement of B with respect to $\mathcal{W}_X$ (written as $A = \sim B$) if and only if A' is the negation of B'. Notice also that the proposition which corresponds to $\emptyset$ is known to be false and the proposition which corresponds to $\mathcal{W}_X$ is known to be true. If A is a proper, non-empty subset of $\mathcal{W}_X$, we shall call the proposition which corresponds to A as *contingent*. Henceforth, we will simple refer to a proposition by its corresponding subset. The set of all subsets of $\mathcal{W}_X$ will be denoted by $2^{\mathcal{W}_X}$.

In an epistemic state for X, some propositions are believed to be true (or simply, *believed*), some are believed to be false (or simply, *disbelieved*) and the remaining are neither believed nor disbelieved. Logical consistency requires that these beliefs satisfy certain conditions (axioms). A set of five axioms defining a consistent epistemic state is as follows.

Definition 1. An epistemic state is said to be *consistent* if the following five axioms are satisfied:

A1. For any proposition A, exactly one of the following conditions holds:
   (i) A is believed;
   (ii) A is disbelieved;
   (iii) A is neither believed nor disbelieved.

A2. $\mathcal{W}_X$ is (always) believed.

A3. A is believed if and only if $\sim A$ is disbelieved.

A4. If A is believed and $B \supseteq A$, then B is believed.

A5. If A and B are believed, then $A \cap B$ is believed. ■

Some simple consequences of axioms A1 to A5 are as follows.

Proposition 1. The following conditions always hold in any consistent epistemic state:

A6. $\emptyset$ is (always) disbelieved.

A7. If A is disbelieved and $B \subseteq A$, then B is disbelieved.

A8. If A and B are disbelieved, then $A \cup B$ is disbelieved.

A9. If A is not believed, then this does not necessarily imply that ~A must be believed; it is possible that ~A is also not believed. And if A is not disbelieved, then this does not necessarily imply that ~A must be disbelieved; it is possible that ~A is also not disbelieved.

A10. If $\mathcal{B}$ denotes the set of all believed propositions, then $\cap\mathcal{B} \neq \emptyset$.

A11. If $\mathcal{B}$ denotes the set of all believed propositions, then $A \in \mathcal{B}$ whenever $A \supseteq_{*} (\cap\mathcal{B}')$ for some $\mathcal{B}' \subseteq \mathcal{B}$. ■

It is clear from axiom A1 and A3 that we can specify a consistent epistemic state simply by listing all propositions which are believed. Then axiom A3 tells us exactly which propositions are disbelieved. And from axiom A1, the remaining propositions are neither believed nor disbelieved.

Let $\mathcal{B}$ denote the set of all propositions (subsets) that are believed, and let $\mathcal{D}$ denote the set of all propositions that are disbelieved in some consistent epistemic state. Theorem 1 gives a characterization of $\mathcal{B}$ and its corollary gives a characterization of $\mathcal{D}$.

Theorem 1. Suppose $\mathcal{B}$ denotes the set of all propositions that are believed in some epistemic state for X. Then the epistemic state is consistent if and only if there exists a unique non-empty subset C of $\mathcal{W}_X$ such that $\mathcal{B} = \{A \in 2^{\mathcal{W}_X} \mid A \supseteq C\}$. ■

Corollary to Theorem 1. Suppose $\mathcal{D}$ denotes the set of all propositions that are disbelieved in some epistemic state for X. Then the epistemic state is consistent if and only if there exists a unique proper subset D of $\mathcal{W}_X$ such that $\mathcal{D} = \{A \in 2^{\mathcal{W}_X} \mid A \subseteq D\}$. ■

The characterization of $\mathcal{B}$ in Theorem 1 (but not Theorem 1 itself) is due to Spohn (1988, p. 108). Spohn defines a consistent epistemic state as one that satisfies A1, A3, A10 and A11, and consequently, Theorem 1 follows trivially from A10 and A11. It follows from Theorem 1 that our definition of a consistent epistemic state (axioms A1 to A5) is equivalent to Spohn's (A1, A3, A10 and A11).

Subset C in Theorem 1 is called the *content* of the consistent epistemic state. Note that the content constitutes a complete specification of an epistemic state. Thus another simple corollary of Theorem 1 is that in a frame $\mathcal{W}_X$ consisting of n elements, there are exactly 2^n-1 distinct possible consistent epistemic states (corresponding to each non-empty subset of $\mathcal{W}$ as the content). The content C represents the smallest believed proposition and D = ~C represents the largest disbelieved proposition.

3. VARIABLES, CONFIGURATIONS AND PROPOSITIONS

In this section, we introduce the notation that will be used in the rest of the paper. We use the multivariate framework because, even though it does not generalize readily to the continuous case, it is more intuitive and easier to comprehend than the measurable-subset framework used by Spohn (1988, 1990).

Consider a variable X. The symbol $\mathcal{W}_X$ denotes the set of possible values of X and we assume that one and only one of the elements of $\mathcal{W}_X$ can be the true value of X. We call $\mathcal{W}_X$ the *frame for X*. For example, suppose we are interested in determining whether a person is a pacifist or not. We can construct a variable P whose frame has two elements: p (for pacifist), and ~p (not pacifist).

Let $\mathcal{X}$ denote the set of all variables. In this paper we will be concerned only with the case where $\mathcal{X}$ is finite. We will also assume that all the variables in $\mathcal{X}$ have finite frames.

We will often deal with non-empty subsets of variables in $\mathcal{X}$. Given a non-empty subset g of $\mathcal{X}$, let $\mathcal{W}_g$ denote the Cartesian product of $\mathcal{W}_X$ for X in g, i.e., $\mathcal{W}_g = \times\{\mathcal{W}_X \mid X \in g\}$. We can think of the set $\mathcal{W}_g$ as the set of possible values of the joint variable g. Accordingly, we call $\mathcal{W}_g$ the *frame for g*. Also, we will refer to elements of $\mathcal{W}_g$ as *configurations of g*. We will use this terminology even when g consists of a single variable. Thus we will refer to elements of $\mathcal{W}_X$ as *configurations of X*. We will use lower-case, bold-faced letters such as **x**, **y**, etc. to denote configurations. Also, if **x** is a configuration of g and **y** is a configuration of h and $g \cap h = \varnothing$, then (**x**,**y**) will denote a configuration of $g \cup h$.

Projection of configurations simply means dropping extra coordinates; if (r,~q,~p) is a configuration of {R,Q,P}, for example, then the projection of (r,~q,~p) to {R,P} is simply (r,~p), which is a configuration of {R,P}. If g and h are sets of variables, $h \subseteq g$, and **x** is a configuration of g, then we will let $\mathbf{x}^{\downarrow h}$ denote the projection of **x** to h.

By extension of a subset of a frame to a subset of a larger frame, we mean a cylinder set extension. If g and h are sets of variables, $h \subseteq g$, and A is a subset of $\mathcal{W}_h$, then the *extension of A to g* is $A \times \mathcal{W}_{g-h}$. We will let $A^{\uparrow g}$ denote the extension of A to g. For example, consider three variables R, P, Q with frames $\mathcal{W}_R = \{r, \sim r\}$, $\mathcal{W}_P = \{p, \sim p\}$, and $\mathcal{W}_Q = \{q, \sim q\}$, respectively. Then the extension of {(r,~p), (~r,p)} (which is a subset of $\mathcal{W}_{\{R,P\}}$) to {R,P,Q} is {(r,~p,q), (r,~p,~q), (~r,p,q), (~r,p,~q)}. Note that the propositions corresponding to A and $A^{\uparrow g}$ are logically equivalent.

We will denote the set of all natural numbers by $\mathbb{N}$ and the set of integers by $\mathbb{Z}$; $\mathbb{N} = \{0, 1, 2, ...\}$ and $\mathbb{Z} = \{ ..., -2, -1, 0, 1, 2, ... \}$. The extended set of natural number consists of the set of all natural numbers to which a symbol, $+\infty$, has been added with the following properties. If $x \in \mathbb{N}$, then (i) $x < +\infty$, and (ii) $x + +\infty = +\infty$. The extended natural number set is denoted by $\mathbb{N}^+$. Similarly, we will let $\mathbb{Z}^+$ denote the extended set of integers consisting of set of all integers to which two symbols $+\infty$ and $-\infty$ have been added with the following properties. If $x \in \mathbb{Z}$, then

(i) $-\infty < x < +\infty$, (ii) $x + +\infty = +\infty$, and (iii) $x + -\infty = -\infty$.

4. DISBELIEF FUNCTIONS

The basic functional representation of an epistemic state in Spohn's theory is called an ordinal conditional function in (Spohn 1988, p. 115) and a natural conditional function in (Spohn 1990, p. 150). We simply call this function a disbelief function.

Definition 2 (Spohn 1990, p. 150). A *disbelief function* for g is a function $\delta: 2^{\mathcal{W}_g} \rightarrow \mathbb{N}^+$ such that

(D0). $\delta(\{\mathbf{w}\}) \in \mathbb{N}$ for all $\mathbf{w} \in \mathcal{W}_g$;

(D1). There exists a configuration $\mathbf{w} \in \mathcal{W}_g$ for which $\delta(\{\mathbf{w}\}) = 0$;

(D2). For any $A \in (2^{\mathcal{W}_g} - \{\varnothing\})$, $\delta(A) = \text{MIN}\{\delta(\{\mathbf{w}\}) \mid \mathbf{w} \in A\}$; and

(D3). $\delta(\varnothing) = +\infty$. ■

Note that although a disbelief function is defined for the set of all subsets of $\mathcal{W}_g$, it is completely specified by its values for each singleton subset of $\mathcal{W}_g$.

A disbelief function is a complete representation of a consistent epistemic state. To see what propositions are believed in state δ, consider the subset $C = \{\mathbf{w} \in \mathcal{W}_g \mid \delta(\{\mathbf{w}\}) = 0\}$. By condition D1 in Definition 2, C is always nonempty. C represents the content of the epistemic state δ, i.e., A is believed in state δ iff $A \supseteq C$. Thus A is believed in state δ iff $\delta(\sim A) > 0$; A is disbelieved in state δ iff $\delta(A) > 0$; and A is neither believed nor disbelieved in state δ iff $\delta(A) = \delta(\sim A) = 0$.

A disbelief function consists of more than a representation of a consistent epistemic state. It also includes degrees of belief and disbelief. If $\delta(A) > 0$, then $\delta(A)$ can be interpreted as the degree of disbelief in proposition A, i.e., A is more disbelieved than B if $\delta(A) > \delta(B) > 0$. And if $\delta(\sim A) > 0$, then $\delta(\sim A)$ can be interpreted as the degree of belief for A, i.e., A is more believed than B if $\delta(\sim A) > \delta(\sim B) > 0$.

Consider the following disbelief function for g: $\delta(\{\mathbf{w}\}) = 0$ for all $\mathbf{w} \in \mathcal{W}_g$. This means that the only proposition that is disbelieved is $\varnothing$ and the only proposition that is believed is $\mathcal{W}_g$. We shall call such a disbelief function *vacuous*. It represents a state of complete ignorance.

Proposition 2 (Spohn 1988, p. 115). Suppose δ is a disbelief function for g. Then

(D4). For each $A \in 2^{\mathcal{W}_g}$, either $\delta(A) = 0$ or $\delta(\sim A) = 0$ or both.

(D5). For each $A, B \in 2^{\mathcal{W}_g}$, $\delta(A \cup B) = \text{MIN}\{\delta(A), \delta(B)\}$ ■

Marginalization. Suppose δ is a disbelief function for g. Suppose $h \subseteq g$. We may be interested only in propositions regarding variables in h. In this case, we would like to marginalize δ to h. The following definition of marginalization is motivated by the fact that each proposition $A \in 2^{\mathcal{W}_h}$ about variables in h can be regarded as a proposition $A^{\uparrow g} \in 2^{\mathcal{W}_g}$ about variables in g.

Definition 3. Suppose δ is a disbelief function for g and suppose $h \subseteq g$. The *marginal of δ for h*, denoted by $\delta^{\downarrow h}$, is a disbelief function for h given as follows:

$$\delta^{\downarrow h}(A) = \delta(A^{\uparrow g}) = \mathrm{MIN}\{\delta(\{(x,y)\}) \mid x \in A, y \in \mathcal{W}_{g-h}\} \tag{4.1}$$

for all $A \in 2^{\mathcal{W}_h}$. ■

In particular, if A is a singleton subset, i.e., $A = \{x\}$ for some $x \in \mathcal{W}_h$, then (4.1) simplifies to:

$$\delta^{\downarrow h}(\{x\}) = \mathrm{MIN}\{\delta(\{(x,y)\}) \mid y \in \mathcal{W}_{g-h}\}.$$

Note that if δ is a disbelief function for g and $k \subseteq h \subseteq g$, then $(\delta^{\downarrow h})^{\downarrow k} = \delta^{\downarrow k}$. In words, if we regard marginalization as reduction of δ by deletion of variables, then the order in which the variables are deleted makes no difference in the final answer.

Example 1. (*Disbelief function and marginalization*) We would like to determine whether a stranger (about whom we know nothing about) is a pacifist or not depending on whether he is a Republican or not and whether he is a Quaker or not. Consider three variables R, Q and P. R has two possible values: r (for Republican), ~r (not Republican), Q has two values: q (Quaker) and ~q (not Quaker) and P has two possible values: p (pacifist) and ~p (not pacifist). Our belief that most Republicans are not pacifists and that most Quakers are pacifists is represented by the disbelief function δ for {R,Q,P} shown in Table 1 (the construction of this disbelief function will be explained later in Section 9).

Note that the marginal of δ for R is the vacuous disbelief function for R, i.e., $\delta^{\downarrow\{R\}}(\{r\}) = \delta^{\downarrow\{R\}}(\{\sim r\}) = 0$. Thus in epistemic state δ, we neither believe he is a republican nor that he is not a republican. Similarly, notice that the marginals of δ for {Q} and {P} are also vacuous. The marginal of δ for {R, P} is as follows:

$\delta^{\downarrow\{R,P\}}(\{r,p\}) = 1$, $\delta^{\downarrow\{R,P\}}(\{r,\sim p\}) = \delta^{\downarrow\{R,P\}}(\{\sim r,p\}) = \delta^{\downarrow\{R,P\}}(\{\sim r,\sim p\}) = 0$.

Thus we disbelieve pacifist Republicans (to degree 1). Similarly, notice that the marginal of δ for {Q, P} is as follows:

$\delta^{\downarrow\{Q,P\}}(\{q,\sim p\}) = 2$, $\delta^{\downarrow\{Q,P\}}(\{q,p\}) = \delta^{\downarrow\{Q,P\}}(\{\sim q,p\}) = \delta^{\downarrow\{Q,P\}}(\{\sim q,\sim p\}) = 0$.

Thus we disbelieve non-pacifist Quakers (to degree 2). Finally note that the marginal of δ for {R, Q} is as follows:

$\delta^{\downarrow\{R,Q\}}(\{r,q\}) = 1$, $\delta^{\downarrow\{R,Q\}}(\{r,\sim q\}) = \delta^{\downarrow\{R,Q\}}(\{\sim r,q\}) = \delta^{\downarrow\{R,Q\}}(\{\sim r,\sim q\}) = 0$.

Thus we disbelieve republican Quakers (to degree 1). ■

Table 1. A disbelief function for {R,Q,P}.

w	δ({w})
(r, q, p)	1
(r, q, ~p)	2
(r, ~q, p)	1
(r, ~q, ~p)	0
(~r, q, p)	0
(~r, q, ~p)	2
(~r, ~q, p)	0
(~r, ~q, ~p)	0

5. SPOHNIAN BELIEF FUNCTIONS

As we saw in the last section, a disbelief function models degrees of disbeliefs for disbelieved propositions directly whereas it models degrees of beliefs for believed propositions only indirectly; if A is a believed proposition, then the degree of belief for A is $\delta(\sim A)$. Can we model both beliefs and disbeliefs directly? The answer is yes. We call such a representation a (Spohnian) belief function.

Definition 4 (Spohn 1988, p. 116). A *(Spohnian) belief function* for g is a function $\beta: 2^{\mathcal{W}_g} \rightarrow \mathbb{Z}^+$ such that

$$\beta(A) = \begin{cases} -\delta(A) & \text{if } \delta(A) > 0 \\ \delta(\sim A) & \text{if } \delta(A) = 0 \end{cases}$$

for all $A \in 2^{\mathcal{W}_g}$ where δ is some disbelief function for g. ■

The number $\beta(A)$ is interpreted as the degree of belief in proposition A. A is believed iff $\beta(A) > 0$, A is disbelieved iff $\beta(A) < 0$, and A is neither believed nor disbelieved iff $\beta(A) = 0$. Furthermore, if $\beta(A) > \beta(B) > 0$, then A is more believed than B, and if $\beta(A) < \beta(B) < 0$, then A is more disbelieved than B. Propositions $\varnothing$ and $\mathcal{W}_g$ are the extreme cases. $\varnothing$ is the most disbelieved proposition ($\beta(\varnothing) = -\infty$) and $\mathcal{W}_g$ is the most believed proposition ($\beta(\mathcal{W}_g) = +\infty$).

Example 2. (*Spohnian belief function*) Consider the disbelief function δ for {R,Q,P} from Example 1. The corresponding belief function β is as shown in Table 2. ■

Table 2. The belief function β corresponding to disbelief function δ.

w	$\delta(\{w\})$	$\beta(\{w\})$
(r, q, p)	1	−1
(r, q, ~p)	2	−2
(r, ~q, p)	1	−1
(r, ~q, ~p)	0	0
(~r, q, p)	0	0
(~r, q, ~p)	2	−2
(~r, ~q, p)	0	0
(~r, ~q, ~p)	0	0

6. REVISION OF DISBELIEFS

In this section, first we state Spohn's A,α-conditionalization rule for modifying a disbelief function in light of new information. Then we describe four properties of this rule. Finally, we describe the general λ-conditionalization rule.

Definition 5 (Spohn 1988, p. 117). Suppose δ is a disbelief function for g representing our initial epistemic state. Suppose we learn something about contingent proposition A (or ~A) which consequently leads us to believe A to degree α (or, equivalently, disbelieve ~A to degree α), where $\alpha \in \mathbb{N}$. The resulting epistemic state called the *A,α-conditionalization of δ*, denoted by disbelief function $\delta_{A,\alpha}$, is defined as follows:

$$\delta_{A,\alpha}(\{w\}) = \begin{cases} \delta(\{w\}) - \delta(A) & \text{if } w \in A \\ \delta(\{w\}) + \alpha - \delta(\sim A) & \text{if } w \notin A \end{cases}$$

for all $w \in \mathcal{W}_g$. ■

Spohn (1988) describes four properties of this rule. Let β and $\beta_{A,\alpha}$ denote the belief functions corresponding to δ and $\delta_{A,\alpha}$, respectively. First note that $\delta_{A,\alpha}(A) = 0$ and $\delta_{A,\alpha}(\sim A) = \alpha$. Therefore, $\beta_{A,\alpha}(A) = \alpha$. What this means is that if $\beta(A) < \alpha$, then what we have learned about A (or ~A) increases our belief in A. However, if $\beta(A) > \alpha$, then what we have learnt about A decreases our belief in A. Finally, if $\beta(A) = \alpha$, then the beliefs remain unchanged after revision, i.e., $\delta_{A,\alpha} = \delta$ (Spohn 1988, p. 118).

Second, since learning about A (or ~A) does not discriminate between propositions contained in A, the relative degrees of disbelief of these propositions is unchanged, i.e., $\delta(E) - \delta(E') = \delta_{A,\alpha}(E) - \delta_{A,\alpha}(E')$ for all $E,E' \subseteq A$. Also, since learning about A does not discriminate between propositions contained in ~A, the relative degrees of disbelief of these propositions is also unchanged. What has changed is that the degrees of disbelief of propositions contained in ~A have shifted upwards relative to propositions contained in A (Spohn 1988, p. 117).

Before we continue with the properties of the belief revision rule, let us illustrate the rule with an example.

Example 3. (*A,α-conditionalization rule*) Consider the story in Example 1. Suppose our initial epistemic state is as given by δ in Table 1. Suppose after a brief conversation with the stranger, we now believe that the person is a republican to degree 3. In this case A = {(r,q,p), (r,q,~p), (r,~q,p), (r,~q,~p)}, $\alpha = 3$, $\delta(A) = 0$, and $\delta(\sim A) = 0$. Then the revised disbelief function, denoted say by δ', is as shown in Table 3 below. Note that as per our epistemic state δ', we now believe that the person is not a pacifist (to degree 1) and not a Quaker (to degree 1).

Suppose after further conversation with the stranger, we now believe that the person is a Quaker to degree 3. In this case, A = {(r,q,p), (r,q,~p), (~r,q,p), (~r,q,~p)}, $\alpha = 3$, $\delta'(A) = 1$, and $\delta'(\sim A) = 0$. The revised disbelief function is denoted by δ" and is also shown in Table 3 below. As per the epistemic state δ", we now believe that the person is a pacifist (to degree 1), a republican (to degree 2) and a Quaker (to degree 3). ■

Table 3. Spohn's rule for belief revision

w	δ({w})	δ'({w})	δ"({w})
(r, q, p)	1	1	0
(r, q, ~p)	2	2	1
(r, ~q, p)	1	1	4
(r,~q, ~p)	0	0	3
(~r, q, p)	0	3	2
(~r, q, ~p)	2	5	4
(~r, ~q, p)	0	3	6
(~r,~q, ~p)	0	3	6

Third, A,α-conditionalization is reversible: Suppose δ is a disbelief function with corresponding belief function β, suppose A is a contingent proposition, and suppose $\beta(A) = \gamma$. Then $(\delta_{A,\alpha})_{A,\gamma} = (\delta_{\sim A,\alpha})_{A,\gamma} = \delta$ (Spohn 1988, p. 118). In words, suppose the initial belief of proposition A is γ in epistemic state δ. Suppose we learn something about A that leads us to believe A to degree α. The resulting epistemic state is given by $\delta_{A,\alpha}$. Suppose that we learn something more about proposition A that has the effect of negating what we learned previously, i.e., we now believe A to degree γ again. Then the resulting epistemic state $(\delta_{A,\alpha})_{A,\gamma}$ is the same as the epistemic state δ that we started with initially.

Fourth, A,α-conditionalization is partially commutative: If δ is a disbelief function, and A and B are contingent propositions such that $\delta(A \cap B) = \delta(A \cap \sim B) = \delta(\sim A \cap B) = 0$, then $(\delta_{A,\alpha})_{B,\gamma} = (\delta_{B,\gamma})_{A,\alpha}$ for all α, $\beta \in \mathbb{N}$ (Spohn 1988, p. 118). Clearly, A,α-conditionalization is not always commutative.

In Definition 5, a belief revision rule was stated in terms of a single proposition A that was believed to degree α. Spohn (1990, p. 152) has generalized this definition to the case where the information gained may concern more than a single proposition. Spohn calls this general belief revision rule λ-conditionalization where λ is the marginal of the resulting disbelief function for some subset h of variables. Here is a formal definition.

Definition 6 (Spohn 1990, p. 152). Suppose δ is a disbelief function for g representing our initial epistemic state. Suppose we learn something about variables in set h which consequently leads us to an epistemic state represented by a disbelief function δ_λ for $g \cup h$ such that $\delta_\lambda^{\downarrow h} = \lambda$ where λ is a disbelief function for h. Then the epistemic state δ_λ, called the *λ-conditionalization of δ*, is defined as follows:

$$\delta_\lambda(\{(\mathbf{w},\mathbf{u},\mathbf{v})\}) = \delta(\{(\mathbf{w},\mathbf{u})\}) + \lambda(\{(\mathbf{u},\mathbf{v})\}) - \delta(\{\mathbf{u}\}^{\uparrow g})$$

for all $\mathbf{w} \in \mathcal{W}_{g-h}$, $\mathbf{u} \in \mathcal{W}_{g \cap h}$, $\mathbf{v} \in \mathcal{W}_{h-g}$. ■

It is easy to show that δ_λ is indeed a disbelief function. Furthermore, it is also obvious that $\delta_\lambda^{\downarrow h} = \lambda$. Notice that A,$\alpha$-conditionalization is a special case of λ-conditionalization where λ is a disbelief function for g such that $\lambda(\{\mathbf{w}\}) = 0$ if $\mathbf{w} \in A$ and $\lambda(\{\mathbf{w}\}) = \alpha$ if $\mathbf{w} \notin A$. As noted by Spohn (1990, p. 152), λ-conditionalization is the analogue of Jeffrey's rule in probability theory (Jeffrey 1983, Ch. 11).

7. BELIEF REVISION AS A RULE OF COMBINATION

In this section, first, we describe a rule of combination. Second, we demonstrate that Spohn's rule for modifying a disbelief function in light of new information can be expressed in terms of this rule of combination.

Definition 7 (A Rule of Combination). Suppose δ_1 and δ_2 are disbelief functions for g_1 and g_2, respectively. The *combination of δ_1 and δ_2*, denoted by $\delta_1 \oplus \delta_2$, is a disbelief function for $g_1 \cup g_2$ defined as follows:

$$(\delta_1 \oplus \delta_2)(\{\mathbf{w}\}) = \delta_1(\{\mathbf{w}^{\downarrow g_1}\}) + \delta_2(\{\mathbf{w}^{\downarrow g_2}\}) - K \qquad (7.1)$$

for all $\mathbf{w} \in \mathcal{W}_{g_1 \cup g_2}$, where

$$K = \text{MIN}\{\delta_1(\{\mathbf{w}^{\downarrow g_1}\}) + \delta_2(\{\mathbf{w}^{\downarrow g_2}\}) \mid \mathbf{w} \in \mathcal{W}_{g_1 \cup g_2}\}.$$ ■

K is a normalization factor that ensures that $\delta_1 \oplus \delta_2$ is a disbelief function. Thus, combination consists of pointwise addition followed by normalization.

We will now show that Spohn's rule for belief revision as described in the previous section can be expressed in terms of the rule of combination.

Theorem 2. Suppose δ_i is an initial disbelief function for g, suppose β_i is the corresponding belief function, and suppose $h \subseteq g$. Suppose we learn something about some contingent proposition A of h. Let δ_f denote the revised disbelief function for g and let β_f denote the corresponding belief function. Suppose $\beta_f^{\downarrow h}(A) = \alpha$ where $\alpha \in \mathbb{N}$, i.e., after revising our beliefs, we believe proposition A to degree α. Then, depending on the value of $\beta_i^{\downarrow h}(A)$, there exists an appropriate disbelief function δ_Δ for h such that $\delta_f = \delta_i \oplus \delta_\Delta$. ■

In the next section we will show that our rule of combination and Spohn's λ-conditionalization are mathematically equivalent.

8. A RULE OF SUBTRACTION

In this section, first we define a rule of subtraction for disbelief functions. Second, we show that we can always recover the incremental disbelief function from the final and initial disbelief functions. Third, we show the mathematical equivalence between Spohn's λ-conditionalization and our rule of combination.

Spohn's belief revision rules were described in terms of the initial disbelief function δ_i and characteristics of the final disbelief function (proposition A and its degree of belief α in A,α-conditionalization, and disbelief function λ in λ-conditionalization). On the other hand, the rule of combination describes the final disbelief function in terms of the initial disbelief function and the incremental disbelief function representing the evidence. If we are given the initial and the final disbelief function, can we always recover the incremental disbelief function? The answer is yes and is stated as Theorem 3 below. First we need a definition.

Definition 8 (A Rule of Subtraction). Suppose δ_1 is a disbelief function for g, suppose δ_2 is a disbelief function for h and suppose h⊆g. Then the subtraction of δ_2 from δ_1, denoted by $\delta_1-\delta_2$, is a disbelief function on g given by

$$(\delta_1-\delta_2)(\{w\}) = \delta_1(\{w\}) - \delta_2(\{w^{\downarrow h}\}) - K$$

for all $w \in \mathcal{W}_g$, where K is a normalization constant given by

$$K = \text{MIN}\{\delta_1(\{w\}) - \delta_2(\{w^{\downarrow h}\}) \mid w \in \mathcal{W}_g\}$$ ■

It is clear from the definition of the normalization constant K that $\delta_1-\delta_2$ is a disbelief function. The next theorem states that we can recover the incremental disbelief function from the initial and final disbelief functions.

Theorem 3. Suppose δ_i and δ_Δ are disbelief function for g and h, respectively. Then $((\delta_i \oplus \delta_\Delta)-\delta_i)^{\downarrow h} = \delta_\Delta$. ■

Corollary to Theorem 3. Suppose δ is a disbelief function for g, suppose δ_Δ is a disbelief function for h, and suppose h⊆g. Then $(\delta-\delta_\Delta)\oplus\delta_\Delta = \delta$.

The property of disbelief functions of being always able to recover the addendum from the sum is unique to this theory and is not shared either by probability theory or by the Dempster-Shafer theory of belief functions. This property is useful for two reasons. First, in cases where it is easier to describe evidence by reference to the final disbelief function, we can always recover the incremental belief function that represents just the evidence. Second, this property is useful in non-monotonic reasoning as it allows us to retract the conclusion of an earlier inference without influencing conclusions drawn using other means (see e.g., Ginsberg 1984).

We now have the necessary tools to demonstrate that Spohn's λ-conditionalization and our rule of combination are mathematically equivalent.

Theorem 4. Suppose δ is a disbelief function for g and λ is a disbelief function for h. Let δ_λ denote the λ-conditionalization of δ. Then there exists a disbelief function δ_Δ for h such that $\delta\oplus\delta_\Delta = \delta_\lambda$. Conversely, suppose δ is a disbelief function for g and δ_Δ is a disbelief function for h. Then there exists a disbelief function λ for h such that $\delta_\lambda = \delta\oplus\delta_\Delta$. ■

9. PROPERTIES OF THE COMBINATION RULE

In this section, we discuss several important properties of the combination rule described in Section 7. First, we state some elementary properties of the rule of combination. Second, we describe how the rule of combination can be used to construct a disbelief function for many variables from independent disbelief functions each of which only involves a small number of variables. We sketch what we mean by independent disbelief functions. Third, we show that Spohn's theory of disbelief functions fits in the same abstract framework as that for the theory of probability and the Dempster-Shafer theory of belief functions.

First, we state some elementary properties of the rule of combination.

Proposition 3. The rule of combination described in (7.1) has the following properties:

(C1, Commutativity): $\delta_1\oplus\delta_2 = \delta_2\oplus\delta_1$.

(C2, Associativity): $(\delta_1\oplus\delta_2)\oplus\delta_3 = \delta_1\oplus(\delta_2\oplus\delta_3)$.

(C3): If δ_1 is vacuous, then $\delta_1\oplus\delta_2 = \delta_2$.

(C4): In general, $\delta_1\oplus\delta_1 \neq \delta_1$. The disbelief function $\delta_1\oplus\delta_1$ disbelieves the same propositions as δ_1, but it will do so with twice the degree, as it were. ■

Note that although Spohn's belief revision rule is partially commutative, the rule of combination is always commutative. There is no conflict here. Spohn's belief revision rule is described in terms of the initial epistemic state and some features of the final epistemic state. On the other hand, the rule of combination describes belief revision in terms of the initial epistemic state and the epistemic state representing the evidence. Thus commutativity for Spohn's rule and commutativity for the rule of combination are two different relations.

Second, the rule of combination is valid not only for belief revision but also for the construction of an initial disbelief function for many variables when this information is given in terms of independent disbelief functions each of which is for small number of variables. What do we mean by independent disbelief functions? Here, we will just sketch an answer by analogy with the theory of probability and the Dempster-Shafer theory of belief functions. A complete answer merits a separate paper.

By independent beliefs, we mean the same as in the theory of probability and the Dempster-Shafer theory of belief functions. In probability theory, beliefs are represented by functions called *potentials* and the rule of combination is pointwise multiplication (see, e.g., Shenoy and Shafer 1988, 1990). However, combining two potentials gives us meaningful results only when the potentials being combined are independent. For example, suppose we have two variables X and Y with frames $\{x, \sim x\}$ and $\{y, \sim y\}$, respectively. Consider a potential $p_1 = (p(x), p(\sim x))$ for X, representing a (prior) probability distribution for X. Consider another potential $p_2 = (p(y|x), p(\sim y|x), p(y|\sim x), p(\sim y|\sim x))$ for $\{X,Y\}$, representing the conditional distributions for Y given X. In this case, the potentials p_1 and p_2 are independent. Combining these by pointwise multiplication, denoted by $\otimes$, gives us the joint potential $p_1 \otimes p_2$ for $\{X, Y\}$ as follows: $(p(y|x)p(x), p(\sim y|x)p(x), p(y|\sim x)p(\sim x), p(\sim y|\sim x)p(\sim x))$. We recognize this potential as the joint probability distribution for $\{X, Y\}$. Consider another potential $p_3 = (p(y), p(\sim y))$ for Y representing a probability distribution of Y. In general, p_1 and p_3 are not independent. If we combine these two potentials, the result $p_1 \otimes p_3 = (p(x)p(y), p(x)p(\sim y), p(\sim x)p(y), p(\sim x)p(\sim y))$ does not necessarily representing pooling of evidence. We know from probability theory that $p_1 \otimes p_3$ represents combination of evidence if and only if X and Y are probabilistically independent, i.e., if and only if p_1 and p_3 are independent potentials.

In the Dempster-Shafer theory, Dempster's rule for combining belief functions represents pooling of evidence only when the belief functions being combined are independent. Shafer (1984, 1987) has described in detail precisely what is meant by independent belief functions in terms of canonical examples for belief functions. In fact, most of the examples that Pearl (1988, pp. 447-450) describes to demonstrate that application of Dempster's rule gives non-intuitive results do so precisely because the belief functions being combined are not independent.

Spohn's theory of disbelief functions is closely analogous to the theory of probability (see Spohn 1990, pp. 152–154 for a comparison of his theory with probability theory). Analogous to the concept of probabilistic conditional independence, Spohn (1988, pp. 120-125; 1990, p. 152) has described conditional independence for disbelief functions. As in the probabilistic case, Hunter (1988) has shown that the conditional independence relation for disbelief functions forms a "graphoid" (see e.g., (Geiger and Pearl 1990) or (Verma and Pearl 1988) for definition of a graphoid). Using the notion of conditional independence, we can define when two disbelief functions are independent by direct analogy with the theory of probability.

Example 4. (*Construction of an initial disbelief function*) Consider two sets of beliefs as follows:

1. Most Republicans are not pacifists.
2. Most Quakers are pacifists.

Suppose further that these two sets of beliefs are independent. Then, if δ_1 is a disbelief representation of the first set of belief and δ_2 is a disbelief representation of the second set of belief, then $\delta_1 \oplus \delta_2$ will represent the aggregation of these two sets of beliefs. In particular, suppose δ_1 is a disbelief function for $\{R,P\}$ as follows:

$$\delta_1(\{(r,p)\}) = 1,\ \delta_1(\{(r,\sim p)\}) = \delta_1(\{(\sim r,p)\}) = \delta_1(\{(\sim r,\sim p)\}) = 0$$

(i.e., we disbelieve pacifist republicans), and suppose δ_2 is a disbelief function for $\{Q,P\}$ as follows:

$$\delta_2(\{(q,\sim p)\}) = 2,\ \delta_2(\{(q,p)\}) = \delta_2(\{(\sim q,p)\}) = \delta_2(\{(\sim q,\sim p)\}) = 0$$

(i.e., we disbelieve non-pacifist Quakers). Then the disbelief function $\delta_1 \oplus \delta_2 = \delta$, say, as shown in Table 1, represents the aggregate belief. Note that there is no belief revision going on here. Of course, Theorem 4 tells us that we can mathematically describe the aggregation of δ_1 and δ_2 using λ-conditionalization (λ is the same as δ_2 for this example). But in general, it is neither practical nor intuitive. ■

Third, as per our reformulation, Spohn's theory of epistemic beliefs shares the essential abstract features of probability theory and the Dempster-Shafer theory of belief functions as described in Shenoy and Shafer (1988, 1990). These features are (i) a functional representation of knowledge (or beliefs), (ii) a rule of marginalization, and (iii) a rule of combination. In Shenoy and Shafer (1988, 1990), we also described three axioms for the marginalization and combination rules that enable one to use local computations in the calculation of the marginals of a joint disbelief function with explicitly having to compute the joint disbelief function. These three axioms are as follows (stated in the notation of disbelief functions):

(L1, Commutativity and Associativity of Combination). Suppose δ_1, δ_2 and δ_3 are disbelief functions for g, h, and k respectively. Then, $\delta_1 \oplus \delta_2 = \delta_2 \oplus \delta_1$, and $(\delta_1 \oplus \delta_2) \oplus \delta_3 = \delta_1 \oplus (\delta_2 \oplus \delta_3)$.

(L2, Consonance of Marginalization). Suppose δ is a disbelief function for g and suppose $k \subseteq h \subseteq g$. Then $(\delta^{\downarrow h})^{\downarrow k} = \delta^{\downarrow k}$.

(L3, Distributivity of Marginalization over Combination). Suppose δ_1 and δ_2 are disbelief functions for g and h, respectively. Then $(\delta_1 \oplus \delta_2)^{\downarrow g} = \delta_1 \oplus (\delta_2{}^{\downarrow(g \cap h)})$.

We have already shown that axioms L1 and L2 are valid for disbelief functions. Theorem 5 below states that axioms L3 is also satisfied.

Theorem 5. Suppose δ_1 and δ_2 are disbelief functions for g and h, respectively. Then $(\delta_1 \oplus \delta_2)^{\downarrow g} = \delta_1 \oplus (\delta_2{}^{\downarrow(g \cap h)})$. ■

Since all three axioms required for local computation of marginals are satisfied, the scheme described in Shenoy and Shafer (1990) can be used for belief updating. Hunter (1990) describes an analogous scheme for belief revision.

10. DISCUSSION

In Shenoy (1989), we describe a valuation-based language for representing and reasoning with knowledge. In such a language, knowledge is represented by functions called valuations and inferences are made from the knowledge-base using two operators called combination and marginalization. Combination corresponds to aggregation of knowledge and marginalization corresponds to crystallization of knowledge. Conceptually, all the valuations are combined to obtain what is called the *joint valuation*. The marginals of the joint valuation are then found for each variable. If combination and marginalization operators satisfy three axioms, then the marginals of the joint valuation can be found using local computation without actually computing the joint valuation. In Shenoy and Shafer (1988, 1990), we show that Bayesian probability theory and Dempster-Shafer's theory of belief functions fit in the abstract framework of valuation-based languages. In this paper, we have shown that Spohn's theory of epistemic beliefs also fits in this abstract framework. One implication of this is that we have a ready-made algorithm for propagating disbelief functions that uses only local computation. Another implication is that we now have a better understanding of the sense in which Spohn's theory differs from probability theory and Dempster-Shafer's theory of belief functions (in the functional representation of knowledge, and rules of combination and marginalization), and the sense in which it is similar to these alternative theories of uncertain reasoning (the abstract features of the axiomatic framework).

ACKNOWLEDGEMENTS

This work was supported in part by the National Science Foundation under grant IRI-8902444 and in part by the Peat Marwick Main Foundation under grant ROA-88-146. The author has profited from discussions with Ken Cogger and Glenn Shafer and correspondence with Daniel Hunter, Judea Pearl, Philippe Smets, and Wolfgang Spohn. I am especially grateful to Wolfgang Spohn for extensive comments on an earlier draft including the suggestion of the statement of Theorem 4. Due to page limitations, proofs are omitted. Proofs of all results can be found in Shenoy (1989b).

REFERENCES

Dempster, A. P. (1968), A generalization of Bayesian inference, *Journal of Royal Statistical Society*, Series B, **30**, 205-247.

Dubois, D. and Prade, H. (1988), Possibility Theory: An Approach to Computerized Processing of Uncertainty, Plenum Publishing Company, New York.

Dubois, D. and Prade, H. (1990), Epistemic entrenchment and possibilistic logic, unpublished manuscript.

Gardenfors, P. (1988), *Knowledge in Flux: Modeling the Dynamics of Epistemic States*, MIT Press, Cambridge, MA.

Geiger, D. and Pearl, J. (1990), On the logic of causal models, *Uncertainty in Artificial Intelligence 4*, Shachter, R., Levitt, T., Lemmer, J and Kanal, L., eds., 3–14, North-Holland.

Ginsberg, M. L. (1984), Non-monotonic reasoning using Dempster's rule, *Proceedings of the Third National Conference on Artificial Intelligence (AAAI-84)*, 126-129, Austin, TX.

Hunter, D. (1988), Graphoids, semi-graphoids, and ordinal conditional functions, unpublished manuscript.

Hunter, D. (1990), Parallel belief revision, *Uncertainty in Artificial Intelligence 4*, Shachter, R., Levitt, T., Lemmer, J and Kanal, L., eds., 241–252, North-Holland.

Jeffrey, R. C. (1983), *The Logic of Decision*, 2nd edition, Chicago University Press

Pearl, J. (1988), *Probabilistic Reasoning in Intelligent Systems: Networks of Plausible Inference*, Morgan-Kaufmann.

Pearl, J. (1989). Probabilistic semantics for nonmonotonic reasoning: A survey, *Proceedings of the First International Conference on Principles of Knowledge Representation and Reasoning*, Toronto, Canada, pp. 505-516.

Shafer, G. (1976), *A Mathematical Theory of Evidence*, Princeton University Press, Princeton, NJ.

Shafer, G. (1984), The problem of dependent evidence, School of Business Working Paper No. 164, University of Kansas, Lawrence, KS.

Shafer, G. (1987), Belief functions and possibility measures, in *Analysis of Fuzzy Information*, volume I: Mathematics and Logic, Bezdek, J. C. (ed.), 51–84, CRC Press.

Shenoy, P. P. (1989), A valuation-based language for expert systems, *International Journal of Approximate Reasoning*, **3**(5), 359-416.

Shenoy, P. P. (1989b), On Spohn's rule for revision of beliefs, School of Business Working Paper No. 213, University of Kansas, Lawrence, KS. To appear in *International Journal of Approximate Reasoning* in 1991.

Shenoy, P. P. and Shafer, G. (1988), An axiomatic framework for Bayesian and belief-function propagation, *Proceedings of the Fourth Workshop on Uncertainty in Artificial Intelligence*, 307-314, Minneapolis, MN.

Shenoy, P. P. and Shafer, G. (1990), Axioms for probability and belief-function propagation, *Uncertainty in Artificial Intelligence 4*, Shachter, R., Levitt, T., Lemmer, J and Kanal, L., eds., 169–198, North-Holland.

Spohn, W. (1988), Ordinal conditional functions: A dynamic theory of epistemic states, in Harper, W. L. and Skyrms, B. (eds.), *Causation in Decision, Belief Change, and Statistics*, **II**, 105-134, D. Reidel Publishing Company.

Spohn, W. (1990), A general non-probabilistic theory of inductive reasoning, *Uncertainty in Artificial Intelligence 4*, Shachter, R., Levitt, T., Lemmer, J and Kanal, L., eds., 149–158, North-Holland.

Verma, T. and Pearl, J. (1990), Causal networks: Semantics and expressiveness, *Uncertainty in Artificial Intelligence 4*, Shachter, R., Levitt, T., Lemmer, J and Kanal, L., eds., 69–78, North-Holland.

Zadeh, L. A. (1978), Fuzzy sets as a basis for a theory of possibility, *Fuzzy Sets and Systems*, **1**, 3-28.

FAST ALGORITHMS FOR DEMPSTER-SHAFER THEORY

Robert KENNES and Philippe SMETS*
IRIDIA, Université Libre de Bruxelles
Av. F. D. Roosevelt 50 - CP 194/6
B-1050 Brussels, Belgium

Abstract: The runtime of the usual algorithms computing the transformation of a basic belief assignment into its associated belief function and conversely is an exponential function of the cardinality of (the domain of) the basic belief assignment. In this paper, new algorithms with a polynomial runtime are presented. These algorithms appear to be optimal in the class of the so-called M-algorithms.

Keywords: Dempster-Shafer theory, Möbius transformation, graph, computational efficiency.

0. INTRODUCTION

This paper grew out of a natural question related to the implementation of Dempster-Shafer theory of evidence: how is it possible to compute the transformation of a mass distribution into its associated belief function in the most 'intelligent' way, i.e. by performing a minimum number of additions? In this paper, we present exact algorithms transforming a mass distribution (m) into its belief function (bel_m) and vice versa. The input of the algorithms is an array containing all the masses, the output is the same array containing the values of the belief function. These algorithms need no extra memory space during computation and compute all values at once. The algorithms have been designed on the model of the Fast Fourier Transform and they indeed appear to be optimal in runtime. Because of lack of space, the proof of the optimality, based on graph considerations, is not given here. All proofs are given in [Kennes 90]. A proof of the optimality of some Fast Fourier Transforms, also based on graph considerations, can be found in [Papadimitriou 79].

Using the ideas presented in the current paper, a whole family of fast algorithms transforming mass distributions, belief functions, plausibility functions, and commonality functions into each other may be designed [Kennes 90].

Independently of us, H. M. Thoma, a student of A. Dempster, also discovered these fast algorithms. However, he writes [Thoma 1989, p. 132] : 'We have not proven explicitly that the fast algorithm requires a minimal number of additions among all possible computing schemes. However, this is shown quite convincingly by the graphical explanation given above.'

---

* The following text presents some research results of the Belgian National incentive-program for fundamental research in artificial intelligence initiated by the Belgian State, Prime Minister's Office, Science Policy Programming. The scientific responsibility is assumed by the authors. These researches have been partially supported by the projects ARCHON and DRUMS which are funded by grants from the Commission of the European Communities under the ESPRIT II-Program, P-2256 and Basic Research Project 3085.

Others studies of the computational complexity related to Dempster-Shafer Theory have been done by P. Orponen (1990) and by G.M. Provan (1990). It should be noted - in order to prevent some misunderstandings - that their point of view is somewhat different from ours.

1. BASIC BELIEF ASSIGNMENTS AND THEIR BELIEF FUNCTIONS

Let Ω be a non empty finite set and $\wp\Omega$ be its power set equipped with the inclusion relation. In Dempster-Shafer theory of evidence - the standard reference of which is [Shafer 76], see also [Smets 88] - a *basic belief assignment* (bba) on Ω, also called a *mass distribution*, is any function

$$m : \wp\Omega \to [0\ 1] \quad \text{such that} \quad \sum_{X \in \wp\Omega} m(X) = 1$$

m(X) is called the *mass* of X. Most often it is also required that the mass of the empty set be 0. Anyway, any basic belief assignment m determines its *belief function* $bel_m: \wp\Omega \to [0\ 1]$ defined, in all cases, by

$$\forall A \in \wp\Omega : bel_m(A) = \sum_{X \in \wp A - \{\emptyset\}} m(X) = \sum_{X \subseteq A, X \neq \emptyset} m(X)$$

$bel_m(A)$ is called the *belief* of A induced by the bba m. If, more generally, we consider all functions $m: \wp\Omega \to \mathbf{R}$, the previous formula defines the functional

$$\mathbf{R}^{\wp\Omega} \to \mathbf{R}^{\wp\Omega} : m \to bel_m$$

where $\mathbf{R}^{\wp\Omega}$ denotes, as usual, the set of functions from $\wp\Omega$ to the set of real numbers. The notation bel_m, although rather explicit, does not do justice to the most important protagonist of the formula, that is, the binary relation $G = \{(X,Y) \in \wp\Omega \times \wp\Omega \mid X \neq \emptyset, X \subseteq Y\}$. The above functional, induced by the relation G, will be called the Möbius transformation of G or, less exactly, of the boolean lattice $(\wp\Omega, \subseteq)$.

2. HOW GRAPHS OPERATE ON FUNCTIONS

Let us first quickly set out the notation. Let S and T be finite sets. A subset G of the cartesian product $S \times T$ is called a *(directed) graph* from S to T. We will indifferently write $G: S \to T$ or $G \subseteq S \times T$. Throughout this paper all graphs are finite. The elements of G will sometimes be called *arrows of G* instead of *ordered pairs of G*. The *source* of (s,t) is s: source(s,t) = s. The *target* of (s,t) is t: target(s,t) = t. The *identity graph* on S is $1_S = \{(s,s) \mid s \in S\}$. When no confusion is possible, the same symbol will denote a binary relation and the set of ordered pairs it determines on a particular set. Explicitly, if R is a binary relation, the graph $\{(s,t) \in S \times S \mid sRt\}$ determined by R on the set S is also denoted by R. According to this identification the two statements $(s,t) \in G$ and sGt are equivalent.
SET denotes the category of sets. Replacing the functions by the relations and restricting the sets to the finite sets we get the category denoted (confusingly!) by **FGRAPH** of which the objects are the *finite sets* and the arrows are the *graphs* $G: S \to T$ together with the usual *composition of graphs.*

Any arrow of the category **FGRAPH**, i.e. any graph, determines its *Möbius transformation.* More explicitly:

Definition 1. The graph G: S $\rightarrow$ T determines the functional (i.e. a second order function):

$$\mathbf{M}^G: \mathbf{R}^S \rightarrow \mathbf{R}^T: f \rightarrow \mathbf{M}^G(f) = f^G$$

defined by

$$\forall t \in T : f^G(t) = \sum_{sGt} f(s) = \sum_{s \in G^{-1}(t)} f(s)$$

The functional $\mathbf{M}^G$ will be called the *Möbius transformation of* G.

An extensive reference to (a special case of) Möbius transformations is [Aigern 79, chap. IV], where it is called *sum function* .
The transformation M^G may be interpreted as follows. It transforms any *mass distribution* on S into a *mass distribution* on T in the following manner: all the masses f(s) are dragged along (or transferred by) the arrows (s,t)$\in$ G to the targets and then are added together at each target. M^G may also be viewed as a discrete analogue of the indefinite integral in calculus replacing the relation $\leq$ by the relation G.
We will say, equivalently: G *computes* M^G
or G is a *M-algorithm* of M^G
or M^G is the *Möbius transformation* of G
(or *induced by* G or *determined by* G or *associated with* G).

For obvious reasons, which will appear later, the graph G will be called the *obvious* M-algorithm computing M^G.
Theorem 1. The map **M** from the category **FGRAPH** to the category **SET**,
M: FGRAPH $\rightarrow$ SET: (S $\rightarrow$ M(S) = $\mathbf{R}^S$, G $\rightarrow$ M(G) = $\mathbf{M}^G$) satisfies:
(1) $\mathbf{M}(1_S) = 1_{\mathbf{M}(S)}$
(2) in general: $\mathbf{M}(G_2 \circ G_1) \neq \mathbf{M}(G_2) \circ \mathbf{M}(G_1)$
(3) For every graphs G, H: S $\rightarrow$ T G = H <u>iff</u> **M**(G) = **M**(H).

Comments: 1. Because of (2) **M** is not a functor **FGRAPH $\rightarrow$ SET**, nevertheless by slightly modifying the category **FGRAPH**, **M** will become a functor as it will be shown in the next section. This will be a key fact for our concern because this new functor **M** will provide us with a criterion for the equality $\mathbf{M}(G_2 \circ G_1)$ = $\mathbf{M}(G_2) \circ \mathbf{M}(G_1)$. 2. The faithfulness of **M** means that a Möbius transformation is induced by exactly one graph.

Let us now consider the following finite sequence of (*queueing*) graphs:

$$S_0 \xrightarrow{G_1} S_1 \xrightarrow{G_2} \ldots \longrightarrow S_i \xrightarrow{G_{i+1}} S_{i+1} \longrightarrow \ldots \xrightarrow{G_n} S_n$$

where the sets $S_0, S_1, \ldots, S_n$ are finite but not necessarily pairewise disjoint. The previous sequence of graphs will be called a *M-algorithm (of graphs)* (of length n). It determines the composite of

$$\mathbf{R}^{S_0} \xrightarrow{\mathbf{M}^{G_1}} \mathbf{R}^{S_1} \xrightarrow{\mathbf{M}^{G_2}} \ldots \longrightarrow \mathbf{R}^{S_i} \xrightarrow{\mathbf{M}^{G_{i+1}}} \mathbf{R}^{S_{i+1}} \longrightarrow \ldots \xrightarrow{\mathbf{M}^{G_n}} \mathbf{R}^{S_n}$$

That is, it determines (*computes*) $M^{G_n} \circ M^{G_{n-1}} \circ \ldots \circ M^{G_1}$.
It should be stressed that $M^{G_n} \circ M^{G_{n-1}} \circ \ldots \circ M^{G_1}$ is not always equal to $\mathbf{M}^{G_n \circ G_{n-1} \circ \ldots \circ G_1}$.

Let us recall that the *composite* of a sequence G

$$S = S_0 \xrightarrow{G_1} S_1 \xrightarrow{G_2} \cdots \xrightarrow{G_n} S_n = T$$

denoted by $C(G): S \to T$

is the graph $C(G) = G_n \circ \ldots \circ G_1 : S \to T$

3. THE MÖBIUS FUNCTOR: HOW WEIGHTED GRAPHS OPERATE ON FUNCTIONS

All the concepts introduced in the previous section can be trivially extended to the case of weighted graphs.
Definition 2. A *weighted graph* $\alpha: S \to T$ from S to T is any function:

$$\alpha: S \times T \to \mathbf{R} : (s,t) \to \alpha(s,t).$$

[Aigner 79] is an extensive reference to weighted graphs or *incidence functions* as they are called there. Actually, α can also be viewed as a matrix $(\alpha(s,t))_{(s,t) \in S \times T}$.
The product of weigthed graphs is defined as follows:

Definition 3. If $\alpha: S \to T$ and $\beta : T \to U$ are two weighted graphs, their *product* $\alpha * \beta$ is defined by

$$\alpha * \beta\ (s,u) = \sum_{t \in T} \alpha(s,t) \cdot \beta(t,u)$$

This is also the product of the matrices α and β.
Each set S defines its identity weighted graph δ_S, the *Kronecker function of S* [Aigner 79 p.140], which is the characteristic function of 1_S as a subset of $S \times S$. The δ_S's are the identities of the product. As a consequence, we get the category **WGRAPH**, the objects of which are the *finite sets* and the arrows are the *weighted graphs* $\alpha: S \to T$ together with the product of weighted graphs. The Möbius transformation of a weighted graph is defined in the same manner as the Möbius transformation of a graph.
Definition 4. The weighted graph $\alpha: S \to T$ determines the following functional

$$M^{\alpha} : \mathbf{R}^S \to \mathbf{R}^T : f \to M^{\alpha}(f) = f^{\alpha}$$

defined by

$$\forall t \in T : f^{\alpha}(t) = \sum_{s \in S} f(s) \cdot \alpha(s,t)$$

M^{α} will be called the *Möbius transformation* of α.
The last expression can be viewed as the product of the column-vector $(f(s))_{s \in S}$ by the matrix $(\alpha(s,t))_{(s,t) \in S \times T}$. M^{α} may be viewed as a discrete analogue of the Stieltjes indefinite integral in calculus and admits the same interpretation as M^G where the $\alpha(s,t)$ are scaling factors.
Now, **M** turns out to be a functor. (Note that the symbol **M** is used to denote two different maps!).
Theorem 2. The map **M** : **WGRAPH** $\to$ **SET**: $(S \to M(S) = \mathbf{R}^S, \alpha \to M(\alpha) = \mathbf{M}^{\alpha})$ from the category of weighted graphs to the category of sets is a functor called the *Möbius functor*. The Möbius functor is faithful.

As in the case of graphs, we say that the following sequence of (*queueing*) weighted graphs

$$S_0 \xrightarrow{\alpha_1} S_1 \xrightarrow{\alpha_2} \ldots \longrightarrow S_i \xrightarrow{\alpha_{i+1}} S_{i+1} \longrightarrow \ldots \xrightarrow{\alpha_n} S_n$$

is a *M-algorithm* (of weighted graphs) (of length n) *computing* the composite of

$$\mathbf{R}^{S_0} \xrightarrow{M^{\alpha_1}} \mathbf{R}^{S_1} \xrightarrow{M^{\alpha_2}} \ldots \longrightarrow \mathbf{R}^{S_i} \xrightarrow{M^{\alpha_{i+1}}} \mathbf{R}^{S_{i+1}} \longrightarrow \ldots \xrightarrow{M^{\alpha_n}} \mathbf{R}^{S_n}$$

There is an obvious link between the two categories **FGRAPH** and **WGRAPH**. Indeed, each graph $G \subseteq S \times T$ determines its weighted graph ζ_G, *the zeta-function of G* [Aigner 79 p.140], which is the characteristic function (except for its codomain) of G as a subset of S×T:

$$\zeta_G: S\times T \to \mathbf{R} : (s,t) \to \zeta_G(s,t)$$

$$\zeta_G(s,t) = 1 \text{ iff } (s,t)\in G \quad \text{and} \quad \zeta_G(s,t) = 0 \text{ iff } (s,t)\notin G.$$

It is trivial to see that: $M^{\zeta_G} = M^G$

which means: $\forall f\in \mathbf{R}^S\ \forall t\in T: \sum_{s\in S} f(s).\zeta_G(s,t) = \sum_{s\in G^{-1}(t)} f(s)$

Lemma. The sequence G $\quad S = S_0 \xrightarrow{G_1} S_1 \xrightarrow{G_2} \ldots \xrightarrow{G_n} S_n = T$

computes the Möbius transformation of C(G) iff

$$\zeta_{C(G)} = \zeta_{G_1} * \zeta_{G_2} * \ldots * \zeta_{G_n}$$

The meaning of the equality $\zeta_{C(G)} = \zeta_{G_1} * \zeta_{G_2} * \ldots * \zeta_{G_n}$ is examined in the following section.

4. GRAPHS DECOMPOSITIONS

The need for decomposing graphs (with respect to *) will become clear in the sequel when it will be shown that a decomposition of a graph may decrease the *computational complexity* of the graph. In fact, the basic idea to get a fast (algorithm for computing) Möbius transformation is to decompose the inclusion relation of $\wp\Omega$.

Definition 5. Let G be the following sequence of graphs

$$S = S_0 \xrightarrow{G_1} S_1 \xrightarrow{G_2} \ldots \xrightarrow{G_n} S_n = T \quad (n \geq 1)$$

a *path u of G* is a n-uple $(g_1,\ldots,g_n)\in G_1\times\ldots\times G_n$ such that $\forall i\in \{1,\ldots,n-1\}$: target($g_i$) = source($g_{i+1}$). The *source* of u is the source of g_1, the *target* of u is the target of g_n. The set of paths of G with source in $X\subseteq S$ and with target in $Y\subseteq T$ is denoted by $P_G(X,Y)$. For the sake of brevity $P_G(\{s\},\{t\})$ is replaced by $P_G(s,t)$. The following map:

$P_G(S,T) \to G_n \circ \ldots \circ G_1: (g_1,\ldots,g_n) \to g_n \circ \ldots \circ g_1 = g_1 \ldots g_n$ = (source(g_1),target(g_n)) defines the usual *composition* or *product* of arrows. Note that the inverse map provides for each arrow (s,t) the set of its *factorizations* of the form $g_1 \ldots g_n$ where $g_i\in G_i$.

We can now state the fundamental theorem which gives a characterization of sequences of graphs computing the Möbius transformation of a graph:

Theorem 3. Let G be the sequence $S = S_0 \xrightarrow{G_1} S_1 \xrightarrow{G_2} \ldots \xrightarrow{G_n} S_n = T$

(1) G computes the Möbius transformation of a graph

iff (2) G computes the Möbius transformation of the graph $C(G): S \to T$

iff (3) $P_G(S,T) \to C(G) : (g_1,\ldots,g_n) \to g_1\ldots g_n$ is a bijection.

To say that $P_G(S,T) \to C(G): (g_1,\ldots,g_n) \to g_1\ldots g_n$ is a bijection is equivalent to saying that every arrow g of C(G) has a unique factorization of the form $g_1\ldots g_n$ where $g_i \in G_i$.

5. THE FAST MÖBIUS TRANSFORMATIONS OF $(\wp\Omega,\subseteq)$

In this section, we study a special case, namely the case of the Möbius transformation of $(\wp\Omega,\subseteq)$.
Let us first recall the notion of Hasse graph of a partial order relation.

Definition 6. If $(P,\leq)$ is a partially ordered set, then the *reflexive Hasse graph* of $\leq$ is $\mathbf{H}(\leq) = \{(a,b) \in P\times P \mid a\leq b \text{ and } \forall x\in P: a<x \Rightarrow b\leq x\}: P \to P$. The *non-reflexive Hasse* graph of $(P,\leq)$ is $\mathbf{H}(<) = \{(a,b) \in P\times P \mid a<b \text{ and } \forall x\in P: a<x \Rightarrow b\leq x\}: P \to P$. The transitive closure of $\mathbf{H}(\leq): P \to P$ is known to be $\leq: P \to P$, and furthermore it is part of the folklore that $\mathbf{H}(\leq)$ is the smallest (with respect to the inclusion relation) subgraph of $\leq$ such that its transitive closure is $\mathbf{H}(\leq)$. Thus, $\mathbf{H}(\leq)$ is characterized by the two properties: (1) $T(\mathbf{H}(\leq)) = \leq$, (2) $T(G) = \leq \Rightarrow \mathbf{H}(\leq) \subseteq G$.

We recall that the *obvious* M-algorithm of M^G is the graph G itself. In this section other M-algorithms for computing M^G are described. These M-algorithms, called the 'fast Möbius transformations' will be shown to be optimal for a specific cost function.

Theorem 4. If $\Omega = \{a_1,a_2,\ldots,a_n\}$, then the following M-algorithm H of length n :

$$\wp\Omega \xrightarrow{H_1} \wp\Omega \xrightarrow{H_2} \ldots \xrightarrow{H_n} \wp\Omega$$

where $H_i = \{(X,Y) \in \wp\Omega\times\wp\Omega \mid Y=X \text{ or } Y=X\cup\{a_i\}\}$

computes the Möbius transformation of $(\wp\Omega,\subseteq)$.

Note the fundamental fact: $\mathbf{U}(H) = \mathbf{H}(\subseteq)$, ($\mathbf{U}(H)$ = union of the H_i). This is a reason for calling these M-algorithms the *Hasse* M-algorithms of (the Möbius transformation of) $(\wp\Omega,\subseteq)$. We also call them the fast Möbius transformations of G.

The same result holds if we take into account the condition $X\neq\emptyset$.

Example. If $\Omega = \{a,b,c\}$, there are 6=3! different Hasse M-algorithms on $(\wp\Omega,\subseteq)$. Each total order of the set Ω determines a Hasse M-algorithm. Here are two of them:

$$\wp\Omega \xrightarrow{H_a} \wp\Omega \xrightarrow{H_b} \wp\Omega \xrightarrow{H_c} \wp\Omega$$

$$\wp\Omega \xrightarrow{H_c} \wp\Omega \xrightarrow{H_b} \wp\Omega \xrightarrow{H_a} \wp\Omega$$

where $H_a = \{(X,Y) \in \wp\Omega\times\wp\Omega \mid Y=X \text{ or } Y=X\cup\{a\}\}$ and similarly for H_b and H_c. The latter M-algorithm may be represented 'vertically' by:

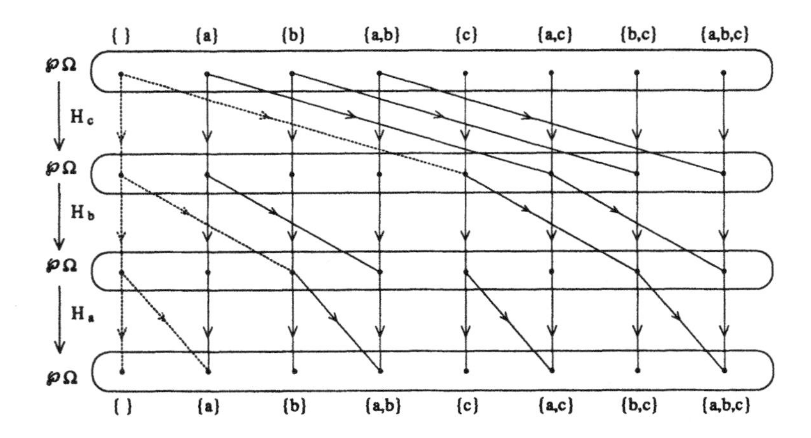

If we take into account the condition X≠Ø, the dotted arrows are not strictly necessary but are put on the diagram for symmetry. Let us now provide, as an example, a 'real life' algorithm computing this transformation. Given a total order on $\Omega = \{a_1,a_2,...,a_n\}$, any subset of Ω is bijectively represented by a n-digit binary numeral. The bijection is realized as follows: the element a_i belongs to the subset A of Ω iff the ith digit of the binary numeral representing A is 'on' (i.e. is 1). Using this coding, it is not difficult to get the following algorithm, which transforms a mass distribution into its belief function:

```
Input: array v[0:2^n -1]          'contains the mass distribution m'
Output: array v[0:2^n -1]         'contains the belief function bel induced by m'
Procedure m-to-bel(v):            'transforms mass into belief'
v[0] ← 0                          'the mass of Ø is not taken into account'
for i ← 1 step 1 until n do
   for j ← 1 step 2 until 2^i do
      for k ← 0 step 1 until 2^(n-i) -1 do
         v[j.2^(n-i) + k] ← v[j.2^(n-i) + k] + v[(j-1).2^(n-i) + k]
      end
   end
end
```

6. THE INVERSE MÖBIUS TRANSFORMATION OF (℘Ω,⊆)

It can easily be shown that the functional **M**(G), determined by a graph G: S → T, is not always injective so there is no hope, in the general case, to be able to 'reverse' (i.e. to get a left inverse of) **M**(G). That means that there does not always exist a functional F: T → S such that: F ∘ **M**(G) = 1 (the identity function on $\mathbf{R}^S$). The problem of the existence of an inverse has a nice solution in the more general setting of weighted graphs. In fact, if ζ_G has an inverse (for the product ∗), then because M is a functor we get: $\zeta_G * (\zeta_G)^{-1} = \zeta_{1S}$ and so we have: $(\mathbf{M}(\zeta_G))^{-1} \circ \mathbf{M}(G) = 1$ (the identity function on $\mathbf{R}^S$).

The Möbius inversion theorem provides a sufficient (but not a necessary!) condition about the graph G in order for the functional $\mathbf{M}(\zeta_G)$ to be a bijection (see [Aigner 79 p.141]). But, for our concern we simply need to find the inverse (with respect to ∗) of the graph $H_i = \{(X,Y) \in \wp\Omega \times \wp\Omega \mid Y=X \text{ or } Y=X \cup \{a_i\}\}$. It can readily be verified, either directly or by using the Möbius inversion theorem, that the following weighted graph is an inverse (hence *the* inverse) of H_i. μ_i: $\wp\Omega \times \wp\Omega \to \mathbf{R}$, defined by: $\forall X \in \wp\Omega$: $\mu_i(X,X) = 1$, $\mu_i(X, X \cup \{a_i\}) = -1$, else $\mu_i(X,Y) = 0$.

Thus we get:

Theorem 5. If $\Omega = \{a_1, a_2, \ldots, a_n\}$, then the following M-algorithm of weighted graphs computes the inverse Möbius transformation of $(\wp\Omega, \subseteq)$:

$$\wp\Omega \xrightarrow{\mu_1} \wp\Omega \xrightarrow{\mu_2} \ldots \xrightarrow{\mu_n} \wp\Omega \text{ where the } \mu_i\text{: } \wp\Omega \times \wp\Omega \to \mathbf{R}$$

are defined by: $\forall X \in \wp\Omega$: $\mu_i(X,X) = 1$, $\mu_i(X, X \cup \{a_i\}) = -1$, else $\mu_i(X,Y) = 0$.

The M-algorithm can be represented vertically by: (The label on a arrow is its weight. In order to simplify the diagram the identity arrows, all having weight 1, have not been represented. All the other non-represented arrows get the weight 0)

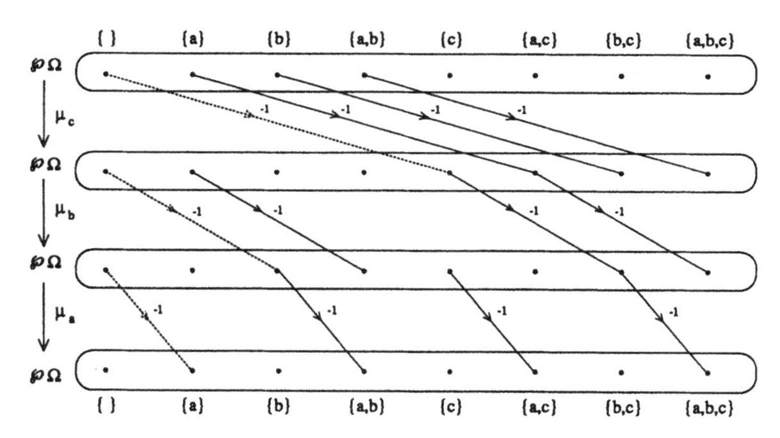

The dotted arrows are unnecessary when the condition $X \neq \emptyset$ is taken into account. Remark: as the number of arrows of weight (-1) in *the* path from X to A is $\#(A-X)$, the mass m(X) moving from X to A is multiplied by $(-1)^{\#(A-X)}$, which provides a new proof of the classical formula transforming bel into m_{bel}.

Here is an algorithm computing the transformation of a belief function into its mass distribution:

```
Input: array v[0:2^n -1]          'contains the belief function bel'
Output: array v[0:2^n -1]         'contains the mass distribution m induced by bel'
Procedure bel-to-m(v):            'transforms belief into mass'
v[0] ← 0                          'the belief of Ø must not be considered'
for i ← 1 step 1 until n do
   for j ← 1 step 2 until 2^i do
      for k ← 0 step 1 until 2^(n-i) -1 do
         v[j.2^(n-i) + k] ← v[j.2^(n-i) + k] – v[(j-1).2^(n-i) + k]
      end
   end
end
v[0] ← 1 – sum of the m(A), for all non-empty set A
```

7. COST OF M-ALGORITHMS OF GRAPHS

We want to count the number of additions performed by a M-algorithm of graphs in the case of the 'worst possible input'. By 'worst possible input' we mean that 'a+b' stands for one addition whatever the values of a and b may be. No multiplication is performed in the case of M-algorithms of *graphs*. The 'cost' function will provide us with the number of additions performed by a M-algorithm. We will define it first for the vertices of a graph, second for graphs, and finally for M-algorithms (sequences) of graphs.

Definition 7. Let $G: S \to T$ be a graph and t be an element of T.

(1) $cost_G(t) = \max\{0, \#G^{-1}(t) - 1\}$

$= \#G^{-1}(t) - 1$ if $G^{-1}(t) \neq \emptyset$, else 0

(2) $cost(G) = \sum_{t \in T} cost_G(t)$

$$= \sum_{t\in T, G^{-1}(t) \neq \emptyset} (\#G^{-1}(t) - 1) = \Big(\sum_{t\in T, G^{-1}(t) \neq \emptyset} \#G^{-1}(t) \Big) - \#G(S) = \#G - \#G(S)$$

(3) If G is a sequence of graphs $S = S_0 \xrightarrow{G_1} S_1 \xrightarrow{G_2} \dots \xrightarrow{G_n} S_n = T$

$$cost(G) = \sum_{i=1}^{n} cost(G_i)$$

Examples:

1. Cost of the obvious M-algorithm G of $M^{\{(X,Y) \in \wp\Omega \times \wp\Omega \,|\, X \neq \emptyset, X \subseteq Y\}}$

   We get : $cost(G) = 3^n - 2^{n+1} + 1$, if $\#\Omega = n$ or $cost(G) = x.(1.5^{\log_2 x} + 1/x - 2)$, if $\#\wp\Omega = x$

The runtime of the obvious algorithm is an '**exponential**' (or better 'superpolynomial') function of $\#\wp\Omega$.

2. Cost of the Hasse M-algorithm H of $M^{\{(X,Y) \in \wp\Omega \times \wp\Omega \,|\, X \neq \emptyset, X \subseteq Y\}}$

   We get: $cost(H) = n\, 2^{n-1} - n$, if $\#\Omega = n$ or $cost(H) = \log_2 x\, .(x/2 - 1)$, if $\#\wp\Omega = x$

The runtime of the fast algorithm is a '**polynomial**' function of $\#\wp\Omega$.

Ratio of the two cost functions:

$\#\Omega$	$\#\wp\Omega$	cost(G)	cost(H)	cost(G)/cost(H)
5	32	180	75	2.4
8	256	6 050	1 016	5.9
10	1 024	57 002	5 110	11.1
12	4 096	523 250	24 564	21.3
15	32 768	14 283 372	245 745	58.1
20	10^6	3.10^9	10^7	332.3

3. Cost of the Hasse M-algorithm of $M^{\subseteq}$

   We get: $cost(H) = n\, 2^{n-1}$, if $\#\Omega = n$

8. OPTIMALITY OF THE FAST MÖBIUS TRANSFORMATIONS

The following theorem provides a lower bound for the cost function of the M-algorithms computing the Möbius transformation of a finite lattice.

Theorem 6. Let $(L,\leq)$ be a finite lattice.
If A is a M-algorithm of graphs computing $\mathbf{M}^{\leq}$ then cost(A) $\geq$ cost($\mathbf{H}(\leq)$).

In fact, a slightly more general result can be proved:

Theorem 7. Let $(P,\leq)$ be a finite partially ordered set in which every upper bounded subset has a least upper bound. If A is a M-algorithm of graphs computing $\mathbf{M}^{\leq}$ then cost(A) $\geq$ cost($\mathbf{H}(\leq)$).

Corollary. (The fast Möbius transformations of $(\wp\Omega,\subseteq)$ are optimal.)
Let $\Omega = \{a_1,a_2,\ldots,a_n\}$ and $(\wp\Omega,\subseteq)$ be the corresponding finite boolean lattice.
If (1) G is a M-algorithm of graphs computing $\mathbf{M}^{\subseteq}$
(2) H is a Hasse M-algorithm of $\mathbf{M}^{\subseteq}$

$$\wp\Omega \xrightarrow{H_1} \wp\Omega \xrightarrow{H_2} \cdots \xrightarrow{H_n} \wp\Omega$$

where $H_i = \{(X,Y)\in \wp\Omega\times\wp\Omega \mid Y=X \text{ or } Y=X\cup\{a_i\}\}$,
then cost(G) $\geq$ cost(H).
If the condition $X \neq \emptyset$ is taken into account, the same result holds.

9. CONCLUSIONS

Algorithms with smaller additive cost than the usual corresponding algorithms have been presented. It can be proved that the presented algorithms are the most efficient in the general case. They unfortunately require an array of length $\#\wp\Omega$. Moreover, in 'practical 'applications, most components of this array get a zero value (in case of a mass distribution), and the fast algorithm does not take into account this extra information. Other algorithms, exact or approximate [Provan 1990], more adapted to those situations, might of course be more efficient.

10. REFERENCES

Aigner M. (1979), *Combinatorial Theory*, Springer-Verlag.

Barr M., Wells C. (1990), *Category Theory for Computing Science*, Prentice Hall.

Kennes R. (1990), *Computational Aspects of the Möbius Transformation of a Graph*, Technical Report, IRIDIA/90-13, Université Libre de Bruxelles, Brussels, Belgium.

Orponen P. (1990), *Dempster's Rule of Combination is #P-complete*, Artificial Intelligence, 44, 245-253.

Papadimitriou C. (1979), *Optimality of the Fast Fourier Transform*, J. ACM 26, 1, 95-102.

Provan G. M. (1990), *A Logic-Based Analysis of Dempster-Shafer Theory*, International Journal of Approximate Reasoning 4,451-495.

Shafer G. (1976), *A Mathematical Theory of Evidence*, Princeton University Press.

Smets P. (1988), *Belief Functions*, In Non Standard Logics for Automated Reasoning, 253-286 (Smets P., Mamdani A., Dubois D. and Prade H. editors), Academic Press.

Thoma H. M. (1989), *Factorization of Belief Functions*, unpublished Ph.D. thesis, Harvard University, Dept. of Statistics.

ON THE COMBINATION OF INFORMATION SOURCES

R. KRUSE and E. SCHWECKE

Institut für Betriebssysteme und Rechnerverbund

TU Braunschweig

Bültenweg 74/75

3300 Braunschweig

West Germany

Abstract: Subject of this paper is a thorough modeling of uncertain knowledge based on the theory of belief functions. In this context it has to be taken into account, that some experts state of mind changes in the light of new information, thus we introduce the notion of an information source. Considering the integration of information sources it turns out that this approach leads to Dempster's well-known rule of combination.

Keywords: belief function, information source, combination, theory of evidence.

1 - INTRODUCTION

In recent years the treatment of uncertainty has been widely accepted as one of the important problems in the range of knowledge based systems. The development of knowledge representation formalisms requires to take into account that the reliability of information may be questionable and thus becomes a matter of degree.

The theory of evidence provides a mathematical framework for the description of this phenemenon. Being a numerical approach it relies on the attachment of numbers to pieces of knowledge according to some experts valuation of the respective sources.

As we will point out in the sequel the formulation of any knowledge representation model has to be done with regard to the fact that the faith one puts in some information is, in essence, determined by the reliability of it's source. So, especially if the integration of different items of information is aspired, the sources and their relationship have to be thoroughly considered.

2 - BELIEF FUNCTIONS

The formal basis of our approach to the representation of knowledge is the set Ω, containing the possible "states of the world". Throughout this paper Ω which we call the universe of discourse or frame of discernment is considered to be a finite set. According to [5] we introduce the notion of a mass distribution in order to specify the attachment of numbers to subsets of Ω.

Definition 1. A mapping $m : 2^\Omega \to [0,1]$ satisfying

(i) $m(\emptyset) = 0$ and

(ii) $\sum_{A:A\subseteq\Omega} m(A) = 1$

is called a *mass distribution.*

From a mass distribution m we can obtain further characterizations of the underlying information in form of belief, plausibility, and commonality functions which we denote by Bel_m, Pl_m and Q_m, respectively.

A mass distribution may change in the light of additional pieces of knowledge. In more formal terms it may be revised with respect to some set $B \subseteq \Omega$, if we learn that with certainty the true state of the world is an element of B.

Definition 2. Let $m : 2^\Omega \to [0,1]$ be a mass distribution and let $B \subseteq \Omega$ be a set with $\mathrm{Pl}_m(B) > 0$. Then we call the mass distribution

$$m_B : 2^\Omega \to [0,1]; \quad m_B(A) = \begin{cases} \frac{1}{c} \cdot \sum_{C:C\cap B=A} m(C) & \text{if } A \subseteq B \\ 0 & \text{otherwise,} \end{cases}$$

where

$$c = \sum_{C:C\cap B\neq\emptyset} m(C),$$

the *revision* of m with respect to B.

From a statistical point of view m_B emerges from a revision of the data underlying m which we conceive to be a random set [3], but *not* from a conditioning of m. On the intuitive level revision means a flow of evidence masses. The masses attached by m to some set A float to $A \cap B$; if $A \cap B = \emptyset$ they are neglected. The final renormalization is done in order to obtain a mass distribution according to Definition 1.

Unfortunately in practice the subject of all knowledge and observations are not the true phenomena but only projections. Thus in general the exact description of some domain would require to consider Ω' being a refinement of the accessible frame of discernnment Ω. The relationship of Ω and Ω' is determined by the respective refinement mapping $\widehat{\Pi} : 2^{\Omega} \to 2^{\Omega'}$, the reverse projection mapping we denote by $\Pi : 2^{\Omega} \to 2^{\Omega'}$. So the mass distributions on Ω we use to encode our valuations are in general projections of refined ones defined on Ω'. The definition of such projection is obvious, we have

$$\Pi(m') : 2^{\Omega} \to [0,1]; \quad \Pi(m')(A) = \sum_{A' : \Pi(A') = A} m'(A').$$

For the same reason considering Ω in general we encounter projections of revisions in Ω' but not revisions originally performed in Ω. In order to obtain a description of such projected revisions in terms of the space Ω, which is the only frame of discernment being accessible for us, we introduce the notion of a specialization (comp. [2],[1],[7]).

Definition 3. Let s, t be two mass distributions defined on Ω. We call s a *specialization* of t ($s \sqsubseteq t$) if and only if there are mass distributions s', t' on a refinement Ω' of Ω, such that $s = \Pi(s')$, $t = \Pi(t')$, and if there is an event $E' \subseteq \Omega'$ such that $s'(B') = t'_{E'}(B')$ for all $B' \subseteq \Omega'$.

From this definition it becomes clear, that the relation $\sqsubseteq$ allows to compare two mass distributions s, t with respect to the specificity of the information they bear. In addition to the definition the next theorem provides an equivalent characterization of the specialization relation which allows to check easily whether $s \sqsubseteq t$ is valid or not.

Theorem 4. Let s, t be two mass distributions on Ω. The following two statements are equivalent

(i) s is a specialization of t ($s \sqsubseteq t$).

(ii) $\forall A \subseteq \Omega : \ (Q_t(A) = 0 \implies Q_s(A) = 0)$.

For a proof see [8]. The verification of Theorem 6 which is the main result of this paper strongly relies on this equivalence.

3 - MODELING KNOWLEDGE

If some expert specifies a mass distribution m on Ω then in general this is not meant to be the final valuation of the problem chiseled in stone but represents his actual state of knowledge. In the light of additional information he is willing to accept modifications of the original mass distribution. Typical admissible modifications are revisions or, in the most general case, specializations. For this reason a complete representation of an expert's knowledge requires also the exact specification of those "updates" of the original mass distribution he considers to be acceptable. This leads to the notion of an information source (see [4]) which we define next.

Definition 5. An *information source* is defined to be a triple $(2^\Omega, m, \mathcal{U}(m))$ where Ω is the frame of discernment, m is a mass distribution defined on Ω and $\mathcal{U}(m)$ is the set of possible updates. In the sequel we assume $\mathcal{U}(m) = \{m' \mid m' \sqsubseteq m\}$.

As we pointed out in the introduction modeling uncertainty requires to consider two spaces, firstly Θ containing the agents or sensors providing statements concerning Ω, and secondly Ω, the evidence space, containing the set of possible states of the world. Note that the elements of the sensor space Θ are human observers or technical devices whereas Ω represents the domain of interest. A mapping $I : \Theta \to 2^\Omega$ (which we can disjunctively extend by defining $I(A) \stackrel{d}{=} \bigcup_{\theta \in A} I(\theta)$) attaches to each sensor $\theta \in \Theta$ its observation $I(\theta) \subseteq \Omega$.

Given a mass distribution n on Θ valuating the sensors being at the expert's disposal, then a mass distribution on Ω is induced by the mapping I if $\sum_{A:I(A)\neq\emptyset} n(A) > 0$. We obtain

$$I[n](B) \stackrel{d}{=} \begin{cases} \dfrac{\sum_{A:I(A)=B} n(A)}{\sum_{A:I(A)\neq\emptyset} n(A)} & \text{if } \exists A \subseteq \Theta : I(A) = B \\ 0 & \text{otherwise} \end{cases}$$

for $B \neq \emptyset$ and $I[n](\emptyset) \stackrel{d}{=} 0$. Starting with the assumption that an information source $(2^\Theta, n, \mathcal{U}(n))$ is given, then by the mapping I a complete information source on Ω is induced.

Theorem 6. Let $(2^\Theta, n, \mathcal{U}(n))$ be an information source. If a point valued mapping $I : \Theta \to \Omega$ together with its disjunctive extension is given, then we have

$$I[\mathcal{U}(n)] = \mathcal{U}(I[n]);$$

for a mapping $I : \Theta \to 2^\Omega$ together with its disjunctive extension, where $\sum_{A:I(A)\neq\emptyset} n(A) > 0$, it holds

$$I[\mathcal{U}(n)] \subseteq \mathcal{U}(I[n]).$$

Proof. As a simple consequence of Theorem 4 for two mass distributions s, t on Ω we derive

$$(\forall C \subseteq \Omega : Q_s(C) > 0 \implies Q_t(C) > 0) \iff (\forall A \subseteq \Omega\ \exists B \supseteq A : s(A) > 0 \implies t(B) > 0).$$

We have to show

$$I[\{n' | n' \sqsubseteq n\}] = \{m' | m' \sqsubseteq I[n]\}.$$

"$\subseteq$" Let $n' \sqsubseteq n$. Consider $M \subseteq \Omega$ where

$$0 < I[n'](M) = \frac{\sum_{N:I(N)=M} n'(N)}{\sum_{N:I(N)\neq\emptyset} n'(N)}.$$

Thus there is $N \subseteq \Omega$ with $I(N) = M$ and $n'(N) > 0$. Since $n' \sqsubseteq n$ we obtain that there is $K \subseteq \Theta$ with $K \supseteq N$ and $n(K) > 0$. This implies $I[n](I(K)) > 0$, where $I(K) \supseteq I(N) = M$. So we have $I[n'] \sqsubseteq I[n]$.

Note that this part of the proof holds for mappings $I : \Theta \to \Omega$ as well as for mappings $I : \Theta \to 2^{\Omega}$.

"$\supseteq$" Let $m' \sqsubseteq I[n]$. Consider $M \subseteq \Omega$ where $m'(M) > 0$. There is $\widetilde{M} \supseteq M$ with $I[n](\widetilde{M}) > 0$. Thus there is $\widetilde{N} \subseteq \Theta$ such that $I(\widetilde{N}) = \widetilde{M}$ and $n(\widetilde{N}) > 0$. Choose $\widetilde{M}, \widetilde{N}$.

Define $N' \stackrel{d}{=} \widetilde{N} \cap I^{-1}(M)$. $I(N') = I(\widetilde{N} \cap I^{-1}(M)) \subseteq I(\widetilde{N}) \cap I(I^{-1}(M)) \subseteq \widetilde{M} \cap M = M$. Let $\omega \in M$, this implies $\omega \in \widetilde{M}$. Thus there is $\theta \in \widetilde{N}$ with $I(\theta) = \omega$. It follows $\theta \in I^{-1}(M)$ and $\omega \in I(N')$. So we obtain $M = I(N')$.

Define $n' : 2^{\Theta} \to [0,1]$; $n'(A) \stackrel{d}{=} m'(A)$ if $\exists M \subseteq \Omega : N' = A$, $n'(A) \stackrel{d}{=} 0$ otherwise. n' is well defined, since $N_1' = N_2' \implies M_1 = I(N_1') = I(N_2') = M_2$. n' is a mass distribution, since $\sum_{A:A\subseteq\Theta} n'(A) = \sum_{M:M\subseteq\Omega} m'(M) = 1$. Since $\widetilde{N} \supseteq N'$ and $n(\widetilde{N}) > 0$ we have $n' \sqsubseteq n$.
We obtain

$$I[n'](M) = \frac{\sum_{A:I(A)=M} n'(A)}{\sum_{A:I(A)\neq\emptyset} n'(A)} = \frac{m'(M)}{\sum_{A:I(A)\neq\emptyset} n'(A)} = m'(M),$$

therefore there is $n' \sqsubseteq n$ such that $I[n'] = m'$.

As the above theorem shows, it is reasonable to use the information source $\left(2^{\Omega}, I[n], \mathcal{U}(I[n])\right)$ for the valuation of events in Ω.

4 - THE COMBINATION OF INFORMATION SOURCES

The essential problem arising with the integration of different pieces of knowledge, i.e. the fusion of sensor data, is the combination of information sources. This cannot be done only in terms of the evidence space but the origin of any information, that is the respective sensor space, has to be taken into account. Moreover (meta) knowledge about the relationship of the different sensor spaces is needed.

Let us start with two sensor spaces Θ', Θ'' and two information sources $(2^{\Theta'}, n', \mathcal{U}(n'))$, $(2^{\Theta''}, n'', \mathcal{U}(n''))$. In addition we have two mappings $I' : \Theta' \to 2^{\Omega}$, $I'' : \Theta'' \to 2^{\Omega}$. The integration of these structures requires the construction of a common sensor space. If we assume the sets Θ' and Θ'' to be disjoint and the mass distributions n' and n'' to be independent (referring to the concept of stochastical independence of random sets), we derive a reasonable joint sensor space by the cartesian product of Θ' and Θ''. We obtain the joint information source $(2^{\Theta' \times \Theta''}, n' \otimes n'', \mathcal{U}(n' \otimes n''))$, where $(n' \otimes n'')(B) \stackrel{d}{=} n'(A') \cdot A''(A'')$, if there are $A' \subseteq \Theta'$, $A'' \subseteq \Theta''$ with $B = A' \times A''$, and

$(n' \otimes n'')(B) \stackrel{\mathrm{d}}{=} 0$ otherwise. Next we have to construct a mapping $I : \Theta' \times \Theta'' \to 2^{\Omega}$. According to some meta expert's decision this mapping may be defined conjunctively or disjunctively, i.e.

$$I(\theta', \theta'') \stackrel{\mathrm{d}}{=} I'(\theta') \cap I''(\theta'')$$

or

$$I(\theta', \theta'') \stackrel{\mathrm{d}}{=} I'(\theta') \cup I''(\theta'').$$

The mappings emerging from a conjunctive or disjunctive combination we denote by $I' \cap I''$ or $I' \cup I''$, respectively. For the induced information source on Ω we obtain in the case of a conjunctive combination

$$(a) \qquad Q_{(I' \cap I'')[n' \otimes n'']}(A) = \frac{1}{c} \cdot Q_{I'[n']}(A) \cdot Q_{I''[n'']}(A),$$

$A \subseteq \Omega$, where

$$c = \sum_{\substack{B, C \subseteq \Omega: \\ B \cap C \neq \emptyset}} I'[n'](B) \cdot I''[n''](C)$$

and, on the other hand, in the case of disjunctive combination

$$(b) \qquad \mathrm{Bel}_{(I' \cup I'')[n' \otimes n'']}(A) = \mathrm{Bel}_{I'[n']}(A) \cdot \mathrm{Bel}_{I''[n'']}(A),$$

$A \subseteq \Omega$. For the verification of these equalities the mass distributions $(I' \cap I'')[n' \otimes n'']$ and $(I' \cup I'')[n' \otimes n'']$ have to be considered. In a straightforward way one obtains

$$(a) \qquad (I' \cap I'')[n' \otimes n''](M) = \frac{\sum_{M', M'': M' \cap M'' = M} I'[n'](M') \cdot I''[n''](M'')}{\sum_{M', M'': M' \cap M'' \neq \emptyset} I'[n'](M') \cdot I''[n''](M'')} = (I'[n'] \oplus I''[n''])(M)$$

where $\oplus$ denotes the well known combination and

$$(b) \qquad (I' \cup I'')[n' \otimes n''](M) = \sum_{M', M'': M' \cup M'' = M} I'[n'](M') \cdot I''[n''](M'').$$

The respective induced information sources on Ω are

$$\left(2^{\Omega}, (I' \cap I'')[n' \otimes n''], \mathcal{U}((I' \cap I'')[n' \otimes n''])\right)$$

and

$$\left(2^{\Omega}, (I' \cup I'')[n' \otimes n''], \mathcal{U}((I' \cup I'')[n' \otimes n''])\right).$$

The above considerations provide a well founded justification of known results on belief functions [6]. The aim of our approach is to clarify the assumptions underlying these results, and to translate them into the context of knowledge representation. It turns out that the classical operations like Dempster's rule of combination are very useful if the assumptions their application requires are satisfied.

5 - References

[1] DUBOIS, D. and PRADE, H., 1986, A set theoretic view of belief functions, logical operations and approximations by fuzzy sets, Int. J. of General Systems 12, 193–226.

[2] DUBOIS, D. and PRADE, H., to appear, Consonant approximations of belief functions, Int. J. of Approximate Reasoning.

[3] GOODMAN, J.R. and NGUYEN, H.C., 1985, Uncertainty models for knowledge based systems, North Holland, Amsterdam.

[4] KRUSE, R. SCHWECKE, E. and HEINSOHN, J., to appear, Uncertainty and vagueness in knowledge based systems: Numerical methods, Springer.

[5] SHAFER, G., 1976, A mathematical theory of evidence, Princeton.

[6] SMETS, P., 1989, personal communications, DRUMS-First workshop at Najac.

[7] YAGER, R.R., 1986, The entailment principle for Dempster/Shafer granules, Int. J. of Intelligent Systems 1, 247–262.

[8] KRUSE, R. and SCHWECKE, E., to appear, Specialization – A new concept for uncertainty handling with belief functions, Int. J. of General Systems.

APPLICATION ASPECTS OF QUALITATIVE CONDITIONAL INDEPENDENCE

Marcus Spies
IBM Germany — Scientific Center
Institute for Knowledge-based Systems
Schloßstr. 70
D - 7000 Stuttgart 1

FAW - University of Ulm
Helmholtzstr. 16
D - 7900 Ulm

Abstract

Axiomatic properties of qualitative conditional independence are compared to those of a Bayesian belief network approach, and judged as to their applicational relevance. It is found that qualitative conditional independence uses weaker axioms and has a clear interpretation in terms of the algebra of non-first normal form relations, and that it can be extended to the recently defined conditional event reasoning.

Keywords: Conditional Independence, Dempster/Shafer theory, Multivariate Models, Database Schemes, Propagation of Evidence

1 Introduction

A key notion for the application of Dempster/Shafer theory (see [15], abbreviated DS theory in the sequel) is that of qualitative independence (Q- independence). It allows to arrange partitions of a given frame of discernment in structures like qualitative Markov trees (see [16,18]). Besides this organizing function w.r.t. knowledge engineering, such structures considerably reduce the computational burden of applying Dempster's combination rule in multivariate models. In this paper, two general kinds of such models will be mentioned.

The first is based on the fact that Q- independence is closely related to multivalued dependencies known from the desgin theory of non-first normal form databases (see [12,3,5,4]). This relationship was mentioned in [16] and, independently, the equivalence of MVDs and conditional Q- independence was proved in [20]. Basically, multivalued dependencies (MVDs) allow to formulate what has been called "heterarchies", i.e. multiple interacting hierarchies. The multivariate models that can be derived from this notion are studied in [8,10,16,18,7].

Now, Q- independence is also related to conditional independence in probability theory. Here, conditional independence became a major practical tool for establishing the structure of probabilistic knowledge bases on which Bayesian inference is performed. This is developed and summarized in [13,9].

Indeed, Q- independence is a generalization of probabilistic independence. The same holds for the conditional dependencies defined qualitatively vs. quantitatively. It is therefore natural to ask whether this generalization has any consequences as for the axiomatic foundations. An axiom system for MVDs has been found in [1], the axiomatic foundations of probabilistic conditional independence are reviewed extensively in [13]. Finally, an axiom system for propagation of probabilities and belief functions is given in [18], based on [17]. However, up to now, the relationship between the axiomatic foundations of the different independence notions has not attracted attention. The present paper is intended to illustrate just how the notions of quantitative independence and qualitative independence are related. I have chosen to do this in a rather informal way because the rather "dry" matter of axiomatic foundations is often not clearly understood as to its applicational relevance. In a companion paper [21], the formal derivation of the results presented here is given.

The second general structure that is based on dependence notions uses conditional events (see [11]). It is shown that conditional events lead to an interesting special case of Q-independences. This allows a very succinct formulation of the problem of combination of evidence in expert systems.

2 Properties of Independence

An important functional aspect of knowledge-based systems is reasoning about properties and their relationships. From elementary predicate calculus, the relationship between properties and partitions is clear. Often various properties are not hierarchical. They represent different views on a set of objects (the universe). If these views are compatible with each other, Q-independence between the partitions corresponding to the properties hold. Let us denote attributes with a finite set of values by capital letters $W, X, ...$, and their elements by small letters $x, y, ...$. The attributes will equivalently be referred to as variables, their values will also be called states. The cartesian product of attributes X and Y is written, like in database theory, as XY with generic elements $xy \in XY$. A relation R_{XY} is, as usual, a subset of XY. Then, Q- independence of two attributes simply means that for all $y_1, y_2 \in Y$,

$$\{x|xy_1 \in R_{XY}\} = \{x|xy_2 \in R_{XY}\},$$

and for all $x_1, x_2 \in X$,

$$\{y|x_1y \in R_{XY}\} = \{y|x_2y \in R_{XY}\},$$

Obviously, this implies that $R_{XY} = XY$ if R is non-empty. In most applications with discrete variables, this kind of independence is unrealistic. It is therefore extended to conditional Q-independence. Conditional Q- independence requires that Q- independence only holds for subsets of variables ***given*** fixed states of of some (usually other) variables.

Example 1: Let X denote a set of producers, Y a set of articles, and Z a set of dealers. $xy \in R_{XY}$ if producer x produces y, $yz \in R_{YZ}$ if dealer z offers y. If some article y produced by x is offered by all dealers that offer y (or, conversely, if a dealer offers each article that he/she has available such that all producers of the article are represented), then we can say that *given* a particular article, its producers and its dealers are Q- independent. This is likely to hold, for instance, for farms (X), some milk products (Y), and supermarkets (Z).

In this case, R_{XYZ} is the relational join (see [3]) of R_{XY} and R_{YZ}. Looking at the example it is obvious that one can prove the following theorem, where sets P_i of variables are considered:

Theorem 1 *Given a finite collection of finite-valued variables P_i, $i \in I$, the conditional Q-independence of any P_B, P_C ($B \subset I$, $C \subset I$), given P_A, $A \subset I$, holds if and only if the multivalued dependency (denoted $\rightarrow\!\!\rightarrow$) $P_A \rightarrow\!\!\rightarrow P_B|P_C$ holds.*

PROOF: [20]

Theorem 1 has a very important consequence. Since a sound and complete set of axioms exists for MVDs (see [1]), they simply can be reformulated to give a sound and complete set of axioms for conditional Q-independence.

Probabilistic conditional independence, on the other hand, is a special kind of conditional Q- indepdence where we require that a distributional product property holds. This notion is excellently discussed w.r.t. knowledge-based systems in [13]. I briefly mention the definition of probabilistic conditional independence in a probability space $(XYZ, \sigma(XYZ), p)$, where $\sigma(XYZ)$ is a σ-algebra over the universe XYZ. X and Z are said to be conditionally independent given Y, if for all $x \in X, y \in Y, z \in Z$ the following holds : If $p(y) > 0$, then

$$p(xz|y) = p(x|y)p(z|y)$$

which is equivalent to

$$p(x|zy) = p(x|y),$$

if we assume $p(zy) > 0$. The first formula establishes the product property, the second establishes the intuitive meaning that "Z tells me nothing about X, once I know Y", as stated by Pearl in [13].

We are now prepared to answer the principal question of this paper: What is the relationship between conditional Q- independence (QCI) and probabilistic conditional independence (PCI)? Is there any significant difference as to their axiomatic foundations? The answer is *yes*, as the following theorem states. It can be established using the axioms of [13] and those stated in [1].

Theorem 2 *Conditional Q-independence inference rules (MVD-axioms) are weaker than inference rules of probabilistic conditional independence rules (PCI-axioms). As a consequence, if a set of MVDs is given, the MVD axioms will allow to infer a set of consequent MVDs, and this set is a subset of the MVDs that will be inferred using the axioms of PCI. Specifically, the decomposition and intersection properties of PCI inferences do not apply to MVDs in general.*

PROOF: [21]

Instead of restating the proof I wish to explain the central points of this theorem. The most tangible difference between QCI and PCI lies in the *decomposition* property. If we describe QCIs by multivalued dependencies, it is possible to have the following MVD on four attributes X, Y, W, Z: $Z \rightarrow\!\!\rightarrow X|YW$. This states that, given a value of attribute Z, sets of corresponding X-values are independent of sets of corresponding pairs of values from attributes Y, W.

Example 2: In a big city, at each station of the underground, ($z \in Z$), there are some office towers ($x \in X$). Specific subsets of the passengers use specific combinations of underground lines ($y \in Y$) and bus lines ($w \in W$) to come to work or to reach the districts of their homes. Note that it is not assumed that the bus lines w and underground lines y stop at the given station. It is only assumed that for each pair (w, y) at least either bus w or underground line y stops there. In this setting we can assume that, given a particular station, people from any office tower sufficiently near to the station will use any of the available combinations of underground and bus lines. Thus, more formally, office towers are independent from combinations of transport lines, given any particular station. This is the pattern $Z \twoheadrightarrow X|YW$. However, from this pattern it cannot be concluded that $Z \twoheadrightarrow X|Y|W$. Given a station, office towers and bus lines and underground lines need not be mutually independent. Some combinations of underground and bus lines do not exist, i.e., there is no way to get to any district using an underground line and a bus line that do not have at least one station in common.

Thus, in our QCI-framework, decomposition of right hand sides of QCIs need not hold. On the other hand, for probabilistic conditional independence, decomposition must always hold. This is, we always have

$$Z \twoheadrightarrow X|YW \Rightarrow_{Prob} Z \twoheadrightarrow X|Y|Z.$$

The reason why decomposition must hold here is simply that any context variable in the consequent part of a conditional probability formula can be summed out:

If we have

$$p(xyw|z) = p(x|z)p(yw|z)$$

it follows that

$$p(xy|z) = \sum_{w \in W} p(xyw|z) = \sum_{w \in W} p(x|z)p(yw|z),$$

which equals

$$p(x|z) \sum_{w \in W} p(yw|z) = p(x|z)p(y|z).$$

It is interesting to note that PCI implies the converse property with weaker premises. In PCIs, we have

$$ZW \twoheadrightarrow X|Y, ZY \twoheadrightarrow X|W \Rightarrow_{Prob} Z \twoheadrightarrow X|WY$$

This is the so-called *intersection* property in Pearl [13]. Again, this is a stronger property than MVD axioms allow to satisfy. In order to keep things readable, I give one more informal example:

Example 3: Assume you enter a huge bar after ten hours of reading papers like this one. Immediately, you will notice the following MVDs. Given a beverage $z \in Z$ and a waiter $w \in W$, there is a set of tables $x \in X$ this waiter serves and a set of barkeepers $y \in Y$ to which the waiter shouts the number of orders of the beverage. Thus, $ZW \twoheadrightarrow X|Y$. On the other hand, you observe that, given a beverage poured by a barkeeper, the tables are independent of the waiter(s). This is true if there is either exactly one waiter per table or if each table has its own disjoint set of waiters. Thus, we have $ZY \twoheadrightarrow X|W$. Assuming either case, it is readily concluded that given a particular beverage tables will usually not be independent of cooperating pairs of waiters and barkeepers. This is so, because tables at which the same beverage is consumed can be served by different waiters exclusively. To clarify this point, let us introduce the following tiny database:

Beverage	*Waiter*	*Table-No.*	*Barkeeper*
beer	Sally	{15, 13 }	{Jack, Henry }
beer	Jim	{12, 14 }	{Joe, Sue }

It is easily verified that this is a relation with the multivalued dependencies stated before. Notice that this relation is not equivalent to the relation depicted in the following table. Under the intersection property, however, this should be the case.

Beverage	*Table-No.*	*Waiter*	*Barkeeper*
beer	{12, 13, 14, 15 }	Sally	Jack
		Sally	Henry
		Jim	Joe
		Jim	Sue

Thus, the intersection property need not hold for conditional Q- independencies. I briefly sketch why intersection must hold in probabilistic conditional independence. The two conditions $ZW \twoheadrightarrow X|Y$ and $ZY \twoheadrightarrow X|W$ amount to having

$$p(x|yzw) = p(x|zw) = p(x|zy)$$

for all $x \in X$, $y \in Y$, $w \in W$, $z \in Z$. If p is strictly positive, this implies $p(x|zwy) = p(x|z)$. Expanding $p(xyw|z)$ we have

$$p(xywz|z) = \frac{p(xzyw)}{p(zyw)} \frac{p(zyw)}{p(z)} = p(x|yzw)p(yw|z) = p(x|z)p(yw|z),$$

which completes the derivation of the intersection property.

The reason for the fact that a quantitative independence notion based on classical probabilities must use less conservative inference rules than a merely qualitative independence structure employs is that quantitative independence can be expressed arithmetically by multiplication, and multiplication is associative. Q-independence, on the other hand, is expressable in terms of MVDs, and MVDs are not arbitrarily decomposable into associative subset-dependencies. This is what makes them intuitively complex and axiomatically economical.

It is important to note that both inequivalences found to exist between QCIs and PCIs vanish if we add one more inference rule to the axiomatic system for QCIs (or MVDs). This inference rule states that each MVD remains valid under projection of its right-hand side attributes. That is, using this rule, we can infer from $X \twoheadrightarrow YZ$ in R_{XYZ} that $X \twoheadrightarrow Y$ in the projection R_{XY}. If we allow this rule to operate on a given set of MVDs, it will find all MVDs that are "embedded" in the given MVDs via projection. Therefore, the inferred MVDs are called embedded multivalued dependencies (EMVDs). They hold in a restricted context of a projected relation.

The interesting point about EMVDs is just what happens if we require them to hold in **any** context, that is, not only in a projection of a given relation. As shown before, the decomposition property in PCI *assumes* any such EMVD (which is a conditional independence of a subset of variables) to hold in any context, i.e., no matter which other variables are taken into account. This is the essence of the stronger inference properties in PCIs as compared to those of "simple" MVDs. As a consequence, we may state the following

Corollary 1 *Probabilistic conditional independence uses EMVD-inference rules not contained in the MVD-axioms.*

PROOF: [21]

It has already been noted in [3] that EMVDs can appear in a relation that do not result from a set of given MVDs together with the MVD axioms [1] and EMVD inference rules. Later, it was proved in [14] that it is *impossible to state a set of sound and complete inference rules for EMVDs.*

Since the axioms for PCI actually use inference rules of EMVDs (as noted in [13], p. 88), they are subject to the incompleteness result from [14]. This seems to allow a negative answer to the "completeness conjecture" stated in [13] for axioms of PCI.

Corollary 2 (Incompleteness Conjecture) *There is no complete formal theory of inference rules for EMVDs, see [14]. Hence, in view of theorem 2 and corollary 1, there is no complete formal theory of probabilistic conditional independence.*

The consequences and detailed proofs of the related facts are in [21]. If the conjecture is true, the main consequence is that reasoning about dependencies is never complete if we use PCIs.

Next, I wish to clarify the consequences of this axiomatic difference between the two approaches for propagation of evidence. As mentioned before, propagation of evidence, given as belief functions or as probabilities, is an emerging computational paradigm for a new generation of expert systems (see [9,13,16,17,7]).

In this approach independence assumptions are summarized in a graphical structure whose most general form is a hypertree or a directed acyclic triangulated graph. The variables of the multivariate model appear as nodes of the triangulated graph. Variables from each triangle form a hyperedge of the corresponding hypertree. The assumption of independence is summarized in the d-separation criterion in [13]. For belief functions, usually a hypertree representation is taken and the independence assumptions correspond to conventional graph separation, see [18]. The hypertree is then called Markov tree, or, as I prefer, qualitative Markov tree (QMT).

If belief functions are to be propagated in QMTs, the underlying independence assumptions are those of QCI, not PCI. Note, however, that the use of Dempster's rule (see [15]) implies locally using PCI for random sets as well. As for the independence assumptions underlying the construction of the QMT, we can conclude from the theorems presented that they are weaker than PCI-assumptions. *In QMTs, one assumes only the Q-independence of the coarsest common refinements of partitions from subtrees that appear after deletion of one node, given the partition at the deleted node. Now, this Q-independence is defined in the context of all relevant variables; "embedded Q-independences" arise only as local consequences of this global context definition.* Therefore, I repeat that there is no use of such a thing as "embedded Q-independence" in evidence theory, and that limitations of inference rules for EMVDs do not apply to Q-independence.

I think the least practical conclusion to be drawn from the developments of this section is that it seems desirable to consider formalisms of evidence propagation that do not rely on a global PCI-assumption. So, one should at least consider to enlarge probabilistic evidence propagation by formalisms that use PCIs only locally or not at all. Moreover, in order to establish complete reasoning about dependencies, one should use the more conservative QCI axioms instead of the PCI axioms. Finally, more work seems desirable on the relationship between normal forms of databases with nested tables (see [12]) and qualitative Markov trees.

3 Conditional Independence and Conditional Objects

One of the apparent problems in using DS theory is the impossibility of specifying conditional information. In DS theory, you can take two informations and condition one on the other, but you cannot specify conditional information directly.

This might be one of the reasons why qualitative Markov trees have been used only quite seldomly. In these trees, two variables can be separated by the joint variable formed by their cartesian product. If one has marginal information on one variable and relational information on the product variable, conditional information concerning the other variable can be deduced. However, this is not what is required in practice. The usual situation is that just conditional and marginal information is available and that relational information remains implicit. Thus, the notion of a stochastic matrix is not easily extendable to a Dempster/Shafer framework.

In the following paragraphs a new approach to conditional evidence is sketched. Here, I give only some very general ideas of the approach. Its relationship to DS theory is extensively treated in [21]. The approach relies on using conditional objects. In order to briefly present them, we switch from the scenario of variables to that of Boolean rings.

The idea behind defining conditional events is to find some purely set-theoretical representation of the relationship "if A, then B", which can serve as a basis for conditional probability. For more information, see [11,2]. Let a, b, x be elements of a Boolean ring $\mathcal{R}$, and let ab denote ring multiplication, which is equivalent to taking $a \sqcap b$ w.r.t. to the ordering of the ring elements. Goodman and Nguyen [11] use the requirement that for any conditional event $[b|a]$ the equality $xa = ab$ must hold, where x is in $[b|a]$. Interpreting the Boolean ring in terms of propositional calculus, this is just the modus ponens. Interpreting the Boolean ring as a σ-algebra, this is the equivalent of the equation $p(b|a)p(a) = p(ab)$. Now, it is clear that many elements of a Boolean ring fulfill $xa = ab$ for fixed a, b. Therefore, one defines conditional objects as a set of ring elements such that the equation defining a conditional event is fulfilled, see [11].

Definition 1

$$[b|a] := \{x|xa = ab\}$$

In the conditional object $[b|a]$, we call a the antecedent, and b the consequent. With a basic result from ring theory, it is proved in [11] that the set of conditional objects for one antecedent partitions the universe of the respective Boolean ring (here I assume a finite Boolean ring. The basic results are much more general).

Now, if we have two partitions of $\mathcal{R}$ corresponding to different antecedents and one partition generated by the conjunction of the antecedents, we have a *formal equivalent of the classical problem of combination of evidence* in expert systems. It is shown in [22] that in this case, Q-independence of the partitions corresponding to the single antecedents given the partition generated by their conjunction holds.

Denoting the partition generated by antecedent x, relative to the ring universe $\mathcal{R}$, as $\mathcal{R}_{/\mathcal{R}x'}$, and conditional Q-independence by the operator $\perp_Q$, we have:

Theorem 3

$$\mathcal{R}_{/\mathcal{R}(xy)'} \perp_Q \mathcal{R}_{/\mathcal{R}x'} \mid \mathcal{R}_{/\mathcal{R}y'}$$

PROOF: [22].

What makes this finding interesting is that Q- independence becomes an analytical property in combination of evidence with conditional objects. That is, you need not verify that it holds for pairs of antecedents. Note this limitation to pairs of antecedents; it shows that the result is not a reformulation of the trivial fact that $XYZ \twoheadrightarrow X|Y|Z$ that holds for any number of variables. As for the valuation of conditional objects that would make this result practically usable, there is a first attempt in [22]. However, I think that some more general valuations than I proposed there are possible; again [21] gives the details.

4 Summary

It has been shown that conditional Q- independence yields a very general dependency structure, which possesses a sound and complete axiomatization. The equivalence to multivalued dependencies seems to offer an elegant way of linking uncertainty management to database design theory. Moreover, probabilistic conditional independence uses inference rules that have been proved in database theory to forbid the formulation of a complete formal theory. It was stressed that QCI-structures can "live" without using such rules. Finally, an interesting reappearance of Q- independence in reasoning with conditional objects was introduced.

Here, Q- independence is, under some circumstances, an analytical property. Some perspectives of integrating the results presented here into the formalism of Dempster/Shafer theory were mentioned. What has still to be studied is the relationship of these results to possibility theory (see [23]). Q- independence together with conditional objects might help in clarifying application conditions of decomposable fuzzy measures.

References

[1] Beeri, C., Fagin, R., Howard, J. (1977): A complete Axiomatization for functional and Multivalued Dependencies in Database Relations. Int. Conf. Mgmt. od Data, ACM, NY, pp. 47-61.

[2] Dubois, D., Prade, H. (1988): Conditioning in Possibility and Evidence Theories - A logical Viewpoint. In: B. Bouchon, L. Saitta, R. Yager (eds.): Uncertainty and Intelligent Systems (Proc. Second IPMU, Urbino 1988), 401-408.

[3] Fagin, R. (1977): Multivalued Dependencies and a new Normal Form for relational Databases. ACM Transactions on Database Systems, 2, pp. 262-278.

[4] Fagin, R. (1983): Degrees of Acyclicity for Hypergraphs and Relational Database Schemes. Journ. ACM, 30 (3), pp. 514 - 550.

[5] Fagin, R. , Mendelzon, A.O., Ullman, J. (1982): A simplified universal Relation Assumption and Its Properties. ACM Transactions on Database Systems, 7 (3), pp. 343 - 360.

[6] Kahneman, D., Slovic, P., Tversky, A. (eds., 1982): Judgment under Uncertainty: Heuristics and Biases. New York, Cambridge University Press.

[7] Kohlas, J., Monney, M. (1990): Propagating Belief Functions through Constraint Systems. FAW, U. Ulm, Tech. Rep. TR-90002.

[8] Kong, A. (1986): Multivariate Belief Functions and graphical Models; Diss., Dept. of Statistics, Harvard University.

[9] Lauritzen, S., Spiegelhalter, D. (1988): Local Computations with Probabilities on Graphical Structures and their Application to Expert Systems. J. R. Statistical Society, 50, 2, pp. 157 - 224.

[10] Mellouli,K. (1987): On the Propagation of beliefs in networks using the Dempster/Shafer theory of evidence; Diss., Sch. of. Business, U. of Kansas.

[11] Goodman, I., Nguyen, H.: Conditional Objects and the Modeling of Uncertainties. In: M. Gupta, T. Yamakawa (eds.): Fuzzy Computing, Noth Holland, Amsterdam, pp. 119-138.

[12] Ozsoyoglu, Z.M., Yuan, L.-Y. (1987): A new normal Form for Nested Relations. ACM Transactions on Database Systems, 12, pp. 111-136.

[13] Pearl, J. (1988): Probabilistic Reasoning in intelligent Systems: Networks of Plausible Inference. Morgan Kaufman, San Mateo, CA.

[14] Sagiv, Y., Walecka, S. (1982): Subset Dependencies and a Completeness Result for a Subclass of Embedded Multivalued Dependencies. J. ACM, 29 (1), pp. 103 117.

[15] Shafer, G. (1976): A mathematical Theory of Evidence. Princeton, Princeton University Press.

[16] Shafer, G., Shenoy, P., Mellouli, K. (1987): Propagating Belief Functions in Qualitative Markov Trees. Int. Journ. of Approximate Reasoning, 1 (4), pp. 349-400.

[17] Shenoy, P. (1989): A valuation-based language for expert systems, Int. Journ. of Approximate Reasoning, 3 (5), pp. 383-411.

[18] Shenoy, P., Shafer, G. (1990): Axioms for Probability and Belief-function Propagation. In: R.D. Shachter, T.S. Levitt, L.N. Kanal, J.F. Lemmer (eds.): Uncertainty in Artificial Intelligence, Vol. 4, pp. 169 - 198.

[19] Smets, P. (1986): Belief Functions and Generalized Bayes Theorem. Preprints of the 2nd IFSA Congress, Gakushuin University, Tokyo, pp. 404-407.

[20] Spies, M. (1988): A Model for the Management of imprecise Queries in relational Databases. in: B. Bouchon, L. Saitta, R. Yager (eds.): Uncertainty and Intelligent Systems. Springer Lecture Notes on Computer Science, vol. 313, Heidelberg; pp. 146-153.

[21] Spies, M. (1991): Evidential Reasoning with Conditional Events *(submitted)*.

[22] Spies, M. (1990): Combination of Evidence with conditional objects and its application to cognitive modeling. In: I. Goodman, H. Nguyen, G., Rogers, M. Gupta (eds.): Conditional Logic in Expert Systems. North Holland (to appear).

[23] Zadeh, L. (1978): Fuzzy Sets as a basis of a theory of possibility, Fuzzy Sets and Systems, 1, pp. 3 - 28.

A STUDY OF PROBABILITIES AND BELIEF FUNCTIONS UNDER CONFLICTING EVIDENCE: COMPARISONS AND NEW METHODS

Mary Deutsch-McLeish[1]
Departments of Computing and Information
Science/Mathematics
University of Guelph
Guelph, Ontario, Canada, N1G 2W1

Abstract

This paper compares the expressions obtained from an analysis of a problem involving conflicting evidence when using Dempster's rule of combination and conditional probabilities. Several results are obtained showing if and when the two methodologies produce the same results. The role played by the normalizing constant is shown to be tied to prior probability of the hypothesis if equality is to occur. This forces further relationships between the conditional probabilities and the prior. Ways of incorporating prior information into the Belief function framework are explored and the results are analyzed. Finally a new method for combining conflicting evidence in a belief function framework is proposed. This method produces results more closely resembling the probabilistic ones.

Keywords:

Belief Functions, Probability Theory, Conflicting Evidence, New Combination Rules.

1. Introduction

There has been considerable recent work concerning the problem of conflicting evidence when using the Dempster combination rule for belief functions [Chatolic, Dubois, Prade, 1987, 1988, 1985, Zadeh, 1986] and about the different models generally [Fagin and Halpern, 1989]. Pearl [1988,1990] has also addressed the problem and about conflicting evidence. The effect of the normalizing constant often distorts the results in a way which produces intuitively incorrect answers. (See Shafer [1979] for the basic definitions about belief functions.)

This paper looks at a very basic problem in the framework proposed by Pearl and examines the conditions under which the two models (belief functions, conditional probabilities) will give the same answer. Two formalisms are considered, one involving two pieces of evidence, one that is weakly for the hypothesis and one strongly for the hypothesis. This is considered to be weakly conflicting evidence. In the case of strongly conflicting evidence, there is strong evidence against and for the hypothesis. In Section 2, Theorems 1, 2 and Corollary 3 provide the conditions for equality of the two approaches under these different types of conflicting evidence. Section 3 discusses the addition of prior information into the probabilistic model and Section 4 provides a new model for handling conflicting evidence for Belief Functions.

[1]This work is supported by NSERC operating grant #A4515.

2. Conflicting Evidence Model

Although other models have been discussed by authors such as Dubois and Prade and Zadeh, this paper is primarily concerned with the model used by [Pearl, 1988] in a discussion really pertaining to non-monotonic reasoning. However the nature of the problem addressed includes the essence of conflicting evidence. Two rules are presented: one strongly confirming a hypothesis and one confirming the negation of the hypothesis. In Pearl's model there is an added connection between the two pieces of evidence. Here we will consider the pieces of evidence to be independent.

As originally presented then, we have two rules r_1: $e_1 \rightarrow \sim h$ with strength m_1 say and r_2: $e_2 \rightarrow h$ with strength m_2. Here $m_1 = 1-\epsilon_1$ and $m_2 = 1-\epsilon_2$, where ϵ_1 and ϵ_2 represent arbitrarily small quantities. In the context of conditional probabilities the rules could be translated to: $P(h|e_1) = \epsilon_1$ and $P(h|e_2) = 1-\epsilon_2$. It can be shown, using the rules of probability that under independence assumptions (both conditional and marginal) $P(h|e_1,e_2) = \frac{P(h|e_1)P(h|e_2)}{P(h)}$ [c.f. Lindley 1985, Dubois and Prade, 1988]. (This formula was also used by Charniak, E., see reference). The first comparison with the belief function model will be made using an interpretation which is not really the correct way to consider the evidence, but follows what has been implied by the conditional probabilities given above. In belief function theory, knowing $m_1 = 1-\epsilon_1$ does not imply a belief in h of ϵ_1. However, this weaker situation of two pieces of evidence, one which strongly confirms the hypothesis and one which very weakly confirms it, can, in real situations, also represent an example of radically differing beliefs in a hypothesis.

2.1 Weakly Conflicting Evidence:

Using a model then where the mass functions are assigned as $m_1\{h\} = \epsilon_1$, $m_1(\theta) = 1-\epsilon_1$ and $m_2\{h\} = 1-\epsilon_2$, $m_2(\theta) = \epsilon_2$, one obtains $m_1 \otimes m_2\{h\} = 1-\epsilon_2+\epsilon_1\epsilon_2$ (as noted in McLeish, 1989, 1990). Here θ consists of the hypothesis h and possibly some other hypotheses. This may be rewritten as $1-\epsilon_2(1-\epsilon_1)$ or $1-P(\sim h|e_1)P(\sim h|e_2)$, which becomes $1-P(\sim h|e_1,e_2)P(\sim h)$. The probability model gives $P(h|e_1,e_2)$ and equating the two expressions one obtains $P(\sim h) = 1$ (provided $P(\sim h|e_1,e_2)$ is not zero).

Theorem 1:

The belief in the hypothesis obtained by combining weakly conflicting evidence using the Dempster combination rule equals the probabilistic result if and only if the prior $P(h) = \frac{\epsilon_1(1-\epsilon_2)}{1+\epsilon_1\epsilon_2-\epsilon_2} \cong \epsilon_1$.

Proof:

This follows from the discussion before the theorem and by noting that

$$\frac{P(h|e_1)P(h|e_2)}{P(h)} = \frac{\epsilon_1(1-\epsilon_2)}{P(h)} = 1-\epsilon_2+\epsilon_1\epsilon_2.$$

Thus, only if the prior of h is very small, for the two expressions be of the same order of magnitude. If the prior is significantly different from zero, the results will also be very different - the probabilistic result will be very small and the mass function value will be close to 1. In the case of an uninformed prior of $\frac{1}{2}$ *the expressions become* $2(\epsilon_1)(1-\epsilon_2) = 1-\epsilon_2+\epsilon_1\epsilon_2$, *which is certainly not possible for the* ϵ_i *assumed small. It also contradicts the requirements that* $P(\sim h)$ *be essentially 1.*

Now $P(h) = P(h|e_1)P(e_1) + P(h|\sim e_1)P(\sim e_1)$. If $P(\sim e_1)$ is very small, then $P(h) \approx P(h|e_1)$, which would ensure the solutions are similar. However there is nothing to assume a requirement like this will be met. Thus, for weakly conflicting evidence, the two approaches will usually provide very different results.

2.2 Strongly Conflicting Evidence:

In this model the evidence for and against the hypothesis are assigned the following values in the belief function model:
$m_1\{\sim h\} = 1-\epsilon_1$, $m_1\{\theta\} = \epsilon_1$, and $m_2\{h\} = 1-\epsilon_2$, $m_2\{\theta\} = \epsilon_2$. As shown in [10], $m_1 \otimes m_2\{h\} = \frac{\epsilon_1(1-\epsilon_2)}{k}$, $m_1 \otimes m_2\{\sim h\} = \frac{\epsilon_2(1-\epsilon_1)}{k}$, $m_1 \otimes m_2\{\theta\} = \frac{\epsilon_1 \epsilon_2}{k}$, where $k = \epsilon_1+\epsilon_2-\epsilon_1\epsilon_2 = 1-m_1 \otimes m_2\{\phi\}$, is the normalizing constant.

Theorem 2

(i) The belief in the hypothesis obtained by combining strongly conflicting evidence using the Dempster combination rule equals that obtained for the probability of the hypothesis, given the evidence, if and only if the prior probability of the hypothesis equals the normalizing constant k (unless $P(h|e_1)$ or $P(h|e_2)$ is identically zero, when the results will both be equal to zero).

(ii) Condition (i) is equivalent to $P(h) = \frac{1}{2-P(h|e_1)} = P(h|e_2)$.

(iii) Some additional results which follow are:

$$P(h) = \frac{P(e_1)}{P(e_1|\sim h)}$$

$$\text{and } O(\sim h) \text{ (odds) } = P(\sim h|e_1).$$

Proof: The first part follows immediately from $m_1 \otimes m_2\{h\} = \frac{\epsilon_1(1-\epsilon_2)}{k}$ and $P(h|e_1,e_2) = \frac{P(h|\epsilon_1)P(h|\epsilon_2)}{P(h)}$.

In order to prove the second and third results, the following observations are necessary: $\epsilon_1+\epsilon_2-\epsilon_1\epsilon_2 = \epsilon_1(1-\epsilon_2)+\epsilon_2 = P(h|e_1)P(h|e_2)+P(\sim h|e_2) = P(h|e_1,e_2)P(h)+P(\sim h|e_2)$. If $P(h)$ equals this quantity, we obtain,

$$P(h)P(\sim h|e_1,e_2) = P(\sim h|e_2)$$

$$\frac{P(h)P(\sim h|e_1)P(\sim h|e_2)}{P(\sim h)} = P(\sim h|e_2),$$

from which $\frac{P(\sim h)}{P(h)} = P(\sim h|e_1)$ or $O(\sim h) = P(\sim h|e_1) = 1-\epsilon_1$.

Continuing with this, one may determine a number of related results using Bayes rule and other basic results of probability theory [6]:

$$P(h) = \frac{P(e_1)}{P(e_1|\sim h)},$$

$$\text{and } P(h) = \frac{1}{2-\epsilon_1}.$$

From this second result as $0 \le \epsilon_1 \le 1$, one determines that $P(h) \ge \frac{1}{2}$. (If ϵ_1 is actually an infinitesimal as originally portrayed, then $P(h)$ will be $O(\frac{1}{2})$.)

Returning to the original expression for k, $\frac{1}{2-\epsilon_1}$ must equal $\epsilon_1+\epsilon_2-\epsilon_1\epsilon_2$. This requires

$$(2-\epsilon_1)\epsilon_1+(2-\epsilon_1)\epsilon_2(1-\epsilon_1) = 1$$

$$\epsilon_2 = \frac{1-(2-\epsilon_1)\epsilon_1}{(1-\epsilon_1)(2-\epsilon_1)} = \frac{1-\epsilon_1}{2-\epsilon_1}$$

Therefore the conditional probabilities are related and it follows immediately that $P(h|e_2) = \frac{1}{2-\epsilon_1} = P(h)$. □

A number of observations follow immediately from this. If ϵ_1 is small (infinitesimal), then the results can only be equal if the prior is approximately $\frac{1}{2}$ and so is $P(h|e_2)$. Thus the amount of conflict cannot be more than a spread of $\frac{1}{2}$ in some sense. In the other interpretation, if the belief opposing the hypothesis is almost 1, then the belief for the hypothesis must be around $\frac{1}{2}$ for the combination rules to give the same result. Other possible scenarios are $m_1\{\sim h\} = \frac{1}{2}$ and $m_2\{h\} = \frac{2}{3}$, provided $P(h) = \frac{2}{3}$.

Graph of Functions

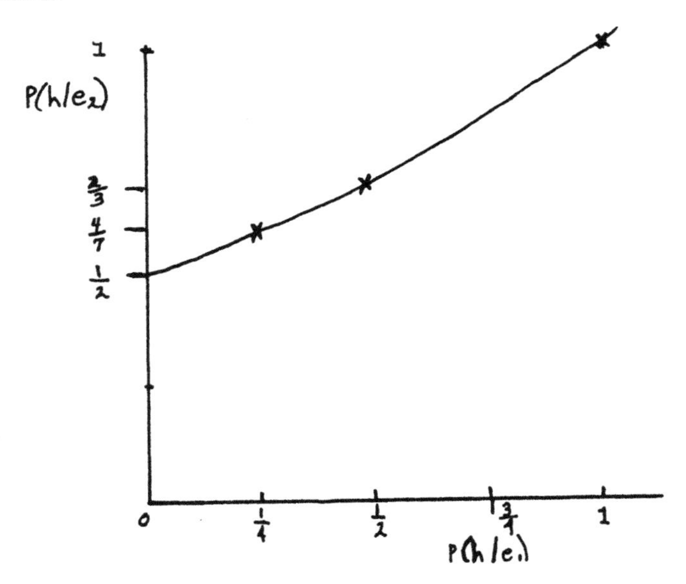

Corollary 2: If the two results are equal and the conditionals are not identically zero, then the combined belief or probability is $P(h|e_1)$.

Proof: This follows from using the result $P(h|e_2) = P(h)$ in the expression for $P(h|e_1,e_2)$. At the extreme, this says that if $\epsilon_1 \approx 0$, (belief against hypothesis is almost 1), and even if the prior and other evidence for the hypothesis is 1/2, then the belief and probability of the combined hypotheses are equal and the value is approximately 0. The results will never be equal for the full extreme: (ie) $\epsilon_1 \approx 0$ and $\epsilon_2 \approx 0$. □

This result is interesting because it is the same as that obtained in [10] for the non-monotonic situation in which you also know that evidence e_2 implies the evidence e_1. Then $P(h|e_1,e_2) = P(h|e_1)$. This corollary states that conditional independence assumption must hold if the probabilistic result is to equal that obtained by combining belief functions under conflicting evidence.

3.1 Addition of Prior Information:

Interesting results can be obtained by including prior information into the belief function model. A number of different ways of doing this in the context of the problems under discussion in this paper are proposed in this section and are compared to the probabilistic results.

There are several alternatives which can be explored. We will first consider constructing belief functions of the form:

(1) $m_3\{h\} = P(h)$
$m_3\{\sim h\} = 1-P(h)$

(2) $m_3\{h\} = P(h)$, $m_3(\theta) = 1-P(h)$

(3) $m_3\{\sim h\} = P(\sim h)$, $m_3(\theta) = P(h)$

These three functions will be combined in turn with $m_1 \otimes m_2$ as given just before Theorem 2. $P(h)$ stands for the prior of the hypothesis.

Theorem 3.

The results of combining prior information about the hypothesis or its negation according to the three belief functions given above result in the following: (all expressions have been normalized)

(1)
$$m_1 \otimes m_2 \otimes m_3\{h\} = \frac{P(h)\epsilon_1}{k = \epsilon_2 P(\sim h) + P(h)\epsilon_1}, \quad m_1 \otimes m_2 \otimes m_3\{\sim h\} = \frac{P(\sim h)\epsilon_2}{k}$$

(2)
$$m_1 \otimes m_2 \otimes m_3\{h\} \approx \frac{\epsilon_1}{\epsilon_1 + \epsilon_2 P(\sim h)}$$

$$m_1 \otimes m_2 \otimes m_3\{\sim h\} \approx \frac{\epsilon_2 P(h)}{\epsilon_1 + \epsilon_2 P(\sim h)}$$

(3)
$$m_1 \otimes m_2 \otimes m_3(h) \approx \frac{P(h)\epsilon_1}{\epsilon_2 P(h) + \epsilon_2}$$

$$m_1 \otimes m_2 \otimes m_3(\sim h) \approx \frac{\epsilon_2}{\epsilon_1 P(h) + \epsilon_2}$$

Proof:

(1) The normalizing constant becomes

$$1 - [\frac{P(h)\epsilon_2(1-\epsilon_1)}{k} + \frac{P(\sim h)\epsilon_1(1-\epsilon_2)}{k}], \quad k = \epsilon_1 + \epsilon_2 - \epsilon_1 \epsilon_2$$

which can be shown to be equivalent to $\epsilon_1 P(h) + \epsilon_2 P(\sim h)$. The expressions for the numerators follow immediately from the application of Dempster's combination rule.

(2) The exact expressions, from Dempster's rule are:

$$m_1 \otimes m_2 \otimes m_3(h) = \frac{\epsilon_1(1-P(\sim h)\epsilon_2)}{\epsilon_1+\epsilon_2 P(\sim h)(1-\epsilon_1)}$$

and $$m_1 \otimes m_2 \otimes m_3(\sim h) = \frac{P(\sim h)\epsilon_2(1-\epsilon_1)}{\epsilon_1+\epsilon_2 P(\sim h)(1-\epsilon_1)}$$

(3) The exact expressions are:

$$m_1 \otimes m_2 \otimes m_3(\sim h) = \frac{\epsilon_2(1-P(h)\epsilon_1)}{k = \epsilon_1 P(h)+\epsilon_2-\epsilon_1\epsilon_2 P(h)}$$

and $$m_1 \otimes m_2 \otimes m_3(h) = \frac{P(h)\epsilon_1(1-\epsilon_2)}{k}.$$

As an example of the calculations for method 2,

(a) $$m_1 \otimes m_2 \otimes m_3(h) = \frac{\epsilon_1(1-\epsilon_2)P(h)}{k = \epsilon_1+\epsilon_2-\epsilon_1\epsilon_2} + P(\sim h)\frac{\epsilon_1(1-\epsilon_2)}{k} + P(h)\frac{\epsilon_1\epsilon_2}{k}$$

But $m(\phi) = \dfrac{P(h)\epsilon_2(1-\epsilon_1)}{k}$. From this the new normalizing constant is

(b) $$\frac{k-P(h)\epsilon_2(1-\epsilon_1)}{k} = \frac{\epsilon_1+\epsilon_2(1-\epsilon_1)P(\sim h)}{k}.$$

From this, the result follows by dividing (a) by (b). □

Comments on these results:

Looking at (2) and (3), the results are similar with the roles of h and $\sim h$ in some sense reversed. The prior belief in $\sim h$ modifies ϵ_2 in method 2. If $P(\sim h)$ is close to 1, the ratios will be similar (same order of magnitude) as $m_1 \otimes m_2$. However if $P(\sim h) \approx 0$ (i.e. $P(h) \approx 1$), then $m' = m_1 \otimes m_2 \otimes m_3$ has $m'\{h\} \approx 1$ and $m'\{\sim h\} \approx 0$, as expected; that is the effect of the prior is very strong. When $P(h) \approx 0$, most of the mass is on θ for m_3 and thus its combination with m_1 and m_2 has little effect.

Exactly the reverse holds for method (3). That is, when $P(h) \approx 0$, $m'\{h\} \approx 0$ and $m'\{\sim h\} \approx 1$, if $P(h) \approx 1$, the results are the same as for $m_1 \otimes m_2$. If $P(h) \approx \frac{1}{2}$, $m'\{h\}$ from (2) becomes $\approx \dfrac{\epsilon_1}{\epsilon_1+\frac{1}{2}\epsilon_2}$ and $m'\{\sim h\} \approx \dfrac{\frac{1}{2}\epsilon_2}{\epsilon_1+\frac{1}{2}\epsilon_2}$. Now results depend on the relative size of the ϵ_i. If $\epsilon_1 = \epsilon_2$, the result is 2/3 and not 1/2 as before. If $\epsilon_1 \to 0$ alone (or ϵ_2), the limiting values remain the same.

In (1), if $P(h) \approx 1$, $m'\{h\} \approx 1$ and $m'\{\sim h\} \approx 0$. If $P(h) \approx 0$, $m'\{h\} \approx 0$ and $m'\{\sim h\} \approx 1$. Thus, the prior can dominate the result in either extreme situation. It is when the prior is 1/2 that the expression reverts to being equivalent to $m_1 \otimes m_2$ alone. This would seem to be the most desirable form - however it is, in a sense, making a strong assumption for Belief function theory that if $m_3\{h\} = P(h)$, then $m_3\{\sim h\} = 1-P(h)$. This is a probabilistic result. If we don't wish to make this strong assumption, the alternatives (2) and (3) favour $P(h)$ or $P(\sim h)$. Which one is used could depend on the size of $P(h)$ - if it is large, (2) could be the preferred formula and if $P(\sim h)$ is large, (3) would be more appropriate. That is, the strong evidence should be put on a specific subset, not θ. This will then have a bigger influence on $m_1 \otimes m_2$ when m' is formed.

Discussion of Equality Conditions after Introduction of the Prior

Conditions for equality of the two methods (Bayesian and Belief function) after introducing prior information, becomes quite tricky. For method (1), one obtains that

$$P(h)^2 + (1-\epsilon_2)(\epsilon_2-\epsilon_1)P(h) - \epsilon_2(1-\epsilon_2) = 0.$$

For small ϵ_i, then $P(h) \approx \sqrt{\epsilon_2}$. That is, the equation reduces to $P(h)^2 + (\epsilon_2-\epsilon_1)P(h) - \epsilon_2 = 0$ which has a positive solution approximately $\sqrt{\epsilon_2}$.

For small ϵ_i, the second method leads to $P(h) \approx \epsilon_1+\epsilon_2$ for equality of the two methods. Thus the prior probability of the hypothesis must be low. However, if the prior is closer to 1, then method (1) produces a result close to 1 also, but the probabilistic value remains small. This is the case for method (2) also. Method (3) would be dependent on the relative size of the ϵ_i's and is not necessarily close to 1 when $P(h) \approx 1$. Thus it could be possible to obtain equality of solutions for other than small priors, depending on ϵ_1 and ϵ_2.

4. A New Combination Rule for Conflicting Evidence

This section introduces a new method of combining conflicting evidence for belief functions. The results are already consistent with a probabilistic interpretation. The method for processing such a system of rules proceeds in two steps. The first step incorporates the rules under one frame of discernment. The second step recombines this resulting evidence in coarser frame. If we suppose that we have two rules of the following form:

$$a_1 \rightarrow b(1-\epsilon_1)$$

$$a_2 \rightarrow \sim b(1-\epsilon_2)$$

Theorem 4.1. Using an interpretation of rules where the frame of discernment Ω becomes the product space $\{(a_1,b),(a_1,\sim b),(a_2,b),(a_2,\sim b)\}$, the following results are obtained after performing Dempster's combination rule:

1) The belief interval for $\{(a_1,b)\} = [1-\epsilon_1, 1]$

2) The belief interval for $\{(a_2,b)\} = [0, \epsilon_1\epsilon_2]$

Proof:

Now the following tables results:

	$\{a_2,b\}^c(1-\epsilon_2)$	$\Omega\ \epsilon_2$
$\Omega\epsilon_1$	$\epsilon_1(1-\epsilon_2)$:$\{(a_1,b)(a_1,\sim b)(a_2,\sim b)\}$	Ω:$\epsilon_1\epsilon_2$
$\{a_1,b\}1-\epsilon_1$	$(1-\epsilon_1)(1-\epsilon_2)$:$\{(a_1,b)\}$	$(1-\epsilon_1)\epsilon_2$:$\{(a_1,b)\}$

From this the belief intervals can be found. This interpretation of rules is similar, but not identical to, the product space interpretation used by Dubois and Prade [1988]. In this model, evidence against the hypothesis is interpreted as they have done, but evidence for the hypothesis is made more specific. Their model, as it stands, does not apply to the conflicting evidence situation because the intersection of the right hand sides of the rules is empty (ϕ), which is not allowed in their model.

If the belief functions are taken from the table of Theorem 4.1 and used as mass functions for the cases b and $\sim b$ separately, one obtains the following theorem:

Theorem 4.2. If the results from the joined (product) space are now used individually to find a truly combined belief in b, the results become:

(1) $Bel\{b\} = \frac{\epsilon_2(1-\epsilon_1)}{(1+\epsilon_2-\epsilon_1\epsilon_2)} = O(\epsilon)$.

(2) $Bel\{b\}^c = \frac{1-\epsilon_1\epsilon_2}{1+\epsilon_2-\epsilon_1\epsilon_2} = 1-O(\epsilon)$,

where ϵ stands for any function of ϵ_1, ϵ_2 which goes to zero as ϵ_1, $\epsilon_2 \rightarrow 0$.

Proof: Now the mass functions to be combined are: $Bel'\{(a_2,b)\}^c = m_1\{b\}^c = 1-\epsilon_1\epsilon_2, m_1(\theta) = \epsilon_1\epsilon_2$, and $Bel'\{(a_1,b)\} = m_2\{b\} = 1-\epsilon_1$, $m_2(\theta) = \epsilon_1$. The resulting $m_1 \otimes m_2 = m'$ becomes:
$m'\{b\}^c = (1-\epsilon_1\epsilon_2)\epsilon_1, m'(\theta) = \epsilon_1^2\epsilon_2$,
$m'\{\phi\} = (1-\epsilon_1\epsilon_2)(1-\epsilon_1), m'\{b\} = (1-\epsilon_1)(\epsilon_1\epsilon_2)$.
After normalizing one obtains the stated results.

Corollary 4.2.

If the prior probability of b is non-infinitesimal, then the above result for $Bel\{b\}$ from Theorem 3 is consistent with the probabilistic result.

Proof. $P(b|a_1 \text{ and } a_2) = \frac{\epsilon_2(1-\epsilon_1)}{P(b)} = O(\epsilon)$ if $P(b) \gg \epsilon_2$. Indeed, if $P(b)$ is taken as the value from the first piece of evidence and we are updating, $P(b|a_1 \text{ and } a_2) = \epsilon_2$. If $P(b)$ is unknown and is assumed to be 1/2, $P(b|a_1 \text{ and } a_2)$ is again $O(\epsilon)$.

Discussion of Theorems 4

The pre-step of combining evidence but carrying the evidence as part of the belief function results in a new function with opposing views contained within it. It can still be difficult to determine the net effect on the outcome b, especially if one piece of evidence is not considered more important than the other. If a combination of the only defined mass functions is then made, results are obtained which are different from simply combining the evidence directly as in J. Pearl. (Recall that these results would give $Bel\{b\} \approx \frac{\epsilon_2}{\epsilon_1+\epsilon_2}$ and $Bel\{b\}^c \approx \frac{\epsilon_1}{\epsilon_1+\epsilon_2}$). They are instead consistent (of same order) with the probabilistic interpretations given in Pearl [1988] and discussed earlier in this paper.

The method actually involves using one frame of reference to translate the uncertain rules to beliefs under a hypothesis of maximizing imprecision and avoiding arbitrariness. This step puts the two rules under one frame of reference and yet maintains their separate identifies within this frame. Only two non-zero beliefs result in this frame and they are opposing as they were to begin with. One subtle change has taken place howeveř, which has a small and yet crucial algebraic effect on the normalizing constant when the coarser frame is used to re-combine (really combine) the evidence.

4.3 Conclusions and Future Work

The paper has provided an in depth study of the similarities and differences between belief function and conditional probability models of combining conflicting evidence. It is shown that there are situations in which the results are indeed the same, even for the ordinary Dempster rule of combination. The effect of introducing prior knowledge into the belief function model in conflicting situations shows how this has a modifying effect on the outcome producing results which 'track' the prior for extreme values (close to 0 or 1). Finally, a method has been presented which combines conflicting evidence in two steps and results in normalized beliefs with orders of magnitude similar to those obtained by probabilistic methods.

Even more information can be discovered from some of the work presented here and only a few observations have been made after each result. Other ways of introducing prior information need to be explored. The normalizing factor could also be distributed perhaps in a different way over the subsets in situations of conflict, to modify its effect.

References

Charniak, E. (1983) "The Bayesian Basis of Common Sense Medical Diagnosis". Proc. of the National Conference on Artificial Intelligence, Wash., D.C., pp. 70-73.

Chatolic, P., Dubois, D., Prade, H. (1987). "An Approach to Approximate Reasoning Based on Dempster Rule of Combination", International Journal of Expert Systems, pp. 67-85.

Dempster, A.P. (1968). "A Generalization of Bayesian Inference", Journal of the Royal Statistical Society, Series B, 30, pp. 205-247.

Dubois, D., Prade, H. (1988). "Representation and Combination of Uncertainty with Belief Functions and Possibility Measurements", Computational Intelligence, Vol. 4, pp. 244-264.

Dubois, D., Prade, H. (1985). "Combination and Propagation of Uncertainty with Belief Functions", IJCAI, 111-113.

Fagin, R., Halpern, J. (1989). "Uncertainty, Belief, and Probability", IJCAI'89 , Proceedings, pp. 1161-1167.

Lindley, D.V. (1985). " Making Decisions", 2nd Edition, John Wiley.

McLeish, M. (March, 1989). "Extreme Probabilities and Non-Monotonic Reasoning in Probabilistic Logic, Dempster-Shafer and Certainty Factor Models", submitted, Journal of Approximate Reasoning.

McLeish, M. (1989). "Further Work on Non-monotonicity in the Framework of Probabilistic Logic with an Extension to Dempster-Shafer Theory", Machine Intelligence and Pattern Recognition, Vol. 8, pp. 23-34.

McLeish, M. (1990). "A Model for Non-Monotonic Reasoning Using Dempster's Rule", Proceedings of the 6th Uncertainty Management Conference, Cambridge, Mass., pp. 518-528.

Pearl, J. (1988). "Probabilistic Reasoning in Intelligent Systems", Morgan Kaufmann.

Pearl, J. (1990). "Reasoning with Belief Functions; an Analysis of Compatibility", Journal of Approximate Reasoning, Vol. 4, pp. 363-389.

Shafer, G. (1979). "A Mathematical Theory of Evidence ", Princeton University Press, Princeton, NJ.

Zadeh, L.A. (1986). "A Simple View of the Dempster Shafer Theory of Evidence and its Implication for the Rule of Combination", A.I. Magazine, 7, pp. 85-90.

PROPAGATING BELIEF FUNCTIONS THROUGH CONSTRAINTS SYSTEMS

Jürg Kohlas and Paul-André Monney
University of Fribourg
Institute for Automation and O.R.
CH-1700 Fribourg (Switzerland)
E-mail kohlas@cfruni51

Abstract

Constraint systems as used in temporal reasoning usually describe uncertainty by constraining variables into given sets. Viewing belief functions as random or uncertain sets, uncertainty in such models is quite naturally and more generally described by belief functions. Here a special class of constraint systems induced by the additive underlying group structure is considered. Belief functions are used to specify uncertain constraints on relations and constraint propagation can be applied to compute belief functions about relations between specified events.

The computations are as usual plagued by combinatorial explosion in the general case. Structural properties of the temporal knowledge base must therefore be exploited. It is explained that there are topological properties of the graph representing the model which can be used to reduce computational complexity. Series-parallel graphs are shown to be particularly simple with respect to computations. They play a role analogous to qualitative Markov trees in multivariate models. These methods are related to the use of reference events - a technique well known in temporal reasoning - to obtain a natural hierarchical structuring of the knowledge base.

Keywords

Belief functions, Dempster-Shafer Theory of Evidence, Temporal Reasoning, Reasoning under Uncertainty, Constraint Propagation.

1. A TEMPORAL MODEL AND RELATED NOTIONS.

Consider a certain number of events i, j, k ... etc. which occur at some time points t_i, t_j, t_k ... on a time axis and let $d_{ij} = t_j - t_i$ represent the duration between events j and i. This imposes the following constraints on the d_{ij} :

$$d_{ik} = d_{ij} + d_{jk} \tag{1}$$

for all triples of events (i, j, k). Of course this implies that $d_{ii} = 0$ and $d_{ji} = - d_{ij}$. Time points are usually represented by real or integer numbers depending on the nature of the events. To fix ideas, suppose that they are all real numbers, $d_{ij} \in \mathbb{R}$.

The values of some variables d_{ij} may not be known exactly. Suppose that it is only known that they are within some subset Θ_{ij} of $\mathbb{R}$: $d_{ij} \in \Theta_{ij}$. These are further constraints to be satisfied by the variables d_{ij}.

The sets Θ_{ij} are often intervals. These data can conveniently be represented by a directed graph G=(V,E) whose vertices are the events i, j, k ... and an arc (i,j) is in E if and only if there exists a constraint $\Theta_{ij} \neq \mathbb{R}$ for variable d_{ij}. This model is at the base of the technique of time maps often used in temporal reasoning [1,2,11]. Note that the results presented in the sequel are not at all limited to this particular temporal model but are much more general. It is shown in [5] that the additive group $(\mathbb{R},+)$ can be replaced by any other group structure. This generalization englobes not only other models of temporal reasoning but also models for geometrical reasoning.

In fact, the problem introduced above corresponds to an ordinary constraint satisfaction problem [7,8] whose solution set S is constituted of all vectors

$$D = (d_{ij})_{(i,j) \in V \times V} \tag{2}$$

such that constraints (1) hold and $d_{ij} \in \Theta_{ij}$ for all $(i,j) \in E$. Such a vector D is called consistent. It may happen that there is no consistent vector and in this case the constraints are said to be contradictory. Moreover, it is not necessarily the overall solution set S which is of interest, but rather its projection to a specified variable d_{rs}. Thereby the question of the relative position of event s with respect to event r is addressed.

2. UNCERTAIN CONSTRAINTS REPRESENTED BY BELIEF FUNCTIONS.

Note that the constraints Θ_{ij} may be uncertain. A body of evidence concerning the value of variable d_{ij} may have many different interpretations $\omega_1, \omega_2 \ldots$, each one leading to a possible constraint $\Theta_{ij}(\omega_1)$, $\Theta_{ij}(\omega_2)$ ... If $p(\omega_h)$ represents the probability that ω_h is the correct interpretation of the body of evidence, then a belief function Bel_{ij} on the possible values of d_{ij} is defined. On the other hand, the so-called structural constraints (1) can be be seen as degenerate belief functions $Bel_{(i,j,k)}$ on $\mathbb{R}^3$.

If stochastic independence between belief functions is assumed, then the Dempster's rule of combination [3,9] can be applied. Unfortunately, all belief functions are not defined on the same frame of discernment. Hence they must first be vacuously extended [9] to the common refinement $\mathbb{R}^n$, where n stands for the number of variables d_{ij}. Let Bel(G) denote the Dempster's rule combination of all these extended belief functions. Here the notation Bel(G) is used to emphasize the dependency of the result on the given model G.

The computation of Bel(G) is a generalized constraint satisfaction problem. In fact, the focal sets of Bel(G) are obtained by solving the constraint satisfaction problems associated to the various samples of focal sets from the belief functions involved in its definition. Once more, it is not necessarily the overall belief function Bel(G) which is of primary interest, but rather its coarsening Bel(G) / (r,s) to some variable d_{rs} [9]. As is to be expected, the determination of this coarsening is plagued by combinatorial explosion so that simplification methods are needed. This is the subject of the following sections.

3. A CONVENIENT MODIFICATION OF THE MODEL.

Shafer, Shenoy, Mellouli [10] considered the problem of computing the coarsening of a belief function resulting from the combination of belief functions defined on different frames of discernment. Unfortunately, the method they propose is not directly applicable to our problem because there are far too many belief functions involved in the definition of Bel(G). More precisely, it is impossible to find a good covering qualitative Markov tree for the hypergraph induced by the model G [6]. This discussion concerning qualitative Markov trees will be continued after two simplification methods are presented in section 4. But before they are exposed, some preparations are needed.

First let us slightly modify the graph G in order to obtain a more uniform structure. Let $\Theta_{ij}(\omega_h)$ be a focal set of Bel_{ij}. Then the opposite set

$$\Theta_{ij}(\omega_h)^{-1} = \{-d_{ij} : d_{ij} \in \Theta_{ij}(\omega_h)\} \tag{3}$$

defines a focal set of a new belief function denoted by Bel_{ij}^{-1}. By definition, the basic probability assignments of Bel_{ij} and Bel_{ij}^{-1} are identical. Let G' = (V,E') be the undirected version of G and for each edge {i,j} in E' define two new belief functions as follows:

$$Bel'_{ij} = Bel_{ij} \oplus Bel_{ji}^{-1}$$
$$Bel'_{ji} = Bel_{ji} \oplus Bel_{ij}^{-1}. \tag{4}$$

If (i,j) is in E but (j,i) is not in E, then define Bel_{ji} as the vacuous belief function having the unique focal set $\mathbb{R}$ [9]. Of course the belief functions in (4) are no longer stochastically independent since they are mutually inverse. This implies that if a focal set is sampled from one belief function then its opposite set must be sampled from the other one. Now, if Bel'(G') denotes the combination of all these inverse belief functions together with all structural belief functions, then the result remains unchanged with respect to the original model.

LEMMA 1: For any model G, Bel(G) = Bel'(G').

The proof of this lemma as well as the proofs of all further theorems stated in this paper can be found in [5]. According to lemma 1, in order to avoid unnecessary complicated notations, it is convenient to consider the undirected graph G' as representing the original model G: to each edge in G is attached a pair of inverse dependent belief functions.

4. MODEL REDUCTIONS.

Suppose there is a vertex $j \in V$, different from r and s, such that in G there are exactly two edges {i,j} and {j,k} incident to j with $i \neq k$. Then the graph G may be replaced by a slightly simpler one, where these two edges are replaced by a new edge {i,k} and node j is removed from V. The new arc (i,k) is labelled by the new belief function

$$Bel_{ik} = Bel_{ij} \oplus Bel_{jk} \oplus Bel_{(i,j,k)}. \tag{5}$$

Note that this combination is contradiction free since focal sets of Bel_{ik} are of the form $\Theta_{ij} + \Theta_{jk}$ if Θ_{ij} and Θ_{jk} are focal sets of Bel_{ij} and Bel_{jk} respectively and if the operation "+" is extended to sets of real numbers. For this reason the belief function in (5) can be written $Bel_{ij} + Bel_{jk}$. The inverse new arc (k,i) is of course labelled by the inverse belief function of Bel_{ik}, which can also be obtained by applying (5) to Bel_{ki}. This operation is called a *series reduction* and represents a first step in the direction of computational simplification. This is the assertion of the following theorem:

THEOREM 1: If G' is the graph and model obtained from G after a series reduction with respect to a vertex $j \neq r,s$, then

$$Bel(G) / (r,s) = Bel(G') / (r,s). \tag{6}$$

A single series reduction alone is surely not yet a big deal, but combined with another type of reduction it may become significant.

A series reduction introduces a new edge which may parallel already existing ones. Parallel edges linking vertices i and k can be replaced by a unique edge {i,k} whose belief function is obtained by combining those on the parallel edges by Dempster's rule. This operation is called *parallel reduction* and represents a second way to reduce and simplify a model.

THEOREM 2: If G' is the graph and model obtained from G after a parallel reduction, then

$$Bel(G) / (r,s) = Bel(G') / (r,s). \tag{7}$$

Note that both series and parallel reductions do not change the nature of the graph representation since a pair of inverse belief functions is again assigned to edges in the reduced model G'. These operations represent *local* simplification procedures which may considerably reduce the computational effort when they are performed repeatedly until no more reduction is possible. This may possibly lead to the trivial graph formed by a single edge between vertices r and s. In this case the final label is precisely Bel(G) / (r,s). Graphs of this sort are said to have a *series-parallel structure* with respect to vertices r and s. In figure

1 it is shown that a graph G may have a series-parallel structure with respect to a pair of vertices r and s (fig. 1 a) whereas the same graph does not possess this structure with respect to another pair of vertices (fig. 1b).

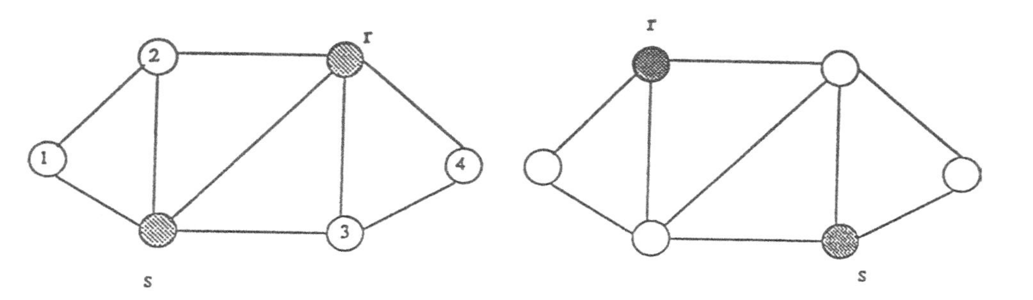

Figure 1

A variable d_{ij} is called *s-p-reducible* if the graph G=(V,E) can be reduced to the only nodes i and j with an edge linking them by successive applications of series and parallel reductions. A graph G is called *series-parallel* if there exists a s-p-reducible variable d_{rs} in G ({r,s} must not necessarily belong to E).

Successive series and parallel reductions for a s-p-reducible variable d_{rs} can be visualized in a bipartite tree $T = (V_1 + V_2, U)$. This tree is constructed as follows:

Each time a series reduction of two edges {i,j}, {j,k} is performed:

- add new vertices to V_1 corresponding to variables d_{ij}, d_{jk} and d_{ik} if they are not already present.
- add a new vertex in V_2 corresponding to the relation $d_{ik} = d_{ij} + d_{jk}$.
- add three edges in U linking the vertices for d_{ij}, d_{jk} and d_{ik} to the vertex for the relation $d_{ik} = d_{ij} + d_{jk}$.

Figure 2 shows this tree for the series-parallel graph of figure 1 a).

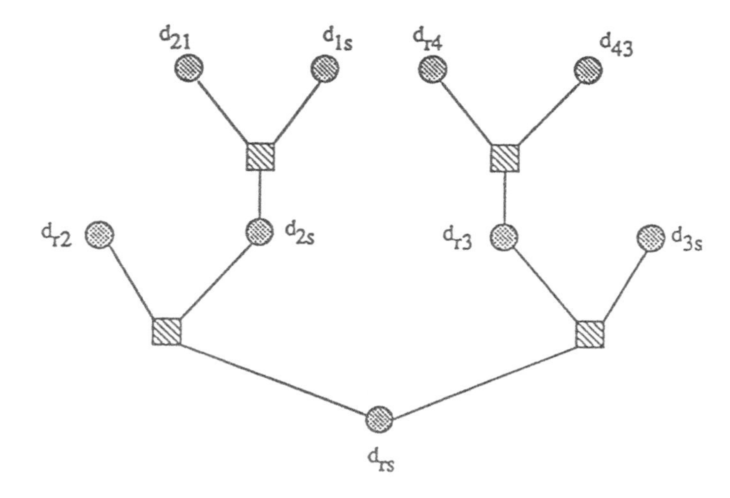

Figure 2

Such a tree represents a *qualitative Markov tree* (this will not be proved here) and the series and parallel reductions correspond exactly to the propagation of belief functions in a qualitative Markov tree to the vertex corresponding to variable d_{rs}. For a discussion of qualitative Markov trees and and the propagation algorithm, see [6] and [10]. Thus series-parallel structures correspond to qualitative Markov trees. If the variable d_{rs} is s-p-reducible, then denote by $T(d_{rs})$ this associated qualitative Markov tree. The following theorem gives properties of series-parallel graphs and their s-p-reducible variables.

THEOREM 3: Let $G=(V,E)$ be a series-parallel graph. Then

(1) G is connected
(2) $T(d_{ij}) = T(d_{kl}) = T$ for all s-p-reducible variables d_{ij} and d_{kl}
(3) A variable d_{ij} is s-p-reducible if and only if it appears as a vertex of V_1 in T.
(4) All variables d_{ij} for which $\{i,j\} \in E$ are s-p-reducible.

Note that if may well happen that a variable d_{ij} is s-p-reducible even though the edge $\{i,j\}$ does not belong to E.

5. MODEL DECOMPOSITIONS.

If a graph G allows no further series or parallel reductions with respect to a fixed pair of vertices r and s, then it is called *irreducible*. Even in this case there may still exist the opportunity to simplify computations. The idea is to *decompose* the model into smaller and hence simpler parts. Just as with reductions, there are two different ways how such a decomposition of a model can arise.

A vertex $k \in V$ is called a *cut-vertex* of G if its elimination from G (together with all the edges incident to it) disconnects the graph G into two or more connected components. Let V_j be the set of vertices in the j-th component and E_j the set of edges in E linking vertices in $V_j + \{k\}$. Then the subgraphs $G_j = (V_j + \{k\}, E_j)$ form what is called a *series partition* of G.

THEOREM 4: Let k be a cut-vertex of the graph G and suppose that vertices r and s belong to the same component G_i of the partition. Then, provided Bel(G) exists

$$Bel(G) / (r,s) = Bel(G_i) / (r,s) \tag{8}$$

Moreover, if Bel(G) does not exist (because of total contradiction between belief functions), then it can be proved that there exists at least one component G_j for which $Bel(G_j)$ does also not exist [5]. According to this theorem, only belief functions pertaining to the component containing vertices r and s are relevant for the computation of Bel(G) / (r,s). All other belief functions can simply be discarded. The next theorem examines the case where r and s belong to different components.

THEOREM 5: Let k be a cut-vertex of the graph G and suppose that the vertex r belongs to the component G_i and s to G_j, $j \neq i$. Then, provided Bel(G) exists

$$Bel(G) / (r,s) = Bel(G_i) / (r,k) + Bel(G_j) / (k,s). \tag{9}$$

The definition of the addition of two belief functions is given in section 4. Only belief functions pertaining to the components containing vertices r and s are relevant for the computation of Bel(G) / (r,s).

Another kind of decomposition arises if the vertices of $V - \{r,s\}$ can be partitioned into two subsets V_1 and V_2 such that there is no edge in E linking a vertex in V_1 to a vertex in V_2. Let E_i denote the edges in E linking vertices within $V_i + \{r,s\}$. The subgraphs $G_i = (V_i + \{r,s\}, E_i)$ are called a *parallel partition* of G. If the edge $\{r,s\}$ itself happens to belong to E, then it will arbitrarily be affected to E_1 and not to E_2.

THEOREM 6: If the graph G allows for a parallel partition G_1, G_2 with respect to the vertices r and s, then

$$Bel(G) / (r,s) = Bel(G_1) / (r,s) \oplus Bel(G_2) / (r,s). \quad (10)$$

Thus parallel components can first be computed independently and then combined by the Dempster's rule.

6. HIERARCHICAL KNOWLEDGE ORGANIZATION.

In temporal reasoning there are often a number of important events (milestones) to be considered and the other events are positioned with respect to these *reference events*. This introduces a natural hierarchical organization in the temporal knowledge base. Reference elements could represent cut-vertices in the graph representing the model and the decomposition results can be applied.

Let R be a set of reference events and G_0 a graph representing relations between them which are defined by belief functions. At any vertex $u \in R$ another graph $G_u = (V_u + \{u\}, E_u)$ is supposed to be attached. The events corresponding to vertices in G_u are linked among themselves and they refer to exactly one reference event. Figure 3 shows such an organization schematically.

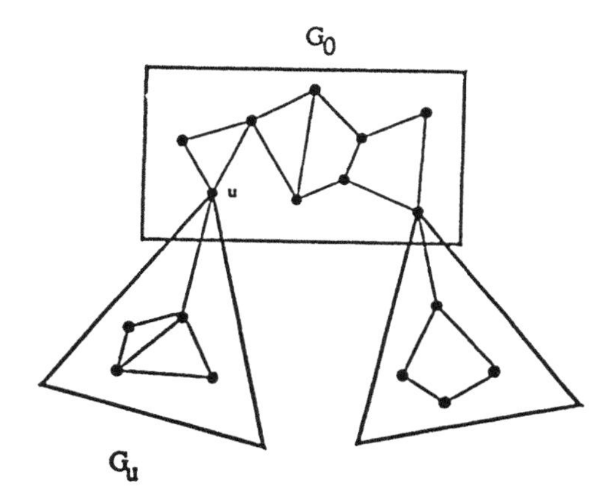

Figure 3

In order to compute Bel(G) / (r,s) in such a structure, different cases must be distinguished:

(1) If r and s are within the same subgraph G_0 or G_u, $u \in R$, then a repeated application of theorem 4 implies that Bel(G) / (r,s) can be computed within the corresponding subgraph:

$$Bel(G) / (r,s) = Bel(G_u) / (r,s) \quad (11)$$

(2) If r belongs to a subgraph G_u and s is a reference event different from u, then theorems 4 and 5 imply that

$$Bel(G)/(r,s) = Bel(G_u)/(r,u) + Bel(G_0)/(u,s) \quad (12)$$

(3) If r and s belong to two different subgraphs G_u and G_v and $\{r,s\} \neq \{u,v\}$, then theorems 4 and 5 imply that

$$Bel(G)/(r,s) = Bel(G_u)/(r,u) + Bel(G_0)/(u,v) + Bel(G_v)/(v,s). \quad (13)$$

In any of these three cases, the hierarchical organization limits the computation on smaller parts of the entire model. It is possible that G_u represents itself a hierarchical organization and the hierarchy may thus extend over several levels. In an extreme case the overall graph is a tree and two vertices are linked by exactly one path. Then Bel(G) / (r,s) is simply obtained by performing series-reductions along the path between r and s. Such a model represents a *pure* series system.

7. IRREDUCIBLE AND UNDECOMPOSABLE MODELS.

The natural question arises how to treat graphs which are both irreducible and undecomposable relative to the methods presented in the previous sections. The exact determination of Bel(G) requires to consider the family of all constraint satisfaction problems represented by all samples of focal sets from belief functions in the model G (see sections 1 and 2). Then these problems are solved using traditional constraints satisfaction algorithms. If it is only a certain coarsening of the overall belief function Bel(G) which is of interest, then local constraint propagation algorithms can be applied. These algorithms are simple, but in general they are not guaranteed to yield the exact solution [1,5]. The following example precisely illustrates this fact.

Suppose that the non-structural constraints are closed intervals on $\mathbb{R}$ of the form $\Theta_{ij} = [l_{ij}, u_{ij}]$ such that either l_{ij} or l_{ji} is positive for all edges $\{i,j\}$ in E. Define the oriented graph G' = (V, E') where an arc (i,j) belongs to E' if and only if the edge $\{i,j\}$ is in E and l_{ij} is positive. Let l denote the length of the longest path between r and s in the graph G' and with arc costs l_{ij}, and let u denote the length of the shortest path between r and s in G' with arc costs u_{ij}. Then it can be shown that in general the interval [l,u] obtained by local constraint propagation strictly contains the projection of the exact solution set S of the associated constraint satisfaction problem to the variable d_{rs} [5].

8. CONCLUSION.

It has been shown here that theory of evidence [6,9] provides a natural framework and methodology for dealing with uncertain, constraint oriented models. Local, efficient and exact computational procedures have been presented. Moreover, similar results as those presented here hold for a larger class of models based on an underlying group structure and also for an interesting class of scheduling models [4].

REFERENCES

[1] Davis E. (1987): Constraint Propagation With Interval Labels. *Artificial Intelligence*, **32**, 281-331.

[2] Dean T. (1985): Temporal Imaginery: An Approach to Reasoning about Time for Planning and Problem Solving. Research Report 433, Yale University, New Haven, CT.

[3] Dempster A.P. (1967): Upper and Lower Probabilities Induced by a Multivalued Mapping. *Annals of Math. Stat.*, **38**, 325-339.

[4] Kohlas J. (1988): Models and Algorithms for Temporal Reasoning III: Combination and Propagation of Belief and Plausibility. Working Paper 158, Institute for Automation and O.R., University of Fribourg, Switzerland.

[5] Kohlas J., Monney P.A. (1989): Propagating Belief Functions Through Constraint Systems. Working Paper 171, Institute for Automation and O.R., University of Fribourg, Switzerland. To appear in *Int. J. of Approximate Reasoning*.

[6] Kohlas J., Monney P.A. (1989): Modeling and Reasoning with Hints. Working Paper 174, Institute for Automation and O.R., University of Fribourg, Switzerland.

[7] Mackworth A.K. (1977): Consistency in Networks of Relations. *Artificial Intelligence*, **8**, 99-118.

[8] Montanari U., Rossi F. (1988): Fundamental Properties of Networks of Constraints: A New Formulation. In: KANAL L., KUMAR V. (Eds.) Search in Artificial Intelligence, Springer, 426 -449.

[9] Shafer G. (1976): A Mathematical Theory of Evidence. Princeton Univ. Press.

[10] Shafer G., Shenoy P., Mellouli K. (1986): Propagating Belief Functions in Qualitative Markov Trees. *Int. J. of Approximate Reasoning*, **1**, 349-400.

[11] Vere S. (1983): Planning in Time: Windows and Durations for Activities and Goals. *IEEE Trans. Patt. Anal. Mach. Intell.*, **5**, 246-267.

UPDATING UNCERTAIN INFORMATION[1]

Serafín MORAL, Luis M. DE CAMPOS
Departamento de Ciencias de la Computacion e I.A.
Facultad de Ciencias, Universidad de Granada
18071 - Granada - SPAIN

Abstract

In this paper, it is considered the concept of conditioning for a family of possible probability distributions. First, the most used definitions are reviewed, in particular, Dempster conditioning, and upper-lower probabilities conditioning. It is shown that the former has a tendency to be too informative, and the last, by the contrary, too uninformative. Another definitions are also considered, as weak and strong conditioning. After, a new concept of conditional information is introduced. It is based on lower-upper probabilities definition, but introduces an estimation of the true probability distribution, by a method analogous to statistical maximum likelihood.

Finally, it is deduced a Bayes formula in which there is no 'a priori' information. This formula is used to combine informations from different sources and its behavior is compared with Dempster formula of combining informations. It is shown that our approach is compatible with operations with fuzzy sets.

Keywords

Theory of Evidence, conditioning, combining informations, Bayes rule.

1. INTRODUCTION

First of all, we are going to consider different kinds of informations that we may have about a variable X taking values on a finite set U. If the value of X is not determined, a probabilistic information may be available, that is a probability measure

$$P:\mathfrak{P}(U) \longrightarrow [0,1],$$

---

[1]This research was supported by the European Economic Community under Project DRUMS (ESPRIT B.R.A. 3085)

or its corresponding probability distribution p.

Our point of view about this kind of informations is objective. That is, P(A) is interpreted as the limit of relative frequencies, considering an infinite sequence of independent values of variable X in the same conditions.

If we do not have an information as precise as a probability, we may have a set of possible probability distributions, $\mathcal{C} = \{P_1,..,P_n\}$. Usually, this set will be a convex set of probabilities $\mathcal{H} = \{\sum_{i=1}^{n} \alpha_i P_i \ / \sum_{i=1}^{n} \alpha_i = 1\}$, where $P_1,...,P_n$ are given probability measures.

With a set of probability measures, $\mathcal{C} = \{P_1,..,P_n\}$, we may associate a convex set of probabilities, precisely its convex hull, $\overline{\mathcal{C}} = \{\sum_{i=1}^{n} \alpha_i P_i \ / \sum_{i=1}^{n} \alpha_i = 1\}$. $\mathcal{C}$ and $\overline{\mathcal{C}}$ can be considered for the same experiment but with different interpretations. For example, assume that we have two urns, U_1 and U_2. U_1 has 99 red balls and one black ball. U_2 has one red ball and 99 black ones. If we pick up randomly a ball from one of the two urns, then for the ball color we have two possible probabilities, $\mathcal{C} = \{P_1,P_2\}$, one for each urn. However, the experiment may be also considered as the selection of one urn and then picking up a ball. In this case, the frequencies of colors will be given by a probability $\alpha P_1+(1-\alpha)P_2$. As α is unknown, we have the convex set $\overline{\mathcal{C}}$ of possible probabilities. We will consider always the second interpretation. However, in the cases where two o more experiments are done with the same unknown probability it will be necessary to distinguish the two situations by introducing the set of possible probabilities as a parameter.

Another important representation of uncertainty are probability envelopes (or lower and upper probabilities). A probability envelope is a pair of ordered fuzzy measures, (l,u), (see [5]), such that there exists a family, $\mathcal{P}$, of probability measures verifying

$$l(A) = \operatorname*{Inf}_{P\in\mathcal{P}} \{P(A)\}$$

$$u(A) = \operatorname*{Sup}_{P\in\mathcal{P}} \{P(A)\}.$$

It is clear that, given a set of probabilities, $\mathcal{C}$, we may associate a probability envelope with it. However, a probability envelope, (l,u), may be defined from different sets of probabilities. But in every case, there is always a maximal family, given by

$$\mathcal{P} = \{ P \ / \ l(A)\leq P(A)\leq u(A),\ \forall\ A\subseteq U \}.$$

If $\mathcal{C}$ is a set of probabilities and we calculate the associated envelope, (l,u), this envelope is equivalent to a maximal family $\mathcal{P}$. In this situation, $\mathcal{C} \subseteq \mathcal{P}$ is always verified. In this sense, we can say that if we transform a set $\mathcal{C}$ on an envelope, then some information is lost (there are more probability measures being possible).

Finally, a pair of belief-plausibility measures (Bel,Pl) will be a probability envelope such that there exists a function

$$m:\mathcal{P}(U) \longrightarrow [0,1]$$

called basic probability assignment and such that

$$Bel(A) = \sum_{B \subseteq A} m(B)$$

$$Pl(A) = \sum_{B \cap A \neq \emptyset} m(B).$$

2. CONDITIONING

The basic operator to update information in Probability Theory is conditioning. In the Theory of Evidence, there is a generalization of this concept, given by Dempster, [2]. According to it the conditional plausibility and belief are,

$$Pl_D(A / B) = \frac{Pl(A \cap B)}{Pl(B)}$$

$$Bel_D(A / B) = \frac{Pl(B) - Pl(\bar{A} \cap B)}{Pl(B)}$$

This formula has been widely used but, in the last years, it has been also severely criticized (see [1],[6]). The main problem is that if we consider the intervals [Bel(A),Pl(A)] as the sets of possible values for the probability of A, then it has a tendency to reduce too much the length of such intervals. It is too informative. We could say that it gets improper information. This formula has a better interpretation under the idea of combining. We differentiate between these two concepts, considering combination as the integration of two informations, given for all possible cases; and, by the other hand, conditioning is considered as a restriction of an information to the cases in which another is verified.

Precisely, in terms of basic probability assignments it can be expressed as

$$m_D(A/B) = \frac{\sum_{E \cap B = A} m(E)}{1 - \sum_{E \cap B = \emptyset} m(E)}$$

That is, the mass of set E, is transferred to the intersection of E with B. It is not considered the possibility of the masses being transferred to $\bar{B}$. From this point of view it may be interpreted better as a rearrangement of information than as a restriction of possibilities. Only the mass assigned to sets with empty intersection with B is eliminated.

Another formula was also proposed by Dempster in [2]. It was defined in terms of families of probability distributions. If $\mathcal{C}$ is a family of probability distributions, then the conditional information given B, is defined as the family of all possible conditional probabilities. That is,

$$\mathcal{C}_B = \{ P(./B) \ / \ P \in \mathcal{C} \}.$$

A pair of belief-plausibility measures is a probability envelope and may be conditioned, by conditioning every probability in its associated family of probabilities, in the following way:

$$Bel_U(A/B) = \text{Inf } \{P(A/B) \ / \ Bel(C) \leq P(C) \leq Pl(C), \forall C \subseteq U\}$$

$$Pl_U(A/B) = \text{Sup}\{P(A/B)\ /\ Bel(C) \leq P(C) \leq Pl(C), \forall C \subseteq U\}.$$

This formula is equivalent to

$$Bel_U(A/B) = \frac{Bel(A \cap B)}{Bel(A \cap B) + Pl(B - A)}$$

$$Pl_U(A/B) = \frac{Pl(A \cap B)}{Pl(A \cap B) + Bel(B - A)}$$

This conditioning will be also called upper and lower probabilities conditioning.

Fagin, Halpern (1989), [4] have shown that the result of conditioning a belief, plausibility measure is a new belief, plausibility measure. The formula (see [7]) has the problem of producing too uninformative (very wide) intervals. For each focal element E, it considers all the possibilities for the transference of its mass: inside or outside of B. These intervals always include the result of Dempster conditioning, in one of interval extremes.

Another formula that has been proposed is strong conditioning (see [9],[3]). It can be explicitly expressed as

$$Bel_S(A/B) = \frac{Bel(A \cap B)}{Bel(B)}$$

$$Pl_S(A/B) = \frac{Bel(B) - Bel(B-A)}{Bel(B)}$$

In terms of focal elements, it may be expressed as

$$m_S(A/B) = \begin{cases} m(A)/Bel(B) & \text{if } A \subseteq B \\ 0 & \text{otherwise} \end{cases}$$

The resulting intervals are also included in upper and lower intervals, precisely in the opposite extreme of Dempster conditioning. In this case, we do a restriction, but to the cases in which B is necessary. Only the focal elements that are included in B, are considered on the conditioning: observe as the normalization is done to Bel(B), the proportion of cases in which we are sure that B is verified. Another point is that it can applied in less cases that former ones. It is necessary that $Bel(B) > 0$. In the others, the condition was only $Pl(B) > 0$.

Another definition is Planchet weak conditioning (see [8]). It may be expressed as

$$Bel_P(A/B) = \frac{Bel(A) - Bel(A-B)}{Pl(B)}$$

$$Pl_P(A/B) = \frac{Pl(A) + Pl(B) - Pl(A \cup B)}{Pl(B)}$$

The corresponding basic probability assignment may be calculated as,

$$m_P(A/B) = \begin{cases} m(A)/Pl(B) & \text{if } A \cap B \neq \emptyset \\ 0 & \text{otherwise} \end{cases}$$

Intuitively, it is a restriction of the possible cases, eliminating the focal elements with empty intersection with B. The difference with Dempster conditioning is that it does not reassign the mass of A to $A \cap B$, in case this intersection is not empty. It remains in the whole set A. This produces some strange results, as for example

$$Bel_P(B/B) = Bel(B)/Pl(B) \leq 1,$$

that is the necessity of B given B may be less than one.

In definitive, we find that the problem of conditioning is not totally solved. There is not a formula with a good behavior in every case.

3. ESTIMATED CONDITIONALS

The purpose of this paper is to determine an intermediate formula of conditioning families of probability distributions. We assume the hypothesis that if we have a family of probabilities $\mathcal{P}$, then this is equivalent to having its convex hull $\mathcal{P}'$. Therefore, we shall give a definition valid for convex sets of probabilities. More particularly, for convex sets with a finite number $\{p_1,..,p_n\}$ of extreme probability distributions.

In Dempster second formula, to condition $\mathcal{P}'$ is enough to calculate the conditional of the extreme probabilities $\{p_1,..,p_n\}$ and then to do the convex hull of $\{p_1(./B),..,p_n(./B)\}$.

The proposed formula, also calculates all the conditional probabilities in a first step, but it considers that knowing that the result is in a set B, then non all the probability distributions have the same possibility. For example, if $P_i(B)$ is very small, it is not easy that given B, p_i be the true probability distribution. Then, each conditional probability, $P_i(./B)$, has an associated possibility equal to $P_i(B)$ normalized by $c(B) = \operatorname{Max}_j \{P_j(B)\}$. That is,

$$\pi(P_i(./B)) = P_i(B)/c(B).$$

We can associate with the possibility distribution π, a pair of necessity-possibility measures (N,Π). Then, the upper and lower conditional informations are calculated doing the Choquet integral of $P_i(A/B)$ with respect to Π and N, respectively:

$$u^*(A/B) = I(\ P_i(A/B)\ /\Pi) = \int_0^1 \Pi(\{\ P_i(./B)\ /\ P_i(A/B) \geq \alpha\ \}).d\alpha$$

$$l^*(A/B) = I(\ P_i(A/B)\ /N) = \int_0^1 N(\{\ P_i(./B)\ /\ P_i(A/B) \geq \alpha\ \}).d\alpha$$

In the case of belief-plausibility pairs, (Bel,Pl), this formula is intermediate between the two former ones:

$$Pl_D(A/B) \leq Pl^*(A/B) \leq Pl_U(A/B)$$

$$Bel_D(A/B) \geq Bel^*(A/B) \geq Bel_U(A/B)$$

In fact, Dempster first formula is equivalent to consider only the probabilities with possibility equal to one. That is to transform π in π_D given by

$$\pi_D(P_i(./B)) = \begin{cases} 1 & \text{if } \pi(P_i(./B))=1 \\ 0 & \text{otherwise} \end{cases}$$

And Dempster second formula is equivalent to consider that the possibility of $p_i(./B)$ is equal to 1 if $p_i(B) > 0$. That is, to transform π in π_U given by

$$\pi_U(P_i(./B)) = \begin{cases} 1 & \text{if } \pi(P_i(./B))>0 \\ 0 & \text{otherwise} \end{cases}$$

Let us see an example. We have the two urns with red and black balls as above. Assume that we pick up two balls with replacement from the same urn. If the events are represented in the following way:

The first ball is black B1,
The first ball is red R1,
The second ball is black B2,
The second ball is red R2,

then we have the following probabilities

$p_1(R1 \cap R2) = 0.9801$ $\qquad$ $p_1(R1 \cap B2) = 0.0099$

$p_1(B1 \cap R2) = 0.0099$ $\qquad$ $p_1(B1 \cap B2) = 0.0001$

$p_2(R1 \cap R2) = 0.0001$ $\qquad$ $p_2(R1 \cap B2) = 0.0099$

$P_2(B1 \cap R2) = 0.0099$ $\qquad$ $p_2(B1 \cap B2) = 0.9801$

Assume that we know that the first ball is red. Then if the conditionals p_1 and p_2 are calculated, the following is obtained,

$p_1(R2/R1) = 0.99$ $\qquad$ $p_1(B2/R1) = 0.01$

$p_2(R2/R1) = 0.01$ $\qquad$ $p_2(B2/R1) = 0.99$

The interval probabilities, given the first red, are the same, [0.01,0.99], for the two colors of the second ball. That is, knowing that the first ball is red says us nothing about the second ball. But, intuitively, we can think that if the first ball is red then the first urn is more possible and then the second ball should also be red with high probability. The problem with this method based on conditioning every probability distribution is that nothing is inferred if we do not have an 'a priori' information. In our method, the following possibilities are assigned,

$\pi(p_1(./R1)) = 1; \quad \pi(p_2(./R1)) = 1/99.$

This produces the following intervals given R1,

R2 [0.9801,0.9900], B2 [0.0100,0.0199],

a much more intuitive result.

Another very well known example is the case of the murder of Mr. Jones (see Ph. Smets,

R. Kennes, [11]). The situation is: It is known that one of three persons, Peter, Paul or Mary, has killed Mr. Jones. The killer was selected tossing a dice. If the result was even, the killer was Mary. If the result was odd, the killer was Peter or Paul, but nothing is known about how the killer was selected between the two possibilities.

This information may be represented by a basic probability assignment:

$$m(\{Mary\}) = 0.5$$
$$m(\{Peter,Paul\}) = 0.5$$

Assume now that we get that Peter is not the killer. The problem is how to update our basic probability assignment, that is, how to calculate the conditional information to the set {Mary, Paul}.

In the case of Dempster rule of conditioning, the following basic probability assignment is obtained,

$$m_D(\{Mary\}) = 0.5$$
$$m_D(\{Paul\}) = 0.5$$

with its corresponding belief-plausibility intervals:

Mary	0.5 - 0.5
Paul	0.5 - 0.5

Here, it is assumed that additional information do not change 'a priori' probabilities, based on dice tossing.

In the case of upper and lower probabilities approach, the solution is

$$m_U(\{Mary\}) = 0.5$$
$$m_U(\{Mary,Paul\}) = 0.5,$$

and the intervals are

Mary	0.5 - 1
Paul	0 - 0.5

It can be observed that intervals are too wide. Dempster conditioning is included as an extreme point of these intervals. Precisely, the values corresponding to transfer the initial mass of the set {Peter,Paul}, 0.5, to the set {Paul}. The other possibility is to transfer this mass to the set {Peter}. The result of Dempster conditioning corresponds to the most possible of the two alternatives, taking into account that Peter is not the killer.

In our approach, the following basic probability assignment is obtained,

$$m^*(\{Mary\}) = 1/2$$
$$m^*(\{Paul\}) = 1/4$$
$$m^*(\{Paul,Mary\}) = 1/4$$

The intervals are,

Mary	0.5 - 0.75

Paul 0.25 - 0.5

These intervals are more informative than in upper-lower probabilities approach and contain Dempster conditioning. The intervals have been shortened at the less possible interval extreme.

Strong conditioning intervals are,

Mary	1 - 1
Paul	0 - 0

That is the less possible extremes of upper-lower intervals. The result is not very intuitive. It is not possible that Paul is the killer, after knowing that Peter is not.

Weak conditioning intervals are,

Mary	0.5 - 0.5
Paul	0 - 0.5
Peter	0 - 0.5

The same intervals, we had before introducing that Peter is not the killer. In fact, the possibility of Peter being the killer is 0.5, after knowing that he is not.

4. BAYES RULE WITHOUT 'A PRIORI' INFORMATION

The model is the following:

- X is a variable taking values on $\{x_1,\dots,x_n\}$
- Y is an observable variable taking values on $\{y_1,\dots,y_m\}$
- We have an information about the conditional probabilities, $p(y_j/x_i)$, consisting in intervals: $p(y_j/x_i) \in [a_{ij},b_{ij}]$.

The problem is: given an observation y_k, What is the conditional information about X.

Ph. Smets (1978), [10], has proposed a more general Bayes rule, in which 'a priori' and conditional informations may be pairs of belief plausibility measures. In general, the results are very different when both models are applicable.

In our case, the set of possible probabilities on X has as extreme probabilities $\{\delta_1,\dots,\delta_n\}$, where $\delta_i(x_i) = 1$, $\delta_i(x_j) = 0$, $j\neq i$.

If we calculate every conditional given y_k, we get $\delta_i(./y_k) = \delta_i$, except when $p(y_k/x_i) = 0$, in which case conditioning is not defined.

That is the possible extreme conditional probabilities are

$$\{\delta_1,\dots,\delta_n\} - \{\delta_i/b_{ik}=0\}.$$

Observe that only the probabilities δ_i with $p(y_k/x_i) = 0$ are eliminated. That would be the conditional using upper and lower probabilities approach. Let us introduce our possibility distribution:

$$\pi(\delta_i,p(./.)) = p(y_k) = \sum_{l=1}^{m} \delta_i(x_l)p(y_k/x_l) = p(y_k/x_i) \in [a_{ik},b_{ik}].$$

Marginalizing on δ_i we get

$$\pi(\delta_i) = \underset{p(./.)}{\mathrm{Max}}(\delta_i, p(./.)) = b_{ik}.$$

Calculating the information induced on X by using Choquet integral, we get a possibility distribution:

$$\pi(x_i) = b_{ik}.$$

That is, $u^*(x_i/y_k) = \pi(x_i) = b_{ik}$.

Let us see an example. Assume that we have an agent, A1, informing us about an unknown element X from a set $\{x_1,x_2,x_3\}$. Consider that this agent has a reliability of α_1 in the sense that with probability α_1 says the true, and with probability $(1-\alpha_1)$ his information has no sense. Assume that the agent says the element is x_1. The induced possibility is

$$\pi(x_1) = 1,\ \pi(x_2) = 1-\alpha_1,\ \pi(x_3) = 1-\alpha_1.$$

that is equivalent to the representation of the Theory of Evidence, the basic probability assignment:

$$m(\{x_1\}) = \alpha_1, \quad m(\{x_1,x_2,x_3\}) = 1-\alpha_1.$$

Assume that another agent, A2, with reliability α_2 says us that X is x_2. If A1 and A2 are conditionally independents given the true value X. Then, the conditionals are

$$p(A1=x_1,A2=x_2/X=x_1) \in [0,1-\alpha_2]$$

$$p(A1=x_1,A2=x_2/X=x_2) \in [0,1-\alpha_1]$$

$$p(A1=x_1,A2=x_2/X=x_3) \in [0,(1-\alpha_1)(1-\alpha_2)].$$

The induced possibility is

$$\pi(x_1) = (1-\alpha_1)/c,$$

$$\pi(x_2) = (1-\alpha_2)/c,$$

$$\pi(x_3) = (1-\alpha_1)(1-\alpha_2)/c$$

where $c = 1-\mathrm{Min}\{\alpha_1,\alpha_2\}$.

This result is different from the combination by Dempster rule of the informations associated with each agent. Our approach is, in general less informative, at element level: $\pi(x_i) \geq Pl(x_i)$.

To end, we want to remark that this combination of possibility distributions is compatible with the intersection of the associated fuzzy sets using the probabilistic operator: $\mu_{A\cap B}(x) = \mu_A(x).\mu_B(x)$.

Another fuzzy operators can be obtained assuming different hypothesis from conditional independence.

Acknowledgments

We are very grateful to Ph. Smets for his useful and valuable comments. We have also frofited from the discussions and conversations with partners in DRUMS Project.

REFERENCES

[1] Campos L.M. de,Lamata M.T.,Moral S.(1989) The concept of conditional fuzzy measure. International Journal of Intelligent Systems 5, 237-246.

[2] Dempster A.P. (1967) Upper and lower probabilities induced by a multivalued mapping. Ann. Math. Statist. 38, 325-339.

[3] Dubois D., Prade H. (1986) On the unicity of Dempster rule of combination. International Journal of Intelligent Systems 1, 133-142.

[4] Fagin R., Halpern J.Y. (1989) Updating beliefs vs. combining beliefs. Unpublished Report.

[5] Lamata M.T., Moral S.(1989) Classification of fuzzy measures. Fuzzy Sets and Systems 33, 243-253.

[6] Pearl J. (1988) Probabilistic Reasoning in Intelligent Systems. Morgan & Kaufman (San Mateo).

[7] Pearl J. (1989) Reasoning with belief functions: a critical assessment. Tech. Rep. R-136. University of California, Los Angeles.

[8] Planchet B. (1989) Credibility and Conditioning. Journal of Theoretical Probability 2, 289-299.

[9] Shafer G. (1976) A Theory of Statistical Evidence. In: Foundations of Probability Theory, Statistical Inference, and Statistical Theories of Science, Vol. II (Harper, Hooker, eds.) 365-436.

[10] Smets Ph. (1978) Un modele mathematico-statistique simulant le procesus du diagnostic medical. Doctoral Dissertation, Universite Libre de Bruxelles. Bruxelles.

[11] Smets Ph., Kennes (1989) The transferable belief model: comparison with bayesian models. Technical Report, IRIDIA-89.

ASSESSING MULTIPLE BELIEFS ACCORDING TO ONE BODY OF EVIDENCE - WHY IT MAY BE NECESSARY, AND HOW WE MIGHT DO IT CORRECTLY

Yen-Teh Hsia*
IRIDIA, Université Libre de Bruxelles
50 av. F. Roosevelt, CP 194/6
1050, Brussels, Belgium

Abstract: Dempster's rule of combination, the main inference mechanism of the Dempster-Shafer theory of belief functions [Shafer 76; Smets 88], requires that the belief functions to be combined must be "independent". This independence assumption is usually understood to be composed of two parts: (1) the uniqueness assumption, which states that each belief function to be combined is based on a unique body of evidence, and (2) the evidential independence (or distinctness) assumption, which states that we must be able to argue about the independence of all these different bodies of evidence. In this paper, we suggest that the uniqueness assumption is not necessary, and that it should be alright if we can come up with "independent specifications" of beliefs.

Keywords: Belief function, Dempster's rule of combination, the uniqueness assumption, the distinctness assumption, specification independence.

1. INTRODUCTION

Dempster's rule of combination, the main inference mechanism of the Dempster-Shafer theory of belief functions [Shafer 76; Smets 88], requires that the belief functions to be combined must be "independent". This independence assumption is usually understood to be composed of two parts: (1) the *uniqueness* assumption, which states that each belief function to be combined is based on a unique body of evidence, and (2) the *evidential independence (or distinctness)* assumption, which states that we must be able to argue about the independence of all these different bodies of evidence.

The uniqueness assumption requires that we always assess one joint belief function (which appropriately formalizes our intuitive belief in the matter) according to a particular body of evidence at hand. This is

* This work was supported in part by the DRUMS project funded by the Commission of the European Communities under the ESPRIT II-Program, Basic Research Project 3085.

no problem if the body of evidence at hand only suggests a few probable relations among few variables. However, if a body of evidence simultaneously suggests various probable relations among many variables, we may have difficulties (sometimes seriously) specifying our belief in the matter using just one joint belief function. This could happen, for example, when we want to express our judgment about the probable relations among many different aspects of a person after interviewing that person. It could also happen when we want to formalize our experience (which may be viewed as *one* body of evidence as a whole) and use the resulting formulation as the knowledge base of an inference system. In cases like this, it is often impractical (sometimes even impossible) for us to express with reasonable confidence our judgment using just one belief function. Instead, it may be more natural and practical *if* we can somehow find a way to express our intuitive belief in the matter using multiple belief functions and then to use Dempster's rule to combine these belief functions. The question, though, is whether this approach is feasible and, if so, how we might do it correctly.

In this paper, we suggest that the uniqueness assumption is not necessary, and that it should be alright if we can come up with "independent specifications" of beliefs. What we mean is that our freedom in specifying any one belief function should *not* be constrained by the specifications of all other belief functions[1].

The remainder of this paper is organized as follows. In Section 2, we settle on a formalism that we will use in this paper, and we also give our version of the well known B-P-F (Bird-Penguin-Fly) problem. In Section 3, we argue for the specification of independent belief functions. In Section 4, we give a technique for specifying independent belief functions. In Section 5, we examine a previously proposed solution to the B-P-F problem and show that it does not consist of independent specifications of belief functions. Finally, Section 6 concludes.

2. THE MULTIVARIATE FORMALISM AND THE B-P-F PROBLEM

Throughout this paper, we assume that the multivariate formalism [Kong 86] is used for specifying belief functions, and we assume that all relevant aspects of the world that we are interested in are already formulated as variables. For example, if we only use three binary variables Bird, Penguin and Fly (or B, P and F) for specifying our knowledge, then the underlying (and perhaps oversimplified) operational assumption should be that these three variables are the only relevant aspects of the world that we need to consider, i.e., we need not take into consideration such things as "equipped with a rocket", "death", "illness", and so forth.

The multivariate formalism is useful for formulating our intuitive belief about the probable relations among different aspects of the world we are interested in. For example, the following belief function

---

[1]This notion of "specification independence" was developed jointly by Philippe Smets and the author.

expresses our intuitive belief about the relation or relations that may exist among (the values of) three binary variables C, D, E: "C=Y or (D=Y and E=Y)" with .6 certainty, **or** "C=Y or D=Y or E=Y" with .3 certainty, **or** "(C=Y or D=Y or E=Y) or (C=N and D=N and E=N)" with .1 certainty. Note that, in this example, the last relation (a tautology that gets the .1 certainty) may be called "the vacuous relation", because it does not really say anything about the relation that we think may exist among C, D, and E.

Belief functions specified using the multivariate formalism may be regarded as being specified on the product space built from the frames[2] of all variables involved. For more details about the multivariate formalism, the reader is referred to [Kong 86; Shafer et al. 87].

In Sections 4 and 5 of this paper, we will also explicitly or implicitly refer to *the B-P-F (Bird-Penguin-Fly) problem* in the following sense. Suppose, for some reason, we need only consider three *and only three* aspects of the world: the aspect as to whether an entity is a bird (i.e., its "birdness", represented as Bird or B), the aspect as to whether *the same entity* is a penguin (i.e., its "penguiness", represented as Penguin or P), and the aspect as to whether this same entity flies (i.e., its "flieness", represented as Fly or F). The problem, then, is to come up with a belief function formulation such that (1) we can deduce a strong belief in the entity's flieness if all we know is that it is a bird (formulated as an additional belief function: "B=Y" with 1.0 certainty), and (2) we can deduce a strong belief in the entity's non-flieness if we also know that the entity is a penguin (formulated as yet another belief function: "P=Y" with 1.0 certainty).

3. INDEPENDENT SPECIFICATIONS OF BELIEF FUNCTIONS

Should we always assess just one (joint) belief function according to one body of evidence? In theory, it is preferable that we do so for each (independent) body of evidence. However, in practice, such a requirement may be unattainable, and we do not think it is necessary to always enforce this uniqueness requirement, especially when it is unnatural to do so.

Our argument is as follows. The applicability of Dempster's rule is fundamentally related to the question as to whether the belief functions to be combined are "independent". Here, we interpret the word 'independence' as what we refer to as *specification independence*; that is, our freedom in specifying any one belief function should *not* be constrained by the specifications of all other belief functions. To see how the specification of a belief function may be constrained by the specification of other belief functions, consider the following example.[3] There are two urns, and each urn contains 100 balls. Urn One has at least 50 white balls in it and urn Two has at least 60 white balls in it. Let Bel_1 be our belief in grabbing a white ball from urn One and Bel_2 be our belief in grabbing a white ball from urn Two. Then

---

[2]The frame of a variable is the set of all possible values of the variable.
[3]This example is due to Philippe Smets.

the specification of Bel_1 must necessarily be constrained by the specification of Bel_2 (and vice versa), as rationality requires that the more white balls we have, the stronger (at least not weaker) our belief in grabbing a white ball ought to be.

When two bodies of evidence are already judged to be "independent" or "distinct" [Shafer 76] in some way and each body of evidence only induces one belief function, the two belief functions are certainly independent, as the specification of any one belief function does not constrain the specification of the other belief function. However, this does not necessarily mean that one belief function per (body of) evidence is the only thing we can do (although we should try). When it is unnatural for us to assess just one belief function according to one body of evidence, we can nevertheless try to specify multiple belief functions, so long as we make sure that these belief functions can indeed be specified independently. The danger, though, is that we might forget to specify some important "pieces" of what we actually believe in and thus only get a partial specification of our belief. But this is the *completeness* issue, an important issue that we should always be aware of, not a prohibitive one.

Therefore, it is only natural that we arrive at the following conclusion. Whenever we formalize (in the belief function framework) our intuitive belief according to the evidence at hand, we should first sort out the independent groups of evidence; that is, we partition the evidence at hand into disjoint groups so that these groups may be viewed as independent in some judgmental sense. Then, for each group (or body) of evidence, we specify one or more belief functions that are "induced" by the evidence. In performing this second step, however, we should try to come up with just one joint belief function first. If we find this first alternative too difficult (because we cannot specify just one belief function with reasonable confidence, etc.), then we try the second alternative: we formulate independent belief functions and use Dempster's rule to combine these belief functions. For this second alternative, we should also make sure that our judgment in the matter is completely specified by the different "pieces" of belief functions.

4. A TECHNIQUE FOR MAKING INDEPENDENT SPECIFICATIONS

The notion of specification independence described in the last section serves as a guidance for specifying multiple belief functions. However, when we are faced with many variables, we may still have problems specifying what our belief is. Where should we start? And how should we proceed? In this section, we offer a technique for specifying independent belief functions. This technique uses a method described in [Smets 78] for specifying independent belief functions. The main spirit of this technique is akin to the idea of Bayesian causal trees [Pearl 86].

Let $\mathcal{A}_0 = \{A_1, A_2, \ldots, A_N\}$ (e.g., {Bird, Penguin, Fly}) be a set of variables. We first specify a set $\mathfrak{C}$ of categorical belief functions for these variables (e.g., $\mathfrak{C}$ = {"if Penguin=Y then Bird=Y and Fly=N" with 1.0 certainty}). Then, we recursively apply the following three steps until the variables contained in $\mathcal{A}_i$ ($i \geq 0$) do not directly "depend on" each other.

Step 1: From $\mathcal{A}_i$ (e.g., $\mathcal{A}_0$), we identify exactly one variable A (e.g., Fly)[4] and also a subset $\mathcal{B}_i$ of $\mathcal{A}_i \backslash \{A\}$ so that A directly "depends on" the valuation of the elements of $\mathcal{B}_i$ (e.g., $\mathcal{B}_0$ = {Bird, Penguin} is a subset of {Bird, Penguin, Fly}\{Fly} and Fly directly "depends" on the valuation of Bird and Penguin).

Step 2: For *each and every* logically possible valuation of the elements of $\mathcal{B}_i$ (e.g., <Bird, Penguin> = <Y, N>; <Bird, Penguin> = <Y, Y>; <Bird, Penguin> = <N, N>), we specify an independent belief function[5] for the valuation of A (e.g., for the valuation <Bird, Penguin> = <Y, N>, we specify the belief: "Fly =Y" with .9 certainty, **or** "(Fly=Y) or (Fly=N)" with .1 certainty). If this specified belief function is non-vacuous and non-categorical, we need to translate it into the following belief function (vacuous belief functions and categorical belief functions are ignored, as they do not affect the result of belief combinations):[6]

"if the-valuation then A's-value-in-ValueSet1" with m_1 certainty, **or**
"if the-valuation then A's-value-in-ValueSet2" with m_2 certainty, **or**
...
"if the-valuation then A's-value-in-ValueSetN" with m_N certainty.

(e.g., the belief function *"Fly =Y" with .9 certainty, or "(Fly=Y) or (Fly=N)" with .1 certainty* is now translated into the belief function *"if (Bird=Y and Penguin=N) then Fly=Y" with .9 certainty, or "if (Bird=Y and Penguin=N) then (Fly=Y or Fly = No)" with .1 certainty.*)

Step 3: Let $\mathcal{A}_{i+1}$ be $\mathcal{A}_i \backslash \{A\}$ (e.g., $\mathcal{A}_1$ = {Bird, Penguin, Fly}\{Fly} = {Bird, Penguin}), and go back to Step 1 if we are not done.

Once we get to $\mathcal{A}_{final}$, we can, if we want, specify a (non-vacuous) belief for each of the variables remaining in $\mathcal{A}_{final}$ (e.g., $\mathcal{A}_{final} = \mathcal{A}_2$ = {Bird}, and we can, if we want, specify a belief function such as: "Bird=Y" with certainty .7, **or** "Bird=Y or Bird=N" with certainty .3; but this is not to be the case in the following example).

The following example shows what we might get from such a belief specification process.

[4]Actually, we can identify more than one variable if we want. Here, just for simplicity, we restrict it to be one.

[5]That is, we make sure that the way A (i.e., Fly) depends on a particular valuation of the elements of $\mathcal{B}_i$ (e.g., <Bird, Penguin> = <Y, N>) is independent of the ways A depends on other valuations of the elements of $\mathcal{B}_i$ (e.g., <Bird, Penguin> = <Y, Y>; <Bird, Penguin> = <N, N>).

[6]This translation is based on the principle of minimum specificity [Dubois and Prade 86].

Example 1. We obtain three belief functions (Penguin $\rightarrow$ Bird $\wedge \neg$Fly) (1);
(Bird $\wedge \neg$Penguin $\rightarrow$ Fly) (.9) or T (.1); and
(Bird $\rightarrow \neg$Penguin) (.95) or T (.05)

from the the categorical belief *"if Penguin=Y then Bird=Y and Fly =N" with 1.0 certainty* and the following "constraints":

valuation	belief about some variable
Bird $\wedge \neg$Penguin	Fly (.9) or T (.1)
Bird $\wedge$ Penguin	$\neg$Fly (1)
$\neg$Bird $\wedge \neg$Penguin	T (1)
$\neg$Bird $\wedge$ Penguin	(logically impossible)
Bird	$\neg$Penguin (.95) or T (.05)
$\neg$Bird	$\neg$Penguin (1)

Once we have specified belief functions according to this technique, we can then use Dempster's rule of combination to combine these belief functions. To use the resulting belief function Bel for reasoning, we simply formulate the context at hand (i.e., available information about the actual situation) as yet another categorical belief function (e.g., "Bird = Y and Penguin = Y" with 1.0 certainty) and combine it with Bel. This, in effect, is equivalent to *conditioning* Bel on the context [Shafer 76]. Thus, for example, if Bel is the belief function resulting from combining the three belief functions in Example 1, then Bel(Fly | Bird) = .9, Bel($\neg$Fly | Penguin) = 1, Bel($\neg$Fly | Bird $\wedge$ Penguin) = 1, etc.

5. A WRONG "SOLUTION" TO THE B-P-F PROBLEM

Before we conclude, we show that a previously proposed solution to the Bird-Penguin-Fly problem (e.g., [Pearl 88] and [Hsia and Shenoy, 89]) is wrong as far as specification independence is concerned. This proposed solution is that we assign a positive value such as .85 to the assertion "if B=Y then F=Y" (the B-F belief)[7], a positive value such as .95 to the assertion "if P=Y then F=N" (the P-NF belief), and the value 1.0 to the assertion "if P=Y then B=Y" (the P-B belief). The problem with this specification of belief functions is that it does not satisfy specification independence.

To show why this is the case, it suffices to show that we do not have total freedom in specifying B-F and P-NF individually once we have specified P-B. This is not very hard. Because once P-B has been specified, we can no longer afford to give a value 1 to "if B=Y then F=Y" while giving a value 1 to "if P=Y then F=N" at the same time (as this will lead to total contradiction when the context is that P = Y). In other words, once we have specified P-B, the specifications of B-F and P-NF become somewhat related, and we no longer have total freedom in specifying them individually.

[7] with the remaining .15 being assigned to a tautology such as "(B=Y or F=Y) or (B=N and F=N)".

6. CONCLUSION

We suggested in this paper that the uniqueness assumption required for using Dempster's rule of combination may be unnecessary, especially if we cannot express with reasonable confidence our judgment using just one belief function. We argued that it is feasible to come up with independent specifications of belief functions and then use Dempster's rule to combine these belief functions; however, we neet to be careful in making such a specification so that the "pieces" we formulate completely describe our intuitive belief in the matter. We gave a technique for making independent specifications of beliefs, and we showed why we think a previously proposed solution to the Bird-Penguin-Fly problem is wrong.

ACKNOWLEDGEMENT

This paper was motivated, in part, by a workshop in Palma de Mallorca, where Enrique Ruspini strongly objected against the wrong "solution" to the Bird-Penguin-Fly problem described in this paper. Discussions with Philippe Smets, Robert Kennes, and Alessandro Saffiotti have helped to greatly enhance the content and presentation of this paper.

REFERENCES

Dubois, D. and Prade, H. (1986). The principle of minimum specificity as a basis for evidential reasoning. In *Uncertainty in Knowledge-Based Systems* (Bouchon and Yager eds.), Springer-Verlag, Berlin, 75-84.

Hsia, Y.-T. and Shenoy, P. P. (1989). An evidential language for expert systems. *Proceedings of the 4th International Symposium on Methodology for Intelligent Systems*, Charlotte, N.C., 9-16.

Kong, A. (1986). Multivariate belief functions and graphical models. Doctoral dissertation, Department of Statistics, Harvard University.

Pearl, J. (1986). Fusion, propagation, and structuring in belief networks. *Artificial Intelligence*, **29**, 241-288.

Pearl, J. (1988). *Probabilistic Reasoning in Intelligent Systems: Networks of Plausible Inference*, Morgan Kaufmann Publishers, Inc., San Mateo, California.

Shafer, G. (1976). *A Mathematical Theory of Evidence*, Princeton University Press.

Shafer, G., Shenoy, P., and Mellouli, K. (1987). "Propagating belief functions in Qualitative Markov Trees", *International Journal of Approximate Reasoning*, **1**, 4, 349 - 400.

Smets, P. (1978). Un modèle mathématico-statistique simulant le processus du diagnostic médical. Doctoral dissertation, Université Libre de Bruxelles, Bruxelles.

Smets, P. (1988). Belief functions. In *Non-Standard Logics for Automated Reasoning* (P. Smets, E. H. Mamdani, D. Dubois and H. Prade eds.). Academic Press, London.

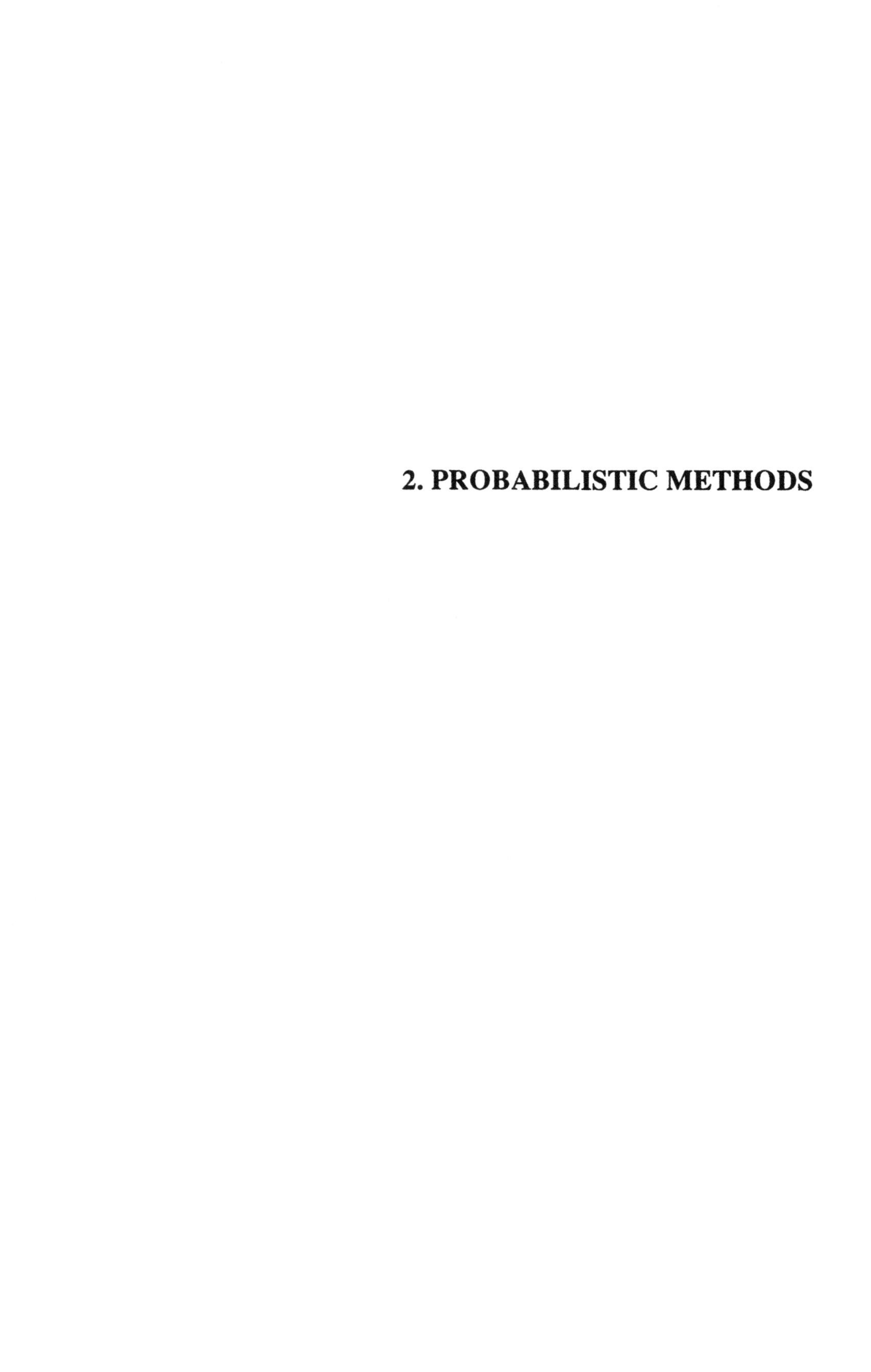

2. PROBABILISTIC METHODS

PROBABILISTIC DEFAULT REASONING

GERHARD PAASS[1]
German National Research Center for Computer Science (GMD)
P.O. Box 1240, D-5205 Sankt Augustin, Germany
email: paass@gmdzi.gmd.de

Abstract

We present an algorithm that is able to integrate uncertain probability statements of different *default levels*. In case of conflict between statements of different levels the statements of the lower levels are ignored. The approach is applicable to inference networks of arbitrary structure including loops and cycles. The simulated annealing algorithm may be used to derive a distribution which best fits to the different statements according to the maximum likelihood principle. In contrast to Pearl's approach to probabilistic default reasoning based on probabilities arbitrarily close to 1 our approach may combine conflicting evidence yielding a compromise between statements of the same default level according to their relative reliability. Between observationally equivalent solutions the maximum entropy criterion is employed to select a distribution with minimal higher order interactions.

1 - INTRODUCTION

In realistic decision problems we have to be able to draw plausible conclusions based on uncertain, incomplete and even inconsistent information. In the realm of logics various nonmonotonic logics have been proposed to tackle this problem. An alternative is probabilistic reasoning where the uncertainty of facts and rules is described by a probability measure. Various formalisms have been developed during the last years [13] and have been studied with respect to their mathematical properties as well as their expressiveness and computational aspects.

The "standard" approaches to probabilistic reasoning start from a consistent set of constraints on the probability measure and derive the desired probabilities of consequences. In this paper we consider constraints which may be *inconsistent*, i.e. there is no probability measure satisfying all constraints simultaneously. If we want to integrate the constraints into a consistent probability distribution, there will always be a deviation for some of the constraints. But which constraints should be violated? We assume that the constraints have been obtained in a sort of inexact *measurement* process from the underlying probability measure. The reliability of each measurement is described by a measurement distribution specifying the random relation between the observed quantities and the corresponding number of the underlying probability measure. Such measurement models are common in statistical and engineering science [3].

If there is a conflict between two different constraints we have two alternatives:

[1]This work is part of the joint project TASSO "Technical Assistance with a System for SOlving unclear problems using inexact knowledge" and was supported by the German Federal Department of Research and Technology, grant ITW8900A7.

- We may try to find a 'compromise' between the constraints. Here the reliability of each constraint can be taken into account. This leads to the minimization of a weighted measure of constraint violations.
- A less reliable constraint is completely discarded if it contradicts a more reliable constraint.

In this paper both approaches are merged. The second alternative up to now is the prevalent way how to deal with inconsistencies in nonmonotonic logics. Current proposals for probabilistic default reasoning (for a discussion see [13, p467ff]) do not allow this mixture and are confined to probabilities 0 and 1.

In the next section the measurement process for probabilities is developed and a best compromise is defined according to the maximum likelihood principle. In the third section different default classes are defined such that a constraint is neglected if it contradicts to a default in a higher class. In the fourth section the approach is applied to standard default problems. In the following the simulated annealing algorithm is employed to generate approximate solutions to the problem. The final section contains a discussion.

2 - UNCERTAIN PROBABILITIES

Suppose the situation in question can be described in terms of n_U different atomic propositional formulae $U_1, \ldots, U_{n_U}$. A term $W_\tau := \tilde{U}_1 \wedge \cdots \wedge \tilde{U}_{n_U}$ with $\tilde{U}_i \in \{U_i, \neg U_i\}$ embodies a comprehensive description of the situation and is called a possible world. The set $\mathcal{W}$ of possible worlds contains $n_q := 2^{n_U}$ elements. A possible world corresponds to a valuation. Define $\mathcal{F}$ as the set of propositions which can be formed by disjunctions of the $W_\tau \in \mathcal{W}$.

We may characterize the uncertainty about the true possible world by a probability measure $p : \mathcal{F} \to [0,1]$, which assigns probability values $q_\tau := p(W_\tau)$ to each W_τ. It is completely defined by the parameter vector $\boldsymbol{q} := (q_1, \ldots, q_{n_q})$ with $q_i \geq 0$, $\sum_{i=1}^{n_q} q_i = 1$. Let $\mathcal{Q}$ be the set of parameters of all probability measures over $\mathcal{F}$. Nilsson [8] shows, that this definition can be extended to a subset of first-order logic.

As [9] and [11] we assume that there are d pieces of evidence $\tilde{\boldsymbol{t}} := (\tilde{t}_1, \ldots, \tilde{t}_d)$ on p collected from possibly different sources: measurement devices, random samples, or expert judgements. Each $\tilde{t}_i$ is related to a measurable function $t_i(\boldsymbol{q})$ of the parameters $\boldsymbol{q}$.

Example: Assume we have the propositions Bi: "*Tweety is a bird*", Fl : "*Tweety can fly*", Pe: "*Tweety is a penguin*". Then $\tilde{t}_i$, for instance, may be equal to ...

- a probability value, e.g. $p(Fl) = 0.8$. This statement can be interpreted as: "The degree of belief that Fl holds is 0.8".
- a value of a conditional probability, e.g. $p(Fl \mid Bi)$=0.95. This statement can be interpreted as a probabilistic rule.
- an indicator I_A for $A \subset \mathcal{Q}$ which takes the value 1 if $\boldsymbol{q} \in A$ and the value zero otherwise. If, for instance, A is the set of all $\boldsymbol{q} \in \mathcal{Q}$ with $p(Fl) \geq 0.95$, then the observation of $I_A(\boldsymbol{q}) = 1$ would imply that $p(Fl) \geq 0.95$, but would not give any evidence on the exact value of p within that interval.

The $\tilde{t}_i$ may be uncertain and incoherent to some extent and are considered as some sort of measurement with an associated 'measurement' distribution $P(\tilde{\boldsymbol{t}} \mid \boldsymbol{q})$. The capital P indicates that

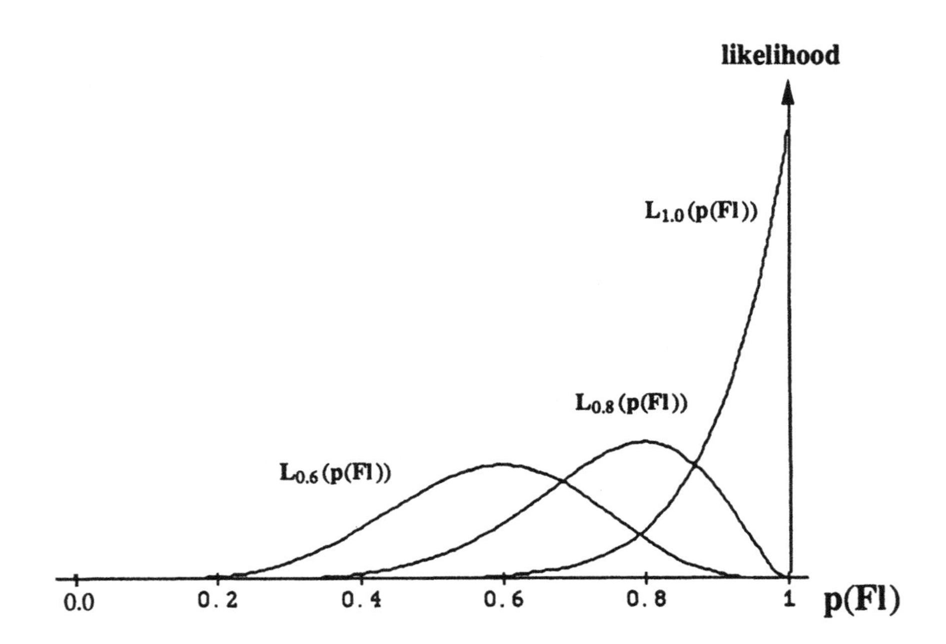

Figure 1: **Binomial Likelihood Functions for Sample Size $n = 10$**

the distribution is a 'second order' probability distribution describing random variables which themselves are probabilities.

We assume that an independent investigator can specify an estimate of $P(\tilde{\boldsymbol{t}} \mid \boldsymbol{q})$ reflecting his subjectively assessed distribution of possible measurements $\tilde{\boldsymbol{t}}$ for each probability vector $\boldsymbol{q} \in \mathcal{Q}$. Note that the investigator does not specify any preference for some specific $\boldsymbol{q}$ as he defines $P(\tilde{\boldsymbol{t}} \mid \boldsymbol{q})$ for every possible value of $\boldsymbol{q}$. The functional form of the measurement distribution is selected according to the type of measurement. For physical measurements often the normal distribution is suitable [6], while the binomial distribution can be used as sampling distribution of counts of binary alternatives [9].

Using the measurement paradigm we may use the procedures developed in statistics for combining uncertain measurements. A prominent approach is the *maximum likelihood* principle [7] which can be shown to exploit the available information in an efficient way. According to this principle a probability distribution $\hat{\boldsymbol{q}}$ with maximal likelihood $P(\tilde{\boldsymbol{t}} \mid \hat{\boldsymbol{q}}) := \max_{\boldsymbol{q} \in \mathcal{Q}} P(\tilde{\boldsymbol{t}} \mid \boldsymbol{q})$ is best compatible with the observations $\tilde{\boldsymbol{t}}$. If the different 'observations' $\tilde{t}_i$ have been obtained independently the likelihood can be factored and the determination of $\hat{\boldsymbol{q}}$ amounts to the maximization of the likelihood function

$$L(\boldsymbol{q}) := P(\tilde{\boldsymbol{t}} \mid \boldsymbol{q}) = \prod_{i=1}^{d} P(\tilde{t}_i \mid \boldsymbol{q}) \tag{1}$$

Note that the set of solutions in general will have more than one element such that $\hat{\boldsymbol{q}}$ is not unique.

If the observed quantities $\tilde{t}_i$ are probability values we should use a measurement distribution which is concentrated on the interval $[0, 1]$. A natural choice is the binomial distribution which describes the distribution of empirical frequencies $\tilde{t}_i$ of white and black balls if we randomly select (with replacement) n_i balls from an urn with a fraction $t_i(\boldsymbol{q})$ of white balls

$$L_{i,\tilde{t}_i}(t_i(q)) := P(\tilde{t}_i \mid t_i(q)) = const * t_i(q)^{n_i\tilde{t}_i}(1 - t_i(q))^{n_i(1-\tilde{t}_i)} \quad (2)$$

Binomial likelihood functions for different observed values $\tilde{t}_i$ of $t_i(q) := p(Fl)$ are plotted in figure 1. Let us first assume an expert has stated $\tilde{t}_i = 0.6$. Then the value of $L_{i,0.6}(p(Fl))$ indicates the 'degree of plausibility' that the observation comes from a distribution, where the true parameter is $p(Fl)$. It can be shown that the likelihood curve takes its maximum for the observed relative frequency $\tilde{t}_i$. Hence $\hat{p}(Fl) = 0.6$ is most compatible with the observation $\tilde{t}_i$. But the likelihood of values in the vincinity of 0.6 is only slightly smaller.

The likelihood approach assumes that the $\tilde{t}_i$-values have been observed from random samples of the distribution. The sample size n_i determines the spread of the distribution around the observed relative frequency $\tilde{t}_i$. A sample of fourfold size will lead to an estimate with halved standard deviation. By evaluating the binomial distribution function we may derive confidence intervals $[a,b]_\alpha$ such that in α percent of the cases the true value will be contained in $[a,b]$. The likelihood functions $L_{i,0.6}(p(Fl))$, $L_{i,0.6}(p(Fl))$, and $L_{i,0.6}(p(Fl))$, for example, have confidence intervals $[0.35, 0.80]_{90}$, $[0.53, 0.92]_{90}$, and $[0.76, 1.0]_{90}$ respectively. Modifying the sample size n_i we may shrink or enlarge these confidence intervals until the measurement has the desired precision.

If two experts independently have assigned the values $\tilde{t}_i = 0.6$ and $\tilde{t}_j = 0.8$ to $p(Fl)$ we may merge the samples giving a new proper random sample with sample size $n_l = n_i + n_j$. The 'observed' relative frequency is given by

$$\tilde{t}_l = \frac{n_i}{n_i+n_j}\tilde{t}_i + \frac{n_j}{n_i+n_j}\tilde{t}_j \quad (3)$$

which is the maximum likelihood estimate if no other observations interfere. The estimate is a convex linear combination of the conflicting measurements and thus a sort of *compromise*. If $n_i = n_j = 10$ we get $\hat{p}(Fl) = 0.7$. The compromise estimate has a higher precision and smaller confidence intervals as it is based on a larger number of sample elements than its subsamples.

If an expert estimates $\tilde{t}_i$=1.0 for $p(Fl)$ and there is no other evidence, then the corresponding likelihood function $L_{i,1.0}(p(Fl))$ takes its maximum for $\hat{p}(Fl) = 1$. Hence we come to the conclusion that Fl holds. Now assume that new evidence arrives and a second expert independently assigns the value $\tilde{t}_j = 0.6$ to $p(Fl)$. Then by (3) we get $\hat{p}(Fl) = 0.8$ if $n_i = n_j$. Therefore in our approach we may derive that a proposition has probability of 0 or 1, i.e. is true or false, and subsequently we may revise this conclusion if new evidence arrives.

3 - DEFAULT REASONING

In the case of conflicting information sometimes a compromise is not desirable. Some pieces of information serve as **defaults** which are to be used only if no other rules or facts apply. Similar to Brewka's extension of the preferred subtheories approach [2] we partition the set $T := \{\tilde{t}_1, \ldots, \tilde{t}_d\}$ of measurements into disjoint *default classes* T_i, $T_1 \cup \cdots \cup T_m = T$. For $i < j$ we assume that the measurements in T_i are only *defaults* with respect to the measurements in T_j. This means that if there is a conflict between some $\tilde{t}_r \in T_i$ and $\tilde{t}_s \in T_j$ the measurement $\tilde{t}_r$ is neglected and only $\tilde{t}_s$ is used.

In principle a default reasoning task can be solved in a stepwise manner. We start with the most reliable default level T_m and form a partial likelihood function corresponding to the elements of T_m

$$L_{(m)}(q) := \prod_{i \in T_m} P(\tilde{t}_i \mid q) \quad (4)$$

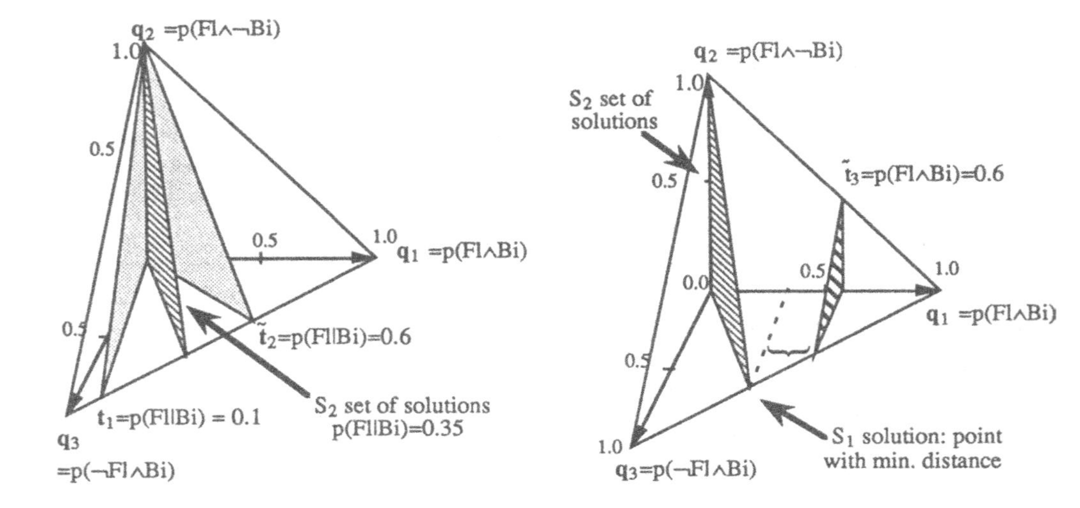

Figure 2: **Stepwise Solution of a Default Problem**

The set of maxima of this function is called $S_m \subset \mathcal{Q}$. If there are conflicting observations $\tilde{t}_i$ and $\tilde{t}_j$ in T_m a compromise is determined according to the relative reliability the measurements. In the case of r independent linear observation functions in T_m the set S_m is a $(n_q - r - 1)$-dimensional simplex.

We subsequently only consider solutions within S_m. We define the partial likelihood function $L_{(m-1)}(q) := \prod_{i \in T_{m-1}} P(\tilde{t}_i \mid q)$ and determine the restricted maximum

$$S_{m-1} := \{q \in S_m \mid L_{(m-1)}(q) = \max_{r \in S_m} L_{(m-1)}(r)\} \tag{5}$$

For conflicting measurements in T_m again a compromise results. In this way the default classes $T_m, \ldots, T_1$ sequentially may be taken into account and we end up with a solution set S_1.

Example:

If we have two atomic propositions Fl and Bi we get four possible worlds with probabilities $q_1 := p(Fl \wedge Bi)$, $q_2 := p(Fl \wedge \neg Bi)$, $q_3 := p(\neg Fl \wedge Bi)$, and $q_4 := p(\neg Fl \wedge \neg Bi)$. As $q_4 = 1 - q_1 - q_2 - q_3$ is determined by the other probabilities we may graphically show the set $\mathcal{Q}$ in 3 dimensions. Assume we have a set T_2 of more reliable defaults with the observations $\tilde{t}_1 = p(Fl \mid Bi) = 0.1$ and $\tilde{t}_2 = p(Fl \mid Bi) = 0.6$ and a second set T_1 with the less reliable default $\tilde{t}_3 = p(Fl \wedge Bi) = 0.6$. Then the set S_2 consists of a compromise between $\tilde{t}_1$ and $\tilde{t}_2$ and is the set of all probability measures with $p(Fl \mid Bi) = 0.35$. This set is shown on the left part of figure 2.

Now we have to select that subset S_1 from S_2 which is most compatible with the observation $\tilde{t}_3 = p(Fl \wedge Bi) = 0.6$ in T_1. This yields the point from S_2 which has the largest value of $q_1 = p(Fl \wedge Bi)$. Hence the desired solution S_1 consists of a single point as shown on the right part of figure 2. If all observations $\tilde{t}_i$ were in the same default class the solution would be located between the set S_2 and the plane $p(Fl \wedge Bi) = 0.6$. The example shows that a default may not be completely neglected even if there are more reliable observations affecting the same variable.

If S_1 contains just one element $\hat{q}$ we are ready. Otherwise we may select solutions $\hat{q}$ from S_1 according to different principles:

- For arbitrary propositions $A, B \in \mathcal{F}$ we may determine upper and lower probabilities

$$p^+(B \mid A) := \max_{q \in S_1} p_q(B \mid A) \qquad p^-(B \mid A) := \min_{q \in S_1} p_q(B \mid A) \tag{6}$$

  These quantities reflect the information that can be deduced from the observed quantities $\tilde{t}_i$ together with the observation model.

- From S_1 we may select a unique $\hat{q}$ according to the *maximum entropy* criterion. In some respect $\hat{q}$ is the element from S_1 which carries the smallest amount of additional information. It corresponds to setting higher order dependencies between the elementary propositions to zero. The maximum entropy criterion can be considered as a weakest default which is applied if no other default interferes [4].

- Finally we may use a ***Bayesian prior distribution*** giving a measure of 'a priori' credibility to each distribution in S_1. This approach is elaborated in [12].

4 - STANDARD DEFAULT PROBLEMS

Tweety

Assume we have the propositions Bi: *"Tweety is a bird"*, Fl : *"Tweety can fly"*, Pe: *"Tweety is a penguin"* and the following knowledge in two default classes:

$$\begin{array}{ll} T_1 := \{p(Fl \mid Bi) = 1\} & \text{(less reliable)} \\ T_2 := \{p(Bi) = 1; \;\; p(\neg Fl \mid Pe) = 1\} & \text{(more reliable)} \end{array}$$

Then the algorithm arrives at $p(Fl) = 1$ and Tweety can fly with certainty. If we add $p(Pe) = 1$ to T_2, the algorithm yields $p(Fl) = 0$. Hence the previous conclusion is revised and it is inferred that Tweety cannot fly.

Now let us consider the proposition Re: *"Tweety is red"*. If the knowledge base contains no rule or fact concerning Re the maximum entropy principle will yield $p(Re) = 0.5$ and Re will be independent from the remaining proposition. If we only have the rule $p(Fl \mid Bi) = a \in [0,1]$ this means

$$\begin{aligned} p(Fl \mid Bi) &= \frac{p(Fl \wedge Bi)}{p(Fl \wedge Bi) + p(\neg Fl \wedge Bi)} = \frac{p(Fl \wedge Bi \wedge Re)}{p(Fl \wedge Bi \wedge Re) + p(\neg Fl \wedge Bi \wedge Re)} \\ &= p(Fl \mid Bi \wedge Re) \end{aligned}$$

as $p(Fl \wedge Bi)p(Re) = p(Fl \wedge Bi \wedge Re)$ because of independence. Therefore we may add irrelevant conditions to the rule without altering the result. If we have $p(Fl \mid Bi) = 1$ we can conclude $p(\neg Fl \wedge Bi) = 0$. Then $p(\neg Fl \wedge Bi \wedge Re) = 0$ and without the independence assumption we arrive at $p(Fl \mid Bi \wedge Re) = 1$.

Nixon

For the propositions Re: *"Nixon is republican"*, Qu: *"Nixon is quaker"* Ha: *"Nixon is hawk"*, and Do: *"Nixon is dove"* we may specify the following knowledge in two default levels:

$$\begin{array}{ll} T_1 := \{p(Ha \mid Re) = 1; \;\; p(Do \mid Qu) = 1\} & \text{(less reliable)} \\ T_2 := \{p(Ha \wedge Do) = 0; \;\; p(Re) = 1\} & \text{(more reliable)} \end{array}$$

Then we get a distribution with $p(Ha) = 1$ and $p(Do) = 0$.

If we add $p(Qu) = 1$ to T_2, then by the knowledge in T_2 we get $p(Re \wedge Qu \wedge (\neg Ha \vee \neg Do)) = 1$. Assume for a moment that in the final distribution $p(Re \wedge Qu \wedge \neg Ha \wedge \neg Do) > 0$. But by changing $\neg Ha$ to Ha we can increase $p(Ha \mid Re)$ without decreasing $p(Do \mid Qu)$. Therefore a solution with $\neg Ha \wedge \neg Do$ is always inferior and consequently $p(Re \wedge Qu \wedge \neg Ha \wedge \neg Do) = 0$. Hence we will arrive at a final distribution

$$p(Re \wedge Qu \wedge Ha \wedge \neg Do) = \eta \qquad p(Re \wedge Qu \wedge \neg Ha \wedge Do) = (1 - \eta)$$

where η is selected according to the relative reliability of the rules in T_1. If they are based on samples of identical size we will get $\eta = 0.5$. Hence the final sample distribution concentrates its mass on two possible worlds: $Re \wedge Qu \wedge Ha \wedge \neg Do$ and $Re \wedge Qu \wedge \neg Ha \wedge Do$. Each possible world corresponds to an extension in the usual default logic.

Exceptions from Exceptions

This example has been introduced by Junker [5] to demonstrate the effect of defaults with several levels. Assume for the propositions Wk: *"P. has to work"*, Ex: *"P. has excuse"*, Il: *"P. is ill"*, and Co: *"P. has a cold"* we have the following knowledge in three default levels:

$$\begin{array}{ll} T_1 = \{p(Wk) = 1\} & \text{(least reliable)} \\ T_2 = \{p(Ex \mid Il) = 1\} & \\ T_3 = \{p(\neg Ex \mid Co) = 1; \;\; p(\neg Wk \mid Ex) = 1\} & \text{(most reliable)} \end{array}$$

This yields a distribution which is concentrated on the possible worlds $Wk \wedge \neg Il \wedge \neg Co \wedge \neg Ex$ and $Wk \wedge \neg Il \wedge Co \wedge \neg Ex$ with equal probability. Consequently $p(Wk) = 1$.

If new evidence $p(Il) = 1$ in T_2 arrives the probability distribution is concentrated on a single possible world $p(\neg Wk \wedge Il \wedge \neg Co \wedge Ex) = 1$ and hence $p(Wk) = 0$. If we subsequently we get the information $p(Co) = 1$ in T_3 producing $p(Wk \wedge Il \wedge Co \wedge \neg Ex) = 1$ and $p(Wk) = 1$. Hence an exception from an exception may restore the original conclusion.

5 - SOLUTION BY SIMULATED ANNEALING

The maximization of the likelihood function 1 and the maximization of entropy on the set S_1 of solutions is a highly nonlinear optimization task. In principle we may solve a series of constrained optimization problems (5), for instance with the Lagrange approach. Here we propose another strategy which avoids constraints during optimization and yields a maximum entropy distribution. Consider again the situation that we have two independent random samples of size n_i and n_j with relative frequencies $\tilde{t}_i$ and $\tilde{t}_j$. Then the compromise between both quantities according to the maximum likelihood approach is given by (3)

$$\hat{t} = \tfrac{n_i}{n_i+n_j}\tilde{t}_i + \tfrac{n_j}{n_i+n_j}\tilde{t}_j$$

and the variance $\sigma_{\hat{t}}^2$ of the estimate $\hat{t}$ is proportional to $1/n_i + n_j$. Now let us increase the sample sizes n_i and n_j with factors $\beta^{\gamma(i)}$ and $\beta^{\gamma(j)}$. Then we get the estimates

$$\begin{aligned} \hat{t}_\beta &= \frac{n_i\beta^{\gamma(i)}}{n_i\beta^{\gamma(i)} + n_j\beta^{\gamma(j)}}\tilde{t}_i + \frac{n_j\beta^{\gamma(j)}}{n_i\beta^{\gamma(i)} + n_j\beta^{\gamma(j)}}\tilde{t}_j \\ &= \frac{n_i}{n_i + n_j\beta^{\gamma(j)-\gamma(i)}}\tilde{t}_i + \frac{n_j\beta^{\gamma(j)-\gamma(i)}}{n_i + n_j\beta^{\gamma(j)-\gamma(i)}}\tilde{t}_j \end{aligned} \tag{7}$$

and the variance $\sigma_{\hat{t}}^2$ of the estimate $\hat{t}$ is proportional to $1/n_i\beta^{\gamma(i)} + n_j\beta^{\gamma(j)}$. Consider the estimate for $\beta \to \infty$

$$\lim_{\beta\to\infty} \hat{t}_\beta = \begin{cases} \tilde{t}_j & \text{if } \gamma(j) > \gamma(i) \\ \frac{n_i}{n_i+n_j}\tilde{t}_i + \frac{n_j}{n_i+n_j}\tilde{t}_j & \text{if } \gamma(j) > \gamma(i) \\ \tilde{t}_i & \text{if } \gamma(j) < \gamma(i) \end{cases} \tag{8}$$

Consequently an observation with a smaller γ-value is neglected while for observations with equal γ-values a compromise results if no other observations interfere. Note that for arbitrary $\gamma(i) > 1$ the sample size $\lim_{\beta\to\infty} n_i \beta^{\gamma(i)}$ approaches infinity and hence all associated variances decrease to zero.

To the different default classes T_i we associate parameters $\gamma(i) > 1$ such that $\gamma(i) > \gamma(j) \iff i > j$. The limiting process is performed using the *simulated annealing* algorithm (SAA). As discussed in the appendix (13) the SAA generates a stationary distribution $\Pr(q \mid \tilde{t})$ over the different probability distributions q which is proportional to their likelihood (1)

$$\Pr(q \mid \tilde{t}) \propto \Pr^c(q) \prod_{i=1}^{d} P(\tilde{t}_i \mid q)^{\beta_i} \tag{9}$$

The term $\Pr^c(q)$ can be interpreted as a prior distribution which gives larger weight to q-vectors with higher entropy. If we replace β_i by $\beta^{\gamma(i)}$ we get

$$\Pr_\beta(q \mid \tilde{t}) \propto \Pr^c(q) \prod_{i=1}^{d} P(\tilde{t}_i \mid q)^{\beta^{\gamma(i)}} \tag{10}$$

If β is gradually increased the algorithm converges to the maximum likelihood solutions (8). The variance around the solution is governed by the variance terms σ_i^2 which approach zero with growing β. For the higher order interactions not affected by the observations fluctuate around 0, which are just the maximum entropy configurations. Consequently higher order interactions are minimized. The resulting values of the individual likelihood terms $P(\tilde{t}_i \mid q)$ in the optimum indicate which observation interfere with each others.

6 - DISCUSSION

We divide the available evidence into disjunct default levels. Statements of lower default level are ignored if they contradict to statements of higher levels. For conflicting statements within a level a compromise is determined taking into account the relative reliability of each statement. From the final set of possible solutions a single probability distribution is selected by the maximum entropy criterion minimizing higher order interactions. The discussion of the standard default problems shows that the approach has many desirable properties of default formalisms.

The algorithm is more expressive than the probabilistic default reasoning approach discussed by Pearl [13, p.467ff]. He assigns probabilities $\geq 1-\epsilon$ or $\leq \epsilon$ with an infinitesimal value $\epsilon > 0$ to defaults. He defines only one additional default level and considers neither probabilities different from 0 or 1 as defaults nor does he allow contradictory statements at the same default level.

Because the determination of the maximum entropy solution seems to be analytically not tractable we use an approximate procedure. The simulated annealing algorithm is a global optimization procedure which has successfully been applied to large combinatorial problems. It can find solutions near the global optimum even if the number of variables is large. Currently this algorithm is implemented in our institute. Numerical experience will show whether it is a serious alternative to other approaches of nonmonotonic reasoning.

Acknowledgement

I would like to thank Ulli Junker and Gerd Brewka for the introduction to the philosophy of default reasoning and many stimulating discussions.

A - APPENDIX: SIMULATED ANNEALING

We approximate q by a random sample $X_\tau := (W_{\tau(1)}, \ldots, W_{\tau(n)})$ of n possible worlds. X_τ corresponds to a distribution $q(X_\tau)$ in $\mathcal{Q}_n := \{q \in \mathcal{Q} \mid q_i \in \{\frac{0}{n}, \frac{1}{n}, \ldots, \frac{n}{n}\}\}$. According to the Law of Large Numbers any distribution $q \in \mathcal{Q}$ can be approximated arbitrary well by a sample, if the sample size n is chosen sufficiently large.

The simulated annealing algorithm [1] can be used to generate a sequence of samples with specific properties. The algorithm starts with an arbitrary X_τ. In an iterative fashion the 'current' sample X_τ is randomly modified to a new sample X_η and subsequently it is checked, whether the modification can be accepted. A modification usually consist of rather small changes, for instance transforming U_i to $\neg U_i$ in one or more possible worlds $W_\tau(j)$. The modification probabilities have to be symmetric, i.e. X_τ is changed to X_η with the same probability as X_η to X_τ. If $L(q) = \prod_{i=1}^{d} P(\tilde{t}_i \mid q(X_\tau))$ is the likelihood (1) of X_τ the modification is accepted with probability

$$P_{acc} := \max\left[1\;,\; \frac{\prod_i P(\tilde{t}_i \mid q(X_\eta))^{\beta_i}}{\prod_i P(\tilde{t}_i \mid q(X_\tau))^{\beta_i}}\right] \tag{11}$$

where the $\beta_i \in \Re$ are control parameters usually having a common value $\beta_i = \beta$. If each X_τ can be transformed into any other X_η by a finite number of modifications, the probability $\Pr(X_\tau \mid \tilde{t})$ of X_τ being generated converges to an unique stationary distribution

$$\Pr(X_\tau \mid \tilde{t}) = c_1 \prod_i P(\tilde{t}_i \mid q(X_\tau))^{\beta_i} \tag{12}$$

as the number of iterations goes to infinity. Here c_1 is a constant normalizing the sum of probabilities to one. If $\beta_i = \beta$ for all i and $\prod_i P(\tilde{t}_i \mid q(X_\tau)) < \prod_i P(\tilde{t}_i \mid q(X_\eta))$ we get $\lim_{\beta \to \infty} \Pr(X_\tau \mid \tilde{t}) / \Pr(X_\eta \mid \tilde{t}) = 0$ and the stationary distribution concentrates more and more around the set of maxima of the likelihood function. The simulated annealing algorithm therefore generates the stationary distribution for a sequence of β-values $\beta(k) = \beta a^k$, $k = 1, 2, \ldots$ with $a > 1$ (e.g. $a = 1.1$). With growing k the stationary distribution is concentrated more and more around the set of maximum likelihood values and in the limit p converges to this set with probability 1.

If we execute the simulated annealing algorithm with constant likelihood function every X_τ appears with equal probability $\Pr^c(X_\tau)$. Consequently we get $\Pr^c(q) = \sum_{X;q(X)=q} \Pr^c(X)$. We define $\Pr(q) := \sum_{X;q(X)=q} \Pr(X)$ and have

$$\begin{aligned}\Pr(q \mid \tilde{t}) &= c_1 \sum_{X;q(X)=q} \prod_i P(\tilde{t}_i \mid q(X))^{\beta_i} = \frac{\sum_{X;q(X)=q} \prod_i P(\tilde{t}_i \mid q(X))^{\beta_i}}{\sum_X \prod_i P(\tilde{t}_i \mid q(X))^{\beta_i}} \\ &= \frac{\sum_{X;q(X)=q} \Pr^c(X) \prod_i P(\tilde{t}_i \mid q(X))^{\beta_i}}{\sum_X \Pr^c(X) \prod_i P(\tilde{t}_i \mid q(X))^{\beta_i}} = \frac{\Pr^c(q) \prod_i P(\tilde{t}_i \mid q)^{\beta_i}}{\sum_u \Pr^c(u) \prod_i P(\tilde{t}_i \mid u)^{\beta_i}}\end{aligned} \tag{13}$$

If we interpret $\Pr^c(q)$ as prior distribution and $P(\tilde{t}_i \mid q)^{\beta_i}$ as a likelihood then the stationary distribution $\Pr(q \mid \tilde{t})$ can be considered as a Bayesian posterior distribution. The prior distribution evolves if we execute the simulated annealing algorithm with constant likelihood. This means every element of the sample is modified independently and takes all possible values with equal probability. Consequently $\Pr^c(q)$ follows a multinomial distribution with sample size n and identical expected values $E(q_i)$. But this is just the maximum entropy distribution for this situation. Whenever a higher order interaction of $\Pr(q \mid \tilde{t})$ is not affected by the observations its distribution is determined by the prior distribution $\Pr^c(q)$ and fluctuates around its maximum entropy value. A rigorous proof is given in [10].

As the possible worlds in the sample have identical weights it is in general not possible to satisfy all constraints exactly. We may partition X_τ into r subsamples $X_{\tau,l}$ of equal size and associate a weight $w_l > 0$, $\sum_{l=1}^{r} w_l = 1$ to each subsample. If there are d or more subsamples it is always possible to generate a the sample fitting exactly to d compatible marginal constraints.

Whenever an element of a subsample is modified the likelihood function imposes constraints on the same statistics as for the whole sample. Therefore within each subsample the higher order interactions α will fluctuate around their maximum entropy value 0. As α is not affected by the constraints the values of α will evolve independently in the different subsamples. Hence the overall α will be a convex combination of the α-s in the subsamples and therefore also will fluctuate around 0. Consequently the maximum entropy property is conserved even if we have subsamples with different weights. Of course the size of the subsamples have to be large enough for the effect to occur.

References

[1] Aarts, E., Korst, J. (1988): *Simulated Annealing and Bòltzmann Machines.* Wiley, Chichester

[2] Brewka, G. (1989): Preferred Subtheories: An Extended Logical Framework for Default Reasoning, *Proc. IJCAI '89*, p.1043-1048

[3] Davis, M.H.A., Vinter, R.B. (1985): *Stochastic Modelling and Control.* Chapman & Hall, London.

[4] Grosof, B. (1988): Non-Monotonicity in Probabilistic Reasoning. In: J.F. Lemmer, L.N. Kanal (eds.) *Uncertainty in Artificial Intelligence 2*, Elsevier, Amsterdam, pp.237-250

[5] Junker, U. (1989): A correct Non-Monotonic ATMS, *Proc. IJCAI '89*, p.1049-1054

[6] Lindley, D.V. (1985): Reconciliation of Discrete Probability Distributions, in Bernardo, J.M., DeGroot, M.H., Lindley, D.V., Smith, A.F.M (eds.) *Bayesian Statistics II*, pp. 375-390, North Holland, Amsterdam

[7] Lloyd, E. (ed.) (1984): *Handbook of Applicable Mathematics. Vol. VI: Statistics.* Wiley, Chichester.

[8] Nilsson, N. (1986): Probabilistic Logic, *Artificial Intelligence*, Vol. 28, p. 71-87

[9] Paass, G. (1988): Probabilistic Logic. In: Smets, P., A. Mamdani, D.Dubois, H.Prade (eds.) *Non-Standard Logics for Automated Reasoning*, Academic press, London, p.213-252

[10] Paass, G. (1988): *Stochastic Generation of a Synthetic Sample from Marginal Information.* Workingpaper 308, GMD, St.Augustin, FRG

[11] Paass, G. (1990): Probabilistic Reasoning and Probabilistic Neural Networks. This volume.

[12] Paass, G. (1989): Second order Probabilities for Uncertain and Conflicting Evidence. *Proc. 6-th Conference on Uncertainty in AI.* Cambridge, Mass. July 1990

[13] Pearl, J. (1988): *Probabilistic Reasoning in Intelligent Systems*, Morgan Kaufmann, San Mateo, Cal.

ON KNOWLEDGE REPRESENTATION IN BELIEF NETWORKS

Bruce Abramson *
Department of Computer Science
University of Southern California
Los Angeles, CA 90089-0782

Abstract

Three focal elements of knowledge-based system design are (i) acquiring information from an expert, (ii) representing the information in a system-usable form, and (iii) using the information to draw inferences about specific problem instances. In the artificial intelligence (AI) literature, the first element is referred to as knowledge acquisition, while the second and third are embodied in a system's knowledge base and inference engine, respectively. AI, however, is not alone in its concern for these issues. Researchers in several of the statistical decision sciences, notably decision analysis (DA), have also investigated them. This paper discusses the use of belief networks—a formalism that lies somewhere between AI and DA—as an overall framework for knowledge-based systems. Unlike previous work, which has concentrated on either the networks' mathematical properties or on their implementation as a specific system, this paper is oriented towards the concerns of general system design. Concrete examples are drawn from one medical system (Pathfinder) and from one financial system (ARCO1), and in particular, from a consideration of their similarities and differences. The design principles abstracted from these systems suggests a powerful, coherent design philosophy guided by the simple thought: *form follows function.*

1 Introduction

Knowledge acquisition, knowledge representation, and inference are common terms in the artificial intelligence (AI) literature; they are the central elements of a knowledge-based (or expert) system. AI, however, is not alone in its concern with these issues. Decision analysis (DA) and other statistical decision sciences have also investigated the elicitation of information from human experts and the formal modeling of this information to infer useful recommendations. The approach traditionally taken by AI has been quite different from that favored by DA. Perhaps the most fundamental distinction between them has been their psychological motivation. Many of the procedures popular in AI are rooted in *descriptive* psychology, or observations about the way that people behave. The prevalent paradigms for expert system design illustrate the point; logic and frames describe commonly encountered

*This research was supported in part by the National Science Foundation under grant IRI-8910173.

situations and production rules attempt to model the problem-solving strategies employed by experts. DA, on the other hand, is an outgrowth of *prescriptive* psychology; its study of tasks, inferences, evaluations, and decisions is geared towards the rational maximization of subjective expected utility (SEU) [vWE86]. Insofar as it relates to human behavior, DA prescribes what decision makers *should* do, rather than describing what they do.

Decision-analytic task analyses often lead to prescriptions that differ systematically from what people would do on their own; these stereotypic deviations from normative behavior have been referred to, alternatively, as *heuristics and biases* [KST82] and *cognitive illusions* [vWE86]. Formally optimal prescriptions growing out of task analyses should have definite implications to the design of knowledge-based systems. A computer, after all, is a psychological *tabula rasa;* it should not suffer from the cognitive illusions that systematically cause humans—even experts—to err. A few recent expert systems have chosen to accentuate the normative over the descriptive—ADRIES [Lev88], IDES [AR87], Pathfinder [HHN90], and ARCO1 [AF90]—to name four.

The entire idea of knowledge-based systems centered around prescriptive ideas is relatively new. One theme that has rapidly gained popularity is the use of *belief networks*—graphical embodiments of hierarchical Bayesian analysis. Bayesian statistics differs from classical statistics in its view of probabilities as orderly opinions rather than as frequencies of occurrence [ELS63] [Che88]. The single most important implication of the Bayesian position to automated system design is that a Bayesian may reasonably ask an expert for an opinion about a rare event; a classical (non-Bayesian) statistician can elicit no such probability in the absence of voluminous data. For this and other less obvious reasons, all existing probability-based systems are Bayesian.

Belief networks are a large family of models; any graphical representation of a problem or a domain in which (i) nodes represent individual variables, items, characteristics, or knowledge sources, (ii) arcs demonstrate influence among the nodes, and (iii) functions associated with the arcs indicate the nature of that influence, qualifies as a member of the family. Prior to the 1980's, the only family member that attracted much attention was the decision tree. Decision trees have been studied in conjunction with virtually every decision problem and statistical technique imaginable; they have been proven themselves useful for some purposes, but not quite sophisticated enough for others. Recent work in AI and DA, however, has uncovered a pair of more powerful belief networks [HBH88]. In DA, *influence diagrams*—deriving their name from the influence indicated by the network's arcs—have been introduced to model complex decision problems [HM84] [Sha86]. Meanwhile, a group of AI researchers have forwarded *Bayes nets* as a mathematically precise method for managing uncertain information in an expert system [Pea88].

The technical distinction between these models is that whereas all nodes in a Bayes net or in a belief network model aleatory (probabilistic) variables and all relationships are defined in terms of conditional dependence, influence diagrams capture determinism, probability, value, and decisions. (Thus, belief networks are a proper subset of influence diagrams). The literature on influence diagrams, however, has generally ignored deterministic relationships, restricted itself to diagrams with a single value node, and focused primarily on discussions of probability and uncertainty—strikingly similar to the settings used for discussions of belief networks. In fact, many of the results derived on these models have been produced by the same researchers, and the distinctions between their uses and capabilities are all-too-often blurred together. Although that blurring is undoubtedly a mistake from a mathematical point of view, its impact on discourse is minimal; the two models are obviously variations on a common theme. The term "belief network" is used throughout this paper to reflect the author's personal preference.

The recent wave of relevant literature has concentrated on but a few topics. First, substantial attention has been paid to the comparative epistemological adequacy of probability theory and alternative systems for representing uncertainty [McL88] [NA90]. Second, a variety of mathematical studies have investigated the inferential and algorithmic power latent in belief networks [Sha86] [Pea88]. Third, systems like those mentioned above have shown that belief networks and related Bayesian models can be implemented in (at least) a few specific domains. Implicit in this literature is a powerful pair of claims:

1. If a belief network accurately captures the domain that it claims to be modeling, then its implications will be powerful and efficient.

2. Useful belief networks can be constructed for at least some interesting domains.

Proof of the first claim can be found in the papers discussing inference and algorithms, proof of the second from the systems already in existence. At this point in its development, then, the belief network is ready to be forwarded as a general model of knowledge representation, at least on par with logic, frames, and rules as an underlying representation for a knowledge-based system. This paper will outline the merits of the model that argue for its inclusion in the canon of knowledge representation, with an emphasis placed on its relationship to the central elements of an expert system: knowledge acquisition, knowledge representation, and inference.

2 Two Belief Network-Based Systems

Knowledge representation can not be discussed sans knowledge. Although most of the early work on belief networks was rather theoretical—their history can be traced from Wright's 1921 study of correlation and causation [Wri21] through Edwards' 1962 ideas about probabilistic information processing [Edw62] to Howard and Matheson's 1981 introduction of influence diagrams [HM84]—they have recently emerged as the bases of several powerful knowledge-based systems in fields as diverse as military reconnaissance [Lev88], automatic lathing [AR87], medical diagnosis [HHN90], and financial forecasting [AF90]. This paper will draw examples from two of these systems: Pathfinder and ARCO1.

Pathfinder, the result of five-plus years of ongoing collaboration between David Heckerman and Eric Horvitz of Stanford University, Bharat Nathwani of the University of Southern California (USC), and several other researchers and programmers, helps pathologists diagnose diseases by guiding their interpretations of microscopic features appearing in a section of tissue [HHN90]. Physicians at community hospitals are rarely experts in all aspects of pathology; they frequently need the help of subspecialists to reach accurate diagnoses. Pathfinder was designed to capture the knowledge of one subspecialty—hematopathology—so that community hospitals could obtain expert input without calling directly on the subspecialists. All tests run on Pathfinder suggest that it has met its primary objective: Pathfinder's diagnoses, given accurately recorded observations, are consistently accurate and on par with those derived by top hematopathologists [HHN90]. (Intellipath, a commercial version of Pathfinder, comes equipped with a laser-disc library of symptomatic cell features, designed to insure that the community hospital's observations are identified accurately).

ARCO1, currently being developed by the author in collaboration with Anthony Finizza, Mikkal Herberg, Peter Jacquette, and Paul Tossetti of the Atlantic Richfield Company (ARCO), models the oil market. It is being designed as an aid to the members of ARCO's corporate planning group who forecast the price of crude oil. The first round of the system was completed in February 1990 [AF90]. Tests are currently being run to assess the accuracy of its forecasts (relative to those of other forecasting tools) and the degree to which it actually helps members of its target audience. Since accuracy and usefulness are the two primary criteria by which the success of a system should be judged, it is too early to make any specific claims about ARCO1's power. It is not, however, too early to discuss the research issues uncovered during its design.

For the sake of this paper, neither system's details are significant[1]. What is important, however, is that much of the basic modeling software used in ARCO1 was borrowed from Pathfinder; Keung-Chi Ng wrote large portions of Pathfinder's code and for all of the modifications necessary for the transition between systems. (Both systems were written in MPW Pascal, an object-oriented language designed for the Mac II). The resultant shift in domain from medical diagnosis to financial forecasting is drastic, and helps highlight many issues in knowledge acquisition, encoding, modeling, and manipulation.

3 Knowledge Acquisition

The knowledge acquisition phases of Pathfinder and ARCO1 were heavily biased towards setup time. The most time-consuming components of knowledge acquisition were the convergence of domain experts and system designers to a common language and a specific approach, and the development of software tools that reflected that convergence. As the use of belief network based-systems matures, development time for software tools will decrease. The education of the system designers and domain experts in each other's languages and approaches, however, will probably always require substantial time and effort.

One of the major breakthroughs in the Pathfinder's development occurred during these preliminaries. The first step towards the design of any system must be the selection of a specific problem and approach. In terms of approach, the decision to use probabilistic modeling and DA elicitation techniques was reached through trial and error. Several representations of the uncertain medical data were attempted; the best performance was obtained with probabilities [Hec88]. Identifying a specific problem, however, was somewhat trickier. Pathfinder's domain contains around sixty diseases and 100 symptoms. Although 160 variables and their value ranges could easily be elicited, discussions of influence and dependence are obviously impractical. This difficulty helped refine the problem from "how can a disease be diagnosed?" to "how can we differentiate between a pair of similar diseases?" This refinement, in turn, led to the creation of the *similarity network*, an important addition to

[1]Discussions of ARCO1's history are based on the author's personal experience. Discussions of Pathfinder's history are based primarily on conversations with David Heckerman.

DA's collection of modeling tools [Hec90]. Pathfinder's similarity network listed all pairs of similar diseases; variables, values, influence, and dependence information was then elicited to differentiate between all similar pairs. These "local" belief networks were then coalesced into a single, coherent, "global" belief network that modeled the entire domain. The information necessary to construct the local networks was elicited rapidly, and in a manner that made the domain expert comfortable and confident with his assessments [HHN90].

In addition to facilitating the design of Pathfinder, similarity networks remove many of the theoretical difficulties that earned probabilistic systems a reputation for being unwieldy and impractical. They allowed elicitation to proceed tractably by applying the time-honored design principle of divide-and-conquer to their domain. By concentrating on diseases pairwise rather than as a set of 63, they reduced the scope of the problem for which probabilities had to be elicited. ARCO1 continued along a similar theme. It made elicitation practical by applying two other common design principles. First, it broke a one-year market model into four quarters, and treated many classes of concepts as subscripted variables whose relationships remained essentially unchanged between quarters. Second, it accepted many of the individual econometric models used by ARCO's forecasters, and specified the corresponding relationships as (deterministic) formulas rather than as lengthy collections of probability distributions. Between them, then, the designers of Pathfinder and ARCO1 built divide-and-conquer, subscripted variables, and algebraic relationships into belief network settings, thereby greatly expanding the scope of DA elicitation techniques. Few of these techniques were ever needed in standard DA settings because the sizes of the problems modeled were relatively small. The incorporation of these design principles allow DA techniques to scale up to models of about 150-plus nodes, if not further.

More specifically, the development of ARCO1 began in 1988 with an analysis of ARCO's forecasting models in light of recent developments in AI and DA [Abr88]. The first phase of this analysis focused on (i) identifying the strengths and weaknesses of the existing models, and (ii) decision analytic problem structuring. Since the current forecasts are based primarily on time series, many of their general characteristics are already well known [Mak86] [Arm88]; they are strong at projecting continued trends, but much weaker at identifying technical aberrations (e.g., turning points, discontinuities, spikes). Another area of difficulty—and one more specific to the oil industry—is that many of the variables affecting the market are essentially political. Whereas most economic variables are numeric, and thus amenable to time series modeling, political considerations tend to be qualitative. As a result, political and economic implications of the market are difficult to combine; political projections are usually used to temper the output of an economic model, and thus rarely given appropriate consideration in the ultimate forecast.

The oil market contains many interesting questions; market models could be used for corporate resource allocation, individual investment strategies, government policies, etc. The need for a highly specific problem, however, directed us towards price forecasting—the first and most basic issue addressed by any model of the market, regardless of its ultimate purpose. The selection of a specific approach was somewhat more difficult than the selection of a specific problem. A one year model of the oil market contains in the neighborhood of 100 nodes, ranging in character from highly precise and public economic data to subjective and occasionally vague interpretations of OPEC's internal politics. Results from the problem structuring phase convinced us that belief networks were appropriate models. In order to capture the domain elegantly, however, the networks had to be extended from the strictly probabilistic character that was appropriate for Pathfinder to more general algebraic models. Although the mathematics behind this shift were trivial, its implications to knowledge representation and inference are not; these topics will be discussed throughout the rest of the paper.

4 Knowledge Representation

The basic procedure for modeling a domain as a belief network can be summarized as:

1. Select a specific problem that is both relevant to at least some members of the target audience and simple enough to be discussed, modeled, and automated efficiently.
2. List all relevant variables.
3. Define the range of values taken on by each variable.
4. Indicate influence among the variables.
5. Specify the (functional or probabilistic) dependence of each variable on all variables that influence it.

The previous section discussed the work that went into selecting appropriate problems for both systems. With the systems' framework set and the design teams comfortable collaborating, actual elicitation and modeling could begin. From that point on, neither system required more than about forty hours of expert time to construct a belief network.

The information necessary to model a domain as a belief network can be broken into four categories: variables, values, influence, and dependence. Variables are simply the different objects within the domain that need to be captured by the model; eliciting them should rarely be difficult. Values are the ranges taken on by each variable; some may be binary, others multi-valued or even continuous. The only difficulties inherent in value elicitation arise from the occasional discretization of continuous ranges. Even in the most difficult cases, however, the domain experts of both Pathfinder and ARCO1 were able to arrive at value sets with which they were comfortable. Influence in a belief network is indicated by an arc; elicitation requires considering the existence (or nonexistence) of direct relationships between pairs of variables. Dependence is indicated by assigning specific relationships to sets of arcs. The value taken on by a variable is dependent on the values taken on by all variables that influence it (i.e., all incident arcs). Pathfinder represented all dependencies as conditional probabilities; ARCO1 extended the concept to include deterministic algebraic dependence, as well. Regardless of the types of allowable functions, however, dependence is the most controversial and difficult information to elicit. Fortunately, the technologies of DA provide many tools for eliciting and refining probability estimates. The power of these elicitation techniques—and the consistency with which they have helped decision-makers overcome cognitive illusions and reach good decisions [vWE86]—are among the greatest benefits accrued by modeling a domain as a belief network rather than as a production system.

Since belief networks lie at the crossroads of AI and DA, they should attempt to capture both fields' strengths while falling prey to the weaknesses of neither. A high-level analysis of DA indicates that one of its greatest assets is past success, while its greatest liability is its expense. Decision analysts have an extensive grab-bag of elicitation techniques; they have used these techniques to help many corporate, military, medical, and other decision makers construct accurate models of decision problems and make "good" decisions [vWE86]. Unfortunately, decision analyses are expensive and time consuming; cost renders DA impractical for all but crucial problems. AI, however, offers an almost complementary set of strengths and weaknesses; expert system technology contains few formal elicitation techniques, but tremendous insights into cost-effective model construction.

One of the most significant contributions of expert systems research has been the conceptual split of knowledge bases from inference engines. Under this division, knowledge bases contain information about a domain, while inference engines process case-dependent information and address specific problems. Thus, representation issues are isolated in the knowledge base; inference is performed by the engine. One of the most attractive features of an expert system's knowledge base is its reusability. This feature allows the design cost to be amortized over the lifetime of the system, thereby greatly extending the range of problems in which decision analytic modeling is practical. Belief networks, like the rule bases of expert systems designed with commercially available shells, are knowledge bases; as such, they offer a forum in which system designers can simultaneously avail themselves of the modeling prowess of DA and the cost effectiveness of AI. Thus, one major benefits provided by belief networks is that they use DA elicitation and modeling techniques to construct domain-level knowledge bases, but amortize cost over the lifetime of the system, a la AI.

Belief networks, however, do more than just facilitate collaboration between researchers in AI and DA. They also offer an environment in which system designers and domain experts can communicate comfortably. Few restrictions are placed on the types of knowledge that can be captured by a network's nodes; the only requirements are that they all have well-defined inputs and outputs, and that the inputs of one node be defined in terms of the outputs of its predecessors. The internal format of each node should be guided by the variable that it models. This practice, which is a marked contrast to the shell-based approach of forcing all information into rules, should increase an expert's willingness to participate in the modeling exercise. Thus, a second merit of belief networks is that each variable in a belief network is encoded in its most natural representation. Analyses leading to the construction of these nodes may be directed by the terminology and approaches with which the domain expert is most comfortable.

Most of the variables in Pathfinder, for example, describe symptoms and diseases. The diseases are collected into a single multi-valued "hypothesis node,"in which each of the 63 possible diseases defines a distinct value. Symptoms are each modeled as nodes with a relatively small set of discrete values (ranging form two values to ten). Functional relationships among these variables are almost exclusively conditional dependencies. In ARCO1, the variables are somewhat more varied; the mix of numeric and symbolic items in a complex politico-economic domain provides a wider variety of variable types than do most medical domains. Functional relationships in ARCO1 are similarly broad; they include an assortment of algebraic equations as well as conditional dependence. The analyses necessary to construct the Pathfinder and ARCO1 belief networks obviously had no overlap. The model construction software, on the other hand, transferred quite nicely from Pathfinder to ARCO1. The only areas that required substantial work were those in which Pathfinder's software assumed that all data would be symbolic, and did not allow ARCO1's heavily numeric data to be entered in the most efficient manner. (The range of possible values taken on by many economic variables, for example, can be described by a minimum, a maximum, and an interval. Since this type of description is rarely meaningful for symbolic variables, the software developed for Pathfinder contained no provisions for specifying variable ranges in this way). Another advantage of belief networks, then, is that the form of a belief network, and the software used to construct it, is essentially independent of both the domain itself and the types of questions typically discussed in the domain.

5 Inference and Information Processing

At the representation level, all belief network-based systems are strikingly similar, regardless of the types of issues that they address. The questions that a system is expected to answer must, however, have a substantial impact on its design. Diagnosis and forecasting, for example, are very different types of questions; medical diagnosis involves observing symptoms and projecting them backward to determine which disease is already present, while forecasting involves projecting forward to predict the future value of some variable. This distinction is completely transparent during the construction of the knowledge base—although it will, of necessity, color the discussions between system designers and domain experts. It is only in the inference engine that these considerations come into play, and it is in the inference engine that differentiation among systems occurs; engines designed to answer different types of questions must be driven by different information processing procedures. This ability to shift drivers can be contrasted to the match-select-act cycle that drives nearly all rule-based engines. It also suggests expanding the notion of an inference engine to that of a *processing engine,* where any procedure that relates a node's outputs to those of its predecessors is a viable driver for a processing engine. Thus, belief network processing engines may be readily tailored to different types of domains.

Most AI research has focused on symbolic, diagnostic domains, such as medicine. Since the questions of interest to diagnosticians are quite different from those that interest forecasters, engines will not transfer easily from diagnostic to forecasting domains. To appreciate this distinction, consider that in a medical setting, influence will generally be indicated from disease to symptom. Evidence, however, tends to be collected as symptoms are observed. In other words, a doctor might specify the probability with which disease A will cause symptom B to appear, or $P(B|A)$. When a specific case is being considered, however, symptom B will be observed, and the posterior probability of disease A, or $P(A|B)$, will be required. Thus, a diagnostic engine must work *against* the arrows in the belief network. Bayes' rule, which relates $P(A|B)$ to $P(B|A)$ is thus the obvious driver for a diagnostic engine.

In a forecast, on the other hand, influence is generally given in terms of either temporal precedence or ease of observation; evidence tends to be accumulated in the same direction, and the processing engine must work *along* the network's arrows. Thus, forecasting engines—such as ARCO1's—must differ substantially from diagnostic engines—like Pathfinder's. These graphical differences reflect a fundamental distinction between diagnoses and forecasts. Consider, for example, a system designed to forecast crude oil prices in the year 2000. As part of the knowledge acquisition phase, prices in 2000 may have been specified in terms of supply and demand in 2000, which, in turn, were dependent on some variables back in 1999, and so on, back to the present. An ideal output for the system would, of course, be a distribution of oil prices in the year 2000, given only the values of the currently observable variables. Unfortunately, this response can rarely be computed for anything but a very simple model. The complexity of multiplying all the relevant distributions is the product of the number of variables, values, and arcs in the network. A Monte Carlo analysis of the years between 1990 and 2000, on the other hand, should yield a reasonable approximation in an acceptable length of time. The simulation proceeds in a straightforward manner: Each rooted node (i.e., a node with no incident arcs) in the network is instantiated according to its specified priors, effectively "removing" the rooted nodes from consideration; all nodes that had depended solely on rooted nodes are now no longer conditioned on anything, and they too may be instantiated. This procedure continues until all variables but the one being forecast have taken on definite values. Although the subject of the forecast could, of course, also be instantiated, the nature of these analyses make it preferable to leave it undetermined, with the instantiation of its inputs describing it probabilistically. Thus, each run through the network yields a scenario describing the future, along with a distribution describing what the target variable (price in 2000) might look like in that future. Although no single scenario is likely to describe the future accurately, the distributions obtained by combining the results of multiple runs should define a meaningful "expected future." Thus, Monte Carlo analyses may replace Bayes' rule as the driving force behind a forecasting engine.

Drivers other than Bayes' rule and Monte Carlo are, of course, possible. In many respects, in fact, these procedures describe opposite ends of a spectrum. In a purely diagnostic system, all information will flow from leaf nodes (those with no outgoing arcs), against the network's arcs, to the rooted nodes. Thus, Bayes' rule is the only driver necessary. In a purely forecasting system, all information will flow from rooted nodes, along the network's arcs, to the leaf nodes; only Monte Carlo is needed. Many settings, however, fall into neither category. When evidence is found in mid-network, mechanisms for simultaneously transmitting information forwards and backwards are necessary; Henrion's *logic sampling* [Hen88] and Pearl's *stochastic simulation* [Pea88] are both reasonable candidates. They are not, however, the only ones. The makeup of the processing engine's driver, like that of individual nodes in a network, should be guided by the conventions of the domain in which the system is attempting to function, and by the types of questions that it is expected to answer.

6 Summary

The purpose of this paper was to forward belief networks as general tools for knowledge representation, with an emphasis placed on their potential as an overall framework for knowledge-based system design; the few examples drawn from existing systems were simply meant to be illustrative. Belief networks are already recognized as precise mathematical models on which many interesting algorithms can be developed. They have also been forwarded as general mechanisms for managing the types of uncertainty with which all AI systems must cope. Pathfinder and ARCO1 provide examples of the usefulness of belief networks as an overall framework for knowledge-based systems. This paper abstracted several principles of system design from specific instances of existing systems. Knowledge acquisition, knowledge representation, and information processing were all discussed; belief networks were shown to suggest clear approaches to all three. The phrase that best summarizes these approaches might well be *form follows function:* specific languages and tasks are selected to maximize the comfort of domain experts, knowledge sources are modeled as individualized nodes (whose only requirements are that they have well-defined inputs and outputs), and processing engines are designed to fit the specific type of questions anticipated within the domain. All of these approaches emerge from the beliefs that systems should be based on task analyses and designed to solve problems, and that the best way to do this is to think about the problem itself rather than the methods employed by human (expert) problem-solvers. Work is currently underway—at USC and elsewhere—to flesh out the underlying design philosophy and to use this philosophy to construct functioning advisory systems [AE89].

References

[Abr88] Bruce Abramson. Towards a Unified Decision Technology: Areas of Common Interest to Artificial Intelligence and Business Forecasting. Technical Report 88-01/CS, University of Southern California, 1988.

[AE89] Bruce Abramson and Ward Edwards. Competent Systems: A Prescriptive Approach to Knowledge-Based Advisors. In *Proceedings of the 1989 IEEE conference on Systems, Man, and Cybernetics*, pages 113–114, 1989.

[AF90] Bruce Abramson and Anthony J. Finizza. Belief Networks for Forecasting the Oil Market. Technical Report 90-06/CS, University of Southern California, 1990. To appear at the Tenth International Symposium on Forecasting, June 24-30, 1990, Delphi, GREECE.

[AR87] A.M. Agogino and A. Rege. IDES: Influence Diagram Based Expert Systems. *Mathematical Modeling*, 8:227–233, 1987.

[Arm88] J. Scott Armstrong. Research Needs in Forecasting. *International Journal of Forecasting*, 4:449–465, 1988.

[Che88] Peter Cheeseman. An Inquiry into Computer Understanding. *Computational Intelligence*, 4(1):58–66, 129–142, 1988.

[Edw62] W. Edwards. Dynamic Decision Theory and Probabilistic Information Processing. *Human Factors*, 4:59–73, 1962.

[ELS63] W. Edwards, H. Lindman, and L.J. Savage. Bayesian Statistical Inference for Psychological Research. *Psychological Review*, 70(3):193–242, May 1963.

[HBH88] E.J. Horvitz, J.S. Breese, and M. Henrion. Decision Theory in Expert Systems and Artificial Intelligence. *International Journal of Approximate Reasoning*, 2:247–302, 1988.

[Hec88] D.E. Heckerman. An Empirical Comparison of Three Inference Methods. In *Proceedings of the fourth Workshop on Uncertainty in Artificial Intelligence*, pages 158–169, 1988.

[Hec90] D.E. Heckerman. *Probabilistic Similarity Networks.* PhD thesis, Stanford University, 1990. In preparation.

[Hen88] Max Henrion. Propagating Uncertainty in Bayesian Networks by Probabilistic Logic Sampling. In Laveen Kanal and John Lemmer, editors, *Uncertainty in Artificial Intelligence 2*, pages 149–163. North Holland, 1988.

[HHN90] D.E. Heckerman, E.J. Horvitz, and B.N. Nathwani. Toward Normative Expert Systems: The Pathfinder Project. Technical Report KSL-90-08, Stanford University, 1990.

[HM84] Ronald A. Howard and James E. Matheson. Influence Diagrams. In Ronald A. Howard and James E. Matheson, editors, *Readings on the Principles and Applications of Decision Analysis, vol. II*, pages 721–762. Strategic Decisions Group, 1984.

[KST82] Daniel Kahneman, Paul Slovic, and Amos Tversky, editors. *Judgement Under Uncertainty: Heuristics and Biases.* Cambridge University Press, 1982.

[Lev88] Tod S. Levitt. Bayesian Inference for Radar Imagery Based Surveillance. In *Uncertainty in Artificial Intelligence 2.* North Holland, 1988.

[Mak86] Spyros Makridakis. The Art and Science of Forecasting. *International Journal of Forecasting*, 2:43–67, 1986.

[McL88] Mary McLeish, editor. *Taking Issue/Forum: An Inquiry into Computer Understanding.* Special issue of Computational Intelligence, 4(1), Feb. 1988, pages 55-142, 1988.

[NA90] Keung-Chi Ng and Bruce Abramson. Uncertainty Management in Expert Systems. *IEEE Expert*, 5(2):29–48, 1990.

[Pea88] Judea Pearl. *Probabilistic Reasoning in Intelligent Systems.* Morgan Kaufmann, 1988.

[Sha86] Ross D. Shachter. Evaluating Influence Diagrams. *Operations Research*, 34(6):871–882, 1986.

[vWE86] Detlof von Winterfeldt and Ward Edwards. *Decision Analysis and Behavioral Research.* Cambridge University Press, 1986.

[Wri21] Sewal Wright. Correlation and Causation. *Journal of Agricultural Research*, 20:557–585, 1921.

STOSS ---- A STOCHASTIC SIMULATION SYSTEM FOR BAYESIAN BELIEF NETWORKS

Zhiyuan Luo and Alex Gammerman

Department of Computer Science, Heriot-Watt University
79 Grassmarket, Edinburgh EH1 2HJ, U.K.

Abstract

This paper describes a computational system, called STOSS (STOchastic Simulation System), using the stochastic simulation method to perform probabilistic reasoning for Bayesian belief networks. The system is then applied to an artificial example in the field of forensic science and the results are compared with the calculations obtained using the Causal Probabilistic Reasoning System (CPRS).

Keywords

Bayesian belief networks, causal models, inference diagrams, legal reasoning, probabilistic reasoning, stochastic simulation.

1. INTRODUCTION

Recent work in theoretical statistics has shown that it is possible to adopt a sound probabilistic approach to uncertain inference using Bayesian belief networks [8, 10, 12]. For example, Lauritzen and Spiegelhalter [8] use a computational technique to perform reasoning in a graph structure. A few different probabilistic reasoning computational models using Lauritzen-Spiegelhalter's technique have been developed, for example, MUNIN [3], and CPRS [2, 5, 6].

Unfortunately, in the worst case, as it has been shown in [4], the complexity of this approach is an exponential function of the number of propositions in the Bayesian belief network, and this makes it difficult to compute probability distributions on large network. Henrion [7] developed a simulation technique called logic sampling, and Pearl suggested [11] an alternative way to perform probabilistic inference by using stochastic simulation. Stochastic simulation is a method of estimating probabilities of a proposition S_i by counting how often the states of the proposition occur in a series of random experiments. In general, given a probability of S_i, say $P(S_i)$, we can generate a sample space for S_i, in which the estimated probability will be "near" to $P(S_i)$ when the sample space is large enough. The sample generation process is called "simulation process". The complexity of this method is a linear function of the number of nodes and edges in the network and the size of sample space [11]; this would allow us to compute probability distributions on large networks. The objective of this paper is to describe a computational system called STOSS using the stochastic simulation method, which may form the basis of an inference mechanism for expert systems. The system is then applied to an artificial example in forensic science and the results are compared with the calculations obtained using the "exact" calculations of CPRS.

2. DESIGN AND IMPLEMENTATION

Let us assume that each proposition has a random number generator. The generator gives a state of the proposition at random with certain restrictions (usually causal relationships and probabilities provided by domain experts). The propositions are processed in turn. According to the approach [11], there are two tasks for each proposition in a simulation process. The first is to calculate the conditional probability of the proposition S_i given the current states of the other propositions. The second is to produce the next state of the proposition S_i. This state is decided by consulting a random number generator which is based on the conditional probabilities calculated in the first step. In the absence of evidence, the initial states of propositions are chosen arbitrarily. When evidence is observed, the states of observed propositions are fixed and the states of unobserved propositions are still assigned arbitrarily. A simulation is conducted again because we have extra restrictions, that is, observed evidence. The updated posterior probability distribution is estimated in the sample space generated.

An algorithm for generating a sample space in the causal model can be described as following. Note that R_{S_i} is the state of all propositions except S_i in the model.

Initializing states of propositions with evidence
WHILE (no query is raised)
  FOR each proposition S_i in the causal model DO
    Compute the conditional probability $P(S_i \mid R_{S_i})$
    Consult a random number generator
    Compare the outcome of the generator and $P(S_i \mid R_{S_i})$
    IF (the outcome of the generator > $P(S_i \mid R_{S_i})$)
     Assign "false" to the state of S_i
    ELSE
     Assign "true" to the state of S_i
     Record the state of proposition S_i
  END
END
Estimate the posterior probability

When we calculate the conditional probability $P(S_i|R_{S_i})$, we simplify the computations by using conditional independence assumptions reflected in the causal model. Let us consider a proposition S_i and its conditional probability $P(S_i|R_{S_i})$:

$$P(S_i|R_{S_i}) = \frac{\prod_{i=1}^{m} P(S_i \mid U_{S_i})}{P(R_{S_i})} \tag{2.1}$$

where m is the total number of nodes in the network and U_{S_i} is a set of parents of node S_i. $P(R_{S_i})$ is a constant independent of S_i. We can rewrite expression (2.1) as:

$$P(S_i|R_{S_i}) = \alpha P(S_i|U_{S_i})\prod_{j=1}^{\beta}P(C_j|U_{C_j})\prod_{k=1}^{\gamma}P(F_k|U_{F_k})$$

where α is a normalizing constant, C_j is a child of S_i and F_k is a node which is neither S_i's parent nor S_i's child. The factor $\prod_{k=1}^{\gamma}P(F_k|U_{F_k})$ in expression above is also a constant independent of S_i. Finally we can get (see Theorem 3.1 in [11]):

$$P(S_i|R_{S_i}) = \alpha P(S_i|U_{S_i})\prod_{j=1}^{\beta}P(C_j|U_{C_j}) \tag{2.2}$$

It is clear that we only need to consider S_i's parents, children and children's parents in order to compute conditional probability $P(S_i|R_{S_i})$ in the causal model.

The architecture developed for CPRS is very flexible and allow us to develop STOSS using basically the same architecture. Basically, the CPRS architecture consists of four main parts: Causal Model Construction, Preprocessor, Enter Evidence and Evidence Propagation. The dependency structure and conditional probabilities are set up in the Causal Model Construction procedure. A computational structure and internal representations for evidence propagation are built up and the initial marginal probability for each node is calculated in Preprocessor. When evidence is observed, it is provided to the system by Enter Evidence procedure. Finally, all the evidence is propagated in Evidence Propagation procedure.

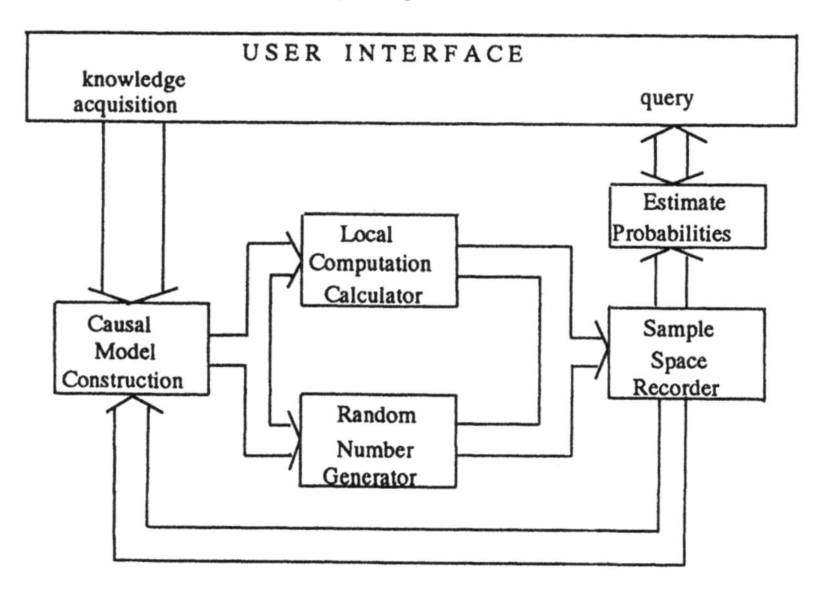

Figure 2.2 Architecture of STOSS System

The STOSS system has four additional modulars. They are *Causal Model Construction*, *Local Computation Calculator*, *Random Number Generator*, *Sample Space Recorder* and *Estimate Probability*. All programs are written in C under UNIX. The architecture of the STOSS is depicted in Figure 2.2 and the function of each modular is described below.

Causal Model Construction. A Bayeain belief network and corresponding probabilities are transferred into the computational model through a *Knowledge Acquisition* procedure contained in User Interface. All this knowledge is changed into internal representation forms and stored in a file called "Causal Model" for the Local Computation Calculator and the Random Number Generator.

Local Computation Calculator calculates the conditional probabilities of a node given current states of other nodes by using formula (2.2). Only the node's parents, children and children's parents are considered in this calculation. The conditional probabilities of the node are sent to the Sample Space Recorder.

Random Number Generator generates a real number between 0 and 1 for the node processed in the Local Computation Calculator. The outcome of the generator goes to the Sample Space Recorder.

Sample Space Recorder gets input from both the Local Computation Calculator and the Random Number Generator. The state of the node, which is determined by comparing the outcome of the Random Number Generator procedure and of the Local Computation Calculator procedure, is recorded in the Sample Space Recorder. This new state is also sent to the Causal Model file for simulating other nodes.

Estimate Probability. When the user raises a query, the current probabilities are estimated from the Sample Space Recorder. The estimated probabilities are presented to the user through the User Interface.

3. APPLYING STOSS TO AN EXAMPLE

In this section, we discuss an experiment in applying STOSS to an example in the field of forensic science described in [1].

> A murder has been committed. There are two suspects, X and Y, who are associates and who say they met the victim, Z, some time before the commission of the crime. If there were an eyewitness to this meeting it would be interesting to know from that witness if the meeting had been cordial or not and, in particular, if there had been a fight. Since X and Y are associates it is feasible that Y may pick up something from X and then deposit it at the scene of the crime. For example, fibres from a jacket of X's may be picked up by some garment of Y's and then be left at the crime scene by Y, thus incriminating X who may, in fact, be perfectly innocent. Such transfer from X to Y may take place if, for example, Y drives X's car frequently. Transfer of evidence may also occur in the other direction, from scene to criminal. For example, a window has been broken at the scene of the crime. Glass may be found in the soles of the shoes of both X and Y but one or other may be glaziers and this will affect the inferences which may be made. Similarly, if Y drives X's car frequently Y may pick up glass from X's car; the glass may or may not have come from the crime scene.

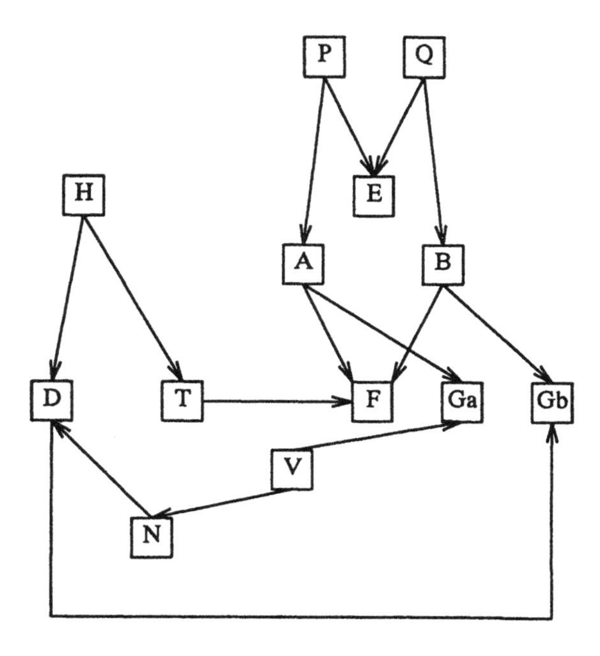

Figure 3.1 Bayesian Belief Network

	Interpretation	Label
E	Eyewitness evidence	S_1
P	X quarrels with Z	S_2
Q	Y quarrels with Z	S_3
B	Y committed the murder	S_4
A	X committed the murder	S_5
H	Y drives X's car frequently	S_6
V	X is a glazier	S_7
T	Y picks up fibres from X's jacket	S_8
N	X leaves glass in his car	S_9
D	Y picks up glass from X's car	S_{10}
F	Fibres evidence	S_{11}
Gb	Glass is found in soles of shoes of Y	S_{12}
Ga	Glass is found in soles of shoes of X	S_{13}

Tab. 3.1 List of Propositions

The Bayesian belief network displayed in Figure 3.1 summarizes different relationships. A "node" in the network represents a proposition, which may be either a piece of evidence, for example that Y drives X's car frequently, or a conclusion, for example that the suspect X committed the crime. A list of the propositions for our example is given in Tab. 3.1. Each proposition may have two states. Upper case letters are used to represent propositions and lower case letters to represent their particular states, for example, t and ¬t ("true" and "false") are two states of proposition T. An arrow between two nodes, say P to E, represents a "causal" relationship and it is said that node P "precedes" node E in the network since there is a directed path between P and E. 36 numbers are necessary to completely specify the model and fictitious probabilities are given below (Tab. 3.2). The task is: given prior probability distributions for nodes P, Q, H, and V, and conditional probabilities of the other nodes on their parents, we wish to assess our revised belief in the guilt or otherwise of X and Y in the light, say, of "fibres evidence".

P(p)=0.05	P(t | h)=0.20	P(d | h, n)=0.40
P(q)=0.05	P(t | ¬h)=0.01	P(d | h, ¬n)=0.01
P(v)=0.02	P(f | a, b, t)=0.80	P(d | ¬h, n)=0.02
P(e | p, q)=0.05	P(f | a, b, ¬t)=0.70	P(d | ¬h, ¬n)=0.01
P(e | p, ¬q)=0.05	P(f | a, ¬b, t)=0.70	P(ga | a, v)=0.80
P(e | ¬p, q)=0.05	P(f | a, ¬b, ¬t)=0.70	P(ga | a, ¬v)=0.70
P(e | ¬p, ¬q)=0.05	P(f | ¬a, b, t)=0.20	P(ga | ¬a, v)=0.75
P(h)=0.70	P(f | ¬a, b, ¬t)=0.03	P(ga | ¬a, ¬v)=0.01
P(a | p)=0.50	P(f | ¬a, ¬b, t)=0.01	P(gb | b, d)=0.70
P(a | ¬p)=0.01	P(f | ¬a, ¬b, ¬t)=0.01	P(gb | b, ¬d)=0.70
P(b | q)=0.50	P(n | v)=0.90	P(gb | ¬b, d)=0.20
P(b | ¬q)=0.01	P(n | ¬v)=0.10	P(gb | ¬b, ¬d)=0.05

Tab 3.2 Prior and Conditional Probabilities Table

sample size	100	500	1000	5000	10000
P(p)	.1161	.0566	.0664	.0471	.0514
P(q)	.0884	.0538	.0601	.0540	.0494
P(h)	.7023	.7060	.7003	.6987	.7011
P(v)	.0454	.0275	.0160	.0203	.0240
P(a)	.1107	.0395	.0570	.0316	.0365
P(e)	.0500	.0500	.0500	.0500	.0500
P(b)	.0749	.0416	.0458	.0395	.0355
P(t)	.1374	.1426	.1423	.1424	.1435
P(ga)	.1359	.0586	.0605	.0459	.0530
P(n)	.1686	.1361	.1090	.1169	.1229
P(f)	.0414	.1070	.0521	.0328	.0368
P(d)	.0564	.0851	.0398	.0415	.0445
P(gb)	.0852	.1198	.0866	.0821	.0788

Tab. 3.3 Initial Belief (STOSS)

In the absence of any evidence, the initial states of nodes in the network are chosen arbitrarily. For demonstration purposes, we choose the sample size as 100, 500, 1000, 5000 and

10000 in turn. The estimated probabilities in these sample spaces are given in Tab. 3.3. When the fibres evidence is confirmed, once again we conduct a simulation starting with the state of proposition F fixed as "true". The estimated probabilities after acceptance of the fibres evidence is displayed in Tab. 3.4.

sample size	100	500	1000	5000	10000
P(p)	.5588	.5573	.5390	.5043	.5119
P(q)	.0953	.0789	.0790	.0801	.0805
P(h)	.7038	.7109	.7084	.7067	.7075
P(v)	.0917	.0361	.0242	.0240	.0220
P(a)	.7233	.7335	.7173	.6786	.6897
P(e)	.0500	.0500	.0500	.0500	.0500
P(b)	.0855	.0724	.0746	.0772	.0769
P(t)	.1679	.1650	.1649	.1651	.1631
P(ga)	.5938	.5534	.5268	.4922	.4979
P(n)	.2049	.1320	.1154	.1192	.1194
P(f)	1.000	1.000	1.000	1.000	1.000
P(a)	.0774	.0485	.0402	.0433	.0436
P(gb)	.1233	.1056	.1044	.1063	.1069

Tab. 3.4 After Acceptance of the Fibres Evidence (STOSS)

The accuracy of the estimated probabilities increases with the sample size of the simulation. If we look at the estimated probability of Gb, P(gb), in Tab. 3.3, it changes from 0.0852 to 0.0788 when the size of sample space increases from 100 to 10000.

P(p)	0.0500
P(q)	0.0500
P(h)	0.7000
P(v)	0.0200
P(a)	0.0345
P(e)	0.0500
P(b)	0.0345
P(t)	0.1430
P(ga)	0.0482
P(n)	0.1160
P(f)	0.0353
P(d)	0.0420
P(gb)	0.0785

Tab. 3.5 Initial Belief (CPRS)

P(p)	.5043
P(q)	.0785
P(h)	.7065
P(v)	.0200
P(a)	.6847
P(e)	.0500
P(b)	.0753
P(t)	.1631
P(ga)	.4884
P(n)	.1160
P(f)	1.000
P(d)	.0423
P(gb)	.1048

Tab. 3.6 After Acceptance of the Fibres Evidence (CPRS)

In [2], a probabilistic reasoning system CPRS which uses Lauritzen-Spiegelhalter's technique [8] was described and implemented. CPRS was one of the first probabilistic reasoning shell systems and has been applied in the field of legal reasoning. It is interesting to compare the performance of the CPRS and STOSS using the same example. The outcome of applying CPRS is shown in Tab. 3.5 and Tab. 3.6.

It is clear that the results of the stochastic simulation system STOSS are "optimal" in the sense that the estimated probability of a proposition using STOSS is "relatively" close to that of the CPRS. For example, the prior probability of Gb, P(gb) in Tab. 3.5 is 0.0785 in the CPRS and that is 0.0788 in the STOSS where the sample size is 10000. Moreover, they remain close after an evidence is propagated. For example, when the "fibres evidence" is observed, our belief in Gb is 0.1048 using CPRS and 0.1069 using STOSS. However, as the example given here is small, there is not much difference in the aspect of computational complexity of two systems.

In general, the complexity of STOSS is a polynomial function of the size of the network, where the size is the sum of the number of nodes and the number of edges in the network [4]. However, our experiments show that the performance of the STOSS may degrade rapidly when there are some extreme conditional probabilities (close to one or zero). The convergence of the algorithm is a problem in this case [9]. Therefore, research for improving the performance of the STOSS with extreme conditional probabilities is needed.

4. CONCLUSION

Stochastic simulation is an alternative way to perform probabilistic inference correctly and is computationally tractable in causal models. A computational system using stochastic simulation is described and implemented. The system has been tested using an artificial example in the field of forensic science. We believe that STOSS provides new opportunities for maintaining uncertain reasoning in complex real problems.

REFERENCES

1. Aitken C.G.G. (1988) In discussion of Lauritzen, S.L. and Spiegelhalter, D.J. Local computations with probabilities on graphical structures and their application to expert systems, J.R. Statist. Soc. (Series B) 50 (1988), 200-201.

2. Aitken C.G.G. and Gammerman A. (1989) Probabilistic reasoning in evidential assessment, Journal of the Forensic Science Society 29 (1989), 303-316.

3. Andreassen S., Woldbye, M., Falck, B. and Andersen, S. (1987) MUNIN - A causal probabilistic network for interpretation of electromyographic findings, Proc. 10th IJCAI 2 (1987), 366-372.

4. Cooper G. (1989) Current research directions in the development of expert systems based on belief networks, Applied Stochastic Models And Data Analysis 5 (1989), 39-52.

5. Gammerman A. and Crabbe W. (1987) Computational models of probabilistic reasoning in expert systems: a causal probabilistic reasoning system, Technical Report 87/16. Computer Science Department, Heriot-Watt University, Edinburgh, U.K.

6. Gammerman A. (1988) In discussion of Lauritzen, S.L. and Spiegelhalter, D.J. Local computations with probabilities on graphical structures and their application to expert systems, J.R. Statist. Soc. (Series B) 50 (1988), 200.

7. Henrion M. (1988) Propagating uncertainty by logic sampling in Bayes' networks, In Uncertainty in Artificial Intelligence 2 (Kanal L.N, Lemmer, J.F. eds.), North-Holland, 149-163.

8. Lauritzen S. and Spiegelhalter D. (1988) Local computations with probabilities on graphical structures and their application to expert systems, J.R. Statist. Soc. (Series B) 50 (1988), 157-224.

9. Luo Z. and Gammerman A. (1989) Probabilistic reasoning using stochastic simulation approach in causal models, Technical Report 89/14. Computer Science Department, Heriot-Watt University, Edinburgh, U.K.

10. Pearl J. (1986) Fusion, propagation, and structuring in belief networks, Artificial Intelligence 29 (1986), 241-288.

11. Pearl J. (1987) Evidential reasoning using stochastic simulation of causal models, Artificial Intelligence 32 (1987), 245-257.

12. Shachter R. (1986) Evaluating influence diagrams, Operations Research 34 (1986), 871-882.

CONDITIONAL EVENTS WITH VAGUE INFORMATION IN EXPERT SYSTEMS

Giulianella Coletti (*) Angelo Gilio (**) Romano Scozzafava (**)

(*) Dipartimento di Matematica, Universita di Perugia, 06100 PERUGIA, Italy.

(**) Dipartimento di Metodi e Modelli Matematici, Universita' "La Sapienza", Via Scarpa 10, 00161 ROMA, Italy.

Abstract

Ad hoc techniques and inference methods used in expert systems are often logically inconsistent. On the other hand, among properties and assertions concerning handling of uncertainty, those which turns out to be well founded can be in general easily deduced from probability laws. Relying on the general concept of event as a proposition and starting from a few conditional events of initial interest, a gradual and coherent assignment of conditional probabilities is possible by resorting to de Finetti's theory of coherent extension of subjective probability. Moreover, even when numerical probabilities can be easily assessed, a more general approach is obtained introducing an ordering among conditional events by means of a coherent *qualitative* probability.

Key-words

Coherence, numerical probability, qualitative probability, uncertainty in expert systems.

1.A STRUCTURE-FREE PROBABILISTIC APPROACH

Many expert systems still rely on *ad hoc* techniques that have little or no theoretical justification.

On the other hand, one of the several fundamental issues concerning Artificial Intelligence is how to deal with uncertainty and vague information in any specific situation of interest, since in many practical cases it is very difficult to build expert systems which have a sufficient degree of flexibility with respect to probability judgements. In fact it is in general mantained that assessments of probabilities require an overall design on the whole set of all possible envisaged situations, and hence that they cannot be easily achieved.

Nevertheless a gradual and coherent assignment of (conditional) probabilities, starting from a few (conditional) events of initial interest, is possible by resorting to de Finetti's theory of coherent extension of subjective probability, as two of us did in Gilio and Scozzafava (1988).

Notice that we rely on the general concept of event as a *proposition* (and probability is looked upon as an ersatz for the lack of information on its actual "value", true or false) so that we can deal also with those uncertain situations in which other approaches need to introduce devices such as "context" or "protocol", or similar stuff. In particular, there is no need to distinguish between hypotheses and evidence.

In order to take into account the need of updating all probability evaluations, all uncertain statements in the expert system are expressed by *conditional* events, which are the natural tool for the Bayesian methodology.

As it is well known, de Finetti's approach (see, for example, de Finetti (1972)) refers to an *arbitrary* set $\mathcal{E}$ of conditional events, *with no underlying structure* (such as Boolean ring, σ-algebra, etc.) and *coherence* of a function P on $\mathcal{E}$ (with range between 0 and 1) entails that P is a *conditional probability distribution*.

The converse in general is not true: sufficient conditions for the coherence of P concern the structure of $\mathcal{E}$, and so the interest for a direct check of coherence is clear.

The study of coherence can be based on mathematical conditions derived by two well known (and equivalent) criteria: bet and penalty methods (see, for example, de Finetti (1970)).

This approach can be exploited using also linear programming tools, as in Bruno and Gilio (1980). See also Nilsson (1986), whose ideas concerning the so-called *probabilistic logic* essentially go back to de Finetti's pioneering papers.

Many other relevant concepts *could be* related to our approach.

We will not deal with questions such as logical operations among conditional events (touched upon also by de Finetti (1970), p.689, and extensively treated, for example, by one of us in Bruno and Gilio (1985), by Dubois and Prade (1988), by Goodman and Nguyen (1988) and by Schay (1968)), neither with approaches based on the logic of conditionals as that by E. Adams (1975).

In particular, we recall that our interpretation of the concept of conditional event $E|H$ (cfr. Sect. 2) is different from that adopted in Dubois and Prade (1988), Goodman and Nguyen (1988) and Schay (1968)), where $E|H$ is informally interpreted as "if H then E" and nevertheless a truth value is given to the case "H true, E false"; moreover also in the case "H false" our interpretation differs from theirs.

We remain within the framework of subjective coherent probability theory (numerical or qualitative): by the way, this approach allows a simple discussion of putative paradoxes in logic or in plausible reasoning (such as those discussed in the book by Pearl (1988), p.447 and p.495, and by one of us, with a different perspective, in Scozzafava (1980) and in Barigelli and Scozzafava (1984).

2. COHERENCE OF NUMERICAL PROBABILITIES

Given a family $\mathcal{F}$ of n conditional events $E_1|H_1,\ldots,E_n|H_n$, let us introduce the *atoms* C_h generated by the 2n events $E_1,H_1,\ldots,E_n,H_n$ and contained in the union of the H_i's. As it is well-known, the atoms are generated by all possible intersections (different from the impossible event $\emptyset$) of the 2n events or their negations.

A sufficient condition for coherence (cfr. Gilio (1990)) is the following: a *prevision point* $\mathcal{P} = (p_1,\ldots,p_n)$ is *coherent* (and so defines a probability distribution on $\mathcal{F}$, with $P(E_i|H_i) = p_i$) if there exists a solution $(\lambda_1,\lambda_2,\ldots,\lambda_r) = \lambda$ of the system

$$\sum_h \lambda_h = 1, \quad \lambda_h > 0 \ (h = 1, \ldots, r), \quad p_i = \sum_h \lambda_h q_{hi} \quad (i = 1, \ldots, n) , \qquad (2.1)$$

where each q_{hi} in (2.1) takes one of the three values $1,0,p_i$ according to whether the corresponding C_h is contained in E_iH_i, or in $E_i^cH_i$, or in H_i^c, respectively. (We denote by AB the intersection of the events A and B).

The q_{hi}'s define the *generalized atoms* Q_h introduced in Gilio (1990), i.e.

$$Q_h = (q_{h1},q_{h2},\ldots,q_{hn}) ,$$

with $h = 1,2,\ldots,r$. In other words, a conditional event $E_i|H_i$ is looked upon as a three-valued event, the third value being its probability p_i (this interpretation is the most natural in terms of bets, while many authors assert that $E_i|H_i$ is *undefined* when H_i is false).

2.1. Example

Let us now show how this procedure works in a particular case. Given three general propositions (events) A, H_1, H_2 and the probability evaluations

$$\alpha = P(A|H_1), \quad \beta = P(A|H_2), \quad \gamma = P(H_2|H_1),$$

we get easily the generalized atoms

$$Q_1 = (1,1,1), \quad Q_2 = (0,0,1), \quad Q_3 = (1,\beta,0),$$

$$Q_4 = (0,\beta,0), \quad Q_5 = (\alpha,1,\gamma), \quad Q_6 = (\alpha,0,\gamma).$$

Then a simple computation allows to write system (2.1) in the form

$$\alpha = \frac{\lambda_1+\lambda_3}{\lambda_1+\lambda_2+\lambda_3+\lambda_4}, \quad \beta = \frac{\lambda_1+\lambda_5}{\lambda_1+\lambda_2+\lambda_5+\lambda_6}, \quad \gamma = \frac{\lambda_1+\lambda_2}{\lambda_1+\lambda_2+\lambda_3+\lambda_4},$$

with the λ_h's all positive. Now, taking, for arbitrary $a,b > 0$,

$$\lambda_1 = \lambda_3 = \lambda_4 = \lambda_6 = \lambda, \qquad \lambda_2 = a\lambda, \qquad \lambda_5 = b\lambda, \tag{2.2}$$

with $\lambda = 1/(4+a+b)$, it follows

$$\alpha = \frac{2}{3+a}, \quad \beta = \frac{1+b}{2+a+b}, \quad \gamma = \frac{1+a}{3+a}.$$

So, given $\varepsilon > 0$, it is possible to have, for example,

$$\alpha < \varepsilon \ , \quad \beta > 1-\varepsilon \ , \quad \gamma > 1-\varepsilon$$

for suitable choices of a and b. This means that the probabilities satisfying these bounds are among the coherent assignments determined by the values (2.2).

Introduce now, besides the previously considered $A|H_1$, $A|H_2$, $H_2|H_1$, the further conditional event $A|H_1H_2$: repeating the same procedure yields for its probability δ the possible values (depending on the parameter a)

$$\delta = P(A|H_1H_2) = \frac{1}{1+a}.$$

On the other hand, if $\alpha < \varepsilon$ and $\gamma > 1-\varepsilon$, since

$$\varepsilon > \alpha = P(A|H_1) \geq P(AH_2|H_1) = P(A|H_1H_2)P(H_2|H_1) > \delta(1-\varepsilon),$$

we have $\delta < \varepsilon/(1-\varepsilon)$.

A possible application of these results is to the so-called *"penguin triangle"* example, as treated for example in Pearl (1988). Consider the events A = "Tweety flies", H_1= "Tweety is a penguin", H_2= "Tweety is a bird", and assume (as lovers of such kind of examples do) that H_1 does not imply H_2. Our conclusions are consistent with those in Pearl's book, pp. 483-484, but in our approach we are able to face, by means of coherence, also the very important and inescapable aspect concerning "the legitimacy" of the conditions $\alpha < \varepsilon$, $\beta > 1-\varepsilon$ and $\gamma > 1-\varepsilon$ when taken as initial assumptions.

Let us now go back to the above sufficient condition of coherence. An important particular case is when $H_1 = H_2 = \ldots = H_n = H_o$: then obviously q_{hi} is 0 or 1 for all values of h and i. Moreover, in this case we have (cfr. Gilio

(1990)) a necessary and sufficient condition for the coherence of $\mathcal{P}$, that is the existence of a solution λ of the system

$$\sum_h \lambda_h = 1, \quad \lambda_h \geq 0 \quad (h = 1, \ldots, r), \qquad p_i = \sum_h \lambda_h q_{hi} \quad (i = 1, \ldots, n) \ . \qquad (2.3)$$

Another particular case is when $H_i H_k = \emptyset$ for all pair i,k (with $i \neq k$). Then all $\mathcal{P}$ such that $0 \leq p_i \leq 1$ (with i=1,...,n) are coherent, with $p_i = 0$ if $E_i H_i = \emptyset$ and $p_i = 1$ if $H_i \subseteq E_i$. For example, given a family $\mathcal{F}$ of n conditional events and a prevision point $\mathcal{P}$, suppose that the events of $\mathcal{F}$ can be clustered in a number of subfamilies $\mathcal{F}_1, \mathcal{F}_2, \ldots, \mathcal{F}_m$, with $\mathcal{F} = \mathcal{F}_1 \cup \ldots \cup \mathcal{F}_m$, $\mathcal{F}_h \cap \mathcal{F}_r = \emptyset$ $(h \neq r)$ such that $H_i H_k = \emptyset$ for $E_i|H_i \in \mathcal{F}_h$, $E_k|H_k \in \mathcal{F}_r$: in this case it is possible to prove that the coherence of $\mathcal{P}$ can be checked in each subfamily.

The evaluation of probabilities may not be a straightforward problem; as a natural consequence the interest toward the use in expert systems of *qualitative probability* is growing: for a recent description, see Coletti (1990).

3. QUALITATIVE PROBABILITY AND COHERENCE

A qualitative probability is a suitable order relation $\geq\cdot$ (on a set of events) expressing the intuitive idea of *"not less probable than"*: we call $\geq\cdot$ *coherent* if it can be represented by a (coherent) numerical probability.

On the other hand, the results in the relevant literature concerning orderings among *conditional* events are in general not suitable for use in expert systems: they usually refer to given well structured sets of events (such as rings or σ-algebras) and conditions for numerical representability are not easily achievable.

We study (cfr. also Coletti, Gilio and Scozzafava (1990)) a suitable model in which vague and varying information is represented by conditional events carrying a conditional qualitative probability, i.e. the expert introduces through qualitative judgements an order relation on a set $\mathcal{F}$ of conditional events not endowed with any particular structure (that is, $\mathcal{F}$ contains only those events strictly related to the problem); moreover at any time the set $\mathcal{F}$ can be extended referring to further conditional events.

This approach is able to check coherence and possibly to give numerical evaluations of conditional probabilities compatible with the order relation.

We list some of our results concerning sufficient conditions for the representability of an ordering $\geq\cdot$ among conditional events by a (coherent) conditional probability P on $\mathcal{F}$ such that

$$E_i|H_i \geq\cdot E_k|H_k \quad \longleftrightarrow \quad P(E_i|H_i) \geq P(E_k|H_k).$$

In the particular case $\mathcal{F} = \{E_1, \ldots, E_n\}$, i.e. $H_1 = \ldots = H_n = \Omega$, where Ω

is the *certain* event, we have

(i) The ordering

$$E_1 \geq\cdot E_2 \geq\cdot \dots \geq\cdot E_n$$

is numerically representable by a coherent point $(p_1,\dots,p_n)$ *such that*

$$p_1 \geq p_2 \geq \dots \geq p_n ,$$

if there exist n-1 *atoms* $C_1,\dots,C_{n-1}$ *(generated by the* n *given events) such that*

$$C_1 \subseteq E_1 E_2^c, \quad C_2 \subseteq E_1 E_2 E_3^c, \dots, C_{n-1} \subseteq E_1 \dots E_n^c .$$

Proof - Putting

$$x_{hr} = \begin{cases} 1, & C_h \subseteq E_r \\ 0, & C_h \subseteq E_r^c \end{cases}, \quad h = 1,2,\dots,n-1; \; r = 1,2,\dots,n \quad ,$$

the atoms generated by $\mathcal{F}$ can be represented as binary vectors

$$C_h = (x_{h1}, x_{h2}, \dots, x_{hn})$$

and each event E_r by the vector

$$V_r = (x_{1r}, x_{2r}, \dots, x_{sr}) .$$

Let s be the number of atoms: by a classical result of de Finetti (1970), a prevision point $\mathcal{P} = (p_1, p_2, \dots, p_n)$ is coherent if and only if a probability distribution $W = (w_1, w_2, \dots, w_s)$, with $w_h = P(C_h)$, exists on the atoms C_h, such that $\mathcal{P} = \sum_{h=1}^{s} w_h C_h$. In the present case, we can take $\mathcal{P}$ in this form, with n-1 in place of s, and

$$\sum_{r=1}^{n-1} w_r = 1 , \qquad w_k \geq \sum_{r=1}^{k-1} w_r = s_{k-1} , \qquad k = 2,3,\dots,n-1 .$$

It follows

$$p_1 = 1 , \; p_2 = 1-w_1 , \text{ and } \; 1-s_k \leq p_{k+1} \leq 1-w_k \leq 1-s_{k-1} \leq p_k \leq 1-w_{k-1}$$

for $k = 3,4,\dots,n$; therefore $p_1 \geq p_2 \geq \dots \geq p_n$.

3.1. Example

If $n = 3$ and $E_2 E_1 \neq \emptyset$, while the other two intersections are $E_3 E_1 = E_3 E_2 = \emptyset$, then $C_1 = E_1 E_2^c$, $C_2 = E_1 E_2 E_3^c$, and we can take $w_1 + w_2 = p_1 = 1/2$, with $w_1 = 1/6$ and $w_2 = 1/3$, $P(E_1^c E_2^c E_3^c) = 1/4$, so that

$$p_1 = 1/2 \geq p_2 = 1/3 \geq p_3 = 1/4 .$$

Another possible assignment is to take $w_1 + w_2 = 1$: then

$$p_1 = 1 \geq p_2 = w_2 \geq p_3 = 0 .$$

(ii) If all the following events

$$C_r = E_1^c \ldots E_{r-1}^c E_r E_{r+1}^c \ldots E_n^c \ , \quad r=1,2,\ldots,n \ ,$$

are atoms, i.e. are all different from the impossible event $\emptyset$, then every ordering of $E_1, E_2, \ldots, E_n$ *is numerically representable by a coherent point* $(p_1,\ldots,p_n)$.

Proof - Let $E_1 \geq\cdot E_2 \geq\cdot \ldots \geq\cdot E_n$ be the given ordering. Then, taking

$$\mathcal{P} = \sum_{r=1}^{n} w_r C_r \ ,$$

with $\sum_{r=1}^{n} w_r = 1$, $w_r \geq w_{r+1}$, $r = 1, 2, \ldots, n-1$, it follows

$$p_r = w_r \geq w_{r+1} = p_{r+1} \ , \quad r = 1, 2, \ldots, n-1 \ .$$

The coherence of any other ordering can be proved by a suitable permutation of the indexes.

In particular, the conditions which entail the above conclusion are obviously satisfied when the number of atoms equals its maximum value 2^n. In this case also the conditions under *(i)* hold. Notice the further interesting particular case in which the events $E_1, E_2, \ldots, E_n$ constitute a *partition* of the certain event Ω.

The computational tool needed to apply the above conditions is a suitable binary matrix corresponding to the set of atoms.

In the general case in which $\mathcal{F}$ is a family of n conditional events, if $\mathcal{F}$ contains $E_i|H_i$ such that $E_i H_i = \emptyset$, and $E_k|H_k$ such that $H_k \subseteq E_k$, then for *every* ordering

$$E_1|H_1 \geq\cdot E_2|H_2 \geq\cdot \ldots \geq\cdot E_n|H_n \qquad (3.1)$$

one necessarily has $i=n$ and $k=1$.

Let $\mathcal{C}$ be the set of atoms generated by the $2n$ events $E_1,\ldots,E_n,H_1,\ldots,H_n$ and contained in the union H_o of the H_i's. Then

(iii) The ordering (3.1) *is numerically representable by means of a coherent point* $(p_1,\ldots,p_n)$ *if among the atoms belonging to* $\mathcal{C}$ *there are* n-1 *such that*

$$C_r \subseteq E_1 H_1 E_2 H_2 \ldots E_r H_r E_{r+1}^c H_{r+1} \ , \qquad r = 1,2,\ldots,n-1 \ .$$

Proof - Put $P(C_h|H_o) = w_h$ for each $C_h \in \mathcal{C}$ and take

$$\sum_{r=1}^{n-1} w_r = 1 \ , \quad w_k \geq s_{k-1} = \sum_{r=1}^{k-1} w_r \ , \quad k = 2, 3, \ldots, n-1 \ .$$

Since $P(E_1 H_1|H_o) = P(H_1|H_o) = 1$, then $P(E_1|H_1) = 1$. Now it is, for each r, $H_r H_o = H_r$, so that, by a well-known probability theorem,

$$P(E_r H_r|H_o) = P(E_r|H_r)P(H_r|H_o) \ .$$

It follows, for $r \geq 2$,

$$P(E^c_{r+1}|H_{r+1}) = P(E^c_{r+1}H_{r+1}|H_o)/P(H_{r+1}|H_o) \geq P(E^c_{r+1}H_{r+1}|H_o) \geq w_r ,$$

and so $p_{r+1} = P(E_{r+1}|H_{r+1}) \leq 1-w_r \leq 1-s_{r-1}$. On the other hand, since $C_k \subseteq E_rH_r$ for $k \geq r$ and $P(E_r|H_r) \geq P(E_rH_r|H_o)$, one has $p_r \geq \sum_{k=r}^{n-1} w_k = 1-s_{r-1}$. Hence $p_r \geq p_{r+1}$ and the ordering is coherent.

3.2. Example

A doctor has to make a diagnosis: let E_1, E_2, E_3 denote some (possibly not exclusive) hypotheses concerning the disease of the patient and H_o the initial "information" of the doctor. The above conditions are met if $E_1H_oE_2^c$ and $E_1H_oE_2E_3^c$ are both different from $\emptyset$: a necessary requirement is that not every patient having disease E_1 has also disease E_2 and that not every patient having both diseases E_1 and E_2 has also disease E_3. Then it is possible to assign a coherent ordering $E_1|H_o \geq\cdot E_2|H_o \geq\cdot E_3|H_o$ (say); depending on this, the doctor chooses to perform a particular test, from which an evidence K_1 is obtained. Now, on the basis of the new information $H_1 = H_oK_1$, if analogous conditions as before are satisfied with H_1 in place of H_o, the doctor can introduce a new coherent ordering among the given conditional events $E_1|H_1$, $E_2|H_1$, $E_3|H_1$; and so on.

(iv) If $H_iH_k = \emptyset$ *for all pair* i,k $(i \neq k)$ *and* $E_iH_i \neq \emptyset$, $E_iH_i \neq H_i$ *for all* i, *then every ordering of the n conditional events* $E_1|H_1$, $E_2|H_2$, ... , $E_n|H_n$ *is numerically representable by a coherent point* $(p_1,...,p_n)$.

Proof - Since the H_i's are mutually exclusive, the assignment

$$P(H_1|H_o) = P(H_2|H_o) = ... = P(H_n|H_o) = 1/n$$

is coherent. So, given the ordering $E_1|H_1 \geq\cdot E_2|H_2 \geq\cdot ... \geq\cdot E_n|H_n$, evaluating

$$P(E_iH_i|H_o) = (r-i)/nr , \quad r \geq n ,$$

gives $p_i = P(E_i|H_i) = (r-i)/r$ $(i=1,2,...,n)$. The coherence of any other ordering can be proved by a suitable permutation of the indexes.

REFERENCES

Adams E. (1975), The Logic of Conditionals, Dordrecht, D. Reidel.

Barigelli B., Scozzafava R. (1984), Remarks on the role of conditional probability in data exploration, Statistics and Probability Letters 2, 15-18.

Bruno G., Gilio A. (1980), Applicazione del metodo del simplesso al teorema fondamentale per le probabilita' nella concezione soggettiva, Statistica 40, 337-344.

Bruno G., Gilio A. (1985), Confronto fra eventi condizionati di probabilita' nulla nell'inferenza statistica bayesiana, Rivista di Matematica per le Scienze Economiche e Sociali 8, 141-152.

Coletti G. (1990), Coherent qualitative probability, Journal of Mathematical Psychology 34, 297-310.

Coletti G., Gilio A., Scozzafava R. (1990), Coherent qualitative probability and uncertainty in Artificial Intelligence, 8th International Congress of Cybernetics and Systems, New York (to appear).

De Finetti B. (1970), Teoria delle probabilita', Torino, Einaudi.

De Finetti B. (1972), Probability, Induction and Statistics, New York, Wiley.

Dubois D., Prade H. (1988), Conditioning in possibility and evidence theories - A logical viewpoint, in Uncertainty and Intelligent Systems (Bouchon-Meunier B., Saitta L., Yager R.R., eds.), Lecture Notes in Computer Science # 313, Springer Verlag, 401-408.

Gilio A. (1990), Criterio di penalizzazione e condizioni di coerenza nella valutazione soggettiva della probabilita', Bollettino Unione Matematica Italiana (7) 4-B, 645-660.

Gilio A., Scozzafava R. (1988), Le probabilita' condizionate coerenti nei sistemi esperti, in Atti delle giornate AIRO su Ricerca Operativa e Intelligenza Artificiale, Pisa, Centro di Ricerca IBM, 317-330.

Goodman I.R., Nguyen H.T. (1988), Conditional objects and the modeling of uncertainties, in Fuzzy Computing (Gupta M., Yamakawa T., eds.), Amsterdam, North Holland, 119-138.

Nilsson, N.J. (1986), Probabilistic Logic, Artificial Intelligence 28, 71-87.

Pearl J. (1988), Probabilistic Reasoning in Intelligent Systems: Networks of Plausible Inference, San Mateo (California, USA), Morgan Kaufmann.

Schay G. (1968), An Algebra of Conditional Events, Journal of Mathematical Analysis and Applications 24, 334 - 344.

Scozzafava R. (1980), Bayesian Inference and Inductive 'Logic', Scientia, 115, 47-53.

ON REPRESENTATION OF SOURCE RELIABILITY IN WEIGHT OF EVIDENCE

Daniel E. O'Leary
Graduate School of Business, University of Southern California
Los Angeles, CA 90089-1421

Abstract

Developers of artificial intelligence-based systems have made frequent use of likelihood ratios. Those ratios have been used to represent the uncertainty associated with events and hypotheses on rules in expert systems and they have been used to establish rankings of resulting diagnoses in other systems. This paper discusses the representation of source reliability through those likelihood ratios, for use in artificial intelligence systems, such as expert systems, influence diagrams and other systems that employ a Bayesian-based approach to the representation of uncertainty to assess the weight of evidence.

This paper presents a means by which that reliability can be captured using a likelihood ratio format. Reliability can have a substantial impact on the value of the likelihood ratio. One example presented in the paper results in about a 60% decrease in the value of the likelihood ratio, with only a 10% decrease in reliability. Then this paper investigates the impact of accounting for reliability in the likelihood ratios. A monotonic property is established for the reliability embedded likelihood ratios. That property provides insight both into the behavior of weights on rules in systems that employ such an approach and into the use of likelihood ratios to rank diagnoses.

Then the impact of reliability-adjusted likelihood ratios is examined for their impact on rank ordering of the ratio. It is found that in some cases where likelihood ratios are examined in terms of the same evidence that reliability does not change the rankings. However, if the likelihood ratios are developed for comparison across different evidence, then the rankings do not remain the same. As a result, accounting for reliability of evidence can be critical to the ultimate success of systems employing this approach.

Keywords

Uncertainty Representation, Influence Diagrams, Expert Systems, Artificial Intelligence, Source Reliability.

1. INTRODUCTION

There are a number of artificial intelligence systems that employ likelihood ratios as a basis of establishing either a strength of evidence, say on a rule in an expert system, or as the weight of evidence to rank alternative diagnoses from an expert system or an influence diagram. For example, both GLADYS (Spiegelhalter and Knill-Jones [1984]), an expert system, and PATHFINDER (Heckerman et.al. [1990]), an influence diagram-based system, employ likelihood ratios to weight evidence. In each case the use of those likelihood ratios is based, in part, on the work of Good [1960], who suggested that likelihood ratios could be used to reflect the extent of "confirmation" or "weight of evidence."

1.1 Reliability

Typically, in expert systems and influence diagrams, information is solicited from the user as the basis of providing evidence to the system. That is, the user provides a "report" of the evidence to the system. That is, the user provides a "report" of the evidence to the system. In addition, in the development of the system, the developer solicits knowledge and estimates of probabilities from the experts in the particular domain. The relationship between the report of the evidence and the actual evidence will be called reliability.

Unfortunately, the evidence on which the likelihood ratios is based is not necessarily perfectly reliable. As noted by Simon [1969], humans have a tendency to "satisfice," leading to less than perfect reliability. In addition, in some cases evidence is statistical (e.g., Spielgelhalter and Knill-Jones [1984], as result, it also is by its very nature less than perfectly reliable.

Unfortunately, currently likelihood ratio use in these systems does not directly account for this lack of reliability. As a result, the purpose of this paper is to integrate a measure of the reliability of the source evidence into the likelihood ratio format. Reliability can have a substantial impact on the weight of evidence. Analytic results are presented that indicate some situations where the reliability model impacts the ranking of diagnoses and some cases where the reliability model has no such impact.

1.2 This Paper

This paper proceeds as follows. Section 2 investigates the use of likelihood ratios on rules in some expert systems and on rankings in some influence diagram systems. Section 3 integrates reliability into those likelihood ratios. Reliability has a major impact on the magnitude of the likelihood ratios. Section 4 examines the impact on the ordering of those likelihood ratios. In some cases the reliability model impacts the ordering, while in other cases there is no such impact. Section 5 provides a brief summary.

2. LIKELIHOOD RATIOS

In expert systems and influence diagram systems, the systems ultimately develop estimates of the probabilities that the evidence ("e") confirms different diagnoses or hypotheses ("h"). Often the likelihood of the individual diagnoses is presented to the system users ranked using likelihood ratios. In addition, in expert systems, knowledge often is represented in rules, using the form

"if e then h."

In some cases these systems employ weights on the rules of a number of different forms, one of which is a likelihood ratio approach. This section discusses both of these uses of likelihood ratios in more detail.

2.1 Notation

In this paper it is assumed that the underlying evidence under consideration is represented using the symbol "e" (or e_1, e_2, etc.) and the hypothesis under consideration is represented using "h" (or h_1, h_2, etc.). The symbol "e#" will be used to represent the report of the evidence e. Finally, p(e) will be used to represent the probability of e.

This paper investigates one particular likelihood ratio designed to provide insight into the confirmation of evidence with respect to a particular hypothesis. In particular, Good [1960] suggested that a measure of the weight of the evidence is $W(e,h) = \log (Z(e,h)) = \log (p(e|h)/p(e|h'))$.

2.2 Uses of Likelihood Ratios to Rank Alternatives

Good's [1960] measure of "weight of evidence" can be used to provide an order to rank the likeliness of alternative diagnoses. As a result, a number of researchers and system developers (Reggia and Perricone [1985], Cooper [1986], and Heckerman et al. [1990] use likelihood ratios of alternative diagnoses and evidence as a means of ranking the likelihood of different disease diagnoses for medical systems. This approach is seen to facilitate the communication of the workings of those systems. The complex interactions of probabilities are described in an easy-to-understand summary.

One of the problems of this approach is noted in Reggia and Perricone [1985, p. 166].

> Providing simpler, more condensed explanations would become even more of an issue were there larger numbers of diseases and symptoms under consideration ... In such a situation the proliferation of relationships between symptoms and diseases ranked higher or lower than the one of interest might prove more confusing to the physician than helpful.

Thus, extensions to a straight ranking of likelihood ratios has been suggested. For example, Reggia and Perricone suggest that only the top ranked diseases could be displayed to the physician. Another approach that they suggest is that a certain threshold must be exceeded. As seen below, both these suggestions are impacted by the introduction of reliability into the model. Reliability impacts the likelihood ratio value and it can change the ordering of diagnoses, depending on characteristics of the evidence.

2.3 Use of Likelihood Ratios as Weights on Rules

Spiegelhalter and Knill-Jones [1984] developed a system where that same likelihood ratio is used as the basis of weights on rules in an expert system. The reliability notions discussed in this paper can be applied to their system, and can be extended to other systems of uncertainty, such as that discussed by Shortliffe and Buchanan [1985]. However, the primary focus of this paper is on the impact on alternative rankings. Some extensions to those weighting systems are discussed in O'Leary [1990].

3. REPRESENTATION OF RELIABILITY

Reliability can be introduced into a likelihood ratio format by replacing the underlying evidence, e, with what actually occurs, a report of that evidence, e#. As a result, we are interested in RW = W (e#, h) and RZ = Z(e#, h). The following lemma establishes the basic model that is discussed throughout the remainder of the paper.

Lemma 1

$$Z(e\#,h) = [p(e\#|h \text{ and } e)\, p(e|h) + p(e\#|h \text{ and } e')\, p(e'|h)] / [p(e\#|h' \text{ and } e)\, p(e|h') + p(e\#|h' \text{ and } e')\, p(e'|h')]$$

Proof--Based in part on Schum and DuCharme [1971]

$$Z(e\#,h) = p(e\#|h)/p(e\#|h')$$

$$= [p(e\# \text{ and } h)/p(h)] \;/\; [p(e\# \text{ and } h')/p(h')]$$

$$= [[p(e\# \text{ and } e \text{ and } h) + p(e\# \text{ and } e' \text{ and } h)] \;/\; p(h)] /$$
$$[[p(e\# \text{ and } e \text{ and } h') + p(e\# \text{ and } e' \text{ and } h')] \;/\; p(h')]$$

$$= [[p(e\#|e \text{ and } h)\; p(e|h)\; p(h) + p(e\#|e' \text{ and } h)\; p(e'|h)\; p(h)]/p(h)] /$$
$$[[p(e\#|e \text{ and } h')\; p(e|h')p(h') + p(e\#|e' \text{ and } h')\; p(e'|h')p(h)] \;/p(h)]$$

$$= [p(e\#|e \text{ and } h)\; p(e|h) + p(e\#|e' \text{ and } h)\; p(e'|h)] /$$
$$[p(e\#|e \text{ and } h')\; p(e|h') + p(e\#|e' \text{ and } h')\; p(e'|h')]$$

The factors in $p(e\#|h)$ that relate to $p(e\#|e \text{ and } h)$ and $p(e\#|e' \text{ and } h)$ contain the information that relates to the reliability of the report, i.e., the relationship between the report of the evidence and the actual evidence. If $p(e\#|e \text{ and } h) = p(e\#|e \text{ and } h') = 1$ and $p(e\#|e' \text{ and } h) = p(e\#|e' \text{ and } h') = 0$ then the model would be perfectly reliable and reduce to the model in the previous section.

3.1 Symmetric and Asymmetric Reliability

In this paper it will be assumed that the reported value does not depend on the hypothesis, h. This is a reasonable assertion since in most cases we would anticipate that the reliability of the medium would be the concern. In these cases $p(e\#|e \text{ and } .)$ and $p(e\#|e' \text{ and } .)$ reduce to $p(e\#|e)$ and $p(e\#|e')$. These are what Schum and DuCharme [1971] refer to as the reliability of the reported evidence or the credibility of the source of evidence.

There are two important cases that allow us to study the impact of reliability. The first case will assume that the report of the evidence is symmetric, so that $p(e\#|e) = p(e'\#|e') = r$. The second will assume that $p(e\#|e) = n$ and $p(e'\#|e') = 1-n$.

Theorem 2 -- Symmetric Case

If $p(e\#|e)$ is symmetric then

$$RZ = [r * p(e|h) + (1-r) * p(e'|h)]/[r * p(e|h') + (1-r) * p(e'|h')]$$

Theorem 3 -- Asymmetric Case

If $p(e\#|e)$ is asymmetric then

$$RZ = [m * p(e|h) + (1-n) * p(e'|h)]/[m * p(e|h') + (1-n) * p(e'|h')]$$

3.2 Impact of Reliability on Likelihood Ratios

The impact of reliability is substantial. For example, assume that $p(e|h) = .90$ and $p(e|h') = .009$. This yields the following for the symmetric case:

r	1.0	0.9	0.8	0.7	0.6	0.5
RW	2.0	0.88	0.56	0.34	0.16	0.0

In this example, simply moving from perfect reliability to a .9 reliability level, yields a decrease in RW of 56%. Thus, if reliability is integrated into the ranking of alternative diagnoses, then using a cut-off value can have a substantial impact on which diagnoses are presented to the user.

3.3 Properties of the Reliability Model

There are a number of properties of the reliability model that are outside the direct scope of this paper. Some of these are summarized in O'Leary [1990]. For example, the reliability model has the desirable characteristic that the weight of evidence is zero when the level of reliability is completely uncertain, i.e., .5. Further, as noted above under perfect reliability have a monotonicity property. In addition, in same cases the models that include reliability have a monotonicity property. In one situation, discussed below, this property basically says that as reliability decreases so does RW. This is an important finding since it says that the weights of evidence, assuming perfect reliability, overstate the weights that include a realistic assessment of the reliability.

The monotonicity results for one case are summarized in the following theorem for the case of symmetric reliability. Further monotonicity results are presented in O'Leary [1990].

3.4 Monotonicity Theorem

In some cases it can be shown that RW is monotonically decreasing or increasing. Although this theorem is concerned with symmetric reliability (Theorem 4), it can be extended to more general cases, such as the asymmetric case (Theorem 5).

Theorem 4

A. If $r'' \geq r'$, $W(e,h) \geq 0$ and there is symmetric reliability, then RW at $r'' \geq$ RW at r'.

B. If $r'' \geq r'$, $W(e,h) \leq 0$ and there is symmetric reliability, then RW at $r'' \leq$ RW at r'.

Proof

A. If $W(e,h) \geq 0$ then $p(e|h) \geq p(e|h')$. Assume that RW at $r'' <$ RW at r', using RZ as follows.

$$[r'' * p(e|h) + (1-r'') * p(e'|h)]/[r'' * p(e|h') + (1-r'') * p(e'|h')] <$$
$$[r' * p(e|h) + (1-r') * p(e'|h)]/[r' * p(e|h') + (1-r') * p(e'|h')]$$

This is equivalent to

$$[r'' * p(e|h) + (1-r'') * p(e'|h)] * [r' * p(e|h') + (1-r') * p(e'|h')] <$$
$$[r' * p(e|h) + (1-r') * p(e'|h)] * [r'' * p(e|h') + (1-r'') * p(e'|h')]$$

After much simplification, we find that

$$p(e|h) * (r''-r') < p(e|h') * (r''-r'),$$

But since $p(e|h) > p(e|h')$ and $r'' > r'$, this is a contradiction.

B. Similar to part A.

Theorem 5

A. If $m''/n'' \geq m'/n'$, $W(h,e) \geq 0$ and there is asymmetric reliability, then RW at m'' and $n'' \geq$ RW at m'' and n''.

B. If $m''/n'' \geq m'/n'$, $W(h,e) \geq 0$ and there is asymmetric reliability, then RW at m'' and $n'' \geq$ RW at m'' and n''.

4. THE IMPACT OF ORDERING OF ALTERNATIVES

The process of accounting for reliability suggests that at least two different errors may arise in the use of likelihood ratios to rank outcomes from an expert systems or influence diagrams. First, in the case of the use of a cut-off value to rank the diagnosis, by not accounting for reliability, the monotonicity property indicates that the computed likelihood value exceeds its value that would occur if we were to account for reliability. Second, by accounting for reliability, the ordering of diagnoses could change. If that were the case then accounting for reliability would be a critical step, since without it, the rankings would be different.

4.1 Different Diagnoses and Same Evidence

If the reliability of the evidence is the same for each of the different pieces or types of evidence then we might expect that the reliability model would have no impact on the rankings, since the same evidence is used to evaluate the hypotheses. The following theorems substantiate that hypothesis, by indicating that the ranking of the alternatives would be independent of reliability measures in both the cases of symmetric and asymmetric reliability. Thus, if the same evidence impacts each of the hypotheses equally, it does not appear to be beneficial to account for reliability, if the concern is only with the ranking of the alternatives.

Theorem 6 -- Symmetric Probabilities

Assume that $r \geq .5$, $Z(e,h_1) \geq Z(e,h_2)$. Let $p(e\#|h_1) + p(e\#|h'_2) \geq p(e\#|h'_1) + p(e\#|h_2)$. For any value of r, $RZ(e,h_1) \geq RZ(e,h_2)$.

Proof

From theorem 2,

$$RZ = [(2r-1) * p(e|h_1) + (1-r)]/[(2r-1) * p(e|h'_1) + (1-r)]$$

Thus, the proof is dependent on noting that the reliability likelihood ratios in theorem 2 takes the form of a set of fractions a/b and c/d, where $a/b \geq c/d$ without reliability, where $a=p(e|h_1)$, $b=p(e|h'_1)$, $c=p(e|h_2)$, and $d=p(e|h'_2)$. The introduction of reliability has the effects of multiplying by a parameter and adding a constant to the numerators and denominators of those fractions.

The condition $Z(e,h_1) \geq Z(e,h_2)$ implies that $a/b \geq c/d$, and thus, that $a*d \geq b*c$. The condition $(p(e\#|h_1) + p(e\#|h'_2) \geq p(e\#|h'_1) + p(e\#|h_2)$ also is required to ensure that the original ordering remains.

Theorem 7 -- Asymmetric Probabilities

Assume that $m \geq n$ and $Z(e,h_1) \geq Z(e,h_2)$. Let $p(e\#|h_1) + p(e\#|h'_2) \geq p(e\#|h'_1) + p(e\#|h_2)$. For any values of m and n, $RZ(e,h_1) \geq RZ(e,h_2)$.

Proof

Similar to theorem 6.

4.2 Different Diagnoses and Different Evidence

On the other hand, if we assume that there is different evidence impacting the different hypotheses then that indicates that there is more than just $p(e\#|e)$. Instead we must consider $p(e\#_1|e_1)$, $p(e\#_2|e_2)$, etc. Thus, unless those reliability probabilities are the same, the rankings will not be the same. As a result, if we assume that there is more evidence detail than e and e', then, in general, even if we are given the ordering of likeliness of the diagnoses without accounting for reliability, it is impossible to predict the ordering when reliability is accounted for, without actually developing reliability-based orderings. This same result holds even if we are considering the impact of different pieces of evidence on the same hypothesis. Thus, reliability must be a part of any system with multiple pieces of evidence leading to different hypotheses or even the same hypotheses.

5. SUMMARY

This paper has introduced the notion of reliability into the ranking of alternative diagnoses produced by expert systems or influence diagrams. The introduction of reliability had an impact on those alternatives presented to the user, if the alternatives must exceed a certain threshold value. In particular, under certain conditions, both symmetric and asymmetric models of reliability were found to be monotonically decreasing in reliability. As a result, when perfect reliability is assumed, but less than perfect reliability actually is the case, the likelihood ratio is larger than it would be under the less than perfect reliability. Thus, it is possible that a case would be presented to a user when it should not be given that the system properly considered reliability.

Second, by accounting for reliability, the order of the alternatives may be different. As a result, the user of the system may be presented with the wrong order of diagnoses, unless the actual reliability of the evidence is taken into consideration.

The potential occurrence of either of these situations suggests the importance of employing reliability in the ranking of outputs from such systems.

REFERENCES

Cooper, G. (1986) A Diagnostic method that uses causal knowledge and linear programming in the application of Bayes' formula. In Computer Methods and Programs in Biomedicine, 22 (1986), 223-237.

Good, I. (1960) Weight of evidence, corroboration, explanatory power, information and the utility of experiments. In Journal of the Royal Statistical Society, B, 22 (1960), 319-331.

Heckerman, D., Horvitz, E., Nathwani, B. (1990) Toward normative expert systems: The pathfinder project. Unpublished paper, University of Southern California School of Medicine and Stanford University School of Medicine, Draft.

O'Leary, D. (1990) On the representation of source reliability through weights on rules. In IPMU Conference Proceedings, Paris.

Reggia, J. and Perricone, B. (1985) Answer justification in medical decision support systems based on Bayesian classification. In Computers in Biology and Medicine, 15 (1985), 161-167.

Schum, D., and DuCharme, W. Comments on the relationship between the impact and the reliability of evidence. In Organizational Behavior and Human Performance, 6, 111-131.

Shortliffe, E. and Buchanan, B. (1985) A model of inexact reasoning in medicine," in rule-based expert systems. Addison-Wesley (B. Buchanan and E. Shortliffe, eds.).

Simon, H. The sciences of the artificial, Cambridge, MA, MIT Press.

Spielgelhalter, D. and Knill-Jones, R. (1984) Statistical knowledge-based approaches to clinical decision support systems, with an application in gastroenterology. In Journal of the Royal Statistical Society, 147 (1984), 35-77.

3. FUZZY SETS

AN ABSTRACT MECHANISM FOR HANDLING UNCERTAINTY

T.P.Martin and J.F.Baldwin
A. I. Group, Advanced Computing Research Centre,
University of Bristol, Bristol BS8 1TR, United Kingdom

ABSTRACT
Fril is an AI language incorporating a powerful mechanism for handling uncertainty in knowledge-based applications. It is implemented as a compiler producing code for an abstract machine. In this paper, we outline the features of the abstract machine used to handle uncertainty and illustrate their operation by reference to a simple example.

KEYWORDS : Support Logic, Uncertainty, Fril, Implementation, Logic Programming, Fuzzy, Probability.

1. INTRODUCTION

Many AI applications have used Prolog, partly because it is suited to symbolic processing, partly because it can be executed as efficiently as conventional (procedural) languages, and partly because of its underlying theoretical foundations. However it lacks any representation of uncertainty, which is a major consideration in most current knowledge-based systems. Support logic programming[1] extends the logic programming approach to handle generalised probabilities in a clean and consistent way, yielding an AI language in which the management of uncertainty has a central role. Fril[2] is a practical implementation of support logic programming, which has been used for a wide variety of applications. It is a compiled language, based on an underlying abstract machine, described in more detailed in [3]. In this paper we outline the mechanism used to handle uncertainty. Owing to the brevity of this paper, we assume a familiarity with Fril and the Warren Abstract Machine (WAM) [4], a von-Neumann oriented instruction set reflecting the features particularly associated with Prolog programs. This has been used in most successful Prolog compilers.

2. SUPPORT LOGIC PROGRAMMING

In support logic programming, uncertainty can appear in the value of a parameter, in the relation between objects, and in drawing a conclusion from a set of conditions. Fril models the first form by allowing pos-

sibility distributions to appear as terms, and the second and third respectively by means of support pairs associated with facts and rules. To handle possibility distributions, syntactic unification must be extended to semantic unification [5]; various other operations can also be extended, such as arithmetic functions [6]. A support pair defines an interval containing a point-value probability. A support pair associated with a fact represents an interval containing the probability of that fact being true - the lower and upper limits give (respectively) the minimum value and maximum value of the probability (alternatively known as the necessary and possible support for the fact). The support *against* the fact being true is related to the support *for* the fact being true, since if its probability of being true is in the interval [n p], its probability of being false is in the interval [1-p 1-n] The difference between the upper and lower supports is a measure of the uncertainty associated with the fact. Baldwin[5] discusses support pairs in greater detail.
We note the following special cases[1]:

((p)) : (1 1) a proposition 'p' that is known to be true for certain
((p)) : (0 0) a proposition 'p' that is known to be false for certain
((p)) : (0 1) a proposition 'p' whose truth is completely uncertain

We also note that a goal that does not match any clause in the knowledge base is assumed to be uncertain, ie have a support of (0 1) rather than being false (0 0). This corresponds to an open-world assumption.
A rule has two support pairs associated with it, for example
((p)(q)) : (s1 s2)
where s1 and s2 are support pairs. The interpretation of a rule is similar to logic programming - the head is supported to some degree, computed from the supports for the body and rule. The conditional probability (p|q) lies in the interval defined by s1, and the conditional probability (p|¬q) lies in the interval defined by s2. If s2 is omitted, it is taken to be (0 1) ie completely uncertain, and if s1 is also omitted it defaults to (1 1). Thus the following support logic rules are equivalent:
((a X) (b X)) : ((1 1) (0 1))
((a X) (b X)) : (1 1)
((a X) (b X))

The support for the conclusion is related to the support for the body by a generalisation of the theorem of total probabilities (see below). Although the body of the rule shown above is a single goal, the approach is equally applicable to a conjunction of goals.

3. Execution of Support Logic Programs

The facts and rules in a knowledge base, together with a query, define an *and/or* tree whose root node is the query. The descendants of *or*-nodes are *and*-nodes, and vice versa. Execution of a support logic program is a full search over the *and/or* tree defined by the knowledge base and query, with calculation of supports for each solution. The methods used to combine support pairs are generalisations of operations on probabilities. For goals b_1 and b_2 with supports (n_1 p_1) and (n_2 p_2), the conjunction (b_1 AND b_2) has a

1. The standard Fril syntax is used in this paper. Variables and constants are represented by upper case and lower case letters respectively.

support of $(n_3\ p_3)$, where

$n_3 = n_1 \times n_2$ and $p_3 = p_1 \times p_2$.

(assuming independence). From two proof paths giving
$p : (n_1\ p_1)$ and $p : (n_2\ p_2)$, we conclude $p:(n_3\ p_3)$ where $n_3 = \max(n_1\ n_2)$ and $p_3 = \min(p_1\ p_2)$, subject to $n_3 \leq p_3$.

Other combinations are allowed[1], but any operation of the form

$$f:\ S^n \rightarrow S, \quad n \geq 0$$

can be incorporated into the abstract machine framework.

To calculate the support for a conclusion, the entire *and/or* tree is searched so that all proof paths may be taken into account (see Figure 1). The support for a conclusion (*or*-node) is calculated from the supports for each descendant *and*-node using the conditional and combination rules. Similarly, the support for an *and*-node is calculated from the supports of its descendant *or*-nodes, using the conjunction rule. This contrasts with the situation in logic programming, where only one proof path is necessary to establish a conclusion, and depth search with backtracking is used. In general, such a mechanism is not appropriate for support logic programs.

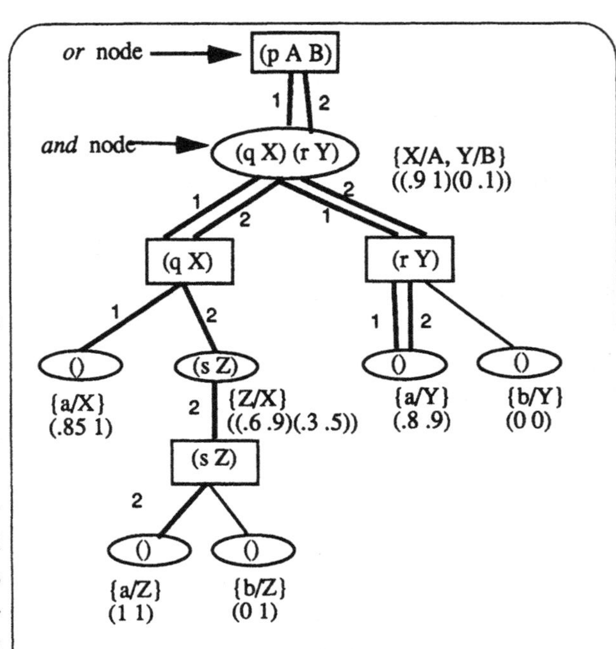

The search tree for the support logic knowledge base

((p X Y) (q X)(r Y)) : ((.9 1)(0 .1))

((q a)) : (0.85 1)

((q Z) (s Z)) : ((.6 .9)(.3 .5))

((r a)) : (.8 .9) ((r b)) : (0 0)

((s a)) : (1 1) ((s b)) : (0 1)

and query qs ((p A B)) is shown. The proof paths for the solution (p a a) are emboldened and labelled 1 and 2 respectively. Support for (p a a) is calculated from the supports associated with the *and*-nodes on the two proof paths.

Figure 1 - Example of a Support Logic Program

The evaluation of support starts at the leaf nodes of the search tree, and moves towards the root. If the goal at an *or*-node is non-ground (ie contains variables), it may have a number of solutions. The set of solutions and supports can be used to replace the sub-tree below the *or*-node with an equivalent set of leaf nodes. This set represents the extension of the relation defined by the clause set for the *or*-node predicate. The set of solutions to an *and*-node can be found by joining its descendant relations on common variables, and projecting the result onto the variables in the *and*-node; by repeating this process from the leaves of the tree towards the root, the extension of the query goal will be produced, giving a set of solutions with associated supports. This process is illustrated in Figures 2 and 3, and a more detailed treatment is given in [3].

Evaluating relations in this manner can be efficiently implemented, as described in the next section, but clearly any measures that reduce temporary storage space will be advantageous. One very important saving arises from the operations used as default combinations, namely product, intersection, conditional, and complement. The support for a conclusion can be written as a nested expression involving these operators; at each appearance of an intersection operator, a temporary relation is created. Since the other operators distribute over intersection [3], the expression can be rewritten so that only one temporary relation is required, at the root node. For example, in Figure 1 there are two proof paths for the conclusion (q a), yielding supports of (.6 .9) and (.85 1), and a temporary relation would be formed as in Figure 2. The support for the conclusion (p a a) is calculated from this intersection and the support (.8 .9) for (r a); since other proof paths may also provide support for (p a a) we must intersect those with the value from this path. Taking S to be the set of support pairs and representing the conjunction, intersection, and conditional operators as

$\text{conj} : S^n \to S,$

$\text{inter} : S^n \to S,$ and

$\text{cond} : S \times (S{\times}S) \to S$ respectively, we can write the support for the conclusion (p a a) as

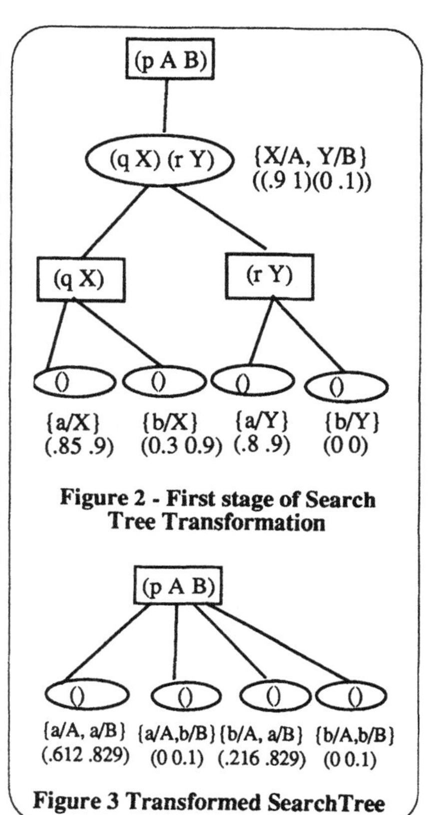

Figure 2 - First stage of Search Tree Transformation

Figure 3 Transformed SearchTree

inter(cond(conj(inter((.6 .9), (.85 1)), (.8 .9)), ((.9 1)(0 .1))), ... expressions from other proof paths)

Rewriting this as

inter(cond(conj((.6 .9), (.8 .9)) , ((.9 1)(0 .1))),
    cond(conj((.85 1), (.8 .9)) , ((.9 1)(0 .1)))   ... expressions from other proof paths)

yields the same result with only one intersection. In this case, there are no other proof paths for (p a a); however, this is not known until the tree has been searched. Any *and/or* tree or subtree involving only the default operators requires a single temporary relation, at the root node. This process is similar to 'setof' in Prolog, with a slight overhead due to calculating supports. Since most support logic programs use only the default operators, there is very little penalty in finding supports compared to simply finding all solutions. However, it should be noted that there is an important difference between the mechanisms used by a Prolog all-solutions predicate such as 'setof', and the search for proof paths required by the support logic evaluation. In Prolog-style execution, a branch can be abandoned as soon as a failure node is found, with execution continuing from the most recent choicepoint. In support logic, however, a failure node is assumed to be unknown, and has a support of (0 1). Generally, this gives a support of (0 1) to the parent *and*-node, since conj((n p), (0 1)) = (0 1) where (n p) is the support for the rest of the conjunction. This is not true when one of the other goals in the body has a support of (0 0) as conj((*n p*), (0 0)) = (0 0) for any support pair *(n p)*. Thus a branch can only be abandoned when a support of (0 0) is found, not at a failure node. This problem arises when there are solutions present that would not be found by a Prolog-style depth search, for example

((p X) (q X)(r X)) : ((0.9 1)(0 0.1))

((q a)) : (0.85 1)

((r a)) : ((0.8 0.9)

((r b)) : (0 0)

has a solution (p b) : (0 0.1), but this would be missed by a depth search as there is no clause ((q b)). The solution is, however, found by the more thorough search outlined above in which the extension of each relation is created. Thus the optimised search can only be performed in cases where it is known for certain that no solutions will be missed. If the code is known to be static (ie it can be guaranteed that no changes are made after the code has been loaded), then it can be fully analysed at compile time and the optimisation applied; otherwise, the more general method must be used.

A consequence of the full search required for support logic is that the *and/or* tree must be finite[2]. This is normally the case for support logic programs, although some care is required in writing recursive rules. A knowledge base of supported clauses can represent uncertain declarative knowledge and reasoning, but purely procedural code is complicated by the open world semantics and hampered by breadth search (most procedural code has a single proof path and is less efficient when other paths are examined) and the constraint that the *and/or* tree must be finite. Thus, support logic programming alone does not form a practical language for all purposes. Since we are dealing with a variant of logic programming, it seems obvious to incorporate procedural code by allowing Prolog-style execution in addition to support logic.

4. The Abstract Fril Machine

For flexibility, support logic and Prolog should be integrated as tightly as possible, and clauses should be usable in either execution mode. To conserve space, it is impractical to keep two versions of the compiled code. The abstract Fril machine is a modification of the WAM to give an efficient support logic programming system. It consists of a set of registers, data areas, and machine instructions. Clauses (possibly involving supports) are compiled into sequences of the instructions, and can be executed by a software emulator or hardware implementation of the abstract machine. The Prolog-like aspects of the machine are derived from (although not identical to) the WAM; additional support logic operations have been integrated into the design. Note that the machine assumes that an address can be given for code etc, but this does not preclude the possibility of a virtual addressing mechanism at a lower level.

Because Fril has a slightly different syntax to Edinburgh Prolog, it is necessary to store the arity of the current call in a register. This is primarily because Fril's list-based syntax allows the construction of goals and clause heads with unspecified arity, such as *(pred a X / Y)* . These can match clauses or goals with different arities, which is not permitted in Edinburgh Prolog.

A register indicates the current search mode, which can be *depth* (Prolog execution), or *breadth* (using dempster or intersection to combine solutions). Various flags are also stored in this register, including one to indicate that a semantic unification has taken place in the current inference step, one to indicate that a

2. In fact, it is possible to handle an infinite tree provided that the infinite branch occurs to the right of a node with support (0 0). This is a comparatively rare case, and is ignored

support has been pushed, one to show which method is used to combine solutions, and one to indicate whether a *call* or *execute* instruction was used to enter the current procedure. The latter is needed to determine action when support instructions have been executed.
The WAM uses three stacks, the control (or environment) stack, the copystack (or heap), and the trail. These are used respectively to store environment and choicepoints, global data, and information about how to reset the copystack on backtracking. In the Fril abstract machine, an extra stack, the support stack, is used to store supports until they are required. A register is needed to indicate the support stack top and an additional field is stored in a choicepoint record to reset the support stack on backtracking. Although the information kept on the support stack could be stored in a choicepoint, this would be inefficient in the case of non-support logic queries. Further fields are needed to record the arity and search mode associated with a choicepoint. An environment is stored in the same manner as in the WAM, ie it contains a continuation environment and code pointer, together with relevant environment variables.
The support stack is used to store instructions for calculating solutions and their supports. Each instruction is stored as a support frame containing the following elements:

operation	- how to combine this support with another
flags	- how to combine different proof paths, information on continuation
support type	- indicates whether there are variables in the support
support	- actual values of support (or address if variable)
solutions	- used to store the bindings of variables relevant to this goal.

The set of solutions can be stored in the knowledge base, or as a vector of suitably indexed cells on the copystack, or in any other appropriate storage location. Conceptually it can be regarded as a relation, in which each column corresponds to a heap location (ie to a global variable), and each row corresponds to a solution to the goal, ie a set of bindings for the global variables. In some cases it is possible that different solutions will not bind precisely the same set of variables; the columns held in the relation must represent the union of all sets of global variables bound by the solutions. Each frame on the support stack represents a machine instruction, which is evaluated when a*proceed* instruction is executed. Thus a support logic program actually creates a stack of instructions, which are executed to produce the overall solution.

4.1 Support Instructions :Various instructions are used to construct and evaluate the support stack. These are treated as null operations if the system is not finding supports:
set_mode(M) sets the current search mode to M, corresponding to intersection or dempster. This is copied into a support frame by the *try_me_else* instruction (see section 4.4).
push_support(S, C) takes a support, and an operator code specifying the combination type - conditional, conjunction, disjunction, negation. This code is a machine instruction, executed during evaluation of the support stack. The support and operator are copied into the top frame on the support stack. A choicepoint is created **after** support is pushed, so the frame is retained after all solutions to this goal have been found.
push_var_support(S, C) is used if the support contains variables. This stores information in the support frame, showing where supports can be found, but is otherwise similar to *push_support*.
Additionally, various instructions are used when the support stack is evaluated. These include:
join - joins two or more relations on their common columns to produce a support relation representing the

conjunction of goals in the body. Conceptually, this is a relational join or cross product, with calculation of supports on each solution; at the machine level, the operation is performed by a sequence of *put* and *get* instructions. Due to limitations of space, the precise mechanism can not be covered in detail in this paper, but is discussed further in [3] and illustrated by example in Figure 4. After a *join* instruction has been performed, it can be discarded from the support stack.

proj either copies a relation, or projects it onto one involving fewer columns. Support for each element of the new relation is calculated from the rule support and the support for corresponding elements in the existing relation(s). Where appropriate, solutions are combined using the intersection or dempster rules. Frequently, it is possible to perform this operation "in place", by modification of an existing relation.

4.2 Procedural Instructions: There are a number of procedural instructions whose behaviour is modified. The *call* and *execute* instructions set the arity register and the call flag, to indicate which was executed. In the WAM, *call* is used when there are further goals in the body, and *execute* is used for the last goal - the difference between them is that *execute* does not have to save a continuation.The *proceed* instruction is changed when supports are being evaluated. It indicates that a success node of the *and-or* tree has been found, and passes control to the top frame on the support stack so that support can be evaluated. This is completed when evaluation reaches the support stack reset of the current choicepoint. Execution continues by backtracking, ie resetting all stacks to the state stored in the current choicepoint and redoing the goal. In the depth search case, execution simply continues from the continuation pointer stored in the current environment.

Another major change is in the *fail* instruction. The control stack is popped back to the last choicepoint, and other stacks are reset in accordance with the information held in the choicepoint. If there are no more matching clauses, then all solutions to the current goal have been found and the support instruction in the associated support frame is set to *conj* (this is only significant if it was originally *cond*). If the call flag stored in the choicepoint indicates that the procedure was *call*-ed, the choicepoint is popped and the next goal is executed. On the other hand, if the procedure was *execute*-d, there is no next goal and control passes to the next most recent choicepoint. Note that the behaviour on finding no more matching clauses only occurs when a choicepoint is not popped by a *trust_me_else fail* instruction, that is when supports are being found. In the standard depth search, *fail* behaves in the normal fashion.

4.3 Indexing Instructions : these are responsible for selecting possible clauses as candidates for unification. The instructions can be classified as "try_me_else", "retry_me_else", and "trust_me_else", and appear respectively before the first, intermediate, and last clauses in a definition. Further instructions are available for more sophisticated indexing, eg when the first argument is known, but these are omitted from this account. The indexing operations are slightly extended when in breadth-search.

try_me_else instructions are responsible for creating choicepoints, and also the necessary structure for saving solutions in breadth search. A support frame is pushed, with support initialised to ((1 1)(0 1)) and combination operator given by the current search mode (either intersection or dempster). *try_me_else fail* is a null operation when in depth search, but creates a choicepoint in breadth search.

retry_me_else instructions set the saved address of an alternative clause, and when in breadth search, also reinitialise the support to ((1 1)(0 1)).

trust_me_else fail pops the choicepoint off the control stack in depth-search; in breadth search, it reinitialises the support to ((1 1)(0 1)) but leaves the choicepoint on the stack, since support evaluation requires information on stack resets from this choicepoint. The address of the next clause is set to *fail*. In breadth search, the choicepoint is removed by a *fail* instruction when control returns to this point.

4.4 Data Representation: The Fril machine includes possibility distributions as data types and the unification instructions permit rapid manipulation of these terms. A Fril machine term is a value and a tag giving the object type (variable, integer etc). Each possibility distribution is represented by a unique value calculated from its elements, in a manner similar to hash-coding for strings. The highest and lowest elements can be extracted from the value (Elements from non-numerical domains are ordered according to their internal representations.). Three unification instructions involve possibility distributions:

- *put_pdist(R, I)* which places a possibility distribution tag and reference value I, into argument register R,
- *get_pdist(R, I)*, which unifies the value in register R with the possibility distribution referenced by I,
- *unify_pdist(I)*, which unifies a possibility distribution in a list with its corresponding term.

In addition, instructions such as *get_const, get_int, get_float* must be extended to allow for the fact that the corresponding argument may be a possibility distribution, in which case semantic unification is used to determine whether or not there is a match. In many cases unification need not access the possibility distribution - binding to variables is straightforward, and non-matching unification with atomic elements or possibility distributions can be detected by comparison with the highest or lowest elements. If semantic unification is invoked, a support is calculated for an implicit rule [5] and pushed onto the support stack with a conditional operator.

5. Example

The clauses in Figure 1 would be compiled as follows. In order to simplify the illustration, we have omitted some of the indexing code which would normally appear at the start of each code segment. Comments are shown on the right, and relate to support logic execution.

```
p1:
set_mode inter                          ((p A B)(q A)(r B)) : ((.9 1)(0 .1))
try_me_else fail                        creates choicepoint (dummy) and support frame
allocate 2                              saves environment, continuation
push_support ((.9 1)(0 .1)) <cond>      fills support frame
call q, 1, 2
put_var A1, Y2                          puts variable 2 (B) in reg 1
deallocate                              reset continuation, discard environment
execute r, 1
q1:
set_mode inter                          ((q a)) : (.85 1)
try_me_else q2                          creates choicepoint and support frame
get_const A1, a                         unify a with argument 1
push_support (.85 1) <conj>             fills in support frame
```

Snapshots of the support stack at various points during the execution are shown. Goals and bindings are also included for illustrative purposes. An asterisk * in the "next" field indicates that a dummy choicepoint is present.

(a) prior to executing the *proceed* instruction after q1

goal	bindings	next	call/exec	support stack	solutions
(p A B)	{X/A, Y/B}	*	c	proj .9 1 0 .1	
(q X)	{a/X}	q2	c	join .85 1 0 1	

(b) the *proceed* instruction evaluates the support and the solution {a/A} : (.85 1) is stored. Execution continues from the most recent choicepoint, namely (q X), and (s Z) is called after the support frame has been overwritten with the conditional support for this rule. Two solutions are found to (s Z)

goal	bindings	next	call/exec	support stack	solutions
(p A B)	{X/A, Y/B}	*	c	proj .9 1 0 .1	
(q X)	{Z/X}	*	c	proj .6 .9 .3 .5	{a/A} : (.85 1)
(s Z)	{a/Z}	s2	e	join <not relevant>	{a/A} : (1 1) {b/A} : (0 1)

(c) these are compounded with the conditional support for q, and the result is combined with the existing solutions for (q X) giving

goal	bindings	next	call/exec	support stack	solutions
(p A B)	{X/A, Y/B}	*	c	proj .9 1 0 .1	
(q X)		-	c	join <not relevant>	{a/A}: (.85 .9) {b/A} : (.3 .9)

(d) Execution continues with the goal (r Y), which has two solutions:

goal	bindings	next	call/exec	support stack	solutions
(p A B)	{X/A, Y/B}	*	c	proj .9 1 0 .1	
(q X)		-	c	join <not relevant>	{a/A} : (.85 .9) {b/A} : (.3 .9)
(r Y)		*	e	join <not relevant>	{a/B} : (.8 .9) {b/B} : (0 0)

The two temporary relations are joined (in this case, by forming their cross product), leaving the solutions {a/A, b/B} : (0 .1), {a/A, a/B} : (.612 .829), {b/A, a/B} : (.216 .829), {b/A, b/B} : (0 .1)

Figure 4 - Development of Support Stack

proceed	causes evaluations of support stack
q2 :	((q Z)(s Z)) : ((.6 .9)(.3 .5))
trust_me_else fail	doesn't remove choicepoint in breadth search
push_support ((.6 .9)(.3 .5)) *<cond>*	
execute s	
r1:	
set_mode inter	((r a)) : (.8 .9)
try_me_else r2	creates choicepoint, support frame
get_const A1, a	unify a with argument 1
push_support (.8 .9) *<conj>*	
proceed	
r2:	((r b)) : (0 0)
trust_me_else fail	

get_const A1, b	unify b with argument 1
push_support (0 0) <*conj*>	
proceed	
s1:	
set_mode inter	((s a)) : (1 1)
try_me_else s2	creates choicepoint, support frame
get_const A1, a	unify a with argument 1
push_support (1 1) <*conj*>	
proceed	
s2:	((s b)) : (0 1)
trust_me_else fail	
get_const A1, b	unify b with argument 1
push_support (0 0) <*conj*>	
proceed	

The support stack for the query qs((p A B)) develops as shown in Figure 4. The root node of the tree is a Fril built-in predicate "supp_query". For clarity, this is not shown.

6. IMPROVEMENTS AND FURTHER DEVELOPMENTS

Work is currently in progress aimed at integrating Baldwin's iterative assignment method[7] into the general Fril framework. This requires certain extensions to the abstract machine, as rules and non-ground facts are used to generate *a priori* assignments over sets of solutions, updated by the assignments for ground solutions. Further work is required on aspects of this problem before the design of an extended Fril abstract machine is finalised.

REFERENCES

[1] **BALDWIN J.F.** (1987) "Support Logic Programming", in Fuzzy Sets - Theory and Applications (Jones A.I. *et al*, eds) Reidel, Dordrecht-Boston, 133-151.

[2] **BALDWIN J.F, MARTIN T.P, PILSWORTH B.W.** (1988) FRIL Manual (2nd Edition) Fril Systems Ltd, Bristol ITeC, St.Annes, Bristol BS4 4AB, United Kingdom.

[3] **MARTIN T.P., BALDWIN J.F.** (1990) "The Fril Language and its Implementation", ITRC Report, University of Bristol; to be published

[4] **WARREN D.H.D.** (1983) "An Abstract Prolog Instruction Set", Tech.Note 903 SRI International.

[5] **BALDWIN J.F.** (1990) "Computational Models of Uncertainty Reasoning in Expert Systems and AI", J.Comp.Math.Applic. 9 (1990), 105-119

[6] **MARTIN T.P,BALDWIN J.F.** (1989) "Fuzzy Sets in Fril", Fuzzy Systems & Signals, AMSE Monographs (Series A) 41-50

[7] **BALDWIN J.F.** (1990) "Combining Evidences for Evidential Reasoning", Int.J.Intelligent Systems (to appear).

A METHOD TO BUILD MEMBERSHIP FUNCTIONS
Application to numerical/symbolic interface building

BOBROWICZ O. - CHOULET C. - HAURAT A*.
SANDOZ F. - TEBAA M.

Equipe Logiciels pour la Productique
Institut de Productique - LAB
15, impasse des Saint Martin
25000 Besançon (France) Tél : (33) 81 88 53 44

* Laboratoire Logiciels pour la Productique
Université de Savoie
41, avenue de la Plaine
BP 806
74016 Annecy Cedex (France) Tél : (33) 50 57 40 70

During the last years, the development of knowledge based systems has taken an increasing expansion in the control of complex industrial processes. The control knowledge is often vague, and it is necessary to create a tool to define the words of the language that are used by the expert when this one is reasoning. We propose a method to acquire and to represent the vague knowledge of the expert.

This method uses membership functions which are coming from the fuzzy set theory. This method is defined in a semantic definition module that allows to build the membership function of three dependent predicates which are defined on the same universe of discourse. Thus, the membership functions permit to obtain a representation which is conformable to the meaning that the expert gives to each predicate that corresponds to the terms of his domain. We determine the parameters of these functions from two kinds of knowledge : first, the knowledge about semantic links joining the three predicates we want to represent on the same universe of discourse, and next, the expert knowledge (heuristics) about the meaning the expert wants to give to each predicate.

The semantic definition module is used in a decision making system assisting the operators which are controlling the process during disturbing conditions. This module is a "numeric / symbolic" interface that, from each numerical datum issuing from the process, interprets it in symbolic information. Then, these symbolic information is in the formalism of the decision making system. The collected information is in harmony with the knowledge of the operators, and leads to a best exploitation of the decision making system.

Keywords : vague knowledge, fuzzy predicate, membership function, semantic definition module, decision making system, "numeric / symbolic" interface.

1 - INTRODUCTION

Complex industrial processes are often very automated and, in normal working conditions, computers ensure the process control without any human intervention. On the other hand, in a change of working conditions or in disturbing conditions, operators are absolutely indispensable to ensure decision tasks.

We propose here a decision making system to assist operators which are controlling the process during disturbing conditions. According to the Rasmussen model [6], the reasoning approach of the operators can be decomposed in four steps :

- to research and to detect the disturbing condition by considering the characteristic parameters of the process
- to analyse the observed situation in order to identify the gravity of the disturbing condition and to find its causes
- to define a corrective strategy according to the general situation of the process
- to apply the corrective actions provided by the last reasoning step

The information used by an operator who is reasoning is vague and uncertain. In this method, we consider only the data processing of vague information, but not uncertain information. Indeed, the expert's knowledge is expressed with fuzzy rules such as :

" if the temperature is high then the pressure is low"
" if the temperature is low then increase the combustible"

The analysis of the observed situation ensures an interpretation of the numerical parameters that are coming from the process sensors. These numerical values are interpreted in symbolic values. For a parameter, the operator uses only three nuances to characterize this one. We will use the fuzzy logic introduced by Zadeh to manage such information. The fuzzy logic defines membership functions to characterize the vague words of the expert's language. We propose here a method to build membership functions based on the expert's knowledge.

2 - METHOD TO BUILD MEMBERSHIP FUNCTIONS

2.1. Recall

Fuzzy logic is a theoretic basis used to manage vague information. A fuzzy subset introduced by Zadeh represents the meaning of a fuzzy word. A fuzzy subset A defined on U is characterized by the membership function [7] :

$\mu_A : U \text{ ------> } [0, 1]$
$u \text{ ------> } \mu_A(u)$ where $\mu_A(u)$ is the grade of membership of U in A.

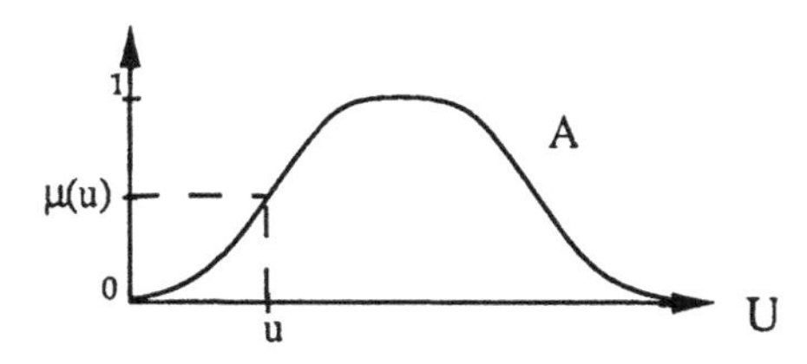

The set U is called the universe of discourse. The membership in a set is also defined by the grades of membership whose values are between 0 and 1. The transition from membership to nonmembership is gradual. A predicate A may be compared with a fuzzy subset. Thus, in order to represent a predicate or more precisely to give a meaning to a predicate, we must determine a membership function μ_A defined on a universe of discourse U.

2.2. Choice of a membership function

We want to represent for each controlled parameter three dependent predicates defined on the same universe of discourse. To present this method, we will represent the predicates : "low", "normal", "high" defined on the universe of discourse of the temperatures.

We have chosen to represent each predicate by a membership function μ which has been defined by [5]:

$$\forall x \in U, \quad \mu(x) = 1/(1+a^b(x-c)^b) \qquad (1)$$

It is continuous on U and takes its values in [0, 1].

The parameters a, b, c characterize the function μ, with :

$\mu(c) = 1$
$\mu(c - 1/a) = \mu(c + 1/a) = 0.5$
b characterizes the shape of the function (a bell-shaped curve more or less flat)

This function is interesting because it offers the possibility to choose some crossing points of the function. We will be able to define different types of curves : left half bell-shaped curves and right half bell-shaped curves.

This study induces us to determine the numerical values of each parameter a, b, c in order to obtain the membership functions which characterize in the best way the predicates whose we want to give a meaning.

2.3. Properties of these functions

Let's take for universe of discourse $U = [x_{min}, x_{max}]$.
The parameters a, b, c have the characteristics :

. $c \in U$ with $\mu(c) = 1$; c is a value of U which membership in the fuzzy subset is total.

. $a \in R^*$ in order to define the two points x_{int1} and x_{int2} such as :

$$\mu(x_{int1}) = \mu(c - 1/a) = 0.5$$
$$\text{and } \mu(x_{int2}) = \mu(c + 1/a) = 0.5$$

. $b \in N^*$, $b \geq 2$ and characterizes the aspect of the function. The more b is high, the more the derived function increases at the point x_{int1} (decreases at x_{int2}). b will be used to increase the interval of values whose have a grade of membership near 1.

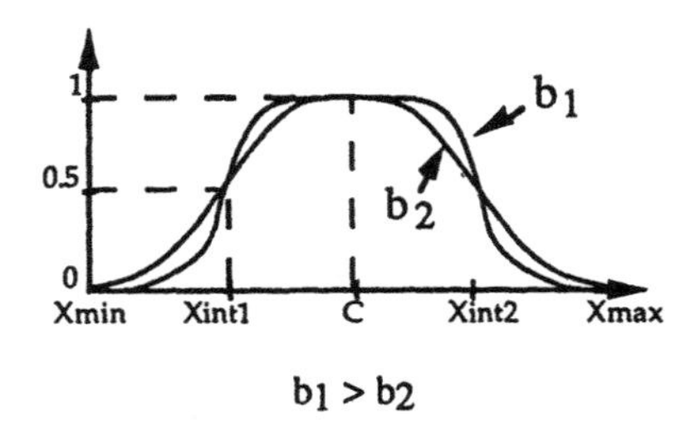

$b_1 > b_2$

This function offers two particular cases.

If $c = x_{min}$, we have a right half bell-shaped curve :

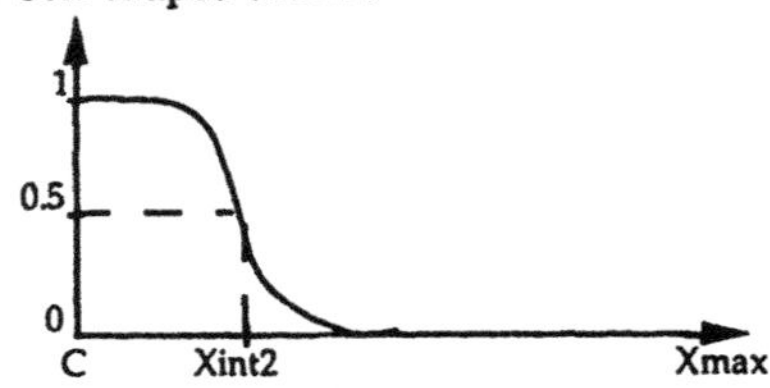

If $c = x_{max}$, we have a left half bell-shaped curve :

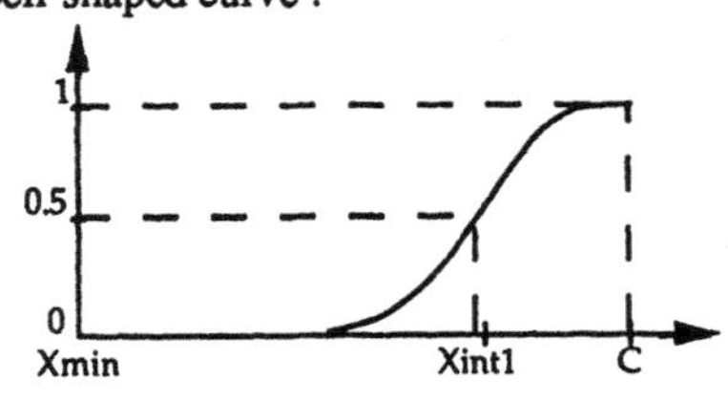

Remark : We can define non symmetrical bell-shaped curves on the x = c axis, by association of a left half bell-shaped curve and a right half bell-shaped curve on the respective intervals $[x_{min}, c]$ and $[c, x_{max}]$. The resulting function is defined by intervals.

2.4. Knowledge used to build membership functions

For each function the three parameters a, b, c are conjointly determined from two kinds of knowledge [3] :

. Semantic links joining the three predicates we want to represent on the same universe of discourse.
. Expert's knowledge indicating the expert's meaning of each predicate.

2.4.1. The semantic links

The semantic links are common sense knowledge and define mutual exclusion between predicates. They specify :

. If a value of U is considered as totally "normal", it can't be "low", neither "high":

$$\forall x \in U, \quad \mu_{normal}(u) = 1 \Rightarrow \mu_{low}(u) \approx 0 \text{ and } \mu_{high}(u) \approx 0$$

. If a value of U is considered as totally "low" (respectively "high"), it can't be "normal" :

$$\forall x \in U, \quad \mu_{low}(u) = 1 \Rightarrow \mu_{normal}(u) \approx 0$$

$$\forall x \in U, \quad \mu_{high}(u) = 1 \Rightarrow \mu_{normal}(u) \approx 0$$

Thus, the three membership functions have some necessary crossing points :

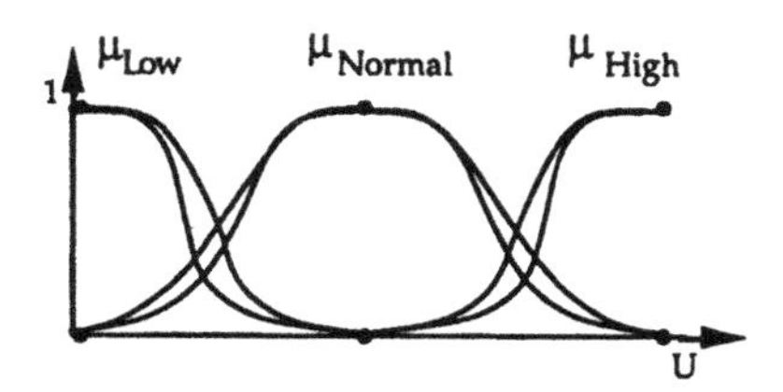

2.4.2. The expert's knowledge

The expert's knowledge is about :

- the values of the universe of discourse U. They correspond to the numerical possible values that the controlled parameter can take :

$$U = [x_{min}, x_{max}]$$

- the values of U that are totally "normal". That means the universe values which have a grade of membership in the fuzzy subset "normal" equal or near 1.

$$\forall c_n \in [c_1, c_2], \; \mu_{normal}(c_n) = 1$$

- the universe value of U from which the grade of membership in the fuzzy subset "low" is higher than the grade of membership in the fuzzy subset "normal".

$$\exists x_{int1} \in U \text{ with } x_{min} \leq x_{int1} \leq c$$

$$\mu_{low}(x_{int1}) = \mu_{normal}(x_{int1}) = 0.5$$

- the universe value of U from which the grade of membership in the fuzzy subset "high" is more important than the grade of membership in the fuzzy subset "normal".

$$\exists x_{int2} \in U \text{ with } c \leq x_{int2} \leq x_{max}$$

$$\mu_{high}(x_{int2}) = \mu_{normal}(x_{int2}) = 0.5$$

We suppose that the functions which characterize the fuzzy subsets "low" and "high" reach the value 1 on the limit of the interval $[x_{min}, x_{max}]$.

2.4.3. Determination of the parameters a, b, c

Notation :

a_l, b_l, c_l , are the parameters of the membership function representing the predicate "low"

a_h, b_h, c_h, are the parameters of the membership function representing the predicate "high"

a_{n1}, b_{n1}, c_{n1}, are the parameters of the left half shaped-curve function representing the left part of the predicate "normal"

a_{n2}, b_{n2}, c_{n2}, are the parameters of the right half shaped-curve function representing the right part of the predicate "normal"

For each membership function, the parameters a and c are computed according to the expert's knowledge :

$c_{n1} = c1$

$c_{n2} = c2$

$c_l = x_{min}$

$c_h = x_{max}$

$x_{int1} = c_{n1} - 1/a_{n1} \Rightarrow a_{n1} = 1/(c_{n1} - x_{int1})$

$x_{int2} = c_{n2} + 1/a_{n2} \Rightarrow a_{n2} = 1/(x_{int2} + c_{n1})$

$x_{min} + 1/a_l = x_{int1}$ and $c_{n1} - 1/a_{nl} = x_{int1} \Rightarrow a_l = 1/(c_{n1} - x_{min} - 1/a_{n1})$

$x_{max} - 1/a_h = x_{int2}$ and $c_{n2} + 1/a_{n2} = x_{int1} \Rightarrow a_h = 1/(x_{max} - c_{n2} - 1/a_{n2})$

The parameter b are obtained by successive approximations in order to respect the semantic links.

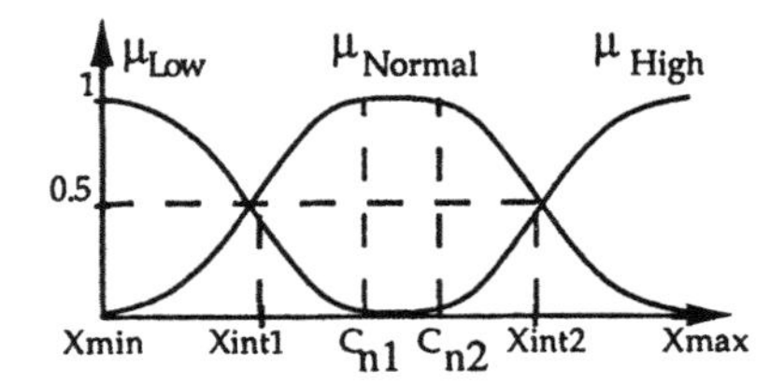

2.5. Semantic definition module

The presented method has been implemented in a *semantic definition module*. This module allows interactive modification made by the expert concerning his knowledge about the meaning of the predicates and also the graphic representation of the resulting membership functions. This modification can be done at the time of a change of process context. The meaning of the predicates and particulary the meaning of the predicate "normal" which is used as reference to control the process can be adjusted.

3 - APPLICATION TO THE PROCESS CONTROL

The information about the process state is a numerical information coming from sensors. But the expert's knowledge is expressed with vague predicates. The decision making system transcribes this knowledge [1]. It will be also necessary to interpret the numerical information to express it in the formalism of the system.

3.1. Perception module

The situation is interpreted by assigning to each numerical value a grade of membership in the three fuzzy subsets defined on the considered universe of discourse.

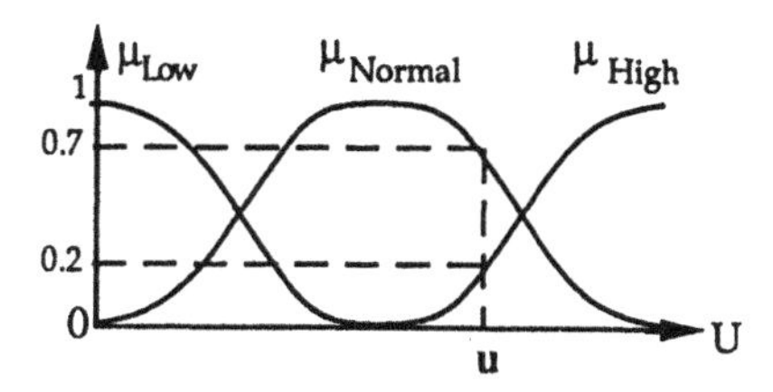

For each numerical information, the *perception module* generates three facts associated with factors, by using the three membership functions relating to this information :

"X is low" CF c_1, "X is normal" CF c_2,
"X is high" CF c_3

Remark : Nearer of 1 is c1, lower is the variable X. If c_1 is near of 0 and c_2 near of 1, then X is slightly low but quiet normal and x isn't high.

The semantic links are defined to have at least one factor equal to 0. The perception module can interpret the situation like the expert does. It is used as a numerical/symbolic interface between the process and the expert system which provides vague information expressed by facts associated with factors.

3.2. Expert System

From the observed situation, the expert system must :

- diagnose the situation in order to find the causes of the observed disturbing condition (there is a disturbing condition as soon as the factor associated to the predicates "low" or "high" isn't equal to 0).

- propose corrective actions to restore the process in a normal state.

The reasoning is described with production rules like "if X is A then Y is B" where A and B are vague predicates and their meaning is given by the semantic definition module (cf 2.5.).

Example : "if the temperature is low then increase the combustible".

If there is the fact "the temperature is low CF c" provided by the perception module in the fact base, the rule will be inferred. The generated fact will be "increase the combustible CF c' " where c' is computed from c (in this example c' is equal to c).

So, we need only one rule to represent the action to do for each nuance of the premise (from "very low" with c equal to 1 to "slightly low" with c near 0).

As initially we have numerical values (precise values), we don't need a specific fuzzy logic tool to manage fuzzy information, so the interest of this method is to simplify the information processing of the inferences.

3.3. Action module

The *action module* converts the factors associated to the actions proposed by the expert system into numerical values which can be immediately applied to the process. This transformation is carried out by the inverse function of the functions (1). As the inverse function can't be found directly, we dispose of an approached function defined by intervals.

3.4. Application

An industrial application of our work has been carried out for the Champagnole S.A Cement in a Cifre convention [2]. The system is operational from February 89.

The perception module treats about 50 numerical values representing a global view of the process situation. There are 12 principal possible actions. Considering the technic used, the number of rules is not very important (around 200 rules to find the causes of the disturbing conditions and 150 to search the associated corrective actions).

The expert system is developed in Guru [4] and the combination of the factors associated to the facts is assured by Guru's algebras. The factor of the premise is transferred to the conclusion of the rule because there aren't any factors associated to the conclusions of the rules. If more than one rule give the same conclusion with different factors, the resulting factors will be obtain with the algebra "maximum".

4 - CONCLUSION

The knowledge of the expert system concerning the interpretation of the observed situation can be applied for each process context because we have regrouped the knowledge in the perception module. Using factors associated to facts, we can also treat vague knowledge without using any fuzzy production rules.

The semantic definition module was built to make uniform the knowledge of different experts. In practice, this uniformity induces a regular process control.

As the formalism used is easily understood by the experts, this making decision system will be used to train beginner operators.

5 - REFERENCES

[1] BOBROWICZ O. - 11 Janvier 1991
La représentation et l'utilisation de connaissances imprécises pour l'aide à la conduite de procédé
Thèse de l'Université de Franche-Comté, Spécialité Automatique- Informatique, France

[2] BOBROWICZ O. - HAURAT A.- Juin 1988
Un système expert d'aide à la conduite d'un four en cimenterie - Acquisition des connaissances basée sur une structuration de l'ensemble des informations du système de production.
IASTED International Conference, Genève , Suisse

[3] CHOULET C. Septembre 1988
La logique floue ou comment représenter des connaissances imprécises en vue de leur captage.
Rapport de DEA Informatique, Automatique et Robotique, Université de Franche-Comté, France

[4] GURU - 1988
Manuel de référence GURU V1.1 - Micro Data Base System

[5] KICKERT W.J.M. - VAN NAUTA LEMKE H.H. - 1976
Application of a Fuzzy Controller in a Warm Water Plant.
Automatica, Vol. 12, pp 301-308, Pergamon Press

[6] RASMUSSEN J. - 1985
The Role of Hierarchical Knowledge Representation in Decision making and System Management
IEEE Transactions on Systems, Man, and Cybernetics Vol. SMC-15, N°2 - March/April 1985

[7] ZADEH L.A. - 1965
Fuzzy Sets.
Information and Control, n° 8, pp 338-353

ALGEBRAIC ANALYSIS OF FUZZY INDISCERNIBILITY

Jian-Ming Gao and Akira Nakamura

Department of Applied Mathematics, Hiroshima University
Higashi-Hiroshima, 724, Japan

Abstract First, this paper investigates a model of the database with fuzzy information and generalizes a class of fuzzy indiscernibility relations from the model. Next, this paper is focused on algebraic analysis of the fuzzy indiscernibility, i.e., defining an algebraic structure based on the fuzzy indiscernibility; showing the representation theorem and the center of a given algebra.

Keywords Fuzzy information system, fuzzy indiscernibility, representation theorem, algebra of fuzzy indiscernibility, center of an algbra.

1. INTRODUCTION

In 1980, Pawlak [9] proposed a mathematical framework, i.e., the so called information system, to formalize the knowledge representation. Later, several researchers introduced fuzzy concepts into the information system based on the work of Pawlak, e.g., fuzzy rough sets [3, 4, 6], rough fuzzy sets [3] and a fuzzy modal logic [7]. In addition, the notion of fuzziness in indiscernibility was suggested by Farinas del Cerro and Prade [4].

The motivation of this approach is stimulated by the difficulty of checking the indiscernibility of objects when attributes of the objects are ambiguous and imprecise. For example, assume that person p1 is very tall and person p2 is extremely tall; the indiscernibility of p1 and p2 with respect to the height can be described by not only " yes " and " no ", but also " fairly " and " hardly " .... To give such a description, we adopt a class of fuzzy relations, called *fuzzy indiscernibility relations*, and model them by a class of algebraic structures.

This paper is devoted to presenting a model of a database with fuzzy information which we call a *fuzzy information system* in correlation with Pawlak's information system; defining distinctively the notion of *fuzzy indiscernibility* of objects in the fuzzy information system and to giving an algebraic approach to it.

2. BASIC NOTIONS

To discuss the above mentioned problem, we define the following notions.

The following operations on the interval [0,1] of real numbers from 0 to 1 follow from [2], [5]: for p, $q \in [0,1]$,

$p \wedge q = \max(0, p+q-1)$,

$p \vee q = \min(1, p+q)$,

$p \to q = \min(1, 1-p+q)$.

$(p \to q) \wedge (q \to p)$ is abbreviated to $p \leftrightarrow q$.

Lemma 2.1.

1) $p \vee q = q \vee p$; $p \wedge q = q \wedge p$.
2) $(p \vee q) \vee r = p \vee (q \vee r)$; $(p \wedge q) \wedge r = p \wedge (q \wedge r)$.
3) $p \vee q = (1-p) \to q$.
4) $(p \to q) \to q = \min(p,q)$.
5) $1-(((1-p) \to (1-q)) \to (1-q)) = \max(p,q)$.
6) $p \leftrightarrow q = 1-|p-q|$.

Proof. The proofs of 1)~5) are immediate.

$$\begin{aligned} 6)\ p \leftrightarrow q &= \max(0, \min(1,1-p+q)+\min(1,1-q+p)-1) \\ &= \min(\min(1,1-p+q), \min(1,1-q+p)) \\ &= \min(1-p+q, 1-q+p) \\ &= 1-|p-q|. \end{aligned}$$ //

A fuzzy information system is the quadruple $I=(\mathcal{OB},\mathcal{FA},[0,1],g)$ where $\mathcal{OB}$ is an ordinary set of objects; $\mathcal{FA}$ is a finite set of fuzzy attributes of the objects, e.g., tall; $g: \mathcal{OB} \times \mathcal{FA} \to [0,1]$ is a mapping which describes fuzzy predicates (e.g., an object has some fuzzy attribute) by the real numbers in [0,1]. In fact, a fuzzy information system can be regarded as a database with fuzzy information. Intuitively, any fuzzy information system can be illustrated by the following table:

	a_1	a_2	...	a_n
o_1	$v_{1,1}$	$v_{1,2}$	...	$v_{1,n}$
o_2	$v_{2,1}$	$v_{2,2}$	...	$v_{2,n}$
⋮				
o_m	$v_{m,1}$	$v_{m,2}$	...	$v_{m,n}$
⋮				

That is, in the considered fuzzy information system, $\mathcal{OB} = \{o_1, o_2, ..., o_m, ...\}$, $\mathcal{FA} = \{a_1, a_2, ..., a_n\}$ and $g(o_i,a_j) = v_{i,j} \in [0,1]$ for i = 1, 2, ..., m, ..., j = 1, 2, ..., n, where m and n are finite; usually, the value $v_{i,j}$ is used to represent the grade of fuzziness of the statement: the object o_i has the fuzzy attribute a_j.

Given a fuzzy information system I, the presentation of fuzzy indiscernibility in I is provided as follows: for $A \subseteq \mathcal{FA}$, a *fuzzy indiscernibility relation* $\tilde{A}(X)$ is a fuzzy subset of $X \times X \subseteq \mathcal{OB} \times \mathcal{OB}$ characterized by the membership function μ:

if $A \neq \emptyset$, then $\mu_{\tilde{A}(X)}(o_1,o_2) = \bigwedge_{a \in A} (g(o_1,a) \leftrightarrow g(o_2,a))$ for $o_1, o_2 \in X$, if $A=\emptyset$, then $\mu_{\widetilde{A}(X)}(o_1,o_2)=1$ for $o_1, o_2 \in X$.

$\mu_{\widetilde{A}(X)}(o_1,o_2)$ means the grade of fuzzy indiscernibility of the objects o_1, o_2 with respect to the set A of fuzzy attributes. The commutativity and the associativity (cf. Lemma 2.1) of $\wedge$ are the guarantee of consistency of the membership function μ. It is easily seen that if $A \neq \emptyset$, then $\mu_{\widetilde{A}(X)}(o_1,o_2)=\bigwedge_{a \in A} \mu_{\widetilde{\{a\}}(X)}(o_1,o_2)$.

Example 2.2. Consider the fuzzy information system represented by the following table:

	tall	fat
p_1	0.3	0.6
p_2	0.5	0.6
p_3	0.3	0.4
p_4	0.8	0.7
p_5	0.7	0.9

$\widetilde{\{tall\}}(\mathcal{OB})$ is characterized by

$$\mu_{\widetilde{\{tall\}}(\mathcal{OB})}(p_i,p_i) = g(p_i,tall) \leftrightarrow g(p_i,tall) = 1 \ (\ i = 1, 2, ..., 5\ ),$$
$$\mu_{\widetilde{\{tall\}}(\mathcal{OB})}(p_1,p_2) = g(p_1,tall) \leftrightarrow g(p_2,tall) = 0.3 \leftrightarrow 0.5 = 0.8,$$
$$\mu_{\widetilde{\{tall\}}(\mathcal{OB})}(p_1,p_3) = 0.3 \leftrightarrow 0.3 = 1,$$
$$\mu_{\widetilde{\{tall\}}(\mathcal{OB})}(p_1,p_4) = 0.3 \leftrightarrow 0.8 = 0.5,$$
$$\mu_{\widetilde{\{tall\}}(\mathcal{OB})}(p_1,p_5) = 0.3 \leftrightarrow 0.7 = 0.6,$$
$$\vdots$$
$$\mu_{\widetilde{\{tall\}}(\mathcal{OB})}(p_4,p_5) = 0.8 \leftrightarrow 0.7 = 0.9;$$

$\widetilde{\{fat\}}(\mathcal{OB})$ by

$$\mu_{\widetilde{\{fat\}}(\mathcal{OB})}(p_i,p_i) = 1 \ (\ i = 1, 2, ..., 5\ ),$$
$$\mu_{\widetilde{\{fat\}}(\mathcal{OB})}(p_1,p_2) = 1,$$
$$\vdots$$
$$\mu_{\widetilde{\{fat\}}(\mathcal{OB})}(p_4,p_5) = 0.8;$$

$\widetilde{\{tall,fat\}}(\mathcal{OB})$ by

$$\mu_{\widetilde{\{tall,fat\}}(\mathcal{OB})}(p_i,p_i) = 1 \ (\ i = 1, 2, ..., 5\ ),$$
$$\mu_{\widetilde{\{tall,fat\}}(\mathcal{OB})}(p_1,p_2) = \mu_{\widetilde{\{tall\}}(\mathcal{OB})}(p_1,p_2) \wedge \mu_{\widetilde{\{fat\}}(\mathcal{OB})}(p_1,p_2) = 0.8 \wedge 1 = 0.8,$$
$$\vdots$$
$$\mu_{\widetilde{\{tall,fat\}}(\mathcal{OB})}(p_4,p_5) = \mu_{\widetilde{\{tall\}}(\mathcal{OB})}(p_4,p_5) \wedge \mu_{\widetilde{\{fat\}}(\mathcal{OB})}(p_4,p_5) = 0.9 \wedge 0.8 = 0.7.$$

Lemma 2.3.

1) $\mu_{\widetilde{A}(X)}(o,o)=1$.

2) $\mu_{\widetilde{A}(X)}(o1,o2)=\mu_{\widetilde{A}(X)}(o2,o1)$.

3) $\max\{\mu_{\widetilde{A}(X)}(o1,o) \wedge \mu_{\widetilde{A}(X)}(o,o2): o \in X\} \leq \mu_{\widetilde{A}(X)}(o1,o2)$.

Proof. The only interesting case is 3) (3) is called max-transitivity of the fuzzy relation $\widetilde{A}(X)$). Let us show it by induction on the cardinality $|A|$ of A.

Basis step. Let $A = \{a\}$ and let o be the object satisfying the equation: $\mu_{\widetilde{A}(X)}(o1,o) \wedge \mu_{\widetilde{A}(X)}(o,o2)=\max\{\mu_{\widetilde{A}(X)}(o1,o') \wedge \mu_{\widetilde{A}(X)}(o',o2): o' \in X\}$. Then it is easily known that

$$\mu_{\widetilde{\{a\}}(X)}(o1,o) \wedge \mu_{\widetilde{\{a\}}(X)}(o,o2)$$
$$= \max(0, \mu_{\widetilde{\{a\}}(X)}(o1,o)+\mu_{\widetilde{\{a\}}(X)}(o,o2)-1)$$
$$= \max(0, 1-|g(o1,a)-g(o,a)|-|g(o,a)-g(o2,a)|),$$ and that

$\mu_{\widetilde{\{a\}}(X)}(o1,o2) = 1-|g(o1,a)-g(o2,a)| \geq 0$, i.e., $|g(o1,a)-g(o2,a)| \leq 1$.

If $\min(g(o1,a),g(o2,a)) \leq g(o,a) \leq \max(g(o1,a),g(o2,a))$, then

$|g(o1,a)-g(o,a)|+|g(o,a)-g(o2,a)| = |g(o1,a)-g(o2,a)|$; otherwise

$|g(o1,a)-g(o,a)|+|g(o,a)-g(o2,a)| > |g(o1,a)-g(o2,a)|$. But $\mu_{\widetilde{\{a\}}(X)}(o1,o2)=1-|g(o1,a)-g(o2,a)|$.

It follows that $\mu_{\widetilde{\{a\}}(X)}(o1,o)\wedge\mu_{\widetilde{\{a\}}(X)}(o,o2) \le \mu_{\widetilde{\{a\}}(X)}(o1,o2)$.

Induction step. Assume that when $|A| = k$, 3) holds. Suppose $b \notin A$ and let o be the object satisfying the equation:

$\mu_{\widetilde{A\cup\{b\}}(X)}(o1,o)\wedge\mu_{\widetilde{A\cup\{b\}}(X)}(o,o2)=\max\{\mu_{\widetilde{A\cup\{b\}}(X)}(o1,o')\wedge\mu_{\widetilde{A\cup\{b\}}(X)}(o',o2): o'\in X\}$. Then,

$$\alpha = \mu_{\widetilde{A\cup\{b\}}(X)}(o1,o)\wedge\mu_{\widetilde{A\cup\{b\}}(X)}(o,o2)$$
$$= \mu_{\widetilde{A}(X)}(o1,o)\wedge\mu_{\widetilde{\{b\}}(X)}(o1,o)\wedge\mu_{\widetilde{A}(X)}(o,o2)\wedge\mu_{\widetilde{\{b\}}(X)}(o,o2).$$ If

$\mu_{\widetilde{A}(X)}(o1,o)\wedge\mu_{\widetilde{\{b\}}(X)}(o1,o)\wedge\mu_{\widetilde{A}(X)}(o,o2) = 0$, or $\mu_{\widetilde{\{b\}}(X)}(o,o2) = 0$, then, the proposition holds without further proof. Suppose not. Then,

$$\alpha = \mu_{\widetilde{A}(X)}(o1,o)\wedge\mu_{\widetilde{\{b\}}(X)}(o1,o)\wedge\mu_{\widetilde{A}(X)}(o,o2)\wedge\mu_{\widetilde{\{b\}}(X)}(o,o2)$$
$$= \max(0, \mu_{\widetilde{A}(X)}(o1,o)+\mu_{\widetilde{\{b\}}(X)}(o1,o)+\mu_{\widetilde{A}(X)}(o,o2)+\mu_{\widetilde{\{b\}}(X)}(o,o2)-3)$$
$$= \max(0, \mu_{\widetilde{A}(X)}(o1,o)+\mu_{\widetilde{A}(X)}(o,o2)+\mu_{\widetilde{\{b\}}(X)}(o1,o)+\mu_{\widetilde{\{b\}}(X)}(o,o2)-3).$$

Similarly, $\beta = \mu_{\widetilde{A\cup\{b\}}(X)}(o1,o2)$

$$= \mu_{\widetilde{A}(X)}(o1,o2)\wedge\mu_{\widetilde{\{b\}}(X)}(o1,o2)$$
$$= \max(0, \mu_{\widetilde{A}(X)}(o1,o2)+\mu_{\widetilde{\{b\}}(X)}(o1,o2)-1).$$

By the induction hypothesis, $\mu_{\widetilde{A}(X)}(o1,o)\wedge\mu_{\widetilde{A}(X)}(o,o2) \le \mu_{\widetilde{A}(X)}(o1,o2)$, i.e.,

$$\max(0, \mu_{\widetilde{A}(X)}(o1,o)+\mu_{\widetilde{A}(X)}(o,o2)-1) \le \mu_{\widetilde{A}(X)}(o1,o2);$$ and

$\mu_{\widetilde{\{b\}}(X)}(o1,o)\wedge\mu_{\widetilde{\{b\}}(X)}(o,o2) \le \mu_{\widetilde{\{b\}}(X)}(o1,o2)$, i.e.,

$$\max(0, \mu_{\widetilde{\{b\}}(X)}(o1,o)+\mu_{\widetilde{\{b\}}(X)}(o,o2)-1) \le \mu_{\widetilde{\{b\}}(X)}(o1,o2).$$

Immediately, we have $\alpha \le \beta$. Thus, we are done. //

The relations $\subseteq$ and $=$ between fuzzy indiscernibility relations in a fuzzy information system are defined as follows:

1) $\widetilde{A}(X)\subseteq\widetilde{B}(X)$ iff $\mu_{\widetilde{A}(X)}(o1,o2)\le\mu_{\widetilde{B}(X)}(o1,o2)$ for all $o1, o2 \in X$;

2) $\widetilde{A}(X)=\widetilde{B}(X)$ iff $\mu_{\widetilde{A}(X)}(o1,o2)=\mu_{\widetilde{B}(X)}(o1,o2)$ for all $o1, o2 \in X$.

<u>Lemma 2.4.</u> 1) If $A\subseteq B$, then $\widetilde{B}(X)\subseteq\widetilde{A}(X)$; but the converse does not hold.

2) If $A=B$, then $\widetilde{A}(X)=\widetilde{B}(X)$; but the converse does not hold.

<u>Proof.</u> 1) Let $B = A\cup C$ where $\emptyset\subseteq C\subseteq\mathcal{FA}$ and $A\cap C = \emptyset$. Then

$$\mu_{\widetilde{B}(X)}(o1,o2) = \mu_{\widetilde{A}(X)}(o1,o2)\wedge\mu_{\widetilde{C}(X)}(o1,o2) = \max(0, \mu_{\widetilde{A}(X)}(o1,o2)+\mu_{\widetilde{C}(X)}(o1,o2)-1).$$

If $\mu_{\widetilde{B}(X)}(o1,o2) = 0$, then $\mu_{\widetilde{B}(X)}(o1,o2) \le \mu_{\widetilde{A}(X)}(o1,o2)$; otherwise, $\mu_{\widetilde{B}(X)}(o1,o2) \le \mu_{\widetilde{A}(X)}(o1,o2)$ as well since $\mu_{\widetilde{C}(X)}(o1,o2) \le 1$.

On the other hand, for example, in Example 2.2, although $\widetilde{\{tall\}}(\{p_1,p_2\}) = \widetilde{\{tall,fat\}}(\{p_1,p_2\})$, $\{tall,fat\}\not\subseteq\{fat\}$.

2) It is obvious that if $A = B$, then $\widetilde{A}(X) = \widetilde{B}(X)$; by obversing the equation: $\widetilde{\{tall\}}(\{p_1,p_2\}) = \widetilde{\{tall,fat\}}(\{p_1,p_2\})$ in Example 2.2, it is shown that the converse does not hold.//

Let $FI(X,A)=\{\widetilde{B}(X): B\subseteq A\}$ for $X\subseteq\mathcal{OB}$ and $A\subseteq\mathcal{FA}$.

We introduce operations $\neg$, $\cap$ and $\Rightarrow$ on the set FI(X,A) for a fuzzy information system:

1) For $B\subseteq A$ $\neg\widetilde{B}(X)=\widetilde{A-B}(X)$, called (X,A)-complement of $\widetilde{B}(X)$;

2) $\widetilde{B}(X)\cap\widetilde{C}(X) = \widetilde{B\cup C}(X)$;

3) $\widetilde{B}(X)\Rightarrow\widetilde{C}(X)=\sup\{\widetilde{D}(X)$: $D\subseteq A$ and $\widetilde{B}(X)\cap\widetilde{D}(X))\leq\widetilde{C}(X)\}$. That is, $\widetilde{B}(X)\Rightarrow\widetilde{C}(X)$ is the least upper bound, or supremum of $\{\widetilde{D}(X)$: $D\subseteq A$ and $\widetilde{B}(X)\cap\widetilde{D}(X))\leq\widetilde{C}(X)\}$ in the sense of the relation $\leq$.

It should be noted that $\cap$ is a symmetric operation.

Further, to express definability of objects in a fuzzy information system, we classify the fuzzy attributes by using the fuzzy indiscernibility as follows:

1) A set A of fuzzy attributes is X-*dependent* in a fuzzy information system I if there is a proper subset C of A such that $\widetilde{A-C}(X)=\widetilde{A}(X)$;
2) A is X-*independent* in I if A is not X-dependent in I;
3) A is X-*superfluous* in the set B of fuzzy attributes iff $\widetilde{B-A}(X)=\widetilde{B}(X)$.

In the example mentioned above, {tall,fat} is $\mathcal{OB}$-independent, $\{p_1,p_2\}$-dependent and $\{p_1,p_3\}$-dependent.

The properties of the dependencies of fuzzy attributes in a fuzzy information system will be investigated in an algebraic way. Given a fuzzy information system I, we should always bear in mind the following fact.

<u>Lemma 2.5.</u>

1) The relation $=$ on FI(X,A) is not a congruence with respect to the operation $\neg$ on FI(X,A).
2) The relation $=$ on FI(X,A) is not a congruence with respect to the operation $\cap$ on FI(X,A).
3) The relation $=$ on FI(X,A) is not a congruence with respect to the operation $\Rightarrow$ on FI(X,A).

<u>Proof.</u> 1) Suppose $\mathcal{FA}=\{a, b, c, d\}$, $X=\{o1, o2\}$, $\mu_{\widetilde{\{a\}}(X)}(o1,o2)=0.4$, $\mu_{\widetilde{\{b\}}(X)}(o1,o2)=0.5$, $\mu_{\widetilde{\{c\}}(X)}(o1,o2)=0.3$ and $\mu_{\widetilde{\{d\}}(X)}(o1,o2)=0.7$. Let $B=\{a, b\}$ and $B'=\{b, c\}$. Then,

$$
\begin{aligned}
&\mu_{\widetilde{B}(X)}(o1,o2)\\
&=\mu_{\widetilde{\{a\}}(X)}(o1,o2)\wedge\mu_{\widetilde{\{b\}}(X)}(o1,o2)\\
&=0.4\wedge0.5\\
&=0,\text{ and}\\
&\mu_{\widetilde{B}'(X)}(o1,o2)\\
&=\mu_{\widetilde{\{b\}}(X)}(o1,o2)\wedge\mu_{\widetilde{\{c\}}(X)}(o1,o2)\\
&=0.5\wedge0.3\\
&=0.
\end{aligned}
$$

It is clear that $\widetilde{B}(X)=\widetilde{B}'(X)$. But $\neg\widetilde{B}(X)\neq\neg\widetilde{B}'(X)$ for FI(X,$\mathcal{FA}$) since

$$
\begin{aligned}
&\mu_{\neg\widetilde{B}(X)}(o1,o2)\\
&=\mu_{\widetilde{\{c,d\}}(X)}(o1,o2)\\
&=\mu_{\widetilde{\{c\}}(X)}(o1,o2)\wedge\mu_{\widetilde{\{d\}}(X)}(o1,o2)\\
&=0.3\wedge0.7\\
&=0,\text{ and}\\
&\mu_{\neg\widetilde{B}'(X)}(o1,o2)\\
&=\mu_{\widetilde{\{a,d\}}(X)}(o1,o2)\\
&=\mu_{\widetilde{\{a\}}(X)}(o1,o2)\wedge\mu_{\widetilde{\{d\}}(X)}(o1,o2)\\
&=0.4\wedge0.7\\
&=0.1.
\end{aligned}
$$

2) Let us give an example to illustrate that "=" on FI(X,A) is not a congruence with respect to $\cap$. Given the following fuzzy information system

	a	b
o1	0.3	0.5
o2	0.5	0.3

$\widetilde{\{a\}}(\{o1,o2\}) = \widetilde{\{b\}}(\{o1,o2\})$ since for all $o_i, o_j \in \{o1,o2\}$

$\mu_{\widetilde{\{a\}}(\{o1,o2\})}(o_i,o_j) = \mu_{\widetilde{\{b\}}(\{o1,o2\})}(o_i,o_j)$. However,

$\widetilde{\{a\}}(\{o1,o2\}) \cap \widetilde{\{a\}}(\{o1,o2\}) = \widetilde{\{a\}}(\{o1,o2\}) \neq \widetilde{\{a\}}(\{o1,o2\}) \cap \widetilde{\{b\}}(\{o1,o2\})$ since

$\mu_{\widetilde{\{a\}}(\{o1,o2\})}(o1,o2) = 0.3 \leftrightarrow 0.5 = 0.8$, but

$$\begin{aligned}
&\mu_{\widetilde{\{a\}}(\{o1,o2\}) \cap \widetilde{\{b\}}(\{o1,o2\})}(o1,o2) \\
&= \mu_{\widetilde{\{a\}}(\{o1,o2\})}(o1,o2) \wedge \mu_{\widetilde{\{b\}}(\{o1,o2\})}(o1,o2) \\
&= (0.3 \leftrightarrow 0.5) \wedge (0.5 \leftrightarrow 0.3) \\
&= 0.8 \wedge 0.8 \\
&= 0.6.
\end{aligned}$$

3) Consider the following table:

	a	b	c	d
o1	0.3	0.7	0.8	0.2
o2	0.5	0.8	0.7	0.5

That is, $\mu_{\widetilde{\{a\}}(\{o1,o2\})}(o1,o2)=0.8$, $\mu_{\widetilde{\{b\}}(\{o1,o2\})}(o1,o2)=0.9$, $\mu_{\widetilde{\{c\}}(\{o1,o2\})}(o1,o2) = 0.9$, and $\mu_{\widetilde{\{d\}}(\{o1,o2\})}(o1,o2)=0.7$; $\widetilde{\{a\}}(\{o1,o2\}) = \widetilde{\{b,c\}}(\{o1,o2\})$. Thus, $\widetilde{\{a\}}(\{o1,o2\}) \Rightarrow \widetilde{\{d\}}(\{o1,o2\}) = \widetilde{\emptyset}(\{o1,o2\})$, $\widetilde{\{b,c\}}(\{o1,o2\}) \Rightarrow \widetilde{\{d\}}(\{o1,o2\}) = \widetilde{\{a\}}(\{o1,o2\})$, but $\widetilde{\{a\}}(\{o1,o2\}) \neq \widetilde{\emptyset}(\{o1,o2\})$. //

3. FIS-ALGEBRAS OF FUZZY INDISCERNIBILITY FOR A FUZZY INFORMATION SYSTEM

In the following, we will assume that $I=(\mathcal{OB}, \mathcal{FA}, [0,1], g)$ is a fixed fuzzy information system. In this section, we define a class of algebras of fuzzy indiscernibility for I and provide some ideas on algebraic analysis of fuzzy indiscernibility in the fuzzy information system I. By the definition given above, it is easily known that for $X \subseteq \mathcal{OB}$ and $A \subseteq \mathcal{FA}$ FI(X,A) is closed under $\neg$, $\cap$ and $\Rightarrow$. Clearly, for any $A \subseteq \mathcal{FA}$ $\widetilde{A}(X) \cap \widetilde{\emptyset}(X) = \widetilde{A}(X)$. Let us denote $\widetilde{\emptyset}(X)$ by T_X.

<u>Definition 3.1.</u> By an FIS-algebra we mean the structure $(\mathrm{FI}(X,A), \neg, \cap, \Rightarrow, T_X)$, denoted by **FI**(X,A), where T_X is called the unit element with respect to X.

<u>Theorem 3.2.</u> Any FIS-algebra **FI**(X,A) obeys the following.

1) $\widetilde{B}(X) \Rightarrow \widetilde{B}(X) = T_X$.
2) $\neg\widetilde{B}(X) = \widetilde{A-B}(X)$.
3) $\widetilde{B}(X) \cap T_X = \widetilde{B}(X)$.
4) $\widetilde{B \cup C}(X) = \widetilde{B}(X) \cap \widetilde{C}(X)$.
5) $\widetilde{B}(X) \cap \widetilde{D}(X) \subseteq \widetilde{C}(X)$ iff $\widetilde{D}(X) \subseteq \widetilde{B}(X) \Rightarrow \widetilde{C}(X)$.
6) $(\widetilde{B}(X) \cap \neg\widetilde{B}(X)) \Rightarrow \widetilde{C}(X) = T_X$.
7) $\widetilde{B}(X) \cap \neg\widetilde{B}(X) = \neg(\widetilde{B}(X) \Rightarrow \widetilde{B}(X))$.

<u>Proof.</u> By simple verifications. //

As consequences of the above, the following elementary properties for any FIS-algebra **FI**(X,A) hold:

<u>Theorem 3.3.</u>

1) $\neg(\widetilde{B}(X)\Rightarrow\widetilde{B}(X))\Rightarrow\widetilde{C}(X)=T_X$.
2) $\widetilde{B}(X)\Rightarrow\widetilde{C}(X)=T_X$ iff $\widetilde{B}(X)\subseteq\widetilde{C}(X)$.
3) $\widetilde{C}(X)\subseteq\widetilde{B}(X)\Rightarrow\widetilde{C}(X)$.
4) $(\widetilde{B}(X)\Rightarrow\widetilde{C}(X))\cap\widetilde{C}(X)\subseteq\widetilde{C}(X)$ and $\widetilde{B}(X)\cap\widetilde{C}(X)\subseteq\widetilde{B}(X)\cap(\widetilde{B}(X)\Rightarrow\widetilde{C}(X))$.
5) $\widetilde{B}(X)\cap\widetilde{C}(X)\subseteq\widetilde{B}(X)$ and $\widetilde{B}(X)\cap\widetilde{C}(X)\subseteq\widetilde{C}(X)$.

Proof. 1) Since $\widetilde{B}(X)\Rightarrow\widetilde{B}(X)=\widetilde{\varnothing}(X)$, $\neg(\widetilde{B}(X)\Rightarrow\widetilde{B}(X))=\widetilde{A}(X)$. But, by Lemma 2.4(1) for any $C\subseteq A$ $\widetilde{A}(X)\cap\widetilde{\varnothing}(X)\subseteq\widetilde{C}(X)$ and hence $\widetilde{A}(X)\Rightarrow\widetilde{C}(X)=T_X$.

2) This proof is an application of Theorem 3.2(5). Suppose $\widetilde{B}(X)\subseteq\widetilde{C}(X)$. Then, $\widetilde{B}(X)\cap\widetilde{\varnothing}(X)\subseteq\widetilde{C}(X)$, i.e., $\widetilde{\varnothing}(X)\subseteq\widetilde{B}(X)\Rightarrow\widetilde{C}(X)$, i.e., $\widetilde{B}(X)\Rightarrow\widetilde{C}(X)=T_X$. Conversely, if $\widetilde{B}(X)\Rightarrow\widetilde{C}(X)=T_X$, then $\widetilde{B}(X)\cap\widetilde{\varnothing}(X)\subseteq\widetilde{C}(X)$, i.e., $\widetilde{B}(X)\subseteq\widetilde{C}(X)$.

3) This follows from Theorem 3.2(5).

The others are easy. //

$\widetilde{B}(X)$ is called the *restriction* of $\widetilde{C}(Y)$ into X if $X\subseteq Y$ and for $o1, o2\in X$ $\mu_{\widetilde{B}(X)}(o1,o2)=\mu_{\widetilde{C}(Y)}(o1,o2)$. **FI**(X,A) is said to be *extensive* to **FI**(Y,B) (symbolically, **FI**(X,A)⊢**FI**(Y,B)) if there is a homomorphism π: FI(X,A)→FI(Y,B) which is one-to-one and for every $\widetilde{C}(X)$ in FI(X,A) $\widetilde{C}(X)$ is the restriction of $\pi(\widetilde{C}(X))$ into X.

Theorem 3.4.
1) All **FIS**-algebras for the fuzzy information system *I* form a partial-ordering under ⊢.
2) If **FI**(X,A)⊢**FI**(Y,A) and B ($\subseteq$A) is Y-dependent, then B is also X-dependent; but the converse does not hold.
3) If **FI**(X,A)⊢**FI**(Y,A) and B ($\subseteq$A) is X-independent, then, B is also Y-independent; but the converse does not hold.
4) If **FI**(X,A)⊢**FI**(Y,A) and B ($\subseteq$A) is Y-superfluous in A, then, B is also X-superfluous in A; but the converse does not hold.

Proof. 1) It is easy to show that ⊢ is reflexive and transitive. Let us show the anti-symmetry of ⊢ only: if **FI**(X,A)⊢**FI**(Y,B) and **FI**(Y,B)⊢**FI**(X,A), then **FI**(X,A)=**FI**(Y,B). Suppose **FI**(X,A)⊢**FI**(Y,B) and **FI**(Y,B)⊢**FI**(X,A). Then $X\subseteq Y$ and $Y\subseteq X$, i.e., X=Y and so for any $\widetilde{A}'(X)\in$ FI(X,A) and any $\widetilde{B}'(Y)\in$ FI(Y,B) $\widetilde{A}'(X)=\pi(\widetilde{A}'(X))$ and $\widetilde{B}'(Y)=\pi'(\widetilde{B}'(Y))$ where π is the homomorphism from FI(X,A) to FI(Y,B) and π' is the homomorphism from FI(Y,B) to FI(X,A). But since the homomorphisms π and π' are one-to-one, FI(X,A)=FI(Y,B) and then **FI**(X,A)=**FI**(Y,B).

2) Since $\widetilde{B}(X)$ is the restriction of $\widetilde{B}(Y)$ into X, if there is a proper subset C of B such that $\widetilde{B-C}(Y)=\widetilde{B}(Y)$, then $\widetilde{B-C}(X)=\widetilde{B}(X)$. The converse does not hold since $\widetilde{B}(Y)$ is not the restriction of $\widetilde{B}(X)$.

The proofs of 3) and 4) are similar to that of 2). //

Given an **FIS**-algebra **FI**(X,A) for a fuzzy information system, define that
Um(FI(X,A))=∪{FI(X,B): $B\subseteq A$ and B is not X-superfluous in A} and
Um(FI(X,A))=∅ if A=∅.

We call Um(FI(X,A)) universe of the subalgebra of **FI**(X,A) with maximal X-independent set of fuzzy attributes.

Theorem 3.5. 1) Um(Um(FI(X,A)))=Um(FI(X,A)).
2) $A \subseteq B$ implies $Um(FI(X,A)) \subseteq Um(FI(X,B))$.
3) $Um(FI(X,A)) \cup Um(FI(X,B)) = Um(FI(X,A \cup B))$.
4) $Um(FI(X,A \cap B)) \subseteq Um(FI(X,A)) \cap Um(FI(X,B))$.
Proof. It is immediate from the definition of Um. //

Theorem 3.6. If A is X-independent, then Um(FI(X,A))=FI(X,A).
Proof. Since A is X-independent, any proper subset of A is not X-superfluous in A. //

4. REPRESENTATION THEOREM FOR FIS-ALGEBRA

The aim of this section is to show a correspondence of an FIS-algebra to a normal algebraic structure, i.e., the representation of an FIS-algebra. Negoita and Ralescu [8] have presented the representation theorems for several fuzzy concepts. To do so, we want to construct an algebra which is isomorphic to the given FIS-algebra.

For $\alpha \in [0,1]$, $\tilde{A}(X)_\alpha = \{(o,o') \in X \times X: \mu_{\tilde{A}(X)}(o,o') \geq \alpha\}$. It should be noted that $\tilde{A}(X)_\alpha$ is not a fuzzy set. By Zadeh [10], $\tilde{A}(X)_\alpha$ is called an α -level-set of $\tilde{A}(X)$. Let $FL(X,A) = \{\{\tilde{B}(X)_\alpha\}_{\alpha \in [0,1]}: B \subseteq A\}$.

The relation $\prec$ on FL(X,A) is defined by
$\{\tilde{B}(X)_\alpha\}_{\alpha \in [0,1]} \prec \{\tilde{C}(X)_\alpha\}_{\alpha \in [0,1]}$ iff for $\alpha \in [0,1]$ $\tilde{B}(X)_\alpha \subseteq \tilde{C}(X)_\alpha$.

Lemma 4.1. For an FIS-algebra FI(X,A), the following conditions are equivalent:
1) $\tilde{B}(X) \subseteq \tilde{C}(X)$.
2) $\{\tilde{B}(X)_\alpha\}_{\alpha \in [0,1]} \prec \{\tilde{C}(X)_\alpha\}_{\alpha \in [0,1]}$.
Proof. $\tilde{B}(X) \subseteq \tilde{C}(X)$ iff for o1, o2 $\in$ X $\mu_{\tilde{B}(X)}(o1,o2) \leq \mu_{\tilde{C}(X)}(o1,o2)$ iff for $\alpha \in [0,1]$ and for o1, o2 $\in$ X $\mu_{\tilde{B}(X)}(o1,o2) \geq \alpha$ implies $\mu_{\tilde{C}(X)}(o1,o2) \geq \alpha$ iff for $\alpha \in [0,1]$ $\tilde{B}(X)_\alpha \subseteq \tilde{C}(X)_\alpha$ iff 2) holds. //

Let us give the definitions of operations ©, $\otimes$ and Θ on FL(X,A) as follows:
1) $\{\tilde{B}(X)_\alpha\}^{©}_{\alpha \in [0,1]} = \{(\neg\tilde{B}(X))_\alpha\}_{\alpha \in [0,1]}$.
2) $\{\tilde{B}(X)_\alpha\}_{\alpha \in [0,1]} \otimes \{\tilde{C}(X)_\alpha\}_{\alpha \in [0,1]} = \{(\tilde{B}(X) \cap \tilde{C}(X))_\alpha\}_{\alpha \in [0,1]}$.
3) $\{\tilde{B}(X)_\alpha\}_{\alpha \in [0,1]} \Theta \{\tilde{C}(X)_\alpha\}_{\alpha \in [0,1]} =$
$\sup\{\{\tilde{D}(X)_\alpha\}_{\alpha \in [0,1]}: D \subseteq A \text{ and } \{\tilde{B}(X)_\alpha\}_{\alpha \in [0,1]} \otimes \{\tilde{D}(X)_\alpha\}_{\alpha \in [0,1]} \prec \{\tilde{C}(X)_\alpha\}_{\alpha \in [0,1]}\}$.

3) means that $\{\tilde{B}(X)_\alpha\}_{\alpha \in [0,1]} \Theta \{\tilde{C}(X)_\alpha\}_{\alpha \in [0,1]}$ is the least upper bound, or supremum of the set $\{\{\tilde{D}(X)_\alpha\}_{\alpha \in [0,1]}: D \subseteq A \text{ and } \{\tilde{B}(X)_\alpha\}_{\alpha \in [0,1]} \otimes \{\tilde{D}(X)_\alpha\}_{\alpha \in [0,1]} \prec \{\tilde{C}(X)_\alpha\}_{\alpha \in [0,1]}\}$ in the sense of $\prec$.

Let $I_X = \{X \times X\}$. Then, $I_X \in FL(X,A)$ since $I_X = \{\tilde{\varnothing}(X)_\alpha\}_{\alpha \in [0,1]}$. FL(X,A) is closed under ©, $\otimes$ and Θ.

Theorem 4.2 (Representation Theorem). The algebra (FI(X,A), $\neg$, $\cap$, $\Rightarrow$, T_X) and the algebra (FL(X,A), ©, $\otimes$, Θ, I_X} are isomorphic.

Proof. Define τ: $\mathbf{FI(X,A)} \rightarrow \mathbf{FL(X,A)}$ by $\tau(\widetilde{B}(X))=\{\widetilde{B}(X)_\alpha\}_{\alpha\in[0,1]}$. Let us show the theorem as follows.

1) τ is onto. This is obvious.

2) τ is one-to-one: if $\widetilde{B}(X)$, $\widetilde{C}(X) \in FI(X,A)$ and $\tau(\widetilde{B}(X))=\tau(\widetilde{C}(X))$, then $\widetilde{B}(X)=\widetilde{C}(X)$. Suppose $\tau(\widetilde{B}(X))=\tau(\widetilde{C}(X))$, i.e., for $\alpha \in [0,1]$ $\widetilde{B}(X)_\alpha=\widetilde{C}(X)_\alpha$. Then, if $\beta=\mu_{\widetilde{B}(X)}(o1,o2)$, then $\widetilde{B}(X)_\beta=\widetilde{C}(X)_\beta$. It follows that $\mu_{\widetilde{C}(X)}(o1,o2)\geq\mu_{\widetilde{B}(X)}(o1,o2)$. Similarly, we can obtain $\mu_{\widetilde{B}(X)}(o1,o2)\geq\mu_{\widetilde{C}(X)}(o1,o2)$. Therefore $\widetilde{B}(X)=\widetilde{C}(X)$.

3) The next question is how to prove that for any $\widetilde{B}(X)$, $\widetilde{C}(X)\in FI(X,A)$ $\tau(\neg\widetilde{B}(X))=\tau(\widetilde{B}(X))^{©}$, $\tau(\widetilde{B}(X)\cap\widetilde{C}(X))=\tau(\widetilde{B}(X))\otimes\tau(\widetilde{C}(X))$ and $\tau(\widetilde{B}(X)\Rightarrow\widetilde{C}(X))=\tau(\widetilde{B}(X))\Theta\tau(\widetilde{C}(X))$. It is immediate that

$$\begin{aligned}
\tau(\neg\widetilde{B}(X)) &= \{(\neg\widetilde{B}(X))_\alpha\}_{\alpha\in[0,1]} \\
&= \{\widetilde{B}(X)_\alpha\}^{©}_{\alpha\in[0,1]} \\
&= (\tau(\widetilde{B}(X)))^{©},
\end{aligned}$$

$$\begin{aligned}
\tau(\widetilde{B}(X)\cap\widetilde{C}(X)) &= \{(\widetilde{B}(X)\cap\widetilde{C}(X))_\alpha\}_{\alpha\in[0,1]} \\
&= \{\widetilde{B}(X)_\alpha\}_{\alpha\in[0,1]}\otimes\{\widetilde{C}(X)_\alpha\}_{\alpha\in[0,1]} \\
&= \tau(\widetilde{B}(X))\otimes\tau(\widetilde{C}(X)), \text{ and}
\end{aligned}$$

$$\begin{aligned}
\tau(\widetilde{B}(X)\Rightarrow\widetilde{C}(X)) &= \{(\widetilde{B}(X)\Rightarrow\widetilde{C}(X))_\alpha\}_{\alpha\in[0,1]} \\
&= \{(\sup\{\widetilde{D}(X): D\subseteq A \text{ and } \widetilde{B}(X)\cap\widetilde{D}(X)\leq\widetilde{C}(X)\})_\alpha\}_{\alpha\in[0,1]} \\
&= \sup\{\{\widetilde{D}(X)_\alpha\}_{\alpha\in[0,1]}: D\subseteq A \text{ and } \widetilde{B}(X)\cap\widetilde{D}(X)\leq\widetilde{C}(X)\} \\
&= \sup\{\{\widetilde{D}(X)_\alpha\}_{\alpha\in[0,1]}: D\subseteq A \text{ and } \\
&\qquad \{(\widetilde{B}(X)\cap\widetilde{D}(X))_\alpha\}_{\alpha\in[0,1]}\prec\{\widetilde{C}(X)_\alpha\}_{\alpha\in[0,1]}\} \\
&= \sup\{\{\widetilde{D}(X)_\alpha\}_{\alpha\in[0,1]}: D\subseteq A \text{ and } \\
&\qquad \{\widetilde{B}(X)_\alpha\}_{\alpha\in[0,1]}\otimes\{\widetilde{D}(X)_\alpha\}_{\alpha\in[0,1]}\prec\{\widetilde{C}(X)_\alpha\}_{\alpha\in[0,1]}\} \\
&= \tau(\widetilde{B}(X))\Theta\tau(\widetilde{C}(X)).
\end{aligned}$$
//

5. THE CENTER OF A GIVEN FIS-ALGEBRA

In this section, we apply a definition of the center of an arbitary algebra given by Freese and McKenzie (cf. [1]) into this paper and answer the question how to calculate the center of an FIS-algebra.

To generalize further the algebraic properties of fuzzy indiscernibility and the operations on an FIS-algebra, we introduce the notion of *term*.

Let V be a set of variables. The set T(V) of terms for FIS-algebras is the least set satisfying the following conditions:

1) $V\subseteq T(V)$;

2) if $t_1,t_2\in T(V)$, then $\neg t_1, t_1\cap t_2, t_1\Rightarrow t_2\in T(V)$.

For a term t, we will write $t(x_1,x_2,\ldots,x_n)$ where $x_1,x_2,\ldots,x_n$ are all the variables occurring in the term t.

Given a term $t(x_1,x_2,\ldots,x_n)$ and an FIS-algebra **FI(X,A)**, we define a valuation function $t^{\mathbf{FI(X,A)}}$ assigning t into FI(X,A):

1) for $t \in V$, $t^{FI(X,A)} \in FI(X,A)$;
2) $(\neg t)^{FI(X,A)} = \neg(t^{FI(X,A)})$;
3) $(t_1 \cap t_2)^{FI(X,A)} = t_1^{FI(X,A)} \cap t_2^{FI(X,A)}$;
4) $(t_1 \Rightarrow t_2)FI(X,A) = t_1^{FI(X,A)} \Rightarrow t_2^{FI(X,A)}$.

For brevity, we write $t(x_1^{FI(X,A)}, x_2^{FI(X,A)}, \ldots, x_n^{FI(X,A)})$ for $(t(x_1, x_2, \ldots, x_n))^{FI(X,A)}$.

Let **FI**(X,A) be an FIS-algebra. Then, the center of **FI**(X,A) is the binary relation **Z**(**FI**(X,A)) such that

$$\mathbf{Z}(\mathbf{FI}(X,A)) = \{(\widetilde{B}(X), \widetilde{B}'(X)) \in FI(X,A) \times FI(X,A) : \text{for any } t(x_1, x_2, \ldots, x_n) \in T(V)\ t(\widetilde{B}(X), \Gamma) = t(\widetilde{B}(X), \Delta) \text{ iff } t(\widetilde{B}'(X), \Gamma) = t(\widetilde{B}'(X), \Delta) \text{ where }, \Gamma, \Delta \in FI(X,A)^{n-1}\}.$$

Lemma 5.1. 1) **Z**(**FI**(X,A)) is an equivalence relation on **FI**(X,A).
2) **Z**(**FI**(X,A)) is a congruence on **FI**(X,A).
Proof. cf.[1]. //

For an FIS-algebra FI(X,A) if B and B' ($\subseteq$A) satisfy one of the following;
1) B=B',
2) if $B \cap B' = B$, then B'-B is X-superfluous in B',
3) if $B \cap B' = B'$, then B-B' is X-superfluous in B,
4) B and B' are X-superfluous in $B \cup B'$,
then, $\widetilde{B}(X) = \widetilde{B}'(X)$; let us denote it by $B(X) \equiv_A B'(X)$.

Lemma 5.2. If $B(X) \equiv_A B'(X)$, then for any $t(x_1, x_2, \ldots, x_n) \in T(V)$ and $\Gamma \in FI(X,A)^{n-1}$ $t(\widetilde{B}(X), \Gamma) \equiv_A t(\widetilde{B}'(X), \Gamma)$.
Proof. This can be easily shown by induction on the structure of t. //

Let us give a precise and intuitive explanation of the center of an FIS-algebra, i.e., actually calculating the center of a given FIS-algebra.

Theorem 5.3. $\mathbf{Z}(\mathbf{FI}(X,A)) = \{(\widetilde{B}(X), \widetilde{B}'(X)) \in FI(X,A) \times FI(X,A) : \widetilde{B}(X) \equiv_A \widetilde{B}'(X)\}$.
Proof. Let $\mathbf{Z}' = \{(\widetilde{B}(X), \widetilde{B}'(X)) \in FI(X,A) \times FI(X,A)$:
1) B=B'; or
2) if $B \cap B' = B$, then B'-B is X-superfluous in B'; or
3) if $B \cap B' = B'$, then B-B' is X-superfluous in B; or
4) B and B' are X-superfluous in $B \cup B'\}$. Then, we must show that $\mathbf{Z}(\mathbf{FI}(X,A)) = \mathbf{Z}'$. First, by applying Lemma 5.2 and Lemma 5.1, we have $\mathbf{Z}' \subseteq \mathbf{Z}(\mathbf{FI}(X,A))$. So, it suffices to show that $\mathbf{Z}(\mathbf{FI}(X,A)) \subseteq \mathbf{Z}'$. Suppose $(\widetilde{B}(X), \widetilde{B}'(X)) \in \mathbf{Z}(\mathbf{FI}(X,A))$ and $(\widetilde{B}(X), \widetilde{B}'(X)) \notin \mathbf{Z}'$. Then,
1) $B \neq B'$; and
2) $B \cap B' = B$, and B'-B is not X-superfluous in B'; and
3) $B \cap B' = B'$, and B-B' is not X-superfluous in B; and

4) B or B' is not X-superfluous in $B \cup B'$.
$\widetilde{B}(X) \cap \widetilde{B}(X) = \widetilde{B}(X) \cap \widetilde{\varnothing}(X)$, but $\widetilde{B}'(X) \cap \widetilde{B}(X) \neq \widetilde{B}'(X) \cap \widetilde{\varnothing}(X)$. That is, $(\widetilde{B}(X), \widetilde{B}'(X)) \notin$ **Z(FI(X,A))**, a contradiction. //

6. FUTURE WORKS

As future works,we will consider the following topics:

1) Algebraic properties of the fuzzy indiscernibility in the case where instead of the above defined the operation $\cap$ on FI(X,A), $\widetilde{B}(X) \cap \widetilde{C}(X)$ is defined by
$\mu_{\widetilde{B}(X) \cap \widetilde{C}(X)}(o1,o2) = \mu_{\widetilde{B}(X)}(o1,o2) \wedge \mu_{\widetilde{C}(X)}(o1,o2)$ for all o1, o2 $\in$ X.
2) Correspondence of this algebraic study to fuzzy logical systems.

REFERENCES

[1] S. Burris and H. P. Sankappanavar, A Course in Universal Algebra, Springer-Verlag, 1980.

[2] D. Dubois and H. Prade, Fuzzy Sets and Systems: Theory and Applications, Academic press, 1980.

[3] D. Dubois and H. Prade, Rough fuzzy sets and fuzzy rough sets, Proceedings of Intern. Conf. Fuzzy Set in Informatics, Moscow, (1988)20-23.

[4] L. Farinas del Cerro and H. Prade, Rough sets, twofold fuzzy sets and modal logic. Fuzziness in indiscernibility and partial information, The Mathematics of Fuzzy Systems, edited by A. D. Nola etc., Verlag TUV Rheinland, 1986.

[5] R. Giles, Lukasiewicz logic and fuzzy sets, *International J. on Man-Machine Studies*, Vol.8, (1976) 313-327.

[6] A. Nakamura, Fuzzy rough sets, Note on *Multiple-Valued Logic in Japan*, Vol.9, No.8, 1988.

[7] A. Nakamura and J.M. Gao, A logic for fuzzy data analysis, to appear in *Fuzzy Sets and Systems*.

[8] C. V. Negoita and D. A. Ralescu, Representation theorems for fuzzy concepts, *Kybernetes*, Vol.4 (1975)169-174.

[9] Z. Pawlak, Information systems-theoretical foundations, *Information Systems* 6, (1981)205-218.

[10] L. A. Zadeh, Similarity relations and fuzzy orderings, *Information Sciences* 3 (1971)177-200.

ON MODELLING FUZZY PREFERENCE RELATIONS

Sergei Ovchinnikov
Mathematics Department
San Francisco State University
San Francisco, CA 94132, U.S.A.

Abstract

A theory of a fuzzy weak preference relation based on multiple-valued logic is developed. The transitivity property of fuzzy strict preference and indifference relations associated with a fuzzy weak preference relation is established.

Keywords: Fuzzy Binary Relations

1. INTRODUCTION

Preference relations play an important role in decision-making, especially, in the context of measurement theory (see, for example, [2]). Let $A = \{ a, b, c, \dots \}$ be a finite set of alternatives. We can suppose that, for each pair of alternatives x and y, you do one of three things: you (strictly) prefer a to b, you (strictly) prefer a to b, or you are indifferent between a and b. Let us say that you *weakly prefer* a to b if either you (strictly) prefer a to b or you are indifferent between a and b. We denote the binary relation of weak preference relation on the set A by R, the binary relation of strict preference on A by P, and the binary relation of indifference on A by I. It is reasonable to assume that the relation R satisfies the following properties:

1) $aRb \vee bRa$,

2) $(aRb \wedge bRc) \rightarrow aRc$,

for all a, b, and c in A. (We use standard notations for logical connectives "and", "or", "implies", and negation "not".) Therefore, R is 1) strongly complete and 2) transitive binary relation on A. Note, that strong completeness implies reflexivity, i.e.,

3) aRa, for all $a \in A$.

(In general, these properties, especially transitivity, are questionable in some application areas; but the model introduced above is considered to be a fairly good first approximation in many practical situations.)

Binary relations of indifference and strict preference are defined in terms of the binary relation of weak preference R as follows. The indifference relation I is defined by

$$aIb \quad \text{iff} \quad aRb \wedge bRa,$$

for all x and y in A, and a strict preference relation P is defined by

$$aPb \quad \text{iff} \quad aRb \wedge \sim bRa,$$

for all a and b in A.

In classical theories binary relations P, I, and R are considered as models for intuitive concepts of strict preference, indifference and preference-or-indifference (weak preference).

The following theorem is a classical result in the theory of preference relations.

Theorem. Suppose R is a weak preference relation on A. Then
i) I is a symmetric and transitive binary relation on A, i.e.,

$$aIb \rightarrow bIa,$$

and

$$(aIb \wedge bIc) \rightarrow aIc,$$

for all a, b, and c in A.
ii) P is an antisymmetric and transitive binary relation on A, i.e.,

$$aPb \rightarrow \sim bPa,$$

and

$$(aPb \wedge bPc) \rightarrow aPc,$$

for all a, b, and c in A.

In many instances, classical binary relations are not adequate models for intuitive concepts of preference and indifference. For example, an individual could prefer a to b to a certain 'degree' or a probability of preference is introduced. To resolve these difficulties a notion of a fuzzy binary relation is introduced. A fuzzy binary relation R on A is a function $R: A \times A \rightarrow [0,1]$; in other words, R is a function of two variables on A with codomain $[0,1]$.

In this paper, we present an approach to the theory of fuzzy preference relations based on multiple-valued logic. The paper is organized as follows:

In Section 2 triangular norms (t-norms), conorms, and negation functions are introduced. These functions are used as models for logical connectives.

In Section 3 we introduce a theory for preference and indifference relations based on multiple-valued logic approach. The classical results about preference and indifference relations are generalized in Section 4 for Lukasiewicz-like t-norms and in Section 5 for strict t-norms.

Some of the results presented in Section 4 appeared in our earlier publications [3] and [4]. We include these results for completeness sake, since these publications are hardly available to a general reader.

2. t-NORMS AND NEGATION FUNCTIONS

This section contains a brief overview of t-norms, t-conorms, and negation functions. Proofs and relevant references are found in [5] and [7].

A *t–norm* T is defined as a function $T: [0,1] \times [0,1] \rightarrow [0,1]$ satisfying the following properties:

(i) $T(x,1) = x$,

(ii) $T(x,y) \leq T(z,u)$, if $x \leq z$ and $y \leq u$,

(iii) $T(x,y) = T(y,x)$, and

(iv) $T(x,T(y,z)) = T(T(x,y),z)$,

for all x,y,z and u in [0,1].

An *Archimedean* t–norm is a t–norm satisfying

(v) $T(x,x) < x$, for all x in (0,1).

Let $\mathbf{R}^+ = [0,+\infty]$. A t–norm is continuous and Archimedean if and only if

$$T(x,y) = f(g(x) + g(y))$$

where

a) g is a continuous and strictly decreasing function from [0,1] to $\mathbf{R}^+$, such that $g(1) = 0$,

b) f is a continuous from $\mathbf{R}^+$ onto [0,1], such that $f(x) = g^{-1}(x)$ on $[0, g(0)]$, and $f(x) = 0$ for all $x \geq g(0)$, or, equivalently,

$$T(x,y) = g^{-1}(\min(g(x)+g(y)), g(0))$$

We say that a t-norm T has *zero divisors* if it satisfies

(vi) $T(x,y) = 0$ for some positive x and y.

A t-norm with no zero divisors is called a *strict* t-norm. A strict t-norm is, actually, a strictly montone function in two variables.

If T is a continuous Archimedean t–norm, then it has zero divisors if and only if $g(0) < +\infty$. A 'canonical' example of a t–norm with zero divisors is given by the Lukasiewicz t–norm

$$W(x,y) = \max\{x + y - 1, 0\}.$$

A 'canonical' example of a strict t-norm is given by the product t-norm

$$\Pi(x,y) = x \cdot y.$$

We call a strictly increasing function ϕ form the unit interval onto itself an ***automorphism*** of the unit interval. Any automorphism of the unit interval is a continuous function satisfying boundary conditions $\phi(0) = 0$ and $\phi(1) = 1$.

Theorem 2.1. A t–norm T is a continuous Archimedean t–norm with zero divisors if and only if there exists an automorphism ϕ of the unit interval [0,1] , such that

$$T(x,y) = W^{\phi}(x,y) = \phi^{-1}(W(\phi(x), \phi(y))).$$

Theorem 2.2. A t–norm T is a strict t-norm if and only if there exists an automorphism ϕ of the unit interval [0,1] , such that

$$T(x,y) = \Pi^{\phi}(x,y) = \phi^{-1}(\phi(x) \cdot \phi(y)).$$

Automorphisms ϕ are called ***generators*** of continuous Archimedean t-norms.

A ***negation*** N is defined as a strictly decreasing function N: $[0,1] \to [0,1]$ satisfying

$$N(N(x)) = x, \text{ for all } x \in [0,1].$$

Thus defined negation is a continuous function satisfying boundary conditions $N(0) = 1$ and $N(1) = 0$.

A standard example of a negation is given by $N(x) = 1 - x$.

Theorem 2.3. N is a negation function if and only if there exists an automorphism ϕ of the unit interval, such that

$$N(x) = N^{\phi}(x) = \phi^{-1}(1 - \phi(x)).$$

The automorphism ϕ from Theorem 2.2 is a *generator* of N.

Suppose a t–norm T and a negation function N are given. Then the function S given by

$$S(x,y) = N(T(N(x),N(y)))$$

is the *t–conorm* of T. For example, if N is the standard negation $N(x) = 1 - x$, the t–conorm of W is given by

$$W^*(x,y) = \min\{x + y, 1\}.$$

A triple $< T, S, N >$ is called a De Morgan triple in fuzzy set theory [1]. In this paper we assume that T is a continuous Archimedean t–norm and T and N are generated by the same automorphism ϕ. Therefore, the automorphism ϕ may be regarded as a 'parameter' in our models. For a given ϕ, elements of the De Morgan triple have the following representations:

$$T(x,y) = \phi^{-1}(\max\{\phi(x) + \phi(y) - 1, 0\}),$$

$$S(x,y) = \phi^{-1}(\min\{\phi(x) + \phi(y), 1\}),$$

$$N(x) = \phi^{-1}(1 - \phi(x)),$$

in the case of Lukasiewicz-like t-norms, and

$$T(x,y) = \phi^{-1}(\phi(x) \cdot \phi(y)),$$

$$S(x,y) = \phi^{-1}(\phi(x) + \phi(y) - \phi(x) \cdot \phi(y)),$$

$$N(x) = \phi^{-1}(1 - \phi(x)),$$

in the case of strict t-norms.

3. MODELS FOR PREFERENCE AND INDIFFERENCE RELATIONS

Let L be the usual first order language with identity based on the following symbols:

a) variables – x, y, z, u, w, ... ;
b) n-ary function symbols and the n-ary predicate symbols;
c) propositional connectives $\sim$, $\wedge$;
d) quantifiers $\forall$ and $\exists$.

The only *logical* predicate symbol is $=$. All other function and predicate symbols are *nonlogical* symbols. We shall be concerned in this paper with languages without function symbols. Formulas in L are constructed in the usual way. We write $F(x,y,\dots,w)$ to indicate that free variables of the formula F are among the variables x, y, ... , w. Propositional connectives $\vee$ and $\rightarrow$ are defined by means of $\sim$ and $\wedge$ as follows:

$A \vee B$ is an abbrevation of $\sim((\sim A) \wedge (\sim B))$;

$\mathbf{A} \rightarrow \mathbf{B}$ is an abbrevation of $(\sim \mathbf{A}) \vee \mathbf{B}$.

Suppose that a De Morgan triple $<T, S, N>$ generated by ϕ is given. We use the following interpretations for propositional connectives:

$$\sim \alpha = N(\alpha) \quad \text{and}$$

$$\alpha \wedge \beta = T(\alpha, \beta),$$

for all α, $\beta \in [0,1]$.
Then, in accordance with our definitions,

$$\alpha \vee \beta = N(T(N(\alpha), N(\beta))) = S(\alpha, \beta)$$

and

$$\alpha \rightarrow \beta = S(N(\alpha), \beta) = N(T(\alpha, N(\beta))),$$

for all α, $\beta \in [0,1]$.

Since t-norms and conorms are associative functions, we will not use parenthesis in expressions like $\alpha \wedge \beta \wedge \gamma$ or $\alpha \vee \beta \vee \gamma$. We shall also use commutative and distributive properties of operations $\wedge$ and $\vee$ (these properties can be easy established using representations for functions in De Morgan triples).

The function $\forall$ is defined for any non-empty set $X \subseteq [0,1]$ by $\forall X = \mathit{inf} X$.

A *model* for L consists of the following things: i) A nonempty set A, called the universe; ii) For each nonlogical n-ary predicate symbol **R** of L, an n-ary value function R from A to [0, 1]. (Note, that we consider only first-order languages without function symbols.)

For each formula $\mathbf{F}(x, y, \ldots, w)$ and interpretation of x, y, ... , w as elements a, b, ... , m in A, we define a value function $F(a, b, \ldots, m)$ with codomain $[0,1]$ by induction on the formulas using interpretations for propositional connectives introduced above. For instance, if **R** is a binary predicate symbol, then a model for **R** is a nonempty set A and a function R(a,b) on A with values in [0, 1], i.e. a model for a binary predicate symbol is a fuzzy binary relation on a nonemty set A. If F(x,y) is a formula such that F(a, b) is defined for all a and b in A, then define

$(\sim F)(a,b) = \sim (F(a,b))$,

$(F \wedge G)(a,b) = F(a,b) \wedge G(a,b)$,

$(\forall y\, F)(a) = \forall \{F(a,b) \mid b \in A\} = \inf \{F(a,b) \mid b \in A\}$, etc.

We say that $F(a, b, \ldots, m)$ is *true* if $F(a, b, \ldots, m) = 1$ (designated value is 1, cf. [6]).

In this paper we investigate properties of models for binary predicates. Suppose **R** is a binary predicate symbol. The following properties can be considered as nonlogical axioms of L:

$(\forall x)(\forall y)\, \mathbf{R}(x,y) \vee \mathbf{R}(y,x)$	*strong completeness*
$(\forall x)(\forall y)\, \mathbf{R}(x,y) \rightarrow \mathbf{R}(y,x)$	*symmetry*
$(\forall x)(\forall y)\, \mathbf{R}(x,y) \rightarrow (\sim \mathbf{R}(y,x))$	*antisymmetry*
$(\forall x)(\forall y)(\forall z)\, (\mathbf{R}(x,y) \wedge \mathbf{R}(y,z)) \rightarrow \mathbf{R}(x,z)$	*transitivity*

We have obvious interpretations for these properties:

$R(a,b) \vee R(b,a) \; = \; 1$ *strong completeness*

$R(a,b) \rightarrow R(b,a) \; = \; 1$ *symmetry*

$R(a,b) \rightarrow (\sim R(b,a)) \; = \; 1$ *antisymmetry*

$(R(a,b) \wedge R(b,c)) \rightarrow R(a,c) \; = \; 1$ *transitivity*

for all a, b, and c in A.

We call R(a, b) a *fuzzy weak preference relation* on the set A if it satisfies conditions of strong completeness and transitivity. We define a *fuzzy indifference relation* I on A by

$$I(a,b) = R(a,b) \wedge R(b,a),$$

for all a and b in A, and a *fuzzy strict preference relation* P on A by

$$P(a,b) = R(a,b) \wedge (\sim R(b,a)),$$

for all a and b in A (compare with the definitions for binary relations in Introduction).

4. MAIN THEOREMS: THE CASE OF LUKASIEWICZ-LIKE t-NORMS

In this section we investigate properties of preference and indifference relations assuming the following interpretations for logical connectives:

$$\alpha \wedge \beta \; = \; T(\alpha,\beta) \; = \; \phi^{-1}(\max\{\phi(\alpha)+\phi(\beta)-1, \, 0\}),$$

$$\alpha \vee \beta \; = \; S(\alpha,\beta) \; = \; \phi^{-1}(\min\{\phi(\alpha)+\phi(\beta), \, 1\}),$$

$$\sim \alpha \; = \; N(\alpha) \; = \; \phi^{-1}(1-\phi(\alpha)),$$

$$\alpha \rightarrow \beta \; = \; N(T(\alpha, N(\beta))) \; = \; \phi^{-1}(\min\{\phi(\beta)-\phi(\alpha)+1, \, 1\}),$$

where ϕ is an automorphism of the unit interval.

The following lemma establishes an important property of functions $\wedge$ and $\sim$.

Lemma 4.1. $\alpha \wedge (\sim \beta) = 0$ iff $\alpha \leq \beta$.

Proof. By definition,

$$\alpha \wedge (\sim \beta) = T(\alpha, N(\beta)) = \phi^{-1}(\max\{\phi(\alpha) + \phi(N(\beta)) - 1, 0\}) =$$

$$\phi^{-1}(\max\{\phi(\alpha) + 1 - \phi(\beta) - 1, 0\}) = \phi^{-1}(\max\{\phi(\alpha) - \phi(\beta), 0\}).$$

The assertion of the lemma follows immediately.

□

In particular, $\alpha \wedge (\sim \alpha) = 0$, for all $\alpha \in [0,1]$, or, equivalently, $\alpha \vee (\sim \alpha) = 1$, for all $\alpha \in [0,1]$, which is the Law of Excluded Middle.

We will need different representations for the basic properties of fuzzy binary relations. They are established in the following lemma.

Lemma 4.2. Listed below are equivalents of the respective properties

i) $R(a,b) \vee R(b,a) \; = \; 1$ *strong completeness*

ii) $R(a,b) = R(b,a)$ *symmetry*

iii) $R(a,b) \wedge R(b,a) = 0$ *antisymmetry*

iv) $R(a,b) \wedge R(b,c) \leq R(a,c)$ *transitivity*

for all a, b, and c in A.

Proof. i) Obvious.

ii) $1 = R(a,b) \rightarrow R(b,a) = (\sim R(a,b)) \vee R(b,a) = \sim(R(a,b) \wedge (\sim R(b,a)))$ is equivalent to $R(a,b) \wedge (\sim R(b,a)) = 0$. By Lemma 4.1, it is equivalent to $R(a,b) \leq R(b,a)$. Similarly, $R(a,b) \leq R(b,a)$.

iii) $1 = R(a,b) \rightarrow (\sim R(b,a)) = (\sim R(a,b)) \vee (\sim R(b,a)) = \sim(R(a,b) \wedge R(b,a))$ is equivalent to $R(a,b) \wedge R(b,a) = 0$.

iv) $1 = [(R(a,b) \wedge R(b,c)) \rightarrow R(a,c)] = \sim(R(a,b) \wedge R(b,c) \wedge (\sim R(a,c)))$ is equivalent to $R(a,b) \wedge R(b,c) \wedge (\sim R(a,c)) = 0$. By Lemma 4.1, it is equivalent to $R(a,b) \wedge R(b,c) \leq R(a,c)$.

□

Suppose that R is a fuzzy weak preference relation on A, and I and P are fuzzy indifference and strict preference relations associated with R. In this section we prove basic properties of these relations.

Theorem 4.3. I is a symmetric and transitive fuzzy binary relation.

Proof. i) $I(a,b) = R(a,b) \wedge R(b,a) = R(b,a) \wedge R(a,b) = I(a,b)$. By Lemma 4.2, I is symmetric.

ii) By transitivity of R,

$I(a,b) \wedge I(b,c) = R(a,b) \wedge R(b,a) \wedge R(b,c) \wedge R(c,b) = [R(a,b) \wedge R(b,c)] \wedge [R(c,b) \wedge R(b,a)]$
$\leq$
$R(a,c) \wedge R(c,a) = I(a,b)$. By Lemma 4.2, I is transitive. □

Theorem 4.4. P is antisymmetric and transitive fuzzy binary relation.

Proof. i) $P(a,b) \wedge P(b,a) = R(a,b) \wedge (\sim R(b,a)) \wedge R(b,a) \wedge (\sim R(a,b)) = 0$, by Lemma 4.1. Therefore, P is antisymmetric.

ii) $P(a,b) = R(a,b) \wedge (\sim R(b,a)) = \phi^{-1}(\max\{\phi(R(a,b)) + 1 - \phi(R(b,a)) - 1, 0\}) =$

$$= \phi^{-1}(\max\{\phi(R(a,b)) - \phi(R(b,a)), 0\}) .$$

If $P(a,b) > 0$, then $\phi(P(a,b)) = \phi(R(a,b)) - \phi(R(b,a))$.

To prove transitivity we need to prove that $P(a,b) \wedge P(b,c) \leq P(a,c)$. It suffices to consider the case when $P(a,b) > 0$ and $P(b,c) > 0$. Then we have to prove the following:

$$\phi^{-1}(\max\{\phi(P(a,b)) + \phi(P(b,c)) - 1, 0\}) \leq P(a,c),$$

or, equivalently,

$$\phi(R(a,b)) - \phi(R(b,a)) + \phi(R(b,c)) - \phi(R(c,b)) \leq$$
$$\leq 1 + \phi(R(a,c)) - \phi(R(c,a)). \tag{4.1}$$

By transitivity of R, $R(b,c) \wedge R(c,a) \leq R(b,a)$, or, equivalently,

$$\phi^{-1}(\max\{\phi(R(b,c)) + \phi(R(c,a)) - 1, 0\}) \leq R(b,a),$$

$$\phi(R(b,c)) + \phi(R(c,a)) \leq 1 + \phi(R(b,a)),$$

or

$$\phi(R(b,c)) \leq 1 + \phi(R(b,a)) - \phi(R(c,a). \tag{4.2}$$

Since R is a complete fuzzy relation, $R(b,c) \vee R(c,b) = 1$, or, equivalently,

$$\phi(R(b,c)) + \phi(R(c,b)) \geq 1.$$

Substituting (4.2) into the last inequality, we obtain

$$1 \leq \phi(R(b,c)) + \phi(R(c,b)) \leq 1 + \phi(R(b,a) - \phi(R(c, a)) + \phi(R(c,b),$$

implying

$$\phi(R(b,a)) + \phi(R(c,b)) \geq \phi(R(c,a)). \tag{4.3}$$

By transitivity of R

$$\phi(R(a,b)) + \phi(R(b,c)) \leq 1 + \phi(R(a,c)).$$

Subtracting (4.3) from the last inequality, we obtain (4.1).

□

Actually, P satisfies a stronger property than antisymmetry:

$\min\{P(a,b), P(b,a)\} = 0$ *strong antisymmetry*

for all a and b in A.

Indeed, if $R(a,b) \leq R(b,a)$, then $P(a,b) = R(a,b) \wedge (\sim R(b,a)) = 0$, by Lemma 4.1. Otherwise, $P(b,a) = 0$, by the same argument.

The existence of non-trivial (i.e., trully fuzzy) fuzzy weak orderings is established in the following example.

Example 4.5. Suppose ϕ is the identity automorphism of the unit interval. Let A be a three element set. We define a fuzzy binary relation R on A by

$$R = \begin{bmatrix} 1 & 1 & 1 \\ \alpha & 1 & 1 \\ \gamma & \beta & 1 \end{bmatrix}$$

where α, β, and γ are in (0, 1). It easy to verify that R is a weak ordering if and only if

$$\max\{\alpha+\beta-1, 0\} \leq \gamma \leq \min\{\alpha, \beta\}.$$

For any given α and β there are infinitely many values of γ satisfying this inequality. The strict preference relation P and indifference relation I are given by

$$P = \begin{bmatrix} 0 & 1-\alpha & 1-\gamma \\ 0 & 0 & 1-\beta \\ 0 & 0 & 0 \end{bmatrix} \quad \text{and} \quad I = \begin{bmatrix} 1 & \alpha & \gamma \\ \alpha & 1 & \beta \\ \gamma & \beta & 1 \end{bmatrix}.$$

5. MAIN THEOREMS: THE CASE OF PRODUCT-LIKE t-NORMS

Theorems 4.3 and 4.4 establish properties of indifference and strict preference relations that are exactly the same as in the classical case. Example 4.5 shows that there exist trully fuzzy (non-classical) binary relations of weak preference and associated with them fuzzy binary relations of indifference and strict preference in the case of Lukasiewicz-like t-norms. This is not true in the case of strict (product-like) t-norms, as Proposition 5.1 shows.

Suppose that basic logical connectives are modelled by elements of De Morgan triple based on a strict t-norm. Then

$$\alpha \wedge \beta = T(\alpha,\beta) = \phi^{-1}(\phi(\alpha)\cdot\phi(\beta)),$$

$$\alpha \vee \beta = S(\alpha,\beta) = \phi^{-1}(\phi(\alpha)+\phi(\beta)-\phi(\alpha)\cdot\phi(\beta)),$$

$$\sim\alpha = N(\alpha) = \phi^{-1}(1-\phi(\alpha)),$$

$$\alpha \rightarrow \beta = N(T(\alpha, N(\beta))) = \phi^{-1}(1-\phi(\alpha)+\phi(\alpha)\cdot\phi(\beta)),$$

for some automorphism ϕ of the unit interval.

Proposition 5.1. Let R be a fuzzy weak preference relation on A. Then R is a classical binary relation.

Proof. Let $\alpha = R(a,a)$ for some a in A. Then

$$1 = \alpha \vee \alpha = \phi^{-1}(\phi(\alpha)+\phi(\alpha)-\phi(\alpha)\cdot\phi(\alpha)) = \phi^{-1}(2\phi(\alpha)-(\phi(\alpha))^2)$$

or, equivalently,

$$(\phi(\alpha))^2 - 2\phi(a) + 1 = (\phi(\alpha) - 1)^2 = 0.$$

Therefore, $R(a,a) = 1$ for all a in A.

Let now $\alpha = R(a,b)$ for some a and b in A. We have, by transitivity of R,

$$(R(a,b) \wedge R(b,b)) \rightarrow R(a,b) = 1,$$

or, equivalently,

$$\alpha \rightarrow \alpha = 1.$$

Therefore, $1-\phi(\alpha)+(\phi(\alpha))^2 = 1$, i.e., $\phi(\alpha)\cdot[1-\phi(\alpha)] = 0$. We conclude, that either $R(a,b) = 1$ or $R(a,b) = 0$, for any a and b in A.

□

Remark. Let α be the value of function R for some given a and b in A. Condition $\alpha \rightarrow \alpha = 1$ is equivalent to the condition $\alpha \wedge (\sim\alpha) = 0$ which is the Law of Excluded Middle. We noted in the previous section that this condition is satisfied for all values of α in the case of Lukasiewicz-like t-norms. In the case of strict t-norms, this condition is satisfied only for Boolean values of α.

Although there are no trully fuzzy weak preferences in the case of strict t-norms, it is easy to construct examples of fuzzy binary relations satisfying properties derived in Lemma 4.2. We shall employ the following definitions in the rest of this section:

$R(a,b) \vee R(b,a) = 1$ *strong completeness*

$R(a,b) = R(b,a)$ *symmetry*

$$R(a,b) \wedge R(b,c) \leq R(a,c) \qquad \textit{transitivity}$$

for all a, b, and c in A. Fuzzy binary relations of weak preference are again defined as strongly complete and transitive relations.

Theorem 5.2. Let R be a fuzzy weak preference relation on A. A fuzzy binary relation of indifference I associated with R is a symmetric and transitive fuzzy binary relation.

Proof is identical to the proof of Theorem 4.3.

Lemma 5.3. The following are equivalent forms of the respective properties:

$$\max\{R(a,b), R(b,a)\} = 1 \qquad \textit{strong completeness}$$

$$\min\{R(a,b), R(b,a)\} = 0 \qquad \textit{antisymmetry}$$

for all a, b, and c in A.

Proof. i) Suppose R is a strongly complete fuzzy binary relation. Let $\alpha = R(a,b)$ and $\beta = R(b,a)$. Then,

$$\phi^{-1}(\phi(\alpha) + \phi(\beta) - \phi(\alpha)\phi(\beta)) = 1,$$

which is equivalent to

$$[\phi(\alpha) - 1] \cdot [\phi(\beta) - 1] = 0.$$

Therefore, either $R(a,b) = 1$ or $R(b,a) = 1$.

ii) If R is an antisymmetric fuzzy binary relation, then $\phi(R(a,b)) \cdot \phi(R(b,a)) = 0$, i.e., either $R(a,b) = 0$ or $R(b,a) = 0$.

□

Theorem 5.4. Let R be a fuzzy weak preference relation on A. A fuzzy binary relation of strict preference P associated with R is antisymmetric and transitive fuzzy binary relation.

Proof. By definition of P,

$$P(a,b) = R(a,b) \wedge (\sim R(b,a)) = \phi^{-1}\{\phi(R(a,b)) \cdot [1 - \phi(R(b,a))]\}.$$

By Lemma 5.3,

$$\phi(P(a,b)) = 1 - \phi(R(b,a)).$$

Since R is a strongly complete fuzzy binary relation, either $P(a,b) = 0$ or $P(b,a) = 0$. Again by Lemma 5.3, P is antisymmetric.

To prove transitivity of P we have to prove that

$$\phi(P(a,b)) \cdot \phi(P(b,c)) \leq \phi(P(a,c)),$$

for all a, b, and c in A. It suffices to consider the case when $P(a,b) > 0$ and $P(b,c) > 0$. Then, by strong completeness of R, $R(a,b) = R(b,c) = 1$. By transitivity of R,

$$\phi(R(c,a)) = \phi(R(c,a)) \cdot \phi(R(a,b)) \leq \phi(R(c,b)),$$
$$\phi(R(c,a)) = \phi(R(b,c)) \cdot \phi(R(c,a)) \leq \phi(R(b,a)),$$

and

$$\phi(R(c,b)) \cdot \phi(R(b,a)) \leq \phi(R(c,a)),$$

for all a, b, and c in A. We have, by previous inequalities,

$$\phi(P(a,b)) \cdot \phi(P(b,c)) = [1 - \phi(R(b,a))] \cdot [1 - \phi(R(c,b))] =$$

$$= 1 - \phi(R(b,a)) - \phi(R(c,b)) + \phi(R(c,b)) \cdot \phi(R(b,a)) \leq$$

$$\leq 1-\phi(R(c,a))-\phi(R(c,a)+\phi(R(c,a)) = 1-\phi(R(c,a)) = \phi(P(a,c)).$$

□

Note, that, like in the case of Lukasiewicz-like t-norms, P is again a strongly antisymmetric fuzzy binary relation, i.e., $\min\{P(a,b), P(b,a)\} = 0$ for all a and b in A.

6. CONCLUSION

Models for weak preference, strict preference, and indifference relations has been developed in the framework of formal multiple-valued logic. We have shown that there are plenty of trully fuzzy weak preferences and associated with them fuzzy strict preferences and fuzzy indifference relations in the case of Lukasiewicz-like t-norms. Classical results about these relations has been established.

On the other hand, it has been shown that there are no fuzzy models for preference and indifference relations in the framework of formal multiple-valued logic if interpretations for logical connectives are given in terms of strict t-norms. This is basically due to the fact that the Law of Excluded Middle is not valid in this case. We have suggested a minimal departure from the rigid framework of multiple-valued logic to develop a theory of fuzzy preference and indifference relations in the case of strict t-norms and proved results similar to the classical ones.

REFERENCES

[1] E.P. Klement, Operations on fuzzy sets and fuzzy numbers related to triangular norms, in: Proc. of the 11th ISMVL, University of Oklahoma, 1981, 218-225.

[2] D.H. Krantz, R.D. Luce, P.Suppes, and A. Tversky, Foundations of Measurement (Academic Press, New York, 1971).

[3] S. Ovchinnikov, Modelling valued preference relations, in: Proc. of the 19th ISMVL, IEEE Computer Society Press, 1989, 82-87.

[4] S. Ovchinnikov, On modelling fuzzy transitive relations, in: Proc. of the European Congress on System Studies, Lausanne, October 3-6, 1989, vol. 1, 413-420.

[5] S. Ovchinnikov and M. Roubens, On strict preference relations, Fuzzy Sets and Systems, to appear.

[6] N. Recher, Many-Valued Logic (McGraw Hill, New York, 1969).

[7] B. Schweizer and A. Sclar, Probabilistic Metric Spaces (North-Holland, Amsterdam, 1983).

CONCEPTUAL CONECTIVITY ANALYSIS BY MEANS OF FUZZY PARTITIONS

Aguilar-Martin, J.[(1)], Martín, M.[(2)(3)] and Piera, N.[(1)(2)]

[(1)]L.A.A.S.-C.N.R.S. 7, avenue du Colonel Roche, 31077 TOULOUSE cedex, France.
[(2)]C.E.A.B.-C.S.I.C. Camí de Santa Bàrbara s/n , BLANES, Spain.
[(3)]Departament L.S.I. of U.P.C., Facultat d'Informàtica, Pau Gargallo 5, BARCELONA, 08028. Spain.

Abstract: In this paper we present a set of tools to analyze concepts that describe and explain a set of observations. Due to the inherent vagueness of concepts, that makes hard to decide in a dichotomic base weather an observation is, or is not, a good example for a concept, we consider the concepts associated to fuzzy subsets. Then we study the adequation and coverage of a collection of fuzzy sets to describe a set of observations. In the same way, once a set of concepts has been acepted to describe a set of objects, we study how the concepts are related by means of the observations, and reciprocally, how the objects are related by the concepts. Finally, a short description of the computer program COCOA is given.

Keywords: Classifications, Data Analysis, Machine Learning, Fuzzy Sets.

1 - INTRODUCTION

To look for a family of concepts that describes or explains a given set of observations is one of the principal tasks in inductive learning and a fundamental proces for the observationals sciences. This task consists on looking for meaningfull collections of subsets of the observations space E. It is not surprising, then, the big interest showed by diverses thecniques like Cluster Analysis, Numerical Taxonomy, Multivariate Data Analysis or Machine Learning are concerned with the conceptual construction task and therefore related with inductive learning.

Most of the classical methods enhance the classification process and do not consider neither the possible connections between classes in a partition nor how the observation set is described by means of the obtained partition. In this work we give theoric criteria that allow to the analysis of this two aspects: a first objective is to give an adequation criteria for determine the description degre of a concepts family to a given space of observations, the second objective is to describe the relations or connexions between the concepts of a family. The results of this research have given rise to a computer package called COCOA (COnceptual COnnectivity Analysis).

2- ADEQUATION AND COVERING LEVELS OF A FUZZY CONCEPT SYSTEM

Inherent vagueness of concepts makes hard to decide in a dichotomic base weather an observation is, or is not, a good exemple for a concept. Therefore concepts are better associated to fuzzy subsets than to crisps ones. The finite set of observations **E** or ***Observed World*** will be considered as the suport of fuzzy partitions, or families of fuzzy subsets.

Definition 2.1: A family **F** of fuzzy subsets of **E**, will be called ***Fuzzy Concepts System*** (FCS). In the particular case that any subset of **F** is crisp, **F** will be named ***Strict Concept System*** (SCS).

Not all the posibles FCS of a given world **E** are adequate to describe it. Let us examine the following elementary example where **E**={cat, dog, tiger, parrot, lyon} is the set of observations, **F**={pet, savage, narine, feline} is the set of concepts that constitute the FCS, and the membership of the observations to the concepts are given in the table 2.1.

	pet	savage	marine	feline
cat	0.95	0.05	0.00	1.00
dog	0.90	0.10	0.00	0.00
tiger	0.00	1.00	0.00	1.00
parrot	0.40	0.50	0.00	0.00
lion	0.00	1.00	0.00	0.00

Table 2.1. Example of a fuzzy partition

It is clear that the concept "marine" is not adequate to describe or explain our world, because there aren't observations that belongs to this concept. Then, it seems important try to evaluate the descriptive contribution of the set of concepts to the set of observations. It will be done studying first the adequation of each concept.

Definition 2.2: The ***adequation degree*** of a concept C to the observed world E is defined by:

$$a_C = \max\{ \mu_C(x) ; x \in E\}.$$

In the example, $a_{marine} = 0$ and $a_{felin} = 1$. The descriptive contribution of a concept to an univers depends on the existence of observations that can be described by it, and it do not depends on the number of this objects. For instance, even if there were not any tiger in E, the feline concept would remain meaningfully to describe the world, because it remains at least one object (cat) that is a good example of the concept feline. This is the reason why in the previous definition the maximum value of membership was used.

By means of the adequation degre of all the concepts, the adequation of the Fuzzy Concept System can be defined as follows:

Definition 2.3: The ***global adequation***, denoted by a_F, of a Fuzzy Concept System F to a observed world E is:

$$a_F = \min\{ a_C ; C \in F\}.$$

The value of a_C and a_F belongs to [0,1], and if F is an SCS then a_F belongs to {0,1}. It will be said that an FCS F is ***completely adequated*** if $a_F = 1$, and that F is a_F-***completly adequated*** if for each concept C, $a_C = a_F$ (see figure 2.1). So it is easy to see that a system completly adequated is 1-equaly-adequated, because for any concept $1 = a_F \leq a_C \leq 1$.

It seems suitable to work with FCS's of E such that their global adequation is greater than a threshold $\eta > 0$. Notice that a FCS F' that fullfils $a_{F'} \geq \eta$ can be built from a given FCS F, if at least exists a concept C of F such that $a_C \geq \eta$. In fact it is only nesesary that $F' = \{C \in F ; a_C \geq \eta\}$. Relating to the example given above, the FCS has $a_F = 0$. But if the concept "marine" is extracted from F, a new set of concepts (FCS) F' is obtained such that $a_{F'} = 0.95$, that is, with a very high adequation degree.

Once the minimum adequation of F is guaranted, it remains to study wheather it is suficient for describe E, that is, if each observation is described or covered by at least one concept of F.

Definition 2.4: The *covering level* of F to an object x, denoted by c_x, is defined in the following way:

$$c_x = \max\{ \mu_C(x) \ ; C \in F \},$$

and the *global covering level* of F to E is

$$c_F = \min\{ c_x \ ; x \in E \}.$$

It will be said that F is a c_f-*covering* of E, if $c_F=1$ then F is a *covering* of E, and in that case any object of the universe is fully described by some concept of the FCS. An FCS F is c_F-*equally-covered,* if for each object of E $c_x=c_F$.

Proposition 1.1: For any FCS F, a_F and c_F satisfy the following properties:

1) $a_F=c_F$ if and only if, F is a_F-equaly-adequated and c_F-equally-covered (see figure 2.3).
2) If $a_F < c_F$ then there exists a concept C that is absorbed by the rest of the concepts, i.e., for each observation x of E exists a concept D of F such that $\mu_D(x) > \mu_C(x)$ (see figure 2.4).

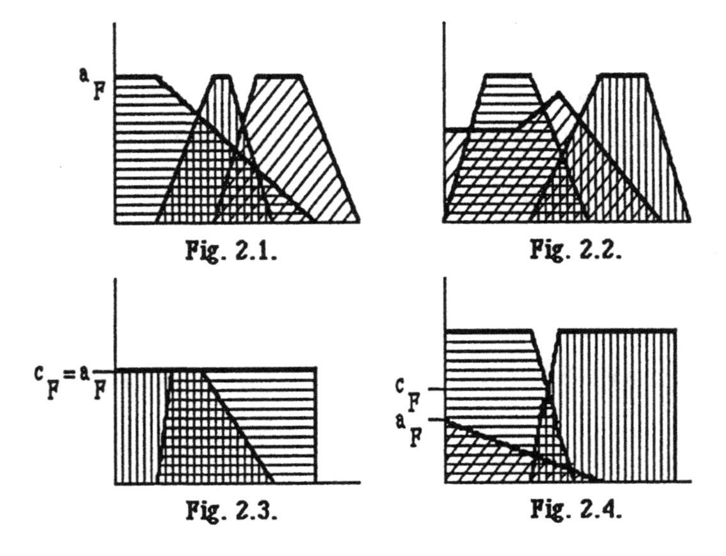

Fig. 2.1 to 2.4. Examples of coverage and adequation values

3- CONNECTIVITY BETWEEN CONCEPTS

The objective of this section is to give the theoric fundaments for the study of the relationships between the concepts of a FCS. This study can help for the interpretation of the partition obtained by an authomatic method. From now on, the FCS considered have an adequation degree greater than η.

3.1- α-exigence

The modelisation of concepts by means of fuzzy sets deals with the problem of deciding if an object x is an good for concept C. To solve this problem, the minimum value α required to the membership of x to C, $\mu_C(x)$, must be known in order to consider x as a good example of C. This carries out to consider the crisp subset of E, α-C, which is obtained by means of the α-cut of C; that is, $\alpha\text{-}C = \{ x \in E; \mu_C(x) \geq \alpha\}$. Therefore, an SCS S_α, is constructed from an α-cut in **F**: $S_\alpha = \{\alpha\text{-}C ; C \in \mathbf{F}\}$.

It is obvious that for having a good description of E using S_α, the S_α-adequation must be at least η. The following theorem gives the condition that guarantees this situation.

Theorem 3.1: S_α-adequation $\geq \eta$ if and only if $a_F \geq \alpha$.

Proof: Assume that C is the concept such that $a_c = a_F$. So $\mu_C(x) \leq a_F$ for any x of E. If $\alpha > a_F$, then $\alpha\text{-}C = \varnothing$, and so $a_{S\alpha} = 0 < \eta$.

Reciprocally for any concept C there exist x of E such that $\mu_C(x) \geq a_F$, so if $a_F \geq \alpha$, then $\alpha\text{-}C \neq \varnothing$, for any concept of **F** and this implies $a_{S\alpha} = 1$.

This theorem can be interpreted by saying that the global adequation degree must be greater or equal to the exigency level.

Since S_α may not be a covering of E, there exist objects in E wich are not good examples of any concept. The following theorem gives the condition for that situation.

Theorem 3.2: S_α is a covering of E if , and only if, the global covering level of **F** is greater or equal to the α-exigence.

Proof: Assume that $a_F \geq \alpha$, then for any x of E, there exist a concept C such that $\mu_C(x) \geq \alpha$, and this implies that x belongs to α-C. The reciprocal is straightforward.

Notice that to impose a high exigence degree, may imply that there exists a great number of objects whitout covering. But, to required that S_α be a covering, may involve a exigence level too low. In the example, when F'={pet, savage, feline}, it appears that a_F=0.95 and c_F =0.5. Then, to obtain an SCS S from F' such that a_S=1, the exigence level must be less or equal than 0.5; whereas to impose 0.8 for this level implies that only one element is not covered by S. It seems necessary to found an equilibrium between covering and exigence levels.

3.2- Connectivity measures

Connection between concepts depends on the number of good examples that they share, and this number varies with the exigency level, and so connectivity is related with α.

Definition 3.1: The α-*connectivity* of C_i to C_j, $\Psi_\alpha(C_i,C_j)$, is defined, in a non symmetric way, by:

$$\Psi_\alpha (C_i,C_j) = \mathrm{Card}\{\alpha\text{-}C_i \cap \alpha\text{-}C_j\} / \mathrm{Card}\{\alpha\text{-}C_i\}.$$

Definition 3.2: The kxk symmetric matrix $A=(a_{ij})$, such that the element a_{ij} is the $\mathrm{Card}\{\alpha\text{-}C_i \cap \alpha\text{-}C_j\}$ will be called ***α-connectivity matrix***.

Via the connectivity between concepts, an indicator of the global connectivity of a FCS with respect to a measured world can be defines using the number of actual links existing between concepts.

Definition 3.3: The ***global α-connectivity*** of Fwith respect to E is the sum for i<j of the a_{ij} of matrix A. The ***potential α-connectivity*** is the maximum possible number of these links. Finaly the ***connectivity function*** $\Gamma(\alpha)$, is defined as the ratio between the global and the potential α-connectivity.

The matrix $A = [a_{ij}]$ is symmetric and diagonal dominant, i.e. $a_{ii} \geq a_{ij}$. The ***connectivity function*** $\Gamma(\alpha)$, is a globally decreasing function, i.e. if $\Gamma(\alpha^*)=0$ then for $\alpha>\alpha^*$ there do not exist any object belonging to more than one α-C of S_α, that means that all the concepts are disjoint and define a crisp sub-partition of E.

4 - SIMILARITY BETWEEN OBJECTS

It is clear that the objects of a given Universe are described by the available concepts, and so the notion of similarity has to be context dependent, as it depends of the set of concepts. Therefore, two objects are similars if they can be described by, more or less, the same concepts. Assume that an SCS F is given to describe E.

Definition 4.1: Given a strict concept system F, two objects $x,y \in E$ are F-*similar*, $x \approx_F y$, if there exist $C \in F$ such that x and y belong to C. The number of concepts that share two objects x,y of E it is called ***extensive similariry degree*** S(x,y).

This binary relation between objects is reflexive, symmetric but it is not transitive. If the set of concepts F is a FCS, the similarity between objects of E must be evaluated by means of the exigency level, and gives rise the notion of α-similarity.

Definition 4.2: Two objects x,y of E are α-*similars*, $x \approx_\alpha y$, if they belong to the same α-concept, i.e., if it exists at least one concept C such that $\mu_C(x) \geq \alpha$ and $\mu_C(y) \geq \alpha$. The ***intensive degree of similarity***, s(x,y), between objects of E is the maximum value of α, such that they are α-similar.

It is clear that the relation of similarity $\approx_\alpha$ is always symmetric, it is reflexive only if S_α is a coverage, and it is not in general transitive.

5 - COMPUTER IMPLEMENTATION

A interactional computer program, called COCOA, (COnceptual COnnectivity Analysis), has been implemented in the Pascal language. From a FCS F, the program evaluates the global adequation to a given data base, or observed world E, and asks the user about the minimum level η of global adequation desired. It is based on the described theory and gives a tools for the interpretation of data using the results of previous classification. It can also be used for the analysis of a fuzzy partition obtained as the result of a weighted query or a imprecise observer.

The connectivity function Γ(α) is displayed so that a value of the exigency α can be chosen and a graphic representation of the α-connections between concepts is also displayed. Another options of COCOA, is to see by screen a graphical representation of α-similarity between objects or the connectivity between concepts.

6. CONCLUSIONS

Constructing concepts that describe, or explain, a set of observations is one of the tasks of inductive learning and it consists on analysing collections of subsets of the observation space E. Inherent vagueness of concepts makes hard to decide in a dichotomic base weather an observation is, or not is, a good example for a concept. Therefore concepts are associated to fuzzy subsets, and the set of observations will be considered as the suport of fuzzy partitions, or families of fuzzy subsets.

Adequation between a FCS and an observed world give a foundament for automatic interpretation of data or situations. Connectivity between concepts, that uses the level of adequation and similarity between objects, is based on the number of concepts shared by each object. This two notions can be displayed by the computer program COCOA, for critical values of α, exigency level, and yield a conversational aid for decision making or diagnostic in front of multiple and multidimensional measurements.

REFERENCES

Aguilar-Martin, J. & Piera, N., (1986). "Les connectifs mixtes: de nouveaux operateurs d'association des variables dans la classification automatique avec apprentissage"; Data Analysis and Informatics, edited by E. Diday Elsevier Science Pub. (pp. 253-265).

Guenoche, A., (1989). "Generalization of conceptual classification: indices and algorithm"; Data Analysis, Learning Symbolic and Numeric Knowledge, edited by E. Diday, INRIA, Nova Science Publishers,NY. (pp.503-510).

Lebbe, L., Lerman, I.C, Nicolas, J., Peter, P. & Vignes, R. (1989)."Conceptual Clustering in Biology: applications and perspectives"; Data Analysis, Learning Symbolic and Numeric Knowledge, edited by E. Diday, INRIA, Nova Science Publishers, NY. (pp. 443-452).

Martín, M. (1991). "Conceptual Connectivity Analysis. Une application pour l'analyse de descriptions conceptuelles floues" LAAS (CNRS) Technical Report. To appear during spring of 1991.

Wille, R. (1989). "Knowledge acquisition by methods of formal concept analysis"; Data Analysis, Learning Symbolic and Numeric Knowledge, edited by E. Diday, INRIA, Nova Science Publishers, NY. (pp.365-380).

Zadeh, L.A. (1965). "Fuzzy Sets"; Information and control n°8 (pp. 338-353).

TRANSITIVE SOLUTIONS OF RELATIONAL EQUATIONS ON FINITE SETS AND LINEAR LATTICES

ANTONIO DI NOLA*, WALDEMAR KOLODZIEJCZYK** AND SALVATORE SESSA*

* Istituto Matematico, Facoltà di Architettura, Università di Napoli, via Monteoliveto 3, 80134 Napoli, Italy.

** Institute of Production Engineering and Management, Technical University of Wroclaw, 50372 Wroclaw, Poland.

Abstract. The set of solutions of relational equations over a finite referential space and with values from a linear lattice is considered. We determine in this set the greatest max-min transitive solution and the related minimal ones. Further, we investigate for the determination of particular max-min transitive solutions, namely those having Schein rank equal to 1. Related properties of convergence of fuzzy systems represented by the involved relations are also given.

Key words: Finite matrix equation, max-min transitive matrix, Schein rank of a matrix.

1. **Introduction.** The theory of the matrix equations via relations defined on ordered algebraic structures ([1], [13]) is a generalization of the classical theory of the Boolean equations [15].

Luce [13] solved the equation:

$$R \circ A = B, \tag{1.1}$$

where "o" is the max-min composition, A and B are Boolean matrices and R is unknown. Rudeanu [15] pointed out that the results of Luce hold for matrices valued on a complete Brouwerian lattice **L**.

In this case, Sanchez [16] was able to determine the greatest solution. Di Nola and Lettieri [3] extended the result of Sanchez for matrix equations assigned on complete right-residuated lattices. In this paper, from now on, we assume that $\mathbf{L}\equiv(L,\wedge,\vee,\leq)$ is a linear lattice (not necessarily complete) with universal bounds 0 and 1. Let A,B,R be matrices defined on referential finite sets and assuming values in **L**, $\mathfrak{R}=\mathfrak{R}(A,B)\neq\emptyset$ be the set of all the solutions R of the Eq. (1.1). In this case, the determination of the minimal elements of $\mathfrak{R}$ (here minimality is understood with respect to the partial ordering induced in $\mathfrak{R}$ from the total ordering of L) is studied by Di Nola [2]. The minimal elements of $\mathfrak{R}$ are determined by Sanchez [17] if A and B are (row-)vectors (fuzzy sets). Under this last hypothesis, therefore we have that if $A,B:\mathbf{X}\to\mathbf{L}$ are vectors and $R:\mathbf{X}\times\mathbf{X}\to\mathbf{L}$ is a matrix defined on the finite set $\mathbf{X}=\{x_1,\ldots,x_n\}$ and valued on **L**, the Eq. (1.1) assumes the following expression for any $j\in\mathbf{I}_n=\{1,2,\ldots,n\}$:

$$\bigvee_{i=1}^{n}(R_{ij}\wedge A_i)=B_j, \tag{1.2}$$

where, for brevity of notation, we put $R(x_i,x_j)=R_{ij}$, $A(x_i)=A_i, B(x_j)=B_j$ for any $R\in\mathfrak{R}$, $i,j\in\mathbf{I}_n$. In this paper, we originate investigations of transitive solutions $R\in\mathfrak{R}$ of the Eq. (1.2). The transitivity is one of the most interesting properties for finite square Boolean matrices [10, p. 83]. This notion is easily extended to finite square matrices defined on **L**, i.e. R is max-min transitive iff $R_{ij}\geq R_{ih}\wedge R_{hj}$ for any $h,i,j\in\mathbf{I}_n$.

Transitive matrices were already applied in various fields of reasearch, e.g. clustering [5], information retrieval [20], fuzzy orderings [22], preference relations [14]. Namely, we can look upon Eq. (1.2) as the description of the system R with regard to the relation between the assigned input A and the output B. R represents a trasformation of the input and output. Alternatively, Eq. (1.2) describes a finite-state system with a (fuzzy) transition matrix R unknown. We are also just interested to determine such matrix R with a given type of transitivity, namely the greatest max-min transitive element of $\mathfrak{R}$ and the greatest max-min transitive element of $\mathfrak{R}$ with Schein rank equal to 1. All these results are successively applied in order to find the corresponding greatest elements of the set $\mathfrak{S}=\mathfrak{S}(Q,T)$ of all the solutions S of the following composite matrix equation, more general than Eq. (1.2):

$$\bigvee_{j=1}^{n} (Q_{ij} \wedge S_{jk}) = T_{ik}, \quad S \circ Q = T, \tag{1.3}$$

where $Q,T:\mathbf{X}\times\mathbf{X}\rightarrow\mathbf{L}$ are assigned matrices, $S:\mathbf{X}\times\mathbf{X}\rightarrow\mathbf{L}$ is unknown, $Q_{ij}=Q(x_i,x_j)$, $S_{jk}=S(x_j,x_k)$, $T_{ik}=T(x_i,x_k)$ for any $i,j,k\in\mathbf{I}_n$. It is easily seen that the max-min composition (1.3) is associative, i.e. if $P,Q,R:\mathbf{X}\times\mathbf{X}\rightarrow\mathbf{L}$, then

$$(P \circ Q) \circ R = P \circ (Q \circ R), \tag{1.4}$$

$P\leq R$ means $P_{ij}\leq R_{ij}$ in $\mathbf{L}$ for any $i,j\in\mathbf{I}_n$, $P<R$ means $P\leq R$ and $P\neq R$, $(P\wedge R)$, $(P\vee R)$: $\mathbf{X}\times\mathbf{X}\rightarrow\mathbf{L}$ are matrices pointwise defined as $(P\wedge R)_{ij} = P_{ij}\wedge R_{ij}$, $(P\vee R)_{ij} = P_{ij}\vee R_{ij}$ for any $i,j\in\mathbf{I}_n$, similarly it is defined the infimum and the supremum of any finite set of matrices, Q^{-1} denotes the transpose of Q, further we put $R^{k+1}=R^k \circ R$ for any positive integer k, $R^1=R$. Thus R is max-min transitive iff $R^2 = R \circ R\leq R$ and it is easy to see that

$$\text{if } P\leq R, \text{ then } (P \circ Q) \leq (R \circ Q), \tag{1.5}$$

$$(P \vee R) \circ Q = (P \circ Q) \vee (R \circ Q). \tag{1.6}$$

We recall that $\mathfrak{R}$ and $\mathfrak{S}$ are upper-semilattices [18], i.e. if $R,R'\in\mathfrak{R}$ (resp. $\mathfrak{S}$) then $(R\vee R')\in\mathfrak{R}$ (resp. $\mathfrak{S}$), but $\mathfrak{R}$ and $\mathfrak{S}$ are not generally meet-semilattices (cfr. Ex. 2.3 of [5]).

Minimal transitive elements in $\mathfrak{R}$ and $\mathfrak{S}$ are characterized too and properties of convergence of systems represented by the above relations are given in the last Section.

2. Max-min transitive solutions of Eq. (1.2). Here we devote our attention only on the Eq. (1.2). Let $A,B:\mathbf{X}\rightarrow\mathbf{L}$ be assigned and following Sanchez ([16], [17]), we define two matrices $(A\sigma B),(A\alpha B):\mathbf{X}\times\mathbf{X}\rightarrow\mathbf{L}$ as

$$(A\sigma B)_{ij} = A_i\sigma B_j = \begin{cases} 0 & \text{if } A_i<B_j, \\ B_j & \text{if } A_i\geq B_j, \end{cases}$$

$$(A\alpha B)_{ij} = A_i\alpha B_j = \begin{cases} 1 & \text{if } A_i\leq B_j, \\ B_j & \text{if } A_i>B_j, \end{cases}$$

for any $i,j\in\mathbf{I}_n$.

REMARK 2.1. *Note that* "α" *is the well known relative pseudocomplementation in* $\mathbf{L}$ *and it is evident that* $a\alpha b\geq b$ *for any* $a,b\in\mathbf{L}$.

Let $\mathfrak{L}=\mathfrak{L}(A,B)$ be the finite set of matrices $L:\mathbf{X}\times\mathbf{X}\rightarrow\mathbf{L}$ such that $L_{ij}=0$ for any $i\in\mathbf{I}_n$ if $B_j=0$ and if $B_j\neq 0$, it is $L_{ij}=0$ for any $i\in\mathbf{I}_n$ except for a unique entry (h,j), if exists, of the j-th column of $(A\sigma B)$ for which it is $(A\sigma B)_{hj}=B_j$ (i.e. by choosing a nonzero element, if exists, in the j-th column of $(A\sigma B)$ for which $B_j\neq 0$ and by putting 0 everywhere along the same column).

Then the following results of Sanchez [17] hold:

THEOREM 2.2. (i) $\mathfrak{R}\neq\emptyset$ *iff* $(A\alpha B)\in\mathfrak{R}$. *Further*, $(A\alpha B)\geq R$ *for any* $R\in\mathfrak{R}$.

(ii) $\mathfrak{R}\neq\emptyset$ *iff* $\bigvee_{i=1}^{n} A_i \geq \bigvee_{j=1}^{n} B_j$.

(iii) $\mathfrak{R}\neq\emptyset$ *iff* $(A\sigma B)\in\mathfrak{R}$. *Further*, $\mathfrak{L}$ *is the set of all the minimal elements of* $\mathfrak{R}$ *and*

$$(A\sigma B) = \bigvee_{L\in\mathfrak{L}} L. \tag{2.1}$$

(iv) *If* $\mathfrak{R}\neq\emptyset$, *for any* $R\in\mathfrak{R}$ *there exists an element* $L\in\mathfrak{L}$ *such that* $L\leq R$.

Of course, L minimal in $\mathfrak{R}$ means that if $R \leq L$ for some $R \in \mathfrak{R}$, then $R=L$. Later we shall use frequently the following easy Lemma, which says that $\mathfrak{R}$ (*resp.* $\mathfrak{S}$) is convex:

LEMMA 2.3. *If* $R', R'' \in \mathfrak{R}$ *(resp.* $\mathfrak{S}$*) and* $R: X \times X \to L$ *is such that* $R' \leq R \leq R''$, *then* $R \in \mathfrak{R}$ *(resp.* $\mathfrak{S}$*).*

As usually denoted in literature [5], let $\mathfrak{R}_\wedge$ be the set of all max-min transitive matrices $R: X \times X \to L$ and let $\mathfrak{T} = \mathfrak{T}(A,B) = \mathfrak{R} \cap \mathfrak{R}_\wedge$ be the set of all the max-min transitive solutions of the Eq. (1.2). If $\mathfrak{R} \neq \emptyset$, we define the following matrix $W: X \times X \to L$ by setting for any $i,j \in \mathbf{I}_n$:

$$W_{ij} = W(x_i, x_j) = \begin{cases} B_j & \text{if } B_i > B_j, \\ A_i \alpha B_j & \text{otherwise.} \end{cases} \tag{2.2}$$

Then the following result holds:

THEOREM 2.4. $\mathfrak{R} \neq \emptyset$ *iff* $W \in \mathfrak{T}$. *Further,* $W \geq R$ *for any* $R \in \mathfrak{T}$.

Proof. Of course, we prove only the nontrivial implication. Let $\mathfrak{R} \neq \emptyset$, then $(A\alpha B) \in \mathfrak{R}$ by Thm. 2.2 (i) and $(A\sigma B) \in \mathfrak{R}$ by Thm. 2.2 (iii). From (2.2), we deduce that

$$W_{ij} = B_j < 1 \text{ iff either } B_i > B_j \text{ or } B_i \leq B_j < A_i, \tag{2.3}$$

$$W_{ij} = 1 \text{ iff } A_i \vee B_i \leq B_j. \tag{2.4}$$

Thus $(A\sigma B) \leq W \leq (A\alpha B)$, i.e. $W \in \mathfrak{R}$ by Lemma 2.3. Assume that $W \notin \mathfrak{R}_\wedge$ and this should imply

$$W_{ij} < W_{ih} \wedge W_{hj} \tag{2.5}$$

for some $h,i,j \in \mathbf{I}_n$. Then $W_{ij} = B_j < 1$ and hence $W_{hj} = 1$ otherwise $W_{hj} = B_j < 1$ and (2.5) should give $B_j = W_{ij} < W_{ih} \wedge B_j \leq B_j$, a contradiction. Thus $B_h < 1$ since, by (2.4),

$$B_h \leq B_j < 1. \tag{2.6}$$

If $W_{ih} = B_h$, then (2.5) should imply $B_j < B_h \wedge 1 = B_h$, a contradiction to (2.6). Then $W_{ih} = 1$, i.e. $A_i \vee B_i \leq B_h$ by (2.4) and hence $A_i \vee B_i \leq B_j$ by (2.6). This should mean that $W_{ij} = 1$ by (2.4), a contradiction to $W_{ij} = B_j < 1$. Then $W \in \mathfrak{R}_\wedge$, i.e. $\mathfrak{T} \neq \emptyset$ since $W \in \mathfrak{T}$.

Now let $R \in \mathfrak{T}$ such that $R \nleq W$, i.e. $R_{ij} > W_{ij}$ for some $i,j \in \mathbf{I}_n$. Then $W_{ij} = B_j < 1$ and hence either $B_i > B_j$ or $A_i > B_j$ by (2.3). If $A_i > B_j$, then, by Thm. 2.2(i), $B_j = A_i \alpha B_j \geq R_{ij} > B_j$, a contradiction. Thus $B_i > B_j$ and since $R \in \mathfrak{R}$, then there exists an element $h \in \mathbf{I}_n$ such that $A_h \wedge R_{hi} = B_i$, which should imply, since $R \in \mathfrak{R}_\wedge$, that $B_j \geq A_h \wedge R_{hj} \geq A_h \wedge R_{hi} \wedge R_{ij} = B_i \wedge R_{ij} > B_j$, a contradiction. This proves that $R \leq W$ for any $R \in \mathfrak{T}$ and the theorem is completely proved. ∎

For any vector $A: X \to L$, let $\mathbf{I}(A) = \{k \in \mathbf{I}_n : A_k = \vee_{i=1}^{n} A_i\}$. Then we state the following:

THEOREM 2.5. *If* $\mathfrak{R} \neq \emptyset$, *then* $\mathfrak{L} \cap \mathfrak{R}_\wedge \neq \emptyset$.

Proof. Let $k \in \mathbf{I}(A)$. Since $\mathfrak{R} \neq \emptyset$, $A_k \geq B_j$ for any $j \in \mathbf{I}_n$ by Thm. 2.2(ii), hence $(A\sigma B)_{kj} = A_k \sigma B_j = B_j$ for any $j \in \mathbf{I}_n$. For any $k \in \mathbf{I}(A)$, let $L^{(k)}: X \times X \to L$ be the matrix defined as

$$L^{(k)}(x_i, x_j) = \begin{cases} B_j & \text{if } i=k, \\ 0 & \text{otherwise,} \end{cases} \tag{2.7}$$

for any $i,j \in \mathbf{I}_n$. Of course, $L^{(k)} \in \mathfrak{L} \cap \mathfrak{R}_\wedge$ for any $k \in \mathbf{I}(A)$, i.e. $\mathfrak{L} \cap \mathfrak{R}_\wedge \neq \emptyset$. ∎

In other words, Thm. 2.5 assures the existence of max-min transitive minimal elements of $\mathfrak{R}$, but it does not give a complete characterization of such elements. In the successive Thm. 2.7 we give a full characterization of the minimal elements of $\mathfrak{T}$ by using the following well known definition ([9], [10], [18]):

DEFINITION 2.6. *Let* $R: X \times X \to L$. *The matrix* $\overline{R} = R \vee R^2 \vee \ldots \vee R^n$ *(the sup of all distinct powers of* R*) is called the max-min transitive closure of* R.

Observing that $\overline{R}$ is the smallest max-min transitive relation including R, we can prove the following:

THEOREM 2.7. *Let* $\mathfrak{R}\neq\emptyset$ *and* $\overline{\mathfrak{L}}=\{\overline{L}:L\in\mathfrak{L}\}$. *Then* $\overline{\mathfrak{L}}\subseteq\mathfrak{T}$, $\overline{\mathfrak{L}}$ *has minimal elements, which are all the minimal elements of* $\mathfrak{T}$ (*i.e.,* L' *is minimal in* $\overline{\mathfrak{L}}$ *iff* L' *is minimal in* $\mathfrak{T}$).

Proof. By Thm. 2.4, $W\in\mathfrak{R}$ since $\mathfrak{R}\neq\emptyset$. By (2.1), we have $L\leq(A\sigma B)$ for any $L\in\mathfrak{L}$ and $(A\sigma B)\leq W$ (cfr. proof of Thm. 2.4), thus $L\leq\overline{L}\leq W$, i.e. $\overline{L}\in\mathfrak{R}$ for any $L\in\mathfrak{L}$ by Lemma 2.3, i.e. $\overline{\mathfrak{L}}\subseteq\mathfrak{T}$. Evidently $\overline{\mathfrak{L}}$ has minimal elements since it is a finite set. Let $L'\in\overline{\mathfrak{L}}$ be minimal in $\overline{\mathfrak{L}}$ and $R\in\mathfrak{T}$ be such that $R\leq L'$. By Thm. 2.2 (iv), there exists an element $L\in\mathfrak{L}$ such that $L\leq R$ and hence $\overline{L}\leq R\leq L'$, where $\overline{L}\in\overline{\mathfrak{L}}$ and since L' is minimal in $\overline{\mathfrak{L}}$, we have $\overline{L}=L'$ which implies $R=L'$, i.e. L' is minimal in $\mathfrak{T}$. Similarly, it is proved that the minimality of an element $R\in\mathfrak{T}$ implies that R is in $\overline{\mathfrak{L}}$ and it is evidently minimal in $\overline{\mathfrak{L}}$. ■

3. **Cross-vectors of Eq. (1.2).** Let $R:X\times X\to L$ be a matrix. Extending a well known concept for Boolean matrices [10], we give the following:

DEFINITION 3.1. R *is a cross-vector if there exist two vectors* $U,V:X\to L$ *such that* $R_{ij}=U_i\wedge V_j$ *for any* $i,j\in I_n$. *In symbols, we write*

$$R = U\times V. \tag{3.1}$$

In ([4], [5]) such relations are called "decomposable" and Definition 3.1 is given for matrices valued on a complete lattice, not necessarily linear. However, we have the following simple characterization ([4], [5]):

THEOREM 3.2. *Let* L *be a linear lattice (not necessarily complete).*
(i) *A matrix* $R:X\times X\to L$ *is a cross-vector iff* $R=U^{\sim}\times V^{\sim}$, *where* $U^{\sim},V^{\sim}:X\to L$ *are vectors defined by* $U_i^{\sim}=\vee_{j=1}^{n}R_{ij}$ and $V_j^{\sim}=\vee_{i=1}^{n}R_{ij}$ *for any* $i,j\in I_n$.
(ii) $U^{\sim}$ *and* $V^{\sim}$ *are the smallest vectors satisfying* (3.1), *i.e. if* $R=U\times V$, *then* $U^{\sim}\leq U$ *and* $V^{\sim}\leq V$.

REMARK 3.3. *In other words, Thm.* 3.2 *says that* ([4], [5]) *a matrix* $R:X\times X\to L$, *where* L *is a linear lattice (not necessarily complete), is a cross-vector iff for any* $i,j,\in I_n$, R_{ij} *is either the sup of the i-th row or the sup of the j-th column.*

Following Di Nola and Sessa [6], we recall the following:

DEFINITION 3.4. *Let* L *be a linear lattice. The Schein rank* $\rho(R)$ *of a matrix* $R:X\times X\to L$ *is the least number of cross-vectors whose sup is* R.

It is easy to see that $\rho(R)\leq n$ for any $R:X\times X\to L$. If $L=\{0,1\}$, Definition 3.4 is an extension of the analogous concept given [10] for Boolean matrices. If L is a semiring, see Kim and Roush [11].

Let $\mathfrak{R}^{(\rho)}$ be the set of the matrices $R:X\times X\to L$ having Schein rank $\rho=\rho(R)$ and hence a cross-vector belongs to $\mathfrak{R}^{(1)}$, further it is known that $\mathfrak{R}^{(1)}\subset\mathfrak{R}_{\wedge}$ ([5], [12]). Let $\mathfrak{D}=\mathfrak{D}(A,B)=\mathfrak{R}\cap\mathfrak{R}^{(1)}$ be the set of all the cross-vectors, solutions R of the Eq. (1.2). If $\mathfrak{R}\neq\emptyset$, we define the following matrix $D:X\times X\to L$ by setting for any $i,j,\in I_n$:

$$D_{ij}=\begin{cases} A_i\alpha B_j & \textit{if } j\in I(B),\\ B_j & \textit{otherwise,}\end{cases} \tag{3.2}$$

where I(B) is defined as in Sec. 2. Then we have the following result:

THEOREM 3.5. $\mathfrak{R}\neq\emptyset$ *iff* $D\in\mathfrak{D}$. *Further,* $D\geq R$ *for any* $R\in\mathfrak{D}$ *and* $D\leq W$.

Proof. We prove only the nontrivial implication. Let $\mathfrak{R}\neq\emptyset$, then $(A\alpha B)\in\mathfrak{R}$ by Thm. 2.2(i) and $(A\sigma B)\in\mathfrak{R}$ by Thm. 2.2(iii). By Lemma 2.3, $D\in\mathfrak{R}$ since $(A\sigma B)\leq D\leq(A\alpha B)$. First of all, we observe that for any $i\in I_n$:

$$\bigvee_{h=1}^{n} D_{ih}=\left[\bigvee_{h\in I(B)} D_{ih}\right]\vee\left[\bigvee_{h\notin I(B)} D_{ih}\right]=\left[\bigvee_{h\in I(B)} D_{ih}\right]\vee\left[\bigvee_{h\notin I(B)} B_h\right]=A_i\alpha B_{h'}$$

for some $h'\in I(B)$ since, by Remark 2.1, $A_i\alpha B_{h'}\geq B_{h'}=\vee_{h=1}^{n}B_h>\vee_{h\notin I(B)}B_h$. Hence, if $j\in I(B)$, we deduce for any $i\in I_n$:

$$\left[\bigvee_{h=1}^{n} D_{ih}\right] \wedge \left[\bigvee_{t=1}^{n} D_{tj}\right] = (A_i\alpha B_{h'}) \wedge \left[\bigvee_{t=1}^{n} (A_t\alpha B_j)\right] = A_i\alpha B_{h'} = A_i\alpha B_j = D_{ij}$$

since $B_{h'}=B_j$. If $j\notin I(B)$, we have for any $i\in I_n$:

$$\left[\bigvee_{h=1}^{n} D_{ih}\right] \wedge \left[\bigvee_{t=1}^{n} D_{tj}\right] = (A_i\alpha B_{h'}) \wedge B_j = B_j = D_{ij}$$

since $A_i\alpha B_{h'}\geq B_{h'}>B_j$ by Remark 2.1.

Thus $D\in\mathfrak{R}^{(1)}$ (cfr. Remark 3.3), i.e. $\mathfrak{D}\neq\emptyset$ since $D\in\mathfrak{D}$. Assume now that there exists an element $R\in\mathfrak{D}$ such that $R_{ij}>D_{ij}$ for some $i,j\in I_n$. Now $j\notin I(B)$ otherwise, by Thm. 2.2(i), $A_i\alpha B_j\geq R_{ij}>D_{ij}=A_i\alpha B_j$, a contradiction. Hence, $R_{ij}>D_{ij}=B_j$ and let $h\in I(B)$. Since $R\in\mathfrak{R}$, there exists an element $k\in I_n$ such that $B_h=A_k\wedge R_{kh}$, so $A_k\geq B_h>B_j$ and then

$$R_{kj} \leq A_k\alpha B_j = B_j < R_{ij}. \tag{3.3}$$

On the other hand, we also have

$$R_{kj} \leq B_j < B_h \leq R_{kh}. \tag{3.4}$$

Thus (3.3) and (3.4) imply that R_{kj} is neither the supremum of the j-th column of R nor the supremum of the i-th row of R, a contradiction to the fact that $R\in\mathfrak{R}^{(1)}$ (cfr. Remark 3.3).

This proves that D is the greatest element of $\mathfrak{D}$. Further, $D\leq W$ since $W\in\mathfrak{T}$ by Thm. 2.3 and $\mathfrak{D}\subseteq\mathfrak{T}$. This concludes the proof. ■

Now we study the minimal elements of $\mathfrak{D}$, but we first need to define the following set:

$$\Gamma(A) = \{i\in I_n : A_i\geq \bigvee_{h=1}^{n} B_h\}.$$

If $\mathfrak{R}\neq\emptyset$, then $\Gamma(A)\supseteq I(A)$ by Thm. 2.2(ii). Let $k\in\Gamma(A)$ and, of course, $A_k\geq B_j$ for any $j\in I_n$. For any $k\in\Gamma(A)$, then we can consider the matrix $L^{(k)}\in\mathfrak{L}$ defined from formula (2.7).

Let $\Gamma=\Gamma(A,B)$ be the set of such matrices and it is immediately seen that $\Gamma\subseteq\mathfrak{D}$ since for any $k\in\Gamma(A)$, it suffices to define the vector $V^{(k)}:X\to L$ by setting

$$V^{(k)}(x_i) = \begin{cases} \bigvee_{j=1}^{n} B_j & \textit{if } i=k, \\ 0 & \textit{otherwise}, \end{cases} \tag{3.5}$$

for any $i\in I_n$ in order to have $L^{(k)}=V^{(k)}\times B$. Moreover, we show the following:

THEOREM 3.6. *If $\mathfrak{R}\neq\emptyset$, then Γ is the set of all the minimal elements of $\mathfrak{D}$.*

Proof. We note that $\mathfrak{D}\neq\emptyset$ by Thm. 3.5 since $\mathfrak{R}\neq\emptyset$. It suffices to prove that for any $R\in\mathfrak{D}$, there exists some element $k\in\Gamma(A)$ such that $L^{(k)}\leq R$. Suppose not. Then there exists some element $R\in\mathfrak{D}$ such that for any $k\in\Gamma(A)$, there exist some $i,j\in I_n$ such that $L^{(k)}(x_i,x_j)>R_{ij}$. We have $i=k$ otherwise $0=L^{(k)}(x_i,x_j)>R_{ij}\geq 0$, a contradiction. Hence, $B_j=L^{(k)}(x_k,x_j)>R_{kj}$ and since $R\in\mathfrak{R}$, then there exists some $h\in I_n$ such that $B_j=A_h\wedge R_{hj}$. This implies that $R_{hj}\geq B_j>R_{kj}$, i.e. R_{kj} is not the greatest element in the j-th column of R. Since $R\in\mathfrak{R}^{(1)}$, we have necessarily

$$R_{kj} = \bigvee_{t=1}^{n} R_{kt}$$

for any $k\in\Gamma(A)$. In particular, it follows that

$$B_j > R_{kj} \geq R_{kt} \tag{3.6}$$

for any $t\in I(B)$ and $k\in\Gamma(A)$. Choose any arbitrary $t\in I(B)$ and let $L\in\mathfrak{L}$, by Thm. 2.2(iv), such that $L\leq R$ and $L_{st}=A_s\sigma B_t=B_t\geq B_j>0$ for some $s\in I_n$ along the t-th column of L. Now $A_s\geq B_t$, hence $s\in\Gamma(A)$ and then, by (3.6), $B_j>R_{sj}\geq R_{st}\geq L_{st}=B_t\geq B_j$, a contradiction. Since $\Gamma\subseteq\mathfrak{L}\cap\mathfrak{D}$, an element of Γ is necessarily minimal in $\mathfrak{D}$. To conclude the proof, it remains to show that a minimal element of $\mathfrak{D}$ is in Γ. Indeed, let R be minimal in $\mathfrak{D}$ and, by the above property, let $k\in\Gamma(A)$ be such that $L^{(k)}\leq R$. Since R is minimal in $\mathfrak{D}$, we have $L^{(k)}=R$ and hence $R\in\Gamma$. ■

4. **Max-min transitive solutions of Eq. (1.3).** In order to deal with Eq. (1.3), we need to recall some results of Sanchez [16], Di Nola [2] and Sessa [19] (for more details, see also [5]).

Assigned two matrices $Q,T:X\times X\rightarrow L$, for any $i\in I_n$ define the (row-)vectors $Q_i,T_i:\{x_i\}\times X\rightarrow L$ by setting $Q_i(x_i,x_j)=Q_{ij}$ and $T_i(x_i,x_k)=T_{ik}$ for any $j,k\in I_n$. As Di Nola [2] pointed out, the study of Eq. (1.3) is equivalent to consider the following system of equations of type (1.2):

$$S \circ Q_i = T_i, \qquad i = 1,2,\ldots,n. \tag{4.1}$$

Denoting by $\mathfrak{R}_i=\mathfrak{R}_i(Q_i,T_i)$ the set of solutions of the i-th Eq. (4.1), we obviously have that

$$\mathfrak{S} = \bigcap_{i=1}^{n} \mathfrak{R}_i. \tag{4.2}$$

THEOREM 4.1 [16]. *$\mathfrak{S}\neq\emptyset$ iff the matrix $S^*=(Q^{-1}\alpha T):X\times X\rightarrow L$ defined pointwise as*

$$S_{jk}^* = S^*(x_j,x_k) = \bigwedge_{i=1}^{n} (Q_{ij}\alpha T_{ik})$$

for any $j,k\in I_n$, belongs to $\mathfrak{S}$. Further, $S^*\geq S$ for any $S\in\mathfrak{S}$.

THEOREM 4.2 [19]. *If $\mathfrak{S}\neq\emptyset$, then $S^* = \wedge_{i=1}^{n} (Q_i\alpha T_i)$.*

THEOREM 4.3 [2]. (i) *Let $\mathfrak{S}\neq\emptyset$. Then the set $\mathfrak{M}_i=\{$minimal elements $L_i\in\mathfrak{R}_i$: $L_i\leq S^*\}$ is nonempty for any $i\in I_n$. Further, the finite set $\mathfrak{M} = \{M:X\times X\rightarrow L: M = \vee_{i=1}^{n} L_i,\ L_i\in\mathfrak{R}_i,\ i\in I_n\}$ is a subset of $\mathfrak{S}$ and has minimal elements, which are all the minimal elements of $\mathfrak{S}$ (i.e., M is minimal in $\mathfrak{M}$ iff M is minimal in $\mathfrak{S}$).*

(ii) *Let $\mathfrak{S}\neq\emptyset$. For any $S\in\mathfrak{S}$, there exists an element $M\in\mathfrak{M}$ such that $M\leq S$.*

(iii) *$\mathfrak{S}\neq\emptyset$ iff the matrix $\Sigma:X\times X\rightarrow L$ defined as*

$$\Sigma = \left[\bigvee_{i=1}^{n} (Q_i\sigma T_i)\right] \wedge S^*$$

belongs to $\mathfrak{S}$. Further, $M\leq\Sigma$ for any $M\in\mathfrak{M}$.

After these preliminaries, we recall the following well known result:

LEMMA 4.4. *Let $R_i\in\mathfrak{R}_\wedge$ for any $i\in I_n$. Then the matrix $R^* = \wedge_{i=1}^{n} R_i$ belongs to $\mathfrak{R}_\wedge$.*

Now we study the set $\mathfrak{T}^*=\mathfrak{T}^*(Q,T)=\mathfrak{S}\cap\mathfrak{R}_\wedge$ of all the max-min transitive solutions of the Eq. (1.3) (from now on, the asterisque on certain symbols distinguishes the same symbols introduced in the previous Sections for the Eq. (1.2)). In accordance to the symbology of Sec. 2, we denote by $\mathfrak{T}_i=\mathfrak{T}_i(Q_i,T_i)=\mathfrak{R}_i\wedge\mathfrak{R}_\wedge$ for any $i\in I_n$, the set of all the max-min transitive solutions of the i-th Eq. (4.1). In account of (4.2), it is evident that

$$\mathfrak{T}^* = \bigcap_{i=1}^{n} \mathfrak{T}_i. \tag{4.3}$$

If $\mathfrak{R}_i\neq\emptyset$ for any $i\in I_n$ (then $\mathfrak{T}_i\neq\emptyset$ for any $i\in I_n$ by Thm. 2.4), we can consider the greatest max-min transitive solution W_i of the i-th Eq. (4.1) and put down,

$$W^* = \bigwedge_{i=1}^{n} W_i. \tag{4.4}$$

Then the following result holds:

THEOREM 4.5. *$\mathfrak{T}^*\neq\emptyset$ iff $W^*\in\mathfrak{T}^*$. Further, $W^*\geq S$ for any $S\in\mathfrak{T}^*$.*

Proof. Of course, we are interested only to the nontrivial implication. Let $\mathfrak{T}^*\neq\emptyset$. Then $\mathfrak{S}\neq\emptyset$ and $\mathfrak{T}_i\neq\emptyset$ for any $i\in I_n$ by (4.3), thus $S^*\in\mathfrak{S}$ and $W_i\in\mathfrak{T}_i$ by Thm. 4.1 and Thm. 2.4, respectively. Let S be an arbitrary element of $\mathfrak{T}^*$, then $S\in\mathfrak{T}_i$ for any $i\in I_n$ by (4.3), hence $S\leq W_i$ for any $i\in I_n$ by Thm. 2.4, i.e. $S\leq W^*$ by

(4.4). On the other hand, $W_i \leq (Q_i \alpha T_i)$ for any $i \in \mathbf{I}_n$ by Thm. 2.2(i), i.e. $W^* \leq S^*$ by Thm. 4.2. Hence $S \leq W^* \leq S^*$ where $S, S^* \in \mathfrak{S}$, i.e. $W^* \in \mathfrak{S}$ by Lemma 2.3 and $W^* \in \mathfrak{R}_\wedge$ by Lemma 4.4. Thus $W^* \in \mathfrak{T}^*$. ■

In Thm. 2.4, we have already seen that, concerning Eq. (1.2), $\mathfrak{T} \neq \emptyset$ iff $\mathfrak{R} \neq \emptyset$. For Eq. (1.3), unfortunately $\mathfrak{T}^*$ could be empty even if $\mathfrak{S} \neq \emptyset$ as it is proved with easy examples.

Now we deal with the determination of the minimal elements of $\mathfrak{T}^*$. If $\mathfrak{S} \neq \emptyset$, we define the finite set $\overline{\mathfrak{M}} = \{ \overline{M} : M \in \mathfrak{M} \}$, where $\mathfrak{M}$ is defined in the statement of Thm. 4.3(i) and $\overline{M}$ is the max-min transitive closure of a matrix $M \in \mathfrak{M}$ (cfr. Def. 2.6). By using Thm. 4.3(i), (ii), it is proved, as in Thm. 2.7, that

THEOREM 4.6. *Let* $\mathfrak{T}^* \neq \emptyset$. *Then* $\overline{\mathfrak{M}} \subseteq \mathfrak{T}^*$, $\overline{\mathfrak{M}}$ *has minimal elements, which are all the minimal elements of* $\mathfrak{T}^*$ (i.e. M *is minimal in* $\overline{\mathfrak{M}}$ *iff* M *is minimal in* $\mathfrak{T}^*$).

5. **Cross-vectors of Eq. (1.3).** We first point out two simple facts, to be used in the sequel.

LEMMA 5.1. *If* $R_i \in \mathfrak{R}^{(1)}$ *for any* $i \in \mathbf{I}_n$, *then the matrix* $R^* = \wedge_{i=1}^{n} R_i$ *belongs to* $\mathfrak{R}^{(1)}$.

Proof. It suffices to observe that if $R_i = U_i \times V_i$ for any $i \in \mathbf{I}_n$, where $U_i, V_i : X \to L$ are vectors, then $R^* = U \times V$, being $U, V : X \to L$ vectors defined by $U = \wedge_{i=1}^{n} U_i$ and $V = \wedge_{i=1}^{n} V_i$. ■

The following result is due to Kolodziejczyk ([5], [12]):

LEMMA 5.2. *If* $S \in \mathfrak{R}^{(1)}$, *then* $(S \circ Q) \in \mathfrak{R}^{(1)}$ *for any matrix* $Q : X \times X \to L$.

Let $\mathfrak{D}^* = \mathfrak{S} \cap \mathfrak{R}^{(1)}$, be the set of all the cross-vectors, solutions S of the Eq. (1.3). Note that $\mathfrak{D}^* \subseteq \mathfrak{T}^*$ since $\mathfrak{R}^{(1)} \subset \mathfrak{R}_\wedge$ (cfr. Sec. 3). Denoting by $\mathfrak{D}_i = \mathfrak{D}_i(Q_i, T_i) = \mathfrak{R}_i \cap \mathfrak{R}^{(1)}$ for any $i \in \mathbf{I}_n$ the set of all the cross-vectors, solutions S of the i-th Eq. (4.1) and taking in account (4.2), we also have that

$$\mathfrak{D}^* = \bigcap_{i=1}^{n} \mathfrak{D}_i. \tag{5.1}$$

If $\mathfrak{R}_i \neq \emptyset$ for any $i \in \mathbf{I}_n$, let $D_i : X \times X \to L$, relatively to the Eq. (4.1), the matrix as defined from formula (3.2) and put down,

$$D^* = \bigwedge_{i=1}^{n} D_i. \tag{5.2}$$

The following result holds (cfr. Thm. 3.5):

THEOREM 5.3. $\mathfrak{D}^* \neq \emptyset$ *iff* $\mathfrak{S} \neq \emptyset$ *and* $T \in \mathfrak{R}^{(1)}$. *Further,* $D^* \geq S$ *for any* $S \in \mathfrak{D}^*$ *and* $D^* \leq W^*$.

Proof. Let $\mathfrak{D}^* \neq \emptyset$ and S be an arbitrary element of $\mathfrak{D}^*$. Then $S \in \mathfrak{R}^{(1)}$, which implies, by Lemma 5.2, that $S \circ R = T \in \mathfrak{R}^{(1)}$ since $S \in \mathfrak{S}$. Vice versa, let $\mathfrak{S} \neq \emptyset$ and $T \in \mathfrak{R}^{(1)}$. Then, by (4.2), $\mathfrak{R}_i \neq \emptyset$ for any $i \in \mathbf{I}_n$, hence $\mathfrak{D}_i \neq \emptyset$ for any $i \in \mathbf{I}_n$ and

$$D_i \leq W_i \leq (Q_i \alpha T_i), \qquad i = 1,2,\ldots,n, \tag{5.3}$$

by Thm. 3.5 and Thm. 2.2(i), where $D_i \in \mathfrak{D}_i$. Thus the matrix D^*, given by (5.2), is such that $D^* \leq S^*$ by Thm. 4.2. Now we show that $\Sigma \leq D^*$, where Σ is the matrix defined in the statement of Thm. 4.3(iii). Suppose that

$$\Sigma_{jk} > D_{jk}^* \tag{5.4}$$

for some $j,k \in \mathbf{I}_n$ and put, for semplicity,

$$D_{jk}^* = \bigwedge_{i=1}^{n} D_i(x_j, x_k) = D_h(x_j, x_k)$$

for some $h \in \mathbf{I}_n$. If $D_{jk}^* = D_h(x_j, x_k) = Q_{hj} \alpha T_{hk}$ (cfr. formula (3.2)), then (5.4) and Thm. 4.1 should imply $Q_{hj} \alpha T_{hk} \geq S_{jk}^* \geq \Sigma_{jk} > D_{jk}^* = Q_{hj} \alpha T_{hk}$, a contradiction. Hence

$$D_{jk}^* = T_{hk} \tag{5.5}$$

and

$$T_{hk} < \bigvee_{i=1}^{n} T_{hi} \tag{5.6}$$

by (3.1). On the other hand, we should deduce also from (5.5):

$$\bigvee_{i=1}^{n} T_{ik} \geq \bigvee_{i=1}^{n} (Q_{ij}\sigma T_{ik}) = \bigvee_{i=1}^{n} (Q_i\sigma T_i)(x_j,x_k) \geq \Sigma_{jk} > D_{jk}^* = T_{hk}. \tag{5.7}$$

Then (5.6) and (5.7) (cfr. Remark 3.3) contradict the fact that D^* belongs to $\mathfrak{R}^{(1)}$ by Lemma 5.1.

Then $\Sigma \leq D^* \leq S^*$ and hence $D^* \in \mathfrak{S}$ by Lemma 2.3. This means that $D^* \in \mathfrak{D}^*$, i.e. $\mathfrak{D}^* \neq \emptyset$.Let S be an arbitrary element of $\mathfrak{D}^*$. By (5.1), $S \in \mathfrak{D}_i$ for any $i \in \mathbf{I}_n$ and, by Thm. 3.5, $S \leq D_i$ for any $i \in \mathbf{I}_n$, i.e. $S \leq D^*$ by (5.2). Further, (4.4) and (5.3) imply that $D^* \leq W^*$ (note that $\mathfrak{T}^* \neq \emptyset$) and this concludes the proof. ■

Now we discuss about the minimal elements of $\mathfrak{D}^*$. Assuming $\mathfrak{D}^* \neq \emptyset$, then $\mathfrak{D}_i \neq \emptyset$ for any $i \in \mathbf{I}_n$ (by (5.1)), i.e. $\mathfrak{R}_i \neq \emptyset$ for any $i \in \mathbf{I}_n$ and hence $\mathfrak{D}_i$ has minimal elements by Thm. 3.6. In analogy with the symbolics of Sec. 3, we denote by $\Gamma_i = \Gamma_i(Q_i,T_i)$ the set of all the minimal cross-vectors solutions $L_i^{(k_i)}$ (belonging to $\mathfrak{D}_i$) of each Eq. (4.1), where

$$k_i \in \Gamma(Q_i) = \{j \in \mathbf{I}_n : Q_{ij} \geq \bigvee_{k=1}^{n} T_{ik}\}$$

and $L_i^{(k_i)}: \mathbf{X} \times \mathbf{X} \to \mathbf{L}$ is a matrix (suitably modifying the symbology of formula (2.7)) defined as

$$L_i^{(k_i)}(x_p,x_q) = \begin{cases} T_{iq} & \text{if } p=k_i, \\ 0 & \text{otherwise}, \end{cases} \tag{5.8}$$

for any $p,q \in \mathbf{I}_n$.

Since $D^* \in \mathfrak{D}^*$, then $D^* \in \mathfrak{D}_i$ for any $i \in \mathbf{I}_n$ by (5.1) and from the proof of Thm. 3.6, there certainly exists some $k_i \in \Gamma(Q_i)$ such that $L_i^{(k_i)} \leq D^*$ for any $i \in \mathbf{I}_n$, i.e.

$$\bigvee_{i=1}^{n} L_i^{(k_i)} \leq D^* \leq S^*. \tag{5.9}$$

Thus the set

$$\Gamma_i^* = \Gamma_i^*(Q_i,T_i) = \{L_i^{(k_i)} \in \Gamma_i : L_i^{(k_i)} \leq D^*, k_i \in \Gamma(Q_i)\}$$

is nonempty for any $i \in \mathbf{I}_n$ and let

$$\Gamma^* = \Gamma^*(Q,T) = \{M: \mathbf{X} \times \mathbf{X} \to \mathbf{L}, M = \bigvee_{i=1} L_i^{(k_i)}, L_i^{(k_i)} \in \Gamma_i^*, i \in \mathbf{I}_n\}. \tag{5.10}$$

Lemma 5.1 assures that $\mathfrak{R}^{(1)}$ is a meet-semilattice, but it is not in general an upper-semilattice as easy examples show. However we have tha following easy Lemma to be used in the successive theorem:

LEMMA 5.4. *If* $R_i \in \mathfrak{R}^{(1)}$ *for any* $i \in \mathbf{I}_n$ *are such that* $R_i = U_i \times V$ *for any* $i \in \mathbf{I}_n$, *where* $U_i, V: \mathbf{X} \to \mathbf{L}$ *are vectors, then*

$$M = \bigvee_{i=1}^{n} R_i$$

belongs to $\mathfrak{R}^{(1)}$.

Proof. It suffices to observe that $M = U \times V$, where $U: \mathbf{X} \to \mathbf{L}$ is a vector given by $U = \vee_{i=1}^{n} U_i$. ■

THEOREM 5.5. *If* $\mathfrak{S} \neq \emptyset$ *and* $T \in \mathfrak{R}^{(1)}$, *then* Γ^* *is the set of all the minimal elements of* $\mathfrak{D}^*$.

Proof. Note that $\mathfrak{D}^* \neq \emptyset$ by Thm. 5.3. By (5.9) and (5.10), we have $\Gamma^* \subseteq \mathfrak{M}$, where $\mathfrak{M}$ is the set defined in the statement of Thm. 4.3(i). We prove that an arbitrary element $M \in \Gamma^*$ is minimal in $\mathfrak{M}$.

Indeed, let $N \in \mathfrak{M}$ be such that $N \leq M$ and assume that $0 \leq N_{pq} < M_{pq}$ for some $p,q \in \mathbf{I}_n$. We put for simplicity:

$$M_{pq} = \left[\bigvee_{i=1}^{n} L_i^{(k_i)}\right](x_p,x_q) = L_h^{(k_h)}(x_p,x_q)$$

for some $h \in \mathbf{I}_n$, where $k_h \in \Gamma(Q_h)$. By (5.8), we should have $k_h = p$ since $M_{pq} > 0$ and hence, by definition of the set $\mathfrak{M}$ (cfr. Thm. 4.3(i)),

$$Q_{hp} \,\sigma\, T_{hq} \le N_{pq} < M_{pq} = L_h^{(p)}(x_p,x_q) = T_{hq}.$$

This should imply

$$Q_{hp} < T_{hq} \le \bigvee_{s=1}^{n} T_{hs},$$

a contradiction to the fact $p=k_h \in \Gamma(Q_h)$. Therefore N=M and M is minimal in $\mathfrak{M}$, hence M is minimal in $\mathfrak{S}$ by Thm. 4.3(i) and then in $\mathfrak{D}^*$, since (3.5) (suitably modified) and Lemma 5.4 assure that $\Gamma^* \subset \mathfrak{R}^{(1)}$, i.e. $\Gamma^* \subseteq \mathfrak{M} \cap \mathfrak{R}^{(1)} \subseteq \mathfrak{S} \cap \mathfrak{R}^{(1)} = \mathfrak{D}^*$.

In order to complete the proof, it remains to prove that a minimal element S of $\mathfrak{D}^*$ is necessarily in Γ^*. Indeed, $S \in \mathfrak{D}_i$ for any $i \in I_n$ (by (5.1)) and let $L_i^{(k_i)} \in \Gamma_i$ (cfr. proof of Thm. 3.6) such that $L_i^{(k_i)} \le S$ for any $i \in I_n$. Since $S \le D^*$, then $L_i^{(k_i)}$ belongs to Γ_i^* and hence the matrix

$$M = \bigvee_{i=1}^{n} L_i^{(k_i)}$$

is in $\Gamma^* \subseteq \mathfrak{D}^*$. It is such that $M \le S$ and because of the minimality of S in $\mathfrak{D}^*$, it follows that $S=M \in \Gamma^*$. ■

A more convenient characterization of the minimal elements of $\mathfrak{D}^*$ is given in the below result, but we premise the following:

DEFINITION 5.6. *Let* L *be a linear lattice (not necessarily complete). For a matrix* $R: X \times X \to L$, *the matrix* $R \in \mathfrak{R}^{(1)}$ *defined by* $R^{\sim} = U^{\sim} \times V^{\sim}$, *where* $U^{\sim}$ *and* $V^{\sim}$ *are vectors as defined in the statement of Thm.* 3.2(i), *is called the cross-vector closure of* R.

Clearly $R \le R^{\sim}$, $R=R^{\sim}$ iff $R \in \mathfrak{R}^{(1)}$ and $R^{\sim}$ is the smallest cross-vector including R by Thm. 3.2(ii). Then, by using Theorem 4.3(ii) and (iii), we can give the following theorem whose proof is similar to that of Thm. 4.6:

THEOREM 5.7. *Let* $\mathfrak{S} \neq \emptyset$, $T \in \mathfrak{R}^{(1)}$ *and* $\mathfrak{M}^{\sim} = \{M^{\sim} : M \in \mathfrak{M}\}$, *where* $\mathfrak{M}$ *is defined in the statement of Thm.* 4.3(i). *Then* $\mathfrak{M}^{\sim} \subseteq \mathfrak{D}^*$, $\mathfrak{M}^{\sim}$ *has minimal elements, which are all the minimal elements of* $\mathfrak{D}^*$ (i.e., $M^{\sim}$ *is minimal in* $\mathfrak{M}^{\sim}$ *iff* $M^{\sim}$ *is minimal in* $\mathfrak{D}^*$).

We conclude this Sec. pointing out the following result concerning the Schein rank of any solution of the Eq. (1.3):

THEOREM 5.9. *Let* $\mathfrak{S} \neq \emptyset$ *and* $\rho(T)=\lambda(\le n)$. *Then* $\rho(S) \ge \lambda$ *for any* $S \in \mathfrak{S}$.
Proof. If $\mu=\rho(S)<\lambda$, let

$$S = \bigvee_{i=1}^{n} S^{(t)}$$

be a decomposition of S in the sup of μ cross-vectors $S^{(t)} \in \mathfrak{R}^{(1)}$, $t=1,2,\ldots,\mu$. Hence, by property (1.6),

$$T = S \circ Q = \left[\bigvee_{t=1}^{\mu} S^{(t)}\right] \circ Q = \bigvee_{t=1}^{\mu} (S^{(t)} \circ Q),$$

where $(S^{(t)} \circ Q) \in \mathfrak{R}^{(1)}$ by Lemma 5.2 for any $t=1,2,\ldots,\mu$. This should imply $\rho(T) \le \mu < \lambda$, a contradiction. ■

6. **Concluding comments.** The above results are motivated by the following general question:

"Assume that Eq. (1.2) describes a finite-state system with a transition matrix R, let A and B be an input and an output, respectively. By applying k times the transition, let $B^{(k)}$ the resulting output given by $B^{(k)}=R^{(k)} \circ A$ (of course, we assume $B^{(1)}=B$). If A and B are assigned, determine a solution $R \in \mathfrak{R}$ which gives the necessary speed of the convergence of R (R is convergent if $R^{k+1}=R^k$ for some integer k)."

It is well known that powers of fuzzy matrices are convergent or oscillate with finite period [21]. For Boolean matrices, see [10, p. 183]. If $R \in \mathfrak{D}$, then $B^{(2)}=R^2 \circ A=B^{(3)}$ since $R^2=R^3$ ([5], [12]). If $R \in \mathfrak{T}$, then $B^{(n)}=R^n \circ A=R^{n+1} \circ A=B^{(n+1)}$ since $R^n=R^{n+1}$ [8]. Of course, similar considerations can be made on the Eq. (1.3). In other words, we can say that the more general transitivity between input and output of the system is used, the slower the convergence is.

REFERENCES

[1] Blyth T.S. (1964) *Matrices over ordered algebraic systems*, J. London Math. Soc. 39, 427-432.

[2] Di Nola A. (1985) *Relational equations in totally ordered lattices and their complete resolution*, J. Math. Anal. Appl. 107, 148-155.

[3] Di Nola A and Lettieri A. (1989) *Relation equations in residuated lattices*, Rend. Circolo Matem. Palermo 38, 246-256.

[4] Di Nola A., Pedrycz W. and Sessa S. (1985) *When is a fuzzy relation decomposable in two fuzzy sets?*, Fuzzy Sets and Systems 16, 87-90.

[5] Di Nola A., Pedrycz W., Sanchez E. and Sessa S. 1989 *Fuzzy Relation Equations and Their Applications to Knowledge Engineering*, Kluwer Academic Publishers, Dodrecht.

[6] Di Nola A. and Sessa S. (1989) *On the Schein rank of matrices over linear lattices*, Lin. Alg. Appl. 118, 155-158.

[7] Dubois D. and Prade H. 1980 *Fuzzy Sets and Systems: Theory and Applications*, Academic Press, New York.

[8] Hashimoto H. (1983) *Convergence of powers of a fuzzy transitive matrix*, Fuzzy Sets and Systems 9, 153-160.

[9] Kaufmann A. 1975 *Introduction to the Theory of Fuzzy Subsets*, Vol. I, Academic Press, New York.

[10] Kim K.H. 1982 *Boolean Matrix Theory and Applications*, Marcel Dekker, New York.

[11] Kim K.H. and Roush F. (1980)*Generalized fuzzy matrices*, Fuzzy Sets and Systems 4, 293-315.

[12] Kolodziejczyk W. (1988) *Decomposition problem of fuzzy relations: further results*, Internat. J. Gen. Syst. 14, 307-315.

[13] Luce R.D. (1952) *A note on Boolean matrix theory*, Proc. Amer. Math. Soc. 3, 382-388.

[14] Ovchinnikov S.V. (1981) *Structure of fuzzy binary relations*, Fuzzy Sets and Systems 6, 169-195.

[15] Rudeanu S. 1974 *Boolean Functions and Equations*, North-Holland, Amsterdam.

[16] Sanchez E. (1976) *Resolution of composite fuzzy relation equations*, Inform. and Control 30, 38-48.

[17] Sanchez E. (1977) *Solutions in composite fuzzy relation equations: Application to medical diagnosis in Brouwerian logic*, in: *Fuzzy Automata and Decision Processes* (M.M. Gupta, G.N. Saridis and B.R. Gaines, Eds.), North-Holland, New York, 221-234.

[18] Sanchez E. (1981) *Eigen fuzzy sets and fuzzy relations*, J. Math. Anal. Appl. 81, 399-421.

[19] Sessa S. (1984) *Some results in the setting of fuzzy relation equations theory*, Fuzzy Sets and Systems 14, 281-297.

[20] Tahani V. (1976) *A fuzzy model of document retrieval systems*, Inform. Processing Manag. 12, 177-187.

[21] Thomason M.G. (1977) *Convergence of powers of a fuzzy matrix*, J. Math. Anal. Appl. 57, 476-480.

[22] Zadeh L.A. (1971) *Similarity relations and fuzzy orderings*, Inform. Sci. 3, 177-200.

GENERALIZED CARDINAL NUMBERS AND THEIR ORDERING

Maciej Wygralak

A. Mickiewicz University, Institute of Mathematics
Matejki 48/49, PL - 60-769 Poznań

Abstract. In this paper a general theory of power for hardly characterizable objects as well as related generalized cardinal numbers are presented. The attention is focused on the questions of order. Łukasiewicz logic is uded as a supporting logic. The theory refers both to fuzzy sets and twofold fuzzy sets, partial sets, rough sets, etc.

Keywords: Łukasiewicz logic, Hardly characterizable objects, Equipotency, Powers, Generalized cardinal numbers, Order, Operations.

1. INTRODUCTION

The purpose of investigations initiated by the author some years .go was to build a general theory of cardinality concerning not only 'uzzy sets but simultaneously referring to some wider class of vague bjects called here hardly characterizable objects (HCH-objects, in hort; [11]). By an HCH-object in some infinite universal set $\mathcal{U}$ we ean a part of $\mathcal{U}$ which is maybe vaguely defined and needs not to be set itself, i.e. which cannot be mathematically described using he classical notion of a (sub)set (cf. semisets). However, we assume hat each HCH-object can be wholly described by means of one or two membership) functions $\mathcal{U}\to\mathcal{L}$ ($\mathcal{L}$ - suitable lattice). This way sets, uzzy sets just as twofold fuzzy sets, partial sets, rough sets, fuzzy ets with t-norms and φ-operators, intuitionistic fuzzy sets, L-fuzzy ets, Heyting algebra valued sets, etc., become particular cases of CH-objects whereas fuzziness can be considered as a kind of difficulty in characterizing some groups of elements from $\mathcal{U}$. So, the introduced definition of HCH-objects is rather informal. Nevertheless, it suffices in this discussion.

If $A: \mathcal{U} \to \mathcal{L}$, then obj(A) denotes the HCH-object 'embedded' in $\mathcal{U}$ and characterized by A. Moreover, $[\![x \in_m \text{obj}(A)]\!] := A(x)$, where $[\![e]\!]$ denotes the truth value of an expression e, := stands for 'equals by definition' and $\in_m$ denotes many-valued membership predicate.

In this paper we discuss problems related to ordering HCH-objects with respect to their powers. We put $\mathcal{L}=\mathcal{I}$, where $\mathcal{I}:=[0,1]$, and use Łukasiewicz logic; the supports $(\text{supp}(\text{obj}(A)):=\text{supp}(A):=\{x\in\mathcal{U} : A(x)>0\}$ can be quite arbitrary (some results for finite supports have been presented in [8,10] while the case $\mathcal{L}$=frame is discussed in [14]). Construction and basic properties of the generalized cardinal numbers related to HCH-objects are presented in detail in [12] and will be selectively recalled in Section 2. Problems of operations are considered in [13]. Proofs have been omitted here because of the limit of space allocated to the papers.

Throughout this discussion capitals in script denote sets (as usual $\emptyset$ symbolizes the empty set). Other capitals denote (membership) functions; in particular, $E:=1_\emptyset$ and $U:=1_\mathcal{U}$, where $1_\mathcal{D}$ is the ordinary characteristic function of $\mathcal{D}\subset\mathcal{U}$. Small letters i,j,...,p,q,r denote both finite and transfinite numbers. Small Greek letters with or without subscripts denote the generalized cardinal numbers. Moreover, let

$$GP(\mathcal{D}):=\mathcal{I}^{\mathcal{D}},\ GP:=GP(\mathcal{U}),$$
$$PS(\mathcal{D}):=\{0,1\}^{\mathcal{D}},\ PS:=PS(\mathcal{U}),$$
$$P_i(\mathcal{D}):=\{\mathcal{Y}\subset\mathcal{D} : \text{card}\,\mathcal{Y}=i\},\ P_i:=P_i(\mathcal{U}).$$

Other symbols will be defined in the sequel of this paper.

2. EQUIPOTENT HCH-OBJECTS AND GENERALIZED CARDINALS

If we try to construct gcn's (generalized cardinal numbers) via many-valued bijections, resulting theories become essentially dependent on the chosen definition of such bijection and seem not to be directed towards applications. Respective calculations are of high complexity even for fuzzy sets with small finite supports (see [1],[2], [4]). So, in this theory we use quite distinct approximative approach in which we try to make a good use of the already existing cardinals and apply some axiomatic method. Moreover, we assume that our information about any membership function can be imprecise or incomplete. For this reason we first approximate each membership function, say A, by means of two other functions f(A) and g(A). In other words, we approximate obj(A) by obj(f(A)) and obj(g(A)). More precisely, we assume that $f,g\colon GP \to GP$, where either f=g=id (id - the identity function) or at least one of the functions f and g is a function to $PS \subset GP$, i.e. obj(f(A)) or obj(g(A)) has to be a set. Moreover, we formulate the following natural postulates:

(A1) $\forall A\in GP\colon f(A) \subset A \subset g(A)$,
(A2) $\forall A,B\in GP\ \forall x,y\in \mathcal{U}\colon A(x) \leqslant B(y) \Rightarrow f(A)(x) \leqslant f(B)(y)\ \&\ g(A)(x) \leqslant g(B)(y)$,
(A3) $\forall A\in PS\colon f(A), g(A) \in PS$.

The family of all the pairs (f,g) fulfilling the above given requirements (excluding the trivial pair (E,U)) is denoted by $\mathcal{F}$. Let $A_t := \{x\in\mathcal{U}\colon A(x) \geqslant t\}$.

THEOREM 2.1. For each $(f,g)\in\mathcal{F}$ and $A\in GP$ we have
a) $f(A) \supset 1_{A_1}$ or $f \equiv E$.
b) $g(A) \subset 1_{supp(A)}$ or $g \equiv U$.
c) If $f\colon GP \to PS$, then $f(A) = 1_{A_1}$ or $f \equiv E$.
d) If $g\colon GP \to PS$, then $g(A) = 1_{supp(A)}$ or $g \equiv U$.

We are now ready to introduce the notion of equipotency of HCH-objects.

DEFINITION 2.2. We say that two HCH-objects obj(A) and obj(B) are equipotent (or: are of the same power) with respect to $(f,g)\in\mathcal{F}$ and we write $A \sim_{f,g} B$ or simply $A \sim B$ (when (f,g) is fixed) iff

$$\bigwedge\{t\colon \operatorname{card} f(A)_t \leqslant i\} = \bigwedge\{t\colon \operatorname{card} f(B)_t \leqslant i\}$$
and
$$\bigvee\{t\colon \operatorname{card} g(A)_t \geqslant i\} = \bigvee\{t\colon \operatorname{card} g(B)_t \geqslant i\}$$

for each cardinal number i.

Of course, $\sim_{f,g}$ is always an equivalence relation. We notice that the equipotency condition is a weakened form of the following one: $\forall t\in(0,1)\colon \operatorname{card} f(A)_t = \operatorname{card} f(B)_t\ \&\ \operatorname{card} g(A)_t = \operatorname{card} g(B)_t$.

Now, we are going to define an operator which will be used to generate gcn's. Let $GCN\colon GP \times GP \to GP(CN)$, where CN denotes the set of all the cardinals which are not greater than $\operatorname{card}\mathcal{U}$ and

$GCN(F,G)(i):=[\![\exists_m \mathcal{Y}\in P_i :\ obj(F)\subset obj(1_{\mathcal{Y}})]\!]\wedge[\![\exists_m \mathcal{X}\in P_i :\ obj(1_{\mathcal{X}})\subset obj(G)]\!]$

provided that $F\subset G$ ($\exists_m$ - many-valued existential quantifier). Let $f_i(A):=\bigvee\{t:\ card\, f(A)_t \geqslant i\}$, $g_i(A):=\bigvee\{t:\ card\, g(A)_t \geqslant i\}$, $a_i:=\bigvee\{t:\ card\, A_t \geqslant i\}$ and let i^+ denote the successor of i; in the same way one defines, say, $f_i(B)$, $g_i(B)$ and b_i. So, $A\sim_{f,g}B$ iff $g_i(A)=g_i(B)$ and $f_{i^+}(A)=f_{i^+}(B)$ for each $i\in CN$.

THEOREM 2.3. $GCN(f(A),g(A))(i)=g_i(A)\wedge 1-f_{i^+}(A)$ for each $(f,g)\in\mathcal{F}$, $A\in GP$ and $i\in CN$.

The following theorem describes so-called decomposition property.

THEOREM 2.4. $GCN(f(A),g(A))=GCN(E,g(A))\cap GCN(f(A),U)$ for each $A\in GP$ and $(f,g)\in\mathcal{F}$.

The key-stone of the outlined construction from [12] is however

THEOREM 2.5. For each $(f,g)\in\mathcal{F}$ and $A,B\in GP$ the following equivalence is fulfilled: $GCN(f(A),g(A))=GCN(f(B),g(B))$ iff $A\sim_{f,g}B$.

So, the values of the operator GCN fulfill the axiomatic definition of cardinals proposed by A.Tarski ([6]) and will be called just gcn's. They will be denoted by small letters from the Greek alphabet equipped sometimes with the indexing pair (f,g) emphasizing which approximating functions have been used. So, what describes powers of HCH-objects in $\mathcal{U}$ are in fact HCH-objects in CN characterized by functions $CN\to\mathcal{I}$ or $CN\to\mathcal{L}$ in general case. There are good reasons (see [12]) to accept such a form of description (see [7] for a review of early approaches). If $GCN(f(A),g(A))=\alpha\in GP(CN)$, we say that the power of obj(A) equals α with respect to $(f,g)\in\mathcal{F}$ and we write then $Gcard_{f,g}(A)=\alpha$. If (f,g) is fixed, we write simply $Gcard(A)=\alpha$. α is called finite when supp(obj(A)) is finite, else α is called transfinite gcn. So, we have $Gcard_{f,g}(A)(i)=\alpha(i)=g_i(A)\wedge 1-f_{i^+}(A)$ and $\alpha=\beta$ iff $\alpha(i)=\beta(i)$ for each $i\in CN$.

Now, we are going to present some comments, examples and selected basic properties. Let $GCN_{f,g}:=\{\alpha\in GP(CN):\ Gcard_{f,g}(D)=\alpha$ for a $D\in GP\}$.

(a) Choosing from $\mathcal{F}$ pairs (E,id), $(1_{(\cdot)_1},id)$ and (id,id) we get gcn's proposed respectively by L.A.Zadeh, D.Dubois, and by the author ([7]).

(b) If $(f,g)\neq(id,id)$, then each $\alpha\in GCN_{f,g}$ is normal and simultaneously is equal to the cardinality of a twofold fuzzy set. Pairs (f,g) such that both f and g are functions into PS are suitable for rough sets (see [9]). If $(f,g)=(1_{(\cdot)_1}, 1_{supp(\cdot)})$, we get partial cardinal numbers of D.Klaua ([3]).

(c) For each $(f,g)\in\mathcal{F}$ each $\alpha\in GCN_{f,g}$ is convex.

(d) For each $(f,g)\in\mathcal{F}$ there exists a bijection Ψ such that for each $\mathcal{D}\subset\mathcal{U}$ we have $Gcard_{f,g}(1_{\mathcal{D}})=\Psi(card\mathcal{D})$.

(e) It is possible that $A\sim_{f,g}B$ while $A\not\sim_{e,h}B$. This is not so surprising because we deal here with different pairs of approximating functions, i.e. we use in essence two distinct criteria and priorities in evaluating powers of obj(A) and obj(B).

(f) So, the theory generates a wide family of different types of gcn's having sometimes common and sometimes distinct properties. We can choose from this 'bag' such type that is most appropriate.

(g) Let $card\mathcal{D}=q$ and $Gcard_{f,g}(1_{\mathcal{D}})=\beta_{f,g}$. It appears that $\beta_{f,g}$ equals $1_{\{q\}}$, $1_{\{i\in CN:\ i\leqslant q\}}$, $1_{\{i\in CN:\ i\geqslant q\}}$ if respectively $f\not\equiv E$ & $g\not\equiv U$, $f\equiv E$, $g\equiv U$. This suggests some approximate informal interpretation of $Gcard_{f,g}(A)(i)$, $A\in GP$, namely: it can be considered to be the degree to which obj(A) has (respectively) exactly, at least, at most i elements.

(h) Let us use some special notation. In this example we put $\mathcal{U}=\mathbb{R}$ (so, $CN=\{i:\ i\leqslant c\}$) and accept only for our convenience the Continuum Hypothesis. The notation $\alpha=(v_0,v_1,\dots,v_r,(v)|\,w_1,w_2)$ will mean that $\alpha(i)=v_i$ for each $i\leqslant r$ (r-finite), $\alpha(i)=v$ for each finite $i>r$, $\alpha(\aleph_0)=w_1$ and $\alpha(c)=w_2$. Let

$$A(x)=\begin{cases}1-x & \text{if } 0\leqslant x\leqslant 1\\ 0 & \text{otherwise}\end{cases},\qquad C(x)=\begin{cases}1-1/x & \text{if } x=2,3,4,\dots\\ 0 & \text{otherwise}\end{cases},$$

$$D(x)=\begin{cases}1 & \text{if } x=0,1\\ 0.9 & \text{if } x=2\\ 0.7 & \text{if } x=3\\ 0.3 & \text{if } 4\leqslant x\leqslant 5\\ 0 & \text{otherwise}\end{cases},\qquad S(x)=\begin{cases}1 & \text{if } x=2,3,4\\ 0.2 & \text{if } x=5\\ 0.3 & \text{if } x=6\\ 0.9 & \text{if } x=7\\ 0.6 & \text{if } x=8\\ 0 & \text{otherwise}\end{cases}.$$

Suppose that $Gcard_{f,g}(A)=\alpha_{f,g}$, $Gcard_{f,g}(C)=\gamma_{f,g}$, $Gcard_{f,g}(D)=\delta_{f,g}$, and $Gcard_{f,g}(S)=\sigma_{f,g}$. Then we get for instance the following values of the generalized cardinals:

(f,g)	$\alpha_{f,g}$	$\gamma_{f,g}$	$\delta_{f,g}$	$\sigma_{f,g}$
(id,id)	$((0)\|0,1)$	$((0)\|1,0)$	$(0,0,.1,.3,.7,(.3)\|.3,.3)$	$(0,0,0,.1,.4,.6,.3,.2,(0)\|0,0)$
(E,id)	$((1)\|1,1)$	$((1)\|1,0)$	$(1,1,1,.9,.7,(.3)\|.3,.3)$	$(1,1,1,1,.9,.6,.3,.2,(0)\|0,0)$
$(1_{(\cdot)_1},id)$	$(0,(1)\|1,1)$	$((1)\|1,0)$	$(0,0,1,.9,.7,(.3)\|.3,.3)$	$(0,0,0,1,.9,.6,.3,.2,(0)\|0,0$
$(id,1_{supp(\cdot)})$	$((0)\|0,1)$	$((0)\|1,0)$	$(0,0,.1,.3,(.7)\|.7,1)$	$(0,0,0,.1,.4,.7,.8,1,(0)\|0,$
$(1_{(\cdot)_1},1_{supp(\cdot)})$	$(0,(1)\|1,1)$	$((1)\|1,0)$	$(0,0,(1)\|1,1)$	$(0,0,0,1,1,1,1,1,(0)\|0,0$

Worth commenting is for instance the value $\alpha_{f,g}$ for f=g=id. We have $\alpha_{f,g}\in PS(CN)$ although $A\notin PS$. This is because A attains continuum of values lying as near to 1 as one likes.

(i) In general, for each $A\in GP$ we get ($m:=\text{card}\,A_1$, $n:=\text{card}\,supp(A)$):

$$Gcard_{f,g}(A)(i) = a_i \wedge 1-a_{i+} \quad \text{if} \quad (f,g)=(id,id),$$

$$Gcard_{f,g}(A)(i) = g_i(A) \quad \text{if} \quad f\equiv E,$$

$$Gcard_{f,g}(A)(i) = \left.\begin{array}{lll} 0 & \text{if} & i<m \\ 1 & \text{if} & i=m \\ g_i(A) & \text{if} & i>m \end{array}\right\} \quad \& \quad f(\cdot)=1_{(\cdot)_1},$$

$$Gcard_{f,g}(A)(i) = 1-f_{i+}(A) \quad \text{if} \quad g\equiv U,$$

$$Gcard_{f,g}(A)(i) = \left.\begin{array}{lll} 1-f_{i+}(A) & \text{if} & i<n \\ 1 & \text{if} & i=n \\ 0 & \text{if} & i>n \end{array}\right\} \quad \& \quad g(\cdot)=1_{supp(\cdot)}.$$

3. ORDERING AND COMPARING GENERALIZED CARDINALS

The inclusion $\subset$ cannot be used as ordering relation for powers and gcn's because $A\subset B$ implies $Gcard_{f,g}(A)\subset Gcard_{f,g}(B)$ iff $f\equiv E$. Thus, we have to define the order in another way.

DEFINITION 3.1. (a) $Gcard_{f,g}(A)\leqslant Gcard_{f,g}(B)$ iff $g_i(A)\leqslant g_i(B)$ and $f_{i+}(A)\leqslant f_{i+}(B)$ for each $i\in CN$.
(b) $Gcard_{f,g}(A) < Gcard_{f,g}(B)$ iff $Gcard_{f,g}(A)\leqslant Gcard_{f,g}(B)$ & $A\not\sim_{f,g}B$.
(c) $\alpha_{f,g}<\beta_{f,g}$ iff for $A,B\in GP$ such that $Gcard_{f,g}(A)=\alpha_{f,g}$ and $Gcard_{f,g}(B)=\beta_{f,g}$ we have $Gcard_{f,g}(A)<Gcard_{f,g}(B)$.
(d) $\alpha_{f,g}\leqslant\beta_{f,g}$ iff $\alpha_{f,g}<\beta_{f,g}$ or $\alpha_{f,g}=\beta_{f,g}$.

If (f,g) is fixed, we shall write simply $Gcard(A)\leqslant Gcard(B)$, $\alpha\leqslant\beta$, $Gcard(A)<Gcard(B)$ or $\alpha<\beta$. Let us list first properties of $\leqslant$ and $<$ for powers and gcn's.

a) $<$ implies $\leqslant$ and excludes both $>$ and $=$.

b) $A \subset B$ implies $Gcard(A) \leqslant Gcard(B)$.

c) $Gcard(A) \leqslant Gcard(B)$ & $Gcard(B) \leqslant Gcard(A)$ iff $A \sim B$.

d) $A \subset B \subset C$ and $A \sim C$ imply $A \sim B \sim C$.

e) $\leqslant$ is partial order relation; however, if (f,g) equals $(1_{(\cdot)_1}, U)$ or $(E, 1_{supp(\cdot)})$, then $\leqslant$ is linear order relation.

e at once notice that (c) and (d) are Cantor-Bernstein-like theorems.

HEOREM 3.2. If there exists $B^* \subset B$ such that $A \sim B^*$, then we have that $Gcard(A) \leqslant Gcard(B)$.

REMARK 3.3. The inverse implication holds for HCH-objects with finite supports. So, then the theory can be constructed in another, say, more elegant way.

THEOREM 3.4. For each $(f,g) \in \mathcal{F}$ and $A,B \in GP$, $Gcard_{f,g}(A) \leqslant Gcard_{f,g}(B)$ iff $Gcard_{E,id}(g(A)) \leqslant Gcard_{E,id}(g(B))$ and $Gcard_{id,U}(f(A)) \leqslant Gcard_{id,U}(f(B))$.

One can easily give an example of $A,B \in GP$ and pairs $(f,g),(\bar{f},\bar{g}),(f',g')$ such that $Gcard_{f,g}(A) < Gcard_{f,g}(B)$ and $Gcard_{\bar{f},\bar{g}}(B) < Gcard_{\bar{f},\bar{g}}(A)$ while $Gcard_{f',g'}(A)$ and $Gcard_{f',g'}(B)$ are incomparable with respect to (f',g'). It happens if for instance $\mathcal{U}=\mathcal{I}$, $A(x):=(1$ if x is rational, else $0)$, $B(x):=(2x$ if $x \leqslant 0.5$, else $-2x+2)$, and $(f,g)=(E,1_{supp(\cdot)})$, $(\bar{f},\bar{g})=(1_{(\cdot)_1},U)$, $(f',g')=(1_{(\cdot)_1},1_{supp(\cdot)})$. This is what we see also in our common life when a few persons try to compare some vague things with respect to some vague attributes and using (what is quite natural) different priority hierarchies they get often quite different results.

THEOREM 3.5. Let $GCN^*_{f,g}:=\{\alpha \in GCN_{f,g}: Gcard_{f,g}(D)=\alpha$ for some $D \in PS\}$. Then $(CN, \leqslant)$ and $(GCN^*_{f,g}, \leqslant)$ are isomorphic for each $(f,g) \in \mathcal{F}$.

As concerns some characterization of $\alpha \leqslant \beta$, we have ($\alpha \leqslant \beta$ iff $\alpha \subset \beta$) for $f \equiv E$ and ($\alpha \leqslant \beta$ iff $\beta \subset \alpha$) for $g \equiv U$. This and Theorem 3.4 are tools for easy characterizing $\alpha \leqslant \beta$ for arbitrary $(f,g) \in \mathcal{F}$.

Now, we like to present some more remarks about the structure of $GCN_{f,g}$. Let $\sigma:=Gcard_{f,g}(E)$ and $\omega_{f,g}:=Gcard_{f,g}(U)$.

THEOREM 3.6. For each $(f,g) \in \mathcal{F}$ the gcn $\sigma_{f,g}$ ($\omega_{f,g}$, resp.) is the least (greatest, resp) and simultaneously unique minimal (unique maximal, resp.) element in $(GCN_{f,g}, \leqslant)$.

It appears that for each $(f,g)\in\mathcal{F}$ and $\alpha,\beta\in GCN_{f,g}$ $(\alpha,\beta\neq\sigma_{f,g},\omega_{f,g})$ α and β always have common (but not immediate, in general) successor and predecessor. For some pairs (f,g) one can say a bit more about $GCN_{f,g}$, for instance $(GCN_{E,id},\cap,\cup,\sigma_{E,id},\omega_{E,id})$ and $(GCN_{id,U},\cup,\cap,\sigma_{id,U},\omega_{id,U})$ are complete distributive lattices (see also (e)).

Finishing this brief discussion about order, let us mention that in the presented theory, which for $\mathcal{L}=\mathcal{T}$ is based on the infinite-valued Łukasiewicz logic, counterparts of the Continuum Hypothesis and Generalized Continuum Hypothesis are not true for HCH-objects. Really, let $\mathcal{U}=\mathcal{T}$, $(f,g)=(E,id)$. Moreover, let $\mathcal{A}$ denote the set of rationals in $\mathcal{U}$, $B(x):=(1$ if x is rational, else $0.5)$. So, card$\mathcal{A}=\aleph_0$, card$\mathcal{U}=\mathfrak{c}$ while $Gcard(1_{\mathcal{A}}) < Gcard(B) < Gcard(U)$.

4. ORDER AND OPERATIONS

Let us mention here some properties related simultaneously to questions of order and operations on the generalized cardinal numbers. Using the modified extension principle (see e.g. [13]; cf. [5]) one can easily define $\alpha+\beta$, $\alpha\cdot\beta$, $\alpha-\beta$ and α^{β} for $\alpha,\beta\in GCN_{f,g}$ and $(f,g)\in\mathcal{F}$. Let $\varsigma_{f,g}:= Gcard_{f,g}(1_{\{z\}})$, where z denotes quite arbitrary element of $\mathcal{U}$. Moreover, let $\sigma:=\sigma_{f,g}$ and $\varsigma:=\varsigma_{f,g}$ if (f,g) is fixed. One can easily check that for each $(f,g)\in\mathcal{F}$ and $\alpha,\beta,\gamma\in GCN_{f,g}$ we obtain the following:

(a) $\alpha+\beta=\beta+\alpha$,
$\alpha\cdot\beta=\beta\cdot\alpha$ (commutativity).

(b) $\alpha+(\beta+\gamma)=(\alpha+\beta)+\gamma$,
$\alpha\cdot(\beta\cdot\gamma)=(\alpha\cdot\beta)\cdot\gamma$ (associativity).

(c) $\alpha+\sigma=\alpha$,
$\alpha\cdot\varsigma=\alpha$ (neutral elements).

(d) $\alpha\cdot(\beta+\gamma)=\alpha\cdot\beta+\alpha\cdot\gamma$ (distributivity).

So, $(GCN_{f,g},+,\cdot,\sigma_{f,g},\varsigma_{f,g})$ forms for each $(f,g)\in\mathcal{F}$ a commutative semiring with zero and unity. Moreover, for each $(f,g)\in\mathcal{F}$ and $A,B\in GP$ the following laws are fulfilled:

(e) $Gcard(A)+Gcard(B) = Gcard(A\cap B)+Gcard(A\cup B)$ (valuation property).
So, $Gcard(A)+Gcard(B) = Gcard(A\cup B)$ if $A\cap B=E$.

(f) $Gcard(A)\cdot Gcard(B) = Gcard(A\times B)$ (Cartesian product rule).

So, addition and multiplication of generalized cardinal numbers are well-defined operations. Finally, the following laws are fulfilled for each $(f,g)\in\mathcal{F}$ and $\alpha,\beta,\gamma\in GCN_{f,g}$:

(g) If $\alpha\leq\beta$ and $\gamma\leq\delta$, then $\alpha+\gamma\leq\beta+\delta$ and $\alpha\cdot\gamma\leq\beta\cdot\delta$ (side-by-side addition and multiplication of inequalities between gcn's). Thus, if $\alpha\leq\beta$, then $\alpha+\gamma\leq\beta+\gamma$ and $\alpha\cdot\gamma\leq\beta\cdot\gamma$ for each γ.

(h) Suppose that α is finite. Then $\alpha+\beta=\alpha+\gamma$ implies $\beta=\gamma$. Similarly, $\alpha\cdot\beta=\alpha\cdot\gamma$ implies $\beta=\gamma$, too.

So, the resulting algebra of generalized cardinal numbers is similar to the algebra of cardinal numbers what is very convenient from the viewpoint of applications. Nevertheless, there are some essential differences (see e.g. [12],[13]). For instance, $\alpha<\gamma$ does not imply in general case that there exists such β that $\alpha+\beta=\gamma$.

REFERENCES

1. BLANCHARD N. (1981). Theories cardinale et ordinale des ensembles flous, Thesis, Claude-Bernard University, Lyon.
2. GOTTWALD S. (1969). Konstruktion von Zahlbereichen und die Grundlagen der Inhaltstheorie in einer mehrwertigen Mengenlehre, Thesis, Leipzig.
3. KLAUA D. (1969). Partielle Mengen und Zahlen, Monatsber. Deut. Akad. Wiss. Berlin, 11, 585-599.
4. KLAUA D. (1972). Zum Kardinalzahlbegriff in der mehrwertigen Mengenlehre. In Theory of Sets and Topology (Asser G., Flachsmeyer J., Rinow W., Eds.), Deut. Verlag Wiss., 313-325.
5. ŠOSTAK A. (1989). Fuzzy cardinals and cardinalities of fuzzy sets. In Algebra and Discrete Mathematics, Latvian State University, 137-144 (in Russian).
6. TARSKI A. (1924). Sur quelques théorémes qui équivalent a l'axiome de choix, Fund. Math., 5, 147-154.
7. WYGRALAK M. (1986). Fuzzy cardinals based on the generalized equality of fuzzy subsets, Fuzzy Sets and Systems, 18, 143-158.
8. WYGRALAK M. (1988). Fuzzy sets, twofold fuzzy sets and their cardinality. In Proc. 1st Joint IFSA-EC & EURO-WG Workshop on Progress in Fuzzy Sets in Europe (Kacprzyk J., Straszak A., Eds.), Warsaw, 369-374.
9. WYGRALAK M., On the power of a rough set, Fasc. Math., in print.
10. WYGRALAK M., Fuzzy sets - their powers and cardinal numbers. In Proc. 10th Inter. Seminar on Fuzzy Set Theory, Linz, 1988, in print.
11. WYGRALAK M., On HCH-objects and the proper choice of basic operations, In Proc. 3rd Polish Symp. IFM, Poznań, 1989, in print.

12. WYGRALAK M., Powers and generalized cardinal numbers for HCH-objects - basic notions, Math. Pann., submitted.

13. WYGRALAK M., Powers and generalized cardinal numbers for HCH-objects - operations, Fuzzy Sets and Systems, submitted.

14. WYGRALAK M., Generalized cardinal numbers for lattice valued HCH-objects, in preparation.

AN INTERVAL-BASED APPROACH FOR WORKING WITH FUZZY NUMBERS

ANTONIO GONZALEZ and MARIA-AMPARO VILA
Dpto. de Ciencias de la Computacion e I.A.
Facultad de Ciencias, Universidad de Granada,
18071-Granada (Spain).

Abstract

An interval-based approach for working with fuzzy numbers is presented. This approach is based on the use of a ranking function calling the average value. This ranking function can be used in different processes associated to fuzzy numbers. Processes as ranking methods, indifference relations, operations on fuzzy numbers and distance measures, has been considered. An interpretation for each one of these processes, as an equivalent process on real intervals representing the mean value of the fuzzy numbers involved, has been obtained. Moreover, from this interpretation the above processes may be visualized through an useful graphic representation.

Keywords

Fuzzy Numbers; expectation; interval analysis; ranking function.

1. INTRODUCTION.

Fuzzy numbers can be introduced in order to model imprecise situations involving real numbers. Recently, different methods for ranking fuzzy numbers have been described. Most of these methods are defined from a ranking function which maps each fuzzy numbers into an ordered set. A particular ranking function evaluated in $\mathbb{R}$, called the Average Value (A.V.), was defined in [2].

An interesting relation between the A.V. and the notion of expectation for fuzzy numbers was studied in [6]. From this relation, we defined a generalized mean value. Now, the aim of this work is to show as the A.V. could be used as an useful tool for solving different problems associated to fuzzy numbers. The approach is based on the substitution of fuzzy numbers by real intervals, representing its mean values.

We have applied this interval-based approach to four particular problems on fuzzy numbers: ranking fuzzy numbers, indifference relations, operations and distance measures.

In these problems, we have obtained that the use of the A.V., for any of the above processes, is equivalent to a similar process on mean values, that is, ranking fuzzy numbers with the A.V. is equivalent to rank its mean values, indifference between fuzzy numbers with the A.V. is equivalent to calculate the indifference between its mean values, and so on.

In conclusion, the A.V. induces processes on real intervals, and when we take as real intervals the mean values of fuzzy numbers, the processes generated are the initial models by taking directly the fuzzy numbers and the A.V. The advantage of this approach is we can interpret the process used as an equivalent process on mean values.

2. THE AVERAGE VALUE.

The following definition of fuzzy number will be used in this work:

Definition 2.1. A fuzzy subset A, of the real line, with membership function $\mu_A(\bullet)$ is said to be a **Fuzzy Number** iff:

i) $\forall\alpha\in[0,1]$, $A_\alpha=\{x\in\mathbb{R} \;/\; \mu_A(x)\geq\alpha\}$ (α-cut of A) is a convex set.

ii) $\mu_A(\bullet)$ is an upper semicontinuous function.

iii) $Supp(A)=\{x\in\mathbb{R} \;/\; \mu_A(x)>0\}$ is a bounded set of $\mathbb{R}$.

We will use $\tilde{\mathbb{R}}$ to denote the set of fuzzy numbers, and α_A to denote the height of the fuzzy quantity A. To facilitate the notation, we will use capital letters to represent fuzzy numbers.

The Average Value was introduced to help in the ordering of fuzzy numbers and defined by means of a process of integration of a parametric function $f_A^\lambda(\alpha)$ representing the position of every α-cut in the real line, and through a subjective assignation of weights related to the relative importance of α levels.

Definition 2.2 (Campos and Gonzalez [2]). Let $Y\subseteq[0,1]$, the **Average Value** (A.V.) of a fuzzy number A with respect to an additive measure S on [0,1] and $\lambda\in[0,1]$ is the value

$$V_S^\lambda(A)=\int_Y f_A^\lambda(\alpha)dS(\alpha)$$

where $f_A^\lambda(\alpha)=\lambda b_\alpha+(1-\lambda)a_\alpha$ for $\alpha\leq\alpha_A$ and 0 otherwise, with $A_\alpha=[a_\alpha,b_\alpha]$ and α_A the height of A.

The A.V. was defined by means of several parameters. The hope is that if a ranking function has enough degrees of freedom to modify its own performance, it would adapt itself toward the decision-maker's preferences. A practical choice of the additive measure can be taken by using the following Stieltjes measure on [0,1]

$$S((a,b])=b^r-a^r, \quad \forall a,b\in[0,1].$$

In this case, the A.V. depends on two parameters λ and r, and it can be writen by

$$V_r^{\lambda}(A)=r\int_0^1 f_A^{\lambda}(\alpha)\alpha^{r-1}d\alpha.$$

We gave a practical interpretation of these elements in [2] and [3], and we showed that when certain parameters are used the A.V. coincides with other comparison indices.

Initially, the A.V. was defined in order to generate a comparison between fuzzy numbers, that is,

$$A\leq_{\lambda}B \Leftrightarrow V_s^{\lambda}(A)\leq V_s^{\lambda}(B), \quad \forall A,B\in\tilde{\mathbb{R}}$$

This relation is a crisp preorder on $\tilde{\mathbb{R}}$, and an order relation on the quotient set generated by the equivalence relation:

$$A\simeq_{\lambda}B \Leftrightarrow V_s^{\lambda}(A)=V_s^{\lambda}(B), \quad \forall A,B\in\tilde{\mathbb{R}}$$

In a previous paper [2] we interpreted the parameter λ as an optimism-pessimism degree, which must be selected by the decision-maker. Accordingly, this choice must be obtained in some way. If the parameter λ is not known "a priori", then it is always possible to calculate the **Dominance Region** of B over A, defined by:

$$R(A,B)=\{\lambda\in[0,1]\ /\ V_s^{\lambda}(A)\leq V_s^{\lambda}(B)\}$$

The computation of this region may be used in the orientation for ranking A and B. This region, provides an initial overall view of the decision, and facilitates the choice of an optimistic or pessimistic approach. Moreover, it tells the decision-maker the sensitivity of the parameter λ, i.e., whether or not a small change of the parameter modifies the final solution.

The dominance region is a subinterval of the unit interval as we proved in [3]. So, R(A,B) is an interval into which we put the different parameter λ, such that we accept $A\leq_{\lambda}B$.

From the dominance region we define the set $I(A,B)=R(A,B)\cap R(B,A)$. This set contains the parameters where the fuzzy numbers A and B are indifferent. In [3] we proved that I(A,B) is either $\emptyset$ or [0,1] or a single point.

The aim of this work is to establish a more general use of the A.V. As we will see later, this ranking function allows us to manipulate fuzzy numbers in a practical way through a substitution between the process generated from the A.V. (e.g. process=ranking, indifference,...) and an equivalent process on real intervals associated to the original fuzzy numbers. These intervals will be the mean values of fuzzy numbers.

3. MEAN VALUES FOR FUZZY NUMBERS.

An interesting relation between the A.V. defined above and the notion of expectation for fuzzy numbers was studied in [6]. The mean value of a fuzzy number was defined by Dubois and Prade [5] as an interval whose bounds are upper and lower expectations, that is,

$$E(A)=[E_*(A),E^*(A)] \quad \text{with} \quad E_*(A)=\int_{-\infty}^{+\infty} x\,dF^*(x) \quad \text{and} \quad E^*(A)=\int_{-\infty}^{+\infty} x\,dF_*(x)$$

where

$F^*(x)=\sup\{\mu_A(r) \ / \ r\leq x\}$ and $F_*(x)=\inf\{1-\mu_A(r) \ / \ r>x\}$.

Obviously, the upper and lower expectations can be written as Choquet's integrals [4] with respect to possibility and necessity measures:

$$E^*(A)=\int x d\Pi \quad \text{and} \quad E_*(A)=\int x dN,$$

where Π and N are respectively the possibility and necessity measures associated with the fuzzy number A.

Thus, by considering the set $\tilde{\mathbb{R}}^*$ of normalized fuzzy numbers, with a continuous and strictly increasing membership function before the modal values and strictly decreasing function after the modal values and the Lebesgue measure L, then the interval $[V_L^0(A),V_L^1(A)]$ coincides with the mean value of a fuzzy number defined by Dubois and Prade, that is,

$$E_*(A)=V_L^0(A),\ E^*(A)=V_L^1(A) \text{ and } E(A)=\{V_L^\lambda(A) \ / \ \lambda\in[0,1]\},\quad \forall A\in\tilde{\mathbb{R}}^*.$$

From this result we defined the following generalized mean value.

Definition 3.1 (Gonzalez [6]). The **S-mean value** of a fuzzy number A, denoted $E_S(A)$, is a set of numbers defined by:

$$E_S(A)=\{V_S^\lambda(A) \ / \ \lambda\in[0,1]\}.$$

Obviously, the S-mean value can also be defined by $E_S(A)=[V_S^0(A),V_S^1(A)]$ and if S is taken as the Lebesgue measure then $E_L(A)$ coincides with the mean value E(A).

Moreover, by taking $\mathcal{C}$ as the set of real closed intervals, then the function

$$E_S:\tilde{\mathbb{R}}\longrightarrow\mathcal{C}$$
$$A\longrightarrow E_S(A)$$

allows us to represent easily fuzzy numbers and relations between them, through the substitution of fuzzy numbers by real intervals. This substitution allows us to give a graphics representation of fuzzy numbers and visualize the order and indifference relations on fuzzy numbers, as well as, the distance measures defined through the Average Value.

This graphic representation is based on a representation of real closed intervals given by Moore [7].

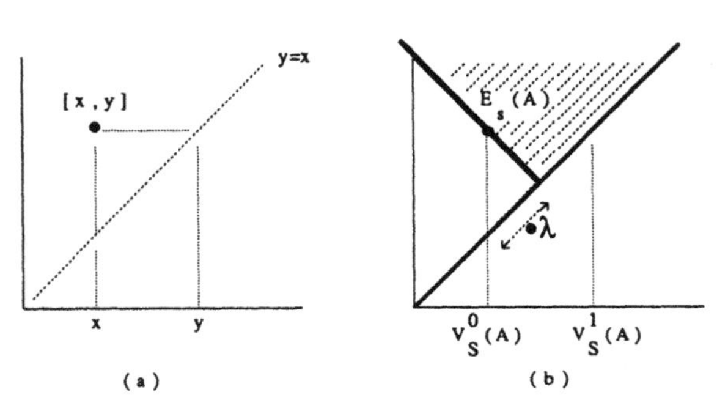

figure 1. (a) Graphic representation of $\mathcal{C}$.
(b) Representations associated to A.

Thus, we can represent the fuzzy number A as figure 1b and visualize the order and indifference relations on fuzzy numbers. The shaded region in figure 1b represents the set of fuzzy numbers that are greater than a concrete fuzzy number A, that is, $\{B / A \lesssim_\lambda B\}$, and the bold line represents the set of fuzzy numbers that are indifferent with A, that is, $\{B / A \propto_\lambda B\}$. In figure 1b, symbol $\bullet\lambda$ represents a number proportional to λ (the exact lenght of this segment is $\sqrt{2}(V_S^1(A)-V_S^0(A))\lambda$).

4. AN INTERVAL-BASED APPROACH.

Until now, the A.V. has been only used as a device for ranking fuzzy numbers. The aim of this work is to show as the A.V. could be used as an useful tool for solving different problems associated to fuzzy numbers, that is, as the A.V. allows us to replace fuzzy numbers by real intervals (representing its mean values) for solving some processes. We have studied the following problems:

4.1. Ranking fuzzy numbers.

By considering the following interval relation:

$$[a,b] \lesssim_\lambda [c,d] \Leftrightarrow \lambda b+(1-\lambda)a \leq \lambda d+(1-\lambda)c$$

with $a,b,c,d \in \mathbb{R}$, the ranking method generated by $V_S^\lambda(\bullet)$ is equivalent to use the mean value as follows:

$$A \lesssim_\lambda B \Leftrightarrow E_S(A) \lesssim_\lambda E_S(B).$$

Thus, the ranking method generated by the A.V. is based on the choice of an appropriate mean value for each fuzzy number (E_S) and an interval relation ($\lesssim_\lambda$) with which to compare the intervals defined by the mean value of each fuzzy number.

Moreover, we can say that $\lesssim_\lambda$ is also a **weak comparison** and the relation

$$A \leq B \Leftrightarrow E_S(A) \lesssim_\lambda E_S(B) \ \forall \lambda \in [0,1]$$

can be defined as a **strong comparison** between fuzzy numbers. Obviously, this last relation ranks all average values independent of the parameter λ, and by defining the following interval relation

$$[a,b] \leq [c,d] \Leftrightarrow a \leq c \text{ and } b \leq d$$

with $a,b,c,d \in \mathbb{R}$, then the strong order can be writen as

$$A \leq B \Leftrightarrow E_S(A) \leq E_S(B).$$

Obviously, there is a relation between dominance region and strong and weak comparison, in the following sense:

Strong comparison $\Leftrightarrow R(A,B)=[0,1]$,

Weak comparison $\Leftrightarrow \lambda \in R(A,B)$.

4.2. Indifference between fuzzy numbers.

By considering the following interval relation:

$$[a,b]\simeq_\lambda[c,d] \Leftrightarrow \lambda b+(1-\lambda)a=\lambda d+(1-\lambda)c$$

with $a,b,c,d\in\mathbb{R}$, the indifference relation generated by $V_S^\lambda(\bullet)$ is equivalent to using the mean value as follows:

$$A\simeq_\lambda B \Leftrightarrow E_S(A)\simeq_\lambda E_S(B).$$

We call to this relation **weak indifference**, since as can be easily verified

$$A\simeq_\lambda B \;\Rightarrow E_S(A)\subseteq E_S(B) \text{ or } E_S(A)\supseteq E_S(B).$$

The opposite implication is only true for a sole parameter λ,

$$E_S(A)\subseteq E_S(B) \Rightarrow \exists\lambda^*\in[0,1] \;/\; A\simeq_{\lambda^*}B$$

and a similar result is obtained for $E_S(A)\supseteq E_S(B)$. Thus, we can also defined a **stronger indifference** relation that is based on the equality of the mean values for each fuzzy numbers

$$A\cong B \Leftrightarrow E_S(A)=E_S(B).$$

Obviously, the weak indifference is relationed with the weak comparison whereas the strong indifference is relationed with the strong comparison.

Thus, the indifference relations generated by the A.V. are based on the choice of E_S as mean value and alternative relations ($\simeq_\lambda$ or $=$) for representing the "equality" between real intervals.

On the other hand, the relation $\leq_\lambda$ defines a total order on the quotient set generated by the equivalence relation $\simeq$, by contrast, the relation $\leq$ is only a partial order on the quotient set generated by the equivalence relation $\cong$.

An alternative to the indifference relations defined above, is the **ε-indifference**

$$A\cong_\varepsilon B \Leftrightarrow |V_S^\lambda(A)-V_S^\lambda(B)|\leq\varepsilon \quad \forall\lambda\in[0,1].$$

The indifference relations verify the following implications, as can be easily proved:

strong indifference $\Rightarrow$ weak indifference $\forall\lambda\in[0,1]$.
($\not\Leftarrow$)
strong indifference $\Rightarrow$ ε-indifference $\forall\varepsilon>0$.
($\not\Leftarrow$)
ε-indifference, $\varepsilon\neq 0$ $\not\Rightarrow$ weak indifference.
weak indifference $\not\Rightarrow$ ε-indifference $\forall\varepsilon>0$.

The last implication is wrong for all the numbers ε, but is true for an infinite set of ε. In a more general context we show that for an infinite set of ε, $A\cong_\varepsilon B$ even if A and B are not weakly indifference, but perhaps these ε are so large to be considered in a practical problem. If use $W_S(A)$ to denote the width of the general mean value

$W_S(A)=V_S^1(A)-V_S^0(A)$, $\forall A\in\tilde{\mathbb{R}}$, then

$\forall A,B\in\tilde{\mathbb{R}}$ A is ε-indifference to B if $\varepsilon\geq|W_S(A)-W_S(B)|+|V_S^0(A)-V_S^0(B)|$.

The strong and the weak indifference are equivalence relation, but the ε-indifference is only a reflexive and symmetric relation, since the transitive relation is not in general verified, that is,

$\exists A,B,C\in\tilde{\mathbb{R}}$ / $A\cong_\varepsilon B$ and $B\cong_\varepsilon C$ but $A\not\cong_\varepsilon C$.

The shaded region in figure 2 represents the set of fuzzy numbers that are ε-indifferent with the fuzzy number A, that is, $\{B \,/\, A\cong_\varepsilon B\}$.

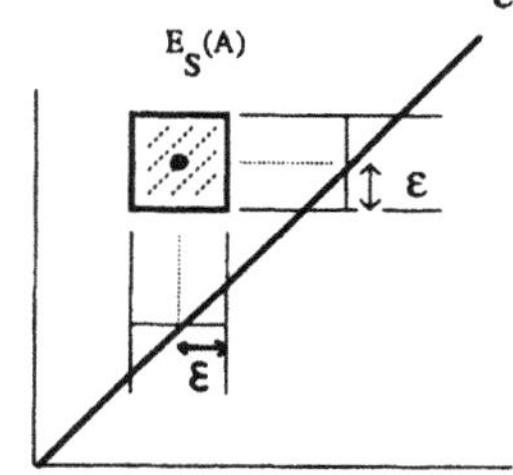

figure 2. The set of fuzzy numbers ε-indifferent with A.

The relation between the difference types of indifference and the indifference regions is:

Strong indifference $\Leftrightarrow$ $I(A,B)=[0,1]$,

Weak indifference $\Leftrightarrow$ $\lambda\in I(A,B)$.

4.3. Operations between fuzzy numbers.

With respect to operations between fuzzy numbers the following properties can be easily proved

$$E_S(A\oplus B)=E_S(A)+E_S(B)$$
$$E_S(rA)=rE_S(A)$$

where $A\oplus B$ and rA are operations between fuzzy numbers and $E_S(A)+E_S(B)$ and $rE_S(A)$ are operations between real intervals.

4.4. Distance measures between fuzzy numbers.

We now propose defining distance measures between fuzzy numbers coherent with the order and indifference generated by the A.V.

Thus, by considering the indifference relation $\simeq_\lambda$ we define the following function

$$d_\lambda:\tilde{\mathbb{R}}/\simeq_\lambda\times\tilde{\mathbb{R}}/\simeq_\lambda\longrightarrow\mathbb{R}_0^+ \qquad d_\lambda(A,B)=|V_S^\lambda(B)-V_S^\lambda(A)|,\ \forall A,B\in\tilde{\mathbb{R}}/\simeq_\lambda.$$

We demonstrated in [3] that d_λ is a distance on $\tilde{\mathbb{R}}/\simeq_\lambda$ and a pseudodistance on $\tilde{\mathbb{R}}$. Moroever, as $V_S^{t\lambda+(1-t)\mu}(A)=tV_S^\lambda(A)+(1-t)V_S^\mu(A)$, $\forall\lambda,\mu,t\in[0,1]$, then the following inequality can easily be proved:

$$d_\lambda(A,B) \leq \lambda d_1(A,B)+(1-\lambda)d_0(A,B). \qquad (1)$$

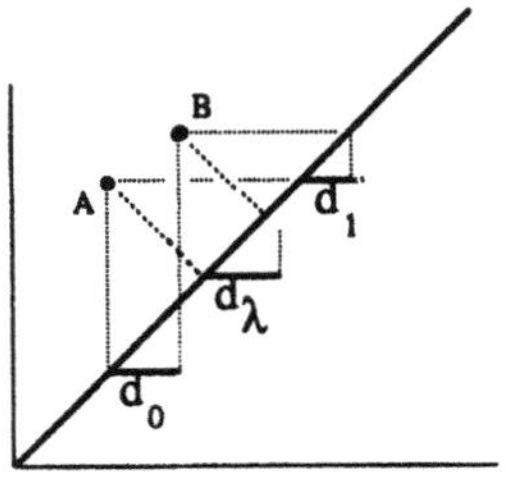

figure 3. The bold line represents d_λ.

By considering the maximun of the second part of this inequality

$$\max_\lambda \{\lambda d_1(A,B)+(1-\lambda)d_0(A,B)\}=\max\{d_0(A,B),d_1(A,B)\}$$

we get

$$d_\lambda(A,B) \leq \max\{d_0(A,B),d_1(A,B)\}. \qquad (2)$$

By other hand, the distance measure can be calculate as

$$d_\lambda(A,B)=\begin{cases} V_s^\lambda(B)-V_s^\lambda(A) & \text{si } \lambda\in R(A,B) \\ 0 & \text{si } \lambda\in I(A,B) \\ V_s^\lambda(A)-V_s^\lambda(B) & \text{si } \lambda\in R(B,A) \end{cases}$$

where $R(\bullet,\bullet)$ and $I(\bullet,\bullet)$ are the dominance and indifference relations, respectively.

The distance measure d_λ is associated to the weak indifference $\simeq_\lambda$. From the d_λ, we can now define distance measures associated to the strong indifference $\cong$. Thus, we define the following functions:

$$d_*:\tilde{\mathbb{R}}/\cong \times \tilde{\mathbb{R}}/\cong \longrightarrow \mathbb{R}_0^+ \qquad d_\infty:\tilde{\mathbb{R}}/\cong \times \tilde{\mathbb{R}}/\cong \longrightarrow \mathbb{R}_0^+$$

$$d_*(A,B)=\int_0^1 d_\lambda(A,B)d\lambda \quad \text{and} \quad d_\infty(A,B)=\sup_\lambda d_\lambda(A,B)$$

with $A,B\in \tilde{\mathbb{R}}/\cong$. In [3] we checked that d_* and d_∞ are distances measures on $\tilde{\mathbb{R}}/\cong$ and pseudistances on $\tilde{\mathbb{R}}$.

Using the inequality (1) we obtain

$$d_*(A,B) \leq \frac{d_0(A,B)+d_1(A,B)}{2} \qquad (3)$$

and using the inequality (2)

$$d_\infty(A,B)=\max\{d_0(A,B),d_1(A,B)\}. \qquad (4)$$

This last equality characterize the distance measure d_∞. By contrast (3) is not enough to know d_*. For this, we consider the following two cases:

-When the dominance region coincides with the interval unity, $R(A,B)=[0,1]$, then

$$d_*(A,B)=\frac{d_0(A,B)+d_1(A,B)}{2}.$$

In this case, the upper bound obtained by (1) for d_*, is reached.

-The dominance region is $R(A,B)=[0,\lambda_{AB}]$ with $\lambda_{AB}\in[0,1)$. In this case λ_{AB} must be the indifference point

$$\lambda_{AB}=\frac{V_S^0(B)-V_S^0(A)}{W_S(A)-W_S(B)},$$

and therefore

$$d_*(A,B)=\frac{d_0^2(A,B)+d_1^2(A,B)}{2(d_0(A,B)+d_1(A,B))}.$$

For other cases, we can use the symmetric property of the distance measure, and finally we obtain the following identity:

$$d_*(A,B)=\begin{cases}\dfrac{d_0(A,B)+d_1(A,B)}{2} & \text{if } R(A,B)=[0,1] \text{ or } \emptyset\\[2ex] \dfrac{d_0^2(A,B)+d_1^2(A,B)}{2(d_0(A,B)+d_1(A,B))} & \text{otherwise.}\end{cases}$$

The following implication can be easily proved, and it shows a relation between the ε-indifference defined in section 4.2. and the distance measure d_∞.

$$\forall A,B\in\tilde{\mathbb{R}}\quad A\cong_\varepsilon B \Leftrightarrow d_\infty(A,B)\le\varepsilon.$$

By using this last equivalence, figure 2 represents also the neighbourhood of the fuzzy number A with the distance measure d_∞, $C_\varepsilon(A)=\{B\in\tilde{\mathbb{R}}/d_\infty(A,B)\le\varepsilon\}$.

Until now, we have defined three distance measures between fuzzy numbers by using the A.V. and the indifference relations considered. From now, we are interested to investigate the relation between these distance measures and associated measures on real intervals.

By using (4), the following equality can be easily obtained

$$d_\infty(A,B)=d^1(E_S(A),E_S(B))$$

where d^1 is the Hausdorff metric on the real intervals and $E_S(\bullet)$ is the mean value of a fuzzy number.

Thus, the A.V. allows the substitution of fuzzy numbers by real intervals again. In this case (when the distance measure d_∞ is used) this substitution generates a schema where the interval distance measure is known. By contrast, d_λ and d_* can give the same situation but with interval distance measures defined by:

$$A_1=[a_1,a_2],\ A_2=[b_1,b_2]$$

$$d^2(A_1,A_2)=\begin{cases}\dfrac{(b_1-a_1)^2+(b_2-a_2)^2}{2(|b_1-a_1|+|b_2-a_2|)} & \text{if } a_1\le b_1\le b_2\le a_2 \text{ or } b_1\le a_1\le a_2\le b_2\\[2ex] \dfrac{|b_1-a_1|+|b_2-a_2|}{2} & \text{otherwise}\end{cases}$$

$$d^3(A_1,A_2)=|\lambda(b_2-b_1-a_2+a_1)+b_1-a_1|$$

obviously d^2 is a distance measure on the set $\mathcal{C}$ of real closed intervals and d^3 is a distance measure on $\mathcal{C}/\simeq_\lambda$. The following identity can be easily proved

$$d_*(A,B)=d^2(E_S(A),E_S(B)) \text{ and } d_\infty(A,B)=d^3(E_S(A),E_S(B)),$$

and the three distance measures defined through the A.V. on fuzzy numbers can be obtained from associated distance measures on the mean values of the fuzzy numbers. Moreover, the associated distances are the original measures but restricted to real intervals.

5. CONCLUDING REMARK.

In summary, the A.V. allows us to generate different processes on fuzzy numbers, and for each one fuzzy numbers have been replaced by real intervals representing its mean value. That is, let P be the process generated from the A.V. on two fuzzy numbers A and B, then

$$P(A,B)=P^*(E_S(A),E_S(B))$$

with P^* an associated process to P on real intervals. P^* can be obtained, in the examples studied, by restricting P to real intervals.

This substitution has two interesting properties, the first one is that it allows us to interpret any process on fuzzy numbers using the A.V. as an equivalent process on mean values (e.g. ranking fuzzy numbers ⟶ ranking its mean values, indifference between fuzzy numbers ⟶ indifference between mean values, and so on). The second one is that it allows us to visualize any process on fuzzy numbers using the Average Value through an useful graphic representation.

REFERENCES

[1] Bortolan, G. and Degani, R. (1985) A review of some methods for ranking fuzzy subsets, Fuzzy Sets and Systems 15, 1-19.

[2] Campos, L.M. and Gonzalez A. (1989) A subjective approach for ranking fuzzy numbers, Fuzzy Sets and Systems 29, 145-153.

[3] Campos, L. and Gonzalez, A. Further contributions to the study of the average value for ranking fuzzy numbers, to appear in International Journal of Approximate Reasoning.

[4] Choquet, G. (1953) Theory of capacities, Ann. Inst. Fourier 5, 263-272.

[5] Dubois D. and Prade H. (1987) The mean value of a fuzzy number, Fuzzy Sets and Systems 24, 279-300.

[6] Gonzalez, A. (1990) A study of the ranking function approach through mean values, Fuzzy Sets and Systems 35, 29-41.

[7] Moore, R. (1966) Interval analysis, Prentice Hall, Englewood Cliffs, N.J.

4. NON-MONOTONIC REASONING

BEYOND SPECIFICITY*

HENRY E. KYBURG, JR.
University of Rochester
Rochester, N.Y. 14627

Abstract

A number of writers have suggested that specificity can be called upon to adjudicate competing default inferences. In the foundations of statistics, specificity is one of several ways to adjudicate the claims of competing reference classes. This suggests that in default inferences also other principles than specificity may be needed. This paper gives examples substantiating this suggestion, and provides formulations of the few other principles needed.

Keywords

Specificity, Probability, Dominance, Inheritance, Reference Class, Ordering Defaults

1. INTRODUCTION

It has been suggested (Poole, 1985; Touretzky, 1984; Touretzky et al, 1987; Neufeld and Pool, 1988; Bacchus, 1988; Etherington, 1987) that considerations suggested by probability theory may throw light on non-monotonic inference. The problems of non-monotonic infernce and probability do seem to be very close to each other; it is the purpose of this paper to explore that relation further. It may seem that we are using probabilistic considerations to throw further obscurity on non-monotonicity. To avoid this impression, we shall first present intuitive cases of non-monotonic inference, without reference to probability, and only subsequently point out the connections.

The general nature of the problem we are considering is the following: We have a set of premises in our body of knowledge or knowledge base, from which we would ordinarily expect to be able to infer a certain statement *S*. But there is another set of statements, that may equally well be regarded as being part of our knowledge base, in that same situation, from which we could infer the denial of *S*. In many cases, what we suppose ourselves to know in the first place entails that these conclusion upsetting statements are part of our knowledge. (Note: this is not just a matter of not being able to infer *S*, but a matter of being able to infer the denial of *S*.)

The classical case is that of Tweety the penguin. We want to infer that Tweety does not fly, even though we know at the same time (ipso facto, we might even say!) that Tweety is a bird and that typically birds fly. By themselves, these facts would warrant the opposite conclusion, namely: Tweety flies.

One approach to this problem, suggested in various forms by the authors cited above, is to observe that when these two possible arguments clash, we prefer the argument with the "most specific" premises. In this case that specificity picks out the argument that classifies Tweety as a penguin rather than the argument that merly classifies Tweety as a bird. This situation is represented by figure 1.

Similar considerations have been called upon to resolve conflicts in Hierarchical nets (Touretzky et al, 1987) and in systems of Defeasible Reasoning (Nute, 1989).

Historically it is worth noting that the idea of specificity goes back at least to Reichenbach (Reichenbach, 1949). The term "specificity" was employed by Hempel (Hempel, 1968), and there has been some philosophical discussion of the notion (for example, Kyburg, 1970). It is also worth observing that exceptions -- even singleton exceptions -- are conveniently handled by reference to specificity. If Tweety is the only penguin in the world who can fly, and we happen to know it, then the facts that Tweety is in {Tweety} (or has the property of being Tweety), and that all the members of {Tweety} (or everything with the property of being Tweety) flie(s), lead us to the conclusion that Tweety flies after all.

Connections have been made to circumscription: There can be abnormal birds (penguins), and then there can be abnormal penguins (who fly), and then perhaps some of these flying penguins are abnormal, ...

We may present the problem more formally this way: We have a knowledge base containing premises

$$P_1, P_2, \ldots P_n$$

On this basis we want to obtain the conclusion C. But if our knowledge base contains the P_i, it also contains

$$R_1, R_2, \ldots, R_k$$

either because they are implied by the P_i or because (like "Birds typically fly") they represent natural assumptions. But given the R_i in our knowledge base, in the absence of the P_i, we would conclude the denial of C, $\sim C$.

2. SIMPLE SPECIFICITY

Specificity resolves conflicts of the form illustrated in figure 1.

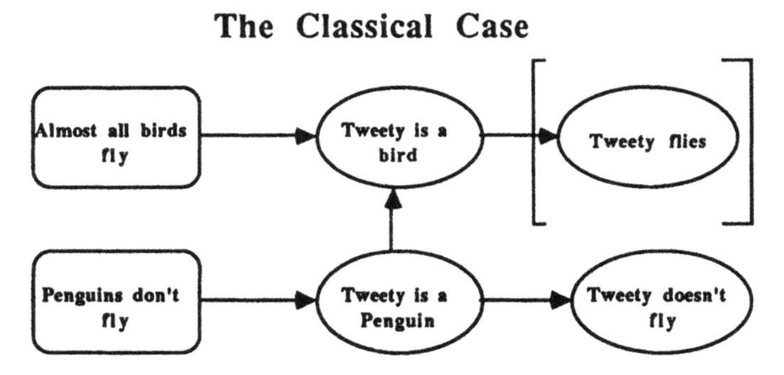

The first inference is defeated by the second, since part of our knowledge includes the fact that penguins are a subset of birds.

figure 1

But even minor variations call for something more. Even in the case of Tweety, this can be easily seen. If we know that Tweety is a penguin, and that penguins don't typically fly, then we also know

(1) "{Tweety} is a subset of birds."

And in general we know that

(2) "Typically subsets of birds are also subsets of the set of flying objects."

From which it is natural to infer that Tweety flies.

Now this may be thought to be a strange and unnatural way of expressing our knowledge about the ability of birds to fly. No doubt. But it represents a fairly straight-forward logical translation of the R-premises of the first example. (If one didn't like the set of which Tweety is the only member, one could talk about the property of being Tweety.)

True, our knowledge base cannot be closed under logical implication, but it seems artificial to rule out any particular forms of inference as illegitimate. In particular, the move from considering a set of objects to considering a set of sets of objects is exactly the move which is required for Bayesian inference, as will be seen shortly.

So it is not unreasonable to suppose that (1) and (2) are in our knowledge base. What prevents the inference to "{Tweety} is a subset of the flyers?"

In the original form of Tweety's problem, specificity did it. That won't work here, since penguins are not a subset of the set of sets of birds. (Sets of anything are abstract objects; penguins aren't.) But we can employ almost the same principle here, as illustrated in figure 2.

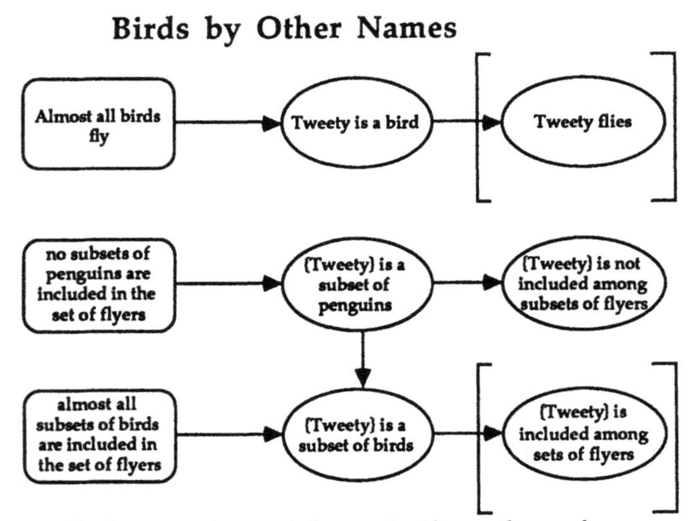

The first pattern gives way to the second, not because the second refers to a subset of the birds referred to in the first (it doesn't), but because the first pattern is reflected in the third, and the third is defeated by the second according to the subset principle.

figure 2

Here is a rough statement of the specificity principle that takes care of the two cases we have considered so far:

PRINCIPLE 1: Generalized Specificity

IF "A's are typically C' s" and "x is an A"are in our knowledge base KB, and
"B's are typically D's" and "y is a B" are in KB, and
"x is a C if and only if y is not a D" is in KB,
THEN the first inference (to "x is aC") is to be prefered to the second
(to "$y$is not aD")whenever we also know that there is a subset B^* of B such that
"y is in B^*" and"Typically members ofB^* are not members of D" are in KB.

This rule applies to the first example, with penguin for $B^* = A =$ Penguin, bird for B , flies for C, and $x = y =$ Tweety. It applies to the second example with sets of penguins for B^* and sets of non–fliers for D , and {Tweety} for y.

3. BAYESIAN CASES

The "artificial" form of the Tweety example is rather Baroque, and unlikely to arise except in the mind of a perverse logician or a perfectly logical computer program. We could keep the argument simple and intuitive in form of the principle if we were to regiment our language adequately,

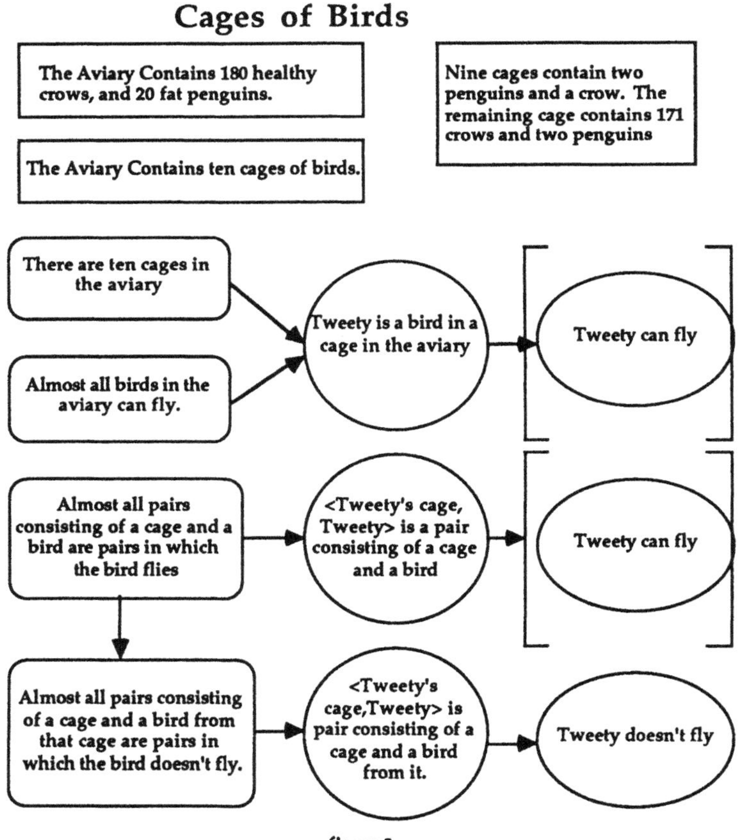

figure 3

though this would leave open the question of where in the regimentation significant power is lost. The second counter–example to simple specificity, represented in figure 3, is much more natural, and in can no way be construed as a matter of "specificity." It reflects the conflict between straight–forward statistics and Bayesian statistics.

In the classical approach to statistics, what is called "direct inference" leads from knowledge of the frequency (or measure) in a reference set, to an assignment of probability to a specific case. This is not entirely uncontroversial, but people — even those who deny the validity of the move in general — frequently act on its basis. The Bayesian approach considers several possible reference classes, and takes the assignment of probability to the specific case to be a weighted average of the frequencies among the several possible reference classes. This if a chip is to be drawn from three bags, one of which contains 10% white chips, and two of which contain 60% white chips, the Bayesian probability to be assigned to the proposition that the draw results in a white chip is to be $1/3*0.10 + 2/3*0.60$.

As we shall see, this assignment of probability does admit of a classical reconstruction. Suppose that we know that a room contains ten cages, nine of which contain one healthy sparrow and two fat penguins, and the tenth cage containing 171 sparrows and only two penguins. We select a cage, and then a bird (whom we call Tweety) from the cage. Typically, this bird will not be a flier. But note: we also know that the selected bird is a bird in the room, and typically birds in that room do fly.

Observe that there is no subset (no specification) of the set of birds in the room to which we know that Tweety, the selected bird, belongs, and in which the typical bird is a non–flier. "For all we know," Tweety was selected from the tenth cage. But "for all we know" Tweety might have been selected from the fifth cage.

It is no answer to say that "most of the time" the selected bird will have come fron one of the cages other than the tenth, whatever "most of the time" means here. Perhaps we are talking of a unique experiment. In any event, we are not talking about statistics, but about inference.

We are often told that "typicality" should not be cashed out in terms of actual frequencies. But the typical bird in the room flys, just as does the typical bird in the world at large. If we are just told that Tweety is a bird in the room, we correctly infer that she flies. It is only when we are told *how* Tweety is selected that we reverse our judgement: typically birds selected in that way don't fly. But what principles govern this reversal?

A general form of an appropriate rule is this:

SECOND PRINCIPLE: The Bayesian Principle

IF "A's are typically C's" and "x is an A" are in KB, and
"B's are typically D's" and "y is a B" are in KB, and
"x is a C if and only if y is not a D" is in KB,
THEN the first inference is to be prefered to the second whenever we can find a cross product $B^* \times B$, a pair $\langle z, y \rangle$, and a predicate of pairs, D^*, of which the following are known to be true in KB:

> $\langle z, y \rangle$ is D^* if and only if y is D ;
> $\langle z, y \rangle$ is in $B^* \Diamond B$;
> our knowledge about $B^* \times B$ and D^* matches our knowledge about B and D ;
> and, finally, there is a subset E of $B^* \times B$ such that $\langle z, y \rangle$ belongs to it, and our knowledge about E and D^* matches our knowledge about A and C .

In the case at hand, C is the property of pairs consisting of a cage and a bird that holds when the bird can't fly. B^* is the set of cages. $E = A$ is the subset of $B^* \times B$ that satisfies the condition that the bird member of the pair comes from the cage member of the pair.

Two remarks on this construction should be made. First, it is easy to see that the first principle — the generalized specificity principle — can be derived from the Bayesian principle. (Let B be arbitrary.) Second, not only does the converse not hold, but it could only be *made* to hold by requiring that our initial statistical knowledge included all tuples that might *possibly* turn out to be relevant. This is not a condition that it is plausible to demand. On the other hand, the use of weighted frequencies or measures is typical of any probabilistic reasoning (even the most classical), so that we really do need principles to guide us in the case of conflict.

4. SUBSAMPLES

The final form of inference to be considered is a bit more specialized than the preceding two, but again cannot be explained in terms of specificity. The situation is illustrated in figure 4.

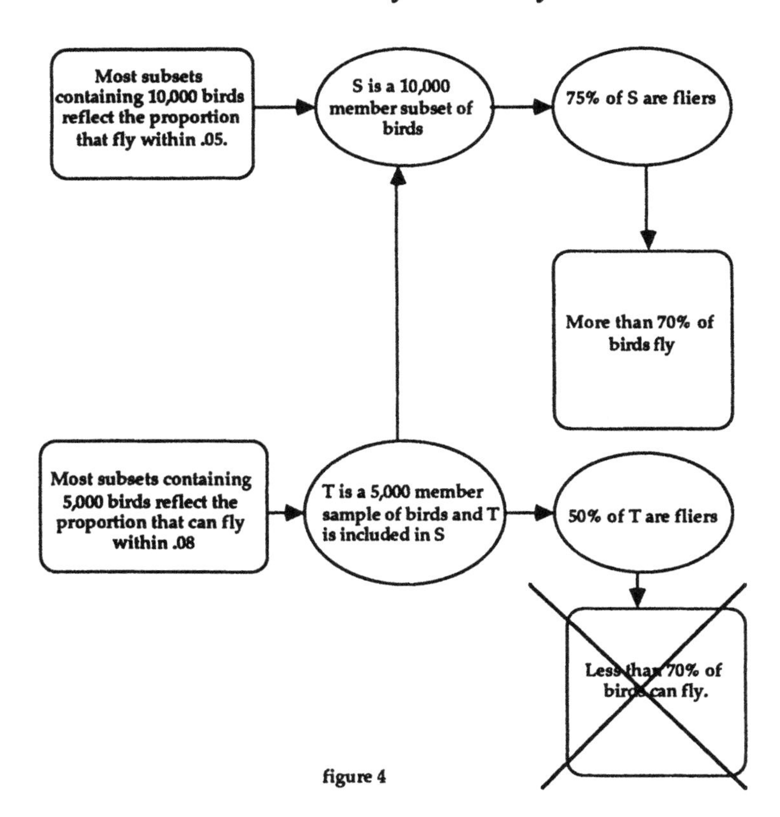

figure 4

Suppose we have examined a sample of 10,000 birds, and observed that 75% of them are fliers. Under the right conditions, it is reasonable for us to conclude that more than 70% of birds fly, since under the right conditions samples of 10,000 typically represent the populations from which they are drawn.

Therefore we also have in our knowledge base knowledge of a sample of 5,000 birds, of which only 50% are fliers. Under the right conditions, it would be reasonable to conclude that less than 70% of birds fly, since samples of 5,000 typically represent the populations from which they are drawn.

The original sample sample is "typical" if and only if the subsample is not "typical." We have two arguments just parallel to the two arguments with which we started: Tweety is a typical bird (and so flies); Tweety is a typical penguin (and so does not fly). Either inference alone seems perfectly reasonable, and in fact we want to be able to make either, considered by itself. It is only when we have the combination of the premises of the two inferences that we want to be able to inhibit one of them.

As we have told the story, if the "right conditions" have been met for the first inference, we want to be able to show that the "right conditions" cannot be met for the second inference. We want to prefer the first inference. Again, "specificity" doesn't help; the generic description of the large sample is no more or less "specific" than the generic description of the small sample.

But the fact that the sample of the second inference is included in the sample of the first inference provides a reasonable criterion.

We can state the rule as follows:

PRINCIPLE THREE: The Subsample Principle

IF"*A*'s are typically *C*'s" and "*x* is an *A*" are in KB, and
"*B*'s are typically *D*'s" and "*y* is a *B*" are in KB, and
"*x* is a *C* if and only if *y* is not a *D*" is in KB,
THEN
the first inference is to be prefered to the second if it is possible to construe the two inferences as statistical inferences differing only in that the sample on which the second inference is based a subset of the sample on which the first inference is based.

The particular case under discussion corresponds to this rule in an obvious way, taking *A* to be the set of large samples, *B* the set of smaller samples, *C* the set of representative samples, and *D* the set of highly atypical samples. Note that the large sample is representative just in case the small sample is highly atypical.

5. DISCUSSION

Are there other forms of competition among plausible inferences than those mentioned here? I think not. My evidence is that the three forms of preferences just illustrated (actually, two, since "specificity" can be construed as a special case of the cross product construction) are the only forms that seem to be necessary in accounting for the choice of a reference class in an evidential probability theory. See (Kyburg, 1983; Kyburg, 1988) for more details. There may be other structures that should be taken account of that have not yet been noticed. But for the moment, at least, these three correspond to our most clear–cut intuitions.

It may be argued that "typicality" and "frequency" are different ideas, and may not be subject to the same rules and constraints. This may be so, but there is some reason to think that they cannot conflict too severely. In fact, in the case of any conflict, it is difficult to see why frequency should not take precedence. When it is argued that *A*'s are "typically" *C*'s, even though most *A*'s are not *C*'s, we may often find an underlying assumption, such as that *in the long run* most A's will turn out to be C's, or that in the long run most A's would turn out to be C's, other things being equal.

In the first place it is clear that we cannot claim that A's are typically C's, when most naturally occuring A's aren't C's — at least not without a long and fairly complicated story. We might suppose that x is an A , that most A's are C's, and at the same time that x is a B , while B's are typically not C's. The upshot of this knowledge would depend on the relation between A and B . If it is one of those relations addressed above, then the considerations adumbrated there should determine the outcome rather than a contest between "typicality" and "frequency".

Another problem that may seem worrisome is whether the process of specification, Bayesianization, or sample expansion will always end. Specification and sample expansion clearly do: it is no big deal to constrain our references to finite sets; and no finite set can admit of arbitrarily fine specification. Before long we must come to that most specific class: that class of which the object in question is the only member. Having come to this class, and been left, therefore, with a totally indeterminate frequency, Analogously, our samples can only be so big. There is a largest, and we may assume that our knowledge base knows about it. The problematic case concerns cross product formation. But even this case cannot lead us on indefinitely; the complexity of any compound experiment must be bounded.

Can the three rules come into conflict? The answer is "No," since the rules work negatively: they do not specify that such and such an argument is a good one, but merely that if there are two conflicting arguments that have the forms described, one of then can not be accepted. This is not to say that the other can. But since (I am proposing) these are all the sources of conflict we can have, at least one argument will remain undefeated.

We thus conclude, tentatively, that the enumerated considerations are all the ones that are relevant to the adjudication of the claims of competing non-monotonic inferences. We also conclude, definitively, that specificity, even if construed quite broadly, doesn't do the trick. And we conclude, quite generally, that however "typicality" is construed, statistical considerations can thrown light on inferences that depend on typicality.

* Research on which this work was based was supported in part under a contract from the U. S. Army Signal Warfare Center.

References:

Bacchus, Fahiem:(1988) "A Heterogeneous Inheritance System Based on Probabilities," mimeo.

Etherington, David W.: (1987) "More on Inheritance Hierarchies with Exceptions: Default Theories and Inferential Distance," AAAI-87, Morgan Kaufman, Los Altos, 352-357.

Hempel, Carl G.:(1968) "Maximal Specificity and Lawlikeness in Probabilisitic Explanation," *Philosophy of Science* **35**, 116-133.

Kyburg, Henry E., Jr.: (1988) "Probabilistic Inference and Non-Monotonic Inference," Shachter, Ross, and Levitt, Todd (eds): The Fourth Workshop on Uncertainty in Artificial Intelligence, pp 229-236.

Kyburg, Henry E., Jr.: (1983) "The Reference Class," Philosophy of Science 50, pp 374-397.

Kyburg, Henry E.,Jr.: (1970) "More on Maximal Specificity," *Philosophy of Science* **37**, pp.295-300.

Neufeld, Eric, and Poole, David: (1988) "Probabilistic Semantics and Defaults," Shachter, Ross, and Levitt, Todd (eds): The Fourth Workshop on Uncertainty in Artificial Intelligence, 1988, pp 275-282.

Nute, Donald (1989) "Defeasible Logic and Inheritance Hierarchies with Exceptions."

Poole, David L.: (1985) "On the Comparison of Theories: Preferring the Most Specific Explanation," IJCAI 85, Morgan Kaufmann, Los Altos, 144-147.

Reichenbach, Hans (1949) *The Theory of Probability*, University of California Press, Berkeley and Los Angles.

Touretzky, D. S., Horty, J. F., and Thomason, R. H.: (1987) "A Clash of Intuitions: the Current State of Non-Monotonic Multiple Inheritance Systems," IJCAI-1987, 476-482.

Touretzky, D. S.: (1984) "Implicit Ordering of Defaults in Inheritance Systems," AAAI-84, 322-325.

ABOUT THE LOGICAL INTERPRETATION OF AMBIGUOUS INHERITANCE HIERARCHIES

E. Grégoire

Unité d'Informatique
Univ. de Louvain
Place Sainte Barbe 2
B-1348 Louvain-la-Neuve
Belgium

UMIACS
Univ. of Maryland
A.V. Williams Bldg
College Park MD20742
USA

Abstract

In this paper, we present a new approach to the logical formalization of ambiguous inheritance hierarchies. This approach involves the translation of inheritance nets into their corresponding hierarchical and stratified logic programs. It allows one to express the semantics of these nets in terms of most major nonmonotonic logics. A theorem is established showing that this approach can be applied to a broad class of inheritance theories dealing with acyclic nets and giving rise to unique extensions.

Keywords: knowledge representation, inheritance, nonmonotonic logics, default reasoning, logic programming

1 Introduction

An important form of uncertainty in knowledge-based systems concerns the conflicting information in ambiguous inheritance hierarchies. Since a knowledge engineer may choose between diverging principles and policies to define what can be inferred from an ambiguous inheritance net (see e.g. [Touretzky et al. 87]), it is of prime importance to provide him with a description and an intelligible semantics of the different available inheritance theories. Logic has long been advocated as a powerful, rigorous and well-understood formalism to express this semantics, which is often poorly understood or expressed in uncommon or less rigorous formalisms [Hayes 79]. Since the reasoning involved in inheritance nets can be defeasible, the logic we need here must be nonmonotonic.

In this paper, we address the issue of providing a semantics for a broad class of inheritance theories, using nonmonotonic logics. To this end, we generalize the approach that has been developed in [Grégoire 89a,89b] with respect to the inheritance theory described in [Horty et al. 87,90]. We show that it can be extended to apply to a broad class of inheritance theories giving rise to unique extensions, i.e. unique maximal sets of inferrable assertions. This unifying approach involves the translation of inheritance nets into their equivalent hierarchical and stratified logic programs. Since the declarative semantics of such programs can be expressed in terms of most major nonmonotonic logics, it follows that the inheritance nets under consideration can be provided with such a semantics too.

First, let us describe the class of inheritance theories we consider in this paper and review the concept of hierarchical and stratified logic programs.

2 A broad class of inheritance theories

The inheritance nets that we are considering here are *acyclic* nets consisting of nodes connected by means of defeasible or non-defeasible positive and negative inheritance links. Positive links are of the form 'x IS-A y' (written $x \rightarrow y$), whose intuitive meaning is '(an) x is (normally) a y'. Negative links are of the form 'x IS-NOT-A y' (written $x \not\rightarrow y$), whose intuitive meaning is '(an) x is (normally) not a y'. A path is defined as a sequence of joined links where only the last link can be negative. It is well known that a topological sort of an acyclic net Γ defines a total order $\mathcal{T}$ between the nodes of Γ in such a way that $\mathcal{T}(x) < \mathcal{T}(y)$ if and only if node x precedes node y with respect to this topological sort of Γ.

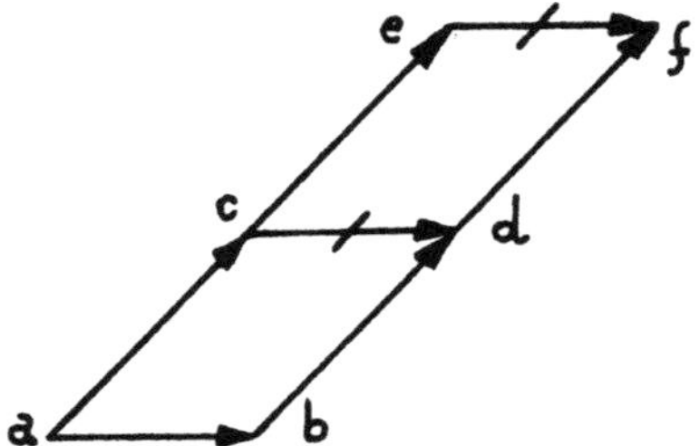

Figure 1. A (simple) ambiguous inheritance net.

In this paper, we consider a broad class of inheritance theories, called *univocal inheritance theories*. These theories deal with acyclic nets, give rise to unique extensions and allow one to infer only positive and negative assertions of the form 'x IS-A y' and 'x IS-NOT-A y' where x and y are nodes in the considered net such that $\mathcal{T}(x) < \mathcal{T}(y)$ for any topological order $\mathcal{T}$ of the nodes in the net. The most interesting theories of this family are the *skeptical theories of inheritance*. Credulous theories of inheritance that always select one extension among the different possible ones also belong to this family.

Let us illustrate the concept of skeptical inheritance by means of a basic example. In the net of Figure 1, the presence of the path $a \rightarrow b \rightarrow d$ gives some tentative evidence that one should be allowed to infer 'a IS-A d' (i.e., '(an) a is (normally) a d') since this path expresses the idea that (an) a is (normally) a b and that a b is (normally) a d. On the other hand, the presence of the path $a \rightarrow c \not\rightarrow d$ gives some tentative evidence that one should be allowed to infer 'a IS-NOT-A d' (i.e., '(an) a is (normally) not a d'). Since we do not want to infer this contradictory information at the same time and since there is no reason to prefer one path over the other one, a *skeptical inheritance reasoner* will conclude neither 'a IS-A d' nor 'a IS-NOT-A d'. Actually, it will avoid inferring any assertion about the relationship between a and d. Different skeptical theories can be defined depending on the way we decide to propagate this form of skepticism inside inheritance nets. For example, in the net of Figure 1, we can decide to propagate the ambiguity about '(an) a being a d' upwards to all nodes y that are accessible from d by means of a path (here, to node f) and avoid inferring either 'a IS-A y' or 'a IS-NOT-A y' [Grégoire 89c] [Stein 89]. Instead, according to the skeptical inheritance theory described in [Horty et al. 87,90], we must accept to infer 'a IS-NOT-A f' since, in this theory, we decide that the path $a \rightarrow b \rightarrow d \rightarrow f$ is not to be taken into account because its segment $a \rightarrow b \rightarrow d$ has been classified ambiguous. In [Touretzky et al. 87], [Grégoire 89c] and [Stein 89], several skeptical inheritance theories have been described according to the 'level' of skepticism we want the theory to obey, according to the exact form of preference for 'more specific information' we want the theory to implement, and according on whether inheritance is understood as a generalization or specialization process.

3 Hierarchical and stratified logic programs

Let us consider finite sets of object, function and predicate symbols with at least one object symbol and at least one predicate symbol. By definition, an *atom* P is an atomic formula formed in the usual way from these symbols and from object variables whereas a *literal* is an atom P or its negation $\neg P$. A *clause* is a formula of the form

$$\neg P_1 \wedge ... \wedge \neg P_n \wedge Q_1 \wedge ... \wedge Q_m \supset R$$

where $R, P_1, ..., P_n, Q_1, ..., Q_m$ are atoms and m and n are greater or equal to 0. $\neg P_1 \wedge ... \wedge \neg P_n \wedge Q_1 \wedge ... \wedge Q_m$ and R are called the *body* and the *head* of the clause, respectively. A *logic program*, or program, consists of the conjunction of the universal closure of a finite set of clauses.

A logic program Π is *stratified* [Przymusinski 88a] [van Gelder 88] [Apt et al. 88] if and only if it is possible to decompose the set $\mathcal{S}$ of all predicates occurring in Π into disjoint sets $\mathcal{S}_1, ..., \mathcal{S}_r$ called *strata*, so that for every clause, we have that

$$\forall i \in [1, n] \; : \; stratum(P_i) < stratum(R)$$

$$\forall j \in [1, m] \; : \; stratum(Q_j) \leq stratum(R)$$

where $stratum(Z) = i$, if the predicate symbol Z belongs to $\mathcal{S}_i$. Any particular decomposition $\mathcal{S}_1, ..., \mathcal{S}_r$ of $\mathcal{S}$ satisfying the above conditions is called a *stratification* of Π. A stratification can be viewed as giving priority levels between predicates, and thus between strata. A positive (rspt. negated) literal can only occur in the body of a clause when it refers to a predicate belonging to a stratum of a (rspt. strictly) higher priority than the stratum to which belongs the predicate occurring in the head of the clause. Note that under this convention, $stratum(P) < stratum(Q)$ means that P belongs to a higher stratum than Q. Intuitively, when one makes use of negation in a stratified logic program, one can only refer to an *already known* predicate, i.e. a predicate which has been defined in a higher-priority stratum. A *hierarchical logic program* Π is a stratified logic program where all (positive and negative) literals belonging to the body of any clause $\mathcal{A}$ refer to predicates belonging to strictly higher-priority strata than the stratum to which belongs the predicate occurring in the head of $\mathcal{A}$.

It is well-known that the declarative semantics of a stratified logic program Π can be expressed by means of its *perfect Herbrand model*, i.e. the unique Herbrand model of Π that results from a prioritized minimization of the predicates occurring in Π, where the scale of priorities corresponds to the priority levels given to these predicates by any stratification of Π (minimizing predicates belonging to higher-priority strata first) [Przymusinski 88a,88b,88c]. It is also well-known that this semantics can be re-expressed as a fixed-point semantics [Apt et al. 88] and in terms of most major nonmonotonic logics [Bidoit and Froidevaux 87] [Gelfond 87] [Przymusinski 88a,88c] [Lifschitz 88].

4 A translation-oriented approach to the semantics of inheritance nets

In this section, we show that, given any univocal inheritance theory, any acyclic inheritance net Γ can be translated into an *equivalent* hierarchical logic program Π and thus be provided with a semantics in terms of most major nonmonotonic logics.

In the following theorems, x and y are nodes in Γ such that $\mathcal{T}(x) < \mathcal{T}(y)$ for any topological order $\mathcal{T}$ of the nodes in Γ. $\Gamma \models x \rightarrow y$ (rspt. $\Gamma \models x \not\rightarrow y$) means that the inheritance theory under consideration allows one to infer $x \rightarrow y$ (rspt. $x \not\rightarrow y$) from Γ. Atoms of the forms Z and $\overline{Z}$ will be introduced in the logic program to correspond with node z in Γ; intuitively, atoms of the form $\overline{Z}$ will be introduced to express the concept 'not being a z'. (Π, X) represents the logic program Π where an additional clause (a ground atom) X is introduced. When Π is ground and stratified, $\Pi \models_{\mathcal{M}} Z$ means that the atom Z is *true* in the unique perfect model $\mathcal{M}$ of Π. This is noted $Z \equiv_{\mathcal{M}}$ *true.*

Theorem 1.
Given any univocal inheritance theory, there exists a translation schema that associates a hierarchical logic program Π *with each acyclic net* Γ *in such a way that*

- $\Gamma \models x \rightarrow y$ iff $(\Pi, X) \models_{\mathcal{M}} Y$
- $\Gamma \models x \not\rightarrow y$ iff $(\Pi, X) \models_{\mathcal{M}} \overline{Y}$

The proof of this theorem is given in the appendix; it makes use of a translation-schema that can be used to provide any acyclic net with a nonmonotonic logic semantics. It is general enough to apply to *any* univocal inheritance theory, even the most complex and unintelligible ones. Therefore, it does not take into account any rule that would be used by a specific inheritance theory to define what is to be inferred from a net. Indeed, it would always be possible to define another theory that would not obey this rule. It should thus be clear that a specific and optimized translation-schema, which takes into account *how* assertions can be inferred in the inheritance theory, should be devised with respect to each considered theory. In [Grégoire 89a,89b], such a specific and optimized translation-schema is established for the elaborate skeptical theory described in [Horty et al. 87,90]. The real importance of the translation-schema given in this paper is that it allows us to prove general feasibility results.

Let us now present the semantical results that one may establish for each theory by means of this general translation-schema or by means of a more specific one.
Since (Π, X) is ground hierarchical, we immediately derive (see e.g. [Shepherdson 88]) that the semantics of Γ can be expressed by means of predicate completion [Clark 78] (Theorem 3) and that SLDNF-resolution can be used to query (Π, X) (Theorem 2). In other words, Π is a Prolog program corresponding to Γ. Thanks to well-known results allowing one to express the declarative semantics of hierarchical and stratified logic programs in terms of most major nonmonotonic logics, this translation-schema allows one to express the semantics of inheritance nets in terms of most major nonmonotonic logics. For example, from Theorem 1 and from results by [Bidoit and Froidevaux 87], [Gelfond 87], [Przymusinski 88a] and [Lifschitz 88], we directly derive Theorems 4,5,6 expressing the semantics of Γ in terms of default logic [Reiter 80], autoepistemic logic [Moore 85] and prioritized circumscription [McCarthy 86].

Theorem 2.

$$\Gamma \models x \rightarrow y \text{ iff } (\Pi, X) \models_{SLDNF} Y$$
$$\Gamma \models x \not\rightarrow y \text{ iff } (\Pi, X) \models_{SLDNF} \overline{Y}$$

Theorem 3.

$$\Gamma \models x \rightarrow y \text{ iff } comp((\Pi, X)) \models Y$$
$$\Gamma \models x \not\rightarrow y \text{ iff } comp((\Pi, X)) \models \overline{Y}$$

Theorem 4.

$\Gamma \models x \rightarrow y$ iff Y is *true* in the unique minimal model associated with the unique extension of the corresponding default theory $<\mathcal{D}, \mathcal{T}>$

$\Gamma \models x \not\rightarrow y$ iff $\overline{Y}$ is *true* in the unique minimal model associated with the unique extension of the corresponding default theory $<\mathcal{D}, \mathcal{T}>$

(where $\mathcal{T}$ is set of all the positive clauses of (Π, X) and the set $\mathcal{D}$ of default rules is obtained by creating a default rule of the form $P_1 \wedge ... \wedge P_n : M(\neg Q_1), ..., M(\neg Q_m)/R$ for each clause of the form $P_1 \wedge ... \wedge P_n \wedge \neg Q_1 \wedge ... \wedge \neg Q_m \supset R$ in Π.)

Theorem 5.

$$\Gamma \models x \rightarrow y \text{ iff } Y \text{ belongs to the unique stable expansion of } I[(\Pi, X)]$$
$$\Gamma \models x \not\rightarrow y \text{ iff } \overline{Y} \text{ belongs to the unique stable expansion of } I[(\Pi, X)]$$

(where $I[(\Pi, X)]$ is obtained by replacing each negative literal of the form $\neg Y$ in Π by a corresponding negative autoepistemic literal $\neg LY$ ("*Y is not believed*").)

Theorem 6.

$$\Gamma \models x \rightarrow y \text{ iff } Y \text{ is } \textit{true} \text{ in the unique model of } CIRC((\Pi, X); \mathcal{S}_1 > ... > \mathcal{S}_r)$$
$$\Gamma \models x \not\rightarrow y \text{ iff } \overline{Y} \text{ is } \textit{true} \text{ in the unique model of } CIRC((\Pi, X); \mathcal{S}_1 > ... > \mathcal{S}_r)$$

(where $\mathcal{S}_1 ... \mathcal{S}_r$ is a stratification of (Π, X) induced by a topological order $\mathcal{T}$ of Γ.)

Let us illustrate this approach by applying it to the inheritance theory described in [Horty et al. 87,90]. To this end, let us directly apply the corresponding specific translation-schema described in [Grégoire 89a,89b] to the net illustrated in Figure 1. We obtain the logic program $\Pi = \{A \supset B, A \supset C, B \wedge \neg A \supset D, C \wedge \neg A \supset \overline{D}, D \supset F, C \supset E, E \supset \overline{F}\}$. A, B, C, D, E, F are ground atoms associated with the nodes a, b, c, d, e and f, respectively. $\overline{D}$ and $\overline{F}$ are additional predicates that are associated with the statements '... is (normally) not a d', and ' ... is (normally) not a f', respectively.
Thanks to theorem 4, we can express the semantics of Γ in terms of default logic. We obtain that Γ allows one to infer $x \rightarrow y$ (rspt. $x \not\rightarrow y$) iff Y (rspt. $\overline{Y}$) is *true* in the unique minimal model associated with the unique extension of the default theory $< \mathcal{D}, \mathcal{T} >$, where the set of defaults $\mathcal{D}$ is given by $\{B\ \mathrm{M}\neg A/D,\ C\ \mathrm{M}\neg A/\overline{D}\}$ and $\mathcal{T}$ results from the union of $\{X\}$ and of the set of positive clauses of the program Π.

5 Conclusion

In this paper we have presented a new approach to the semantics for ambiguous inheritance nets using nonmonotonic logics. This approach can be applied for a broad class of inheritance theories. This approach allows one to express this semantics in terms of most major nonmonotonic logics and derive Prolog programs that correspond to the nets.

Acknowledgements

I would like to thank J. Horty, P. Siegel and M. Sintzoff for valuable discussions on the subject of this paper. This research has been supported by the Belgian SPPS under grants RFO/AI/15 and RFO/AI/15/2 and has been completed during a stay at UMIACS (University of Maryland Institute for Advanced Computer Studies).

Appendix.
Translation-schema and proof of Theorem 1.

Translation-Schema.
Initially, the logic program Π is empty. Let $\mathcal{T}$ be a topological order of the nodes in Γ. Let us examine the nodes y of Γ according to the increasing value of $\mathcal{T}(y)$. For each node y, according to the decreasing order of $\mathcal{T}(x)$, we successively consider each node x such that $\mathcal{T}(x) < \mathcal{T}(y)$.
When $\Gamma \models x \to y$ (rspt. $\Gamma \models x \not\to y$), we introduce in Π a new ground clause $X \wedge \neg T_1 \wedge ... \wedge \neg T_l \supset Y$ (rspt. $X \wedge \neg T_1 \wedge ... \wedge \neg T_l \supset \overline{Y}$) where $\overline{Y}$ is an additional predicate and where the set $\{T_i | 1 \leq i \leq l\}$ corresponds to the set $\{t_i | 1 \leq i \leq l\}$ of all nodes t_i such that $\mathcal{T}(t_i) < \mathcal{T}(x)$ and such that $\Gamma \not\models t_i \to y$ (rspt. $\Gamma \not\models t_i \not\to y$).

In order to prove Theorem 1, we need to establish several lemmas. Theorem 1 is a direct consequence of Lemmas 2 and 4.

Lemma 1.
We can number the predicates occurring in Π according to the total order $\mathcal{T}$ of the nodes of Γ, in such a way that R_j (and $\overline{R_j}$) $< R_{j+k}$ (and $\overline{R_{j+k}}$), where $k \geq 1$ and where R_i and $\overline{R_i}$ correspond to node r_i and are assigned the same number. Π contains only clauses of the form

1. $R_j \supset R_{j+k}$
2. $R_j \supset \overline{R_{j+k}}$
3. $R_j \wedge \bigwedge_k \neg R_{j-k} \supset R_{j+l}$
4. $R_j \wedge \bigwedge_k \neg R_{j-k} \supset \overline{R_{j+l}}$

where $k \geq 1$ and $l \geq 1$.

Proof. Obvious. □

Lemma 2. The program (Π, X) is hierarchical.

Proof.
This is straightforward thanks to the structure of the program Π. One of the stratifications of Π is given by the ordering of the predicates R and $\overline{R}$ described in Lemma 1 and where atoms of the form R and $\overline{R}$ belong to the same stratum. □

Since (Π, X) is stratified and ground, its declarative semantics can be expressed by means of its unique *perfect model* [Przymusinski 88a,88b,88c], noted $\mathcal{M}$. $Z \equiv_{\mathcal{M}}$ *true* (rspt. *false*) means that the literal Z is *true* (rspt. *false*) in $\mathcal{M}$.

Lemma 3.

$T_i \equiv_{\mathcal{M}}$ *false* for every ground atom T_i such that $\mathcal{T}(t_i) < \mathcal{T}(x)$.

Proof.
In the unique perfect model $\mathcal{M}$ of a ground stratified logic program, atoms belonging to higher-priority strata are minimized first, i.e. their truth values are *false* in that model. Recall that a

stratification of (Π, X) is given by the ordering of predicates corresponding to the ordering $\mathcal{T}$ of the corresponding nodes in Γ.
By definition, X must be *true* in $\mathcal{M}$ because $\mathcal{M}$ is the perfect model of (Π, X). According to the structure of the program Π, whenever $\mathcal{M}$ remains a model of (Π, X), all atoms T_i such that $\mathcal{T}(t_i) < \mathcal{T}(x)$ must be minimized with respect to $\mathcal{M}$, since their predicates belong to the highest strata (in particular, to strata higher than the stratum to which X belongs). Accordingly, let us show that these T_i's are *false* in $\mathcal{M}$.
To this end, let us show how to construct a model $\mathcal{N}$ of (Π, X) such that $X \equiv_{\mathcal{N}}$ *true* and such that $T_i \equiv_{\mathcal{N}}$ *false* for any atom T_i such that $\mathcal{T}(t_i) < \mathcal{T}(x)$. It is straightforward to see that any such a $\mathcal{N}$ satisfies any clause belonging to (Π_1, X) where Π_1 denotes the subset of Π consisting of all clauses whose positive literal in the body is one of the T_i's. Indeed, the body of any clause of Π_1 is *false* in $\mathcal{N}$. Thus, at this point, the truth values of the literals occurring in the consequents of these clauses can be chosen freely (except the truth values of the T_i's and X which are already known to be *false* and *true*, respectively).
To be a model of (Π, X), $\mathcal{N}$ must also satisfy the clauses belonging to $(\Pi, X) \setminus \Pi_1$.
In particular, it must satisfy all clauses having the positive literal X in their bodies. These bodies are always *true* in $\mathcal{N}$ since the other literals in the bodies are negated literals $\neg T_i$'s. Therefore, the consequent of any of these clauses must be *true* in $\mathcal{N}$. Since these consequents do not refer to any of the T_i's or X, this is achieved by assigning the truth value *true* to positive literals occurring in these consequents.
Any other clause of (Π, X) has a consequent whose literals are referring to predicates Z or $\overline{Z}$ such that $\mathcal{T}(t_i) < \mathcal{T}(x) < \mathcal{T}(z)$. Thanks to the hierarchical structure of Π, we can easily prove by induction that truth values can be assigned to these atoms Z and $\overline{Z}$ in such a way that $\mathcal{N}$ is also a model of these clauses and that already assigned truth values are not modified. □

Let us prove that the hierarchical logic program Π is 'equivalent' to the considered net Γ.

Lemma 4.

1. $\Gamma \models x \rightarrow y$ iff $(\Pi, X) \models_{\mathcal{M}} Y$
2. $\Gamma \models x \not\rightarrow y$ iff $(\Pi, X) \models_{\mathcal{M}} \overline{Y}$

when the logic program Π and the ground atoms X, Y and $\overline{Y}$ are obtained by means of the translation-schema described above.

Proof.
1. $\Longrightarrow$
Assume $\Gamma \models x \rightarrow y$ (rspt. $\Gamma \models x \not\rightarrow y$). Let us show that assuming that Y (rspt. $\overline{Y}$) $\equiv_{\mathcal{M}}$ *false* yields a contradiction.
According to our translation-schema, since $\Gamma \models x \rightarrow y$ (rspt. $\Gamma \models x \not\rightarrow y$), there exists a clause of the form $X \wedge \neg T_1 \wedge ... \wedge \neg T_n \supset Y$ (rspt. $X \wedge \neg T_1 \wedge ... \wedge \neg T_n \supset \overline{Y}$) in Π. Therefore, in order to have Y (rspt. $\overline{Y}$) $\equiv_{\mathcal{M}}$ *false*, we must have $\exists\, i \in [1, n] : T_i \equiv_{\mathcal{M}}$ *true*. According to Lemma 1, such a T_i must be such that $\mathcal{T}(t_i) < \mathcal{T}(x)$. This result contradicts Lemma 3 stating that $T_j \equiv_{\mathcal{M}}$ *false* for any ground atom T_j such that $\mathcal{T}(t_j) < \mathcal{T}(x)$. □

2. $\Longleftarrow$
Let us show that assuming both $(\Pi, X) \models_{\mathcal{M}} Y$ (rspt. $(\Pi, X) \models_{\mathcal{M}} \overline{Y}$) and $\Gamma \not\models x \rightarrow y$ (rspt. $\Gamma \not\models x \not\rightarrow y$) yields a contradiction.

- When there exists a clause of the form $X \wedge \neg T_1 \wedge \ldots \wedge \neg T_n \supset Y$ (rspt. $X \wedge \neg T_1 \wedge \ldots \wedge \neg T_n \supset \overline{Y}$) in Π, according to our translation-schema, we have $\Gamma \models x \rightarrow y$ (rspt. $\Gamma \models x \not\rightarrow y$). This result contradicts the initial assumption.

- Let us show that we can also yield a contradiction when such a clause does not exist in Π. To this end, let us discover the set of atoms that must be ***true*** in $\mathcal{M}$ and show that Y (rspt. $\overline{Y}$) does not belong to that set. Such a result will contradict the initial assumption $(\Pi, X) \models_{\mathcal{M}} Y$ (rspt. $(\Pi, X) \models_{\mathcal{M}} \overline{Y}$), i.e. Y (rspt. $\overline{Y}$) $\equiv_{\mathcal{M}}$ ***true***.
First, let us consider all clauses of Π containing the positive literal X in their bodies.
According to Lemma 3, all negative literals occurring in the bodies of these clauses are *true* in $\mathcal{M}$ since they refer to predicates T_i such that $\mathcal{T}(t_i) < \mathcal{T}(x)$. Therefore the consequents X_2 or $\overline{X_2}$ of each of these clauses are *true* in $\mathcal{M}$ too. Let us discover the consequences of this feature on the truth value of other atoms.

  - 1. There is no clause in Π that contains $\overline{X_2}$ in its body (see Lemma 1). According to the hierarchical structure of Π, this means that having $\overline{X_2} \equiv_{\mathcal{M}}$ *true* will not influence the truth value in $\mathcal{M}$ of any other literal X_3 such that $\mathcal{T}(x_2) < \mathcal{T}(x_3)$ (in particular, Y or $\overline{Y}$).
  - 2. Clauses having X_2 as a positive literal in their bodies *can* have the truth value of their consequents X_3 or $\overline{X_3}$ required to be *true* in $\mathcal{M}$.
  We thus iterate the analysis of points 1 and 2 until it yields no literal X_n to consider that is such that $\mathcal{T}(x_n) < \mathcal{T}(y)$.

According to the increasing order $\mathcal{T}(z)$ of their consequent Z or $\overline{Z}$, we initialize and apply the above analysis until all clauses in Π have been considered. Two cases may occur.

- If this procedure ends without having considered a clause having Y (rspt. $\overline{Y}$) as a consequent, this means that there is no reason to require Y (rspt. $\overline{Y}$) to be *true* in $\mathcal{M}$. According to the definition of a perfect model, this atom must be *false* in $\mathcal{M}$. This contradicts the initial assumption $(\Pi, X) \models_{\mathcal{M}} Y$ (rspt. $(\Pi, X) \models_{\mathcal{M}} \overline{Y}$).
- If we eventually consider clauses of the form $X_p \wedge \neg T_1 \wedge \ldots \wedge \neg T_n \supset Y$ (rspt. $X_p \wedge \neg T_1 \wedge \ldots \wedge \neg T_n \supset \overline{Y}$), that means, by the way that we have constructed Π, that, since we have assumed that $\Gamma \not\models x \rightarrow y$ (rspt. $\Gamma \not\models x \not\rightarrow y$), the atom X must occur as a negative literal in the bodies of these clauses and thus that these bodies are *false* in $\mathcal{M}$. Since the consequents of these clauses (i.e., Y (rspt. $\overline{Y}$)) are in no way required to be *true* in $\mathcal{M}$, we must have Y (rspt. $\overline{Y}$) $\equiv_{\mathcal{M}}$ *false*. This result contradicts the initial assumption $(\Pi, X) \models_{\mathcal{M}} Y$ (rspt. $(\Pi, X) \models_{\mathcal{M}} \overline{Y}$). □

References

[Apt et al. 88] K.R. Apt, H. Blair and A. Walker, Towards a theory of declarative knowledge, in: J. Minker (ed.), *Foundations of Deductive Databases and Logic Programming*, Morgan Kaufmann, Los Altos, pp. 89-148, 1988.

[Bidoit and Froidevaux 87] N. Bidoit and C. Froidevaux, Minimalism subsumes default logic and circumscription in stratified logic programming, *Proc. LICS-87*, pp. 89-97, 1987.

[Clark 78] K.L. Clark, Negation as failure, in: H. Gallaire and J. Minker (eds.), *Logic and Data Bases*, Plenum Press, New York, pp. 293-322, 1978.

[Doherty 89] P. Doherty, A correspondence between inheritance hierarchies and preferential entailment, in: Z.W. Ras (ed.), *Methodologies for Intelligent Systems 4*, North Holland, pp. 395-402, 1989.

[Etherington and Reiter 83] D.W. Etherington and R. Reiter, On inheritance hierarchies with exceptions, *Proc. AAAI-83*, pp. 104-108, 1983.

[Etherington 88] D.W. Etherington, *Reasoning with Incomplete Information*, Research Notes in Artificial Intelligence, Pitman, London, 1988.

[Gelfond 87] M. Gelfond, On stratified autoepistemic theories, *Proc. AAAI-87*, pp. 207-211, 1987.

[Gelfond and Przymusinska 89] M. Gelfond and H. Przymusinska, Inheritance reasoning in autoepistemic logic, in: Z.W. Ras (ed.), *Methodologies for Intelligent Systems 4*, North Holland, pp. 419-429, 1989.

[Gelfond et al. 89] M. Gelfond, H. Przymusinska, T. Przymusinski, On the relationship between circumscription and negation as failure, *Artificial Intelligence*, 38, pp. 75-94, 1989.

[Grégoire 89a] E. Grégoire, Reducing inheritance theories to default logic and logic programs, in: J. Hannu and L. Seppo (eds.), *Second Scandinavian Conference on Artificial Intelligence (SCAI-89)*, vol. 2, pp. 1064-1079, Tampere, Finland, June 1989.

[Grégoire 89b] E. Grégoire, Skeptical theories of inheritance and nonmonotonic logics, in: Z.W. Ras (ed.), *Methodologies for Intelligent Systems 4*, North Holland, pp. 430-438, 1989.

[Grégoire 89c] E. Grégoire, *Logiques non monotones, programmes logiques stratifiés et théories sceptiques de l'héritage*, PhD thesis, Unité d'Informatique, Faculté des Sciences Appliquées, Université de Louvain, Louvain-la-Neuve, Belgium, June 1989.

[Grégoire 90a] E. Grégoire, Skeptical inheritance can be more expressive, in: L.C. Aiello (ed.), *Proc. of the 9th Europ. Conf. on Artificial Intelligence (ECAI-90)*, Pitman, pp. 326-332, 1990.

[Grégoire 90b] E. Grégoire, *Logiques non monotones et Intelligence Artificielle*, Hermès, Paris, 1990.

[Hayes 79] P.J. Hayes, The logic of frame, in: D. Metzing (ed.), *Frame Conceptions and Text Understanding*, de Gruyter, Berlin, pp. 46-61, 1979.

[Horty et al. 87,90] J.F. Horty, R.H. Thomason and D.S. Touretzky, A skeptical theory of inheritance in nonmonotonic semantic networks, *Proc. AAAI-87*, pp. 358-363, 1987. Expanded version in: *Artificial Intelligence*, vol. 42, pp. 311-348, 1990.

[Horty and Thomason 88] J.F. Horty and R.H. Thomason, Mixing strict and defeasible inheritance, *Proc. AAAI-88*, pp. 427-432, 1988.

[Lifschitz 88] V. Lifschitz, On the declarative semantics of logic programs with negation, in: J. Minker (ed.), *Foundations of Deductive Databases and Logic Programming*, Morgan Kaufmann, Los Altos, pp. 177-192, 1988.

[McCarthy 86] J. McCarthy, Applications of circumscription to formalizing common-sense knowledge, *Artificial Intelligence*, 28(1), pp. 89-116, 1986.

[Moore 85] R.C. Moore, Semantical considerations on non-monotonic logic, *Artificial Intelligence*, 25(1), pp. 75-94, 1985.

[Przymusinski 88a] T.C. Przymusinski, On the semantics of stratified deductive databases, in: J. Minker (ed.), *Foundations of Deductive Databases and Logic Programming*, Morgan Kaufmann, Los Altos, pp. 193-216, 1988.

[Przymusinski 88b] T.C. Przymusinski, Perfect model semantics, *Proc. Fifth Int. Conf. on Logic Programming*, pp. 1081-1096, 1988.

[Przymusinski 88c] T.C. Przymusinski, On the relationship between logic programming and non-monotonic reasoning, *Proc. AAAI-88*, pp. 444-448, 1988.

[Reiter 80] R. Reiter, A logic for default reasoning, *Artificial Intelligence*, 13(1-2), pp. 81-131, 1980.

[Shepherdson 88] J.C. Shepherdson, Negation in logic programming, in: J. Minker (ed.), *Foundations of Deductive Databases and Logic Programming*, Morgan Kaufmann, Los Altos, pp. 19-88, 1988.

[Stein 89] L.A. Stein, Skeptical inheritance: computing the intersection of credulous extensions, *Proc. IJCAI89*, 1989.

[Thomason and Horty 89] R.H. Thomason and J.F. Horty, Logics for inheritance theory, in: M. Reinfrank et al. (eds.), *Non-Monotonic Reasoning (Proc. of the 2nd Int. Workshop)*, Springer Verlag, LNCS 346, pp. 220-237, 1989.

[Touretzky 86] D.S. Touretzky, *The Mathematics of Inheritance Systems*, Morgan Kaufmann, Los Altos, 1986.

[Touretzky et al. 87] D.S. Touretzky, J.F. Horty and R.H. Thomason, A clash of intuitions: the current state of nonmonotonic multiple inheritance systems, *Proc. IJCAI-87*, pp. 476-482, 1987.

[van Gelder 88] A. van Gelder, Negation as failure using tight derivations for general logic programs, in: J. Minker (ed.), *Foundations of Deductive Databases and Logic Programming*, Morgan Kaufmann, Los Altos, pp. 149-176, 1988.

NONMONOTONIC REASONING AND MODAL LOGIC,

FROM NEGATION AS FAILURE TO DEFAULT LOGIC

Philippe Balbiani
Institut de Recherche en Informatique de Toulouse, université Paul Sabatier,
118 route de Narbonne, F-31062 Toulouse Cedex

abstract : We present a modal characterization of two well-known nonmonotonic formalisms : the negation as failure rule and default logic. The semantics of logic programming with the negation as failure rule is described through the definition of a modal completion. In modal logic K4, this completion characterizes provability in logic programming with respect to SLDNF-resolution while in modal logic Pr (the modal logic of provability) it characterizes unprovability in logic programs.

keywords : nonmonotonic reasoning, logic programming, modal logic

1. PRESENTATION

Negation as failure is a non-monotonic rule of inference. It introduces provability and unprovability features inside a linear resolution-based deduction mechanism. It makes reference to a logic program inside the program itself. See Clark (1978), Jaffar, Lassez and Lloyd (1983) and Lloyd (1984) for deeper investigations in logic programming with the negation as failure rule.

Default logic uses inference rules like A : B / C which allows us to add C to our current knowledge database whenever A belongs to that database and B is consistent with that database. Since something which is consistent with a set of belief is not always consistent with a superset of that set of belief, default logic is a nonmonotonic logic. See Reiter (1980) and Besnard (1989) for deeper investigations in default logic.

The purpose of this report is to study some properties of negation as failure and default logic. It gives a modal translation of logic programs. It shows this translation is sound and complete with respect to the negation as failure rule. It gives a modal translation of default theories and shows the soundness of this translation with respect to the provability in default logic. It studies the relationship between logic programming with negation as failure and default reasoning. It defines two very similar linear resolution-based mechanisms. The first mechanism can be proved to be equivalent to SLDNF-resolution - see Lloyd (1984) for a detailed study of SLDNF-resolution. The second mechanism enlightens us on the notion of provability in a fragment of default logic. It characterizes from a proof theory point of view the relationship between logic programming with negation as failure and default logic.

Though we will only consider ground logic programs, the results of sections 3 and 4 can be extended to the predicate case.

2. MODAL LOGIC

We will use the modal logics $K4^{\pm}$, Pr^{-} and $Pr^{\pm}$. Their language is based on a finite set VAR of propositional variables or atoms - atoms and negation of atoms will be called literals. The language includes the operators $\neg$, $\vee$, $\wedge$ and $\rightarrow$, the modal operators [+] and [-], and the rule : if F is a well-formed formula (wff) then so are [+]F and [-]F.

Modal logic $K4^{\pm}$ possesses the propositional calculus axioms and rules of inference plus the axioms : $[s](A\rightarrow B)\wedge[s]A\rightarrow[s]B$ (s=+,-) and $[s]A\rightarrow[s'][s]A$ (s=+,- and s'=+,-) and the rule : if F is a theorem then so are [+]F and [-]F. Modal logic Pr^{-} possesses the axioms and rules of $K4^{\pm}$ plus the axiom : $[-]([-]A\rightarrow A)\rightarrow[-]A$. Modal logic $Pr^{\pm}$ possesses the axioms and rules of Pr^{-} plus the axiom : $[+]([+]A\rightarrow A)\rightarrow[+]A$. Modal logics Pr^{-} and $Pr^{\pm}$ are variations of the modal logic of provability Pr - see Boolos (1979) or Smorynski (1984) - which is famous because of its relationship with the godelian concept of provability in arithmetic.

The semantics of these modal logics is defined in term of Kripke models. A Kripke model is composed of :

(1) a non-empty set W of possible worlds,
(2) two binary relations R^{+} and R^{-} defined over the members of W and
(3) an application v which associates to any pair (w,A) -w being an element of W and A being a well-formed formula- an element of {0, 1} such that :

(3)(1) $v(w,\neg A)=1-v(w,A)$,
(3)(2) $v(w,A\wedge B)=v(w,A)\times v(w,B)$ and, for any s in {+,-},
(3)(3) $v(w,[s]A)=\min\{\ v(w',A) : wR^{s}w'\ \}$.

A wff A is said to be valid in a model (W,R^+,R^-,v) when $v(w,A)=1$ for every element w of W.

The adequation results - see Balbiani (1991) for detailed proofs - between the axiomatic and the semantics of the modal logics $K4^\pm$, Pr^- and $Pr^\pm$ are as follows :

Theorem : The well-formed formula A
(1) is a theorem of $K4^\pm$ iff it is valid in any model (W,R^+,R^-,v) where R^+ and R^- are mutually transitive, that is to say : $wR^{s'}w''$ whenever $wR^{s}w'$ and $w'R^{s'}w''$,
(2) is a theorem of Pr^- iff it is valid in any model (W,R^+,R^-,v) where W is finite, R^+ and R^- are mutually transitive and R^- is irreflexive, that is to say : there is no element w of W such that wR^-w,
(3) is a theorem of $Pr^\pm$ iff it is valid in any model (W,R^+,R^-,v) where W is finite, R^+ and R^- are mutually transitive and R^+ and R^- are irreflexive.

Following the filtration method detailed in Hughes and Cresswell (1984), it can be showed that modal logics $K4^\pm$, Pr^- and $Pr^\pm$ possess the finite model property, that is to say :

Theorem : For any formula A of $K4^\pm$ (respectively : Pr^- and $Pr^\pm$) such that $\neg A$ is not a theorem of $K4^\pm$ (respectively : Pr^- and $Pr^\pm$), there is a model (W,R^+,R^-,v) of $K4^\pm$ (respectively : Pr^- and $Pr^\pm$) such that W is a finite set of possible worlds and, for some w in W, $v(w,A)=1$.

Consequently, and for obvious reasons that will not be much more detailed, $K4^\pm$, Pr^- and $Pr^\pm$ are decidable.

3. LOGIC PROGRAMMING

3.1. Preliminary definitions

A goal is a set of literals. A clause is a formula like $l_1 \wedge l_2 \wedge \ldots \wedge l_m \rightarrow A$ where A is an atom and each l_i is a literal. A logic program is a set of clauses. For every program P and for every atom A, P(A) will denote the set $\{L : L \rightarrow A \in P\}$ and $P(\neg A)$ will denote the set $\{ L : (\forall l \in L)(\exists L' \in P(A))(-l \in L') \} \cap \{ L : (\forall L' \in P(A))(\exists l \in L)(-l \in L') \}$, where, for every atom B, $-B=\neg B$ et $-\neg B=B$.

Let N be an application which associates to (P,l) - P being a program and l a literal - the program $P'=N(P,l)$.

An N-derivation from the goal L and the program P will be any sequence (L_0,P_0), (L_1,P_1), ..., (L_n,P_n) such that :

$$n\geq 0,\ L_0=L,\ P_0=P \text{ and, for every i ranging from 0 to n-1,}$$
$$(\exists l\in L_i)(\exists L'\in P(l))(L_{i+1}=(L_i\setminus\{l\})\cup L')(P_{i+1}=N(P_i,l)).$$

An N-refutation of L in P will be an N-derivation (L_0,P_0), (L_1,P_1), ..., (L_n,P_n) from L and P such that $L_n=\emptyset$.

N_1 will denote the application which associates P to (P,l). N_1-refutations are of course the refutations that would be obtained in an idealized version of PROLOG. N_2 will denote the application which associates P to (P,A) and $P\setminus A$ to $(P,\neg A)$, where $P\setminus A=P\setminus\{L\rightarrow A: L\in P(A)\}$. If $N=N_1$ then, using Fitting (1985) and Kunen (1987), it can be proved that :

Theorem : L is N_1-refutable in P if and only if (iff) there is an SLDNF-refutation of $P\cup\{L\rightarrow\bot\}$, that is to say iff L succeeds in P with respect to a refutation-based mechanism using the negation as failure rule.

3.2. Translation in modal logic

We translate "L is N_1-refutable in P" into the formula $P^*\rightarrow\wedge_{B\in L}[+]B\wedge\wedge_{\neg B\in L}[-]B$, [+] and [-] being the modal operators of $K4^{\pm}$; P^* being a translation of P in $K4^{\pm}$. In the modal logic $K4^{\pm}$ we have previously defined, $P^*= P_1\wedge P_2$ where

$$P_1=\wedge_{A\in VAR}\,[+](\vee_{L\in P(A)}(\wedge_{B\in L}[+]B\wedge\wedge_{\neg B\in L}[-]B)\rightarrow A),$$
$$P_2=\wedge_{A\in VAR}\,[-](\wedge_{L\in P(A)}(\vee_{B\in L}[-]B\vee\vee_{\neg B\in L}[+]B)\rightarrow A)$$

is both sound and complete with respect to N_1-refutation, that is to say :

Theorem : L is N_1-refutable in P iff $P^*\rightarrow\wedge_{B\in L}[+]B\wedge\wedge_{\neg B\in L}[-]B$ is a theorem of $K4^{\pm}$.

Proof : The proof of the if-part is done by induction on the length of the N_1-refutation of L in P. The proof of the only if-part is done by induction on the depth of the closed tableau of $\{\neg(P^*\rightarrow\wedge_{B\in L}[+]B\wedge\wedge_{\neg B\in L}[-]B)\}$. See Balbiani (1991) for further details.

Examples :
The modal translation of $P=\{\neg A\rightarrow B\}$ is the formula
$P^*=[-]A\wedge[+]([-]A\rightarrow B)\wedge[-]([+]A\rightarrow B)\wedge[-]C$.
$\{\neg A,B,\neg C\}$ is N_1-refutable in P and $P^*\rightarrow[-]A\wedge[+]B\wedge[-]C$ is a theorem of $K4^{\pm}$.

The modal translation of $P=\{\neg B \rightarrow B, \neg A \rightarrow B, \neg C \rightarrow A\}$ is the formula
$P^*=[+]([-]C \rightarrow A) \wedge [-]([+]C \rightarrow A) \wedge [+]([-]B \vee [-]A \rightarrow B) \wedge [-]([+]B \wedge [+]A \rightarrow B) \wedge [-]C$.
$\{A, \neg C\}$ is N_1-refutable in P and $\{B\}$ et $\{\neg B\}$ are not N_1-refutable in P and $P^* \rightarrow [+]A \wedge [-]C$ is a theorem of $K4^{\pm}$ and $P^* \rightarrow [+]B$ and $P^* \rightarrow [-]B$ are not theorems of $K4^{\pm}$.

As for non-provability in logic programs with negation, in $Pr^{\pm}$ P^* is both sound and complete, that is to say :

Theorem : L is not N_1-refutable in P iff $P^* \rightarrow \vee_{B \in L}[-]B \vee \vee_{\neg B \in L}[+]B$ is a theorem of $Pr^{\pm}$.

Note : The previous property is equivalent to :

$$L \text{ is } N_1\text{-refutable in P iff } P^* \wedge \neg(\vee_{B \in L}[-]B \vee \vee_{\neg B \in L}[+]B) \text{ is } Pr^{\pm}\text{-satisfiable}$$

Proof : The if-part is proved by induction on the length of the N_1-refutation of L in P. The only if-part is proved by induction on the depth of the Kripke model of $Pr^{\pm}$ which satisfies $P^* \wedge \neg(\vee_{B \in L}[-]B \vee \vee_{\neg B \in L}[+]B)$. See Balbiani (1991) for further details.

These results characterize the notions of provability and unprovability in logic programs with the negation as failure. They use the fact that the formula $[+]A \vee [-]A$ - which logically translates the alternative between the refutability of A and the refutability of $\neg A$ - is not a tautology of modal logic. It proves - especially for the notion of unprovability - that modal logic is able to represent the deduction characteristics in logic programs with negation. The last result can be explain by the fact that for any goal N_1-refutable in some program there exists a finite derivation tree which possesses an empty leaf and by the fact that Kripke's models of $Pr^{\pm}$ are finite irreflexive trees. This is the point which has led us to translate default theories into the modal logic Pr^-.

4. DEFAULT LOGIC

4.1. Preliminary definitions

A default theory T is composed with a set of Horn clauses Ax(T) and a set D(T) of rules (or defaults rules) like $A_1 \wedge A_2 \wedge ... \wedge A_m : \neg B_1, \neg B_2, ..., \neg B_n / C$ where $m \geq 0$, $n > 0$ and where$A_1, A_2, ..., A_m, B_1, B_2, ..., B_n$ and C are atoms.

An extension of T will be any fixpoint of Γ_T where for every set E of wff, $\Gamma_T(E)$ is the smallest set E' of wff such that $Ax(T) \subseteq E'$, $Th(E')=E'$ - Th(E') being the deductive closure of E' - and : $(\forall\ F{:}\neg B_1\neg B_2...\neg B_n/C \in D(T))$(if $(F\in E')(\forall i=1..n)(B_i \notin E)$ then $C\in E'$).

Examples : The theory ({A}, {A : ¬B / C, true : ¬C / B}) possesses two extensions:Th({A, B}) and Th({A, C}). The theory ({A}, {true : ¬B / C, C : ¬A / B, C: ¬B / D}) possesses only one extension:Th({A, C, D}). The theory ({}, {true : ¬B / B}) possesses no extension.

4.2. Translation in modal logic

T(A) will denote the set $\{L{:}L\rightarrow A\in Ax(T)\} \cup \{\{A_1, A_2, ..., A_m\}\cup\{\neg B_1, \neg B_2, ..., \neg B_n\} : (A_1\wedge A_2\wedge...\wedge A_m{:}\neg B_1\ \neg B_2\ ...\ \neg B_n /A \in D(T))\}$. Let T be a theory which possesses almost one extension.

The formula T^* of Pr^- where

$$T^* = T_1\wedge T_2,$$
$$T_1 = \wedge_{A\in VAR}\ [+](\vee_{L\in T(A)}(\wedge_{B\in L}[+]B\wedge\wedge_{\neg B\in L}[-]B) \rightarrow A) \text{ and}$$
$$T_2 = \wedge_{A\in VAR}\ [-](\wedge_{L\in T(A)}(\vee_{B\in L}[-]B\vee\vee_{\neg B\in L}[+]B) \rightarrow A)$$

is sound for provability in T. That is to say :

Theorem : For every goal L, if there is an extension E of T such that $(L\cap VAR\subseteq E)(\forall B)(\neg B\in L\rightarrow B\notin E)$ then $T^*\rightarrow\wedge_{B\in L}[+]B\wedge\wedge_{\neg B\in L}[-]B$ is a theorem of Pr^-.

Let T=({A}, {A : ¬B / C, true : ¬C / B}). T possesses two extensions:Th({A, B}) and Th({A, C}). We have $T^*=[+]A\wedge[+]([+]A\wedge[-]B\ \rightarrow C)\wedge[+]([-]C\ \rightarrow B)\wedge...$
$...\wedge[-]([-]A\vee[+]B\ \rightarrow C)\wedge[-]([+]C\ \rightarrow B)$ and $T^*\rightarrow[+]A, T^*\rightarrow[+]B, T^*\rightarrow[+]C, T^*\rightarrow[-]B$ and $T^*\rightarrow[-]C$ are theorems of Pr^-. Our translation is not complete however and the previous theory T shows it well. The modal wff $T^*\rightarrow[+]B\wedge[+]C$ is a theorem of Pr^- but B and C does not belong to the same extensions. We believe however that this difficulty could be solved by another kind of translation of default theory in Pr^-.

Let T=({A}, {true : ¬B / C, C : ¬A / B, C : ¬B / D}). T possesses only one extension : Th({A, C, D}). We have :

$$\begin{aligned}T^* = {} & [+]A\wedge[+]([-]B\ \rightarrow C)\wedge[+]([+]C\wedge[-]A\ \rightarrow B)\wedge\\ & [+]([+]C\wedge[-]B\ \rightarrow D)\wedge[-]([+]B\ \rightarrow C)\wedge\\ & [-]([-]C\vee[+]A\ \rightarrow B)\wedge [-]([-]C\vee[+]B\ \rightarrow D)\end{aligned}$$

and $T^*\rightarrow[+]A\wedge[-]B\wedge[+]C\wedge[+]D$ is a theorem of Pr^-.

The modal logic Pr^- expresses the notion of provability in a fragment of default logic. Its main axiom $[-]([-]A \rightarrow A) \rightarrow [-]A$ expresses the uselessness of an axiom like $A \wedge L \rightarrow A$ or a default rule like $A \wedge L{:}\neg B_1, \neg B_2, \ldots, \neg B_n / A$ in a default theory.

In the case where T possesses no extension, the soundness is not assured. Let $T=(\{B\}, \{true : \neg A \ / \ A\})$. T possesses no extension. Its modal translation is $T^*=[+]B \wedge [+]([-]A \rightarrow A) \wedge [-]([+]A \rightarrow A)$. In Pr^-, we have $[-]([-]A \ \rightarrow [+][-]A)$.
Thus $T^* \rightarrow [-]([-]A \ \rightarrow A)$ and $T^* \rightarrow [-]A$, though no extension of T contains A. Our translation is not able to represent the fact that the presence of the default $true{:}\neg A/A$ in a theory can make this theory inconsistant in the sense that it possesses no extension.

5. LOGIC PROGRAMMING AND DEFAULTS

The previous result does not give to our fragment of default logic a complete formalization in modal logic. Now we show an interesting result concerning the relationship between logic programming with negation and default reasoning. Provability in default logic is in fact equivalent to the notion of provability based on N_2-refutation. Semantical relations between the autoepistemic logic of Moore (1985) and stratified logic programs - see Apt, Blair and Walker (1988) for precise investigations inside this particular class of program - has been presented by Gelfond (1987). Bidoit and Froidevaux (1988) has given similar results for default logic. Lifschitz (1988) has studied the relations between stratified programs and circumscription. Our approach will do no restriction concerning the stratifiability of logic programs.

Let T be a theory of default logic. Let $P(T)=\{ L \rightarrow A : (A \in VAR)(L \in T(A)) \}$. An N_2-refutation $(L_0,P_0), (L_1,P_1), \ldots, (L_n,P_n)$ of L in P(T) is said to be total when $\{ A : (A \in VAR)(P(T)(A) \neq \emptyset) \}$ is included in $\{A{:}\{A, \neg A\} \cap (L_0 \cup L_1 \cup \ldots \cup L_n) \neq \emptyset\}$.

The following result establishes the equivalence between the notion of provability in default logic and N_2-refutability in P(T).

Theorem : If there is a total N_2-refutation of L in P(T) then there is an extension E of T such that $(L \cap VAR \subseteq E)(\forall B)(\neg B \in L \rightarrow B \notin E)$ and if there is an extension E of T such that $(L \cap VAR \subseteq E)(\forall B)(\neg B \in L \rightarrow B \notin E)$ then there is an N_2-refutation of L in P(T).

Examples : The theory $T=(\{\}, \{true : \neg A \ / \ B, true : \neg B \ / \ A\})$ possesses two extensions : $Th(\{A\})$ and $Th(\{B\})$. The associated program is $P(T)=\{\neg A \rightarrow B, \neg B \rightarrow A\}$. The sequence $(\{\neg A\}, P(T)), (\{B\}, P(T) \backslash A), (\{\neg A\}, P(T) \backslash A), (\{\}, P(T) \backslash A)$ is a total N_2-

refutation of $\{\neg A\}$ in P(T). As for the goal $\{A, B\}$ - which is not provable in T - it does not possess any N_2-refutation in P(T).
The theory $T=(\{B\}, \{true : \neg A / A\})$ possesses no extension. The associated program is $P(T)=\{\varnothing\rightarrow B, \neg A\rightarrow A\}$. The sequence $(\{B\}, P(T)), (\{\}, P(T))$ is an N_2-refutation of $\{B\}$ in P(T) which is not total, and that corresponds to the fact that $\{B\}$ is not provable in T.

6. PERSPECTIVES AND REFERENCES

We have presented a modal semantic for negation in logic programming. This semantic used a very particular modal logic, namely Pr, which is related to the notion of provability in arithmetic. We believe that non-monotonic formalisms and this modal logic share common specificities and that some of the properties of Pr are those the main non-monotonic logic are searching for. For full first order logic programs with negation, previous results have been extended, see Balbiani (1991).

REFERENCES

Apt K. R., Blair H. A. and Walker A. 1988, Towards a theory of declarative knowledge, in Foundations of Deductive Databases and Logic Programming (J. Minker, Ed.), Morgan Kaufmann Publishers, Los Altos, CA, 89-148.

Balbiani P. (1991), Une caractérisation modale de la sémantique des programmes logiques avec négation, thèse de l'université Paul Sabatier, janvier 1991.

Balbiani P. (1991), A modal semantics for the negation as failure and the closed world assumption rules, to be presented at the eighth Symposium on Theoretical Aspects of Computer Science, Hamburg, february 1991.

Besnard P. 1989, An introduction to default logic, Springer-Verlag.

Bidoit N. and Froidevaux C. 1988, More on stratified default theories, Proc. of the European Conf. on Artificial Intelligence, 492-494.

Boolos G. 1979, The unprovability of consistency, Cambridge University Press.

Chellas B. F. 1980, Modal logic : an introduction, Cambridge University Press.

Clark K. L. 1978, Negation as failure, in Logic and Databases (H.Gallaire and J.Minker, Eds) Plenum Press, New York, 293-322.

Fitting M. R. 1985, A Kripke-Kleene semantics for logic programs, J. of Logic Programming, 2:295-312.

Gelfond M. 1987, On stratified autoepistemic theories, Proc. of the Sixth Nat. Conf. on Artificial Intelligence (AAAI-87).

Hughes G. E. and Cresswell M. J. 1984, A companion to modal logic, London, Methuen.

Jaffar J., Lassez J.-L. and Lloyd J. W. 1983, Completeness of the negation as failure Rule, Proc. of the Eighth Int. Joint Conf. on Artificial Intelligence, Karlsruhe, 500-506.

Kripke S. 1963, Semantical analysis of modal logic, Zeitschrift für Mathematische Logik und Grundlagen der Mathematik, 9:67-96.

Kunen K. 1987, Negation in logic programming, J. of Logic Programming, 4:289-308.

Lifschitz V. 1988, On the declarative semantics of logic programs with negation, in Foundations of Deductive Databases and Logic Programming, (J. Minker, Ed.), 177-192.

Lloyd J. W. Foundations of logic programming, Springer Verlag, 1984.

Moore R. C. 1985, Semantical Considerations on Nonmonotonic Logic, Artificial Intelligence, 25:75-94.

Reiter R. 1980, A logic for default reasoning, Artificial Intelligence, 13:81-132.

Smorynski C. 1984 Modal logic and self reference, Handbook of Philosophical Logic (D. Gabbay et F. Guenthner, Eds), Reidel, 441-496.

ON THE NOTION OF UNCERTAIN BELIEF REVISION SYSTEMS

C. Bernasconi, S. Rivoira and S. Termini
Dipartimento di Matematica, Università degli Studi
06100 Perugia (Italy) [1]

Abstract

The notion of uncertain belief revision systems (UBRS) is introduced as an extension of assumption-based truth maintenance systems (ATMS) to a many valued logic.
In this framework, some results relative to the many-valued implication $p \Rightarrow q$ equivalent to *not*-$p \vee q$ are considered.
Problems arising when different implication functions are defined in many valued logic are then discussed and some steps of a research development are described.

Keywords

intelligent systems, reasoning systems, inference, uncertainty, approximate reasoning, resolution principle, fuzzy resolution principle, fuzzy Prolog, ATMS.

1. MOTIVATIONS

The aim of the present paper is to pave the way for the definition of an uncertain belief revision system (UBRS). The paper will indicate some results which can be usefully applied for the solution of our problem and will also presents some problematic points emerging when we face the question in a general way.

The reason for looking for such a definition are essentially the following two.

First, A.I. is progressively acknowledging the importance of treating inexactitude and approximation with powerful tools and at a general level. We agree with H. Berliner that:

---

[1] This paper has been supported by C.N.R. through P. F. Robotica - contratto di ricerca N° 89.00556.67.
A partial support by: M.U.R.S.T-40%,"Tecniche di ragionamento automatico in sistemi intelligenti" is also acknowledged.

"notions such as fuzziness, ranges, uncertainty, and probability distributions, are not only the equal of symbols, lists, slots, scripts and plans, they far exceed these in importance. Symbolic constructions are just great when their presence can be determined with exactitude, and it is known how to use them. However inexactitude is a necessary fact of life in all other circumstances, which include almost all of what there is. ... The future for A.I. appears very much tied up in these notions" [see Bobrow & Hayes (eds) (1985), page 403].

Secondly, in our view, the notion of Truth Maintenance System is from the start strictly associated with (uncertain) beliefs and so to try to define an extension of these systems to capture the uncertainty present in beliefs, seems to be a useful and natural development of the philosophy behind TMS' and ATMS'

The notion of truth maintenance system grew out of the need of assuring the coherence of a certain set of assumptions and of their consequences also after their subsequent extensions. It is clear that this notion is from the start strictly associated with uncertain beliefs,

From this perspective, then, it seems to be a strong restriction that the mechanisms proposed are strictly modelled as if the assumptions do not suffer of "inexactitude", even if they can be withdrawn when they are incompatible with other assumptions, considered unrenouncible. We think however that, in general, unless we confine to the cases of withdrawal of wrong assumptions, the essentiallly uncertain nature of beliefs should play here a role.

In the present investigation we shall concentrate only on approximate reasoning aspects and shall not consider non monotonic reasoning, for which we refer to the interesting updated state of the art provided by Reinfrank et al (1989) as well as to the classical references (monographic issue of AI, 1980, survey by Reiter, etc.) We shall never deal with circumscription, default etc, since we think that at this preliminary stage of development of the proposal, this would make less clear the analysis regarding approximation aspects.

In the present paper we shall essentially look for many-valued versions of ATMS, seen as a first attempt at providing efficient UBRS. In the following section we shall present a review of ATMS, in section 3 we shall discuss the possibility of extending the definition of ATMS by means of many-valued logics, briefly rewieving some results related to the many-valued extensions of the resolution principle. In section 4 we shall discuss how these basic properties of the ATMS could be modified in order to define an UBRS, while in the concluding section we shall discuss some critical problems related to future developments.

2. A REVIEW OF ATMS.

The best general-purpose mechanisms for dealing with the revision of certain beliefs are currently represented by truth maintenance systems [Reiter,1987]. Following Doyle's TMS [Doyle,1979], many such systems (also called belief revision or reason maintenance systems) have been developed (see [de Kleer,1986b] for a review).

Most of them can be viewed [Brown,1987] as constraint propagation mechanisms which tell a problem solver what things it is currently obliged to believe, given a single set of premises and a set of deduction constraints, some of which may be nonmonotonic.

The Assumption-based TMS [de Kleer,1986a] avoids the restriction that the overall set of premises is contradiction-free, maintaining a disjunctive set of sets of premises and telling the client problem solver what things it is obliged to believe, assuming one or another of the sets of premises.

An assumption-based TMS therefore solces a more complex problem than a justification-based TMS do and it is consequently less efficient in many cases. The main goal of this research is to understand the relations between uncertainty and truth-mantainence from a general point of view as a first step, leaving efficiency to be analized in a second phase. For this reason we chose an ATMS as an early reference system for dealing with the revision of uncertain beliefs. The basic features of such a system will briefly be recalled in the following.

Every datum introduced or derived by the problem solver biunivocally corresponds to an ATMS *node*.

A *justification* j: $x_1, x_2, \ldots x_m \dashrightarrow n$ is a propositional Horn clause asserting that the consequent node n is derivable from the conjunction of the antecedent nodes $x_1, \ldots x_m$.

An *assumption* is a self-justifying node representing the decision of introducing an hypothesis; it is connected to the assumed data through justifications.

An *environment* is a set of logically conjuncted assumptions.

An environment E is inconsistent if falsity can be derived from E and the current set of justifications J : $E \overset{J}{\vdash} \mathbf{F}$.

An ATMS *context* is defined as the set formed by the assumptions of a consistent environment and all the nodes derivable from those assumptions: $C = E \cup N \mid E \overset{J}{\nvdash} \mathbf{F}$, $N = \{ n \mid E \overset{J}{\vdash} n \}$.

The goal of ATMS is to efficiently update the contexts when new assumptions or justifications are provided by the problem solver.

This goal is achieved by associating with every node a description (label) of every context in which the node holds.

More formally, a *label* L_n associated with the node n is defined as the set of all the consistent environments from which n can be derived:

$$L_n = \{ E_i \mid E_i \overset{J}{\nvdash} \mathbf{F} , E_i \overset{J}{\vdash} n \} .$$

Four important properties can be defined for labels:

- a label L_n is *consistent* if all of its environments are consistent:

$$\forall E_i \, [E_i \in L_n \Longrightarrow E_i \overset{J}{\nvdash} \mathbf{F}] \; ;$$

- a label L_n is *sound* if n is derivable from each of its environments:

$$\forall E_i \, [E_i \in L_n \Longrightarrow E_i \overset{J}{\vdash} n] \; ;$$

- a label L_n is *complete* if every consistent environment from which n can be derived is a superset of some environment in L_n:

$$\forall E\,[\,\exists E_i\,[\,E \vdash^{J} n \wedge E_i \in L_n \implies E_i \subset E\,]]\ ;$$

- a label L_n is *minimal* if no environment in L_n is a superset of any other environment in L_n:

$$\forall E_i\,[\,\nexists E_k\,[\,E_i \in L_n \wedge E_k \in L_n \implies E_i \subset E_k\,]]\ .$$

The task of the ATMS is to ensure that each label of each node is consistent, sound, complete and minimal with respect to the current set of justifications. This task is performed by invoking a label-update algorithm every time the problem solver adds a new assumption or a new justification.

Node labels are completely insensitive to the order in which the assumptions, nodes, and justifications are introduced.

The main feature of the ATMS is the coexistence of multiple, mutually contradictory contexts corresponding to different theories. Therefore it is easy to compare different contexts and to switch from one context to another as consequence of belief revisions.

The only effect of discovering a contradiction is the partitioning of the related context into new contradiction free contexts, where mutually contradictory data will never be combined.

In section 4 we shall discuss how these basic properties of the ATMS can be modified in order to define an uncertain belief revision system.

3. ON MANY VALUED INFERENCE TECHNIQUES.

Before discussing how these basic properties of the ATMS can be modified in order to define an uncertain belief revision system (UBRS) let us briefly discuss what can be done to formalize the ideas proposed in the first paragraph by means of the techniques of ATMS.

The first idea is to add a many valued device to the classical logical tools on which the working of the ATMS is based. In this sense then the most straightforward path to be followed is to modify the logical evaluation part in a many-valued way. Since the logical inference rule which is most useful in automatic deduction systems is the resolution principle, it would be good to preserve this principle also in a many-valued setting. Luckly enough some old (Lee, 1972) and new (Mukaidono, 1982 and1989) results are very helpful in this direction. R.C.T. Lee studied in 1972 the problem of extending the resolution principle to fuzzy logic. Let us remember his main result as a basis for a discussion. We refer to his paper for details and extensions. Let the well-formed formulas be defined exactly as in classical (two-valued) logic, let the interval [0, 1] of the real line the set of truth-values, and let [S] denote the truth-value of the formula S.

The truth-values of the propositional connectives be defined by:

$$[\text{not } S] = 1 - [S]$$
$$[R \vee S] = \max\{\,[R], [S]\,\}$$
$$[R \wedge S] = \min\{\,[R], [S]\,\}$$
$$[R \Rightarrow S] = \max\{1 - [R], [S]\,\}$$

Following Lee let us give the following definition of satisfiability in the previous logic: an interpretation I is said to satisfy a formula S if [S] is not less than O.5 under I.

Lee's main result can be stated as follows: *"Let S be a set of clauses whose truth value is not less than 0.5, let the most unreliable clause have the truth value a, and let the truth value of the most reliable clause be equal to b, then we are guaranteed that all the logical consequences obtained by repeatedly applying the resolution principle will have truth value at least equal to a, but never exceeding b".*

This result, as is explicitely stressed by the author works only in the case in which the truth values of all the clauses in S are not less than 0.5. As Lee observes this is not, in fact, a strong limitation since we could consider A or not A, according to the truth value of A and so work always with set of clauses whose truth value is not less than 0.5. With this "trick" Lee's result quoted above can be always applied by remembering the choice done.

The problem has been recently studied again by Mukaidono (1982) who has shown that Lee's results, under suitable conditions, can be extended so as to be applicable to every clause independently of its truth-value. The previous results guarantee the possibility of using efficient inference rules. In the following section we shall outline a definition of UBRS.

4. REVISION OF UNCERTAIN BELIEFS.

An UBRS justification is a pair (j,t_j), where j is an ATMS justification and t_j is the truth-value of the consequent node when it is derived by j.

UBRS assumptions and environments are defined exactly as in ATMS.

An environment E has an inconsistency degree [E] if falsity can be derived with a truth-value [E] from E and the current set of justifications.

The label L_n associated with the node n is defined as the set of all the environments with a suitable inconsistency degree less than 1, from which n can be derived.

The consistency, soundness, completeness and minimality properties of labels can be redefined taking into account the inconsistency degrees of environments and using the truth rules of the many-valued logic adopted by the problem solver.

It is worth noticing that the notion of contradiction plays the major role for the mechanisms of UBRS, while approximate inference is mainly a matter of the problem solver, even if it is crucial for the overall reasoning system. The main difference of UBRS with respect to ATMS is the possibility for a node to be derived from different sets of assumptions which are contradictory at some degree.

The notion of contradiction is not unique and can be changed by the problem solver at different steps of the solution process.

The label-update algorithm of UBRS allows the revision of both the truth-value of any datum and the contradiction degrees of the sets of assumptions from which it can be derived.

5. CONCLUDING REMARKS AND FUTURE PROSPECTS.

Let us stress that the results of Lee and Mukaidono are strongly based on the chosen implication, which extends to the many-valued case the classical identification between *not*-$p \vee q$ and $p \Rightarrow q$, and their choice of defining that a formula S is satisfiable under any interpretation I if the truth value of S is not less than 0.5. What happens if we change the notion of satisfiability or also the same implication? All the developed machinery does not work or, at least, it is not straightforwardly applicable to the most general cases.

In the case in which we would be interested in using different definitions of implication we should then face also the problem of looking for efficient inference strategies. In fact, since the general machinery of many valued logic offer a wide spectrum of possibilities, a fruitful path to be followed is to study the behaviour of a whole family of implication functions with respect to the problems posed by a many-valued extension of an ATMS.

A preliminary study on the most general definition of implication (and also of the other connectives) in many-valued logic is due to Salomaa (1959). Trillas and Valverde (1984) has explicitly provided a characterization of those implication functions which are continuous functions I from [0, 1] x [0, 1] into [0, 1] satisfying the following conditions:

I_1) if $x \le x'$ then $I(x', y) \le I(x, y)$
I_2) if $y \le y'$ then $I(x, y) \le I(x, y')$
I_3) $I(0, y) = 1$
I_4) $I(1, y) = x$
I_5) $I(x, I(y, x)) = I(y, I(x, z))$

One further step forward of our investigation will then be to examine whether the implication functions of Trillas and Valverde's family are easily matchable with an ATMS-like structure. Subsequently one should look again at Salomaa's conditions in order to see whether it is possible to find out explicitely other useful families of implication functions characterized by their defining conditions, and which could then be easily related to the type of uncertainty which is present in a specific situation.

A complete analysis of the different types of uncertainty will also force to consider the problem of the truth-functionality of logical connectives and the (related) problem of the comparison of the previously considered models with probabilistic approaches.

Several basic inference models for different aspects of uncertainty will be considered and compared from the viewpoint of efficiency both of the inference rules and of the revision mechanisms.

REFERENCES.

Bobrow D.G. & Hayes P.J. (eds) (1985), A.I. - Where Are We?, Artificial Intelligence 25, 375-415.

A.L. Brown, D.E. Gaucas, D. Benanav (1987), An Algebraic Foundation for Truth Maintenance, Proc. 10th IJCAI, Milano, 973-980.

Doyle J.(1979), A Truth Maintenance System, Artificial Intelligence 12, pp. 231-272.

de Kleer J. (1986a), An Assumption-based TMS. Artificial Intelligence 28, pp. 127 -162.

de Kleer J. (1986b), Problem solving with the ATMS. Artificial Intelligence 28, pp. 197-224.

Lee R.C.T.(1972), Fuzzy Logic and the Resolution Principle, Journal of A.C.M. 19, 109-119.

Mukaidono M. (1982), Fuzzy Inference of resolution style, in "Fuzzy Sets and Possibility Theory", R.R. Yager (ed), Pergamon Press, New York, 224-231.

Mukaidono M. et al. (1989), Fundamentals of Fuzzy Prolog, International Journal of Approximate Reasoning, 3, 179-193.

Reinfrank M., de Kleer J., M.L. Ginzberg, E. Sandewall (Eds) (1989), Nonmonotonic reasoning, Springer Lecture Notes in A.I., vol. 346.

Reiter R. (1987), Nonmonotonic reasoning, Ann. Rev. Comput. Sci. 2, pp. 147-186.

Salomaa A. (1959), On many valued systems of Logic, Ajatus XXII, 115 - 159.

Trillas E. & Valverde L. (1984), On Implication and Indistinguishability in the Setting of Fuzzy Logic, in Kacprzyk & Yager (eds), Management Decision Support Systems using Fuzzy Sets and Possibility Theory, Verlag TUV.

NON-MONOTONIC REASONING IN A SEMANTIC NETWORK

by Marcel CORI

Université Paris 7, Laboratoire de Linguistique Formelle,
Tour centrale 8ème étage, 2 place Jussieu,
75251 Paris Cedex 05, France

Abstract:

We define semantic networks for representing knowledge as being multilabeled semiordered directed graphs. Node labels allow us to treat the problem of universal and existential quantification. Arc labels are used for treating the "defined classes" and negations.

Question-answering process amounnts to searching for graph morphisms called "enlargements". Such morphisms are also used for making networks consistent.

Precedence rules are given for non-monotonic inferences, i.e. for selecting between enlargements or between "enlargement paths". These rules are suitable for simple hierarchical taxonomies, and also when predicates occur.

Keywords: *knowledge representation, semantic network, graph morphism, non-monotony.*

Semantic networks are very convenient for representing knowledge. Scott Fahlman's NETL [4] is particularly interesting because it is based on a hierarchical net with exceptions. NETL has received first formalizations by Etherington and Reiter [3] and Touretzky [6] who related networks to non-monotonic logic, more precisely to default logic.

We here present an original way of formally defining and handling such a semantic network, using graph theory and graph morphisms. This allows us to treat some problems unsolved by Touretzky, as Fahlman's "defined classes", and existential quantification in predicates. We give rules about non-monotonic reasoning adapted from Touretzky's approach to our more complete networks.

1 NETWORKS FOR REPRESENTING KNOWLEDGE

1.1 Definition

A **network** is a multilabeled semiordered directed graph,

$H = \langle X,\delta,\sigma,\omega,\rho \rangle$.

Each **node** $x \in X$ of the graph is labeled with a **sort** $\sigma(x)$ taken in the set $S = \{\underline{i},\underline{t},\underline{p},\underline{j}\}$: $\underline{i}$ for **individual** nodes (or $\underline{i}$-nodes), $\underline{t}$ for **typical** nodes ($\underline{t}$-nodes), $\underline{p}$ for **predicates** ($\underline{p}$-nodes), and $\underline{j}$, denoting individual objects depending on typical objects, is used for representing existential quantification (j-nodes).

Each node x of a subset X_0 of X also receive a **lexical item** $\omega(x)$.

$\delta : X \rightarrow X^*$ define the successor function. Order between successors of x is only significant for the x such as $\sigma(x) = \underline{p}$.

Each p-node has at most two successors, labeled with i, t or j.

Each t-node has a non limited number of successors, labeled with t, i or j.

i-nodes and j-nodes have no successors.

Each arc is labeled with an element of the set $\Xi = \{\subset, \neq, +, \&\}$: "$\subset$" for ordinary inheritance arcs, "$\neq$" for negative or uncertain inheritance arcs, "+" and "&" for defined classes. In the following we suppose that there is a single arc <x,y> for each $x,y \in X$, and we note $\rho(x,y)$ its label. Ordinary inheritance arc labels may be omited in figures.

We will note $\delta^*(x)$ the set of descendants of a node x.

For example, the following knowledge

the typical dog hates the typical cat,
the typical cat has a father being a cat,
Fido is a dog and Pussy is a cat,
the typical dog and the typical cat are animals

is represented by the network of figure 1.

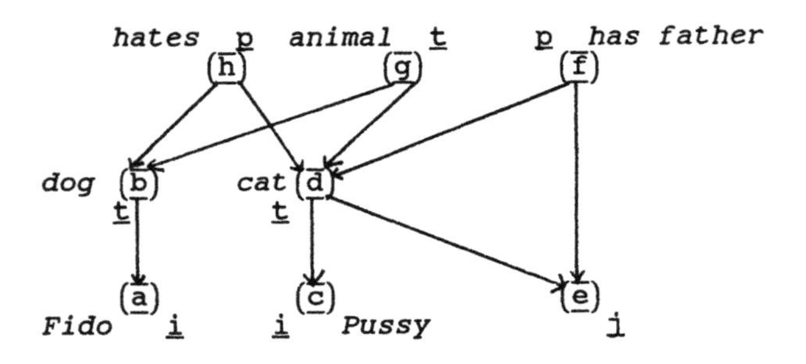

figure 1

Note that node e, labeled with j, represents *the father of the typical cat*, i.e. an existential quantified notion. Each existential quantified notion, and then each j-node x, depends on a universal quantified notion. In the simplest case, this universal quantified notion is represented by a single t-node y and there is a p-node z such as $\delta(z) = yx$ or $\delta(z) = xy$. More generally, a j-node is at least the successor of one p-node.

Thus we have a representation of universal quantification and of existential quantification. Touretzky only allows implicit universal quantification, while KL-ONE [1] allows only the representation of existential quantification in predicates. Hendrix [5] allows both existential quantification and universal quantification, but its "partitioned networks" are objects more complicated than simple graphs. We will see later (section 2.1) that extracting answers to questions remains for us a simple process.

1.2 Defined classes

As Fahlman did, we have to distinguish between arcs denoting definition properties of a typical object (these arcs are labeled with "+") and arcs denoting ordinary inferences (labeled with "⊂"). Then, it is possible to represent knowledge like *purple mushrooms are poisonous* (figure 2), without confusing it with *poisonous mushrooms are purple*.

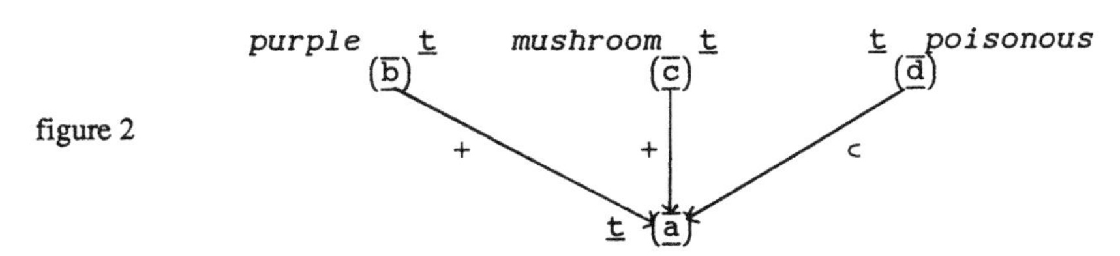

figure 2

Moreover, we use arc label "&" in order to distinguish between adjectives denoting intrinsic properties and adjectives denoting relative properties of an object. For instance, *a small elephant* is not *a small object*. In figure 3 is represented the knowledge *the typical small elephant is kind.*

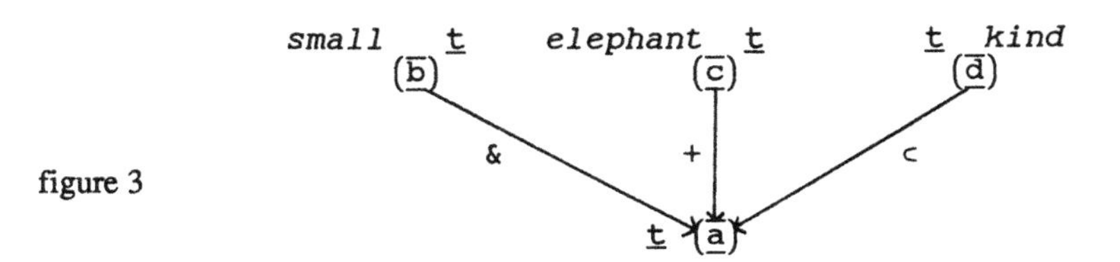

figure 3

Now we can say, if there is a path $<x_0,x_1,...,x_n>$ where

$\forall i \in \{0,1,...,n-1\}\ \sigma(x_i) = \underline{t}$

$\sigma(x_n) \in \{\underline{i},\underline{j},\underline{t}\}$

$\forall j \in \{1,...,n\}\ \rho(x_{j-1},x_j) \in \{\subset,+\}$

that x_n **inherits** all properties of x_0.

It is not the case if one of the arcs is labeled with "&".

Predicates may also be used to define typical objects. So, in figure 4 node f represents *the typical person who owns a plane* and node a represents *the typical man who owns a Boeing.*

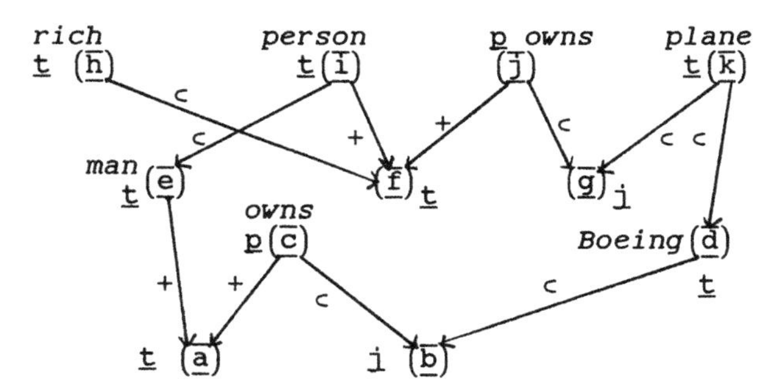

figure 4

We extend this notation for also defining individuals, as in *the man who loves Mary is young* (figure 5).

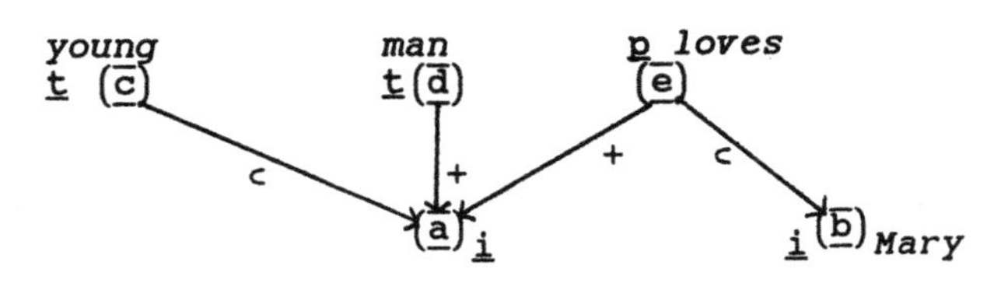

figure 5

Hence, any typical (or individual) object is defined by a subgraph of the network: the **definition subgraph** of a node x.

This subgraph is obtained by recursively building its node set Y(x), as follows:

(1) If $x \in X_0$, then $Y(x) = \{x\}$
(2) Else let be $x_1,...,x_p$ all the X t-nodes such as, for each i, $x \in \delta(x_i)$ and $\rho(x_i,x) \in \{+,\&\}$, and let be $y_1,...,y_q$ the p-nodes such as, for each j, $x \in \delta(y_j)$ and $\rho(y_j,x) =$ "+".

We have:
$Y(x) = \{x\} \cup Y(x_1) ... \cup Y(x_p) \cup Z(y_1,x) \cup ... \cup Z(y_q,x)$

Z(y,x), for y being a p-node and x a y successor, is the set defined by:

(1) if $\delta(y) = x$ then $Z(y,x) = \{y\}$
(2) else let z be the y successor different from x;
(2.1) if $\sigma(z) = \underline{i}$ or $\sigma(z) = \underline{t}$ then $Z(y,x) = \{y\} \cup Y(z)$
(2.2) if $\sigma(z) = \underline{j}$ let $u_1, u_2,..., u_r$ be the X t-nodes such as, for each j, $z \in \delta(u_j)$;
let $v_1, v_2,..., v_s$ be the X p-nodes such as, for each j, $v_j \neq y$ and $z \in \delta(v_j)$. We have:
$Z(y,x) = \{y,z\} \cup Y(u_1) \cup ... \cup Y(u_r) \cup Z(v_1,z) \cup ... \cup Z(v_s,z)$

Example: For the definition subgraph H' of node f (figure 4):
$Y(f) = \{f\} \cup Y(i) \cup Z(j,f)$
$Y(i) = \{i\}$
$Z(j,f) = \{j,g\} \cup Y(k) = \{j,g,k\}$
$Y(f) = \{f,i,j,g,k\}$

This subgraph is shown in figure 6.

figure 6

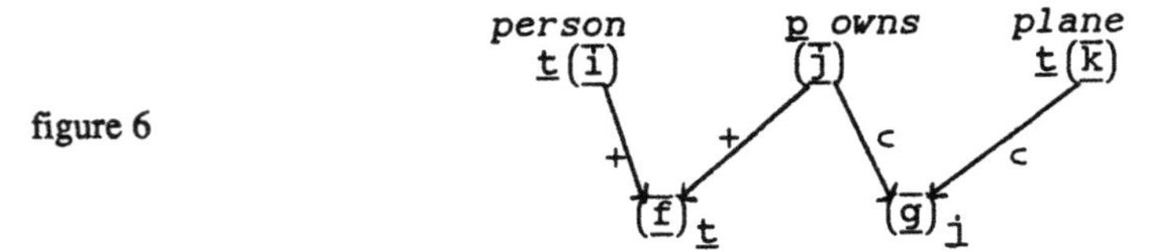

1.3 Negation

There exist two different types of negation: negation over inheritance properties and negation over predicates.

To indicate that x *is not a* y, we create a typical node z labeled with the lexical item *no*, representing the typical object which *is not a* y. z is a y successor such as $\rho(z,x) = "\neq"$.

In network of figure 7 are represented the following negative inheritances:

the typical dolphin is not a fish

the typical fish is not a mammal

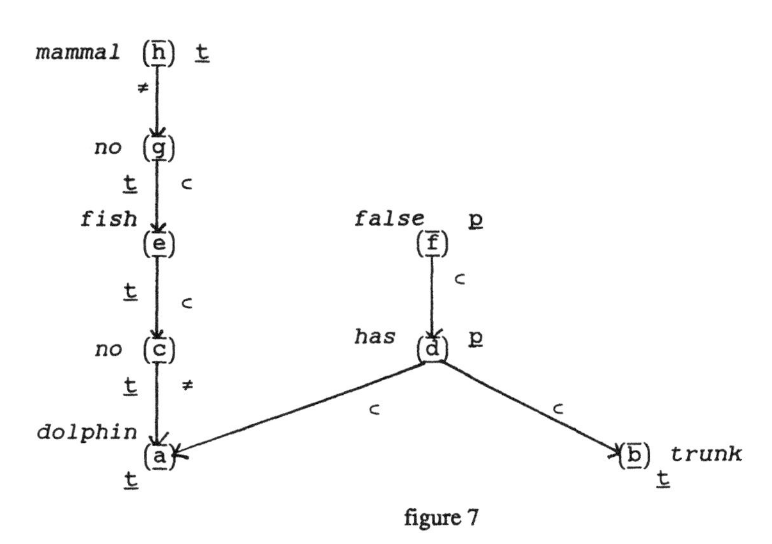

figure 7

To indicate that the predicate node u has *false* as truth-value, we create a predicate node v labeled with the lexical item *false*, such as u is a v successor.

For example in figure 7 is also represented the knowledge:

the typical dolphin hasn't any trunk

This is in fact the representation of the knowledge: *the typical dolphin hasn't the typical trunk.* Then, if we consider any individual *dolphin* x and any individual *trunk* y, by inheritance we know that x *hasn't* y.

Because there is a special node labeled with *no* or *false*, we can generalize this approach by substituting to the lexical items *no* or *false* other items: adverbs like *a little*, *very much*, *perhaps*, or even temporal conditions or numerical items of certainty. In such a case, the answer to a question is not *yes* or *no* and it is not obtained by proving a theorem.

For example, if is recorded the knowledge *John is somewhat intelligent*, or the knowledge *John perhaps loves Mary* (figure 8), it is possible to give an answer to the question *who is intelligent?* or to the question *does John love Mary?* The user of the system will interpret this answer in accordance with his convenience.

2 QUESTION-ANSWERING AND ENLARGEMENTS

2.1 Question-answering

A question is represented by a network $G = \langle X_1,\delta_1,\sigma_1,\omega_1,\rho_1\rangle$ in which $\sigma_1(x)$ is generally not defined for all x. For example, network G of figure 9 represents the question:

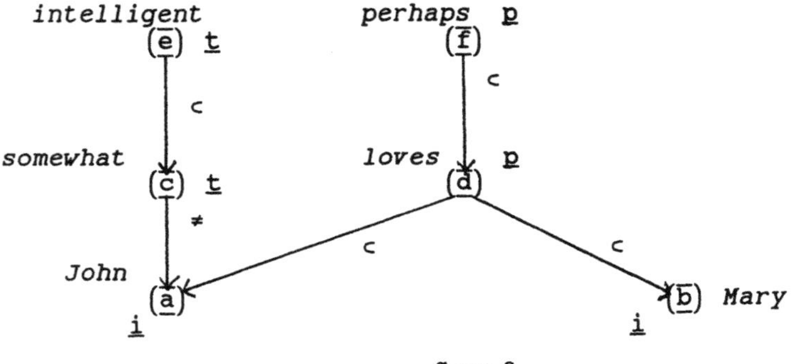

figure 8

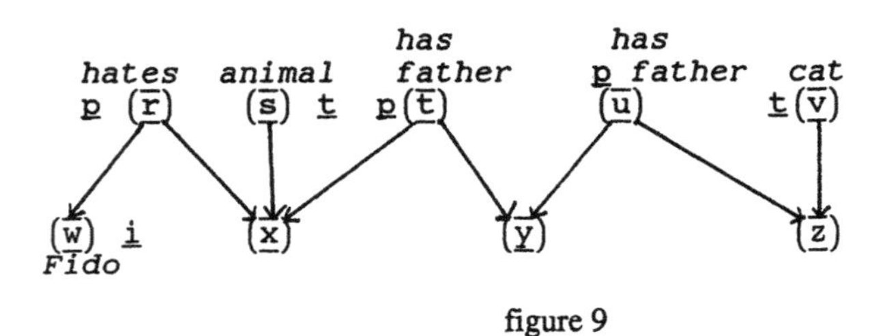

figure 9

does Fido hate an animal having a father whose father is a cat ?

Searching for the answer to a question consists in searching for a graph morphism from G to H representing the recorded knowledge, $\varphi : G \rightarrow H$, called **enlargement**. Such a morphism preserves the sorts and the lexical items, and has the property that to each arc of G corresponds a path of H:

$\forall$ x,y $\in X_1$ $y \in \delta_1(x) \Rightarrow \varphi(y) \in \delta^*(x)$

φ also satisfies the following condition on order: if x is a p-node of G such as $\delta_1(x) = yz$, then there exist y', z' $\in$ X such as $\delta\varphi(x) = y'z'$, $\varphi(y) \in \delta^*(y')$ and $\varphi(z) \in \delta^*(z')$.

The answers to a question are found by a parallel marker bits propagation algorithm. (see [2] for more details)

In the above example related to network H of figure 1, we have three graph morphisms, and therefore three answers to the question.

These morphisms are described by the following array:

r	s	t	u	v	w	y	z	x
h	g	f	f	d	a	e	e	c/d/e

Nodes corresponding to x are labeled with i, t, or j. The three answers may be expressed by:

(1) *Fido hates Pussy*

(2) *Fido hates the typical cat*

(3) *Fido hates the typical cat's father*

Node e of graph H, labeled with j, has different meanings when occuring in an answer.

In (1), it represents *Pussy's father* and *Pussy's grandfather*.

In (2), it represents *the typical cat's father* and *the typical cat's grandfather*.

In (3), it represents *the typical cat's father, the typical cat's grandfather* and *the typical cat's great-grandfather*.

Thus we see how nodes labeled with j may be used in the question-answering process.

2.2 Consistency rule about defined classes

The following consistency rule must be verified:

Rule 1: Let H' be the definition subgraph of a node x. If there is a subgraph H" different from H' and an enlargement φ : H' $\rightarrow$ H" preserving sorts (unless j which may be converted to i), then $\varphi(x)$ has to be a descendant of x.

This rule means that if $\varphi(x)$ has all definition properties of x, it also has all its incidental properties. To find if it is the case amounts to the same process than question-answering.

Example: In figure 4, node a has to be a descendant of node f, else we cannot infer that *the man who owns a Boeing is rich*. This is because there is an enlargement φ between subgraph H' of figure 6 and subgraph H" (figure 10). This enlargement is shown in the following array:

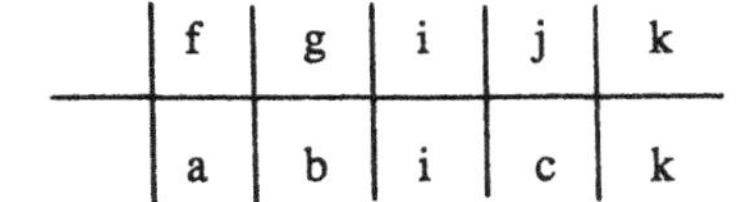

f	g	i	j	k
a	b	i	c	k

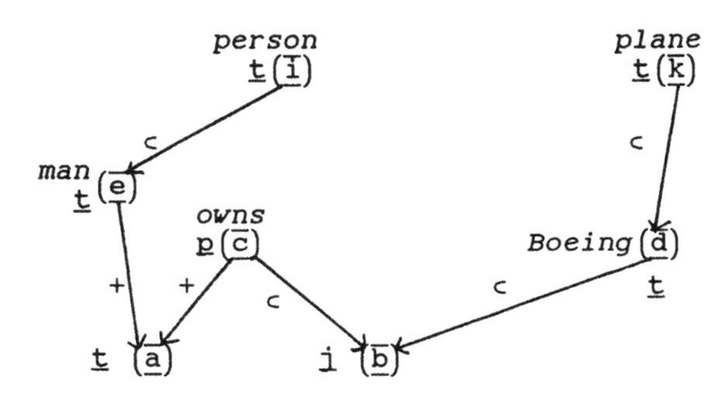

figure 10

3 NON-MONOTONIC REASONING

If there are several answers to a question, it is possible to find out all valid enlargements, and to let the user make a choice. But it is also possible to give precedence rules for automatically chosing one answer.

3.1 Case of hierarchical taxonomies

An enlargement φ from a network G having a single arc <y,x> to a network H is called a **(first type) elementary enlargement**. All paths from $\varphi(y)$ to $\varphi(x)$ are **enlargement paths**.

Consider such a path $<z_1,z_2,...,z_p>$. It is a **yes path** iff for each i $\rho(z_i,z_{i+1}) = "\subset"$. It is a **no path** iff $\rho(z_1,z_2) = "\neq"$ and $\rho(z_i,z_{i+1}) = "\subset"$ for each other i. It is a **no conclusion path** in other case.

There is a **conflict** iff there is both a yes path $<v_1,v_2,...,v_q>$ and a no path $<w_1,w_2,...,w_r>$ with $v_1 = w_1 = \varphi(y)$ and $w_r = v_q = \varphi(x)$. This conflict is solved by the following precedence rule:

Rule 2: v_q *is a* v_1 iff there is a yes-path $<v_1,v_2,...,v_q>$ and for each no-path $<w_1,w_2,...,w_r>$ w_3 is a strict ancestor of v_2.

w_r *is not a* w_1 iff there is a no-path $<w_1,w_2,...,w_r>$ and for each yes-path $<v_1,v_2,...,v_q>$ v_2 is a strict ancestor of w_3.

In any other case, it is impossible to conclude.

Consider for instance the network H of figure 11 and the question *Is Flipper a mammal ?* (network G). We have a single enlargement, but three enlargement paths <f,h,g,c,b,a> , <f,c,b,a> and <f,e,d,b,a> corresponding to <y,x>. First path is a no path, second path is a yes path and third path is a no conclusion path.

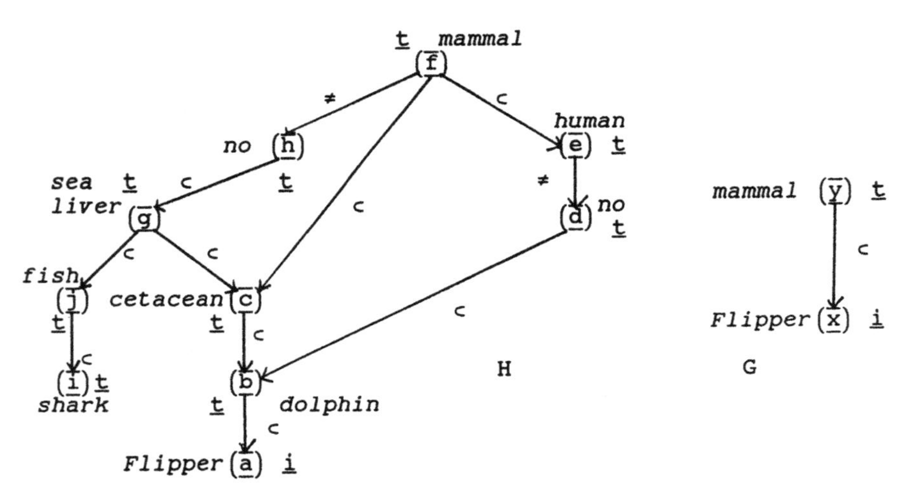

figure 11

By applying rule 2, we conclude that *Flipper is a mammal*, because $c \in \delta^*(g)$ and $c \neq g$.

A **second type elementary enlargement** is such as there are two paths $<v,v_q,...,v_1>$ and $<v,v_{q+1},...,v_r>$ where $v_1 = \varphi(x)$, $v_r = \varphi(y)$ and where $\rho(v,v_q) = "\neq"$ or (exclusively) $\rho(v,v_{q+1}) = "\neq"$, all the other arc labels being equal to "$\subset$".

With such an enlargement, we can infer that x *is not a* y.

For example, in graph of figure 11 we see that *the typical shark is not a human* (or that *the typical human is not a shark*). This is because there are the paths <f,h,g,j,i> and <f,e>.

When there is a conflict between a first and a second type elementary enlargements, first type has always precedence.

For example, in figure 11 we conclude that *the typical dolphin is a sea liver*, although there are the paths <f,c,b> and <f,h,g>.

Conflicts can't occur between second type enlargements.

3.2 Case of knowledge with predicates

Consider the question *does John love a woman ?* It is possible to provide the non-contradictory answers *John loves Mary* and *John doesn't love Ann.* We can even simultaneously accept the two answers *John loves Mary* and *John doesn't love the typical woman.* These answers must not be compared. Nevertheless, the answers *John loves Mary* and *John doesn't love Mary* have to be compared. That is why we give the following definition:

Two enlargements φ and φ' from G to H are **similar** iff for each x of X_1 such as $\sigma_1(x)$ is undefined or $\sigma_1(x) \in \{\underline{i},\underline{t},\underline{j}\}$ $\varphi(x) = \varphi'(x)$.

If φ and φ' are two similar enlargements, let be, for each $\underline{p}$-node $x \in X_1$, $\delta(x) = x_1x_2...x_q$ the sequence of successors of x, $\delta\varphi(x) = y_1y_2...y_q$, and $\delta\varphi'(x) = z_1z_2...z_q$. We can now express the second precedence rule:

Rule 3: φ is preferred to φ' iff for each $\underline{p}$-node x and for each i, $y_i \in \delta^*(z_i)$.

Therefore, we deduce the notion of strict precedence, which allows us to conclude. If there is no strict precedence between two enlargements, we can't conclude.

Consider for example (figure 12) network H representing knowledge and network G representing the question *does Flipper like an animal who has a trunk ?*

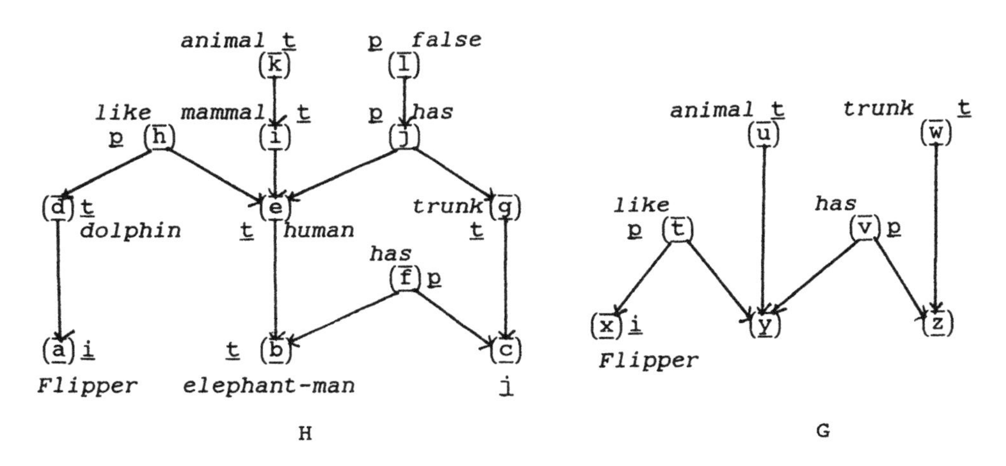

figure 12

There are five enlargements from G to H, corresponding to the following answers:

(1) *Flipper like the typical human, but the typical human hasn't the typical trunk*

(2) *Flipper like the typical human, but the typical human hasn't the trunk of the typical elephant-man*

(3) *Flipper like the typical elephant-man, but the typical elephant-man hasn't the typical trunk*

(4) *Flipper like the typical elephant-man, but the typical elephant-man, being a human, hasn't the trunk of the typical elephant-man*

(5) *Flipper like the typical elephant-man, and the typical elephant-man has a trunk*

The two last enlargements from G to H are similar. They are shown in the following array:

t	u	v	w	x	y	z
h	k	j/f	g	a	b	c

By applying rule 3 we conclude that the enlargement strictly preferred is the one verifying $\varphi(v) = f$, i.e. the last one. Then we have a positive answer to the question.

We can combine precedence rule 3 to rule 2. To each inheritance arc of the question, must correspond a yes path $<v_1,...,v_q>$ of the preferred answer such as v_q *is a* v_1.

Notes

[1] BRACHMAN R. J. and SCHMOLZE J. G., 1984, "An Overview of the KL-ONE Knowledge Representation System", FLAIR Technical Report nł 30, Palo Alto.
[2] CORI M., 1987, "Modèles pour la représentation et l'interrogation de données textuelles et de connaissances", Thèse de doctorat d'Etat ès sciences, Université Paris 7.
[3] ETHERINGTON D.W. and REITER R., 1983,"On Inheritance Hierarchies With Exceptions", Proc AAAI-83, Washington.
[4] FAHLMAN S.E., 1979, *NETL A system for representing and using real-world knowledge*, The MIT Press.
[5] HENDRIX G.G., 1979, "Encoding Knowledge in Partitioned Networks". *Associative networks. Representation and Use of Knowledge by Computers*, Edited by N.V. Findler, Academic Press.
[6] TOURETZKY D. S., 1986, *The Mathematics of Inheritance Systems*, Pitman/Morgan Kaufmann.

INFERENCE IN POSSIBILISTIC HYPERGRAPHS

Didier DUBOIS - Henri PRADE

Institut de Recherche en Informatique de Toulouse
Université Paul Sabatier, 118 route de Narbonne
31062 TOULOUSE Cedex (FRANCE)

Abstract

In order to obviate soundness problems in the local treatment of uncertainty in knowledge-based systems, it has been recently proposed to represent dependencies by means of hypergraphs and Markov trees. It has been shown that a unified algorithmic treatment of uncertainties via local propagation is possible on such structures, both for belief functions and Bayesian probabilities, while preserving the soundness and the completeness of the obtained results. This paper points out that the same analysis applies to approximate reasoning based on possibility theory, and discusses the usefulness of the idempotence property for combining possibility distributions, a property not satisfied in probabilistic reasoning. The second part analyzes a previously proposed technique for handling dependencies, by relating it to the hypergraph approach.

Keywords

Hypergraph, Markov trees, possibility theory, approximate reasoning.

1. INTRODUCTION

Starting with a knowledge base made of pieces of information which may be pervaded with uncertainty or vagueness, an ideal inference mechanism should be able to provide the user with conclusions which are i) sound, ii) as certain and precise as it is possible, iii) obtained by a computation procedure which is as efficient as possible. Obviously, this aim may be more or less easily reached, depending on the modeling of uncertainty which is used and the format of the pieces of knowledge allowed in the framework of this modeling. We may distinguish between local and global computation methods (e.g. Dubois and Prade, 1987). A typical local approach to the treatment of uncertainty is the one used in rule-based expert systems where rules are evaluated and triggered one after the other according to the control procedure of the inference engine. The procedure in that case is computationally efficient but no guarantee exists usually about the optimality in terms of certainty and precision of the conclusion and even in some cases its soundness may be questioned. This is mainly due to problems raised by the necessity of combining partial uncertain conclusions pertaining to the same matter, without being able to take into account implicit

5. NON-STANDARD LOGICS

dependencies or redundancies in the knowledge base. The probabilistic reasoning method advocated by Nilsson (1986) is an example of global approach where the best lower and upper bounds on the probability of a conclusion are obtained by solving a (linear) programming problem whose constraints are directly obtained from the available bounds on the probability of the different granules in the whole knowledge base. In that case the soundness and the optimality of the conclusion can be guaranteed, but the computational cost may be heavy in practice and moreover it is then difficult to provide such an inference system with explanation capabilities.

Local propagation methods have been recently proposed in Bayesian networks (Pearl, 1986), and Markov networks (Lauritzen and Spielgelhalter, 1988). This kind of approach assigns precise probability values to any proposition in a given context, making an extensive use of conditional independence assumptions. Then the soundness and optimality issues coincide in the sense that in these approaches one is interested in getting estimations of the probability values which are as close as possible to the actual values. Both Bayesian and Markov networks aim at representing conditional independence assumptions using oriented graphs and non- oriented graphs respectively. Similarly, Shafer et al.(1987) have proposed to use hypergraphs and Markov trees for representing dependence relations between variables, the uncertainty being modelled in terms of belief functions (Shafer, 1976) and have developed local propagation methods. In a more general framework Shafer and Shenoy (1988, 1990) have studied axioms which propagation and combination operations should satisfy to ensure that local algorithms correctly work in hypergraph representations ; see also (Williams, 1990).

Independently and quite at the same time an approach has been proposed for handling dependencies in approximate reasoning by Chatalic, Dubois and Prade (1987). It is applicable to Shafer's belief functions as well as to Zadeh's (1978) possibility measures. The purpose of this paper is to give an improved presentation of this approach, in the framework of possibility theory, to connect it with the approaches mentioned at the end of the preceding paragraph, and also to relate it to recent extensions of the constraint propagation paradigm for the handling of fuzzy values (Dubois and Prade, 1989a, b ; Yager, 1989). Indeed the updating of the possible ranges of variables linked by a set of constraints has been studied for a long time in Artificial Intelligence and the general procedure originally proposed by Waltz has been refined for many particular cases corresponding to different kinds of relations between the variables ; see Davis (1987) for instance. Let us also mention the work recently developed by Kruse and Schwecke (1988, 1989, 1990) in the possibilistic framework for handling dependencies in (causal) networks.They propose a local propagation algorithm for rule-based systems where rules are expressed as dependence relations, and show that this approach carries over to fuzzy rules following Zadeh(1979).

2. POSSIBILISTIC REASONING AND HYPERGRAPHS

In possibility theory (Zadeh, 1978), what is known about the value of variables or about the existing relations between variables is represented by means of possibility distributions. Namely, a possibility distribution π_X attached to the variable X, is a mapping from the domain $\mathcal{X}$ of X to the interval [0,1] ; $\forall$ x $\in$ $\mathcal{X}$, $\pi_X(x)$ reflects to what extent it is possible that X = x according to the available information. If two possibility distributions π_X and π'_X such that $\pi_X \leq \pi'_X$ are available from two different sources, π_X is said to be more restrictive than π'_X since each possible value for X receives a smaller possibility degree according to π_X than according to π'_X ; π_X then expresses a less uncertain and/or imprecise information. The normalization condition which is usually applied to π_X is $\exists\ x \in \mathcal{X}$, $\pi_X(x) = 1$ which expresses that at least one value in $\mathcal{X}$ is considered as completely possible for X (exhaustiveness of the domain) and which allows that distinct values in $\mathcal{X}$ be simultaneously regarded as completely possible. More generally a possibility distribution $\pi_{X_1,\ldots,X_n}$, defined on a Cartesian product $X_1 \times \ldots \times X_n$, expresses a dependency relation between the variables $X_1, \ldots, X_n$; for instance $\pi_{X_1,\ldots,X_n}(x_1, \cdot, \ldots, \cdot)$ characterizes the fuzzy set of the more or less possible values of the tuple $(X_2,\ldots,X_n)$ when $X_1 = x_1$. Normalization of $\pi_{X_1,\ldots,X_n}$ guarantees the existence of a completely possible interpretation $(x_1,\ldots,x_n)$ compatible with the piece of knowledge expressed by $\pi_{X_1, \ldots, X_n}$.

In this framework a general procedure known as the conjunction/projection method (Zadeh, 1979) can be applied to a knowledge base in order to deduce what can be said about the value of a variable of interest, or the relationship which jointly constrains a set of variables. Namely, let $A_1, \ldots, A_n$ be the fuzzy sets of the possible values of the variables $X_1, \ldots, X_n$ (i.e. $\pi_{X_i} = \mu_{A_i}$) ; if nothing is known about the value of X_i, then $\mu_{A_i}(x_i) = 1$, $\forall\ x_i \in \mathcal{X}_i$. Let $R_1,\ldots, R_m$ be the (fuzzy) relations stated in the knowledge base between variables. R_j is then defined on the Cartesian product of the domains $\mathcal{X}_k$'s of the variables X_k involved in the relationship represented by R_j. Then the conjunction/projection method consists in

i) performing the combination

$$\pi^*_{X_1,\ldots,X_n} = \min(\min_{i=1,n} \mu_{A_i}, \min_{j=1,m} \mu_{R_j}) \qquad (1)$$

which is the *least restrictive* possibility distribution for the tuple $(X_1, \ldots, X_n)$ compatible with all the constraints.

ii) projecting the result $\pi^*_{X_1,\ldots,X_n}$ on the domain(s) of the variable(s) we are interested in.For instance we get for X_i

$$\forall\ x_i \in \mathcal{X}_i,\ \pi_{X_i}(x_i) = \sup_{x_j,\ j=1,n,\ j\neq i} \pi^*_{X_1,\ldots,X_n}(x_1, \ldots, x_n) \qquad (2)$$

$\pi^*_{X_1,\ldots,X_n}$ is generally supposed to be normalized.The lack of normalization of $\pi^*_{X_1,\ldots,X_n}$ would express the inconsistency of the knowledge base ; namely, the quantity $1 - \sup_{x_1,\ldots,x_n} \pi^*_{X_1,\ldots,X_n}$ estimates the degree of inconsistency of the knowledge base. The projection preserves the normalization. So, there always exists an interpretation fully compatible with any conclusion obtained by the combination/projection approach when the knowledge base is consistent.

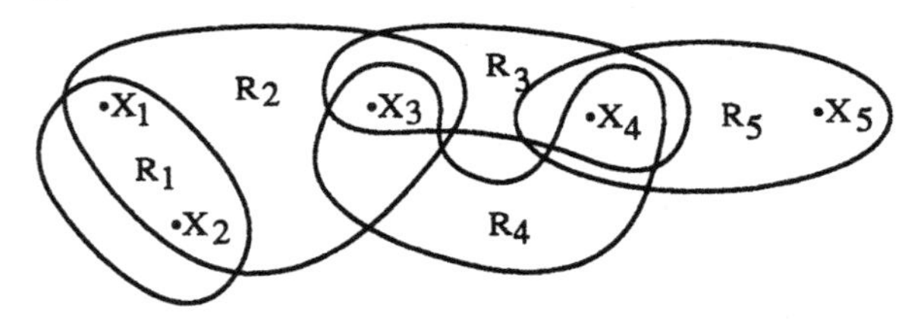

Figure 1

An example of knowledge base is illustrated by the hypergraph on Fig. 1, where variables are nodes and relations are pictured by hyperedges ; an hyperedge, denoted S, corresponds to a subset of variables (nodes) ; however there may be several hyperedges between the same subset of nodes. Then for instance computing the possibility distribution attached to X_3 from the knowledge of A_2 and A_3 (the other A_i's being such that $A_i = \mathcal{X}_i$) leads to the expression, $\forall\, x_3 \in \mathcal{X}_3$

$$\pi_{X_3}(x_3) = \sup_{x_1,x_2,x_4,x_5} \min(\mu_{R_1}(x_1,x_2), \mu_{R_2}(x_1,x_2,x_3), \mu_{R_3}(x_3,x_4), \mu_{R_4}(x_3,x_4), \mu_{R_5}(x_4,x_5), \mu_{A_2}(x_2), \mu_{A_3}(x_3)).$$

Here the hypergraph visualises the decomposition of a possibility distribution into non-interactive components; this feature make approximate reasoning similar to Bayesian networks (decomposition of a joint probability into a product of marginals or conditional probabilities) and belief function techniques (decomposition of a belief function into independent pieces of evidence). From the point of view of constraint propagation, the updating of the variable X_i, from A_i to A'_i, is expressed in this framework by

$$\forall i, \forall x_i,\ \mu_{A'_i}(x_i) = \sup_{x_j,\ j=1,n\ ;\ j\neq i} \min(\min_{k=1,m} \mu_{R_k}(x_1,\ldots,x_n), \min_{j=1,n} \mu_{A_j}(x_j)) \quad (3)$$

$$\leq (\min(\mu_{A_i}(x_i), \min_{k=1,m}[\sup_{x_j,\ j=1,n\ ;\ j\neq i} \min(\mu_{R_k}(x_1,\ldots,x_n), \min_{j=1,n\ ;\ j\neq i} \mu_{A_j}(x_j))]) \quad (4)$$

The inequality (4) expresses that if we take into account each R_k separately in the updating process, we are not sure, even if we iterate the procedure as in the Waltz algorithm, of obtaining the most restrictive possibility distribution for X_i ; however the result provided by the right-hand part of (4) is more easy to compute in general and obviously sound (in the sense that what is obtained is not as restrictive and informative as what could be obtained, but, as such, cannot be arbitrarily restrictive). Note that in case of binary relations, it can be shown that the Waltz procedure (i.e. the separate processing of the R_k's) yields the most accurate result given by (3), provided there is at most one relation R_k between any pair of variables and that there is no cycle in the non-oriented graph whose nodes correspond to variables and edges to binary relations.

More generally, it is not surprizing that the general procedure which consists first in performing the general combination of all the representations of the pieces of information in the knowledge base and then projecting the result, is generally computationally untractable in practice. Then it is natural to try to take advantage of the fact that each relation usually relates only a (small) subset of the variables and to exploit the structural properties enjoyed by the combination and projection operations. Indeed it can be easily checked that the four axioms which are at the basis of the local computation scheme proposed by Shafer and Shenoy (1988a) (see also Williams, 1990), are satisfied in the possibilistic framework. Namely let $I \subset \{1, \ldots, n\}$, R be defined on the Cartesian product of $\mathcal{X}_i$'s where $i \in I$, we have

A0 ("Identity") $\sup_{x_i,\ i \notin I} \mu_R(\ldots, x_i, \ldots) = \mu_R$ (assimilating μ_R with its cylindrical extension on $\mathcal{X}_1 \times \ldots \times \mathcal{X}_n$).

A1 ("Consonance of marginalization") Let $K \subseteq J \subseteq I$, then

$$\sup_{x_i,\ i \notin K} \mu_R(\ldots, x_i, \ldots) = \sup_{x_i, i \notin K} (\sup_{x_i, i \notin J} \mu_R(\ldots, x_i, \ldots))$$

A2 Commutativity and associativity of the combination (obvious with 'min')

A3 ("Distributivity of marginalization over combination") Let S be a relation defined on the Cartesian product of $\mathcal{X}_i$'s where $i \in L \subset \{1, \ldots, n\}$, then

$$\sup_{x_i,\ i \notin I} \min(\mu_R, \mu_S) = \min(\mu_R, \sup_{x_i \notin I \cap L} \mu_S)$$

(for instance, let R and R' be defined on $X_1 \times X_2$ and $X_2 \times X_3$ respectively, then

$$\sup_{x_3} \min(\mu_R(x_1,x_2), \mu_{R'}(x_2,x_3)) = \min(\mu_R(x_1,x_2), \sup_{x_3} \mu_{R'}(x_2,x_3)))$$

A2 and A3 also hold with conjunctive operations other than min (e.g. product, which is used in the Bayesian setting and in evidence theory). These properties ensure that the local treatment of uncertainty is equivalent to the global treatment, provided that the hypergraph can be translated into a Markov tree. A Markov tree is a tree $G = (\mathcal{N},\mathcal{A})$ such that its nodes correspond to hyperedges of the hypergraph, or unions thereof ; there is an edge $(S,S') \in \mathcal{A}$ whenever $S \cap S' \neq \emptyset$, and moreover if S, S' are any two nodes, $S \cap S'$ is contained in all nodes on the (unique) path from S to S' (Markov property). When the Markov tree conditions are fulfilled, $\mathcal{N}$ is called an hypertree. Given a general hypergraph, it is possible to cover it with a hypertree by merging suitable hyperedges. The bigger the hyperedges, the greater the spaces in which global combination projection operations must be done. Hence the problem is to cover hypergraphs with hypertrees having as small hyperedges as possible (see Shafer et al., 1987 ; Shafer and Shenoy, 1990).

In the possibilistic framework the combination is also idempotent. An interesting question is then : how does this property simplify a local treatment of uncertainty ? First, it must be noticed that in possibilistic reasoning the projection operation is the same as for general belief functions. Now, as a consequence of the idempotence of min, performing local inference on a possibilistic hypergraph will never produce wrong results. The reason why this is not so with belief functions is that Dempster rule used on related pieces of evidence can take into account twice the same information, thus leading to arbitrary reinforcement effects Nothing of the like occurs in possibilistic reasoning. The tree structure turns out to be useful for possibilistic reasoning because a local treatment of cyclic structures usually leads to a loss of information through projection steps.

3. AN HYPERTREE GENERATION AND QUESTION ANSWERING PROCEDURE

A procedure for simultaneous treatment of dependencies and question answering has already been proposed by Chatalic, Dubois and Prade (1987) and Chatalic (1986). The procedure makes use of two basic transformations which are to be applied the dependency hypergraph, namely

<u>The (hyper)edge fusion</u>
It aims at merging dependence relations pertaining to the same variables, so that combination may take place before any projection step. A simple example is on Figure 2 : knowing $\pi_X = \mu_{A_1}$, if we want to compute π_{X_2} we cannot "propagate along" R_1 and R_2 separately and combine the results afterwards, without loosing information, as a consequence of (4) (see Dubois, Martin-Clouaire and Prade, 1988) ; namely :

$$\sup_{x_1} \min(\mu_{A_1}(x_1), \mu_{R_1}(x_1,x_2), \mu_{R_2}(x_1,x_2)) \leq \min(\sup_{x_1} \min(\mu_A(x_1), \mu_{R_1}(x_1,x_2)), \sup_{x_1} \min(\mu_A(x_1), \mu_{R_2}(x_1,x_2)))$$

(which is a particular case of (3)-(4)). This operation also applies to any family of hyperedges such that one of them contains the other ones.

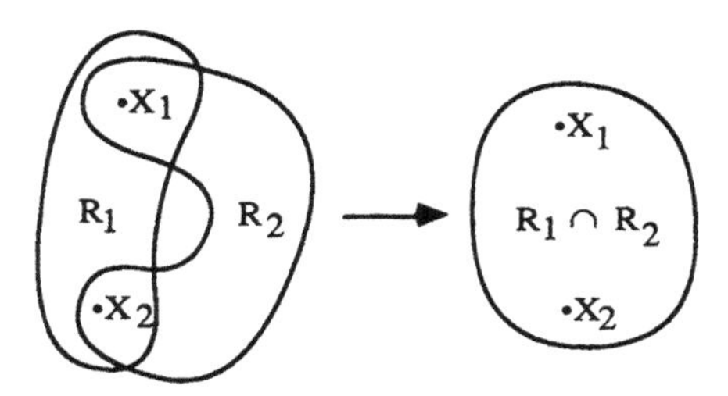

Figure 2

The variable fusion

It aims at destroying cyclic structures by considering two or more variables jointly as a vector-valued variable. It results in the fusion of hyperedges in the original hypergraph and eventually produces a Markov tree structure. A simple example of this operation is on Figure 3. If we want to evaluate X_1, we must evaluate X_2 and X_3 due to R_2 and R_3 ; in turn evaluating X_2 requires the evaluation of X_3 and evaluating X_3 requires the evaluation of X_2 due to R_1. The way out of this tricky situation is to conjointly evaluate X_2 and X_3. Then the hyperedges corresponding to $R_2 \cap R_3$ and R_1 can be fused if needed. Clearly this still applies when there is more than one variable in common between two relations. It is also used when longer cycles are present. Consider the example on Figure 4

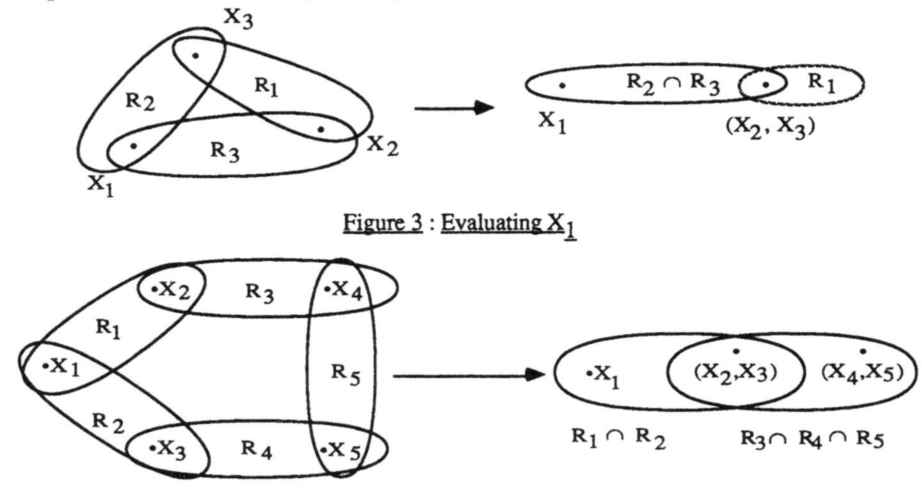

Figure 3 : Evaluating X_1

Figure 4

Starting with X_1, X_2 and X_3 are reached via R_1 and R_2, and then X_4 and X_5. The discovery of the dependence between X_4 and X_5 leads to fusing X_2 and X_3, and also X_4 and X_5.

One basic idea of the procedure presented in the following is that the treatment of the "dependency hypergraphs", induced by the relationships stated in the knowledge base, can depend on the query which is addressed to the inference system. Indeed if we consider the hypergraph in the left part of Fig. 4, in order to evaluate X_1 we can perform a variable fusion as explained above, while for instance for computing π_{X_2} we can perform another variable fusion (i.e. combining R_1 and R_3). The procedure consists in

i) starting with variable(s) to evaluate, identifying the relations in which the variable(s) take(s) part, then finding out the other variables involved in these relations and then iterating this process until we have considered all the dependency relations between variables ;

ii) recursively applying the hyperedge fusion and the variable fusion to the dependency graph thus oriented at step i).

An example of this procedure is shown in Fig. 4, starting with the "dependency hypergraph" shown in Fig. 4.a, the result of step i) is shown in Fig. 4.b and the hypertree finally obtained is in Fig. 4.c. It is easy to see that the obtained hypertree is a covering of the original hyperstructure generally. Details can be found in (Chatalic, 1986 ; Chatalic, Dubois and Prade, 1987).

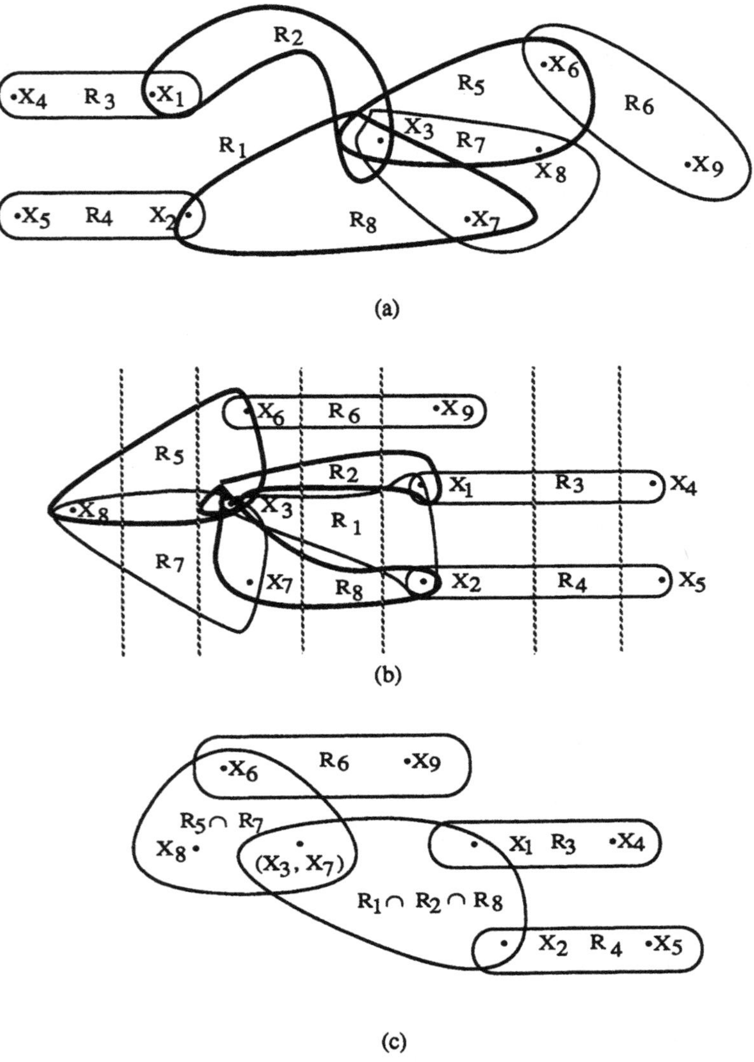

Figure 5

This procedure produces an hypertree which can be directly exploited in order to obtain a sound, optim (in the sense that it yields the most restrictive possibility distribution) evaluation of the variable(s) we are interested in. Moreover the hypertree shows what are the sequence(s) of computations to be performed and what computations can be done separately taking advantage as far as possible of the property expressed by axiom A_2. For instance, in Figure 5.c, we see that when looking for evaluation of X_8, we can evaluate X_6 only on the basis of i) R_6 and X_9 and also of ii) the information we may already have on X_6 (updating process) ; we also see that X_3 and X_7 have to be evaluated conjointly, R_1, R_2 and R_8 have to be used in combination for evaluating this pair of variables and so on. Note also that if we have no information on some variable (e.g. X_4), it is allowed to forget the corresponding branch (here X_1 R_3 X_4) in the evaluation, provided that the involved relation is normalized (i.e. $\sup_{x_4} \mu_{R_3}(x_1,x_4) = 1, \forall\ x_1$). The procedure puts the different computation steps at a level which is as local as possible if we want to be sure to have a result which is both sound and optimal, since in the hyperedge and variable fusions, and thus in the proposed procedure, we are only taking advantage of the properties A0-A3 (but not of idempotence).

Chatalic's procedure mixes up the problem of finding an hyperedge cover and the problem of evaluating a variable. Namely the hypertree in Figure 5 is obtained assuming that X_8 is to be evaluated. When evaluating another variable, another hypertree is produced from 5(a). However it is worth noticing that the hypertree on 5(c) can serve for a sound and complete evaluation of any variable. The above procedure can produce as many hypertrees as involved variables in the network. The best hypertree is again the one that has the smallest hyperedges. However the above procedure is not optimal in that respect. Figure 6 gives a better hypertree cover for the example on Figure 4.

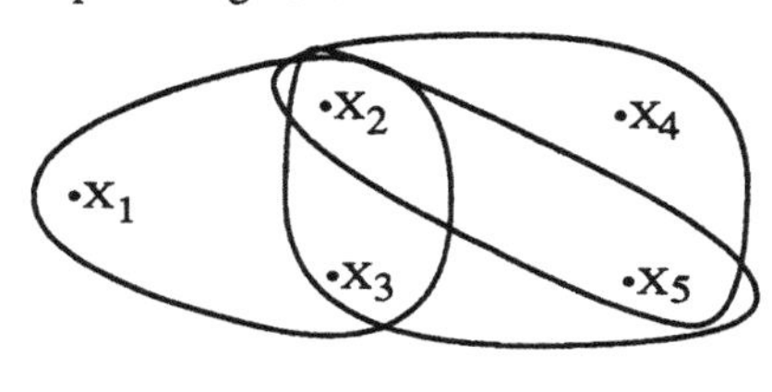

Figure 6

Indeed, the biggest hyperedge of any hypertree cover for the hypergraph, computed via Chatalic's procedure will contain four variables, due to obvious symmetry reasons, while on Figure 6, hyperedges contain only 3 variables.

N.B. : Let R be a fuzzy relation defined on $\mathcal{X}_1 \times \ldots \times \mathcal{X}_k \times \mathcal{X}_{k+1} \times \ldots \times \mathcal{X}_n$ such that $\forall x_1, \ldots, \forall x_k$, $\exists x_{k+1}, \ldots, \exists x_n$ such that $\mu_R(x_1, \ldots, x_k, x_{k+1}, \ldots, x_n) = 1$. $X_1, \ldots, X_k$ are called input variables of R. This condition guarantees that whatever the available information on $X_1, \ldots, X_k$ is, provided that this information is expressed in terms of *normalized* possibility distributions, the result of any projection of the combination of R with this information, will be always normalized. In other words any information on $X_1, \ldots, X_k$ will be consistent with R. Thus if all the relations obtained in the hyperedge fusion process (e.g., R_3, R_4, R_6, $R_1 \cap R_2 \cap R_8$, $R_5 \cap R_7$ in Fig. 5.c) enjoy this property with respect to their input variables, then whatever the available information on the input variables of the hypertree we have built, we are sure to obtained a normalized possibility distribution on the variable(s) we are evaluating (i.e. X_8 in the example of Fig. 5). When some relations do not satisfy the above condition with respect to their input variables, we are in the situation recently considered by Yager and Larsen (1990) of a *potentially* inconsistent knowledge base.

4. CONCLUDING REMARKS AND OTHER RELATED ISSUES

In this paper we have considered that what is available in the knowledge base, are possibility distributions restricting tuples of variables. In practice, the expert knowledge is rather given under the form of rules from which it is possible to build the associated possibility distributions, taking into account the intended meaning of the rules (see Dubois and Prade, 1989c). So, the procedure described here could be applied to fuzzy expert systems, especially the hyperedge fusion for combining parallel rules relating the same variables, rather than combining their conclusions. However it can be shown that the most efficient way of computing with rules in parallel is not to combine their associated possibility distributions explicitly as formally suggested by the procedure we have presented, but generally to consider them jointly and to build from them a new rule adapted to each input value to consider. It gives the optimal result directly (see Dubois, Martin-Clouaire and Prade, 1988). But this point concerns the local computation within a node of a Markov tree representation. It does not question the interest of representing a knowledge base of fuzzy rules via a Markov tree structure.

Based on this framework, it is interesting to investigate computational problems similar to the ones addressed by Pearl (1986) in Bayesian networks : updating the value of X_i from a new piece of data about X_j elsewhere in the network ; re-computing the most possible tuple(s) of values for the variables $X_1 \ldots X_n$. These problems should be solved via local propagation, provided that decomposability assumptions hold for the global fuzzy relation relating the set of variables. This achievement would be a significant step ahead with regard to current fuzzy logic applications (e.g. in fuzzy control), where only one layer of parallel rules is considered. A first attempt has been recently made by Fonck (1990) who proposes a possibilistic counterpart of Pearl's local computation approach. It would correspond here to the particular case where all the variables take their values on binary universes of discourse. Lastly, another worth-considering issue would use the proposed framework for analyzing the consequences of the *approximation* of complex hyperedges by simpler (hyper)edges, by fixing the values of "secondary" variables by means of fuzzy default values, as already discussed in (Dubois and Prade, 1988).

ACKNOWLEDGEMENTS : This work has been partially supported by the P.R.C.-G.D.R. Intelligence Artificielle (CNRS and French Ministry of Research) and by the ESPRIT-BRA project DRUMS. Thanks also to Sandra Sandri for useful comments on a first draft of this paper.

REFERENCES

Chatalic P. (1986) Raisonnement déductif en présence de connaissances imprécises et incertaines : un système basé sur la théorie de Dempster-Shafer. Thèse de Doctorat, Univ. P. Sabatier, Toulouse, France.

Chatalic P., Dubois D., Prade H. (1987) A system for handling relational dependencies in approximate reasoning. Proc. of the 3rd Inter. Expert Systems Conf., London, June 2-4, 495-502.

Davis E. (1987) Constraint propagation with interval labels. Artificial Intelligence, 32, 281-331.

Dubois D., Martin-Clouaire R., Prade H. (1988) Practical computing in fuzzy logic. In : Fuzzy Computing (M.M. Gupta, T. Yamakawa, eds.), North-Holland, 11-34.

Dubois D., Prade H. (1987) A tentative comparison of numerical approximate reasoning methodologies

Int. J. of Man-Machine Studies, 27, 717-728.

Dubois D., Prade H. (1988) On the combination of uncertain or imprecise pieces of information in rule-based systems – A discussion in the framework of possibility theory. Int. J. of Approximate Reasoning, 2, 65-87.

Dubois D., Prade H. (1989a) Fuzzy arithmetic in qualitative reasoning. In : Modeling and Control of Systems in Engineering, Quantum Mechanics, Economics, and Biosciences (Proc. of the Bellman Continuum Workshop 1988, June 13-14, Sophia Antipolis, France) (A. Blaquière, ed.), Springer Verlag, 457-467.

Dubois D., Prade H. (1989b) Processing fuzzy temporal knowledge. IEEE Trans. on Systems, Man and Cybernetics, 19(4), 729-744.

Dubois D., Prade H. (1989c) A typology of fuzzy "if... then..." rules. Proc. of the 3rd IFSA Congress, University of Washington, Seattle, Aug. 6-11, 782-785.

Fonck P. (1990) Building influence networks in the framework of possibility theory. Proc. of the RP2 First Workshop DRUMS (Defeasible Reasoning and Uncertainty Management Systems), Albi, France, April 26-28, 1990.

Kruse R., Schwecke E. (1988) Fuzzy reasoning in a multi-dimensional space of hypotheses. Proc. of 7th Annual Meeting of the North American Fuzzy Information Processing Society (NAFIPS'88), San Francisco State University, June 8-10, 147-151.

Kruse R., Schwecke E. (1989) On the treatment of cyclic dependencies in causal networks. Proc. of the 3rd IFSA Congress, University of Washington, Seattle, USA, Aug. 6-11, 416-419.

Kruse R., Schwecke E. (1990) Fuzzy reasoning in a multidimensional space of hypotheses. Int. J. of Approximate Reasoning, 4, 47-68.

Lauritzen S.L., Spiegelhalter D.J. (1988) Local computations with probabilities on graphical structures and their application to expert systems (with discussion). J. of the Royal Statistical Society, B-50, 157-224.

Nilsson N.J. (1986) Probabilistic logic. Artificial Intelligence, 28, 71-87.

Pearl J. (1986) Fusion, propagation and structuring in belief networks. Artificial Intelligence, 29, 241-288.

Shafer G. (1976) A Mathematical Theory of Evidence. Princeton University Press.

Shafer G., Shenoy P.P. (1988) Local computation in hypertrees. Working Paper n° 201, School of Business, Univ. of Kansas, USA.

Shafer G., Shenoy P.P. (1990) An axiomatic framework for Bayesian and belief-function propagation. In : Uncertainty in Artificial Intelligence, Vol. 4 (R.D. Shachter et al., eds.), North-Holland, Amsterdam, 119-198.

Shafer G., Shenoy P.P., Mellouli K. (1987) Propagating belief functions in qualitative Markov trees. Int. J. Approximate Reasoning, 1, 349-400.

Williams P.M. (1990) An interpretation of Shenoy and Shafer's axioms for local computation. Int. J. of Approximate Reasoning, 4, 225-232.

Yager R.R. (1989) Some extensions of constraint propagation of label sets. Int. J. of Approximate Reasoning, 3, 417-435.

Yager R.R., Larsen H.L. (1990) On discovering potential inconsistencies in validating uncertain knowledge bases by reflecting on the input. In Tech. Report #MII-1001, Machine Intelligence Institute, Iona College, New Rochelle, N.Y..

Zadeh L.A. (1978) Fuzzy sets as a basis for a theory of possibility. Fuzzy Sets and Systems, 1, 3-28.

Zadeh L.A. (1979) A theory of approximate reasoning. In : Machine Intelligence, Vol. 9 (J.E. Hayes, D. Mikulich, eds.), Elsevier, Amsterdam, 149-194.

SEMANTIC EVALUATION IN POSSIBILISTIC LOGIC
APPLICATION TO MIN-MAX DISCRETE OPTIMISATION PROBLEMS

Jérôme LANG
Institut de Recherche en Informatique de Toulouse
Université Paul Sabatier, 118 route de Narbonne
31062 TOULOUSE Cedex (FRANCE)

Abstract

Propositional possibilistic logic is a logic of uncertainty in which the notion of inconsistency is gradual, each interpretation having a compatibility degree with the uncertain available knowledge. We present here an algorithm for the search of the best interpretation of a set of uncertain clauses (i.e., the most compatible with it), which is an extension to possibilistic logic of semantic evaluation (based on the Davis and Putnam procedure). Possibilistic logic is also a general framework for translating discrete "min-max" optimisation problems (some examples of such problems are discussed).

1. FOUNDATIONS OF POSSIBILISTIC LOGIC

Possibilistic logic (see for example (Dubois, Lang and Prade, 1989)) is an uncertainty-handling logic, in which inconsistency is a gradual notion, each interpretation having a compatibility degree with the available uncertain knowledge. We present here an algorithm for the search of the "best" interpretation(s) of a set of uncertain clauses.

In possibilistic logic, we attach to formulas possibility and necessity degrees which measure to what degree they are possible or certain ; we consider only ordinary formulas which are uncertain owing to an incomplete available information ; it differs completely from fuzzy logic handling formulas involving vague predicates which can have an intermediary degree of truth. It differs also from probabilistic logic (which is also a logic of uncertainty) which is not able to represent states of incomplete information (partial or total ignorance). See (Dubois, Lang and Prade, 1990b).

In possibilistic logic LP1 we handle conjunctions of necessity-valued formulae defined as follows : let F be a set of (classical) propositional[1] formulas ; if $\varphi \in F$, then $(\varphi\ \alpha)$ is a possibilistic formula, expressing that $N(\varphi) \geq \alpha$, where $N(\varphi)$ is the necessity degree of φ. The axioms governing the necessity measures are the following :

1 Here we deal only with the propositional case, since the semantic evaluation algorithm works only for propositional logic. For details about first-order possibilistic logic see for example (Dubois, Lang and Prade, 1990a).

(i) $N(\perp) = 0$ (ii) $N(1) = 1$

(iii) $\forall\ \varphi, \psi \in F$, $N(\varphi \wedge \psi) = \min [N(\varphi), N(\psi)]$ (1)

(where $\perp$ and 1 denote the contradiction and the tautology, respectively).

We have always $\min [N(\varphi), N(\neg\varphi)] = 0$; $N(\varphi) = 1$ expresses that φ is certainly true ; but $N(\varphi) = 0$ does not express that φ is false, but only that φ is not at all certain, which is much weaker. In case where we have no information about the truth or the falsity of φ, we have $N(\varphi) = N(\neg\varphi) = 0$.

Clausal form is then defined in propositional possibilistic logic : if c is a classical propositional clause, then $(c\ \alpha)$ is is possibilistic propositional clause. A possibilistic formula can be easily put into clausal form : let $(\varphi\ \alpha)$ be a propositional possibilistic formula, and let $c_1 \wedge c_2 \wedge \ldots \wedge c_n$ be a conjunction of propositional clauses equivalent to φ. Then the conjunction of possibilistic clauses $(c_1\ \alpha) \wedge (c_2\ \alpha) \wedge \ldots \wedge (c_n\ \alpha)$ is equivalent to the initial possibilistic formula $(\varphi\ \alpha)$. (See (Dubois, Lang and Prade, 1990b)). Possibilistic logic LP1 restricted to clauses is called LCP1.

A semantics was defined for LP1 (see for example (Dubois, Lang and Prade, 1989)) : if $(\varphi\ \alpha)$ is a necessity-valued clause, let us call $M(\varphi)$ the set of the (classical) models of φ. Then the models of $(\varphi\ \alpha)$ will be defined by a fuzzy set of interpretations $M(\varphi\ \alpha)$ with a membership function

$$\mu_{M(\varphi\ \alpha)}(I) = 1 \text{ if } I \in M(\varphi) ;$$
$$\mu_{M(\varphi\ \alpha)}(I) = 1 - \alpha \text{ if } I \in M(\neg\varphi). \quad (2)$$

Then the fuzzy set of the models of a conjunction of necessity-valued formulae $\mathcal{F} = \phi_1 \wedge \phi_2 \wedge \ldots \wedge \phi_n$, will be the intersection of the fuzzy sets $M(\phi_i)$, i.e.

$$\mu_{M(\mathcal{F})}(I) = \min_{i=1,\ldots,n} \mu_{M(\phi i)}(I) \quad (3)$$

It measures to what degree the interpretation I is compatible with the knowledge base $\mathcal{F}$. The quantity $1 - \mu_{M(\mathcal{F})}(I)$, measuring the incompatibility degree of I with $\mathcal{F}$, will be noted inc $(\mathcal{F},I)$ and verifies

$$\text{inc}(\mathcal{F},I) = \max \{\text{val}(\varphi), (\varphi\ \text{val}(\varphi)) \in \mathcal{F}, I \models \neg\varphi\} \quad (4)$$

The quantity $\text{Inc}(\mathcal{F}) = \min_I \text{inc}(\mathcal{F},I)$ will be called the inconsistency degree of $\mathcal{F}$. The interpretation(s) minimizing inc$(\mathcal{F},I)$ will be called the best interpretation(s) of $\mathcal{F}$; this (these) is (are) the one(s) which are the least contradictory with the system constitued by $\mathcal{F}$ and the axioms of possibility theory.

In next sections we develop a procedure computing the best interpretation(s) with a set of possibilistic clauses.

Resolution can be generalized to first-order possibilistic logic (Dubois and Prade, 1987) and its completeness for the logic LCP1 is proved in (Dubois, Lang and Prade, 1989).

2. SEMANTIC EVALUATION IN CLASSICAL PROPOSITIONAL LOGIC

Semantic automated deduction methods consist in proving the consistency (resp. the inconsistency) of a set of propositional clauses H by finding a model of H (resp. by showing that there is no model of H). The Davis and Putnam procedure for propositional calculus (Davis and Putnam, 1960) produces a semantic tree of H, i.e. it constructs interpretations by evaluating all atomic propositions one by one, it stops as soon as it finds a model of H or when it has verified that all branches lead to failure nodes. Recently, Jeannicot, Oxusoff and Rauzy (1988) have shown that the Davis and Putnam procedure can be considerably improved by pruning some branches of the tree. In the formulation of the Davis and Putnam procedure we use the notations of (Jeannicot, Oxusoff and Rauzy, 1988) which we briefly recall : let H be a set of propositional clauses ; let p be a literal appearing in H ; then $T_p(H)$ will be the set of clauses obtained from H by deleting all clauses containing p, and by deleting all occurencies of $\neg p$ in the remaining clauses. For example, if H = $\{a \vee b, \neg a \vee b, \neg b \vee c, \neg c \vee a\}$ then $T_{\neg a}(H) = \{b, \neg b \vee c, \neg c\}$ and $T_{\neg a,c}(H) = T_c(T_{\neg a}(H)) = \{b, \bot\}$ $T''_p(H)$ will be the simplification by subsumption of $T_p(H)$. The property H consistent $\Leftrightarrow$ $T''_p(H)$ consistent or $T''_{\neg p}(H)$ consistent gives us a recursive algorithm similar to the Davis and Putnam procedure for testing the consistency of a set of clauses H. To make this algorithm more efficient, Jeannicot, Oxusoff and Rauzy (1988) have shown an original pruning method based on the "model partition theorem".

For some classes of problems, semantic evaluation may be more efficient than resolution ; furthermore it gives a constructive proof of the consistency of a set of clauses by giving one of its models, which cannot be done by resolution.

3. SEMANTIC POSSIBILISTIC TREES

In the following $\mathcal{H}$ is a finite set of <u>propositional</u> clauses of LCP1. As seen in section 2, it may be interesting for some problems to find the best interpretation of $\mathcal{H}$; this cannot be done by resolution, which can only give the inconsistency degree of $\mathcal{H}$.

In a classical semantic tree associated with a set of clauses H, every terminal node corresponds to an interpretation of H and that every non-terminal node correspond to a partial interpretation of H, i.e. an evaluation of a proper subset of the whole set of atoms of H ; if I and I' are two interpretations (partial or total), we say that I ≤ I' if the node corresponding to I is an ancestor of the node corresponding to I'. Then we define a possibilistic semantic tree as a classical tree in which each node has a valuation which is the incompatibility degree of $\mathcal{H}$ with the (partial or total) interpretation corresponding to I. The valuation of terminal node is directly computed by the formula (4) of section 1, and the valuation of a non-terminal node is the minimum of the valuations of its two son nodes. Then it can be shown that the valuation computed for the root of the semantic tree is the inconsistency degree of $\mathcal{H}$ (see Lang (1989)) ; besides, the terminal node(s) having the minimal value correspond(s) to the best interpretation(s) of $\mathcal{H}$.

Example : $\mathcal{H} = \{C_1 : (p\ \ 0.8), C_2 : (\neg p \vee r\ \ 0.4), C_3 : (\neg p \vee q\ \ 0.5), C_4 : (\neg q\ \ 0.2), C_5 : (\neg r\ \ 0.1), C_6 : (\neg p \vee \neg r\ \ 0.3)\}$. The semantic tree associated with $\mathcal{H}$ is :

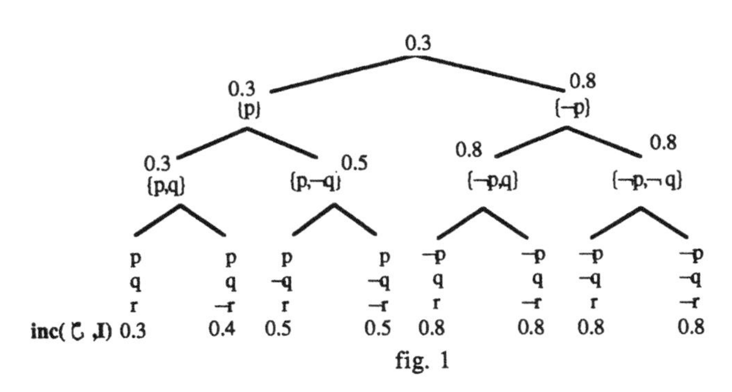

fig. 1

The inconsistency degree of $\mathcal{H}$ is 0.3 and the best interpretation of $\mathcal{H}$ is $\{p,q,r\}$.

4. THE DAVIS AND PUTNAM PROCEDURE IN POSSIBILISTIC LOGIC

The Davis and Putnam procedure is naturally extended to possibilistic logic by exploring the possibilistic semantic tree corresponding to a set of uncertain clauses $\mathcal{H}$. The main difference with the classical Davis and Putnam procedure is that it does not stop once it has found any model of $\mathcal{H}$, but once it has found the best interpretation of $\mathcal{H}$. Exploring the whole semantic tree may be very time-consuming, for there are 2^n terminal nodes, where n is the number of atomic literals appearing in $\mathcal{H}$. To improve the efficiency of semantic evaluation we first show that to a possibilistic semantic tree we can associate a min-max tree in which inconsistencies are taken into account as soon as they appear : each time an empty clause with a degree α appears in $\mathcal{H}$, we take it out as well as all clauses whose degree is $\leq \alpha$. The algorithm corresponding to the exploration of this min-max tree is the following, where $T''_p(\mathcal{H}')$ is the where $T''_p(\mathcal{H}')$ is the simplification by subsumption[2] of $\mathcal{H}'$ and $\mathcal{H}_{\bar{\alpha}}$ is the strong α-cut of $\mathcal{H}$ considered as a fuzzy set, i.e. the subset of clauses of $\mathcal{H}$ whose valuation is strictly greater than α.

[2] The subsumption being defined in possibilistic by : $(c\ \alpha)$ subsumes $(c'\ \beta)$ if and only if c subsumes c' and $\alpha \geq \beta$.

<u>Begin</u>
Associate the root node of the tree to $\mathcal{H}$; this node is a MIN node. Compute recursively **Inc($\mathcal{H}$)** :
<u>if</u> the corresponding node is a MIN node
<u>then</u> <u>if</u> $\mathcal{H} = \emptyset$
    <u>then</u> **Inc($\mathcal{H}$)** $\Leftarrow$ **0**
    <u>else</u> choose an atom p appearing in $\mathcal{H}$
        Inc($\mathcal{H}$) $\Leftarrow$ **min(Inc(T"$_p$($\mathcal{H}$)), Inc(T"$_{\neg p}$($\mathcal{H}$)))**
<u>else</u> {the corresponding node is a MAX node}

    <u>if</u> $\mathcal{H}$ contains an α-empty clause [3]
    <u>then</u> val($\mathcal{H}$.left) $\Leftarrow \alpha$
    <u>else</u> val($\mathcal{H}$.left) $\Leftarrow 0$
        Inc($\mathcal{H}$) $\Leftarrow$ **max(val($\mathcal{H}$.left), Inc($\mathcal{H}_{\bar{\alpha}}$)))**

<u>end</u>
<u>End</u>

This algorithm computes exactly the inconsistency degree of $\mathcal{H}$ and gives the best interpretation of $\mathcal{H}$ (or one of the best ones if there are more than one) .

Then the efficiency of the algorithm is improved by cutting some branches of the tree, firstly by using the alpha-beta pruning, secondly by using a generalized version of the model partition theorem. For more details and proofs see (Lang, 1989, 1991).

The MIN-MAX tree corresponding to the set of clauses of the previous example is shown in fig. 2.

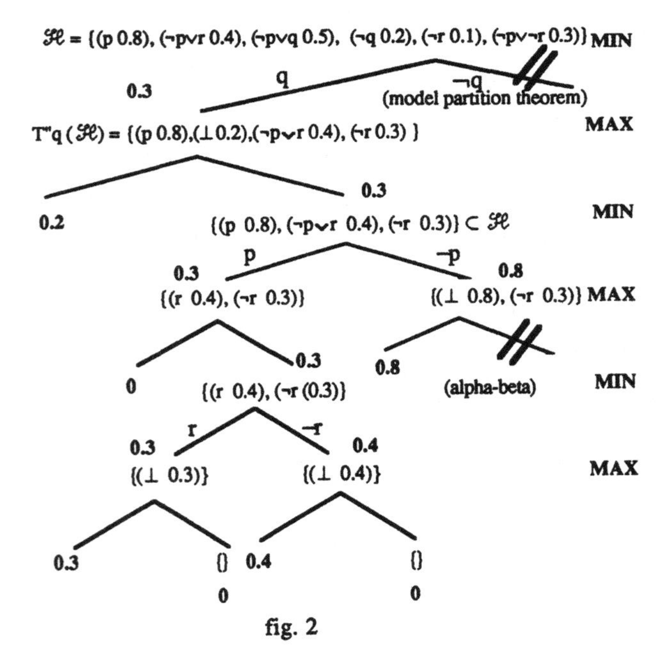

fig. 2

[3] it can be shown that at this step $\mathcal{H}$ contains at most one empty clause.

5. SOME APPLICATIONS OF POSSIBILISTIC SEMANTIC EVALUATION

Classical semantic evaluation enables us to find one or more model(s) of a given set of clauses, and thus to give a solution a logically-formulated problem ; whereas resolution gives only a non constructive answer (by checking the consistency, it answers for the existence or the non-existence of a model ; one can only have the path associated to the proof, and in first-order logic, the list of substitutions done). Examples of applications of semantic evaluation to combinatory problems can be found in (Oxusoff and Rauzy, 1988).

Classical semantic evaluation does not enable us to take account of a priority order upon clauses, and thus does not enable us to define a preferential order upon models. Now, for many problems consisting in finding a solution minimizing a "max-like" function, possibilistic logic enables a translation of the problem and semantic evaluation gives a general method for solving it.

Of course, semantic evaluation is (in the case where we use non-Horn clauses) an exponential algorithm[4] and it is clear that for a given problem, there generally exists a specific algorithm whose complexity is at least as as good as (often better than) the complexity of possibilistic semantic evaluation. Thus, we do not claim to give, for the problems we shall deal with, a more efficient algorithm than those already existing ; however, we think that translation in possibilistic logic is useful, for several reasons :

- the search method is independent of the problem ;
- the pruning properties in the search tree can confer to the algorithm a good average complexity (even polynomial, in some cases) ;
- possibilistic logic enables a richer representation ability in the formulation of a problem (one can specifies complex satisfaction constraints not specifiable in the classical formulation).

5.1. Min-max assignment problems

The min-max assignment problem (also called "bottleneck assignment problem") is formulated as follows : one has n tasks to assign to n machines (one and only one task per machine) ; if machine i is assigned to task j, the resulting cost is a_{ij}. Then the total cost of the global assignment is not the sum, but the maximum of the costs of the elementary assignments. Thus one looks for a permutation $\mathcal{P}$ of {1,2,...,n} such that $\mathrm{Max}_{\{i=1,2,\ldots,n\}}[a_{i,\mathcal{P}(i)}]$ be minimum[5].

The min-max assignment problem associated to the n x n matrix $A = (a_{i,j})$ can be translated and solved in possibilistic logic : first, the coefficients of A are supposed to lay in [0,1] (if it is not the case, we normalize A) ; then, to A one associates the set of clauses EC(A), whose atoms are $\{B_{i,j}, i = 1\ldots n, j = 1\ldots n\}$:

[4] However, the average complexity of semantic evaluation can be polynomial in some particular cases (see for example (Purdom, 1983)).

[5] The min-max assignment problem can be solved in polynomial time using the Ford-Fulkerson algorithm for searching a maximal flow in a network.

$EC(A) = EC1(A) \cup EC2(A)$, with

$$EC1(A) = \{(B_{1,1} \vee B_{1,2} \vee \ldots \vee B_{1,n}\ 1),$$
$$(B_{2,1} \vee B_{2,2} \vee \ldots \vee B_{2,n}\ 1),$$
$$\ldots\ldots\ldots\ldots\ldots\ldots ,$$
$$(B_{n,1} \vee B_{n,2} \vee \ldots \vee B_{n,n}\ 1)\}\quad \text{(at least one task per machine)}$$
$$\cup\ \{(\neg B_{1,1} \vee \neg B_{1,2}\ 1)),$$
$$\ldots\ldots\ldots\ldots ,$$
$$(\neg B_{2,1} \vee \neg B_{2,2}\ 1),$$
$$\ldots\ldots\ldots\ldots ,$$
$$(\neg B_{n,n-1} \vee \neg B_{n,n}\ 1)\}\quad \text{(at most one task per machine)}$$
$$\cup\ \{(B_{1,1} \vee B_{2,1} \vee \ldots \vee B_{n,1}\ 1), \ldots\ldots\}\text{(at least one machine per task)}$$
$$\cup\ \{(\neg B_{1,1} \vee \neg B_{2,1}\ 1), \ldots\ldots\}\quad \text{(at most one machine per task)}^{6};$$

$EC2(A) = \{C_{i,j}, 1 \le i \le n, 1 \le j \le n\}$ avec $C_{i,j} = (\neg B_{i,j}\ a_{i,j})$.

Intuitively, stating that $B_{i,j}$ is true if (i,j) belongs to the assignment, it can be seen that searching for the optimal assignment for A is equivalent to searching the best model of EC(A). Which is expressed more formally by the following theorem :

If $\mathcal{P}(n)$ is the permutation set of $\{1,\ldots,n\}$, then we have $\mathrm{Min}_{P \in P(n)} \max_{i=1,\ldots,n}(a_{i,\mathcal{P}(i)}) = \mathrm{Min}_I \max_{C \in EC(A)}(inc(C,I)) = Inc\ (EC(A))$. (the proof is in (Lang, 1990).

More generally, possibilistic logic can formalize and solve "min-max" discrete optimization problems, i.e. problems formulated as : compute $\mathrm{Min}_{x \in X} \max_{y \in Y(x)} f(x,y)$ (and find the solution, i.e. the value of x giving the optimum).

As it has already be said, such a translation may appear useless since one knows a polynomial algorithm for solving the problem, whereas semantic evaluation is NP-complete. However the average complexity of semantic evaluation may often be good (sometimes polynomial), all the more that it can be improved by using heuristics.

Lastly, the possibilistic formulation of a min-max assignment problem enables the formalization of extra *soft* constraints.

For example, the constraint "exactly one task per machine" can be weakened ; we give two examples of such weakenings (with n = 3) :

- "the machine 3 may, if necessary, can be assigned a second task but certainly not a third one" is translated by replacing the clauses of EC (A) corresponding to the 3rd range by $\{(\neg B_{3,1} \vee \neg B_{3,2}\ \alpha) ; (\neg B_{3,1} \vee \neg B_{3,3}\ \alpha) ; (\neg B_{3,2} \vee \neg B_{3,3}\ \alpha) ; (\neg B_{3,1} \vee \neg B_{3,2} \vee \neg B_{3,3}\ 1)\}$, knowing that the least α, the more permissive the expression "if necessary" ;

---

6 Any three of these four sets of clauses imply the fourth one, so it is useless, for example, translating that there is at most one machine per task ; in this case, EC1 contains $(n^3-n^2+4n)/2$ clauses.

- "the 2nd task may, if necessary, not be done" is translated by replacing the clause $(B_{1,2} \vee B_{2,2} \vee B_{3,2}\ \ 1)$ by $(B_{1,2} \vee B_{2,2} \vee B_{3,2}\ \ \alpha)$.

Complex soft constraints can also be formalized ; for example :

- "if task 3 is assigned to machine 1 then task 2 should rather be assigned to machine 2", translated by $(\neg B_{1,3} \vee B_{2,2}\ \ \alpha)$;
- "it is absolutely forbidden to assign tasks 1 and 2 to the same machine, excepted maybe machine 3", translated by $\{(\neg B_{1,1} \vee \neg B_{1,2}\ \ 1) ; (\neg B_{2,1} \vee \neg B_{2,2}\ \ 1) ; (\neg B_{1,1} \vee \neg B_{1,2}\ \ \alpha)\}$; and even
- "if it rains, task 3 may not be done, if necessary ; but if it does not rain, then it must be done", translated by $\{(\neg rains \vee B_{1,3} \vee B_{2,3} \vee B_{3,3}\ \ \alpha) ; (rains \vee B_{1,3} \vee B_{2,3} \vee B_{3,3}\ \ 1)\}$.

5.2. Algorithms for fuzzy graphs

A fuzzy graph $\tilde{G}$ (see Dubois and Prade, 1980 ; Rosenfeld, 1975) is a pair $(\tilde{V},\tilde{E})$ where $\tilde{V}$ is a fuzzy subset of V (vertice set) and $\tilde{E}$ a fuzzy relation on V (edge set), i.e. a fuzzy subset of $V_x V$ such that $\forall v, v' \in V$, $\mu_{\tilde{E}}(v,v') \leq \min (\mu_{\tilde{V}}(v), \mu_{\tilde{V}}(v'))$. Most of the definitions relative to classical graphs have been generalized to fuzzy graphs. A large part of the problems relative to fuzzy graphs are in the "min-max" problems class, and can be thus formalized and solved with propositional possibilistic logic.

An example of such a problem solved by possibilistic semantic evaluation is the search for a stable with maximal strength and cardinal $\geq$ p. Let $\tilde{G} = (\tilde{V}, \tilde{E})$ be a finite fuzzy graph. $S \subset V$ is a stable with strength α (or α-stable) if and only if, $\forall s, s' \in S$, we have $\mu_{\tilde{E}}(s,s') \leq 1 - \alpha$ (or equivalently, $\tilde{G}_{\overline{1-\alpha}}$ is a classical stable [7]). Let $v_1,\ldots,v_n$ be the vertices of $\tilde{G}$. One wish to find a stable with maximal strength containing at least p vertices. Let $ECS(\tilde{G})$ be the set of possibilistic clauses defined by $ECS(\tilde{G}) = ECS1(\tilde{G}) \cup ECS2(\tilde{G})$ where $ECS1(\tilde{G})$ is the set of clauses (weighted by 1) expressing that least p literals among $\{v_1, v_2, \ldots, v_n\}$ are true (literal v_i is true if and only if vertex v_i is in the stable), and $ECS2(\tilde{G}) = \{(\neg v_i \vee \neg v_j\ \ \mu_{\tilde{E}}(v_i,v_j)), i \neq j, i, j = 1,\ldots,n\}$. Then the positive atoms v_i of the best model obtained by possibilistic semantic evaluation on $ECS(\tilde{G})$ correspond to the vertices componing a stable of maximal strength with cardinal $\geq$ p.

The search for an Hamiltonian path of maximal strength in a fuzzy graph and the couplage of maximal strength in a bipartite fuzzy graph can be expressed in terms of a min-max assignment problem and can thus be solved by possibilistic semantic evaluation.

The search for a path of maximal strength between two points, the computation of the transitive closure of a fuzzy graph, the composition of binary fuzzy relations can also be translated in propositional possibilistic logic, but first-order possibilistic logic (with use of the extended resolution principle) is more adapted to these problems. See (Dubois, Lang and Prade, 1990a ; Lang 1991).

---

[7] In a classical graph, a stable is a subset of vertices such that any two vertices of this subset are not connected.

6. CONCLUSION

The described algorithm has been implemented in Le_Lisp, with pruning methods ; an improvement by using heuristics is currently being implemented. A study of the average complexity of the algorithm is also currently being studied.

Possibilistic logic proved to be a well-adapted general logical framework for representing and solving min-max discrete optimisation problems. The next step is to generalise semantic evaluation from possibilistic logic to a more general class of uncertainty logics (based on decomposable measures) enabling representation of additive (or other) discrete optimisation problems. This would lead to the implementation of a generalized possibilistic logic programming language.

REFERENCES

Davis M., Putnam H. (1960) A computing procedure for quantification theory. J. of the Assoc. for Computing Machinery, 7, 201–215.

Dubois D., Prade H. (1980) Fuzzy Sets and Systems : Theory and Applications. Mathematics in Science and Engineering Series, Vol. 144, Academic Press, New York.

Dubois D., Prade H. (1987) Necessity measures and the resolution principle. IEEE Trans. Systems, Man and Cybernetics, 17, 474–478.

Dubois D., Lang J., Prade H. (1987) Theorem proving under uncertainty — A possibility theory-based approach. Proc. 10th Inter. Joint Conf. on Artificial Intelligence, Milano, 984–986.

Dubois D., Prade H. (1988) (with the collaboration of Farreny H., Martin-Clouaire R., Testemale C. Possibility theory : an Approach to Computerized Processing of Uncertainty, Plenum Press, New York.

Dubois D., Lang J., Prade H. (1989) Automated reasoning using posibilistic logic : semantics, non-monotonocity and variable certainty weights. Proc. 5th Workshop on Uncertainty in Artificial Intelligence, Windsor (Canada), Aug. 18–20, 81–87.

Dubois D., Lang J., Prade H. (1990a) POSLOG : an inference system based on possibilistic logic. Proc. in National Fuzzy Information Processing Society Congress, 1990, to appear.

Dubois D., Lang J., Prade H. (1990b) Fuzzy sets in automated reasoning : logical approaches. Fuzzy Sets and Systems, to appear.

Jeannicot S., Oxusoff L., Rauzy A. (1988) Evaluation sémantique : une propriété de coupure pour rendre efficace la procédure de Davis et Putnam. Revue d'Intelligence Artificielle, 2(1), 41–60.

Lang J. (1990) Evaluation sémantique en logique possibiliste propositionnelle. Research Report IRIT/90–7/R, IRIT, Univ. P. Sabatier, Toulouse, France. Revue d'Intelligence Artificielle, 4(4), 27–48.

Lang J. (1991) "Logique possibiliste : aspects formels, déduction automatique, et applications. Thèse de l'Université P. Sabatier, Toulouse, France.

Oxusoff L., Rauzy A. (1989) L'évaluation sémantique en calcul propositionnel. PhD of the University Aix Marseille II, January 1989.

Purdom P.W. (1983) Search rearrangement backtracking and polynomial average time. Artificial Intelligence, 21, 117–133.

Rosenfeld A. (1975) Fuzzy graphs. In : Fuzzy Sets and their Applications to Cognitive and Decision Processes (L.A. Zadeh, K.S. Fu, K.Tanaka, and M. Shimura, eds), Academic Press, New York 77–95.

FORMALIZING MULTIPLE-VALUED LOGICS AS INSTITUTIONS*

J. AGUSTÍ-CULLELL, F. ESTEVA, P. GARCIA, Ll. GODO.
Centre d'Estudis Avançats de Blanes (CSIC).
17300 BLANES. Girona Spain.

Abstract.

Many of the uncertainty management systems used in the Knowledge Based Systems technology can be considered as the set of mechanisms that a certain underlying multiple-valued logic supplies: certainty values, numeric or linguistic, would be the truth-values of that logic, a knowledge base would be a set of axioms, and the mechanisms of uncertainty combination and propagation would be the inference rules of the deduction system. In this communication we formalize multiple-valued logics inside the institutional framework. We structure multiple-valued logics as families of institutions, each one being indexed by a class of truth-values algebras, in such a way that each morphism between truth–values algebras determines a corresponding morphism of institutions. These institution morphisms are a basic mechanism in modular Expert System languages in order to build uncertainty management systems that deal with different logics in different modules.

Keywords: Multiple-valued Logics, Institution, Entailment System, Truth-Values Algebra

1. INTRODUCTION.

Present research on methods for formal software development tries to comprehend and unify, inside a general theory, the diversity of logic based languages proposed for specification and design in computer science. In particular, the theory of "institutions" (Goguen, Burstall 1983) formalizes the intuitive notion of a logical system and seems useful to understand what a specification is and to unify programming paradigms. As it is known, an Institution is a 4-tuple $<$ SIGN, SEN, MOD, $\models$ $>$, where:

*Research partially supported by the SPES project, CICYT id. 880j382, and by the ESPRIT-II Basic Research Action DRUMS.

- SIGN is a category whose objects are signatures containing a set of sorts and sorted sets of relation and function symbols to build the language, and whose morphisms are signature morphisms, i.e., a mapping between sorts, a mapping between relation symbols and a mapping between function symbols preserving the types of relations and functions respectively.
- SEN is a functor from SIGN to SET, giving the sentences of the language and how these sentences change by signature morphisms.
- MOD is a functor from SIGN to CAT^{op}, giving the models associated to each signature and how they change by signature morphisms.
- $\models$ is a function giving for each signature Σ a satisfaction relation $\models_\Sigma$ between models and sentences of Σ, that must be preserved by signature morphisms, i.e., the following invariance condition, known as the Satisfaction condition,

$$[MOD(\sigma)](M') \models_\Sigma E \quad \text{if, and only if,} \quad M' \models_{\Sigma'} [SEN(\sigma)](E).$$

must hold for all Σ'-model M', for all Σ-sentence E and for all signature morphism σ from Σ to Σ':

In this general framework, any commitment to particular logical systems is avoided by doing constructions once and for all at the more general level of institutions (Harper, Sannella, Tarlecki 1989). In particular, several modularization techniques work in this general setting: there are operations that allow both to put theories together and to write each theory in a different logical system. The possibility of using several logical systems at once is based on the notion of morphism between institutions.

Expert System development languages can be considered as logic based specification languages. Thus the general theory and techniques of the institutional framework could be applied to the design of Expert Systems (ES, for short) languages. Nevertheless, some characteristics of ES programming must be taken into account. Formal software development emphasizes the construction of programs from a precise given formal specification. The starting point of software engineering is the ultimate goal of ES development: the process by which a precise formal specification is obtained starting from an informal, vague, and often incomplete problem solving knowledge of a human expert. Thus, the methodology of ES development is based on the evolution of specifications whose adequacy is checked by running them, i.e. searching the space of proofs the specification potentially defines. As we need to manipulate and change specifications, it would be very useful to have them available in a structured way, as a hierarchy of modules, each one written in a different logical system if needed (Agusti-Cullell, Sierra 1989; Sierra 1989).

Another characteristic of ES is the need to manage uncertainty. Many of the uncertainty management systems which are used in the Knowledge Based Systems technology can be considered as the set of mechanisms that a certain underlying multiple-valued logic supplies: certainty values (numeric or linguistic) would be the truth-values of that logic, a knowledge base would be a set of axioms, and the mechanisms of uncertainty combination and propagation would be the inference rules of the deduction system.

In this communication we formalize multiple-valued logics in the institutional framework. In the second section, truth-values algebras are introduced, and in the third one we structure multiple-valued logics as families of institutions, indexed by a class of truth-values algebras. Finally, in the fourth section we show that each morphism between truth–values algebras determines a corresponding morphism of institutions. These institution morphisms are a basic mechanism in modular Expert System languages i order to build uncertainty management systems that deal with different logics in different modules. It is assumed that the reader is acquainted with the basic notions of category and institutions theory.

2. TRUTH-VALUES ALGEBRAS.

Multiple-valued predicate logics (quantifiers free) are built on a set of predicate and connective symbols and on a set of truth-values (Rescher 1969; Rasiowa 1974). Predicate and connective symbols are used to build sentences and truth-values are used to evaluate these sentences. Usually, the truth-values set is structured as an algebra similar to the algebra of sentences with two distinguished elements ('True', 'False'). Valuations are morphisms from the free algebra of sentences to a truth-values algebra.

The most uniform way to formalize multiple-valued logics as institutions is to consider truth-values as a part of the language signature and, thus, part of the sentences and models as well. Nevertheless, there is a problem with this approach. The class of models is too large, that is, it does not allow us to make a clear distinction between model changes due to symbol changes and model changes due to truth-values structure changes. In order to overcome this problem we propose to keep truth-values as part of the sentences, but not as a part of the language signatures. This leads to building families of institutions indexed by a class of truth-values algebras.

Technically, a truth-values algebra will be given as an object of the category MOD(TV,EQ) of models of a presentation (TV,EQ), where TV is a signature with only one sort V, the set of truth-values, and EQ is a set of axioms the signature operations must fulfil. Moreover, these signature operations correspond to the set of logical connectives. In this way, the class of all possible multiple-valued logical systems over these truth-values algebras can be formalized as a family of institutions parametrized by the corresponding algebras.

Here, there is an example of a general presentation of a truth-values structure which covers most of the systems that have been defined from the original Lukasiewicz multiple–valued systems and, in particular, all those arising from Fuzzy Set Theory.

Example:

Let **TV** be the following signature:

Sorts : V (a set of truth-values)
Opns : True : ——> V
False : ——> V
Neg : V ——> V
And : VxV ——> V
Or : VxV ——> V
Imp : VxV ——> V
Order : VxV ——>{True,False}

Let **EQ** be the following set of axioms:

Neg :
Neg(True) = False
Neg(Neg(x)) = x
Imp(Order(x,y),Order(Neg(y),Neg(x))) = True

And :
Commutative
Associative
And(True,x) = x
And(False,x) = False
Imp(Order(x,y),Order(And(x,z),And(y,z))) = True

Or :
Or(x,y) = Neg(And(Neg(x),Neg(y))) (De Morgan law)

Imp :
Imp(True,x) = x, Imp(False,x) = True
Imp(And(x,y),x) = True
Imp(x,Imp(y,z)) = Imp(y,Imp(x,z))
Imp(Order(x,y),Order(Imp(z,x),Imp(z,y))) = True
Imp(Order(x,y),Order(Imp(y,z),Imp(x,z))) = True

Order :
Order(False,x) = True
Order(x,True) = True
Order(x,x) = True
Order(x,y) = Neg(Order(y,x)), for all $x \neq y$
Imp(and(Order(x,y),Order(y,z)),Order(x,z)) = True

From now on, L will stand for an object of MOD(TV,EQ), and |L| will denote its carrier.

3. L-MULTIPLE-VALUED LOGIC INSTITUTIONS.

In this section, for each truth-values algebra L of MOD(TV,EQ), we are going to describe th L-Multiple-valued Logic institution, $\mathbf{MVL_L}$ for short. This institution will be defined by a 4-tupl ($\mathbf{SIGN}$, $\mathbf{SEN_L}$, $\mathbf{MOD_L}$, $\models_L$), satisfying the usual Satisfaction condition.

Previously, let's assume we have the following objects:

- **Nat**, the set of natural numbers,
- a set **C** of logical connective symbols, along with their arity,
- a function **con_de** from **C** to the set of operations of the signature TV,
- a function **Ary** from **C** to **Nat**, giving the arity of connective symbols,
- a sorted set **X** of variable symbols.

Each of the components of the $\mathbf{MVL_L}$ institution will be described in detail in the following.

*** SIGN.**

The category of signatures **SIGN** have as objects signatures Σ, consisting of a pair (Σ_{sort}, Σ_{rel}), where Σ_{sort} is a set S of sorts and Σ_{rel} is an S*-sorted set of relation symbols, where S* stands fot the union of cartesian products of S. The morphisms of SIGN will be signature morphisms

*** $\mathbf{SEN_L}$.**

At this point we would like to note that if we define sentences and models as in classical logic, the satisfaction relation would become a fuzzy relation between them. In our approach we have used L-sentences, pairs of sentences and subsets of truth-values, which allow us to have a classical satisfaction relation and therefore a proper Institution.

In order to define the functor $\mathbf{SEN_L}$ from SIGN to SET, first of all we need to build classical sentences by means of the functor **SEN** from SIGN to SET, which is defined in the usual way:

1- SEN(Σ), the set of Σ-sentences, consists of:

- atomic Σ-sentences are sentences of type (r,{x_{s_1},...,x_{s_n}}) where r is a (s_1, ..., s_n)-sorted relation symbol of Σ_{rel}, and x_{s_i} is a variable symbol of sort s_i.

- connected Σ-sentences are sentences of type con(c,[E_1,...,E_n]), where c is a logical connective, E_i is a Σ-sentence, and ary(c)=n.

2- Given a signature morphism σ:Σ ——> Σ', SEN(σ) translates Σ-sentences into Σ'-sentences in the following way:

- if E = (r,X), then SEN(σ)(E) = (σ(r),X).

- if E = con(c,[E_1,...,E_n]), then SEN(σ)(E)=con(c,[SEN(σ)(E_1),...,SEN(σ)(E_n)]).

Now, the functor $\mathbf{SEN_L}$ is defined by:

1- $\mathrm{SEN_L}(\Sigma)$, the set of Σ-L-sentences, is the set of pairs (s, SV), where s is a Σ-sentence, and SV is a subset of truth-values of |L|.

2- $SEN_L(\sigma)$ is the functor formed by the pair of functors (SEN(σ), Id.), i.e., $SEN_L(\sigma)(s,CV)$ = = (SEN(σ)(s),SV)

Let be noticed that this idea of introducing subsets of truth-values in our sentences has been largely used in the context of Expert Systems dealing with uncertainty.

*** MOD_L.**

Models in Multiple-valued Logics are defined here as L-fuzzy sets, defining fuzzy relations in the similar way that sets, the models of the classical logic institution, define classical relations where predicate symbols are interpreted.

The functor $\mathbf{MOD_L}$, from SIGN to the category CAT^{op}, gives, for each signature Σ, the category of Σ-L-models $MOD_L(\Sigma)$, and a functor $MOD_L(\sigma) : MOD_L(\Sigma') \longrightarrow MOD_L(\Sigma)$, for each morphism $\sigma: \Sigma \longrightarrow \Sigma'$.

The objects of the category $MOD_L(\Sigma)$ are Σ-L-models, defined as triplets < Σ, Ind, f >, where:

- Σ is a signature of SIGN.
- Ind is a set of individuals.
- f is the interpretation function that associates a relation f(r) to each relation symbol r of Σ_{rel} (actually, f(r) is a function from Ind^n to |L|, being n=ary(r), i.e., f(r) is a |L|-valued fuzzy relation).

In the category $MOD_L(\Sigma)$, there is a morphism from <Σ, Ind, f > to <Σ, Ind', f' > if and only if there exists a mapping m : Ind ——> Ind' such that the following diagram

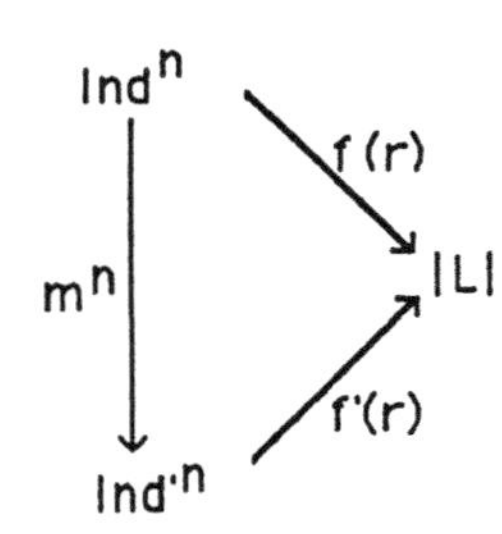

commutes for each r of Σ_{rel}, being n = ary(r). This means that there exists a morphism between two models if there is a renaming mapping between the individual sets such that the interpretation of each relational symbol is preserved.[1]

The functor $MOD_L(\sigma)$ translates Σ'-L-models into Σ-L-models in the following way:

$$MOD_L(\sigma)(< \Sigma', Ind', f' >) = < \Sigma, Ind', f' \circ \sigma >.$$

[1] A weaker condition in order to define these morphisms could be the preservation of the interpretation via a fixed endomorphism of L.

On the other hand, the functor $MOD_L(\sigma)$ is the identity over the morphisms.This definition makes commutative the next diagram:

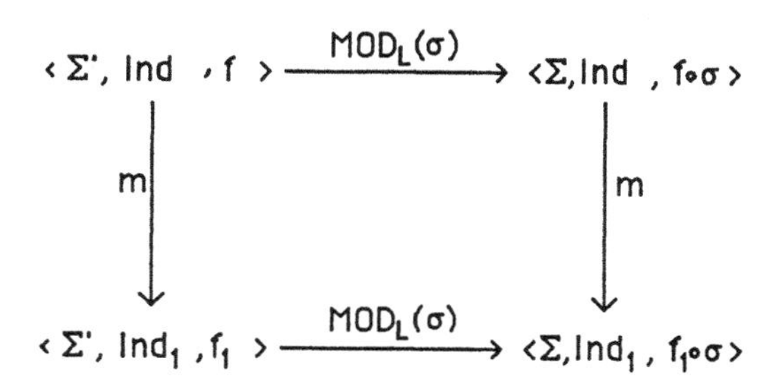

THE SATISFACTION RELATION

The definition of the **Satisfaction Relation** $\models$ requires first of all to define the **satisfaction degree** of a Σ–L-sentence in a Σ–L-model.

Given a model $A = <\Sigma, Ind, f>$ and an assignment function $v : VAR \longrightarrow Ind$, where VAR is a set of variables, the **satisfaction degree** of an Σ-sentence in the model A, under the assignment v, is given by the function d_v: $MOD_L(\Sigma) \times SEN(\Sigma) \longrightarrow |L|$, defined by

$$d_v(A ,(r,X))= f(r)(v(X)), \text{ and}$$
$$d_v(A ,connect(c,[s_1,...,s_n])) = con\text{-}den(c)(d_v(A,s_1),...,d_v(A,s_n)).$$

Now, we define the **satisfaction relation** $\models_\Sigma$ between Σ–L-models and Σ–L-sentences as follows:

Definition: A $\models_\Sigma$(s,SV) iff d_v(A,s) belongs to SV, for all assignment functions v.

The next theorem proves that MVL_L is a proper institution.

Theorem (**Satisfaction relation condition**): The satisfaction relation $\models$ is preserved by the functor $MOD_L(\sigma)$, i.e., given a signature morphism $\sigma : \Sigma\text{----}> \Sigma'$, and an Σ'-L-model A', then A' $\models_\Sigma$ $SEN_L(\sigma)$(s,SV) if and only if $MOD_L(\sigma)(A') \models_\Sigma$ (s,SV).

Proof: The theorem follows from the fact that, for each assignment v, it can be checked that $d_v(A',SEN(\sigma)(s))) = d_v(MOD_L(\sigma)(A') , s)$.

From now on, we will write $\models$ instead of $\models_\Sigma$, if no confusion is possible.

REMARK: Given a signature Σ, the satisfaction relation between L-models and L-sentences can be extended, as usual, to a relation |- between sets of L-sentences and L-sentences in the following way:

$$\Omega \text{ |- } (s,W) \text{ iff } M \models (s,I) \text{ for each } M \text{ belonging to } MOD_L(\Sigma,\Omega),$$

where Ω is set of Σ-L-sentences and $MOD_L(\Sigma,\Omega)$ is the full subcategory of $MOD_L(\Sigma)$ determined by those Σ-L-models that satisfy all the Σ-L-sentences in Ω. The relation |- is an entailment relation (i.e., it is a reflexive, monotonic and transitive relation), and ($SIGN, SEN_L$, |-) is an Entailment System (Meseguer,1989). This framework allows to derive several inference schemes, for which the soundness property is guarantied. For instance, let us consider the truth-values algebra L = [0,1] equipped with a t-norm T and an implication function I, standing for a conjunction and implication connectives respectively. It can be checked (Godo,1990) that the following inference schemes hold for all $a,b,c \in [0,1]$:

$$\{ (p, [a,1]), (q, [b,1]) \} \vdash_L (p \text{ and } q, [T(a,b),1])$$

$$\{ (p, [a,1]), (\text{if } p \text{ then } q, [b,1]) \} \vdash_L (q, [m_I(a,b),1]),$$

where $m_I(x,y) = \mathrm{Inf}\{z \in [0,1] \,/\, I(x,z) \geq y\}$ is known as the Modus Ponens generating function with respect to I, and thus the functional formulation of Modus Ponens in Multiple-valued Logics in (Trillas, Valverde 1985) is recovered and proved its soundness.

4.- MORPHISMS BETWEEN L-INSTITUTIONS.

In this section, a sufficient condition for the existence of morphisms between L-institutions is given. Let L and L' be two truth-values algebras and let $MVL_L = < SIGN, SEN_L, MOD_L, \models_L >$ and $MVL_{L'} = < SIGN, SEN_{L'}, MOD_{L'}, \models_{L'} >$ be their corresponding multiple-valued logic institutions defined as in section 3.

Theorem: For each morphism of algebras $\varphi: L \text{------>} L'$, an institution morphism $M_\varphi: MVL_L \text{------>} MVL_{L'}$ can be defined.

Proof: An institution morphism (Goguen, Burstall 1983) consists of: 1) a functor between signatures, 2) natural transformation between sentences and 3) a natural transformation between models, preserving the satisfaction relation. In our case we define the institution morphism M_φ as follows:

1) The functor between signatures ID : SIGN ——> SIGN is the identity one.

2) The natural transformation t_φ from $SEN_{L'}$ to SEN_L is defined by the set of functors

$$t_\varphi = \{t_{\varphi\Sigma}: SEN_{L'}(\Sigma) \text{------>} SEN_L(\Sigma) \,/\, \Sigma \in SIGN\},$$

where $t_{\varphi\Sigma}((s,SV'))=(s,\varphi^{-1}(SV'))$. This definition makes the following diagram commutative for each signature morphism σ:

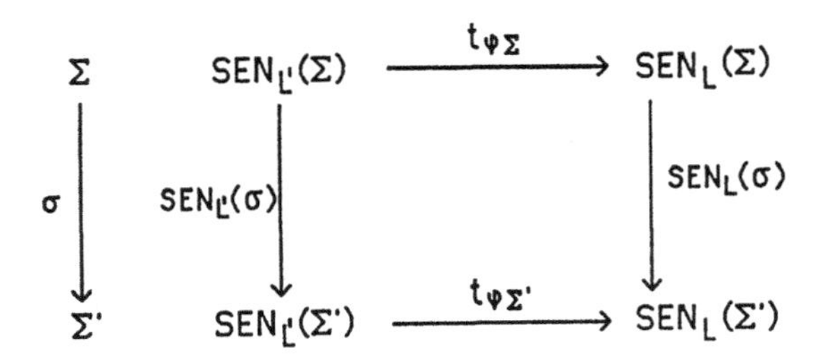

3) The natural transformation m_φ from $MOD_{L'}$ to MOD_L is given by the set of functors $m_\varphi = \{m_{\varphi\Sigma} : MOD_L(\Sigma) \longrightarrow MOD_{L'}(\Sigma) \ / \ \Sigma \in SIGN\}$, where $m_{\varphi\Sigma}(<\Sigma, Ind, f>) = <\Sigma, Ind, \varphi of>$, and it is applied to morphisms in the natural way. With this definition, the following diagrams commutes for each signature morphism σ:

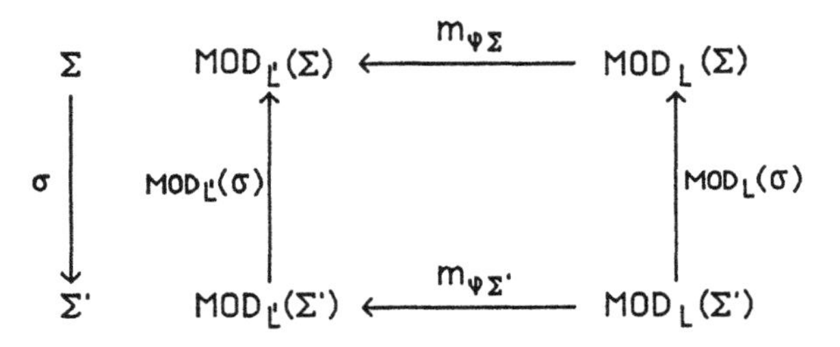

Finally, it is easy to check that $m_{\varphi\Sigma}(A) \models (s,SV')$ if and only if $A \models t_{\varphi\Sigma}(s,SV')$, for each model A belonging to $MOD_L(\Sigma)$ and for each L'-sentence (s,SV') belonging to $SEN_{L'}(\Sigma)$. Thus, the theorem is proved.

Example

Let us consider the following truth-values algebras over the unit interval $L = ([0,1], T_1, I_1, n_1)$ and $L'=([0,1], T_2, I_2, n_2)$, where

$T_1(x,y) = xy$	;	$T_2(x,y) = xy \ / \ (x+y-xy)$
$I_1(x,y) = 1$, if $x \leq y$	;	$I_2(x,y) = 1$, if $x \leq y$
$I_1(x,y) = x/y$, otherwise	;	$I_2(x,y) = xy \ / \ (x-y+xy)$, otherwise
$n_1(x) = \exp(1/\ln x)$	;	$n_2(x) = 1-x$

are their conjunction, implication and negation operations respectively. Straightforward computation shows that the mapping $\varphi:[0,1] \longrightarrow [0,1]$ defined by $\varphi(x) = 1/(1-\ln x)$ is an isomorphism between L and L' (Garcia, Valverde 1989). Taking into account the remark in section 3, it is easy to show that the following entailment relations hold:

$$\{ (p,[a,1]), (q,[b,1]), (\text{if p and q then r}, [c,1]) \} \vdash_L (r,[d,1]), \text{ being } d = T_1(a,T_1(b,c))$$

$$\{ (p,[a',1]), (q,[b',1]), (\text{if p and q then r}, [c',1]) \} \vdash_{L'} (r,[d',1]), \text{ being } d' = T_2(a',T_2(b',c'))$$

The above theorem states that φ changes the entailment relations consistently, that is $\varphi(d)=d'$ whenever $\varphi(a)=a'$, $\varphi(b)=b'$ and $\varphi(c)=c'$.

Acknowledgments.

The authors are indebted to D. Sannella and A. Tarlecki for their useful suggestions and comments.

REFERENCES.

AGUSTI-CULLEL J., SIERRA C. (1989) *'Adding Generic Modules to Flat Rule-Based Languages: a low cost approach'*. Methodologies for Intelligent Systems, pp. 43-52, North Holland.

ALSINA C. ,TRILLAS E., VALVERDE L. (1983)*' On some logical connectives for Fuzzy Set Theory'* Journal of Mathematical Analysis and Applications, 93, pp. 15-26.

GARCIA P., VALVERDE L. (1989) *'Isomorphisms between De Morgan Triplets '* Fuzzy sets and Systems Vol. 30, Nº 1. pp. 27-36. North-Holland

GODO L. (1990) *'Contribució a l'estudi de models d'inferència en els sistemes possibilístics '*. Ph. D Universitat Politècnica de Catalunya. Barcelona.

GODO L., LOPEZ DE MANTARAS R., SIERRA C., VERDAGUER A. (1989) *`MILORD:The architecture and management of linguistically expressed uncertainty'*. Int. Journal of Intelligent Systems Vol. 4 nº 4, pp. 471-501.

GOGUEN J., BURSTALL R.M. (1983) *'Introducing Institutions'* . Proc. Workshop on Logics of Programs, Carnegie-Mellon University. Springer LNCS 64, pp. 221-256.

HARPER R., SANNELLA D., TARLECKI A. (1989) *'Structure and Representation in LF'* Proc.4th IEEE Symposium on Logic of Computer Science.

MESEGUER J. (1989) *'General Logics '*. In H.D. Ebbinghaus et al. (eds) Proc. Logic Colloquium '87 North -Holland.RASIOWA H., (1974) *'An algebraic approach to non-classical logics'* North Holland.

RESCHER N. (1969) *'Many-valued Logic'* Mc Graw-Hill.

SIERRA C. (1989) *' MILORD: Arquitectura multi-nivell per a sistemes experts en classificació"* Ph. D Universitat Politècnica de Catalunya . Barcelona.

EMPIRICAL PLAUSIBLE REASONING BY MULTIPLE-VALUED LOGIC

Paolo BOTTONI *, Luca MARI **, Piero MUSSIO *

* Università degli Studi di Milano - Dipartimento di Fisica, Via Viotti 5, 20133 - Milano - Italy

** IMU - Consiglio Nazionale delle Ricerche, Via Ampére 56, 20131 - Milano - Italy

Abstract

A method to approximate the heuristic reasoning of an expert in judging the behaviour of a system is proposed. The method is based on the use of label functions, mapping the observed value of one attribute into a local judgement, and of multiple-valued logic trees, mapping a set of local judgements into a global one. The method is introduced within the fuzzy set approach, so that the involved approximation can be discussed.

Keywords

Multiple-valued logics, label functions, plausible reasoning, fuzzy set theory, empirical judgement

1. INTRODUCTION

In many experiments, observers describe a situation by several linguistic descriptors and deduce linguistic judgements on it by a heuristic reasoning based on the combination of those descriptors. Linguistic descriptors reflect the observers' model of the situation and their inability to state observations in a numerical way, either for the inadequacy of their measurement procedures [1] and models [2], or for the model complexity [3]. This way, the deduction is neither repeatable in the physical sense nor explicitly controlled and criticizable by other experts: the communication of the experiment results is no longer an intersubjective activity. In order to achieve these fundamental properties of an observation process, in all the quoted cases the rules underlying the heuristic methods used by the expert were investigated so as to formalize them. This study led to recognizing that experts use their own models in the reasoning process and that a correct formalization has to be based on the underlying (often non classical) logic. Moreover, such a formalization is also necessary for the automation of the expert judgement process.

Approaches based on fuzzy set and fuzzy measure theories are widely used to model and formalize plausible judgement processes. These approaches, in fact, allow both the assignment of a meaning to a linguistic descriptor and the evaluation of the uncertainty associated with a given judgement [4]. However,

some problems arise in practice, especially due to the recognized difficulty of assessing the membership functions of the fuzzy sets [5] and to the constraints that the standard fuzzy connectives impose on the modeling of the reasoning process.
It was observed in several experiences in different scientific fields that experts describing the system under observation via a set of linguistic labels are seldom aware of the possibility of defining fuzzy membership functions mapping each label into its meaning and, when aware, seem unable to state those functions. Despite that, given an attribute[1] of the system and an associated suitable set of linguistic labels, experts contextually choose one of the latter to describe the observed situation.
Moreover, since the system can be observed with respect to several different attributes, experts combine the linguistic labels, each one chosen considering a single attribute, into new ones, which synthesize a judgement on the whole system.
In this paper,at first, it is proposed an outline of the judgementprocedure we derived from the observation of the experts within a fuzzy set theory framework- under the hypothesis that the membership functions relating each linguistic label to the corresponding attribute are known. Next, it is shown how this procedure can be empirically approximated -avoiding any a priori specification of the membership functions in the case that the sets of the attributes and of the linguistic labels associated with each attribute are finite. The paper argues that such a judgement procedure can be formalized in terms of a suitable multiple-valued logic, in which the truth values are denoted by the linguistic labels, reflecting the reasoning schemes underlying the expert judgment. It is also suggested that the proposed approximation even provides a means to facilitate the elicitation of knowledge.

2. A FUZZY SET FRAMEWORK FOR THE JUDGEMENT PROCEDURE

In describing the behaviour of a system during an experiment, observers generally exploit a finite set of attributes A. For each attribute $\alpha \in A$, the set P_α of its possible values is defined, called the set of properties. A "description" is a set of "descriptors", each one being a couple <attribute, (associated) property>.
Were the experts able to state the membership functions for the properties, a global judgement about the behaviour of the observed system could be derived from a description via the application of a two step deductionmethod.
In the first step, given an "initial description" D_{init} obtained by a direct observation of the attributes on the system, each property in D_{init} is judged independently of any other, as if it were the result of the only possible observation of the system (for simplicity, we will assume that the observed properties are singletons of P_α, and not, for example, intervals or fuzzy subsets). To this end, a finite set $L_\alpha=\{\lambda_i\}$ of linguistic labels must be defined for each attribute α. Each label λ_i defines a fuzzy subset of P_α characterized by a membership function $\mu_{\lambda_i}:P_\alpha \rightarrow [0,1]$. Each property $p \in P_\alpha$ is associated with each label

[1] We adopt the term "attribute", instead of the more usual "observable" or "(measurable) quantity" [6], order to emphasize that the sets of its possible values need not be metric spaces: they could even unordered sets of labels. In this sense an attribute is measurable even on a nominal scale [7].

$\lambda_i \in L_\alpha$ and the appropriateness of each association is given by the value $\mu_{\lambda_i}(p)$. A way of defining the set of the μ_{λ_i}'s is to formalize the association between labels and properties via a function $\Lambda_\alpha : P_\alpha \rightarrow (L_\alpha \times [0,1])^n$, where n is the cardinality of the set L_α, associating with each property p a n-tuple of couples $(\lambda_i, \mu_{\lambda_i}(p))$ (see Fig.1). In this sense Λ_α can be thought of as either a function mapping properties into fuzzy subsets of L_α, or a one-to-many fuzzy binary relation on $P_\alpha \times L_\alpha$.

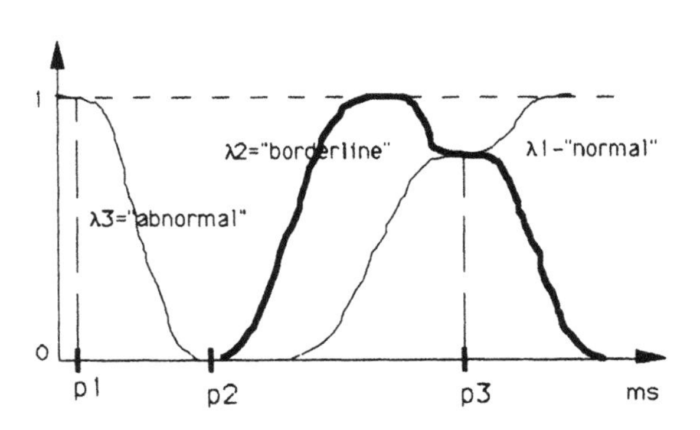

Figure 1
A (non-realistic) $\Lambda_{duration}$ function for the P wave from the example discussed in the text. Ambiguities are exemplified which arise in the judgement of the properties p_2 and p_3 when a synthesis function based on a prevalence criterion is used.

Given a property $p \in P_\alpha$, the derived n-tuple $\Lambda_\alpha(p) = ((\lambda_1, \mu_{\lambda_1}(p)), \ldots, (\lambda_n, \mu_{\lambda_n}(p)))$ can be resumed by stating just one couple (λ_i, m), instead of the whole n-tuple, where $m \in [0,1]$ is the degree of appropriateness assigned to the selected label λ_i in the resuming process. The couple is chosen by means of a "synthesis function" $\sigma_\alpha : (L_\alpha \times [0,1])^n \rightarrow (L_\alpha \times [0,1])$, so that $(\lambda_i, m) = \sigma_\alpha((\lambda_1, \mu_{\lambda_1}(p)), \ldots, (\lambda_n, \mu_{\lambda_n}(p)))$.
As an example, consider a case drawn from electrocardiology [8], in which the attribute "duration" of an observed P wave structure in a tracing can be judged according to $L_\alpha = \{\lambda_1$ = "Normal", λ_2 = "Borderline", λ_3 = "Abnormal"$\}$ in order to state whether the structure is a clue to the presence of a pathological state. Because of the strong correlation among the meanings of these three labels (for example, high normality implies low abnormality; the case of both high normality and high abnormality is not accepted, as being a non self-consistent judgement), it is sufficient to consider the label λ_3, and its associated membership function. In this case, σ_α could then be simply a projector, extracting from $\Lambda_\alpha(p)$ the couple $(\lambda_3, \mu_{\lambda_3}(p))$.
To sum up, for each attribute α the combination of a function Λ_α with a synthesis function σ_α is used to express the interpretation $\sigma_\alpha(\Lambda_\alpha(p))$ of the property p measured on the observed system with respect to the goals of the observation. As a result of the first step, for each α a new descriptor $<\alpha, \sigma_\alpha(\Lambda_\alpha(p))>$ is obtained. The set of these descriptors constitutes the "evaluated description" D_{eval}, which is the input to the next step of the procedure.
In the second step, the properties of D_{eval} are in turn contextually combined into a global judgement, recognizing the fact that the system is a whole. This can be formalized by a "global judgement function" $G: \sigma_{\alpha_1}(\Lambda_{\alpha_1}(P_{\alpha_1})) \times \ldots \times \sigma_{\alpha_n}(\Lambda_{\alpha_n}(P_{\alpha_n})) \rightarrow (L_S \times [0,1])^k$ where L_S is the predefined set of labels experts use in judging the whole situation S, k is the cardinality of L_S and $\sigma_{\alpha_i}(\Lambda_{\alpha_i}(P_{\alpha_i})) \subseteq L_{\alpha_i} \times [0,1]$.

In a way similar to what happens for partial judgements, the global judgement is eventually synthesized into a single label via a synthesis function $\Sigma:(L_S\times[0,1])^k\rightarrow L_S\times[0,1]$.

For example, the properties of a P wave can be studied to obtain a judgement from L_S={"right atrial hypertrophy", "left atrial hypertrophy", "normal", "uncertain"}.

Labels to be combined can take values in different universes: in general if $\alpha_i\neq\alpha_j$, $L_{\alpha_i}\neq L_{\alpha_j}$, so that G cannot be simply modeled by means of the classical fuzzy connectives, such as T-norms or T-conorms [4], also because different attributes can have different relevances to the criteria according to which the behaviour of the system is judged [9].

3. A MODEL OF EMPIRICAL PLAUSIBLE REASONING

In order to face experts' inability to assess the membership functions μ_{λ_i}, an approximate procedure is proposed based on the following observations:

a) experts have a clear vision of the direction of the judgement (or "goal at infinity" [10]). They make this view explicit by defining the set of labels L_S, whose elements represent the judgements considered as appropriate by experts about the behaviour S of the observed system;
b) experts derive the description D_{eval} in a heuristic way, often associating the linguistic labels used in this description with subsets of the possible numerical values;
c) experts generally synthesize the couple $\sigma_\alpha(\Lambda_\alpha(p))$ by a linguistic label, the first term of the couple (e.g. "the duration of the P wave is normal"), so accepting to consider as certain the label obtained by synthesis. Moreover, they are ready to associate a subset of P_α to this term (e.g. "normal duration"→ [80-110] ms) [11]. The boundaries of this subset may be fuzzy;
d) for each observed property $p\in P_\alpha$ experts seem to choose the most plausible linguistic label in L_α. Since this criterion does not guarantee the uniqueness of the selected label, experts seem to implicitly adopt, in place of the synthesis function σ_α, a relation selecting the labels λ_i such that $\mu_{\lambda_i}(p)$ is maximum. For instance, in Fig.1, no label can be associated with the value p_2 by this criterion, whereas both "borderline" and "normal" can be associated with the value p_3;
e) in expressing their linguistic judgements, experts state contextually the subsets of P_α in which each linguistic label prevails, rather than define the most plausible linguistic label λ for each $p\in P_\alpha$;
f) experts express these decisions by indicating the (possibly fuzzy) boundaries of each subset, i.e. the points (or the subsets) in which one linguistic label ceases to prevail, and another begins;
g) if subsets exist in which it is not clear which linguistic label prevails, a new label, usually denoting this uncertainty, is introduced (see Fig.2);
h) when experimental results give some evidences of a wrong or inadequate linguistic classification experts tend to contextually change the boundaries of the defined subsets, rather than discuss the problem in terms of the (never formally expressed) μ_{λ_i}s;
i) only a finite set of labels is used by experts.

The procedure assumes that:

1) the linguistic judgement for each observed property $p\in P_\alpha$ relative to the single attribute α is derived by the prevalence criterion;

2) the linguistic judgement is expressed by a single label, so that if either no or more than one labels are recognized to be applicable to the situation, a new label has to be defined;
3) the linguistic judgements assigned to each attribute are contextually combined, taking into account the relative relevance of each attribute to the behaviour of the whole system.

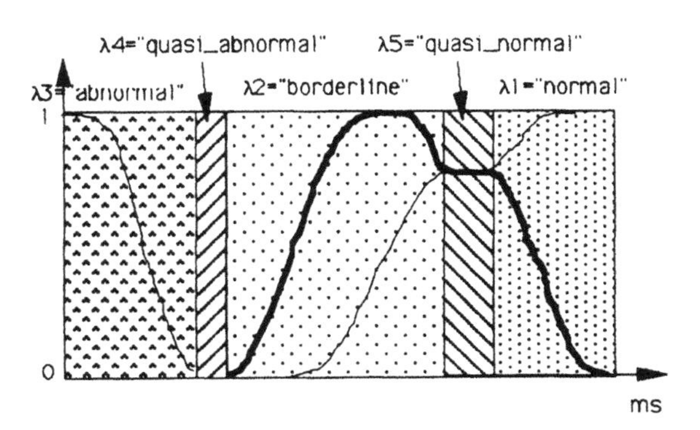

Figure 2
The subsets associated with each label in the proposed approximate procedure.

From these observations and assumptions, the empirical two-step procedure is derived. This procedure approximates the proposed judgement method by substituting the precise definitions of the Λ_α's and σ_α's by a "label function" (L-function), and of G and Σ by a "label combination function".

For each attribute α an L-function $\Lambda'_\alpha : P_\alpha \rightarrow L_\alpha$ is defined which associates the observed property $p \in P_\alpha$ with a label $\lambda = \Lambda'_\alpha(p) \in L_\alpha$. This reflects the empirical assumption that the appropriateness of λ as a judgement on p is greater than that of all the other possible labels. In this sense assigning a label to a property is a contextual process, in that it has to take into account the appropriateness of each possible label for that property.

In the definition of an L-function no assumptions are required on the topological and metric structure of the domain P_α. On the other hand, if an ordering is defined on P_α, the procedure can take advantage of this by indicating only the boundaries of the interval(s) of values to which a label applies.

When Λ'_α is applied to the observed value $p \in P_\alpha$, a descriptor $<\alpha, \lambda>$ is obtained, where $\lambda = \Lambda'_\alpha(p)$, which suggests an interpretation of the situation under study on the basis of the observation of the attribute α alone.

Such an interpretation must not, in general, be taken as definitive and exclusive. Rather, it acts as a constraint on the possible values that the final judgment can take on. The same λ can in fact be suggestive of different judgements, yet be such as to exclude others. Although the interpretation process is brought forward within the world of experts' models for the situation, an underlying logical structure can be recognized which rules the appropriateness of the interpretation.

Indeed, both label functions and label combination functions can be thought of as truth functions. In this sense, indicating with $\mathbf{P}(L_S)$ the power set of L_S, λ can be interpreted as an indicator of the truth value, in a multiple-valued logic, of a sentence of the kind: "the whole observed situation S, on the basis of the observation of the attribute α alone, is interpreted as $\underline{d}$", (shortly "S is $\underline{d}$" | α), where $\underline{d} \in \mathbf{P}(L_S)$ expresses the choice of the possible targets of the judgement procedure. The cardinality of the set $\mathbf{P}(L_S)$ is assumed as the cardinality of the set of truth values. Once $\underline{d}$ is fixed, the associated truth value(s) becomes designated (in the sense of formal logic [12]), and the logic underlying the judgement process can be derived. In the different applications the sets of truth values have therefore different structures, in which at least one value is designated. In the example, if $\underline{d}$ = "atrial hypertrophy" (≡ {"right atrial hypertrophy"}∪{"left atrial hypertrophy"})∈$\mathbf{P}(L_S)$, the elements of L_S {"right atrial hypertrophy"} and {"left atrial hypertrophy"} denote the designated values.

The evaluated description D_{eval} which is obtained by the application of the set of L-functions to the properties of D_{init} is now seen as a set of local interpretations, each one assessing the degree of truth of the sentence "S is $\underline{d}$" | α_i.

The local propositions of D_{eval} are combined interpreting the labels as indicators of sets of truth values and taking into account the mutual relations of the attributes to weigh their relevance to the global judgement of the situation S.

Due to their complexity, relations among the attributes are studied two at a time, thus defining a set of dyadic operators which realize the label combination function. Each operator combines the sets of labels of two attributes to combine the corresponding truth values. Therefore, the overall label combination functio is a multiple-valued logic tree (MVLT), as defined in [13]. The so obtained partial results are agai combined two at a time into intermediate contextual interpretations, so that each combination of loca interpretations is examined until a global interpretation is reached. In practice, the process is made simpler because -on the basis of the user model- only the combinations of attributes meaningful in the experimen are taken into account.

Intermediate and final interpretations are in their turn expressed as couples <α', λ> where α' is a name identifying the set of attributes $\alpha_1,\ldots,\alpha_n$ considered in the judgement process and λ denotes the truth value assumed by the sentence "S is $\underline{d}$" | $\alpha' = \{\alpha_1,\ldots,\alpha_n\}$.

In the example of atrial hypertrophy, left hypertrophy is characterized by tall, pointed P waves of norma duration, whereas in right hypertrophy P waves present a long duration (0.12s or more), normal amplitud and a notch. [14]. This way, a normal duration of the P wave can be present in a normal ECG as well as i a left hypertrophy ECG. On the other hand, a normal amplitude is a sign either of a normal heart or of right atrial hypertrophic one. Only the combination of the three attributes (duration, amplitude, shape) ca solve such ambiguities.

4. CONCLUSIONS

The paper has proposed a method to approximate the heuristic reasonings performed by an expert i judging the state or the behaviour of a system. This method has been developed and tested durin collaboration with experts from several disciplines (astronomy, reliability, cardiology) and aims requiring realizable assumptions about the a priori assessments and automatic reasoning methods reflectin the user models.

The presented hypotheses on experts' plausible judgement formulation have been met by some real cases in our experiences, and the proposed approximate procedure seems to match and explain the empirical procedure followed by the expert in the quoted experiments.

In addition, the whole procedure can be considered as a method for expert's knowledge elicitation. In fact, the empirical model proposed allows situations to be highlighted in which the judgements expressed by the expert are contextually incomplete or incoherent.

5. ACKNOWLEDGEMENTS

We would like to thank Dr. Barbara De Cristofaro of REMCO ITALIA - Cardioline for her suggestions about the cardiological example.

REFERENCES

[1] Accomazzi A., Bordogna G., Mussio P., Rampini A. (1989), An approach to heuristic exploitation of astronomers' knowledge in automatic interpretation of optical pictures, in: Knowledge-based systems in astronomy, F. Murtagh ed., Springer-Verlag, pp.191-212

[2] Brambilla P., Lechi G., Mussio P., What is a remotely sensed tree? A proposal for structural and pseudo-spectral reconnaissance (1978), Proc. Int. Symp. Rem. Sens. Obs. Inven. Earth Res., pp.491-506

[3] Garribba S., Guagnini E., Mussio P. (1986), Uncertainty reduction techniques in an expert system for fault tree construction, in: Uncertainty in knowledge-based systems, B.Bouchon, R.R.Yager eds, Springer-Verlag, pp.252-255

[4] Dubois D., Prade H. (1980), Fuzzy sets and systems, Academic Press

[5] Chandhuri B.B., Dutta Majumder D. (1982), On membership evaluation in fuzzy sets, in: Approximate reasoning in Decision analysis, M.Gupta, E.Sanchez eds., North-Holland, pp.3-11

[6] International vocabulary of basic and general terms in metrology, ISO et al., 1984

[7] Siegel S. (1956), Non-parametric statistics for the behavioural sciences, McGraw Hill

[8] Bortolan G., Degani R. (1988), Linguistic approximation of fuzzy certainty factors in computerized electrocardiography, in: Fuzzy computing, Theory, Hardware and applications, M.M.Gupta, T.Yamakawa eds., North-Holland, pp.243-261

[9] Bonissone P.P. (1982), A fuzzy sets based linguistic approach: theory and applications, in: Approximate reasoning in Decision analysis, M.M.Gupta, E.Sanchez eds., North-Holland, pp.329-339

[10] Kierulf A., Chen K., Nievergelt J., (1990), Smart Game Board and Go Explorer: A study in software and knowledge engineering, Comm. ACM, vol.33, n.2, pp.152-166

[11] Gotts N., Hunter J., Hamlet I., Vincent R. (1989), Qualitative spatio-temporal models of cardiac electrophysiology, University of Aberdeen, AUCS/TR8903

[12] Rescher N. (1969), Many-valued logic, McGraw-Hill

[13] Garribba S., Guagnini E., Mussio P. (1985), Multiple-Valued Logic Trees: meaning and prime implicants, IEEE Trans.on Reliability, vol.R-34, n.5, pp.463-472

[14] Marriott H.J.L. (1977), Practical Electrocardiography, The Williams and Wilkins Co.

TIME, TENSE, AND RELATIVITY REVISITED

Frank D. Anger and Rita V. Rodriguez

Division of Computer Science
The University of West Florida
Pensacola, FL 32514 USA
(904) 474-3022, 3065
fa@ufl.edu

Abstract

Interest in the problem of expressing temporal relations between events in a coherent fashion has undergone a revival due to the creation of data bases and knowledge bases containing time-dependent information and also through the scrutiny of concurrent algorithms and real-time systems. Presented herein is a simple temporal model, designated an F-complex, which develops from a single *future* operator and a single order axiom yet encompasses several of the current proposals for models to systematize reasoning about one or more of the aforementioned areas. The rudimentary F-complex commits to no special ontology of time, giving the advantage of clarifying the properties which are common to most methods of temporal modeling. Concepts of past, future, and temporal precedence are formulated within the posited structure, allowing comparison to the published temporal models of Lamport [11], Allen [1], Milner [14], Rodriguez [17], and others [21]. Specifically, Allen's thirteen linear-time and Rodriguez's eighty-two relativistic *atomic* relations are characterized, as well as the axiomatic scheme of Lamport. The models are treated more thoroughly than in [6]. Furthermore, the main theorem is strengthened.

Key Words: *Temporal Models, Distributed Systems, Tense Logics, Knowledge Bases, Event Complexes, F-Complexes, Relativity.*

1 Introduction

The expression of time in natural language communication has received extensive study through the development of *tense logics* [15,18]. A variety of temporal conundrums dating from antiquity have been analyzed and re-analyzed in the light of different suppositions about the significance of verb tenses and the standard logical connectives [20]. Does *"Dahlia danced and sang all evening"* mean that at every moment during the evening Dahlia was dancing **and** singing, or that at each moment she was either dancing **or** singing, or that these activities were dispersed over the entire evening in some way even though poor Dahlia may have "taken five" a few times? The development of temporal logics from modal logics has allowed the investigation of many different attitudes toward time in a rigorous fashion. **Totally ordered** and **partially ordered** models of the temporal sequence of events can both be expressed in terms of differing axioms imposed on modal operators, as can **discrete** versus **continuous** and **bounded** versus **unbounded** time [21].

Throughout the literature on tense logics and temporal logics, the concepts of **past**, **present**, and **future** play a central role just as they occupy such a position in our own understanding of our existence with respect to other phenomena. While some systems treat past and future symmetrically, others capture the asymmetry we experience from the irreversible flow of time from the past through the present and into the future [20]. In the construction of expert planning systems, intelligent robot control programs, and other AI tools dealing with a changing domain, temporal issues arise either implicitly or explicitly in the way the electronic product models the actual application [8,13]. When the application is one in which relativistic effects are significant (from satellite networks and interplanetary communications to VLSI behavior) or in which a common clock is not available for measuring event order (from distributed systems to military intelligence), the temporal model must be sufficiently sophisticated to allow the expression and analysis of complex non-sequential relations yet simple enough to allow efficient computation.

The authors presented a temporal model [4,5,16,17], based on work in natural language processing [1] and the analysis of distributed systems [11,12], having the expressiveness required for such applications. The approach also developed efficient methods for drawing temporal conclusions from partial information. Although that model makes no reference to past, present, and future, these concepts are easily derivable from it. The intent of the paper at hand is to start with a *future* operator, related to the tense operator G, "It will always be the case that," and show how an equivalent model can be built from this single concept. Rather than modal logic operators, the future operator, F, is an assignment of a set, $F(e)$, of *future* events to each given event e. Depending on the assumptions on the relationship between e, $F(e)$, and the *past*, as defined in the next section, different models arise. The paper concentrates on retrieving the models of Lamport, Allen, Milner, and Rodriguez [11,1,14,17].

2 F-complexes

Start with a set, S, of *events*, and a *future* operator, F: $S \longrightarrow \mathcal{P}(S)$, all subsets of S.

DEFINITION 1: An **F-complex** is a pair (S,F), in which S is a set and F a future operator on S satisfying the single axiom

(E1) $e \in F(a)$ implies $F(e) \subset F(a)$ (proper subset).

Define a *past* operator, P by

$P(a) = \{e | a \in F(e)\}$.

Observe that a precedence relation, $\longrightarrow$, can be defined in (F, S) by

$a \longrightarrow b$ if and only if $b \in F(a)$ (iff a in $P(b)$).

The condition (E1) is equivalent to making $\longrightarrow$ an irreflexive partial order on S. ($a \longrightarrow a$ iff $a \in F(a)$, which implies $F(a) \subset F(a)$, which is impossible. Transitivity is clear.) This simple structure needs more flesh on the bones in order to be able to apply F-complexes to most particular tasks, but a surprising amount can be derived from the order structure alone. For the moment, no further assumptions are made.

DEFINITION 2: If A is a subset of S, define

$F(A) = \{e | e \in F(a)\, \forall a \in A\}$, and

$P(A) = \{e | e \in P(a)\, \forall a \in A\}$.

In particular, $F(S) = P(S) = \emptyset$ and $F(\emptyset) = P(\emptyset) = S$.

$F(A)$ is the intersection of all the sets $F(e)$ for $e \in A$, and, consequently, $F(A) \subseteq F(e)\, \forall e \in A$. Proposition 1 now follows easily.

PROPOSITION 1: If (S, F) is an F-complex, then:

(1) $A \subseteq B$ implies $F(B) \subseteq F(A)$ and $P(B) \subseteq P(A)$.

(2) $A \subseteq PF(A)$ and $A \subseteq FP(A)$ for all subsets A of S.

(3) $FPF = F$ and $PFP = P$.

PROOF:

(1) Immediate.

(2) By the remark preceding the Proposition, $a \in A$ implies $F(A) \subseteq F(a)$. Applying (1) yields $PF(a) \subseteq PF(A)$. But $a \in PF(a)$ iff $\forall b \in F(a)$, $a \in P(b)$ iff $a \in P(b)$ for every b with $a \in P(b)$, which is true. Therefore, $a \in PF(a) \subseteq PF(A)$ implies $a \in PF(A)$. Since this is true for any $a \in A$, conclude that $A \subseteq PF(A)$. That $A \subseteq FP(A)$ is dual.

(3) The set of subsets of S is partially ordered by inclusion, and F and P are order-reversing functions on this partial order satisfying (2). (3) is then a standard result which follows by observing that $A \subseteq PF(A)$ (by (2)) implies $FPF(A) \subseteq F(A)$ (by (1)), while $F(A) \subseteq FP(F(A)) = FPF(A)$ by (2). Thus $FPF = F$.

The intention is to think of $F(A)$ as the "*future* of the collection A of *events*," and of $a \longrightarrow b$ as *a temporally precedes b* or *a happens before b*.

Still without restricting the temporal model, we define a second relation, $a \dashrightarrow b$, to be understood as *a can (causally) affect b*.

DEFINITION 3: In an F-complex (S, F), write $a \dashrightarrow b$ or a **can affect** b to signify $F(b) \subset FP(a)$ (proper subset).

3 System Executions and Event Complexes

Lamport, with two primitive relations, $\longrightarrow$ and $\dashrightarrow$ and four axioms A1 to A4 (actually more which dealt with cardinality and *non-terminating* events), described a **system execution** $(S, \longrightarrow, \dashrightarrow)$. The elements of S were thought of as actual executions of operations in a particular run of a concurrent program. Anger [5] added two more axioms, M1 and M2, to create an axiomatic system which can always be modeled as a set of subsets of a partial order. We will demonstrate that if we use Definition 3 to define the can-affect arrow $\dashrightarrow$, then we can establish all of these six axioms from the one E1; in other words, a relation $\dashrightarrow$ can be derived order-theoretically from the one relation $\longrightarrow$ without the need for further primitives and axioms. After establishing the required properties we discuss the relationship between this approach and Lamport's [11]. As usual, more definitions are required to accomplish our goal.

DEFINITION 4: An **event complex** is a triple $(S, \longrightarrow, \dashrightarrow)$ in which S is a set and the two arrows represent relations on S, named **precede** and **can affect**, respectively, satisfying:

(A1) $\longrightarrow$ is an irreflexive partial order.

(A2) $A \longrightarrow B$ implies that $A \dashrightarrow B$ and $B \not\dashrightarrow A$

(A3) $A \dashrightarrow B \longrightarrow C$ or $A \longrightarrow B \dashrightarrow C$ implies $A \dashrightarrow C$.

(A4) $A \longrightarrow B \dashrightarrow C \longrightarrow D$ implies $A \longrightarrow D$.

(M1) $A \dashrightarrow B \longrightarrow C \dashrightarrow D$ implies $A \dashrightarrow D$.

(M2) $\dashrightarrow$ is reflexive (but not transitive).

These axioms are consistent with the properties exhibited by any collection of subsets of a partially ordered set if $A \longrightarrow B$ is interpreted as *every* $a \in A$ *is less than every* $b \in B$ and $A \dashrightarrow B$ is taken to indicate that *some* $a \in A$ *is less than some* $b \in B$. Intervals on the real line are only a particular case of the given situation corresponding to *global time* (there is a global clock to which every event can be referred) or to the added assumption that, for every pair of events A and B, either $A \longrightarrow B$ or $B \dashrightarrow A$. A collection of subsets of a relativistic space-time continuum is a more interesting case. In this interpretation, $A \longrightarrow B$ means that B is contained in the *future cone* of A, and $A \dashrightarrow B$ that some part of B is in the *future cone* of some part of A. The more relevant image is that of a collection of sets of instructions running on a distributed system, where $A \longrightarrow B$ represents that A is completely done before B begins, and $A \dashrightarrow B$ that some part of A could have successfully sent a signal which could be received by B. Next we show that an F-complex is also an event complex under the given definition of precede and can affect.

THEOREM 1: If (S, F) is an F-complex and if $\longrightarrow$ and $\dashrightarrow$ are defined as in Definitions 1 and 3, then $(S, \longrightarrow, \dashrightarrow)$ is an event complex. Conversely, if $(S, \longrightarrow, \dashrightarrow)$ is an event complex and F is defined by $F(a) = \{b \mid a \longrightarrow b\}$, then (S, F) is an F-complex satisfying $a \dashrightarrow b$ implies $F(b) \subset FP(a)$.

PROOF: Each of the six axioms, A1 to A4 and M1 and M2, must be established.

(A1) is equivalent to (E1).

(A2) $a \longrightarrow b$ means that $F(b) \subset F(a)$ (since $b \in F(a)$, $b \notin F(b)$). But also $F(a) \subseteq FP(a)$ (always) and hence $F(b) \subset FP(a)$. Therefore $a \dashrightarrow b$. Moreover $b \in P(a)$ implies $FP(a) \subseteq F(b)$ (Proposition 1, part (1)), hence $F(b) \not\subset FP(a)$. Therefore, $b \not\dashrightarrow a$.

(A3) Suppose $a \dashrightarrow b \longrightarrow c$. $c \in F(b)$ implies $F(c) \subset F(b)$, while $a \dashrightarrow b$ implies $F(b) \subset FP(a)$. Therefore, $F(c) \subset FP(a)$; i.e. $a \dashrightarrow c$. The other half of (A3), that $a \longrightarrow b \dashrightarrow c$ implies that $a \dashrightarrow c$, is only slightly harder: noting that $b \in F(a)$ implies $FP(b) \subseteq FPF(a) = F(a) \subseteq FP(a)$ and that $b \dashrightarrow c$ implies $F(c) \subset FP(b)$, conclude that $F(c) \subset FP(a)$. Thus $a \dashrightarrow c$

(A4) Suppose $a \longrightarrow b \dashrightarrow c \longrightarrow d$. $a \longrightarrow b$ implies $b \in F(a)$, hence $FP(b) \subseteq F(a)$ (Proposition 1, part (1)). But also $F(c) \subset FP(b)$ and $d \in F(c)$. These three facts yield $d \in F(a)$, hence $a \longrightarrow d$.

(M1) For any a, $F(a) \subset FP(a)$, asserting that $a \dashrightarrow a$.

(M2) Suppose $a \dashrightarrow b \longrightarrow c \dashrightarrow d$. Then $F(b) \subset FP(a)$ and $FP(c) \subseteq F(b)$ (as in A4), and $F(d) \subset FP(c)$. Combining gives $F(d) \subset FP(a)$, implying that $a \dashrightarrow d$.

The converse, that (S, F) is an F-complex, is an immediate result of (A1), Definition 1, and the comments following that definition. Suppose, finally, that $a \dashrightarrow b$ and that $e \in F(b)$. It must be shown that $e \in FP(a)$. If $e' \in P(a)$, then $e' \longrightarrow a \dashrightarrow b \longrightarrow e$, which implies $e' \longrightarrow e$ by (A4). Therefore, $e \in F(e')$. Since the latter is true for any e' in $P(a)$, $e \in FP(a)$, establishing that $F(b) \subseteq FP(a)$. But $F(b) \neq FP(a)$ because $a \in FP(a)$, while (A2) guarantees that $a \notin F(b)$.

A closing remark: In general, in an event complex $(S, \longrightarrow, \dashrightarrow)$, $F(b) \subset FP(a)$ does not imply that $a \dashrightarrow b$. For example, let $(S, \longrightarrow)$ be any partially ordered set, taking $\dashrightarrow$ to be the same as $\longrightarrow$, with the addition of $a \dashrightarrow a$ for all a. The result is an event complex in which $a \dashrightarrow b$ if and only if $b \in F(a)$ or $b = a$, a subrelation of $F(b) \subset FP(a)$.

Table 1: **Classification of relations in an event complex. Each REL is a relation between events A and B**

	REL	ABBR	DEFINITION
1.	affects	- - ->	*A*- - -> *B* (primitive term)
2.	precedes	—>	*A*—> *B* (primitive term)
3.	starts before	*sb*	∃ *C*, *A*- - -> *C*—> *B*
4.	finishes before	*fb*	∃ *C*, *A*—> *C*- - - > *B*
5.	starts later	*sl*	∃ *C*, *B*- - -> *C* & *A*- -/- > *C*
6.	finishes later	*fl*	∃ *C*, *B*—> *C* & *A*—/-> *C*
7.	co-original with	*co*	not *sb* & not *sl*
8.	co-terminal with	*ct*	not *fb* & not *fl*
9.	before	<	∃ *C*, *A*—> *C*—> *B*
10.	meets	*m*	—> & not <
11.	overlaps	*o*	*sb* & *fb* & —/->
12.	during	*d*	*sl* & *fb*
13.	contains	*cn*	*sb* & *fl*
14.	surpasses	*su*	*sl* & *fl* & - - ->
15.	starts	*s*	*co* & *fb*
16.	finishes	*f*	*ct* & *sl*
17.	projects to	*p*	*co* & *ct*
18.	extends	*e*	*co* & *fl*
19.	pre-extends	*pe*	*ct* & *sb*
20.	after	>	∃ *C*, *B*—> *C* & *A*- -/- > *C*
21.	follows	*fol*	- -/- > & not >
22.	inverse REL	RELi	*B* REL *A* (for any REL)
23.	REL1 & REL2i	REL1/REL2i	*A* REL1 *B* & *B* REL2 *A* (for legal pairs)

4 Relations in an F-Complex

Rodriguez and Anger [17] demonstrated that, starting with the axioms of Definition 4, there are 8 possible distinct **atomic** temporal relations which can be defined in terms of **precedes** and **can affect**. Atomic connotes that the relations cannot be written as a proper disjunction of other relations defined in the same way. A relation can be recognized as atomic if its conjunction with any other relation produces itself or the false (empty) relation. A naming scheme is presented in Table 1 allowing the description of the atomic relations, which are all legal relations of the form given in line 23, with each of REL1 and REL2 being one of the relations 9 through 21 of the table. To describe the corresponding atomic relations in an F-complex in terms of the operator F (and P, which is defined in terms of F) is the object of this section.

The plan of attack begins with enumerating all possible relations between two members a and b of S using only the future operator F, the past operator P defined in terms of F, their compositions, set membership ($\in$), and set inclusion ($\subseteq$). Next we form the further relations which can be expressed using set equality, intersection, conjunction (&), and negation, allowing thereby expressions containing $=$, $\neq$, $\subset$, etc.. The closure of the collection of relations under disjunction, conjunction, negation, converse and composition forms a relation algebra [10] whose atoms are the atomic relations we seek.

First consider the possible compositions of F and P: $G_1G_2 \ldots G_n(a)$, where each G_i is either F or P. There are an infinite number of such compositions, but fortunately only ten *different* functions arise. It was already established, for example, that $PFP = P$ and $FPF = F$. Calculating $FF(A)$

if $x \in F(A)$, then $e \in FF(A)$ implies $e \in F(x)$, which, by Definition 1, is contained in $F(A)$. But if $e \in F(x)$ and also $e \in FF(x)$, then e must be in its own future, which is impossible. Therefore, $FF(A) = \emptyset$ if there is any $x \in F(A)$, but $FF(A) = S$ if $F(A) = \emptyset$. The case for PP is similar, making it convenient to define a "characteristic function," $\chi_\emptyset^S$ as

$$\chi_\emptyset^S(A) = \begin{cases} S, \text{ if } A = \emptyset \\ \emptyset, \text{ if } A \neq \emptyset \end{cases}$$

or just $\chi(A)$. Then $FF(A) = \chi F(A)$ and $PP(A) = \chi P(A)$. Moreover, for any non-empty A, $FP(A)$ is non-empty, since $A \subseteq FP(A)$ by Proposition 1, and hence $FFP = \chi$. In short, the only ten different compositions obtainable are:

F, P, FP, PF, χ, χF, χP, $\overline{\chi}$, $\overline{\chi}F$, and $\overline{\chi}P$,

where $\overline{G}$ is the *complement* of G: $\overline{G}(A) = \overline{G(A)}$.

Under the assumption that *no event has an empty future or an empty past*, it can be seen that, for a given element a of S, $FF(a) = \chi F(\{a\}) = \emptyset$ and $PP(a) = \chi P(\{a\}) = \emptyset$; thus there are only four distinct non-trivial sets which can be obtained of the form $G_1G_2 \ldots G_n(a)$, where each G_i is either F or P, to wit: $F(a)$, $FP(a)$, $PF(a)$, and $P(a)$. Of the 32 containment relations which can be written of the form $G(a) \subseteq H(b)$, where G and H represent any of the four given operators, only sixteen can ever hold (provided neither term is empty), namely: $F(a) \subseteq F(b)$, $F(a) \subseteq FP(b)$, $FP(a) \subseteq F(b)$, $FP(a) \subseteq FP(b)$, the four duals obtained by interchanging F and P, and the eight converses obtained from the above by interchanging a and b.

There are eight more relations: $a \in F(b)$, $a \in FP(b)$, $a \in PF(b)$, $a \in P(b)$, and the four converses obtained by interchanging a and b. It will now be shown, however, that there are only eight, rather than 24, relations which can be constructed using $\subseteq$ and $\in$ together with the future and past operators.

PROPOSITION 2:

(1) $a \in F(b)$ iff $FP(a) \subseteq F(b)$ iff $PF(b) \subseteq P(a)$.
(2) $a \in FP(b)$ iff $P(b) \subseteq P(a)$ iff $FP(a) \subseteq FP(b)$.
(3) $b \in PF(a)$ iff $F(a) \subseteq F(b)$ iff $PF(b) \subseteq PF(a)$.
(4) $F(a) \subseteq FP(b)$ iff $P(b) \subseteq PF(a)$.

The relations (1′) through (4′) are the dual relations obtained by exchanging F and P.

(1′) $a \in P(b)$ iff $PF(a) \subseteq P(b)$ iff $FP(b) \subseteq F(a)$.
(2′) $a \in PF(b)$ iff $F(b) \subseteq F(a)$ iff $PF(a) \subseteq PF(b)$.
(3′) $b \in FP(a)$ iff $P(a) \subseteq P(b)$ iff $FP(b) \subseteq FP(a)$.
(4′) $P(a) \subseteq PF(b)$ iff $F(b) \subseteq FP(a)$.

Moreover, these are all the relations which can be written using F, P, $\subseteq$, $\in$, and composition.

PROOF: The relations listed as (1) through (4) and (1′) through (4′), together with the obvious $\in F(b)$ iff $b \in P(a)$, are all the 24 possible relations as discussed previously. Therefore there are, in act, only eight distinct relations if the equivalences stated can be established.

(1) Using Proposition 1, $a \in F(b)$ implies $FP(a) \subseteq FPF(b) = F(b)$, and the converse is also true nce $a \in FP(a)$ is always true, yielding that $FP(a) \subseteq F(b)$ implies $a \in F(b)$. For the other half, $P(a) \subseteq F(b)$ implies $PF(b) \subseteq PFP(a) = P(a)$, which in turn implies $FP(a) \subseteq FPF(b) = F(b)$; ius this half is also *"if and only if."*

(2) $a \in FP(b)$ implies $P(b) = PFP(b) \subseteq P(\{a\}) = P(a)$. Continuing: $P(b) \subseteq P(a)$ implies $FP(a) \subseteq FP(b)$. But $a \in FP(a)$, so $a \in FP(b)$. Since this is a circular implication, all of the implications are "if and only if," and (2) is established.

(3) $b \in PF(a)$ implies $F(a) = FPF(a) \subseteq F(b)$ implies $PF(b) \subseteq PF(a)$, which implies $b \in PF(a)$. Once again, both equivalences are thereby proved.

(4) $F(a) \subseteq FP(b)$ implies $P(b) = PFP(b) \subseteq PF(a)$ implies $F(a) = FPF(a) \subseteq FP(b)$, and (4) follows.

(1') through (4') follow by duality (any set-theoretic formula involving F and P operators which is true for all values of all variables appearing in the formula yields a true formula when all F's are replaced by P's and all P's by F's.)

At this point, therefore, there are only eight distinct relations. None of these relations is atomic; moreover, it is easily shown that the eight are not independent. In fact,

$$(1) \Longrightarrow (2) \Longrightarrow (4) \text{ and}$$
$$(1) \Longrightarrow (3) \Longrightarrow (4),$$

and likewise with the dual relations.

We next consider set equality, intersection, conjunction, and negation. Since for sets X and Y, $X = Y$ iff $(X \subseteq Y)$ & $(Y \subseteq X)$, it is apparent that equals is unnecessary; however, it is convenient to retain its use in the special case of $X = \emptyset$, even though it could be written $X \subseteq \emptyset$. Similarly, since $X \neq Y$ iff $X \not\subseteq Y$ or $Y \not\subseteq X$, $\neq$ will not be needed; notwithstanding, once again $X \neq \emptyset$ will be convenient. Consider intersections: $Z \subseteq X \cap Y$ iff $Z \subseteq X$ & $Z \subseteq Y$, while $X \cap Y \subseteq Z$ iff $X \cap Y \cap \overline{Z} = \emptyset$; consequently, intersection can be restricted to appear in the form

$X_1 \cap \ldots \cap X_n = \emptyset$ or $X_1 \cap \ldots \cap X_n \neq \emptyset$.

Continuing under the assumption that $F(a)$ and $P(a)$ are never empty, using the eight operators F, P, FP, PF, and their complements and intersection, for a given pair of elements a and b we obtain 40 possibly distinct nontrivial sets of the forms $G_1(a) \cap G_2(b)$, $\overline{G_1(a)} \cap G_2(b)$, $G_1(a) \cap \overline{G_2(b)}$ and $\overline{G_1(a)} \cap \overline{G_2(b)}$. Requiring any of these to be empty (or non-empty) yields 80 possible relations, of which 64 are distinct.

As a result of this discussion, it follows that all relations definable in terms of F and P can be expressed using only containment ($\subseteq$), negation, conjunction and disjunction, and relations involving intersections in the particular forms $X_1 \cap \ldots \cap X_n = \emptyset$ or $X_1 \cap \ldots \cap X_n \neq \emptyset$. For example, $F(a) \cap P(b) \neq \emptyset$ is the relation called $a < b$ in Table 1, while $F(b) \not\subseteq F(a)$ is *a finishes later than b*. Of course, by Definition 3, $F(b) \subset FP(a)$ means a- - -> b in this paper; it is equivalent to $F(b) \subseteq FP(a)$ and $FP(a) \not\subseteq F(b)$. The atomic relations can now be obtained as conjunctions of the relations discussed so far. Many conjunctions simply give empty relations, while many lead to the same relation.

Theorem 1 established that an F-complex with - - -> given by Definition 3 satisfies the axioms of an event complex and hence should have the same (or more restricted) set of atomic relations. In order to define the atomic relations in terms of F and the most general **can affect** relation, from Table it is sufficient to define, in terms of F, the relations **sb**, **fb**, **sl**, **fl**, $<$, and $>$, since all others can be defined as conjunctions of these, their complements, and their converses. The following two definitions accomplish the task.

DEFINITION 5: For any element $a \in S$,

$E(a)$ = the *effect* of $a = \{x \mid a$- - -> $x\}$ and $C(a)$ = the *cause* of $a = \{x \mid x$- - -> $a\}$.

In case - - -> is given by Definition 3,

$E(a) = \{x \mid F(x) \subset FP(a)\}$ and $C(a) = \{x \mid F(a) \subset FP(x)\}$.

DEFINITION 6: The relations (compare Table 1) **start before**, **finish before**, **start later**, **finish later**, **before**, and **after** in an F-complex are given by:

a **sb** b	$E(a) \cap P(b) \neq \emptyset$
a **fb** b	$F(a) \cap C(b) \neq \emptyset$
a **sl** b	$E(b) \not\subseteq E(a)$
a **fl** b	$F(b) \not\subseteq F(a)$
$a < b$	$F(a) \cap P(b) \neq \emptyset$
$a > b$	$F(b) \not\subseteq E(a)$

For example, the atomic relation written as **d/cni** (during & contained-in converse) can be written as a **d/cni** b iff a (**sl** & **fb** & **sbi** & **fli**) b iff $E(b) \not\subseteq E(a)$ & $F(a) \cap C(b) \neq \emptyset$ & $E(b) \cap P(a) \neq \emptyset$ & $F(a) \not\subseteq F(b)$.

5 Other Temporal Models

If we assume, with Allen [1], that S consists of the collection of intervals on the real line (really ordered pairs of real numbers with the first less than the second), and that $F(<x,y>) = \{<r,s> \mid y \leq r\}$, then $<r,s> \not\in F(<x,y>)$ implies that $FP(<r,s>) = F(\{<p,q> \mid q \leq r\}) = \{<t,u> \mid r \leq t\} \supset \{<t,u> \mid s \leq t\} = F(<r,s>)$; that is, $<x,y>$ —/-> $<r,s>$ implies $<r,s>$ - - -> $<x,y>$. In this very restricted case, the thirteen atomic relations are obtained as

<:	$F(a) \cap P(b) \neq \emptyset$
meets:	$b \in F(a)$ & $F(a) \cap P(b) = \emptyset$
overlaps:	$b \in FP(a)$ & $a \in PF(b)$
starts:	$P(a) = P(b)$ & $F(a) \cap PF(b) \neq \emptyset$
during:	$P(a) \cap FP(b) \neq \emptyset$ & $F(a) \cap PF(b) \neq \emptyset$
finishes:	$F(a) = F(b)$ & $P(a) \cap FP(b) \neq \emptyset$
equal:	$P(a) = P(b)$ & $F(a) = F(b)$

and the converses of these obtained by interchanging a and b. Note that **equals** is self-converse. It is also apparent from this form of definition that the **dual** of each relation, obtained by interchanging P and F in the definition, is the same as the converse except for **starts** and **finishes**, which are duals of one another but not converses. In the general setting of relativistic time, there are events which are so separated in space and time that they can only be classified as temporally *incomparable* or *concurrent*, while others may appear to different observers to be in different temporal relations to one another.

Another interesting model is that of Milner's **event structures** [14]. An *elementary event structure* is just a partially ordered set $(E, \leq)$, in which the objects of interest are the *left-closed sets*: subsets L of E satisfying $a \in L$ and $b \leq a$ implies $b \in L$. In our context, sets of the form $P(A)$ are left closed. Milner's sets $\lceil a \rceil = \{x \mid x \leq a\}$ would, moreover, correspond to $P(a) \cup \{a\}$ in our notation. Although we do not pursue the lattice-theoretic development of those investigators, translation into F-complexes is possible.

6 Conclusion

By looking at what could be considered a minimal structure for any temporal model, it has been shown that many of the definitions and basic structure of several proposed temporal models can already be captured with a single *future* operator in an **F-complex** . The definitions and reasoning in an F-complex are purely set theoretic, and all derived relations can be described using only first order logic, the Boolean algebra of the subsets of a set S, and the construction of sets $\{x \mid Q(x)\}$ in which Q is itself a predicate expressible in terms of the operator F and the same logical and set-theoretic constructs. The approach presented offers the advantage of isolating those properties which do not depend on the particular temporal ontology, such as discreteness, linearity, boundedness, or the atomicity of events. Moreover, no introduction of special logics is necessary at this level of description. There are limitations, of course. The study does not approach the problem of associating propositions about the events with the underlying temporal structure: the relationship between the truth value of propositions at points and over intervals of time or over intervals and their subintervals is not addressed.

The ability to describe temporal relations in F-complexes was applied to a number of current temporal models, emphasizing the expressive power of this analysis and the simplicity of many of the properties of the other models. Lamport's axiomatic model [11,12] as extended by Anger [5] and Rodriguez [16] was expressed in a straight forward way in terms of F-Complexes. Furthermore, the atomic relations of event complexes [17] were characterized. The utility of knowing the atomic relations pertinent to a given model has been shown in [1,10,17,7], in which constraint-propagation methods based on the composition of atomic relations are used to provide feasible heuristics for testing the consistency of a collection of temporal relations. Moreover, the work of Allen [1] and Winskel [19] was likewise shown to be expressible in the language of F-Complexes.

Acknowledgment: This work was partially supported by a *Florida High Technology and Industry Council* Applied Research Grant.

References

[1] Allen, J. Maintaining Knowledge about Temporal Intervals. *Comm. of ACM 26*, 11 (1983), pp. 832-843.

[2] Allen, J. Towards a General Theory of Action and Time. *Artificial Intelligence 23*, (1984), pp 123-154.

[3] Allen, J. and Hayes, P. A Commonsense Theory of Time. *Proceedings of IJCAI*, Los Angeles CA, 1985, pp. 528-531.

[4] Anger, F., Morris, R., Rodriguez, R., Mata, R. A Temporal Logic for Reasoning in Distributed Systems, *Australian Joint AI Conference*, Adelaide, Australia, Nov 1988, pp. 146-155. (Also in *Lecture Notes in Artificial Intelligence*, Vol. 406, C. Barter and M. Brooks, eds., Springer-Verlag 1990, pp. 177-186.)

[5] Anger, F. On Lamport's Interprocessor Communication Model. *ACM Trans. on Prog. Lang. and Systems 11*, 3 (July 1989), pp. 404-417.

[6] Anger, F. and Rodriguez, R. Time, Tense, and Relativity. *Proceedings of Information Processing and Management of Uncertainty in Knowledge-Based Systems (IPMU)*, Paris, France, (July 1990), pp. 74-78.

[7] Anger, F., Ladkin, P., and Rodriguez, R. Atomic Temporal Interval Relations in Branching Time: Calculation and Application. *Applications of Artificial Intelligence IX, Proceedings of SPIE*, Orlando, (April 1991).

[8] Dean, T. and Boddy, M. Reasoning about Partially Ordered Events. *Artificial Intelligence 36*, 3 (Oct 1988), pp. 375-399.

[9] Ladkin, P. Specification of Time Dependencies and Synthesis of Concurrent Processes. *Ninth ACM Software Engineering Conference*, (1987), pp. 106-115.

[10] Ladkin, P. Satisfying First-Order Constraints about Time Intervals. *Proceedings of the Seventh National Conference on Artificial Intelligence*, St. Paul, MN, (Aug 1988), pp. 512-517.

[11] Lamport, L. The Mutual Exclusion Problem: Part I–A Theory of Interprocess Communication. *Journal ACM 33*, 2 (April 1986), pp. 313-326.

[12] Lamport, L. The mutual exclusion problem: Part II–Statement and Solutions. *Journal ACM 33*, 2 (April 1986), pp. 327-348.

[13] Leban, B., McDonald, D., Forster, D. A Representation for Collections of Temporal Intervals. *Proceedings of the Fifth National Conference on Artificial Intelligence*, Pittsburgh, PA, (July 1986), pp. 367-371.

[14] Milner, R. *A Calculus of Communicating Systems*. Springer Lecture Notes in Comp. Sci., Vol. 92, 1980.

[15] Prior, A. *Past, Present, and Future*. Clarendon Press, Oxford, 1967.

[16] Rodriguez, R., Anger, F. Reasoning in Relativistic Time. Submitted to *Computational Intelligence Journal*. July 1989.

[17] Rodriguez, R., Anger, F., Ford, K. Temporal Reasoning: A Relativistic Model. *International Journal of Intelligent Systems*, to appear.

[18] Rodriguez, R. and Anger, F. Prior's Temporal Legacy in Computer Science. To appear in *The Arthur Prior Memorial Volume*, Oxford University Press, Oxford, 1991.

[19] Winskel, G. An Introduction to Event Structures. *Linear Time, Branching Time and Partial Order in Logics and Models for Concurrency*, G. Goos and J. Hartmanis, eds., Springer-Verlag, New York, 1989, pp. 1-49.

[20] van Benthem, J. *A Manual of Intensional Logic*, 2nd ed. Center for the Study of Language and Information (CSLI), Stanford, CA, 1988.

[21] van Benthem, J. Time, Logic and Computation. *Linear Time, Branching Time and Partial Order in Logics and Models for Concurrency*, G. Goos and J. Hartmanis, eds., Springer-Verlag, New York, 1989, pp. 1-49.

6. INFORMATION

INFORMATION AND THE MIND-BODY PROBLEM

Giuseppe Longo
Dipartimento di Elettrotecnica Elettronica Informatica
Università di Trieste, 34100 Trieste, Italy

Creatura and Pleroma

It was not until the Forties that the existence of the world of information was explicitly recognized, i.e. it was recognized that along with the world of physics, i.e. the world of forces, masses and impacts, there exists the realm of communication, difference, organization and meaning, where the laws which hold sway are quite different from those of physics and sometimes surprising.

Following the Gnostics, Carl Gustav Jung and Gregory Bateson, I shall call Pleroma the world of matter and forces, and Creatura the world of information and structure. In the Pleroma each thing always stands for itself, whereas in the Creatura each thing can stand for another thing, thus becoming a symbol: every thing can bear a meaning, which is not inherent in the thing itself but is given to it by man, i.e. by a meaning-giving and meaning-using being. Meaning, in turn, is strictly related to redundancy, i.e. to repetitive information. Actually it is when we perceive a redundant pattern (e.g. a circle) that we can grasp or understand its form even before we see it all: as it happens with associative memories, we can reconstruct the whole form from a part of it.

In the Creatura there is no conservation law for information and the absence of information can be information. In this world, the world of communication and organization, the letter which you did not write can precipitate an angry reaction because zero is different from one and zero can therefore be a cause, which is obviuosly not true in physics. So in the Creatura the differences are of paramount importance, but for a difference to be effective, i.e. to make a sequence of differences along the communication and causation pathways, we also need a context and an observer able to appreciate such differences. Even sameness can be a cause, because sameness differs from difference! (We now understand why Pleroma is a good word for the world of physics: the Pleroma is "full", i.e. has no room for difference.)

To a large extent, Creatura is the world of living organisms. We organisms, like some of the machines we make, are able to store energy, and this energy we can spend in response to an external stimulus irrespective, to a large extent, of the size of the stimulus. The same amount of sound energy associated with a word can evoke an affectionate reply, a puzzled look or a furious reaction according to the word we pronounce to whom in which language. If you kick a stone, it moves according to the laws of dynamics with the energy it got from your kick. If you kick a dog, it moves (and can attack you) with a "collateral" energy, an energy it gets from its metabolism.

The explanatory principles of Creatura are different from those of Pleroma. These principles are essentially ideas. Ideas are differences, and differences are information. Information can be communicated only by means of a material support, but cannot be reduced to that support. In Pleroma there is no information, although Pleroma is the matrix of all (potential) information and the source of all (potential) differences. The whole energy structure of Pleroma is unable to explain (in an "appropriate" way) an event like a "lecture", i. e. a situation in which someone talks about some subject in front of several people who listen to him.

Remark that the event "lecture" can be described in many ways, i.e. at many levels of description. A quantum-mechanical description is possible (at least in principle), as well as a molecular one. But such descriptions can be of interest only in particular situations and to a limited number of people. Actually most situations in which men find themselves can be described, explained and understood adequately only in terms of information, meaning, and communication. But what makes the usual, common-language description of a lecture and other frequent human interactions so appropriate and easy to convey and to grasp is the long experience that we have of lectures. Such experience absorbs and conceals a lot of complexities and subtelties, which should otherwise be made explicit.

Imagine an Alien from a planet where lectures are not held. He visits our world (Strangeland, to him) and is taken to a lecture room. Here are his impressions:

"I entered a large room, dimly lighted, where many Strangelandians were sitting in comfortable armchairs all of which faced the same direction. As I looked more closely into he vast hall, I noticed that the floor was gently sloping and converging, as it were, to a central point brightly lighted, where a stately handsome Strangelandian was standing. I surmised he might be the City Mayor or even the King, for he was so noticeably distinguished and imposing. When the confusion that had seized me upon entering the room abandoned me and I could perceive the details out of the whole, I was bitterly surprised to see and hear that the majestic Strangelandian, whom I thought of as the King, was uttering strange sounds, shrilling and buzzing and shrieking, from his masticatory apparatus as if he were in great physical discomfort. But to my outmost bewilderment nobody helped or assisted him in the least, quite to the contrary. Actually, looking at the faces of the Stangelandians in the audience, I could see some grinning, some evidently enjoying the scene, some lightheartedly smiling as if the great pain of their Monarch was a source of joy to them, which I took as a sign of disgusting cruelty and I was thus confirmed of the many tales I had heard about the horrible population of Strangeland."

The mental world

Let me now state that Creatura is the mental world. If we adopt this viewpoint, many obscurities and puzzling intricacies concerning the so-called Mind-Body Problem (MBP) are likely to be somewhat clarified.

Mind and pattern as explanatory principles were excluded long ago from scientific biology, but this exclusion probably contributed to maintain a radical and dangerous dichotomy in epistemology, the age-old Cartesian dualism between mind and body. Many of you are well aware of the difficulties that this dualism raises but also the attempts to eliminate the dualism in one way or another have raised intricate problems.

The MBP can be stated in several different ways, each emphasizing a particular side of it. The first, utterly inadequate impression that the MBP evokes is that there is something called "body" to which something called "mind" is attached. The point is that there are two possible ways of conceiving a man: either the machine-man, immersed in the Pleroma and subject to its laws, and the person-man, active in the Creatura, where he performs symbolic and informational activities. These two views seem difficult to conciliate.

However, if we use information, communication and pattern as explanatory (not metaphysical) principles, many of the difficulties of the MBP would seem less severe. The MBP concerns essentially men: each person is a peculiar focal point where Pleroma and Creatura meet, and each such interface is an ideal source of problems, pathologies, illusions and mistakes. But man can neither be reduced to one component (all mind or all body) nor be separated into components: each man is a mind-in-a-body performing many physical and mental activities in many different contexts at the same time, and these activities lend themselves to different kinds of description. Each activity has its own right level of description, which is adequate because it meaningfully explains that activity. So what is needed to grasp the wide territory of mind is not a single all-encompassing and all-embracing approach, but rather a plurality of descriptions, a multitude of approaches and interpretations.

One can go further, and say that mental phenomena do not concern a "mind", considered as some component of man, they concern a man in its full and rich complexity. There are no mental phenomena without a body. On the other hand, to understand such mental phenomena as joy or pain or love, I cannot study only their physiological and neural counterparts, which are certainly there, can be investigated and give a lot of information to those who are interested in that kind of information. To really understand love and pain, I have to study love-in-a-man, or pain-in-a-man. A human, a particular man or woman, expresses love or pain through many and different aspects of his or her person: physical, chemical, physiological, behavioural, communicational, cultural and social. A person is a system, and its mental activities are systemic activities, concerning his or her whole existence.

When we consider the mental activity of a man, we always have to consider this man in a larger context, i.e. in the family, in the society, even in the whole of Pleroma and Creatura. Perhaps this complexity is reducible, but so far the attempts to reduce it have been largely unsuccessful, oscillating between an inadequate dualism and a crude reductionism. This is why the study of mind is so difficult and fascinating: psychological facts are not only linked to a physiological substrate, they are also linked to values, beliefs, laws, ends, all of which belong to the Creatura, although they could not exist and work without the Pleroma. A mind, every mind, is a sort of irreducible microcosm where the "subject" itself is reflected in its whole complexity.

In the last few decades, mind has gradually become less and less an object of idealistic or spiritualistic mysticism and more and more the focus of systematic studies which have begun to confer explanatory powers to it. Information theory, system theory and artificial intelligence have contributed in various degrees to this transition.

I said that information is the hallmark of Creatura and that Creatura is the realm of mind. Mind is characteristic of vast processes, by no means limited to what is traditionally meant by mind. It is not only our individual minds that live and act in the Creatura, there are larger mental processes, the largest of all

being perhaps biological evolution. It is therefore important to state some of the characteristics which, according to Gregory Bateson, are essential to any mental process:
1) A mind is an aggregate of interacting parts
2) The interaction between the parts of a mind is triggered by difference
3) Mental processes require collateral energy, i.e. a stored energy which is not in the mental process
4) Mental processes require circular or more complicated chains of communication or causation.

The systemic and diachronic character of mind and intelligence

Among the characteristics of any mind, its systemic and diachronic nature should be emphasized: on one hand mental processes consist of, and unfold through, communication and interaction, and on the other hand they build up through an evolutive and historical accumulation. Intelligence, too, is systemic and diachronic: actually, intelligence is probably best defined as the autoconservative and evolutive capability of mind.
Each mind is a set of computations (although this word is perhaps inadequate, as we shall see) which tend primarily to maintain the conditions in which those computations can go on (in some cases such computations are adequately called "life"). We use to speak of a person (or machine) as being intelligent, but we often forget that such characteristic can emerge and operate only through interaction and communication. It is not by chance that the Turing test for computers' intelligence is based on a highly formalized communication process.
In the original version of the Turing test for conscious intelligence in a symbol-manipulating machine, the inputs to the machine are questions and remarks typed on sheets of paper, and the outputs are the typewritten answers from the machine. The test is passed by the machine if its responses cannot be discriminated (by a human) from the responses of a real person.
If we adopt the systemic and diachronic viewpoint for mind and intelligence, the seemingly hopeless problems raised by strong AI might be somewhat palliated. Attempts to reproduce human-like intelligence in computers in the last 35 years have met a curious mix of success and failure. Although computers have been able to master such intellectual tasks as chess and calculus, they have yet to attain the skill of a squid in dealing with the actual world. Some argue that it is the very nature of computers that will ever prevent them from acquiring true intelligence.
In my opinion, though, it is not fair to ask whether a machine (or a program) can ever exhibit human-like intelligence, since that machine or program is always to be considered as the product of an evolutionary process (perhaps a second-order evolutionary process) which is deeply immersed in a more or less complex context. It is this spatio-temporal context which should be examined for intelligent behaviour. The machine itself has not undergone a human-like evolution and does not live in a human-like context: how could it, then, exhibit human-like intelligence?
On the other hand, strong AI insists that the essence of this human-like intelligence can be reduced to the manipulation of formal symbols. Hence a symbol manipulating machine could, after all, exhibit human intelligence. (Remark that the intricacies of the issue are increased by the use of undefined words, like "intelligence", "human-like", etc.)

Searle's Chinese room thought experiment

Actually in recent years the long-standing question: Can a machine think? has been replaced by a different one: Can a machine think by virtue of implementing a program? Is the program by itself constitutive of thinking? A certain number of people in AI believe that the answer is yes. They believe that by designing the right inputs and outputs they literally create minds. This is the claim of strong AI: thinking in nothing but the manipulation of formal symbols, and the mind is to the brain as the program is to the circuits.
Now, John Searle strongly objects to this view. He does not want to prove that computers cannot think, or that only biological systems like our brains can think. He wants to refute the strong AI claim that computers must be thinking by virtue of the right programs because that is all there is to thinking.
To this end, Searle describes the following thought experiment. A person who does not understand Chinese at all and to whom Chinese writing looks like so many meaningless squiggles, is locked in a room containing baskets full of Chinese symbols. The person is also given a rule book (in his mother-tongue) for matching Chinese symbols with other Chinese symbols. Understanding the rules does not require understanding the symbols.
Now, from outside the room, people who understand Chinese hand in small bunches of symbols to the person in the room, and in response the person manipulates the symbols according to the rules in the book and hands back more small bunches of Chinese symbols to the people outside.
Searle identifies the rule book with the computer program, their author with the programmer, the person in the room with the computer, while the baskets full of symbols are the data base, the small bunches handed in are questions and the small bunches handed out are the answers. Now assume that the rules are written in such a way that the answers that the man gives to the questions are indistinguishable from those of a Chinese native speaker, so that the man satisfies the Turing test for understanding Chinese. Nevertheless he is still utterly ignorant of this language and can in no way learn the meaning of the symbols he manipulates. Like a computer he manipulates symbols but does not attach any meaning to them. He has the syntax but has no semantics for the symbols he deals with, so he does not really understand what he is doing.
Searle concludes that just manipulating symbols is not by itself enough to guarantee perception, cognition, understanding, thinking and so on. Hence computers, that are symbol-manipulating devices, are not able to think only by virtue of running formal programs.
The artificiality of the situation in the Chinese room conceptual experiment is apparent: you cannot lock a person in a room like that and expect he or she ever come to understand Chinese. It is like asking a boy: multiply 3 by 7 but do not make use of what you know about multiplication. In the real situations where men act, the communication links are much richer, and extend well outside the room, and this is what we should do with computers, if we ever want them to become intelligent: they should be able to communicate much more intensely with the outer world. Therefore they should be endowed with flexible and sensitive sense organs and effectors and with closed-loop communication links. Only in this way they might go from syntax to semantics, i.e. attain meaning.
This is why I believe that the robot reply to Searle's experiment is very appropriate. The robot reply goes this way: "computers would have semantics, and not just syntax, if their inputs and outputs were put in the appropriate causal relation to the rest of the world. Imagine that we put the computer into a robot,

attached television cameras to the robot's head, installed transducers connecting the television messages to the computer and had the computer output operate the robot's arms and legs. Then the whole system would have a semantics."

The problem is: are the extra communication links that have been added essentially different from the previous ones, the difference being such that they can create semantics and not only syntax? Are they not again simply carrying the results of a symbol-manipulating activity? Or, to go to the very heart of the problem: what is the difference between syntax and semantics? I feel inclined to say that semantics is a redundant quasi-isomorphism between two or more syntaxes, one or more among which could in turn be a semantics.

There is more than that, however, because semantics is not a zero-one property for a mind. Some minds can have a richer semantics than others (i.e. the quasi-isomorphism involves more syntaxes). It will only be possible for the man in the room to understand Chinese (and it will certainly be possible in the long run) if he can watch how the outside world behaves in giving him the inputs and how it reacts to his outputs. After all, this is more or less the way children come to understand language.

But we should not forget the important fact that the man in the room is still potentially able to learn Chinese. His philogenetic evolution has endowed him with a brain able to learn a language. Consider the following variations of the Chinese room experiment: put a lobster instead of the man. Put a monkey instead of the man. Put a computer instead of a man.... These candidates are to various degrees unable to learn and understand Chinese, no matter how rich their links with the external world are. (The case of the computer is particularly intriguing, since there is no standard or ideal model for a computer, and the future could bring the most amazing innovations and developments.)

In other words, semantics is not acquired by the particular individual only: in the case of living beings, they have inherited instruments to this end, and it is as if the whole biological chain that gave rise to the individual locked in the Chinese room were at work for helping him to get his semantics. Intelligence and mind are diachronic, evolutionary, cumulative properties.

To better reflect the difference between man, having a semantics, and computers, having only syntax, Searle should perhaps modify his experiment, locking in the room many generations of beings, so that, in the final "person" in the room, there is not even the memory of semantics, not even the potential instruments for acquiring a semantics. Then the symbol shuffling of the man in the room would be even more syntactic, more similar to those of a computer.

In the experiment as it is, intelligence and semantics are destroyed by the cutting of the necessary communication channels that usually link a man to its environment. Here what is left of the environment is highly artificial and limited. The artificiality of the Chinese room is not the obvious and normal artificiality of any thought experiment, it is an essential artificiality, which destroys the very systemic nature of intelligence by severing the necessary communication channels through which information propagates.

Searle certainly proves what he wants to proof, namely that the skilful manipulation of a set of formal rules is not always a sufficient condition for understanding the meaning of that manipulation "out there", in the larger world. But in his experiment this is tautologically true: since the man in the room has no access whatever to the outer world, how could he ever understand what his symbol shuffling cause there?

The Church-Turing thesis

But this is not all in Searle's criticism to strong AI. To me, his most important, and perhaps overlooked, statement is that not all mental processes in the human brain are algorithmic. This is a negation of the famous Church-Turing thesis, which forms the philosophical foundation of strong AI.
Mathematicians use the words "algorithm", "computable", "recursive" or "effective" to denote the mechanical operations that can be performed by the so-called Turing machines. Actually, if a procedure is clear-cut and mechanical enough, it is reasonable to believe that a Turing machine can be found that performs it. (Other theoretical schemes which have been shown to possess a totally equivalent computational power are Church's lambda calculus and a scheme by Emil Post.) The viewpoint emerged that the Turing machine concept defines what is meant by an algorithmic (or effective, or mechanical) procedure: what can be done by an algorithm can also be done by some Turing machine. This is the original Church-Turing thesis, which only refers to mathematical computations. Today the thesis is stated in different ways, and emphasis is given to mathematical computations performed in physical (or biological) systems. This shift is of fundamental importance.
Now, following Douglas Hofstadter, let me list a few versions of the Church-Turing thesis.
CT thesis, standard version: Suppose there is a method that a sentient being follows in order to sort numbers into two classes. Suppose further that this method always yields an answer for a given number. Then: some general recursive function exists which gives exactly the same answer as the sentient being's method does.
CT thesis, isomorphism version: exactly as the previous version, moreover: the mental process and the program implementing the general recursive function are isomorphic, in the sense that on some level there is a correspondence between the steps being carried about in both computer and brain.
CT thesis, reductionist's version: All brain processes are derived from a computable substrate.
CT thesis, AI version: Mental processes of any sort can be simulated by a computer program whose underlying language is of power such that all partial recursive functions can be programmed.
In practice many AI people follow an article of faith closely related to the CT thesis, which runs more or less like this:
AI thesis: As the intelligence of machines evolves, its underlying mechanisms will gradually converge to the mechanisms underlying human intelligence.
Now Searle points out that there are many processes in the world which are not algorithmic, and offers the example of a footprint being made in the sand. The production of the footprint does not follow an algorithm, the footprint is produced because the physical structure of the world requires that a footprint be made when a foot presses the sand.
But this example concerns Pleroma and is not very interesting. Moreover one could object that this process of leaving a footprint in the sand could be mapped onto the Creatura, i.e. could, in principle at least, be described by a set of equations whose solutions are computable by a Turing machine, say. What would be needed is a formal proof that some of the activities that our brain performs are not algorithmic, hence not reproducible by a computer within the frame of today's AI.
(Remark that there is something awkward in looking for a formal proof that not everything we do is formal: perhaps this has to do with the fact that to study the brain we can only rely on our brain...)

Probably the best we can do, in this direction, is to consider Gödel's theorem on the incompleteness of formal systems. I shall not go through the details of this celebrated result, since I am mostly concerned with its implications on the nature of mind. Essentially, what Gödel's theorem states is that in any formal system, provided it is powerful enough, there is a proposition, P, say, which is known to be true but whose truth cannot be proven within the system.

Mathematical insight

This is really remarkable: we can actually establish that P is a true statement, although it is not possible to prove its truth formally. We have somewhat "seen" the truth of P, but P cannot be assigned the truth-value "true" by a formal procedure. It seems that the formalist's notion of truth is incomplete: we need to employ insights from outside the system. Such insights cannot be formalized, and are bound to be outside any sort of algorithmic approach. This holds true for any consistent formal system used for arithmetic, no matter how we enlarge it by adding new axioms to account for the Gödel proposition(s). The type of insight which is involved here enables one to jump out the rigid confinements of any formal system to get a mathematical vision that did not seem to be available before.

At this point, an objection might be raised by the formalist: a proposition such as P above should not worry us, since it is so complicated and exotic. In "normal" mathematics things are very different, and formalistic criteria for truth are perfectly adequate. But this is not the case, since reasonably simple mathematical statements have recently been exhibited which are equivalent to Gödel-type propositions. (E. g., Gregory Chaitin has found a remarkable mathematical proposition of this kind.)

To Penrose all this seems a sufficient basis to believe that in the concept of mathematical truth there is something absolute, to which man-made formal mathematics, in spite of its usefulness, is just a partial, approximate and provisional guide. Moreover, there are several examples of mathematical results of noncomputable nature. Actually, the computable part of mathematics is a rather small province of that realm and it seems that Gödel-like results are much more widespread than one might think: as Chaitin has pointed out, far from being the exception, indecidibility appears almost everywhere in mathematics.

Penrose's "proof" of the non-algorithmic nature of mind goes this way. Assume that mathematicians use algorithms to decide truth. All these algorithms must be equivalent, since one of the striking features of mathematics is its communicability: a mathematical argument that convinces one mathematician will also be able to convince another, as soon as the argument is fully understood, and this applies also to the truth of Gödel-type propositions. This "universal mathematicians'" algorithm cannot be the one used to get mathematical truth, otherwise we could construct its Gödel proposition and know that it is true as well: hence the algorithm is obscure and complicated.

But this is against what we know to be the very characteristic of mathematics, that there is nothing obscure or dogmatic or mysterious in its procedures: on the very contrary, each step in an argument can be reduced to something simple and obvious, which everybody can see and be persuaded of. This proves that there is not an algorithm which mathematicians use to establish mathematical truth.

The strong AI people might object that this is at variance with the machine-like physical structure of brain: after all the brain too is a computer, although made of meat instead of silicon, and like all computers it works by algorithms. It is the algorithm which is all-important, not its embodiment.
But the language in which the algorithm is expressed belongs to the Creatura, so it is connected with a mind which gives it its meaning, although it is embodied in a pleromatic support. This appears to be the strong AI version of the Body-Mind Problem.
All this belongs, so to say, to the destructive part of the argument. We would very much like to have a constructive part as well, i.e. it would be important to exhibit some physical property of the brain (perhaps, as Penrose has suggested, at the quantum level, or at the level of that mysterious transition from the quantum level to the macroscopic level which is involved in measurements and observations) which could support the view that our brains perform some non algorithmic, i.e. non computable, activity. To the best of my knowledge, we do not yet know such a property. The non computable features which we know of physical systems (such as chaotic systems, where predictability is highly impaired) do not seem to be useful in this respect.
Remark that the same physical properties that would introduce noncomputabily in the brain should be present in the physical structure of computers, but here we prevent them to show up from the very beginning, because we design computers to possess an algorithmic nature.
On the other hand, even if the Church-Turing thesis were finally disproved, the AI enterprise would not be dramatically impaired, given the systemic nature of intelligence. For one thing, the algorithmic part of men's mental activity could be strongly enhanced by these would-be algorithmic "intelligent" machines; moreover, there is no in-principle reason why some machines could not be constructed in the future which might be able to perform non algorithmic activities: after all, we perhaps are such machines already.

Bibliography

G. Bateson, Steps to an Ecology of Mind, Chandler Publishing Company, 1972.
G. Bateson, Mind and Nature, E. P. Dutton, 1979.
G. Chaitin, Algorithmic Information Theory, Cambridge University Press, 1987.
G. Chaitin, "Incompleteness Theorems for Random Reals", Advances in Applied Mathematics, 8, 119-146, 1987.
S. Harnad, "Minds, Machines, and Searle", Journal of Experimental and Theoretical Artificial Intelligence, 1, 1, 5-25, 1989.
D. Hofstadter, Gödel, Escher, Bach, Basic Books, 1979.
R. Penrose, The Emperor's New Mind, Oxford University Press, 1989.
J. Searle, "Minds, Brains, and Programs", Behavioral and Brain Sciences, 3, 3, 417-458, 1980.

A GENERAL INFORMATION FOR FUZZY SETS

P. Benvenuti - D. Vivona - M. Divari
Dipartimento di Metodi e Modelli Matematici per le Scienze Applicate
Università di Roma "La Sapienza", V. A. Scarpa, 10 - 00161 Roma

ABSTRACT – *In this paper we introduce a general definition of an information measure for fuzzy sets and we study its main properties. Then we describe a procedure to extend any information measure from a σ-algebra of ordinary sets to the σ-algebra of fuzzy sets of the same space.*

KEY WORDS – *Information theory, Fuzzy sets.*

1. Introduction

In an abstrac set Ω of elements ω, let $\mathcal{C}$ be a σ-algebra of ordinary (or crisp) subsets of Ω, $\mathcal{A}$ the σ-algebra of fuzzy sets [11], whose membership functions $f_A \colon \Omega \longrightarrow [0,1]$ are $\mathcal{C}$-measurable.

Identifying the crisp sets with the corresponding characteristic functions, the σ-algebra $\mathcal{C}$ is contained in $\mathcal{A}$ and moreover $\mathcal{C} = \mathcal{P}(\Omega) \cap \mathcal{A}$.

For fuzzy sets, in [11], the order relation $\subset$ is introduced by putting: $A \subset B \iff f_A \leq f_B$ together with the following operation: $A \cup B \iff f_{A\cup B} = f_A \vee f_B$, $A \cap B \iff f_{A\cap B} = f_A \wedge f_B$ and the complement $A^c \iff f_{A^c} = 1 - f_A$. With respect to these operations the family of fuzzy sets is a lattice with a least element $\emptyset$ ($f_\emptyset = 0$) and a greatest element Ω ($f_\Omega = 1$).

Several Authors introduced some information measures on fuzzy sets; for instance, Nguyen gave a definition, involving also a composition law [7], and Weber came to an extension of informations associated to measures [10].

It is advisable for us to adopt a definition as general as possible, extending the general concept of Kampé De Fériet - Forte [3] related to the σ-algebra of ordinary sets. This definition must share the crucial properties of an information measure: it must be applicable to every family of fuzzy sets and must drop axioms, which express some additional, non essential properties.

2. Certainty and possibility sets

2.1. DEFINITION

With every fuzzy set $A \in \mathcal{A}$, we can associate the sets $m(A)$ and $M(A)$, defined by maps m and $M \colon \mathcal{A} \longrightarrow \mathcal{C}$, where

$$m(A) = \{\omega \in \Omega: f_A(\omega) = 1\} = \max\{C \subset A\,,\, C \in \mathcal{C}\}\,,$$

$$M(A) = \{\omega \in \Omega: f_A(\omega) > 0\} = \min\{C \supset A\,,\, C \in \mathcal{C}\}\,.$$

The sets $m(A)$ and $M(A)$ are called *certainty set* and *possibility set*, respectively [2].

2.2. PROPOSITION

The following immediate properties hold:
— *for sets:*

$$(a) \quad m(A) \subset A \subset M(A) \qquad , \quad \forall\, A \in \mathcal{A}\,,$$

$$(b) \quad m(C) = C = M(C) \qquad , \quad \forall\, C \in \mathcal{C}\,,$$

$$(c) \quad A \subset B \Longrightarrow \begin{cases} m(A) \subset m(B) \\ M(A) \subset M(B) \end{cases} \quad , \quad \forall\, A, B \in \mathcal{A}\,;$$

— *for sequences of sets:*

$$(d) \quad m(\cap A_n) = \cap m(A_n) \quad , \quad \forall\, \{A_n\}_{n \in \mathbf{N}}\,,$$

$$(e) \quad M(\cap A_n) \subset \cap M(A_n) \quad , \quad \forall\, \{A_n\}_{n \in \mathbf{N}}\,,$$

$$(f) \quad m(\cup A_n) \supset \cup m(A_n) \quad , \quad \forall\, \{A_n\}_{n \in \mathbf{N}}\,,$$

$$(f) \quad M(\cup A_n) = \cup M(A_n) \quad , \quad \forall\, \{A_n\}_{n \in \mathbf{N}}\,.$$

3. An information measure for fuzzy sets

3.1. DEFINITION

An *information measure* for fuzzy sets is a map $\hat{J}$, such that:

$$(i_1) \qquad \hat{J}: \mathcal{A} \longrightarrow [0, +\infty]\,,$$

$$(i_2) \qquad A \subset B \Longrightarrow \hat{J}(A) \geq \hat{J}(B)\,,$$

$$(i_3) \qquad \hat{J}(\emptyset) = +\infty \quad , \quad \hat{J}(\Omega) = 0\,.$$

These axioms are the natural extensions of those introduced by Kampé De Fériet and Forte and moreover have the same form, but for their setting in a σ-algebra of fuzzy sets. These axioms have been assumed by Nguyen too, who required however the compositivity law, which seems to us to be less natural.

3.2. PROPOSITION

The following properties hold[1]*:*

$$(a) \qquad \text{if } A \subset B \text{ , then } \hat{J}(A^c) \leq \hat{J}(B^c),$$

$$(b) \qquad \hat{J}(A \cap B) \geq \hat{J}[(A,B;L)] \geq \hat{J}(A \cup B) \qquad \forall L \in \mathcal{A}.$$

3.3. PROPOSITION

The following properties are immediate:

(a) the restriction J of $\hat{J}$ to $\mathcal{C}$ is an information in the sense of Kampé De Fériet,

(b) by introducing inner information *and* outer information, *respectively, as follows*

$$\hat{J}_m = \hat{J}(m(A)) = \inf\{J(C) \colon C \subset A, C \in \mathcal{C}\},$$
$$\hat{J}_M = \hat{J}(M(A)) = \sup\{J(C) \colon C \supset A, C \in \mathcal{C}\},$$

it holds

$$(1) \qquad \hat{J}_m(A) \geq \hat{J}(A) \geq \hat{J}_M(A).$$

$\hat{J}_m$ and $\hat{J}_M$ are informations on $\mathcal{A}$: they are respectively, the maximum and the minimum among all possible informations on $\mathcal{A}$, which agree with J on $\mathcal{C}$,

(c) any convex combination of $\hat{J}_m$ and $\hat{J}_M$:

$$\hat{J}(A) = \lambda \hat{J}_m(A) + (1-\lambda)\hat{J}_M(A) \quad , \quad 0 \leq \lambda \leq 1,$$

is an information measure for which inequalities (1) hold;

(d) the restriction of $\hat{J}_m$ and $\hat{J}_M$ to $\mathcal{C}$ coincide with the restriction J of $\hat{J}$ to $\mathcal{C}$.

In [4] Kampé De Fériet gave the definition of continuity of an information measure J for decreasing (resp. increasing) sequences of crisp sets. In the setting of fuzzy sets we have also these concepts.

3.4. DEFINITION (c.d.s.)

An information $\hat{J}$ is *continuous for decreasing sequences* (c.d.s.) if for any decreasing (or non increasing) sequence $\{A_n\}_{n \in \mathbb{N}}$ of fuzzy sets, it holds:

$$\hat{J}(\cap A_n) = \sup \hat{J}(A_n).$$

[1] $(A,B;L)$ denotes the convex combination of A and B; it is defined by the relation $(A,B;L) = LA + L^cB$, where LA and L^cB are algebraic products in the sense of Zadeh [11].

3.5. DEFINITION (c.i.s.)

An information $\hat{J}$ is *continuous for increasing sequences* (c.i.s.) if for any increasing (or non decrteasing) sequence $\{A_n\}_{n\in\mathbf{N}}$ of fuzzy sets, it holds:

$$\hat{J}(\cup A_n) = \inf \hat{J}(A_n).$$

It is easy to recognize that the properties of continuity of $\hat{J}$ are partially transferred either on inner information $\hat{J}_m$ or on outer information $\hat{J}_M$.

3.6. PROPOSITION

(a) If $\hat{J}$ is c.d.s., then $\hat{J}_m$ is c.d.s.
(b) if $\hat{J}$ is c.i.s., then $\hat{J}_M$ is c.i.s.

PROOF. (a) If the sequence $\{A_n\}_{n\in\mathbf{N}}$ is decreasing, then the sequence $\{m(A_n)\}_{n\in\mathbf{N}}$ is decreasing too, and it converges to the set $\cap m(A_n) = m(\cap A_n)$. Then we have $\hat{J}_m(\cap A_n) = \hat{J}(m(\cap A_n)) = \hat{J}(\cap m(A_n)) = \sup \hat{J}(m(A_n)) = \sup \hat{J}_m(A_n)$; i.e. $\hat{J}_m$ is c.d.s..

(b) If the sequence $\{A_n\}_{n\in\mathbf{N}}$ is increasing, the sequence $\{M(A_n)\}_{n\in\mathbf{N}}$ is increasing too, and it is converging to the set $\cup M(A_n) = M(\cup A_n)$. Then we have $\hat{J}_M(\cup A_n) = \hat{J}(M(\cup A_n)) = \hat{J}(\cup M(A_n)) = \inf \hat{J}_M(A_n)$; i.e. $\hat{J}_M$ is c.i.s..

In the general case, on the contrary, if $\hat{J}$ is c.d.s. $\hat{J}_M$ it is not, and analogously if $\hat{J}$ is c.i.s. $\hat{J}_m$ it is not: relations (e) and (f) of prop. 2.2 are indeed *strictly verified.*

4. Ideals and filter

In previous papers [9,1] it has been shown that with every information measure $J\colon \mathcal{C} \longrightarrow [0,+\infty]$ some collections of crisp sets can be associated, according with the definitions below:

$$\mathcal{N} = \{N \in \mathcal{C}\colon J(C-N) = J(C) = J(C \cup N),\, \forall\, C \in \mathcal{C}\},$$

$$\mathcal{F} = \{F \in \mathcal{C}\colon J(C \cap F) = J(C) = J(C \cup F^c),\, \forall\, C \in \mathcal{C}\},$$

$$\mathcal{I}_\infty = \{C \in \mathcal{C}\colon J(C) = +\infty\},$$

$$\mathcal{I}_0 = \{C \in \mathcal{C}\colon J(C) = 0\}.$$

Analogously, with every information measure on fuzzy sets $\hat{J}\colon \mathcal{A} \longrightarrow [0,+\infty]$ we can consider families defined in this way:

$$\hat{\mathcal{N}} = \{N \in \mathcal{A}\colon \hat{J}(A \cap N^c) = \hat{J}(A) = \hat{J}(A \cup N),\, \forall\, A \in \mathcal{A}\},$$

$$\hat{\mathcal{F}} = \{F \in \mathcal{A}\colon \hat{J}(A \cap F) = \hat{J}(A) = J(A \cup F^c),\, \forall\, A \in \mathcal{A}\},$$

$$\hat{\mathcal{I}}_\infty = \{A \in \mathcal{A}\colon \hat{J}(A) = +\infty\},$$

$$\hat{\mathcal{I}}_0 = \{A \in \mathcal{A}\colon \hat{J}(A) = 0\}.$$

The sets of $\hat{\mathcal{N}}$ are called *negligible sets* and the elements of $\hat{\mathcal{F}}$ are called *essential sets.*

4.1. PROPOSITION

The following properties hold:
(a) $\hat{\mathcal{N}}$ *is an ideal in* $\mathcal{A}$,
(b) $\hat{\mathcal{F}}$ *is the dual filter of* $\hat{\mathcal{N}}$,
(c) $\hat{\mathcal{N}} \subset \hat{\mathcal{I}}_\infty$, $\hat{\mathcal{F}} \subset \hat{\mathcal{I}}_0$,

PROOF. (a) $\hat{\mathcal{N}}$ is a nonempty hereditary family: the empty set belongs to $\hat{\mathcal{N}}$. If $X \in \mathcal{A}$ and $N \in \hat{\mathcal{N}}$ with $X \subset N$, by the monotonicity of $\hat{J}$, we have $\hat{J}(A) \leq \hat{J}(A \cap X^c) \leq \hat{J}(A \cap N^c) = \hat{J}(A)$, $\forall\, A \in \mathcal{A}$; i.e. $X \in \hat{\mathcal{N}}$.

Moreover the family $\hat{\mathcal{N}}$ is stable with respect the finite unions: in fact if $N_1, N_2 \in \hat{\mathcal{N}}$, then $\hat{J}(A \cup (N_1 \cup N_2)) = \hat{J}((A \cup N_1) \cup N_2) = \hat{J}(A \cup N_1) = \hat{J}(A)\ \forall\, A \in \mathcal{A}$, so $N_1 \cup N_2 \in \hat{\mathcal{N}}$.

(b) By definition of $\hat{\mathcal{F}}$ replacing A by A^c, we obtain $\hat{J}(A^c \cap F) = \hat{J}(A^c) = \hat{J}(A^c \cup F^c)$, hence $F^c \in \hat{\mathcal{N}}$.

(c) If $N \in \hat{\mathcal{N}}$ by the definition of $\hat{\mathcal{N}}$ when $A = \emptyset$ we have $\hat{J}(\emptyset \cup N) = \hat{J}(\emptyset) = +\infty$. Similarly it is $\hat{\mathcal{F}} \subset \hat{\mathcal{I}}_0$, taking $A = \Omega$.

4.2. PROPOSITION

Let J *be an information mesure defined on* $\mathcal{C}$ *and* $\hat{J}$ *be one of its extensions on* $\mathcal{A}$. *Putting* $\mathcal{N}$ *and* $\hat{\mathcal{N}}$ *the families of negligible sets related to* J *and* $\hat{J}$, *respectively, the following properties are immediate:*
(a) $\hat{\mathcal{N}} \cap \mathcal{C} \subset \mathcal{N}$, $\hat{\mathcal{F}} \cap \mathcal{C} \subset \mathcal{F}$, $\hat{\mathcal{I}}_\infty \cap \mathcal{C} \subset \mathcal{I}_\infty$, $\hat{\mathcal{I}}_0 \cap \mathcal{C} \subset \mathcal{I}_0$,
(b) if $A \in \hat{\mathcal{N}}$, *then* $m(A) \in \mathcal{N}$ *and if* $A \in \hat{\mathcal{F}}$, *then* $M(A) \in \mathcal{F}$.

REMARK for an information $\hat{J}$ the equality $\hat{\mathcal{N}} \cap \mathcal{C} = \mathcal{N}$ can be either true or false. For example for $\hat{J}(A) = \lambda \hat{J}_m(A) + (1-\lambda)\hat{J}_M(A)$, $0 \leq \lambda \leq 1$, it is true that $\mathcal{N} \subset \hat{\mathcal{N}}$, and so $\hat{\mathcal{N}} \cap \mathcal{C} = \mathcal{N}$. An other significant example will be seen in section 6.

On the contrary, in the following example, the inclusion $\mathcal{N} \subset \hat{\mathcal{N}}$ is false. Let $\Omega = [0, +\infty]$, $\mathcal{C}$ the Borel σ-algebra, and $J(C) = \nu(C^c)\ \forall\, C \in \mathcal{C}$, ν being the Lebesgue measure. For all $A \in \mathcal{A}$, put

$$\hat{J}(A) = \begin{cases} J(M(A)) & \text{if } f_A(\omega) = 1 \text{ for at least } \omega \in \Omega \\ +\infty & \text{if } f_A(\omega) < 1 \text{ for all } \omega \in \Omega. \end{cases}$$

It is easy to see that $\hat{J}$ is an information measure on $\mathcal{A}$ extending J. The family of negligible sets with respect to J is: $\mathcal{N} = \{N \in \mathcal{C}\colon \nu(C^c - N) = \nu(C^c)\ \forall\, C \in \mathcal{C}\} = \{N \in \mathcal{C}\colon \nu(N) = 0\}$. For every $a \in \Omega$ $\{a\} \in \mathcal{N}$. Now we show that $\{a\} \notin \hat{\mathcal{N}}$. In fact let E be the fuzzy set whose membership function is $f_E(\omega) = \frac{1}{2}\ \forall\, \omega \in \Omega$. We have $\hat{J}(E) = +\infty$ and $f_{E \cup \{a\}}(a) = f_E(a) \vee \chi_{\{a\}}(a) = 1$; then $\hat{J}(E \cup \{a\}) = J(M(E \cup \{a\})) = J(M(E) \cup \{a\}) = 0$, as $M(E) = \Omega$. So $\{a\} \notin \hat{\mathcal{N}}$, i.e. $\mathcal{N} \not\subset \hat{\mathcal{N}}$.

5. Fuzzy measure and information

The notion of fuzzy measure was introduced by Sugeno in his Doctoral Thesis [8] for families of ordinary sets and it was studied by other Authors (see [5]).

As it involves only the partial order, it will be extended in natural way to the σ-algebra of fuzzy sets.

5.1. DEFINITION

A *fuzzy measure* $\hat{\mu}$ is a map $\hat{\mu}\colon \mathcal{A} \longrightarrow [0,+\infty]$ such that
(i) $\hat{\mu}(\emptyset) = 0$,
(ii) $A \subset B \implies \hat{\mu}(A) \leq \hat{\mu}(B)$.

Immediately we can remark that the restriction of $\hat{\mu}$ to $\mathcal{C}$ is a fuzzy measure in the sense of Sugeno, in a wider sense for which continuity conditions are not request, and no special value is impose on $\hat{\mu}(\Omega)$.

Fixed a fuzzy mesure $\hat{\mu}$ on $\mathcal{A}$, immediate examples of information measures are the following functions:

$$(2) \qquad \hat{J}(A) = \varphi(\hat{\mu}(A)) \quad \forall\, A \in \mathcal{A}$$

φ being any non increasing function $\varphi\colon [0, \hat{\mu}(\Omega)] \longrightarrow [0,+\infty]$ with $\varphi(0) = +\infty$ and $\varphi(\hat{\mu}(\Omega)) = 0$;

$$(3) \qquad \hat{J}(A) = \hat{\mu}(A^c) \quad \forall\, A \in \mathcal{A},$$

if $\hat{\mu}$ satisfies $\hat{\mu}(\Omega) = +\infty$.

For these information measures some properties of continuity hold: first of all we consider $\hat{J}$ defined by (2).

5.2. PROPOSITION

(a) If $\hat{\mu}$ is continuous from below [6] (i.e. c.i.s) and φ is a left continuous function, then $\hat{J}$ is c.i.s.,

(b) if $\hat{\mu}$ is continuous from above [6] (i.e. c.d.s.) and φ is a right continuous function, then $\hat{J}$ is c.d.s..

Now we consider $\hat{J}$ defined by (3).

5.3. PROPOSITION

(a) If $\hat{\mu}$ is continuous from below, then $\hat{J}$ is c.d.s.,

(b) if $\hat{\mu}$ is continuous from above, then $\hat{J}$ is c.i.s..

In a previous paper [6] Murofushi introduced the definition of null set, and we can extend it to the σ-algebra of fuzzy sets.

5.4. DEFINITION

Fixed a fuzzy measure $\hat{\mu}$, a fuzzy set N is said to be a ***null set with respect to*** $\hat{\mu}$ if it holds $\hat{\mu}(A \cup N) = \hat{\mu}(A) = \hat{\mu}(A \cap N^c)$, $\forall\, A \in \mathcal{A}$.

5.5. PROPOSITION

The ideals of negligible sets associated with informations (2) and (3) coincide with the family of $\hat{\mu}$-null sets.

6. An information measure by means of an integral

In a previous paper [1] we introduced an integral with respect to an information measure J for crisp sets, by putting

$$\int_{\Omega} f\, dJ =: \int_{0}^{+\infty} J(\{f \leq x\})\, dx\,.$$

This integral is defined for all non negative measurable functions; in particular, for characteristic functions χ_C of crisp sets C, we have

$$\int_{\Omega} \chi_C dJ = \int_{0}^{+\infty} J(\{\chi_C \leq x\})\, dx = J(C^c)\,, \quad \forall\, C \in \mathcal{C}\,.$$

This suggests the possibility of extending this kind of information measure to the fuzzy sets [2], in this way:

$$\hat{J}^{\bullet}(A) =: \int_{\Omega} (1 - f_A)\, dJ = \int_{0}^{+\infty} J(\{1 - f_A \leq x\})\, dx\,. \tag{4}$$

As the integrand is non increasing, and lying between $J_m(A^c)$ and $J_M(A^c)$, the integral (4) is meant in elementary sense, and lies between $\hat{J}^{\bullet}(m(A))$ and $\hat{J}^{\bullet}(M(A))$.

For $x \geq 1$ one has $\{1 - f_A \leq x\} = \Omega$, so

$$\hat{J}^{\bullet}(A) = \int_{0}^{1} J(\{1 - f_A \leq x\})\, dx = \int_{0}^{1} J(\{1 - f_A < x\})\, dx\,. \tag{5}$$

6.1. Proposition

The map $\hat{J}^{\bullet}$, defined by (5), is an information measure, which is an extension of the information J to the lattice $\mathcal{A}$.

Proof. The map $\hat{J}^{\bullet}$ satisfies axioms (i_1), (i_2), (i_3). In fact $\hat{J}^{\bullet}\colon \mathcal{A} \longrightarrow [0, +\infty]$; for the monotonicity of the integral [1], if $A \subset B$ then $\hat{J}^{\bullet}(A) \geq \hat{J}^{\bullet}(B)$; $\forall\, C \in \mathcal{A}$ one has $\{1 - \chi_C < x\} = C$ $\forall\, x \in [0, 1]$ hence $\hat{J}^{\bullet}(C) = J(C)$, in particular $\hat{J}^{\bullet}(\Omega) = 0$ and $\hat{J}^{\bullet}(\emptyset) = +\infty$.

If the information measure $J\colon \mathcal{C} \longrightarrow [0, +\infty]$ satisfies one continuity condition, then $\hat{J}^{\bullet}$, under mild conditions, has the same property.

6.2. Proposition

If J is c.d.s., then $\hat{J}^{\bullet}$ is c.d.s..

Proof. The proof follows from monotone convergence theorem, proved in [1] and applied to the (non decreasing) sequence of crisp set $\{1 - f_{A_n} < x\}_{n \in N}$, which converges to the set $\{1 - f_A < x\}$ for all x.

The c.i.s. property of continuity is not exactly transposing from $\hat{J}$ to $\hat{J}^{\bullet}$: only for bounded informations the information's limit equals limit's information.

6.3. PROPOSITION

Let $\{A_n\}_{n\in\mathbf{N}}$ be a increasing sequence of fuzzy sets, converging to a fuzzy set A. If J is c.i.s. and $\hat{J}^*(A_1) < +\infty$, then $\lim\limits_{n\to+\infty} \hat{J}^*(A_n) = \hat{J}^*(A)$.

PROOF. The proof derives from the dominated convergence theorem [1] applied to the (non increasing) sequence $\{1 - f_{A_n} < x\}_{n\in\mathbf{N}}$, which converges to $\{1 - f_A < x\}$.

REMARK It is impossible to avoid the bound hypothesis on $\hat{J}^*(A_n)$ informations because even if J is c.i.s., for every fuzzy set A with $\hat{J}^*(A) < +\infty$, we can construct a sequence $\{A_n\}_{n\in\mathbf{N}}$ of fuzzy sets, $A_n \nearrow A$ such that $\hat{J}^*(A_n) = +\infty \;\forall\, n$. In fact it is sufficient to consider the sequence $\{A_n\}_{n\in\mathbf{N}}$ of fuzzy sets, with $f_{A_n} = \left(1 - \frac{1}{n}\right) \wedge f_A$, and to make use of the following proposition.

6.4. PROPOSITION

If $A \in \mathcal{A}$ satisfies $\{1 - f_A < \delta\} \in \mathcal{I}_\infty$ for some $\delta \in (0,1)$, then $A \in \hat{\mathcal{I}}_\infty$.

PROOF. For $x < \delta$ it is $\{1 - f_A < x\} \subset \{1 - f_A < \delta\}$ and then

$$\hat{J}^*(A) = \int_0^1 J(\{1 - f_A < x\})\, dx = \int_0^\delta J(\{1 - f_A < x\})\, dx + \int_\delta^1 J(\{1 - f_A < x\})\, dx \geq$$

$$\geq \int_0^\delta J(\{1 - f_A < x\})\, dx \geq \int_0^\delta J(\{1 - f_A < \delta\})\, dx = \delta J(\{1 - f_A < \delta\}) = +\infty\,,$$

i.e. $A \in \hat{\mathcal{I}}_\infty$.

7. Equivalent fuzzy sets

By means of the ideal $\mathcal{N}$, defined in section 4, we have an equivalence relation for fuzzy sets.

7.1. DEFINITION

If $A, B \in \mathcal{A}$ we shall say that A and B are *equivalent* ($A \sim B$) if the crisp set $\{f_A \neq f_B\}$ belongs to $\mathcal{N}$ (or equivalenty $\{f_A = f_B\} \in \mathcal{F}$).

Then $\sim$ is an equivalence relation.

7.2. PROPOSITION

Two fuzzy sets A, B are equivalent if and only if one of the following conditions holds:

$$(a) \qquad A^c \sim B^c\,,$$

$$(b) \qquad A \cup D \sim B \cup D \qquad \forall\, D \in \mathcal{A}\,,$$

$$(c) \qquad A \cap D \sim B \cap D \qquad \forall\, D \in \mathcal{A}\,,$$

and then (d): $A \sim B$ if and only if it is $(A, B; D) \sim A \;\forall\, D \in \mathcal{A}$ or $(A, B; D) \sim B \;\forall\, D \in \mathcal{A}$.

PROOF. For (a), (b), (c) the assertion follows from: $\{f_A \neq f_B\} = \{1 - f_A \neq 1 - f_B\}$; $\{f_{A\cup D} \neq f_{B\cup D}\} \subset \{f_A \neq f_B\}$, similarly for $f_{A\cap D}$ and $f_{B\cap D}$.

(d) If $A \sim B$, it holds $\{f_A = f_B\} \in \mathcal{F}$; as $\{f_A f_D + (1-f_D)f_B = f_A f_D + (1-f_D)f_A\} \supset \{f_A = f_B\}$, it follows that $f_{(A,B;D)} = f_A f_D + (1 - f_D)f_B \sim f_A$.

Viceversa if $f_A f_D + (1 - f_D)f_B = f_A \ \forall\ D \in \mathcal{A}$, in particular for $D = \emptyset$ we have $f_B = f_A$ $\forall\ A, B \in \mathcal{A}$.

7.3. THEOREM

If $A, B \in \mathcal{A}$ are equivalent, then $\hat{J}^*(A) = \hat{J}^*(B)$.

PROOF. The proof follows from theorem 3.2 in [1].

The following consequences are immediate:

7.4. COROLLARY

(a) $A \in \mathcal{A},\ A \sim \emptyset \implies M(A) \in \mathcal{N}$,
(b) $A \in \mathcal{A},\ A \sim \Omega \implies m(A) \in \mathcal{F}$.

7.5. THEOREM

The families $\hat{\mathcal{N}}^*$ and $\hat{\mathcal{F}}^*$ defined for the information $\hat{J}^*$ are extensions of $\mathcal{N}$ and $\mathcal{F}$ respectively, i.e. $\mathcal{N} = \hat{\mathcal{N}}^* \cap \mathcal{C}$ and $\mathcal{F} = \hat{\mathcal{F}}^* \cap \mathcal{C}$.

PROOF. The equality will follow if we prove that $\mathcal{N}$ is contained in $\hat{\mathcal{N}}^*$. If $N \in \mathcal{N}$, $f_{A\cap N^c}(\omega) = f_A(\omega) \wedge (1 - \chi_N(\omega)) = 0\ \forall\ \omega \in N$ and $A \in \mathcal{A}$, and $f_{A\cap N^c}(\omega) = f_A(\omega)\ \forall\ \omega \in N^c$ and $A \in \mathcal{A}$. Thus the set $\{\omega \in \Omega\colon f_A(\omega) \wedge (1 - \chi_N(\omega)) \neq f_A\}$ is included in N, and $\{f_{A\cap N^c} \neq f_A\} \in \mathcal{N}$; i.e. $A \cap N^c$ is equivalent to A. Then $\hat{J}^*(A \cap N^c) = \hat{J}^*(A)$ and $N \in \hat{\mathcal{N}}^*$.

The other equality is a consequence of proposition 4.1.

8. Essential convergence

The ideal $\mathcal{N}$ (see section 4) can be used to give a definition of essential convergence.

8.1. DEFINITION

If $\{A_n\}_{n\in\mathbb{N}}$ is a sequence of fuzzy sets, we shall say that A_n is *essentially pointwise convergent* to a fuzzy set A, if it exists a set $N \in \mathcal{N}$ such that $\lim_{n\to\infty} f_{A_n}(\omega) = f_A(\omega)\ \forall\ \omega \in N^c$.

Now we shall give two theorems of essential convergence for a sequence of fuzzy sets.

8.2. THEOREM (Monotone convergence)

Let J be a c.d.s. information measure on $\mathcal{C}$, and $\{A_n\}_{n\in\mathbb{N}}$ a decreasing sequence of fuzzy sets. If $\{A_n\}$ is essentially convergent to $A \in \mathcal{A}$, then $\lim_{n\to+\infty} \hat{J}^*(A_n) = \hat{J}^*(A)$.

PROOF. Let F be an element of $\mathcal{F}$, such that $\lim_{n\to+\infty} f_{A_n}(\omega) = f_A(\omega)$ for all $\omega \in F$. By virtue of theorem 7.3, we have $\hat{J}^*(A_n) = \hat{J}^*(A_nF)$ for all n, and $\hat{J}^*(A) = \hat{J}^*(AF)$.

Now the conclusion follows by theorem 1.3 of [1], applied to the sequence $\{1 - f_{A_nF}\}_{n\in\mathbf{N}}$.

8.3. THEOREM (Dominated Convergence)

Let's assume that J is continuous (i.e. J is c.d.s. and c.i.s.), and let $\{A_n\}_{n\in\mathbf{N}}$ be a sequence of fuzzy sets, essentially convergent to $A \in \mathcal{A}$. If there exists a fuzzy set B, such that $1 - f_{A_n} \leq f_B$ for each n, and satisfying $\hat{J}^*(B) < +\infty$, then $\lim_{n\to+\infty} \hat{J}^*(A_n) = \hat{J}^*(A)$.

PROOF. If $F \in \mathcal{F}$ is as above, then it suffices to apply theorem 1.7 in [1] to the sequence $\{1 - f_{A_nF}\}_{n\in\mathbf{N}}$.

REFERENCES

[1] P. BENVENUTI - D. VIVONA - M. DIVARI: *Sull'integrale nella teoria dell'informazione*, Rend. Mat. VII, **8**, (1988) 31-43.

[2] P. BENVENUTI - D. VIVONA - M. DIVARI: *An integral for fuzzy sets in information theory*, Proceedings of the 8th International Congress of Cybernetics and Systems, New York, June 1990 (to appear).

[3] J. KAMPÉ DE FÉRIET - B. FORTE: *Information et probabilité*, C.R.A.S., Paris, **265**, (1967) 110-114, 142-146, 350-353.

[4] J. KAMPÉ DE FÉRIET - P. BENVENUTI: *Sur une classe d'information*, C.R.A.S., Paris, **269**, (1969) 529-534.

[5] G.J. KLIR: *Where do we stand on measures of uncertainty, ambiguity, fuzziness, and the like?*, F.S.S. **24**, (1987) 141-160.

[6] T. MUROFUSHI: *Two approaches to fuzzy measure theory: integrals based on pseudo-addition and Choquet's integral*, Doctoral Thesis, Tokio Institute of Technology (1987).

[7] H.T. NGUYEN: *On fuzziness and linguistic probabilities*, JMAA **61**, (1977) 658-671.

[8] M. SUGENO: *Theory of fuzzy integrals and its applications*, Doctoral Thesis, Tokio Institute of Technology (1974).

[9] D. VIVONA: *L'informazione integrale*, Quad. **19**, I.M.A., Univ. Roma, (1982).

[10] S. WEBER: *Decomposable measures and measures of information for crisp and fuzzy sets*, Proc. of the I.F.A.C. Symposium, Marseille, july 1983.

[11] L.A. ZADEH: *Fuzzy sets*, Inf. and Control **8**, (1965) 338-353.

INFORMATION THEORY BASED ON FUZZY (POSSIBILISTIC) RULES

ARTHUR RAMER
SCHOOL OF ELECTRICAL ENGINEERING AND COMPUTER SCIENCE
UNIVERSITY OF OKLAHOMA, NORMAN, OK 73019,USA

December 18, 1990

ABSTRACT The paper presents a systematic design of information theory based on fuzzy rules of combining evidence. The theory is developed both for discrete and continuous cases, represented respectively by finite distribution of possibility values and by an arbitrary measurable function on the domain of discourse.

Given a fuzzy set its associated uncertainty is expressed by an information function of the corresponding possibility distribution. Some natural properties of information make such expression of uncertainty essentially unique. A further extension is provided by the notion of information distance between two possibility distributions on the same domain of discourse. Finally, information functions and information distances are used to formulate a possibilistic principle of maximum uncertainty.

KEY WORDS Fuzzy set, information measure, maximum uncertainty, possibility theory.

KEYWORDS Fuzzy information, Information measures, Information theory, Possibility theory

Fuzzy sets and fuzzy systems [D-P88, Y80,Z78] address issues of imprecision and vagueness inherent in various modes of approximate reasoning. Human analysis and inference mechanisms of intelligent systems are but particular domains of such reasoning. On the other hand, behavior of essentially inanimate systems is often described using probability theory. In both theories there is a need to express in quantitative terms notions of uncertainty and information. Having captured these notions formally, their numerical expressions can be utilized in decision making under uncertainty.

Probabilistic techniques for dealing with those issues are known as Shannon's information theory [S48,A-D75]. Entropy is a fundamental expression of uncertainty and the principle of its maximization is a standard decision method [J82,S-J80].

This paper demonstrates that an equally integrated theory can be built on the foundation of possibility theory and fuzzy reasoning it represents. We show how to define measures of information and uncertainty for possibility distributions. We then show that a very natural decision method is obtained; it becomes a possibilistic principle of maximum uncertainty. Our presentation stresses motivation and rationale for various formalisms and includes examples of their use. Detailed proofs will have been presented in a near future [R-L87,R90a,R90c].

Possibility distributions for fuzzy sets

We use a model of possibility theory close to one introduced in [Z78]. Given set X as a domain of discourse we declare a *possibility distribution* as a function $\pi : X \to [0,1]$ such that $\sup_{x \in X} \pi(x) = 1$.

It expresses a (basic) assignment of possibility values $\pi(x)$ to elementary events $x \in X$. To extend it to arbitrary subsets $Y \subset X$ we define a possibility measure $\Pi(Y) = \sup_{x \in Y} \pi(x)$. Thus possibility distributions are closely related to fuzzy membership functions. In particular, given a fuzzy set $\mu : X \to [0,1]$ its α-cuts

$$c_\alpha = \{x : \mu(x) \geq \alpha\}$$

form a nested family of subsets, called the focal subsets of the corresponding possibility distribution. This correspondence [D-P88,Z78] has been the basis of defining joint and marginal possibility values.

We discuss first finite domains of discourse, possibility being expressed as a sequence of values between 0 and 1. Given two domains X and Y and two independent possibility assignments $\pi_1 : X \to [0,1]$, $\pi_2 : Y \to [0,1]$ we define a joint distribution $\pi : X \times Y \to [0,1]$ as

$$(x,y) \mapsto \min(\pi_1(x), \pi_2(y)).$$

One denotes such π as $\pi_1 \otimes \pi_2$.

Given an arbitrary assignment π on a product space $X \times Y$ we define marginal assignments π' on X and π'' on Y as

$$\pi'(x) = \max_{y \in Y} \pi(x,y),$$

$$\pi''(y) = \max_{x \in X} \pi(x,y).$$

It is also convenient to define an *extension* of π from its domain X to a larger set $Y \supset X$. We put

$$\pi^Y(y) = \pi(y), \quad y \in X$$

$$\pi^Y(y) = 0, \quad \text{otherwise.}$$

Lastly, given a permutation s of $\{1, \ldots, n\}$ we can define a permuted possibility assignment $s(\pi)$ by putting

$$s(\pi)(x_i) = \pi(x_{s(i)}).$$

The general case consists of an arbitrary set X as the universe of discourse and an arbitrary function $\mu : X \to [0.1]$. It is often described as a continuous case, as opposed to a discrete one, even though the membership function need not be continuous. It is impractical to allow for a totally arbitrary function, and we restrict our attention to measurable functions. To simplify the presentation we assume the the domain of discourse consists of the unit interval $[0,1]$ and thus $\mu : [0,1 \to [0,1]$. The formulas defining joint, marginal and extended distributions have obvious general formulations. It is also straightforward that those derived distributions are again presented as measurable functions. A permuted distribution is obtained by applying a measurable isomorphic transformation of $[0,1]$ onto itself.

Information functions in possibility theory

This simple structure, outlined above, provides the *possibilistic* context for introducing concepts of information theory [D-P87,H-K82,R-L87].

Let $(X, \mu(x))$ be a fuzzy set corresponding to possibility distribution (p_i). We view the membership function μ as computing a degree of assurance or certainty that an element of X satisfies the property of being a member of that fuzzy set. Therefore, if we were to select a given $x \in X$ as a representative object, we would be only certain to a degree $\mu(x)$ that our selected object satisfies the aforementioned property.

Another perspective is offered by considering element $x \in X$ as an outcome of an experiment—an attempt to find a descriptive object for the property defined by our fuzzy set. Therefore we may wish to quantify the uncertainty associated with using a given possibility distribution as a basis of selection. Thus deciding on a specific selection entails a gain of information. This information consists of specifying the precise outcome rather than having only a fuzzy description of possible outcomes.

We shall attempt such quantification through an assignment of an overall value to the uncertainty inherent in the complete distribution π. We consider such uncertainty value as equivalent to the information that can be gained by selecting a specific event x from the total domain X. Accordingly, we intend to define an information function I which would assign a nonnegative real value to an arbitrary distribution π. Following established principles of information theory, [A-D75,G77] we stipulate that such information function satisfies certain standard properties. Specifically, we require

- *additivity*

$$I(\pi_1 \otimes \pi_2) = I(\pi_1) + I(\pi_2)$$

- *subadditivity*

$$I(\pi) \leq I(\pi') + I(\pi'')$$

- *symmetry*

$$I(s(\pi)) = I(\pi)$$

- *expansibility*

$$I(\pi^Y) = I(\pi)$$

It turns out that these properties essentially characterize the admissible information functions [K-M87,R-L87].

Given $X = \{x_1, \ldots, x_n\}$, let $p_1 \geq p_2 \geq \ldots \geq p_n$ be a descending sequence formed from the values $\pi(x_1), \ldots, \pi(x_n)$. Then, up to a multiplicative constant

$$I(\pi) = \sum_{i=1}^{n-1} (\tau(p_i) - \tau(p_{i+1})) \log i$$

where τ is a nondecreasing mapping of $[0,1]$ onto itself. If we assume some kind of a linear condition (like the branching property in [K-M87]) we obtain a particularly simple expression of information value associated with distribution π. It has been first introduced under the name of U-uncertainty [H-K82]

$$U(\pi) = \sum (p_i - p_{i+1}) \log i.$$

We observe that the distribution which carries the highest uncertainty value consists of assigning possibility 1 (maximal) to all the events in X. Such distribution can be considered the *most uninformed* or a *possibilistic uniform* one.

U-uncertainty serves to define information distance [H-K83, R90a] between two distributions π and ρ defined on the same domain X. If $\pi(x) \leq \rho(x)$, we put

$$g(\pi, \rho) = U(\rho) - U(\pi).$$

For the general case, given π and ρ, we first define their lattice meet and join

$$\pi \wedge \rho : x \mapsto \min(\pi(x), \rho(x)),$$

$$\pi \vee \rho : x \mapsto \max(\pi(x), \rho(x)).$$

We then put

$$G(\pi, \rho) = g(\pi, \pi \vee \rho) + g(\rho, \pi \vee \rho),$$
$$H(\pi, \rho) = g(\pi \wedge \rho, \pi) + g(\pi \wedge \rho, \rho).$$

These functions have several attractive properties; for example G is a metric, while H is additive in both arguments.

Comparison to probability theory

Historically, concepts of information have been tied to probability theory [A-D75,M-R75,G77]. Given a probability distribution $\mathbf{p} = (p_1, \ldots, p_n)$, its associated information is defined as the Shannon entropy [S48]

$$H(\mathbf{p}) = -\sum p_i \log p_i.$$

This function satisfies the aforementioned properties of information—it is additive, subadditive symmetric and expansible. In turn these properties can be used to characterize such probabilistic information uniquely.

A closely related notion is information distance, also called I-divergence [C75] or cross-entropy [S-J80]. Given $(p_1, \ldots, p_n)$ and $(r_1, \ldots, r_n)$ such that $p_i = 0$ whenever $r_i = 0$, we define

$$D(\mathbf{p}, \mathbf{r}) = \sum p_i \log \frac{p_i}{r_i}$$

Although not a metric, D has several useful properties which make suitable to decide on proximity of probability distributions. Both H and D found extensive applications in several domains, like signal processing, optimization, patterns recognition and others. Recently, the entropy function has been applied to the analysis of dependencies in relational databases.

These applications often take form of the principle of *maximum entropy* [S-J80,J-82]. In a search for, as yet unknown, distribution subject to some constraints, we should select one with the largest entropy. This method generalizes to the case of a known prior distribution and a search for an admissible posterior. We should then prefer a distribution that has a minimum information distance to the prior. This principle can be formally justified (and derived) by supposing certain natural decision making conditions. Specifically, if the decision is to be made by minimizing some function of both prior and posterior distributions, and the decision process is consistent, then that function must be the I-divergence between these two distributions [S-J80].

The probabilistic information theory can be extended to continuous (or even measurable) domains and distributions [G77]. Given probability density $f(x)$ on X with measure $\mathcal{M}$, an obvious generalization of the Shannon entropy would be

$$H(f) = -\int_X f(x) \log f(x)\, d\mathcal{M}(x),$$

and similarly for the information distance

$$D(f_1, f_2) = \int_X f_1(x) \log \frac{f_1(x)}{f_2(x)}\, d\mathcal{M}(x).$$

For technical reasons [G77], it is preferable to replace $H(f)$ by the distance between f and the (probabilistic) uniform distribution on X. Now the complete theory of information functions carries on to the continuous case. Moreover, continuous information values can be obtained through approximations by discrete distributions.

Structure of continuous possibility information

We shall attempt to extend previous definitions to arbitrary 'continuous' domains. We develop ur approach first for a special, albeit typical case where X is the unit interval. Now a fuzzy tructure (possibility distribution) is given as a function

$$f : [0,1] \rightarrow [0,1]$$

uch that $\sup_{x \in [0,1]} f(x) = 1$. Although in a variety of practical situations it is sufficient to consider nly continuous functions, a much more general case of an arbitrary measurable function can be reated through the same means.

As a first step the discrete formula

$$U(x) = \sum p_i \nabla \log i$$

uggests forming an expression like

$$\int_0^1 \tilde{f}(x) d \ln x,$$

here $\tilde{f}$ would be some 'descending sorted' equivalent of f, while $d \ln x$ substitutes for $\nabla \log x$. The atter part is well known and simply represents $x^{-1}dx$, however a proper definition of $\tilde{f}$ requires ome work. We clearly want $\tilde{f}$ to be decreasing, or at least nonincreasing. We also would like it o 'stay' above any given value α, $0 \leq \alpha \leq 1$ over about the same space as the original function does. This would assure us that all the α-cuts [D-P88] of f are of the same size (have the ame measure as α-cuts of $\tilde{f}$. Such construction is well known in the literature; we follow here a lassical description [HLP].

We define

$$P(y) = \mathcal{M}\{x : f(x) \geq y\},$$

here $\mathcal{M}$ is a standard measure on $[0,1]$. We then put $\tilde{f}(x) = P^{-1}(y)$. As an illustration let us onsider a couple of examples.

xample 1.

$$f(x) = \begin{cases} 2x, & 0 \leq x \leq 0.5, \\ 2 - 2x, & \text{otherwise.} \end{cases}$$

hen $P(y)$ represents the *combined* length of the intervals where $f(x)$ is $\geq y$. In our case

$$P(y) = 1 - y$$

nd

$$\tilde{f}(x) = 1 - x.$$

is immediate that $\tilde{f}(x) \geq \alpha$ over the set of the same measure as the set where $f(x) \geq \alpha$.

xample 2.

$$f(x) = 4(x - \frac{1}{2})^2 = 4x^2 - 4x + 1.$$

ow $P(y) = 1 - \sqrt{y}$, as

$$\{x : f(x) \geq y\} = [0, \frac{1 - \sqrt{y}}{2}] \cup [\frac{1 + \sqrt{y}}{2}].$$

Therefore $y = (1 - P(y))^2$ and $\tilde{f}(x) = (1 - x)^2$.

Using this definition we can consider

$$\int_0^1 \frac{\tilde{f}(x)}{x} dx$$

as a candidate expression for the value of information. Unfortunately, $\tilde{f}(x)$ is equal to 1 at 0, and the integral above diverges. A solution can be found through a technique that has been used in probability theory. Shannon's entropy [S48]—a customary information function for discrete probability distributions, could be generalized to Boltzmann's entropy

$$\int_0^1 f(x) \log f(x) dx$$

for a continuous probability density f. However the latter expression is less than satisfactory as an information function as it cannot be obtained as a limit of discrete approximations. The solution [G77] is to use instead the information distance between a given density and the uniform one.

We shall use the same approach here. In possibility theory we consider a constant function $f(x) \equiv 1$ as representing a uniform distribution. It is also the most 'uninformed' one—its discrete form clearly attains maximum U-uncertainty. Our final formula becomes

$$I(f) = \int_0^1 \frac{1 - \tilde{f}(x)}{x} dx.$$

This integral is well defined and avoids the annoying singularity at 0. We demonstrate its use on a class of polynomial functions.

Example 3. Let us consider possibility distributions represented by $f(x) = x^n$, $n = 0, 1, \ldots$ First a few complete integrations. Denoting $J_n = I(x^n)$ and remembering that $\tilde{x^n} = (1-x)^n$, we find

$$J_0 = I(1) = \int_0^1 \frac{1-1}{x} dx = 0$$

$$J_1 = I(x) = \int_0^1 \frac{1-(1-x)}{x} dx = 1$$

$$J_2 = I(x^2) = \int_0^1 \frac{1-(1-x)^2}{x} dx = \int_0^1 (2-x)^2 dx = 1\frac{1}{2}$$

To find a general expression for J_n, let us first compute $J_n - J_{n-1}$

$$\int_0^1 \frac{(1-(1-x)^n) - (1-(1-x)^{n-1})}{x} dx =$$

$$\int_0^1 \frac{(1-x)^{n-1} - (1-x)^n}{x} dx = \int_0^1 (1-x)^{n-1} dx = \frac{1}{n}$$

As $J_0 = 0$ we find that $J_n = 1 + \frac{1}{2} + \cdots + \frac{1}{n} = H_n$, the nth harmonic number.

Properties of continuous information measures

We state here without proof (complete proofs will be published separately) that $I(f)$ satisfies all the required conditions of an information function. It is additive, subadditive, expansible symmetric and continuous. We shall demonstrate additivity with an example.

Example 4. We use as an example $f = g = x^\alpha, \quad \alpha \geq 0$. Before the verification that $I(f \otimes g) = 2I(f)$, we indulge first in an explicit integration for $n = 2$. We find, putting $h = f \otimes g, \quad h(x,y) = \min(x^2, y^2)$, that

$$P(z) = \mathcal{M}\{(x,y) : h(x,y) \geq z\} =$$
$$\mathcal{M}\{(x,y) : x^2 \geq z, \quad y^2 \geq z\} =$$
$$(1 - \sqrt{z})^2$$

and therefore

$$\tilde{h}(t) = P^{-1}(z) = (1 - \sqrt{t})^2.$$

Now

$$\int_0^1 \frac{1 - (1 - \sqrt{t})^2}{t} dt =$$
$$\int_0^1 (\frac{2}{\sqrt{t}} - 1) dt = -1 + 2 \int_0^1 t - \frac{1}{2} dt = 3.$$

This agrees with $I(f) = I(g) = 1\frac{1}{2}$.

For the general case, we put $h(x,y) = \min(x^\alpha, y^\alpha)$ and find

$$\tilde{h}(t) = (1 - \sqrt{t})^\alpha.$$

Then

$$I(h) = \int_0^1 \frac{1 - (1 - \sqrt{t})^\alpha}{t} dt,$$

which after the substitution $u = \sqrt{t}$ becomes

$$\int_0^1 \frac{1 - (1 - u)^\alpha}{u^2} \cdot 2u du =$$
$$2 \int_0^1 \frac{1 - (1 - u)^\alpha}{u} du = 2I(f).$$

Using $I(f)$ we can define continuous extensions of information distances g, G and H. Specifically, if $f_1(x) \leq f_2(x), \quad x \in [0,1]$

$$g(f_1, f_2) = \int_0^1 \frac{\tilde{f}_2(x) - \tilde{f}_1(x)}{x} dx.$$

For a general case we put

$$f_1 \wedge f_2 : x \mapsto \min(f_1(x), f_2(x)),$$
$$f_1 \vee f_2 : x \mapsto \max(f_1(x), f_2(x)).$$

Now

$$G(f_1, f_2) = g(f_1, f_1 \vee f_2) + g(f_2, f_1 \vee f_2)$$

and

$$H(f_1, f_2) = g(f_1 \wedge f_2, f_1) + g(f_1 \wedge f_2, f_2).$$

As in the discrete case, G is a metric, while H is additive in both arguments.

Finally, we remark that $I(f)$ can be approximated as a limit of $U(p_n)$, where p_n are discrete distributions approximating f. Already a non-trivial example is offered by a linear function.

Example 5. We select $f(x) = 1 - x$ and approximate it using the values at $\frac{1}{n}, \frac{2}{n}, \ldots 1$. The approximating distributions are $\pi^{(n)} = (\frac{n-1}{n}, \frac{n-2}{n}, \ldots 1)$ and have as their information measures

$$U(\pi^{(n)}) = \sum (p_i - p_{i+1}) \ln i =$$

$$\sum \left(\frac{n-i+1}{n} - \frac{n-i}{n}\right) \ln i =$$

$$\frac{1}{n} \sum \ln i = \frac{1}{n} \ln n!$$

Invoking Stirling's formula, we find

$$\ln n! = n \ln n - n + O(\ln n),$$

and therefore

$$U(\pi^{(n)}) \sim \ln n - 1, \qquad n \to \infty.$$

This gives

$$I(\pi^{(n)}) = \ln n - U(\pi^{(n)}) = 1$$

which agrees with $I(f)$.

Principle of maximum uncertainty

The principle of maximum uncertainty can be stated independently of any specific theory used to capture analytically the notions of information and uncertainty. If the theory of probability is employed we arrive at the principle of maximum entropy. It offers an effective method of statistical inference by selecting a distribution of maximum entropy. Conditions of such selection usually specify a set of constraints, thus defining the set of admissible distributions. The choice from among those, the reasoning continues, should be made without introducing extraneous information, or to be as 'uninformed' as possible. And because the Shannon's entropy is the preferred information function, it leads to selecting the distribution for which the entropy value is maximum. Were they no constraints whatsoever (other than the number of elements), the entropy would be maximized by the uniform distribution. It follows that whenever the entropy of a distribution reaches maximum its information distance from the uniform distribution is minimized. This leads to a generalized criterion—if a certain distribution is specified as a prior information, then a new distribution is selected from among the admissible ones to minimize the information distance from the prior.

In possibility theory such principle would state that, given a prior assignment of possibility values and certain constraints on the posterior, we should select the latter as the closest admissible assignment. The proximity here is expressed through the possibilistic information distance. If there is no known or assumed prior, we should consider the distance from the most 'uninformed' possibility distribution, which is given by assigning a constant value 1 to every element of the domain of discourse. It clearly has the highest value of U-uncertainty; it also agrees with the intuitive perception that, in absence of constraints, every choice should be accorded maximum possibility.

This method of determining the posterior distribution has several desirable properties, which apply both to possibility and probability theories.

- *Invariance*

  The result is not affected by a permutation of the domain of discourse

- *Consistency*

  Refinement of the constraints corresponds to the refinements of the resulting selection

- *Separability*

  Invoking all the constraints simultaneously, or proceeding through a series of posteriors, each one derived on the basis of an additional constraint, will eventually lead to the same result.

The opposite result also holds in both theories—if the selection process is based on maximization of some function of the distribution, and if this process satisfies above postulates, then that function can be selected to be an information measure. A similar situation arises when prior assignment is available—the decision process becomes one of minimizing an information based distance. We can thus consider entropy and I-divergence as natural objective functions in probability theory, while U-uncertainty and its related distances are natural in possibility theory.

An interesting example of applying this method arises in defining *conditional* possibility measures. Given a possibility distribution $\pi = (p_1, \ldots, p_n)$ on X we want to describe the effect of the condition that the domain of discourse is restricted to a subset Y. We thus want to find a distribution $\pi_{|Y}$ on Y, which could be derived from π by constraining its domain. It turns out [R89] that the same definition can be obtained by two independent methods. One of them relies on the principle of minimum information distance.

Example 6.

We shall consider π on X as the given prior and seek ρ on Y as a posterior distribution. To have both π and ρ defined on the same domain, we consider an expansion of ρ to ρ^X, obtained by putting

$$\rho^X : x \mapsto 0, \quad x \in X \setminus Y.$$

Let

$$\pi = (p_1 \geq p_2 \geq \ldots \geq p_n$$

and

$$\rho^X = (r_1 \geq r_2 \geq \ldots \geq r_m, 0, \ldots, 0).$$

We want to minimize their distance

$$g(\pi, \rho^X) = \sum_{i=1}^{m}((r_i - p_i) - (r_{i+1} - p_{i+1})) \log i$$

subject to the feasibility conditions

$$0 \leq r_i \leq 1, \qquad \max_i q_i = 1.$$

The part of the summation between $m+1$ and n is constant, while the initial part can be made 0 if we put

$$r_1 = 0, \qquad r_i = p_i, \quad i = 2, \ldots, m.$$

Now the G-distance between ρ^X and π is minimized if we modify one of the maximum values that π reaches on Y to become 1 and leave all the remaining values intact.

We note that this approach is not confined to the theory of possibility. For example, it can be used in defining conditional *probability* distributions [R90b].

References

A-D75 J.Aczel and Z.Daroczy, On measures of information and their characterization, Academic Press, New York 1975.

A-R87 M. Anvari and G. Rose, Fuzzy relational databases, Analysis of Fuzzy Information, v.II, CRC Press, Boca Raton, FL 1987.

B-P87 B. Buckles and F. Petry, Generalized database and information systems, Analysis of Fuzzy Information, v.II, CRC Press, Boca Raton, FL 1987.

C75 I.Cziszar, I-divergence geometry of probability distributions and minimization problems, Ann.Prob.,3,1975.

D-P88 D. Dubois and H. Prade, Possibility theory, Plenum Press, New York 1988.

D-P87 D. Dubois and H. Prade, Properties of measures of information in evidence and possibility theories, Fuzzy Sets Syst.,2,24,1987.

G77 S. Guiasu, Information Theory and Applications, McGraw Hill, New York 1977.

H-K83 M.Higashi and G. Klir, On the notion of distance representing information closeness, Int. J. Gen. Sys.,9,1983.

H-K82 M. Higashi and G.Klir, Measures of uncertainty and information based on possibility distributions Int.J.Gen.Sys., 8,1982.

HLP34 G.Hardy, J.Littlewood, G.Polya, Inequalities, Cambridge University Press, Cambridge 1934.

J82 E.Jaynes, On the rationale of maximum entropy methods, Proc. IEEE, 70, 1982.

K-M87 G.Klir and M.Mariano, On the uniqueness of possibilistic measure of uncertainty and information, Fuzzy Sets Syst.,2,24,1987.

R90a A.Ramer, Structure of possibilistic information metrics and distances, Int. J. Gen.Sys., 17,1990.

R90b A.Ramer, A note on defining conditional probability, Am.Math.Monthly, 1990.

R90c A.Ramer, Concepts of fuzzy information measures on continuous domains, Int.J.Gen.Sys., 17,1990.

R89 A.Ramer, Conditional possibility measures, Cybern. Syst.,20,1989.

R-L87 A.Ramer and L.Lander, Classification of possibilistic uncertainty and information functions, Fuzzy Sets Syst.,2,24,1987.

S48 C.Shannon, A mathematical theory of communication, Bell. Sys. Tech.J.,27,1948.

S-J80 J.Shore and R.Johnson, Axiomatic derivation of the principle of maximum entropy and the principle of minimum cross-entropy, IEEE Trans. Inf. Theory, IT-26, 1980.

Y80 R.Yager, Aspects of possibilistic uncertainty, Int. Man-Machine Stud.,12,1980.

Z78 L. Zadeh, Fuzzy sets as a basis for a theory of possibility, Fuzzy Sets Syst.,1,3,1978.

MEASURING UNCERTAINTY GIVEN IMPRECISE ATTRIBUTE VALUES

J.M. Morrissey
University of Windsor
Windsor Ontario N9B 3P4
Canada

Abstract

A consistent and useful treatment of missing and imprecise data is required for database systems. One problem is that when imprecise data are present then the semantics of query evaluation are no longer obvious and uncertainty is introduced. It is proposed that the query result consist of two sets of objects: those where there is complete certainty and those where there is some uncertainty. Furthermore, the uncertainty should be measured and used to rank the objects for presentation. *Self-information* and *entropy* are examined as possible measures of uncertainty.

Keywords: databases, imprecise data, uncertainty, self-information, entropy.

1. INTRODUCTION

In conventional database systems real world objects are modelled by selecting representative attributes to which values are assigned. For example, a memo might be modelled and stored as follows:

```
sender(memo_34)   = Jim Smith
receiver(memo_34) = John Brown
subject(memo_34)  = "Next System Dump"
text(memo_34)     = "July 4th."
```

where the representative attributes for the object *memo* are *sender, receiver, subject* and *text*, and the values are as shown above. However, some exceptions arise which require a consistent and useful treatment. Some of the values may be missing or unknown to the system. For example, the sender of a memo may not give a subject heading, or an employee may be reluctant to supply certain personal details for records. The system must cope with such missing information. In many cases the representative attributes will not be applicable to every object. For example, single people will not have the attribute *'name of spouse'*; companies without a fax machine will not have the attribute *'fax number'* and so on. That an attribute is not applicable is very precise information, and it is treated as such in this paper. More frequently, the exact attribute value will be unknown but some imprecise information will be available which is useful and should be recorded. For example, if we have imprecise information about the age of an employee which states that it is in the range 20 to 30, and if a query asks for a list of employees who are near retirement then the system can decide with complete certainty that this person should not be part of the list. Similarly, we may know that the location of a certain meeting must be either London, Paris or Rome. If we need a list of all European meetings then this one can be part of the list even though we do not have complete information about the location. However, such imprecise information introduces uncertainty since query evaluation is no longer obvious. For example, if a query asks for a list of red parts and there is a part whose colour is only known to be either red or blue then should it be listed ? In this paper we address this problem, proposing that the response should consist of two sets of objects: those which should certainly be retrieved, and those where there is some uncertainty. Furthermore, we propose that this uncertainty should be measured and used to rank objects for presentation.

2. BACKGROUND

Codd [1, 2] presents a standard treatment of missing and non-applicable attributes in relational database systems. In each case a null value is stored and 3-valued logic is used for query evaluation. In [3] the treatment is extended so that nulls have been replaced by two special markers. One signifies that the value is *missing but applicable*, the other signifies that the value is *not applicable*. This work does not address the problem of imprecise values and uncertainty. Date [4] proposes a system of default values whereby a default value is associated with each attribute domain and this value is used instead of a null. However, such values confuse the semantics of the data since it is no longer possible to distinguish between missing values and non-applicable attributes. There is no treatment of imprecision or uncertainty. Vassiliou [9, 10] is concerned with three types of information: missing values, non-applicable attributes and inconsistent data arising from an inconsistent database. A special 4-valued logic is introduced for query evaluation and two result lists are produced: a *true* result consisting of all tuples evaluating to true and a *maybe* result consisting of all tuples evaluating to unknown. However, there is no provision for imprecise values and no concept of uncertainty. Lipski [5] provides the most comprehensive treatment of imprecise information in information systems. Imprecise values are allowed and complete semantics for query evaluation are provided. However, the semantics are not obvious and hence may be of limited use in end-user information systems. Although it is recognised that uncertainty is introduced, there are no proposals for estimating it.

3. IMPRECISE VALUES AND UNCERTAINTY

To enable the storage of imprecise data we allow any attribute value to be one of the following:

- A regular precise value. For example,

  ```
  Age(Joe) = 17
  ```

  means that Joe is aged 17.

- A special value denoting that the value is unknown. For example

  ```
  Address(Mary) = nk
  ```

  means that Mary's address is not known to the system.

- A special value denoting that the attribute is not applicable. For example,

  ```
  Spouse(John) = na
  ```

  means that John does not have a spouse.

- An imprecise value. Two types of imprecise value are allowed: a *p-range* or a *p-domain*. A *p-range* is a range of possible values where the actual value lies somewhere in the range specified. For example,

  ```
  Age(Anne) = [20-40]
  ```

  means that Anne's age is somewhere in the range 20 to 40 years. A *p-domain* is a list of possible values, one of which is the actual value. For example,

  ```
  Colour(part-100) = [red, blue, green]
  ```

  means that the colour of part 100 is either red, green or blue.

The value 'unknown' is treated as a p-domain where the list of possible values is the entire domain. The value 'not applicable' is treated as a precise value since it gives complete information about the attribute.

Allowing imprecise values into the database introduces uncertainty. It is no longer obvious how queries should be evaluated nor how the results should be ranked for presentation to the user. For example, given the following data:

```
colour(part-1) = red
colour(part-2) = [red,green]
colour(part-3) = [red, green, blue]
colour(part-4) = pink
colour(part-5) = nk
```

and the query *'list all red parts'*, then clearly part 1 should be part of the list and part 4 should not, but there is uncertainty about the other parts since we can't be sure that they are red, nor can we be sure that they are not red. Intuitively there is less uncertainty about part 2 than part 3 since there are fewer possible colours, but there is a lot of uncertainty about part 5 since we have no information about it.

Such intuitive reasoning is ad-hoc. In [6] the precise semantics of query evaluation are specified so that two sets of objects are formed: those *known* to satisfy the query with absolute certainty and those which *possibly* satisfy the query with varying degrees of uncertainty. This paper is concerned with how this uncertainty may be measured and used to rank objects for presentation.

4. MEASURES OF UNCERTAINTY

Probability Theory is the classic method of handling uncertainty, but other methods include Certainty Theory [8], the Dempster-Shafer Theory of Evidence [7] and Possibility Theory [12]. In this section we examine possible methods based on two concepts in Information Theory: *self-information* and *entropy.*

4.1. Self-Information

The amount of information given by an event, or the self-information, is defined as $-\ log\ P(v)$ where $P(v)$ is the probability of event v occurring. Two 'events' are of interest here:

- The occurrence of a specific attribute value. Measuring the self-information of this event measures the amount of information given by the value. This quantity is denoted I_{given}.
- The occurrence of an attribute value such that we have the necessary and sufficient information required to be certain that the object should be retrieved. Measuring the self-information of this event will give a measure of the minimum amount of information required to be certain that the object should be retrieved. This quantity is denoted $I_{required}$.

Example 4.1

For example, the database contains data about a part which is known to be either red or blue. If only five colours are possible and the probability of red is 0.4, and the probability of blue is 0.1 then the amount of information given by the attribute value, denoted I_{given}, is

```
= -log P(any part is red or blue)
=  - log 0.5
=  0.693
```

This quantity is a measure of the amount of information we have about the colour of this part.

The quantity I_{given} can be calculated for each object where there is some uncertainty and used for ranking. The underlying hypothesis is that a user is most interested in the object which gives the most information, even if the information is imprecise.

It is not always necessary to have complete information to be certain that an object satisfies a query. For example, if a user requires a list of parts which are either red, black or white then he is clearly wants details of a part where the colour is only determined to the extent that it is either black or white. In this example no extra information is required but in some cases we need some information about an object, which is both necessary and sufficient to be certain that it should be retrieved. For example, given the same request and an object which is either red, black or pink then the necessary and sufficient information required in this case is that the object is black or red. By calculating the self-information of this information we can quantify the amount of information required, $I_{required}$, for certainty about the object.

When there is uncertainty $I_{required}$ is always greater than I_{given}, so subtracting the two quantities gives a measure of the amount of extra information required to remove uncertainty. This quantity is denoted by I_{extra}. It can be calculated for each object where there is uncertainty and used to give an alternate ranking. The hypothesis here is that the user is most interested in that object about which the least extra information is required.

Example 4.2

The database contains information about the colour of parts. Any part can only be red, green, pink or blue. Based on customer preferences the probability of any colour is as follows:

red	green	pink	blue
0.4	0.3	0.2	0.1

The information available is:

```
colour(p1) = [red, green]
colour(p2) = [red, green, blue]
colour(p3) = [red, green, pink]
colour(p4) = red
colour(p5) = [red, pink, blue]
colour(p6) = nk
colour(p7) = [red,blue]
```

A query asks for a list of red or green objects and two sets are formed:

```
known red or green: p1, p4
possibly red or green: p2, p3, p5, p6, p7
```

Since I_{given} for each object is as follows:

p2	p3	p5	p6	p7
0.223	0.105	0.356	0.0	0.693

then ranking objects by the amount of information they contain gives the following ordering:

```
possibly red or green: p7, p5, p2, p3, p6
```

Based on the information we have, the necessary and sufficient information for each object is

```
p2: red or green
p3: red or green
p5: red
p6: red or green
p7: red
```

and therefore I_{extra} for each object is

p2	p3	p5	p6	p7
0.133	0.251	0.560	0.916	0.223

So ranking in terms of I_{extra}, the extra information required to remove uncertainty gives the following ranking:

```
possibly red or green: p2, p7, p3, p5, p6
```

Another alternative is to consider the minimum negative information required about each object, $I_{negative}$, which in this example is as follows:

```
p2: not blue
p3: not pink
p5: not blue or pink
p6: not blue or pink
p7: not blue
```

Calculating the self-information in this case gives the following quantities for $I_{negative}$:

p2	p3	p5	p6	p7
0.105	0.233	0.356	0.356	0.105

which gives the partial ordering

```
possibly red or green:  (p2, p7), p3, (p5, p6)
```

The parts *p2* and *p7* are ranked in the same position since in both cases we need the information *'not blue'*.

4.2. Entropy

The entropy of a system S is defined as the average of self-information. That is,

$$H(S) = -\sum_{v=1}^{N} P(v) \, log \, P(v)$$

where $P(v)$ is the probability of event v. Entropy is a direct measure of uncertainty since it is at a maximum when there is complete uncertainty and at a minimum when there is no uncertainty. The following quantities are of interest:

- The entropy of the information given by the attribute value. This is denoted H_{given}.
- The entropy of the necessary and sufficient information required to remove uncertainty. This can be less than complete information, since we can tolerate some uncertainty about the exact value and still be certain that the object satisfies the query. This quantity is denoted $H_{tolerated}$.
- Subtracting the two quantities gives a measure of the amount of uncertainty which must be removed to be certain about the object. This quantity is denoted $H_{to\ be\ removed}$.

Entropy can be used to give two alternate rankings: one based on H_{given} and the other based on $H_{to\ be\ removed}$. In the former the hypothesis is that the user is most interested in that object about which there is the least uncertainty, and in the latter case the hypothesis is that he is most interested in the object where there is the least uncertainty to be removed.

Example 4.3

Given the same data and query as in Example 4.2 we get the following quantities for each part where there is uncertainty:

	$p2$	$p3$	$p5$	$p6$	$p7$
H_{given}	0.9579	1.04958	0.9186	1.2798	0.5967
$H_{tolerated}$	0.7277	0.7277	0.3665	0.7277	0.3665
$H_{to\ be\ removed}$	0.2302	0.3218	0.552	0.552	0.2302

Using H_{given} and $H_{to\ be\ removed}$ gives the following two rankings

```
possibly red or green: p7, p5, p2, p3, p6

possibly red or green: (p2, p7), p3, (p5, p6)
```

The first is based on uncertainty about the attribute value, the second is based on the amount of uncertainty to be removed to have certainty.

5. DISCUSSION

A number of different ways of ranking objects for presentation have been examined. In all cases two sets of objects are formed during query evaluation and the uncertainty is estimated and used for ranking. However, the underlying hypothesis is somewhat different for each method.

With I_{given} the hypothesis is that the user is more interested in the quantity of information given than in the quantity required to have certainty about an object. Thus, in example 4.2, part $p7$ ranks before $p2$ since it gives a greater quantity of information. The method is simple, easy to understand and intuitively obvious, but it does not take into consideration the fact that complete information is not always required for certainty.

The I_{extra} method recognises that incomplete information may be sufficient for certainty and calculates the quantity of extra information required. Thus, for example, $p2$ ranks before $p7$ since the necessary and sufficient information required about $p2$ is 'red or green' while for $p7$ it is 'red'. Less extra information is required about $p2$ and thus it ranks before $p7$.

With $I_{negative}$ we consider the possibilities which must be ruled out to have certainty. In this case $p2$ and $p7$ rank equal since we require the same negative information about both, 'not blue'. Similarly p5 and p6 share the same rank position since the information 'not pink or blue' is required about both. This method is also simple and intuitively satisfying but again it does not take into account the information which is available. For example, more information is available about $p5$ than $p6$ but they are ranked equal.

With H_{given} the hypothesis is that the user is more interested in the absolute uncertainty about the attribute value than in the quantity of uncertainty to be removed. Thus $p7$ ranks before $p2$ since there is less uncertainty about $p7$. In contrast, $H_{to\ be\ removed}$ recognises that some uncertainty as to the exact value can be tolerated and considers the minimum amount of uncertainty which must be removed. Thus $p2$ and $p7$ rank equal since in both cases the same uncertainty, about the colour being blue, must be removed.

All methods provide a useful way of ranking objects for presentation to users, none is the *'correct'* method. Some users may find one method more useful than another. For example casual naive users may prefer the methods based on I_{given}, $I_{negative}$ and $H_{to\ be\ removed}$ since they are obvious, simple and easy to understand. Users with an understanding of information theory could be given a choice of methods.

To use self-information and entropy it is necessary to know the range of possible values and the probability of each occurring. In general this is not possible. One possibility is to allow users to make subjective estimates at data entry time and have the system maintain the probability values automatically. Another solution is to make the closed world assumption which would imply that the only possible values are those in the system and that the probabilities could be calculated from frequency information. However, this may not be realistic as information systems are frequently updated. It is also possible to assume that all values are equally likely but this would not be as useful.

In the case of queries concerning more than one attribute it is necessary to combine measures to get a total measure of uncertainty. Given that we have information measures and if it is assumed that the attributes are independent then it is possible to simply add the individual measures to get a total.

This paper addresses the problem of calculating the uncertainty in order to rank objects in response to a query. An information theoretic solution is proposed, but it is recognised that other approaches, including possibility theory [11, 12, 13] and the Dempster-Shafer theory of evidence [7] may yield significant results. Future work will examine these areas.

References

[1] Codd, E.F. Understanding relations. ACM SIGMOD 7, (1975), 23-28.

[2] Codd, E.F. Extending the database relational model to capture more meaning. ACM TODS 4,4 (Dec. 1979), 394-434.

[3] Codd, E.F. Missing information (applicable and inapplicable) in relational systems. SIGMOD Record 15,4 (Dec. 1986).

[4] Date, C.J. Null values in databases. Proceedings of the 2nd British National Conference on Databases, (1982).

[5] Lipski, W. On semantic issues connected with incomplete information databases. ACM TODS 4,3 (Sept. 1979) 262-296.

[6] Morrissey, J.M. A treatment of Imprecise Data and Uncertainty in Information Systems. Ph.D. Thesis. National University of Ireland. 1987.

[7] Shafer, G. A Mathematical Theory of Evidence. Princeton University Press, Princeton, New Jersey, (1976).

[8] Shortliffe, E.H.; Buchanan, B.G. A model of inexact reasoning in medicine. Mathematical Biosciences, 23, (1975), 351-379.

[9] Vassiliou, Y. Null values in database management: a denotational semantics approach. Proceedings of the 1979 ACM SIGMOD international conference on management of data. Boston, (1979).

[10] Vassiliou, Y. A formal treatment of imperfect information in database management. Ph.D. Thesis. University of Toronto, (Sept. 1980).

[11] Zadeh, L. Fuzzy sets. Information and Control, 8, (1965), 338-353.

[12] Zadeh, L. Fuzzy sets as a basis for a theory of possibility. In: Fuzzy Sets and Systems, North Holland, Amsterdam, (1978).

[13] Zadeh, L. Commonsense knowledge representation based on fuzzy logic. Computer, 16, 10, (1983), 61-65.

DISCRIMINATION BY OPTIMIZING A LOCAL CONSISTENCY CRITERION

A. ZIGHED, D. TOUNISSOUX, J.P. AURAY, C. LARGERON
URA 934 Bt 101
University of Lyon I
43 Bd. du 11 Novembre 1918
69100 VILLEURBANNE

Abstract.

The paper we present here offers a method of pattern recognition which could be considered as a non-parametrical technique based on geometrical procedures. The ideas developed can be found in certain approaches for restoring pictures with sound and in supervised learning algorithms based on relaxation.

Our approach is based on three points:

* The definition of a neighbourhood structure endowed with certain properties.
* Finding a local consistency criterion which we will try to optimize with a view to relabeling
* Adopting a labeling rule for anonymous individuals.

We will conclude by presenting a few experimental results.

Keywords.

Discrimination, Relabeling, Neighbourhood, Geometrical approach, local consistency, relaxation.

1.INTRODUCTION

In a recent article [17], M. Terrenoire and D. Tounissoux suggest a method for restoring pictures with sound based on the optimization of a local homogeneity criterion. Other authors have also developed similar approaches [7], [14]. This paper, while drawing on them, tries to adapt the ideas developed in [17] to problems of classification, also called discrimination. Restoring a picture with sound is related to a discrimination problem, whether it be greyness level or binary. On a binary picture with sound, for example, we attempt to determine the correct state of a pixel (0 or 1). In the same way, for a greyness level picture with sound,we try to find the real greyness level of each pixel.

With discrimination, we have to determine, for each element of a population Π , the state of a so-called endogneous variable. What is special about the restoration of pictures with sound is that the points are in a frame which is a sub-set of ZxZ and the endogenous "greyness level" variable is of ordinal type. But with discrimination, the representation space is $\mathbb{R}^n$ and the endogenous variable can have values within a finite and unarranged set of labels.

The approach we recommend is non-parametric and, what is more, it does not attempt, as do discrimination and certain methods of regression, to establish a mathematical model linking together endogenous and exogenous variables. It is related to relaxation methods and is intended to be mainly descriptive, giving nonetheless a labeling procedure for a point not accounted for in the learning phase.

In this article, we will first make clear what a discrimination problem is and we will then develop the three basic concepts around which our approach is built:

- the first aims at defining a neighbourhood structure in $\mathbb{R}^n$. Instead of classical structures such as "the k-nearest neighbours" or "the neighbourhood in a hypersphere of radius ε", we prefer to use non-parametric neighbourhoods based on geometrical properties.This means using the neighbourhood according to Gabriel's graph [16]. This choice is justified by considerations which are both mathematical and algorithmic.

- the second deals with choosing a procedure for relabeling the points of the learning set. The one we recommend is classical and relatively simple. It consists of relabeling a point according to the relative majority of labels occuring in its neighbourhood.

- the third, which is our main contribution, is to propose a criterion called "local consistency" which we will try to optimize. We consider that neighbouring points ought to have in their neighbourhood "similar" labels. This resemblance will be evaluated using the notion of profile (obtained by counting the labels in the neighbourhood).

We will end this section dealing with methodology with a procedure for identifying individuals not accounted for in the learning phase and an experiment with a real case taken from the medical field.

2. DISCRIMINATION

Out of a sample Ω called the learning sample, taken from a population, it is supposed that the follwing are known:

1°) the p values taken by the p statistical variables $X_1,\ldots,X_j,\ldots,X_p$ called exogenous or explicative variables which will be supposed to be quantitative.

$X(\omega) = (X_1(\omega),\ldots,X_j(\omega),\ldots,X_p(\omega)) \in \mathbb{R}^P$ for any $\omega \in \Omega$

2°) the state of the endogenous variable Y, known as the variable to be explained:

$Y(\omega) \in E = \{y_1,\ldots, y_n\}$ E is called the set of labels. For any $\omega \in \Omega$.

In a problem of discrimination, our objective is to construct a procedure φ i.e. an application of Π in E, via which we would be able to forecast the state of the variable Y for all $\pi \in \Pi - \Omega$ individuals. For this forecasting rule φ to be of any use, it must be easier to determine than Y and in a majority of cases the prediction must be good, i.e:

$$\varphi(\pi) = Y(\pi) \qquad \pi \in \Pi - \Omega.$$

3. BASIC NOTIONS

3.1 p-neighbourhood[1]

Let $p(\omega, \omega')$ be a proposition relative to ω and ω' for any $(\omega,\omega') \in \Omega \times \Pi$. For any $\omega' \in \Pi$ the set of points $\omega \in \Omega$ confirming the property p is noted $V(\omega')$.

---

[1] The word "neighbourhood" is wrongly used here, since the definition we give for it does not conform to the defintiond used in topology; hence the expression "p-neighbourhood".

$$V(\omega') = \{\omega \in \Omega \;/\; p(\omega,\omega') \text{ true}\}$$

and is called p-neighbourhood of ω'.

There are many ways of building the structure of this p-neighbourhood: the k-nearest neighbours [1], [8], [16], the points within a ball of radius ε (ε-neighbour) [17], Delaunay's neighbourhood [2], [20], [21], the sphere of influence, the "lunule" [18], [19] etc...

The local consistency criterion which we will construct in §IV and the calculations it requires in order to be optimized suppose that our neighbourhood structure be symmetrical, i.e. for $\omega \in \Omega$, if ω is the p-neighbour of ω', then ω' is the p-neighbour of ω. This requirement excludes the k-nearest neighbours, the ε-neighbours and all those not possessing this property of symmetry.

In this paper, we shall keep the neighbourhood as it appears in Gabriel's graph [16], [19], since it checks the property of symmetry and, moreover, its construction is relatively simple to programme, compared with Delaunay's triangles for example.

The property p defining the neighbourhood as in Gabriel's graph can be stated thus [16]:

$p(\omega, \omega')$ is true if and only if the open disk of diameter $\|X(\omega) - X(\omega')\|$ containing ω and ω' contains no other point of Ω.

Let us consider the example of 5 points placed in $\mathbb{R}^2$ as shown in the following figure:

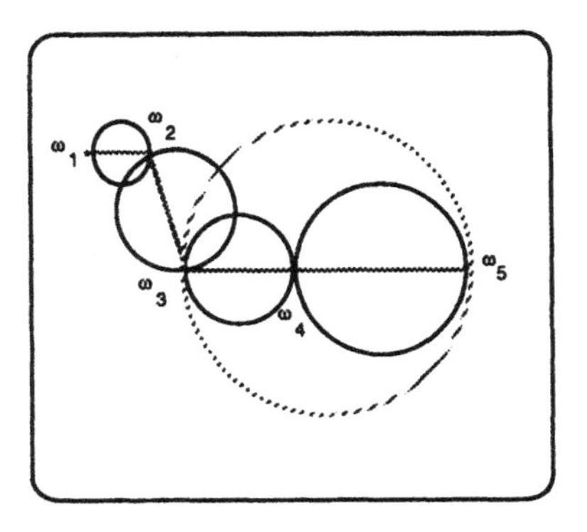

In this figure $V(\omega_1) = \{\omega_2\}$

$V(\omega_2) = \{\omega_1, \omega_3\}$

$V(\omega_3) = \{\omega_2, \omega_4\}$

$V(\omega_4) = \{\omega_3, \omega_5\}$

$V(\omega_5) = \{\omega_4\}$

ω_3 does not belong to $V(\omega_5)$ since the disk of diameter $\|X(\omega_3) - X(\omega_5)\|$ contains the point ω_4.

3.2 Profile

Definition

Let $S_n = \{(\gamma_0, \gamma_1, ..., \gamma_n) \in \mathbb{R}^{n+1} ; \forall \ i = 0,..., n, \gamma_i \geq 0 \text{ and } \Sigma \gamma_i = 1 \}$ the simplex of dimension (n+1), where n is the cardinal of E. Any application ℓ of Ω in S_n is called a profile. If $\omega \in \Omega$, then $\ell(\omega)$ will be a profile of ω. The set of profiles will be notated $\mathcal{L}(\Omega)$.

In the following,we suggest using a profile which will make it possible to characterize the environment of ω, i.e. the points of $V(\omega)$ by means of the proportion of points of $V(\omega)$ labelled y_k. To do this, we construct the profile ℓ in the following way:

We add to $E = \{y_1,...., y_n\}$ an element $y_0 \notin E$ which will be a new modality of Y, attributed to the points for which no attribution decision will be possible. This label y_0 will only be used during relabelling of all the $\omega \in \Omega$.

Let $k = 0,...., n$ $C_k = Y^{-1}(\{y_k\}) = \{ \omega \in \Omega / Y(\omega) = y_k \}$. C_0 is therefore the group of unlabelled points. As at the start all the points are already labelled, we have $C_0 = \emptyset$ and for $k \neq 0$ we have $C_k \neq \emptyset$.

Upon relabelling, on the one hand certain points will keep the same label (we hope this will be the case for the majority) and on the other hand C_0 can remain empty (which is to be hoped for).

The application $\ell : \Omega \longrightarrow \mathbb{R}^{n+1}$ defined by
$\ell(\omega) = (\ell_0(\omega), \ell_1(\omega),..... \ell_k(\omega),.......... \ell_n(\omega)) \in \mathbb{R}^{n+1}$

with

$$\forall k = 0,....,n \quad \ell_k(\omega) = \frac{\text{card}(V(\omega) \cap C_k))}{\text{card}(V(\omega))}$$

is thus a profile which makes the proportion of p-neighbour to ω points which appear in $C_0, C_1,......$ $C_j,......, C_n$ correspond to any $\omega \in \Pi$.

3.3 Relabelling

Relabelling(which concerns only the individuals of the learning sample) of a point $\omega \in \Omega$ must, as we made clear in §III.2 take into account the environment of ω which is summed up by the notion of profile.

Definition:

Any application ε of S_n in $E^* = \{ y_0,......, y_n \}$ is refered to as relabelling.
Given a relabelling ε, the application $\varepsilon \circ \ell$ notated simply as $\varepsilon\ell$ is an application of Ω in E^* which, to any $\omega \in \Omega$ makes it possible to associate an element of E^* called label of ω.

We would, of course, like $\varepsilon\ell$ to be such that $\varepsilon\ell(\omega) = Y(\omega)$ for any ω: this would prove that Y can be reconstituted by combining ε and ℓ. If we could than extend ℓ to the whole of Π, we would have available the means of predicting what $Y(\omega')$ could be for any $\omega' \in \Pi-\Omega$.

It goes without saying that the choice of ε will have to be made depending on that of ℓ. Given the choice made in §III.2, it seems natural to take for ε one or another of the applications ε_1 or ε_2 below.

a) for $\gamma \in S_n$ we take:

* $\varepsilon_1(\gamma) = y_k$ if $k \in \{1,....,n\}$ exists such that $\forall \; i <> k \quad \gamma_i < \gamma_k$.

* $\varepsilon_1(\gamma) = y_0$ if not.

In this way, ω is labelled y_k if and only if $\ell_k(\omega)$ is the greatest of the components of $\ell(\omega)$ (this is the relative majority vote rule).

b) We can imagine a more demanding procedure which would not allow ω to be labelled y_k even if $\ell_k(\omega)$ is the greatest of the components of $\ell(\omega)$, when this component does not seem high enough. In this case, a threshold $s \in]0, 1[$ is chosen and ε_2 is defined, taking for $\gamma \in S_n$:

* $\varepsilon_2(\gamma) = y_k$ if $\forall \; i <> k \quad i, k \in \{1,.....,n\}, \; \gamma_i < \gamma_k$ and $s \le \gamma_k$
* $\varepsilon_2(\gamma) = Y_0$ if $\max \{\gamma_i \; , \; i = i, ...n\} < s$

or if the max. is not unique.

In this way, for one of these choices, any $\omega \in \Omega$ receives the label y_k if the points of C_k are for the majority within the set of p-neighbout of ω points. Other more elaborate relabelling rules are possible, in particular those based on the principle of relaxation [4], [9], [13].

There is, unfortunately, no reason why we should have $Y(\omega) = \varepsilon\ell(\omega)$ for any $\omega \in \Omega$ and consequently our procedure for reconstituting Y is certainly not perfect: we might try to improve it either by modifying ℓ or by modifying ε or by modifying both of them.

Below, we suggest a procedure which aims simply to modify ℓ by optimizing one criterion.

Relabelling algorithm based on a local consistency criterion.

Let ℓ be the profile already constructed in §III.2 and let $\ell' \in \mathcal{L}(\Omega)$ be another profile that we are going to substitute for it. The principle of the ℓ profile modification procedure hinges on two ideas:

a) The notion of fidelity: the profile ℓ is constructed from direct observation of Y. It would therefore be deisrable for the profile ℓ' which we are going to substitute for it not to be too far away from .

b) The notion of compatibility: the p-neighbours of $\omega \in \Omega$ points are in fact points showing, with respect to the exogenous variables $X_1,\ldots, X_j,\ldots X_p$, neighbouring characteristics (this is due to the choice of the property *p* cf.§III.1); as we hope to predict the state of the endogenous variable from exogenous variables, we can only achieve this if closely related exogenous characteristics correspond in general to identical endogenous values. This is a hypothesis which we are taking for granted. In these conditions, for any $\omega \in \Omega$ the p-neighbours of ω must have a profile as close as possible to that of ω. Hence:

Definition.

We shall call ℓ-efficient any profile $\ell' \in \mathcal{L}(\Omega)$ such that

$$\delta(\ell') = \sum_{\omega \in \Omega} \left[\| \ell(\omega) - \ell'(\omega) \|^2 + \frac{1}{\mathrm{card}\,(V(\omega))} \sum_{\omega' \in V(\omega)} \| \ell'(\omega) - \ell'(\omega') \|^2 \right]$$

be a minimum

The first term of the above expression represents a kind of fidelity between the observed profile ℓ and the calculated profile ℓ'. The second term represents a notion of local compatibility which did not necessarily exist beforehand and which we wish to establish by means of this calculation. We are looking for the vectors $\ell'(.)$ which would give minimum $\delta(\ell')$ which represents what we call "local homogeneity".

This problem can easily be solved by a technique of optimization. In fact, the above expression can be written as follows:

$$\delta(\ell') = \sum_{k=1}^{n} \left[\sum_{\omega \in \Omega} \left((\ell_k(\omega) - \ell'_k(\omega)) \right)^2 + \frac{1}{\mathrm{card}(V(\omega))} \sum_{\omega' \in V(\omega)} (\ell'_k(\omega) - \ell'_k(\omega'))^2 \right) \Big]$$

which we will notate as: $\delta(\ell') = \sum_{k=1}^{n} \delta(\ell'_k)$

As the labels are independant, we will optimize according to each of the $\delta(\ell'_k))$, $k=1,\ldots n$. By rewriting the previous equation for a $\omega \in \Omega$, we will have to look for the optimum of the following expression:

$$\delta(\ell'_k) = \sum_{\omega \in \Omega} \Big[(\ell_k(\omega) - \ell'_k(\omega))^2 + \frac{1}{\mathrm{card}\,(V(\omega))} \sum_{\omega_j \in V(\omega)} (\ell'_k(\omega) - \ell'_k(\omega_j))^2 + \sum_{\omega_i \in V(\omega)} \frac{1}{\mathrm{card}\,(V(\omega_i))} (\ell'_k(\omega) - \ell'_k(\omega_i))^2 \Big]$$

This last expresion is only possible on condition that for any $\omega_i \in \Omega$ and any $\omega_j \in \Omega$, if $\omega_i \in V(\omega_j)$ then $\omega_j \in V(\omega_i)$. This condition is confirmed in our case as the neighbourhood structure we have chosen (p-neighbourhood, in accordance with Gabriel's graph cf. §IV.1) possesses this property of symmetry.

The strict convexity of $\delta(\ell'_k)$ can be demonstrated without difficulty. $\ell'(.)$ is a point of $\mathbb{R}^{\mathrm{card}(\Omega)}$.

The minimum will be a point of gradient zero. We will therefore have to determine $\ell'_k(\omega)$ for any ω for k=1,........, n. Or:

$$\frac{\partial\ \delta}{\partial\ \delta(\ell'_k(\omega))} = 2\ell'_k(\omega) - 2\ell_k(\omega) + 2\ell'_k(\omega)\left[1 + \sum_{\omega_i \in V(\omega)} \frac{1}{\text{card}\ (V(\omega_i))}\right]$$

$$- 2 \sum_{\omega_i \in V(\omega)} \ell'_k(\omega_i) \left[\frac{1}{\text{card}(V(\omega))} + \frac{1}{\text{card}\ (V(\omega_i))}\right]$$

Therefore, for any $\omega \in \Omega$ and k=1,....n, we will have:

$$\ell'_k(\omega) = \frac{\ell_k(\omega) + \sum_{\omega_i \in V(\omega)} \ell'_k(\omega_i) \left[\frac{1}{\text{card}(V(\omega))} + \frac{1}{\text{card}(V(\omega_i))}\right]}{2 + \sum_{\omega_i \in V(\omega)} \frac{1}{\text{card}\ (V(\omega_i))}}$$

This fixed point equation can be resolved efficicently using the GAUSS-SEIDEL method [6]. We have demonstrated that the sufficient convergence conditions of the GAUSS-SEIDEL algorithm are confirmed. Thus, the iterative method is convergent, whatever the starting point. In practice, we choose for ℓ^o the ℓ profile which we have observed in the learning sample, which gives the following general iteration:

$$\ell_k^{'(j+1)}(\omega) = \frac{\ell_k(\omega) + \sum_{\omega_i \in V(\omega)} \ell_k^{'(j)}(\omega_i) \left[\frac{1}{\text{card}(V(\omega))} + \frac{1}{\text{card}(V(\omega_i))}\right]}{2 + \sum_{\omega_i \in V(\omega)} \frac{1}{\text{card}\ (V(\omega_i))}}$$

It is shown that for any $\omega \in \Omega$ we have $\sum_{k=1}^{n} \ell'_k(\omega) = 1$ and $\ell'_k(\omega) \geq 0$

for k=1,.....,n: this is why we did not introduce these constraints into the optimization programme.

To relabel the points of the sample Ω we will proceed as follows:

Having available, for each point $\omega \in \Omega$ of an ℓ-efficient profile $\ell'(\omega)$, we relabel the set of points Ω according to one of two procedures described in §III.3. For the examples we are dealing with, we used the second rule, i.e. that a point $\omega_i \in \Omega$ will be relabelled y_k if the component $\ell'_k(\omega)$ is the greatest and if, moreover, it is greater than a threshold s.

5. EXTENSION PROCEDURE

This extension procedure is indispensible if we want to have an attribution rule for a test or anonymous individual.

Having available an ℓ-efficient profile $\ell'(.)$, we associate with it an application $\hat{\ell}'$ of Π in S_n defined thus:

1. For any k=0,.....,n, we note

$$C'_k = \{\omega \in \Omega \;/\; \varepsilon\ell'(\omega) = y_k\}$$

2. For any $\omega \in \Pi$ we put:

$$\hat{\ell}'_k(\omega) = \frac{\mathrm{card}(V(\omega) \cap C'_k))}{\mathrm{card}(V(\omega))}$$

and $\hat{\ell}'(\omega) = (\hat{\ell}'_0(\omega), \hat{\ell}'_1(\omega), \ldots\ldots\ldots, \hat{\ell}'_n(\omega))$

It is certain that $\hat{\ell}'(\omega) \in S_n$ and we will say that $\hat{\ell}'$ is the extended profile associated with ℓ'. In this way, for $\omega \in \Pi - \Omega$ we take $\varepsilon\hat{\ell}'(\omega)$ as the prediction for Y.

Whenever an anonymous individual appears, we will therefore be able to determine its neighbours in Ω and depending on the latter's labels (after relabelling) and depending on the relabelling rule chosen, predict its unknown label.

6. APPLICATION

The example we give comes from the medical field [3], [15]. We have to devise a procedure for assessing the risk of infection in patients suffering from serious burns. Three groups have to be discriminated: those alive exhibiting no complications, those who contracted septicemia and those who died. 174 patients making up the learning sample were chosen. The figure below gives the break-down into three groups: 87 living without complications, 60 living with septicemia and 27 who died as a result of an infection. The x and y axes give age and degree of burns respectively.

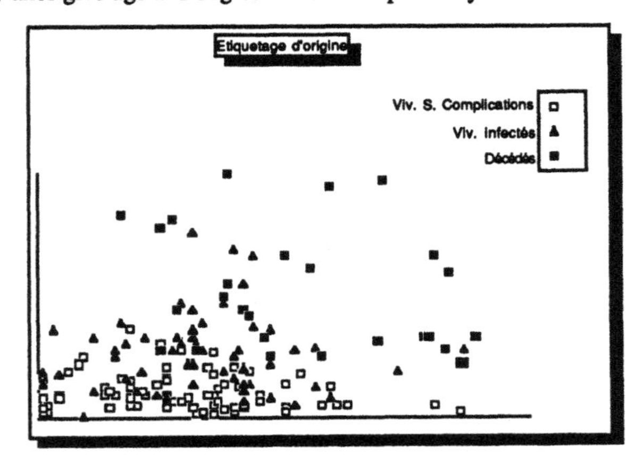

Based simply on initial profiles $\ell(\omega)$, $\omega \in \Omega$, relabelling of the individuals in the learning sample Ω gives a proportion of properly attributed points of 66.6%

It is considered that a point is properly attributed if it has not changed label after relabelling.

ℓ-efficient profile calculation and relabelling each element ω of the learning sample Ω using the new profile, give a noticeably better recognition rate (proportion of individuals who did not change labels after relabelling) of 73.5%

		Relabelling (attribution)			
		Living without complications	Living infected	Deceased	Not attributed Boundary points
Original labelling	Vivant Sans complications	73	14	0	0
	Vivant infectés	20	35	5	0
	Décédés	0	7	20	0

Proportion of properly attributed points: 73.5%

The figure below gives the result of relabelling obtained on the learning sample. We should note that discriminatory analysis has given a recognition rate of 72.4%. Other non-parametric techniques [3] did not give more significant results.

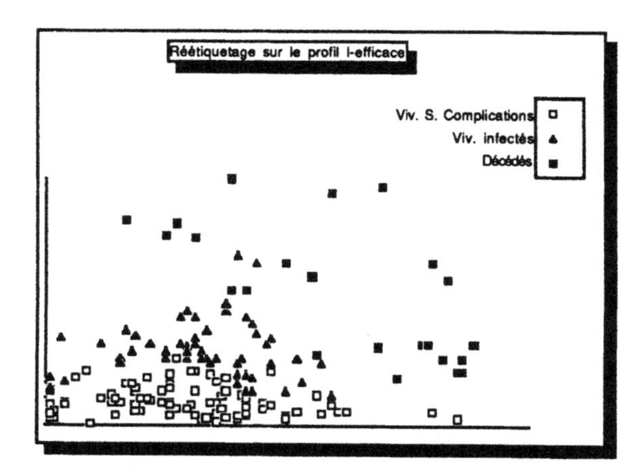

In order to better appreciate the quality of the results given by our method, it would seem wise to refine the experiment. To do this, we adopted the following procedure:

The 174 individuals in the learning sample were apportioned at random into 5 packs, respecting the proportions in each of the 3 groups to be discriminated. We applied our algorithm 5 times, taking each pack in turn as the test sample, the others being put together to make up the learning set. The results obtained are shown in the following table. We also give the results obtained by discriminatory analysis on the same data.

Test N°	Numbers		Proportion properly classified			
			Our approach		Discriminatory analys.	
	Base	Test	Base	Test	Base	Test
1	139	35	75,54%	71,43%	71,74%	71,43%
2	139	35	72,66%	77,14%	71,22%	74,29%
3	140	34	75,71%	67,65%	74,29%	70,59%
4	139	35	75,54%	63,00%	72,66%	74,29%
5	139	35	74,10%	68,57%	56,12%	68,57%
Average recognition rate :			74,71%	69,55%	69,20%	71,83%

From this, we can see that the results are stable. One of the advantages of the suggested algorithm is that, on a configuration where the groups are not linearly seperable, the results remain good, whereas they deteriorate using discriminatory analsis.

7 CONCLUSION

The main idea of this paper is based on the principle that "birds of a feather flock together", which, in our language becomes "neighbouring points in the representation space ought to have the same label".

First, it was necessary to give a precise meaning to the term "neighbours". We.examined many possibilities and in fact we discriminate betweentwo main classes of neighbourhood structure:

1°) That which requires the user to fix a parameter. For example, the k-plus nearest neighbours, the ball of radius ε... The main problem with these structures is that the learning result depends on the correct choice of parameters k, ε,...Their advantge is that they may lead to probabilistic approaches [10], [13].

2°) The neighbourhood structures which we chose are not parametered: they draw on geometric properties linked to the representation space. For example, Voronoi's neighbourhoods, Delaunay's triangles, Gabriel's graph etc... These structures are well-known and have often been used in pattern recognition [19]. The disadvantage is the complexity of their algorithm when in $\mathbb{R}^p$ [16].

Secondly, we look to see if neighbouring individuals have in general the same label. As this is often not the case, we are induced to relabel the set of individuals so as artificially to create this situation. Various procedures are possible, such as relabelling according to the majority in the neighbourhood or relaxation methods. We suggest a criterion enabling relabelling which takes account of the local stucture of each point, and in this sense, our procedure is related to relaxation methods.

We explained in §III.1 the reasons for choosing the neighbourhood using Gabriel's graph. We should point out, however, that this neighbourhood may turn out to be inappropriate in certain layouts. For example, to discriminate between two groups situated along two parallel straight lines, discriminatory analysis can give the better results.

As far as the criterion of homogeneity is concerned, other methods are being studied: they depend on taking not only immediate neighbours of each point into account, but also neighbours of the k-th rank, weighting where necessary fidelity or local compatibility...which opens the way for statistical formalization.

We are currently carrying out a number of tests, real or simulated using various methods: relaxation, k-nearest neighbours, ε-neighbours, discriminatory... Although the results we have so far obtained are not systematically noticeably better than those obtained with the other methods mentionned above, which are the most appropriate to the problem we are dealing with, it is nevertheless true that the approach we suggest has practical advantages. For example, the user does not have to look for the right parameter k or ε, nor the most appropriate procedure. Moreover, our method can adapt to the geometric structure of the patterns being looked for and is not algorithmically complicated, both when looking for neighbourhoods and when optimizing the local homogeneity criterion.

8. BIBLIOGRAPHY

[1] P. Devijver
Selection of prototypes for nearest neighbour classification.
Indian statistical institute golden jubilee
Proc. Int. Conf. on advances in information sciences and technology.1982

[2] P. DEVIJVER & M. DEKESEL,
Computing multidimensional Delaunay tesselation
Report R.464 Philips Research Laboratory. Brussels.,1982

[3] R. Fages, M.Terrenoire, D. Tounissoux, A. Zighed
Non supervised classification tools adapted for supervised classification.
Nato ASI Series Vol. F30 Springer-Verlag.1985

[4] O. D. Faugeras & M. Berthod.
Improving consistency and reduction ambiguity in stochastic labelling: an optimization approach.
IEEE Vol. PAMI-30, N°4, July 1981

[5] K. Fukunaga.
Introduction to statistical pattern recognition.
Academic Press. 1972.

[6] R. GASTINEL, Analyse numérique linéaire
Herman Paris 1966.

[7] X. Guyon & J.F Yao
Analyse discriminante contextuelle.
Actes 5° Jr. Anl. des données et informatique. Tome1 p43-52. 1987

[8] Hossam A. El Gindy, Godfried T. Toussaint
Computing the relative neighbour decomposition of simple polygon.
Computational morphology Ed. G.T. Toussaint
North-Holland 1988.

[9] R.A. Hummel & S.W. Zucker
On the foundation of relaxation labeling
IEEE VOL. PAMI-5, NO. 3, May 1983

[10] J. Illingworth & J. Kittler
Optimisation algorithms in probabilistic relaxation labelling
Pattern recognition Theory and application. Nato series Vol. 30 Springer Verlag.
1987

[11] M. Jambu, M.O. Lebeaux
Classification automatique pour l'analyse des données
Dunod 1978

[12] A.K. Jain
Advances in statistical pattern recognition
Pattern recognition Theory and application. Nato series Vol. 30 Springer Verlag.
1987

[13] J. Kittler
Relaxation labelling
Pattern recognition Theory and application. Nato series Vol. 30 Springer Verlag. 1987

[14] M. Levy
A new theoriktical approach to relaxation, application to edge detection. Actes
IAPR Rome 1988 IEEE

[15] J. Marichy, G. Buffet, A. Zighed, P. Laurent
Early detection of septicemia in burnt patients,
Actes 3rd INt. Conf. Sci. in Health Care, p 505-508
Munich 1984 Ed. Springer-Verlag.

[16] F. P. Preparata & M. I. Shamos.
Computational geometry: an introduction.
Springer-verlag. 1988

[17] M. Terrenoire, D. Tounissoux.
Restauration d'image par optimisation d'un critère d'homogénéité locale.
Proceding of workshop on syntactical and structural pattern recognition.
Pont-à-Mousson. 1988

[18] G. T. Toussaint..
Computational geometry recent relevant to pattern recognition.
Nato Asi Series Vol.F30 Springer-Verlag. 1987.

[19] G. T. Toussaint.
A Graph-theorical primal sketch.
Computational morphology.
E.d Toussaint, North-Holland 1988.

[20] J. I. Toriwaki, S. Yokoi.
Voronoi andrelated neighbors on digitized 2-dimensiona space with application of texture analysis.
Computational morphology.
E.d Toussaint, North-Holland 1988.

[21] D. F. Watson.
Computational the n-dimensional Delaunay tesselation with application to voronoi polytopes.
The computer journal. Vol. 24 N°2 1981.

"MINIMUM LOSS OF INFORMATION AND IMAGE SEGMENTATION"

Bruno Forte and Vladimir Kolbas
Department of Applied Mathematics
University of Waterloo
Waterloo, Ontario, Canada N2L 3G1

Abstract

A new measure of the information loss in image segmentation is derived from a set of natural properties. A similar quantity can be used in the quantization of a continuous real random n-vector. A new method for thresholding the grey-level histogram of a picture is then introduced. The method is based on the natural requirement of minimum information loss.

Keywords

quantization, image segmentation, information loss, entropy.

1. INTRODUCTION

Following the one by T. Pun, several other attempts have been made to use Shannon's entropy and related measures for optimal thresholding of grey-level histograms (for a survey of these entropic methods see [9, 10]). Recently [8] a new (entropic) criterion has been suggested which is based on a simple remark.

The grey level histogram of an image conveys a partial information about the subject of the picture. In the further transition from the original image to its binarized version (the one obtained by thresholding the grey level values) there is an additional loss of information. In selecting the threshold value we should try to minimize this loss of information. The idea of minimizing the loss of information for the choice of the (optimal) threshold value seems to be natural. The problem is to find an equally natural measure for such a loss.

Note that one finds himself in a similar situation when he wants to discretize a continuous probability distribution or to quantize a real valued random variable and in general when one wants to compress data.

The transition we are interested in is that from a digitized image (with a raster of the order of 55,000 pixels and quantization into 256 grey levels) to a binarized one with the same number of pixels of only two possible grey level values (black or white).

If X_d is the real valued random variable defined by the grey level histogram of the digitized image, let $U(X_d)$ represent the measure of uncertainty (about the subject of the image) when the only relevant information is the one provided by the said histogram. Similarly, let X_b be the real valued random variable (with only two possible values) defined by the thresholded histogram and let $U(X_b)$ be the measure of the actual uncertainty when the only relevant information about the same subject is the one provided by the thresholded histogram or equivalently the one provided by the binarized image.

It seems natural to measure the amount of information loss by the (non-negative) difference $U(X_b) - U(X_d)$ as it has been done in [8].

On the other hand, several derivations for the measure of the uncertainty (entropy) of a real valued random vector, given its probability distribution and its range, can be found in the literature [7]. These derivations are from a set of natural properties that have been selected for an axiomatic definition of entropy. For the special application in image processing it seemed to be natural to impose that the measure of uncertainty U exhibit the following properties: additivity, subadditivity, expansibility, weak-symmetry, boundedness, continuity.

2. MEASURE OF THE INFORMATION LOSS

Let $(\Omega, \mathcal{F}, P)$ be a probability space, $I^n = [0,1]^n$ the unit hypercube in $\mathbb{R}^n$, $\mathcal{B}^{(n)}$ the σ-algebra of all Borel subsets of I^n,

$$\psi : \Omega \rightarrow I^n$$

a measurable mapping $(\forall\ B \in \mathcal{B}^{(n)}\ :\ \psi^{-1}(B) \in \mathcal{F})$. Such a mapping defines a real valued random vector $X^{(n)}$ in I^n. We shall assume that the probability measure on $\mathcal{B}^{(n)}$ defined by $P(B) = P(\psi^{-1}(B))$ has a density function f on I^n. We shall denote by m^n the Lebesgue measure on $(I^n, \mathcal{B}^{(n)})$.

A quantization of the random vector $X^{(n)}$ consists of

a) the choice of a (finite) partition of I^n into N sets $(B_1, B_2, \dots, B_N)$ in $\mathcal{B}^{(n)}$.

b) the selection of N vectors (quanta) $b_i \in B_i$, $i = 1, 2, \dots, N$.

The quantified random vector $X^{(n)}$ $(n > 1)$ can be considered as the cartesian product of two random vectors say $Y^{(r)}$ and $Z^{(s)}$ with $r + s = n$. The subadditivity of the measure of uncertainty $U(X^{(n)})$ means that

$$U(X^{(n)}) = U(Y^{(r)} \times Z^{(s)}) \leq U(Y^{(r)}) + U(Z^{(s)})$$

where the equality holds when the two vectors $Y^{(r)}$ and $Z^{(s)}$ are stochastically independent (additivity).

A representation theorem [7] states that if the entropy $U(X^{(n)})$ of the given random n-vector exhibits the specified properties (additivity, subadditivity, expansibility, weak-symmetry, boundedness, continuity) then there exists a function $g^{(n)} : I^n \times [0,1] \rightarrow \mathbb{R}$ such that

$$(*) \qquad U(X^{(n)}) = -\sum_1^N P(B_i) \log[P(B_i)/m^n(B_i)] + \sum_1^N P(B_i) g^{(n)}(b_i, m^n(B_i)) \,.$$

In this formula there is some arbitrariness in the choice of the functions $g^{(n)}$. In [8] such a function has been replaced by a linear function in m^n and in [4] by a quadratic polynomial in the same variable $(g^{(n)} = am^n + b(m^n)^2)$.

It is easy to recognize in

$$-\sum_1^N P(B_i) \log [P(B_i) / m^n(B_i)]$$

the discrete analogue of the continuous Shannon entropy $H(X^{(n)})$ which is associated to a (continuous) random n-vector $X^{(n)}$ with probability density f

$$H(X^{(n)}) = -\int_{I^n} f \log f \, dm^n \, .$$

The non-negative difference

$$-\sum_1^N P(B_i) \log [P(B_i) / m^n(B_i)] - H(f)$$

represents, therefore, the contribution to the amount of information loss due to the discretization (a) of the probability distribution (from f to $(P(B_1), P(B_2), \ldots, P(B_N))$. The remaining sum in $(*)$

$$(**) \qquad \sum_1^N P(B_i) g^{(n)} (b_i, m^n(B_i))$$

represents the contribution to the information loss due to the quantization (b). Note that when $P(B_i) = m^n(B_i)$ (uniform probability distribution) the amount of information loss reduces to

$$\sum_1^N m^n(B_i) g^{(n)} (b_i, m^n(B_i))$$

which shows that $g^{(n)}$ must also be non-negative.

In the general case $(**)$ is the mean value of the random variable $g^{(n)}(b_i, m^n(B_i))$. It seems natural to consider this quantity as the measure of the amount of information that we would lose if b_i occurs. In fact when b_i occurs this means that any of the $b \in B_i$ has occurred with a loss of information compared with what we would have if we knew which one of the b's $\in B_i$ has occurred. The information we receive from the occurrence of b_i is that none of the b's $\in I^n \backslash B_i$ has occurred.

To find an explicit representation for $g^{(1)}$ let us consider the special case $n = 2$. If u in B_1 occurs, none of the $u^c \in I \backslash B_1$ occurred, if $v \in B_2$ occurs none of the $v^c \in I \backslash B_2$ occurred. Thus if $z = (u, v)$ occurs none of the (u^c, v^c) in $I^2 \backslash \{(B_1 \times [0\, 1]) \cup ([0, 1] \times B_2)\}$ occurred (see figure 1).

We shall assume that the loss of information

$$g^{(2)}(z, m^2 [(B_1 \times [0, 1]) \cup ([0, 1] \times B_2)])$$

due to the occurrence of z is the sum of the losses of information due to the occurrence of the two components u and v of z, separately. This yields the functional equation

$$g^{(2)}(u,v,m^{(2)}[(B_1 \times [0,1]) \cup ([0,1] \times B_2)])$$
$$= g^{(1)}(u,m^{(1)}(B_1)) + g^{(1)}(v,m^{(1)}(B_2)),$$

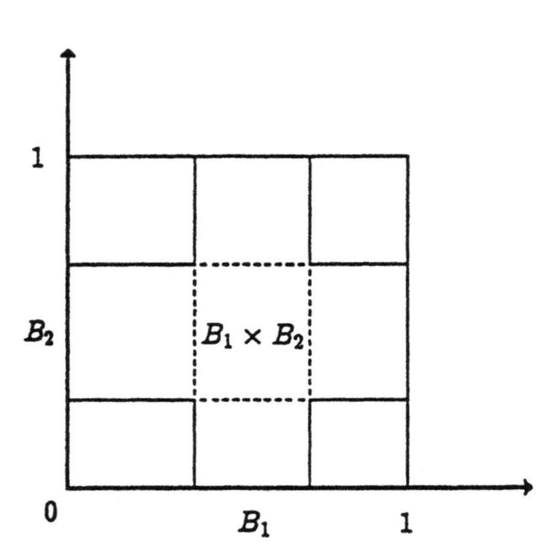

Figure 1

for all $u,v \in [0,1]$, $B_1,B_2 \in \mathcal{B}^{(1)}$. Let $x = m^1(B_1)$, $y = m^1(B_2)$, then

$$m^{(2)}[(B_1 \times [0,1]) \cup ([0,1] \times B_2)] = x + y - xy$$

With fixed u, v, and $x, y \in [0,1)$, define

$$f(1-x-y+xy) := g^{(2)}(u, v, x+y-xy)$$
$$h(1-x) := g^{(1)}(u, x)$$
$$k(1-y) := g^{(1)}(v, y).$$

We have

$$f((1-x)(1-y)) = h(1-x) + k(1-y) \qquad \forall x, y \in [0, 1).$$

As a consequence we have

$$g^{(1)}(u,x) = -\log(1-x) + h(u)$$

where $h(u)$ is again a non-negative but otherwise arbitrary function of u in $[0,1]$.

Without changing the final results, a simplified representation for the loss of information can be derived from a set of axioms that seem to be more natural in the kind of application we are considering in this paper. Let $X_q^{(n)}$ be a quantized random vector, i.e. a finite partition of I^n into N sets $(B_1, B_2, \ldots, B_N)$ in $\mathcal{B}$, a set $(m_1^{(n)}, m_2^{(n)}, \ldots, m_N^{(n)})$ of positive numbers, their measures, and a set $(b_1, b_2, \ldots, b_N)$ with $b_i \in B_i$, $i = 1, 2, \ldots, N$, the quanta. Let L be the measure of the loss of information due to the quantization, transition from continuous to discrete; the following different set of axioms seem to translate natural properties for L. The measure of $m_i^{(n)}$ representing somehow the extent of the set of values that have been replaced with the quantum b_i.

Axiom 1 - The measure of the loss of information is a function

$$L = L_N(m_1^{(n)}, m_2^{(n)}, \ldots, m_N^{(n)},\ p_1, p_2, \ldots, p_N)$$

of the probabilities p_i and the measures $m_i^{(n)}$ of the events B_i, $i = 1, 2, \ldots, N$.

Axiom 2 - If $X_q^{(n)} = Y_q^{(r)} \times Z_q^{(s)}\ (r + s = n)$, where $Y_q^{(r)}$ is discretized with the partition $(C_1, C_2, \ldots, C_M)$ and Z with $(D_1, D_2, \ldots, D_N)$, and $P(C_i \times D_j) = P(C_i)P(D_j)$ (stochastic independence) while $m^{(n)}(C_i \times D_j) = m^{(r)}(C_i)m^{(s)}(D_j)$, then (*additivity*)

$$L(X_q^{(n)}) = L(Y_q^{(r)}) + L(Z_q^{(s)})\,.$$

Axiom 3 - (*branching*) If $D_4 := \{(u,v,x,y) \in I^4 : 0 < u+v < 1, 0 < x+y < 1\}$ there exists a non-positive function $\Delta^{(n)} : D_4 \to \mathbb{R}$ such that for all $m_i^{(n)} \in (0,1)$, $p_i \in (0,1)$ with $\sum_1^N m_i^{(n)} = \sum_1^N p_i = 1$,

$$\begin{aligned} &L_N(m_1^{(n)}, m_2^{(n)}, m_3^{(n)}, \ldots, m_N^{(n)},\ p_1, p_2, \ldots, p_N) - \\ &L_{N-1}(m_1^{(n)} + m_2^{(n)}, m_3^{(n)}, \ldots, m_N^{(n)},\ p_1 + p_2, \ldots, p_N) \\ &= \Delta_4(m_1^{(n)}, m_2^{(n)}, p_1, p_2) \leq 0\,. \end{aligned}$$

Axiom 4 - (*symmetry*)

$$\begin{aligned} &L_N(m_{j_1}^{(n)}, m_{j_2}^{(n)}, \ldots, m_{j_N}^{(n)},\ p_{j_1}, p_{j_2}, \ldots, p_{j_N}) \\ &= L_N(m_1^{(n)}, m_2^{(n)}, \ldots, m_N^{(n)}, p_1, \ldots, p_N) \end{aligned}$$

for all permutations $(j_1, j_2, \ldots j_N)$ of $(1, 2, \ldots, N)$.

By recourse to a recent result of B.R. Ebanks [5] one can show immediately that axioms 1, 3 and 4 imply that there exist functions $b^{(n)} : (0,1) \times (0,1) \to \mathbb{R}$ such that (*sum property*)

$$L_N(m_1^{(n)}, m_2^{(n)}, \dots, m_N^{(n)}, p_1, p_2, \dots, p_N) = \sum_1^N b^{(n)}(m_i^{(n)}, p_i).$$

Using the first part of a representation Theorem by J. Aczél [2] and axiom 2 one obtains that there exist functions $g^{(n)} : (0,1) \to \mathbb{R}$ such that

$$L_N(m_1^{(n)}, m_2^{(n)}, \dots, m_N^{(n)}, p_1, p_2, \dots, p_N) = -c \sum_1^N p_i \ell n p_i / m_i^{(n)} + \sum_1^N p_j g^{(n)}(m_j).$$

At this point we join the proof of the previous representation theorem. Namely, we set $N = n = 2$ and add a special additivity property for the loss of information $g^{(n)}$ and we derive with the above described procedure a representation for the measure of the loss of information that differs from the one found before by just a constant.

3. OPTIMAL THRESHOLD VALUE

In image segmentation $n = 1$ (see, however [1.] where $n = 2$), the quanta are the grey levels a_i of the digitized image (total number 256) and the Borel subsets B_i are the intervals $[a_i, a_{i+1}]$, $m(B_i) = 1/256$, $i = 0, 1, \dots, 255$. The corresponding entropy is

$$U(X_d) = -\sum_0^{255} p_i \log(p_i 256) + \sum_0^{255} p_i h(i/256) - \log(255/256),$$

where $p_i := P([a_i, a_{i+1}])$.

The entropy associated with the binarized histogram when t is the threshold value and $w(t) := \sum_0^t p_i$ is given by

$$\begin{aligned} U(X_b) = & -w(t) \log[w(t)/t] \\ & -[1 - w(t)] \log[(1 - w(t))/(1 - t)] \\ & -w(t) \log(1 - t) - [1 - w(t)] \log t \\ & +w(t) h(0) + [1 - w(t)] h(1), \end{aligned}$$

having chosen as the only possible grey levels (quanta) the black (0) and the white (1). By symmetry argument one can see that $h(0) = h(1)$, thus

$$t \to U(X_b) - U(X_d) =$$

$$
\begin{aligned}
&-w(t)\log[w(t)/t]\\
&-[1-w(t)]\log[(1-w(t))/(1-t)]\\
&-w(t)\log(1-t)\\
&-[1-w(t)]\log t + const.
\end{aligned}
$$

is the function to be minimized.

The validity of this method is supported by experimental results. The details of these results are reported in a separate paper [6]. In the following tables, however, one can find the different threshold values which are obtained for some classical pictures with the present criterion, the previous method (Caselli-Forte [4]) with g^n quadratic, in three cases: $a = 1, \quad b = 0, \quad a = 1, \quad b = \frac{1}{2}, \quad a = 1, \quad b = 1$. A modified method was used. Instead of viewing a digitized image as one class of pixels with grey values between 0 and 255, one considers it as three classes of pixels, namely: the *black pixels*, the *white pixels* and the remaining (*grey*) pixels. Black are considered all those pixels with grey value between 0 and a certain grey value $n_{\min}$, that depends on the quality of the picture. Similarly we consider white all those pixels with grey value between a certain number $n_{\max}$ and 255. Thus the set of grey-levels is divided into three intervals

$$[0, n_{\min}], \quad [n_{\min}+1, n_{\max}-1], \quad [n_{\max}, 255]$$

The idea is that since the binarization by thresholding is not going to change the grey-level of those pixels that are already considered to be black or white, only the pixels with grey-levels in the interval

$$[n_{\min}+1, n_{\max}-1]$$

should be considered. The function that is going to be minimized is then the following

$$
\begin{aligned}
L(t) \quad = \quad & -c\frac{w(t)-w(n_{\min})}{w(n_{\max})-w(n_{\min})}\,\ell n\,\frac{w(t)-w(n_{\min})}{t-n_{\min}}\\
& -c\frac{w(n_{\max})-w(t)}{w(n_{\max})-w(n_{\min})}\,\ell n\,\frac{w(n_{\max})-w(t)}{n_{\max}-t}\\
& +c\,\ell n\,[w(n_{\max})-w(n_{\min})]\\
& -c\frac{w(t)-w(n_{\min})}{w(n_{\max})-w(n_{\min})}\,\ell n\,(n_{\max}-t)\\
& -c\frac{w(n_{\max})-w(t)}{w(n_{\max})-w(n_{\min})}\,\ell n\,(t-n_{\min})\\
= \quad & A\left\{[w(t)-w(n_{\min})]\left[\ell n\frac{w(t)-w(n_{\min})}{t-n_{\min}}+\ell n\,(n_{\max}-t)\right]\right.\\
& \left.+\,[w(n_{\max})-w(t)]\left[\ell n\,\frac{w(n_{\max})-w(t)}{n_{\max}-t}+\ell n\,(t-n_{\min})\right]\right\}\\
& +B
\end{aligned}
$$

where A and B are negative constants. Further details of this method can be found in [6]. The experimental results in the tables refer to the choice $(n_{\min}, n_{\max})$ such that

$$w(n_{\min}) = 0,\ w(n_{\min}+1) > 0,\ w(n_{\max}-1) < 1,\ w(n_{\max}) = 1$$

Threshold Values (Modified Method)

New Method	
1. Cameraman	164
2. Building	63
3. Model	75
4. Robot Arm	125

	Old Methods		
	$a=1, b=0$	$a=1, b=\frac{1}{2}$	$a=1, b=1$
1.	178	174	171
2.	57	57	57
3.	60	67	68
4.	166	166	166

It cannot pass unnoticed that in special cases one obtains the same results as in [4], indeed, this occurs whenever one has

$$-\log(1-t) \cong t + t^2/2\ .$$

As expected, small are the differences in the resulting binarized versions of pictures 1, 2 and 3. For the robot arm the difference is decisive: with the new method the robot is still recognizable, with the others it is totally undistinguishable from the background.

REFERENCES

Abutaleb, A.S. (1989) "Automatic Thresholding Grey-Level Pictures Using Two- Dimensional Entropy", Comput. Vision, Graphics Image Process., 42, pp. 22-31.

Aczél, J. (1978) "A mixed Theory of Information - II: Additive Inset Entropies (of Randomized systems of Events) with Measurable Sum Property", Utilitas Mathematica, Vol. 13, pp. 49-54.

Aczél, J. and Daroczy, Z. (1975) "On measures of Information and their Characterizations", Academic Press, New York-San Francisco-London.

Caselli, R. and Forte, B. (1988) "Thresholding Grey-Level Histograms by Minimum Information Loss", SASIAM internal report.

Ebanks, B.R. (1990) "Branching inset entropies on open domains", Aequationes Mathematicae, 39, pp. 100-113.

Forte, B. and Kolbas, V. "Some Experimental Results in Image Segmentation by Minimum Loss of Information", manuscript.

Forte, B., Ng, C.T. and Lo Schiavo, M. (1984) "Additive and Subadditive Entropies for Discrete Random Vectors", Journal of Comb. Info. and Systems Sc., Vol. 9, No. 4, 207-216.

Forte, B. and Sahoo, P.K., "Minimal Loss of Information and Optimal Thresholds for Digital Images", manuscript.

Sahoo, P.K., Soltani, S., Wong, A.K.C. and Chen, Y.C. (1988) "A Survey of Thresholding Techniques", Computer Vision, Graphics and Image Process., 41, 233-260.

Wong, A.K.C. and Sahoo, P.K. (1989) "A Grey-Level Threshold Selection Method Based on Maximum Entropy Principle", IEEE Trans. on Systems, Man and Cybernetics, Vol. 19, No. 4, July/August 866-871.

7. HYBRID APPROACHES TO UNCERTAINTY

TOWARDS A GENERAL THEORY OF EVIDENTIAL REASONING

J. F. BALDWIN
Engineering Mathematics Department
University of Bristol
Bristol BS8 1TR
England

ABSTRACT
This paper describes a general assignment method for combining evidences in the form of mass assignments over the power set, *P* (X), of a set of labels. It also describes an iterative assignment method for updating an apriori assignment over *P* (X) with evidences E1, ..., Er all expressed as assignments over *P* (X). This can be thought of as an extension of Bayesian updating with uncertain information. A minimum relative information optimisation of the updated assignment relative to the apriori is used. The methods have application to knowledge engineering for processing rules with uncertainties, causal nets and other representations for inference purposes. Solutions to the non-monotonic logic and abduction problems are special cases of this inference process. The methods are not the same as the Dempster Shafer theory of evidential reasoning, [Shafer 1976] but does use Shafer's belief function form of representing uncertainty.

KeyWords
Logic, Logic Programming, Abduction, Non-Monotonic Logic, Bayesian Updating, Causal Nets, Evidential Reasoning, Probability Logic, Fuzzy Sets, Fuzzy Logic, Expert Systems, FRIL.

1. INTRODUCTION

Knowledge Engineering requires a theory of evidential reasoning which incorporates such reasoning methods as logical deduction, induction, abduction, non-monotonic inference, probability logic and fuzzy logic. Most practical problems are resolved using a combination of these methods. Such a theory must be able to handle incompleteness of information and of precision. Sources of information must be distinguished. Information can be derived from general tendencies as probabilistic statements or as specific statements about a unique situation. These specific statements can also to some extent be uncertain. The general information can help to guide the completion of the incompleteness of the specific information. Assignment methods are described in this paper which help in this process. One method, the general assignment method, deals with combining evidences of the same type. The other, the iterative assignment method, allows the relevant general type of information to be updated using the relevant specific information. This is a generalisation of Bayes updating theorem to more general cases including updating with uncertain information. We will illustrate the methods with simple examples and briefly outline the theory to this approach of evidential reasoning. A fuller account of the theory can found in [**Baldwin 1990(a)(b)**]. The methods can be used for uncertainties expressed as fuzzy sets, [**Zadeh 1965**].

2. THE PENGUIN PROBLEM

Whatever form of knowledge representation is used to represent the knowledge of our application, the inference mechanism will involve extracting a set of variables and providing a set of possible instantiations to these variables. Each combination of instantiated variables can be given a label.

Consider the problem

90% of birds can fly -- (1)

All penguins are birds -- (2)

No penguins can fly -- (3)

An object x, which we call Mary, from the set of objects is tested for "bird quality" and tested for "penguin quality" and it is estimated that

$\Pr\{(x \text{ is } b)\} = 0.9$ -- (4)

$\Pr\{(x \text{ is } p)\} = 0.4$ -- (5)

What is the probability that Mary can fly?

Let BPF denote (Mary is B) $\wedge$ (Mary is P) $\wedge$ (Mary can F) where B, P, F can be instantiated to b or $\neg$ b, p or $\neg$ p, f or $\neg$ f respectively. The constants b, p, f stands for bird, penguin, fly respectively .

The eight possible instantiations of BPF are labels. Three of these labels, namely

$$\{\neg bpf, \neg bp\neg f, bpf\}$$

cannot have any probability of occurring using (2) and (3). The possible set of labels is therefore

$$\{\neg b\neg p\neg f, \neg b\neg pf, b\neg p\neg f, b\neg pf, bp\neg f\}$$

To answer the query we assign apriori probabilities to sets of labels as given from (1), (2) and (3). These rules do not give a unique solution for each label. We can assign the following masses for each piece of information expressed as a mass assignment over the power set of the set of labels. This family of assignments will be updated using (4) and (5).

The inference methods described in this paper assume that any query can be answered by extracting a set of labels, S say, and associating an apriori mass assignment over the power set of S, P (S) say, using each relevant rule or fact which applies in general and processing these assignments to give an overall apriori assignment. If further relevant specific facts are available then this apriori assignment is updated using these facts in such a way that the relative information between the updated assignment and the apriori assignment is minimised.

The problem can be expressed as a FRIL program, [**Baldwin et al** 1987].

((fly X) (bird X)) : (0.9 0.9) -- (1)
((bird X) (penguin X)) -- (2)
((fly X) (penguin X)) : (0 0) -- (3)
((penguin Mary)) : (0.4 0.4) -- (4)
((bird Mary)) : (0.9 0.9) -- (5)

A FRIL statement of the form ((a X) (b X)) : ((x1 x2)(y1 y2) is interpreted as Pr{(a x) | (b x)} lies in the interval [x1, x2] and Pr{(a x) | ¬ (b X)} lies in the interval [y1, y2]. (x1 x2) and (y1 y2) are called support pairs. If only the first support pair is given then (y1 y2) is assumed to be (0 1). The body of the rule can contain more than one list and such a rule can have more than two support pairs associated with it corresponding to the various conditional probabilities that can be defined.

The rules (1), (2) and (3) define the family of apriori assignments. (2) and (3) eliminate certain possible labels as discussed above. The labels, li, with their corresponding apriori probabilities, yi, are:

y1 ¬b¬p¬f
y2 ¬b¬pf
y3 b¬p¬f
y4 b¬pf
y5 bp¬f

so that y4 / (y3 + y4 + y5) = 0.9 and y1 + y2 + y3 + y4 + y5 = 1

If we let

k = y3 + y4 + y5 -- (6)

then

y4 = 0.9k -- (7)

and the family of apriori assignments for a given k, $0 < k \leq 1$ is

{b¬pf} : 0.9k
{¬b¬p_} : 1-k
{b_¬f} : 0.1k

determined by combining (6) and (7). The "_" indicates that it can be instantiated to any allowed value.

We can use the general assignment method to combine the two evidences (6) and (7). The solution i obvious in this case but the general assignment method is general and can be automated on a computer. We will use this example to illustrate the method.

This family of assignments are updated using the specific evidences (4) and (5) with the iterativ assignment method as described later.

3. General Assignment Method

Let m1 and m2 be two mass assignments over the power set *P* (X) where X is a set of labels. Evidence and evidence 2 are denoted by (m1, F1) and (m2, F2) respectively, where F1 and F2 are the sets of foca elements of *P* (X) for m1 and m2 respectively. A focal element is an element of *P* (X) with a non zer assignment. An assignment is to be understood in the sense of a mass probability as given by Shafer.

Suppose F1 = {L1k} for k = 1, ..., n1 and F2 = {L2k} for k = 1, ..., n2 then Lij is a subset of X fo which mi(Lij) ≠ 0. mi(Lij) is associated with the whole subset Lij and not to any subset of it.

Let (m F) be the evidence resulting from combining evidence 1 with evidence 2 using the general assignment method. This is denoted by

$(m, F) = (m1, F1) \oplus (m2, F2)$ where

$F = \{L1i \cap L2j \mid m(L1i \cap L2j) \neq 0\}$

$$m(Y) = \sum_{ij\,:\,L1i \cap L2j = Y} m'(L1i \cap L2j) \qquad \text{for any } Y \in F$$

$m'(L1i \cap L2j)$ for $i = 1, \ldots, n1$; $j = 1, \ldots, n2$ satisfies

$$\sum_{j} m'(L1i \cap L2j) = m1(L1i) \qquad \text{for } i = 1, \ldots, n1$$

$$\sum_{i} m'(L1i \cap L2j) = m(L2j) \qquad \text{for } j = 1, \ldots, n2$$

$m'(L1i \cap L2j) = 0$ if $L1i \cap L2j = \emptyset$ the empty set ; for $i = 1,\ldots, n1$; $j = 1, \ldots, n2$

The problem of determining the mass assignment m is an assignment problem.

If there are more than two evidences to combine then they are combined two at a time. For example to combine (m1, F1), (m2, F2), (m3, F3) and (m4, F4) use

$(m, F) = ((((m1, F1) \oplus (m2, F2)) \oplus (m3, F3)) \oplus (m4, F4))$

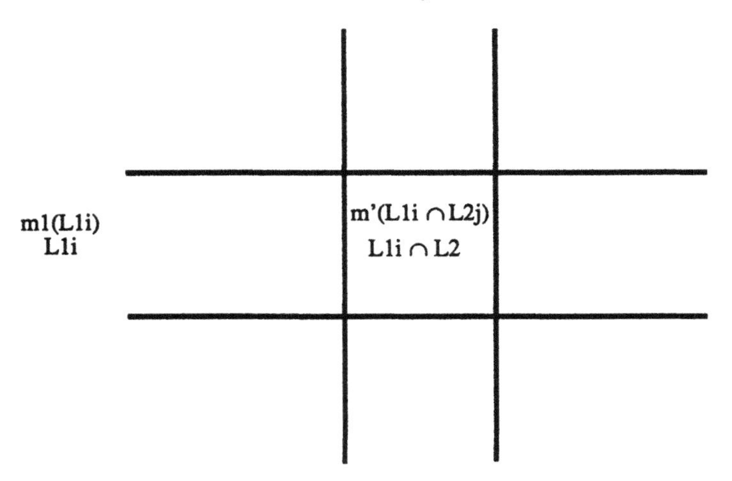

The labels in a cell is the intersection of the subset of labels of evidence 1 associated with the row of the cell and the subset of labels of evidence 2 associated with the column of the cell.
The mass assignment entry in a cell is 0 if the intersection of the subset of labels of evidence 2 associated with the column of the cell and the subset of labels of evidence 1 associated with the row of the cell is empty.
The mass assignment in a cell is associated with the subset of labels in the cell.
The sum of the cell mass assignment entries in a row must equal the mass assignment associated with m1 in that row.
The sum of the cell mass assignment entries in a column must equal the mass assignment associated with m2 in that column.
If there are no loops, where a loop is formed by a movement from a non zero assignment cell to other non zero assignment cells by alternative vertical and horizontal moves returning to the starting point, then the general assignment problem gives a unique solution for the mass assignment cell entries. If a loop exists then it is possible to add and subtract a quantity from the assignment values around the loop without violating the row and column constraints and the solution will not then be unique.

Applying this method to combining (6) and (7) above gives

	0.9k : {b¬pf}	1-0.9k : {¬b¬p_, b_¬f}
k {b¬p_, bp¬f}	{b¬pf} 0.9k	{b_¬f} 0.1k
1 - k {¬b¬p_}	Ø 0	{¬b¬p_} 1-k

4. THE ITERATIVE ASSIGNMENT METHOD

Let each specific evidence Er, r = 1, ..., n be expressed as a mass assignment (mr Fr) where Fr = {Xri}, = 1, ..., nr and mass mri is associated with Xri, a subset of X,
Let the apriori assignment be expressed as a mass assignment (t F) where F = {Ti}, i = 1, ..., m. Ti is a subset of X, and mass ti is associated with Ti.
The intersection of the row subset of labels of the apriori assignment with the column subset of labels of the evidence assignment for a given cell of the tableau is a subset of X. In the case when this intersection is the empty set the mass assignment for that cell is zero, When the intersection is not empty then the mass assignment is the product of the row apriori assignment and the evidence column assignment scaled with the K multiplier for that column. The K multiplier for a column is the inverse of the sum of the

apriori row assignments corresponding to those cells of the tableau in the column which have non-empty label intersections. The update is a mass assignment over the set of subsets of labels in the cells of the tableau. The update can therefore be over a different set of subsets of labels to that of the apriori. The apriori is updated with each evidence Ei in turn and the whole process repeated until convergence of the final assignment is obtained. Convergence will be obtained with a mass assignment over *P* (X) which will correspond to a family of possible probability distributions over X.

We can give a pictorial view of the algorithm:

Evidence Er

Apriori is mass assignment (t T) where T = {T1 ... Tm} where t = t1, ..., tm} where Ti is subset of P(X)	mrk Xrk	- - -	mrq Xrq
tv : Tv	Tv ∩ Xrk Kk.tv.mrk if non null intersection, 0 otherwise		Tv ∩ Xrq Kq.tv.mrq if non null intersection, 0 otherwise
		- - -	
	$1/Kk = \sum_{s : Ts \cap Xrk \neq \emptyset} ts$		$1/Kq = \sum_{s : Ts \cap Xrq \neq \emptyset} ts$

The apriori assignment was also a family of possible probability distributions. For an apriori distribution, the calculation satisfies the minimum relative information principle where the constraints are in terms of beliefs and plausibilities from the evidence assignments. An apriori probability distributions over the label set X, say p, will be updated with the evidences E1, ..., Er to a final probability distribution over X, say p', satisfying the following relative information optimisation problem.

$$\sum_{x \in X} p'(x) \, Ln \, (p'(x) / p(x))$$

is minimised subject to the constraints

$Bel(Y) \leq p'(Y) \leq Pl(Y)$ for all subsets Y of *P* (X)

where Bel(Y) and Pl(Y) are beliefs and plausibilities, as defined by Shafer in [**Shafer 1976**], determined from the mass assignment (mr , Fr).

If the apriori is a mass assignment the updating procedure allocates the row apriori mass for a given cell of the tableau to those apriori elements consistent with the column evidence. Alternatively we can

allocate this row apriori mass among all the apriori row elements according to entropy considerations. This is called "proportional fill in" and will be discussed more fully in a future paper.
The updating tableau can contain loops in a similar manner to the general assignment case. The loop can be treated in exactly the same way as for the general assignment method.
The following schematic shows the updating procedure applied to the penguin problem given above.

0.9k	{b¬pf}	UPDATE USING	{b¬pf}	
0.1k	{b _ ¬f}	Pr{(b)} = 0.9	{b _ ¬f}	
1-k	{¬b¬p _}	TO GIVE UPDATE	{¬b¬p _}	
	{b¬pf}	UPDATE USING	{bp¬f}	
	{b _ ¬f}	Pr{(p)} = 0.4	{b¬pf}	
	{¬b¬p _}	TO GIVE UPDATE	{b¬p¬f}	
			{¬b¬p _}	
	{bp¬f}	UPDATE USING	{bp¬f}	
	{b¬pf}	Pr{(b)} = 0.9	{b¬pf}	I
	{b¬p¬f}	TO GIVE UPDATE	{b¬p¬f}	T
	{¬b¬p _}		{¬b¬p _}	E
				R
				A
	{bp¬f}	UPDATE USING	{bp¬f}	T
	{b¬pf}	Pr{(p)} = 0.4	{b¬pf}	E
	{b¬p¬f}	TO GIVE UPDATE	{b¬p¬f}	
	{¬b¬p _}		{¬b¬p _}	

From the final family of assignments
f(k) : [assignment for {b¬pf}, assignment for {b¬pf} + assignment for {¬b¬p _}]
= [0.45, 0.55]
This final support pair is in actual fact independent of k, so that this is the actual support pair for "f" ie

Pr{(f Mary)} ε [0.45, 0.55]

5 A COMPLETE MODEL FOR THE PENGUIN EXAMPLE

Consider the program
((bird X)) : (0.7 0.7)
((fly X)(bird X)) : (0.9 0.9)
((fly X) (bird X)(penguin X)) : ((0 0)(0.95 0.95)(_ _)(0.1 0.1))
((bird X) (penguin X)) : (1 1)
((fly X)(penguin X)) : (0 0)
((penguin Mary)) : (0.4 0.4)

((bird Mary)) : (0.9 0.9)

This program says that:

the proportion of birds in the relevant population of objects is 70%. 90% of the birds can fly. No object which is a bird and penguin can fly, 95% of birds which are not penguins can fly, 10% of objects which are not birds can fly. All penguins are birds. No penguin can fly.

It also gives specific information about the object Mary, namely that there is a probability of 0.9 that Mary is a bird and a probability of 0.4 that Mary is a penguin.This comes from tests as for the case above. This information allows the following unique distribution over the relevant labels to be constructed:

Apriori

y1 = 0.27	$\neg b \neg p \neg f$
y2 = 0.03	$\neg b \neg pf$
y3 = 0.0332	$b \neg p \neg f$
y4 = 0.63	$b \neg pf$
y5 = 0.0368	$bp \neg f$

using

$y4 / (y3 + y4) = 0.95$; $y2 / (y1 + y2) = 0.1$; $y4 / (y3 + y4 + y5) = 0.9$

$y3 + y4 + y5 = 0.7$; $y1 + y2 + y3 + y4 + y5 = 1$

The iterative assignment update then gives (fly Mary) : (0.485 0.485).

Intuitive solution

In this problem we are presented with two pieces of information:

1. Object Mary came from a population with statistics

$\neg b \neg p \neg f$	0.27
$\neg b \neg pf$	0.03
$b \neg p \neg f$	0.0332
$b \neg pf$	0.63
$bp \neg f$	0.0368

so that

IF object Mary has properties bp then Pr(Mary can fly) = 0

IF object Mary has properties $b \neg p$ then Pr(Mary can fly) = 0.63 / 0.6632 = 0.95

IF object Mary has properties $\neg b \neg p$ then Pr(Mary can fly) = 0.03 / 0.3 = 0.1

2. Object properties

2(a) b : 0.9 ; $\neg$b : 0.1

2(b) p : 0.4 ; $\neg$p : 0.6

2(c) Mary cannot be a penguin and not a bird

Combining 2(a) and 2(b) taking account of 2(c) by allowing only the set of labels

{¬b¬p, b¬p, bp}

using the general assignment method gives

	0.4 p	0.6 ¬p
0.9 b	bp 0.4	b¬p 0.5
0.1 ¬b	¬bp (not allowed) 0	¬b¬p 0.1

giving

bp : 0.4 ; b¬p : 0.5 ; ¬b¬p : 0.1

Expected value of Pr(Mary can fly) = 0.5*0.95 + 0.1*0.1 = 0.485

which is the value given by the iterative assignment method.

We can write

P'r(f) = Pr(f | bp)P'r(bp) + Pr(f | b¬p)P'r(b¬p) + Pr(f | ¬b¬p)P'r(f | ¬b¬p)

where Pr(.) signifies a probability determined from the population statistics, information 1, and P'r(.) signifies a probability determined from the specific information, information 2 and set of possible labels.

Dempster Shafer Solution

We might like to compare the solution given here with that given by the Dempster Shafer theory of evidence. [**Shafer 1976**], for the independent informations 1 and 2 given above.

Combining 2(a) and 2(b) allowing for 2(c) using the Dempster rule gives

bp : 0.375 ; b¬p : 0.5625 ; ¬b¬p : 0.0625

This is then combined with information 1 given above using the Dempster rule of combination to give Pr(Mary can fly) = 0.8783.

The Dempster Shafer result is the same if 2(c) is ignored and simply 2(a) and 2(b) are combined using the Dempster rule giving

bp : 0.36 ; b¬p : 0.54 ; ¬bp : 0.04 ; ¬b¬p : 0.06

and this is combined with information 1.

The solution given here by the Dempster rule of combination is most certainly not in accord with intuition.

6 A SIMPLE RULE SYSTEM

((a X) (b X)(c X)) : ((0.9 0.9)(0 0)) -- (1)

((c X) (d X)) : ((0.85 1)(0 0)) -- (2)

((b Mary)) : (0.8 0.8) -- (3)

((d Mary)) : (0.95 0.95) --- (4)

This is a simple FRIL program. X is a variable and a, b, c, d are predicates. The first two sentences are rules which express general statements about persons and the third and fourth are facts about a specific person Mary.

The iterative assignment method applied to this problem to find the Pr{(a Mary)} by forming the family of apriori assignments and updating this with evidences (3) and (4) gives the same solution as when applied to the decomposed problem:

problem 1

Use the iterative assignment method to solve

((c X) (d X)) : ((0.85 1)(0 0)) --- (5)

((d Mary)) : (0.95 0.95) --- (6)

by forming the family of apriori assignments using (5) and updating this using (6) to give the assignment

(9) below corresponding to Pr{(c Mary)} ε [0.8075, 0.95].

problem 2

Use the iterative assignment method to solve

((a X) (b X)(c X)) : ((0.9 0.9)(0 0)) -- (7)

((b Mary)) : (0.8 0.8) -- (8)

(c Mary) : 0.8075 ; ¬ (c Mary) : 0.05 ; {(c Mary), ¬ (c Mary)} : 0.1425 ----------- (9)

by updating the family of apriori assignments formed using (7) with the specific evidences (8) and (9) to give assignment (a Mary) : 0.6675 ; ¬ (a Mary) : 0.28 ; {(a Mary), ¬ (a Mary)} : 0.0525

corresponding to Pr{(a Mary)} ε [0.6675, 0.72].

Decomposition methods are most important in this approach to evidential reasoning to reduce the computational demands. These are discussed more fully in Baldwin's papers cited above.

7 Conclusions

A theory of evidential reasoning with wide application has been demonstrated.

8. References

BALDWIN J. F. (To appear 1990) "*Combining Evidences for evidential reasoning*", Int. J of Intelligent Systems, pp 1 - 40

BALDWIN J. F. (1990), I.T.R.C Report, University of Bristol, " *Assignment Methods for Evidential Reasoning for Knowledge Engineering*", pp 1 - 45.

BALDWIN J. F., PILSWORTH, B., MARTIN, T., (1987), "FRIL Manual", Fril Systems Ltd, St Anne's House, St Anne's Rd, Bristol BS4 4A, UK

SHAFER G., (1976), "*A mathematical theory of evidence*" , Princeton Univ. Press

ZADEH L., (1965), "*Fuzzy sets*" , *Information and Control*, **8**, pp 338-353.

A PRAGMATIC WAY OUT OF THE MAZE OF UNCERTAINTY MEASURES

Giuseppe Longo
DEEI, University of Trieste, 34100 Trieste, Italy
Andrea Sgarro
Department of Mathematics and Computer Science
University of Udine, 33100 Udine, Italy
and: Department of Mathematical Sciences
University of Trieste, 34100 Trieste, Italy

Abstract. We argue that the construction of a general (non-specific) theory of pragmatic uncertainty measures is a dubious undertaking. To make our point we discuss fractional entropy, a measure introduced in the specific context of cryptography. Two new properties of fractional entropy are proved.
Key-words: entropy, conditional entropy, uncertainty measures.

1 - INTRODUCTION

An appropriate measure of uncertainty appears to be, at least in principle, a convenient tool to be used in uncertainty management. However, the prospective user gets easily (and rightly) scared by the abundance of products offered on the market, some of them being quite mysterious, or even mystifying. The risk is that, as for IQ tests, one puts oneself in a position where one is able to make very accurate numerical measurements, but one does not exactly know what on earth is being measured. (As a general reference to uncertainty measures, we refer to Klir (1987), which is an excellent review of the state-of-the-art up to 1986. The author distinguishes two main types of uncertainty: *vagueness* and *ambiguity*; further, three subtypes of ambiguity are considered: *non-specificity*, *dissonance* and *confusion*. For a more extensive reference cf Klir and Folger (1988).)

A general criticism of uncertainty measures is that they tend to ignore the *semantic* and *pragmatic* aspects of uncertainty. Below we explore the consequences of letting pragmatics into the domain of uncertainty measures; we argue that the construction of a general (abstract) theory of pragmatic uncertainty measures is a more or less hopeless (or even undesirable) undertaking.

The uncertainty measure (dissonance measure according to Klir (1987)) which at present stands in the best position is obviously Shannon entropy, for which uniqueness theorems have been proven. However, in Sgarro (1988a) and Sgarro (1988b) the second author has put forward a new measure of statistical uncertainty (dissonance again!), to be applied in a cryptographic context, and *different* from Shannon entropy. This proposal is less daring than it sounds, as we argue now.

2 - A PRAGMATIC MEASURE OF UNCERTAINTY

The *axiomatic approach* is considered to be the most sophisticated vindication of an uncertainty measure: one puts forward a list of desirable properties, or *axioms*, and then tries to prove that the new measure is the unique solutions to the list of axioms, seen as a sort of system of functional "equations". This approach works quite well for entropy and also for *U-uncertainty,* a much more recent measure of non-specificity (cf Klir (1987)); unfortunately, uniqueness theorems are quite difficult to prove for *meaningful* lists of axioms (however vague, meaningfulness is a paramount requirement to exact for a list of axioms; past experience shows that the temptation to choose ad-hoc axioms to suit the proposed solution is difficult to resist, indeed!). Below we take the less ambitious *operational-heuristic* approach, which anyway works quite well in the case of Shannon's entropy. In this case an operational theorem, e.g. a coding theorem, suggests that a certain functional, whose operational meaning is made clear by that theorem, might be an adequate measure of uncertainty; the properties of the functional are carefully scrutinized: if appropriate, the use of that functional as an uncertainty measure is vindicated a posteriori; if not, the functional is dropped (in Popper's terms we put forward a hypothesis and then try to "falsify" it - of course in the hope that the falsification utterly fails). The operational vindication of entropy as a measure of statistical uncertainty is based on

Shannon's source coding theorem; this theorem deals with binary descriptions (encodings) of information sources which have to be shortest and *faithful*. Statistically, the source is assumed to be stationary and memoryless; at each time instant a random letter X is output (a random experiment X is performed) according to the probability distribution, or probability vector, $P=(p_1,p_2,...,p_K)$. Even if the information source remains the same, however, the *fidelity requirements* of the user may change: consequently, his uncertainty as to source outputs will change, too. (In the limit, if the user is simply not interested in what is going on, his uncertainty will be equal to zero!) Now, "pragmatic" coding theorems are offered by *rate-distortion theory*. In the setting of this theory, the binary descriptions of the information source are converted (in an optimal way) into strings written using the letters of a reproduction alphabet. An allowed distortion level *d* is fixed, and the reproduction is considered faithful if the distortion between the sequence output by the source and the reproducing sequence is at most *d*. If one imposes suitable limitations on the mathematical form of the distortion measure, it turns out that the place of Shannon entropy is taken by a function of *d* called the *rate-distortion function* (Shannon entropy is re-found when the source alphabet coincides with the reproduction alphabet and when no distortion is allowed).

Now, in the cryptographic context considered in Sgarro (1988a), the user was not interested in the precise outcome of the random "experiment" X (in the precise value of the cleartext, it happened) but was quite contented when he could find a 2-element set containing that outcome (more generally, an N-element set with N fixed, $N \le K$; cf the remark below; for N=1 one re-finds entropy). This led to consider a reproduction alphabet whose "letters" were unordered couples of source letters, the distortion between a source letter and an unordered couple being zero iff the former belonged to the latter. The resulting uncertainty measure, S(X) or S(P), set equal to the value of the corresponding rate-distortion function at $d=0$, was called *semientropy* and was proved in Sgarro (1988a) to have the form (logs are to any base ≥ 2):

$$S(P) = H(P) - H(p,1-p) \quad \text{if } p = \max p_i \ge \frac{1}{2}$$

$$S(P) = H(P) - \log 2 \quad \text{else}$$

Heuristically, semientropy has several convenient properties: it is zero iff P has at most 2 non-zero components, it is maximal iff P is uniform, it is concave; weakly concave, however, while entropy is strictly concave (the region where S(P) is linear is described in Sgarro (1988a); for K=3, S(P) as a function of (p_1,p_2) is a sort of wigwam without doors, with a strictly concave dome over it, and with infinite supporting sticks connecting the ground with the seam of the dome). Weak concavity is an important point, since it implies that two random variables can be "non-interactive" even if they are stochastically dependent: this is in violation of one of the axioms uniquely leading to entropy, the "two-step axiom" (the uncertainty in a two-step experiment should be the uncertainty in the first step plus the conditional uncertainty in the second step, given the result of the first).

To make our point, we find it convenient to introduce the *conditional semientropy* S(X|Y) (cf Sgarro (1988a)). Set

$$S(X|Y) = \Sigma_y \text{Prob}\{Y=y\}\ S(X|Y=y)$$

Observe that the conditional entropy H(X|Y) can be introduced in exactly the same way; S(X|Y=y) is semientropy of the conditional distribution of X given Y=y and so S(X|Y) is an "average semi-uncertainty". Since S(X)=S(P) is concave:

$$S(X|Y) \leq S(X)$$

Since S(X|Y=y) does not depend on y when X and Y are independent:

$$X,Y \text{ independent} \quad \text{implies} \quad S(X|Y) = S(X)$$

Unlike in the case of entropy this implication cannot be reversed: even if X and Y are dependent one has S(X|Y)=S(X) whenever the distributions of X given Y=y lie all in a region where S(P) is linear. The fact that there can be such "lost information" is readily interpretable in our pragmatic context, since *side information which only serves to discriminate within 2-element sets is completely useless!* An adequate measure of semi-uncertainty *has* to be weakly concave and so *has* to violate the two-step axiom.

To further make this point, let's pursue the analogy with conditional entropy and set

$$S^*(X|Y) = S(XY) - S(Y)$$

(Basically, we are imposing that the two-step axiom should hold). Unfortunately, this would be quite an inappropriate definition of conditional semientropy. To see this take the case of X and Y both binary, so that S(X)=S(Y)=0, S*(X|Y)=S(XY); then one has S*(X|Y)>0=S(X) whenever XY takes on at least three of its four values with positive probability: however, side information should not increase uncertainty of any form and so this is an unacceptable inequality! Instead S(X|Y)=0=S(X), as it should be. While S(X) is an upper bound to S(X|Y) and not to S*(X|Y), non-negativity is a property both of S(X|Y) and of S*(X|Y). Luckily so, since S(XY)=S(YX)≥S(Y) is an adequateness property which cannot be set aside. A proof of it in the general case of fractional entropy is given in the appendix.

Remark 1. A formula for *fractional entropy* in the general case N≥1 is given in Sgarro (1988a), based on the expression of the rate-distortion function (cf also the Appendix below). Unfortunately, we were able to derive an *explicit* formula only in the (trivial) case N=1 and in the (non-trivial) case N=2 (entropy and semientropy). The implicit formula is however enough to prove that fractional entropy passes the most obvious checks: it is non-negative, weakly concave, zero iff P has at most N non-zero components, maximal iff P is uniform.

Remark 2. In our model the observer "wins the game" when he can put forward a subset of N objects containing the correct object. It might be argued that the observer should be given the chance to put forward subsets of *at most* N objects, rather that subsets of exactly N objects. Then, the reproduction alphabet in the definition of the rate-distortion function should be correspondingly extended. However, intuition wants that the new opportunities given to the observer are useless, so that his "fractional uncertainty" should remain exactly the same. That this is precisely the case will be formally proved in the appendix.

3 - FINAL COMMENT

Quite appropriately, semientropy was introduced in a specific context, and was not derived from the general theory of uncertainty measures. Constructing an all-purpose (abstract) theory such as to cover all the facets of uncertainty is a formidable task, indeed. To the risk of ad-hocness, the judicious user of uncertainty measures still has to tailor problem-oriented measures to suit his own needs. Historically, even Shannon entropy was introduced in the *specific* context of coding! Later all sorts of applications were made of it which led to great achievements as well as to blunt failures. The latter were too easily attributed to "weaknesses" in information theory: a rush towards generalization and diversification ensued which has brought about the present confusion. Using a prefabricated (abstract, possibly obscure) measure is, as the saying goes, "to put the cart before the horse" (the horse being replaced by oxen in the Italian and in the French version).

Acknowledgment. We gratefully acknowledge helpful technical comments made by János Körner.

Appendix. Below N is fixed, $1 \leq N \leq K$; I(X;U) is the mutual information of the two random variables X and U (cf Csiszár and Körner (1982)). For this N fractional entropy is defined as

$$S(X) = \min I(X;U)$$

the minimum being taken with respect to all stochastic matrices U given X s.t. $\Pr\{U=u|X=x\}>0$ implies $x \in u$ (recall that u is a subset of N primary letters). S(X) is a rate-distortion function computed at allowed distortion level 0 (cf above; for rate-distortion theory cf Csiszár and Körner (1982)). Extend now the reproduction alphabet (the range of U) by adding also (some or all) subsets with less than N letters, possibly also the empty set. Let $S'(X) = \min I(X;U')$ be the corresponding rate-distortion function.

Theorem 1. $S'(X) = S(X)$

Since a stochastic matrix U given X can always be seen as a stochastic matrix U' given X by adding zero columns, one has $S'(X) \leq S(X)$. Let f be any deterministic function from the alphabet of U' to the alphabet of U s.t. $u' \subseteq f(u')$, so that $x \in u'$ implies $x \in f(u')$. Assume XU' achieves S'(X). Then by the data-processing lemma (cf Csiszár and Körner (1982)): $S'(X) = I(X;U') \geq I(X;f(U')) \geq \min I(X;U) = S(X)$.

We shall need theorem 1 to prove theorem 2 below.

Theorem 2. $S(XY) = S(YX) \geq S(X)$

The equality is obvious. We prove $S(XY) \geq S(X)$. Assume XYV achieves S(XY). A possible value for V is a subset of N ordered couples; let g(V) be deterministically obtained from V be deleting the Y-components (the X-components are not necessarily distinct, but this is irrelevant in view of theorem 1). By the data-processing lemma and by a basic inequality for mutual information: $S(XY) = I(XY;V) \geq I(XY;g(V)) \geq I(X;g(V)) \geq \min I(X;U) = S(X)$.

REFERENCES

Klir G.J. (1987) Where do we stand on measures of uncertainty, ambiguity, fuzziness, and the like?, Fuzzy Sets and Systems, 24 (1987) 141-160

Klir G.J., Folger T.A. (1988) Fuzzy Sets, Uncertainty, and Information, Prentice-Hall Int.

Sgarro A. (1988a) A measure of semiequivocation, in Advances in Cryptology ed. by Ch. Günther, Lecture Notes in Computer Science 330 (1988) Springer Verlag, 375-387

Sgarro A. (1988b) Information measures from rate-distortion theories, in Uncertainty and Intelligent Systems, ed. by B. Bouchon, L. Saitta and R.R. Yager, Lecture Notes in Computer Science 313 (1988) Springer Verlag, 112-118

Csiszár I., Körner J. (1982) Information Theory, Academic Press

MODELS FOR REASONING WITH MULTITYPE UNCERTAINTY IN EXPERT SYSTEMS

J.C.A. van der Lubbe, E. Backer
Information Theory Group, Dept. of Electrical Engineering,
Delft University of Technology, P.O. Box 5031,
2600 GA Delft, the Netherlands

W. Krijgsman
Lab. for Clinical and Experimental Image Processing, Thoraxcenter,
Erasmus University Rotterdam, Rotterdam, the Netherlands

Abstract

Uncertainty models play an important role within expert systems. However, there are different types of uncertainty (inaccuracy, inexactitude, fuzziness etc.). It can be shown that the various uncertainty models as known from literature in fact are dealing with different types of uncertainty. The type of uncertainty, which is characteristic for the application for which the expert system is used, has a direct impact on the selection of the appropriate uncertainty model within a given application domain.

Problems appear when within an application domain various types of uncertainty should be handled at the same time (multitype uncertainty). In this case a special inference calculus for e.g. the combination of evidences, related to various types of uncertainty, is needed. In this paper two general methods for an inference calculus for multitype uncertainty will be proposed and evaluated.

Keywords

Expert systems, multitype uncertainty, uncertainty classes and types, uncertainty calculi, inference calculi, certainty vector, rule inference, classification inference, rule generation

1. UNCERTAINTY MODELS

A characteristic aspect of human reasoning is that in many cases one has to deal with uncertain or imprecise information. Since expert systems are thought to imitate or represent human subjective reasoning as close as possible, at least with respect to the input-output performance, the study of uncertainty models is inevitable.

With respect to an uncertainty model distinction can be made between on the one hand the uncertainty calculus which deals with the mathematical representation of uncertainty and on the other hand the inference calculus dealing with the combination and propagation of uncertainties through the inference process.

Considering uncertainty it may be concluded that the concept of uncertainty itself is ambiguous. Moreover, in a proposition uncertainties play a role on different levels. Consider for example the proposition "This apple is round". This proposition is troubled with various classes of uncertainty. First the concepts "apple" and "round" are not well-defined; there is what can be called a conceptual uncertainty in their use. Furthermore, even if they are well-defined it is the question how far apples can be tested on the property of roundness: the so-called relational uncertainty. Finally, under the condition that apples can be tested on roundness it should be discussed how far "this apple" is round; the corresponding uncertainty can be denoted as propositional uncertainty.

There is a hierarchical relation between these uncertainties. Whereas the relational uncertainty is influenced by the conceptual uncertainty, the propositional uncertainty is influenced by both conceptual and relational uncertainty.

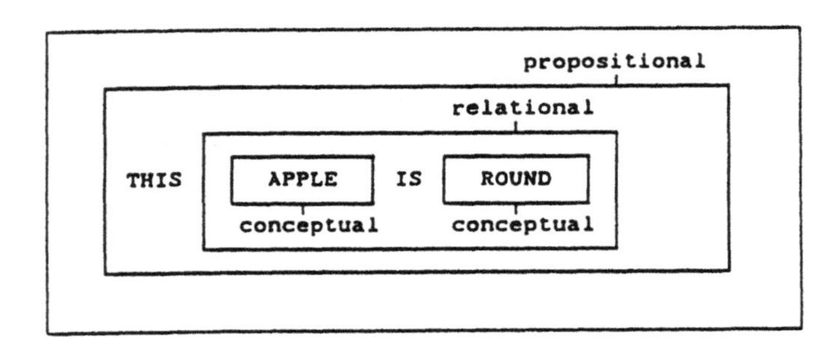

Figure 1. Uncertainty classes.

Here, it should be remarked that the propositional uncertainty is the uncertainty which is usually considered in expert systems. However, also the meaning of propositional uncertainty is far from clear. It can express to what degree a proposition is assumed to be true. It can also express the precision with which its value can be determined on a given scale. In general uncertainty can be related to fuzziness, incredibility, statistical uncertainty, inexactitude, unreliability, indistinctness etc. A proposition is fuzzy if its value cannot be determined with sufficient precision on a given scale. Further, the term incredibility is used to express to what degree a proposition is assumed to be false etc.

On the basis of the various types of uncertainty the various uncertainty models as known in literature (e.g. Shortliffe-Buchanan model (Shortliffe, Buchanan (1975), Buchanan, Shortliffe (1984)), Dempster-Shafer theory (Shafer (1975)), fuzzy theory etc.) can be ordered. For a survey of the wide variety of approaches to the modelling of uncertainty one is referred to Buxton (1989) among others.

Knowledge about the types of uncertainty for a given application or context prescribes which uncertainty models are most appropriate for that given application or context. It is remarked here that control over uncertainty models is part of the knowledge involved from a meta-view. The characteristics of uncertainty in a given application together with knowledge about the uncertainty model assumptions (domain, range, precision, independence assumptions, availability of relevant parameters etc.) can help the system to select the optimal uncertainty model, whereby optimal means that the output corresponds to at least that of a human expert (see Figure 2).

In most applications it is assumed that one has to do with precisely one type of uncertainty. According to this one out of the available uncertainty models is selected. However, in most applications, it is very unlikely that one has to deal with just one type of uncertainty. Often in a given context more than one type of uncertainty is relevant at the same time (multitype uncertainty) (Backer, Van der Lubbe, Krijgsman (1988)).

Compare the medical expert system ESATS in development by Thoraxcenter, Rotterdam and Delft University of Technology (Backer et al (1988)), which is an Expert System for the Analysis of Thallium-201 Scintigrams for the assessment of abnormalities in the heart muscle, from which the presence of coronary artery disease is estimated. The knowledge base consists among others of patient data from certain populations characterized by an average age, typical risk factors, sex, etc. This in fact is a group characterization, and is often given in terms of statistical quantities with respect to a given population. Then, for a particular patient study, the patient is identified in terms of belonging to a certain population. Thus, certain statistical properties are expected. However, as each patient is a unique creature, patient specific features like age, weight, degree of fattening, predisposition, heredity, etc. are also important. These features refer to different aspects of uncertainty, which are not per se statistical. The same holds with respect to the information derived from the scintigrams based on visual image interpretation and quantitative analysis. Therefore, in order to make a diagnosis concerning a patient multitype uncertainty has to be included in the reasoning process.

This implies that in situations as outlined above models are requested for inference procedures in those cases whereby more than one type of uncertainty plays a role. In the following such calculi are proposed.

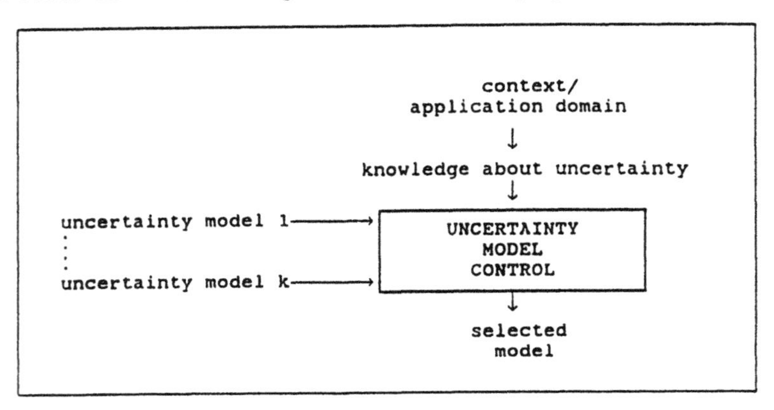

Figure 2. Uncertainty model control.

2. REASONING WITH MULTITYPE UNCERTAINTY

Assume there are k types of uncertainty which are relevant within a given application. The assessment of the degree of uncertainty is usually performed in terms of the so-called certainty factors. Let, according to the k types of uncertainty, the corresponding measurements of the certainty factors be denoted by f1,...,fk. The (un)certainty of e.g. a proposition (or rule) can now be represented by the k-dimensional certainty vector $\underline{f}$ = (f1,...,fk). The main problem is the development of an inference calculus on the basis of certainty vector $\underline{f}$ (see Figure 3). This means a calculus with respect to e.g. the calculation of the certainty via a conclusion on the basis of the certainties of the evidences and the rule, methods for the combination of co-concluding rules, methods in the case of disjunctions or conjunctions of evidences etc.

Two general methods are proposed here; the first one can be considered as an extension of known inference calculi, the other one is closely related to classification methods, as used in the field of pattern recognition.

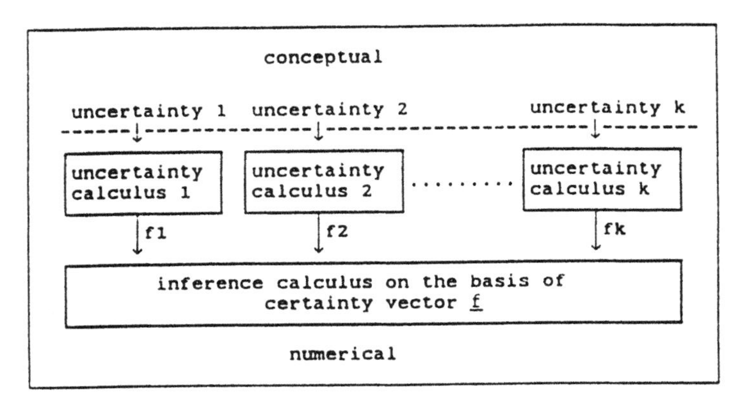

Figure 3. Uncertainty model using multitype uncertainty.

2.1. RULE INFERENCE

In line with the usual uncertainty and inference calculi the following methods in the case of vectors can be considered. See also Figures 4,5,6 and 7 in the case of modus ponens and 2-dimensional certainty vectors.

(i) Let for i = 1,...,k the degree of certainty of an evidence e, according to uncertainty calculus i, be denoted by fi(e).
Analogously fi(e⇒H) and fi(H) are the certainty of rule (e⇒H) and conclusion H, respectively, according to uncertainty calculus i. For each type of uncertainty the certainty of the conclusion can be calculated on the basis of the corresponding certainties of the evidence and the rule and just according to the procedure related to that type of uncertainty. This means that fi(H) is determined on the basis of fi(e) and fi(e⇒H) for each i = 1,...,k.

$$\left.\begin{array}{l} f1(e),\ f1(e\Rightarrow H) \Rightarrow f1(H) \\ f2(e),\ f2(e\Rightarrow H) \Rightarrow f2(H) \\ \\ fk(e),\ fk(e\Rightarrow H) \Rightarrow fk(H) \end{array}\right\} \Rightarrow \underline{f}(H)$$

$$\begin{array}{ll} \uparrow & \uparrow \\ \underline{f}(e) & \underline{f}(e\Rightarrow H) \end{array}$$

Also in the case of conjunctions and disjunctions of evidences or in the case of combination of rules, combination and propagation of certainties is performed with the help of the calculi corresponding to the various types of uncertainty. This method guarantees that the various types of uncertainty play their role during the entire inference process. Thus at the end of the reasoning process the certainty of the final conclusions is still represented as a certainty vector.

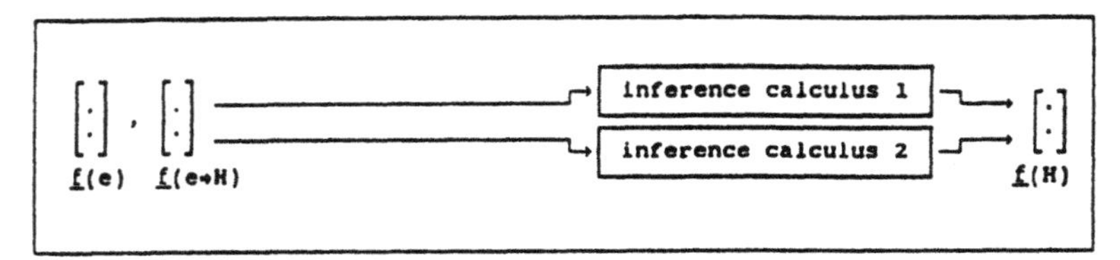

Figure 4. Rule inference: for each type of uncertainty a specific inference calculus.

(ii) In the foregoing method each type of uncertainty is handled by means of its corresponding calculus during the reasoning and inference process. However, the fundamental question is whether each type of uncertainty needs its own inference calculus in order to deal with the combining and propagation of uncertainties? That means that the question is whether the formal procedure of inference is rather independent of the type of uncertainty.

Within this framework there is no clear reason according to which combination of co-concluding rules, of which the certainty is expressed in terms of credibility, should be in another way than in the case that certainty is related to statistical certainty. The same holds with respect to the calculation of the certainty of a conclusion on the basis of the certainties of the evidence and the corresponding rule or for conjunctions and disjunctions of evidences.

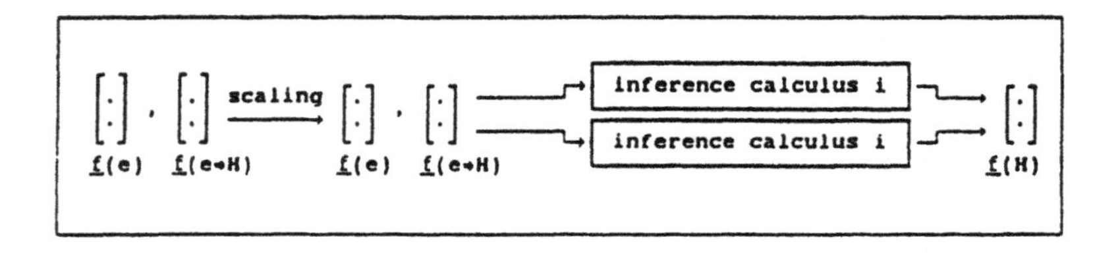

Figure 5. Rule inference: the same inference calculus for each type of uncertainty.

However, using one inference calculus for the various types of uncertainty scaling can be necessary, since the different types of uncertainty do not imply a priori that their measurements have the same range. By scaling all elements of a vector become values on the same interval. In the following it is assumed that the scaling is such that all elements of the certainty vector have values on the interval [-1,1], whereby -1 and 1 correspond to complete negation and confirmation of the proposition, respectively, in the sense of the type of uncertainty. The reason for scaling to [-1,1] instead of e.g. to [0,1] is that the first one implies that the value 0 corresponds to the situation where knowledge concerning the proposition is missing; since most mathematical operations are neutral with respect to zero values, this in fact implies that one is neutral with respect to absence of knowledge.

The selection of the inference calculus depends mainly on the application domain and the underlying reasoning model as implicitly or explicitly used by the human expert. In the following sections we will return to this question. If it is assumed that the formal structure of the Shortliffe-Buchanan model is applicable within the given application domain, this means it is assumed that the expert is reasoning "Shortliffe-Buchananian", then e.g. the modus ponens becomes:

$$\underline{f}(H) = \underline{f}(e) \otimes \underline{f}(e \Rightarrow H),$$

whereby the operation $\otimes$ is the "vector point-product" by means of which corresponding elements of the vectors are multiplied. In fact this correponds to a matrix multiplication of diagonal matrices.

With respect to disjunctions and conjunctions of evidences min- and max-operators can be defined which can be considered as generalizations of the 1-dimensional case. This can be done by taking the minimum or maximum for each certainty (this means for each dimension) or by defining a min- or max-operator which orders vectors in such a way that a vector can be labeled as minimum- or maximum-vector, respectively. With respect to the latter one the following method can be used.

As a matter of fact the reasoning process tries to achieve a situation whereby each element of the certainty vector of the conclusion has value 1. This vector, which can be considered as a unit vector, can be denoted by $\underline{u}$. The maximum-vector can now be defined as that vector which supports $\underline{u}$ the best. This can be measured by considering the projections of the vectors upon $\underline{u}$. Let $(\underline{f}(e1), \underline{f}(e2), \ldots)$ the set of certainty vectors with respect to evidences e1, e2,... and let for all h, $\underline{f}^{\wedge}(eh)$ be the projection of $\underline{f}(eh)$ upon $\underline{u}$. The set $\Psi 1$ is the set of all certainty vectors of which the projections upon $\underline{u}$ have the same direction as $\underline{u}$. $\Psi 2$ is the set consisting of the remaining certainty vectors. Now the algorithm is as follows.

a) If $\Psi 1 \neq \emptyset$ then $\underline{f}(ek)$ is maximum-vector if for all h for which $\underline{f}(eh) \in \Psi 1$:
$|\underline{f}^{\wedge}(eh)| \leq |\underline{f}^{\wedge}(ek)|$.

b) If $\Psi 1 = \emptyset$ then $\underline{f}(ek)$ is maximum-vector if for all h:
$|\underline{f}^{\wedge}(eh)| \geq |\underline{f}^{\wedge}(ek)|$.

This algorithm guarantees that that certainty is a maximum-vector which supports $\underline{u}$ the most (case a) or opposes $\underline{u}$ the least (case b), depending on the positions of the certainty vectors in the k-dimensional space.

On the basis of the operations defined above it is possible to consider methods for the combination of co-concluding rules in a similar way.

(iii) In the foregoing inference on the basis of the separate elements was studied. The question arises whether it is possible to develop an inference calculus which gives a direct mapping from an input vector to an output vector. This means a calculus that e.g. in the case of co-concluding rules the resulting certainty vector computes directly.

Assume that one is interested in the combination of the rules $\underline{f}(e \Rightarrow H)$ and $\underline{f}(e2 \Rightarrow H)$. It is possible to define the resulting vector as the resultant of vector addition of $\underline{f}(e1 \Rightarrow H)$ and $\underline{f}(e2 \Rightarrow H)$ in the vector space. In order to guarantee that all elements of the resulting vector are on interval [-1,1] scaling is necessary. However, which scaling is used, the result is always such that if one of the elements of one of the input vectors is equal to 1, the corresponding element in the resulting vector is unequal to 1, unless all elements of the input vectors corresponding to the element 1 are equal to 0. This is due to the properties of vector additions. However, it is a fundamental question to what extent this property is admissible. In general it is stated in the field of expert systems that if during the reasoning process complete certainty (=1) is achieved this cannot be decreased by other certainty values obtained during the same reasoning process. Each direct operation in the vector space does not have these properties, unless one performs operations on the separate dimensions.

Furthermore, there are also other problems concerning the direct mapping within the vector space. With respect to the modus ponens it is wished to define some vector product for the computation of the certainty vector of a conclusion on the basis of the certainty vectors of evidences and corresponding rule. At this moment there seem no clear alternatives for the vector point-product mentioned above. Therefore the possibility of vectorial mapping is still an open problem.

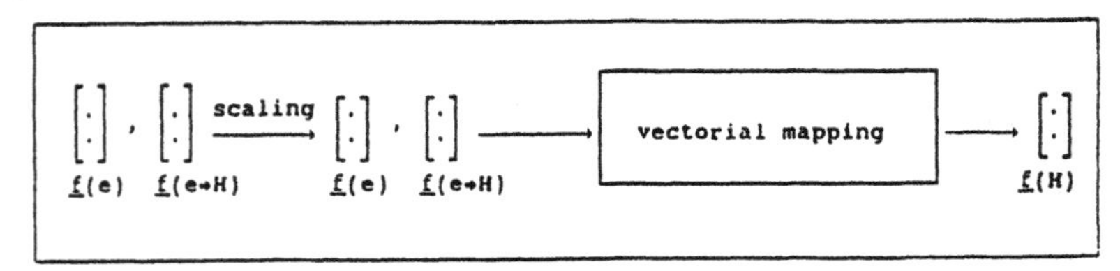

Figure 6. Rule inference on the basis of vectorial mapping.

(iv) Instead of using certainty vectors it is also possible to perform inference on the basis of the so-called overall certainty F , which is some weighted function G of f1,...,fk related to evidence, rule or conclusion. This means $F = G(\underline{f})$. An appropriate candidate for the function $G(\underline{f})$ is the so-called modified R-norm, which is defined by:

$$F = G(\underline{f}) = \left[\frac{1}{k}\sum_{i}(fi+1)^{R}\right]^{1/R} - 1 \qquad (R>0),$$

whereby only these certainty factors are taken into account for which $fi \neq 0$. It can easily be shown that the following inequality holds for the modified R-norm: Min (fi) $\leq F \leq$ Max (fi), and that in the case that for all i: fi = f that then $F = f$. The parameter R gives the possibility to attach stronger weights to some certainty values. For R = 1 the R-norm yields the arithmetic mean, whereby all fi's contribute equally. It can be shown that for $R \Rightarrow \infty$ the modified R-norm F tends to Max (fi). Together with the fact that F is a non-decreasing function of R, this implies that for increasing R the influence of the largest uncertainty value in $\underline{f}$ dominates.

If instead of the k-dimensional uncertainty vector $\underline{f}(.)$ the 1-dimensional certainty number $F(.)$ is used, then the model corresponds to an uncertainty model whereby instead of one type of uncertainty some "averaged" uncertainty is used. The overall certainties $F(e)$ and $F(e \Rightarrow H)$ can now be computed on the basis of $\underline{f}(e)$ and $\underline{f}(e \Rightarrow H)$, respectively, and $F(H)$ directly on the basis of $F(e)$ and $F(e \Rightarrow H)$:

$$\begin{array}{ll} F(e \Rightarrow H), & F(e) \Rightarrow F(H). \\ \uparrow & \uparrow \\ \underline{f}(e \Rightarrow H), & \underline{f}(e) \end{array}$$

The advantage of dealing with overall certainty is reduction of dimensionality, which is attractive from a computational point of view. However, it does not right to the idea that there are different types of uncertainty. Moreover, in fact it corresponds to the inference calculus we usually apply. The only difference is that it has been stated explicitly that an (un)certainty factor is the resultant of various types of (un)certainty.

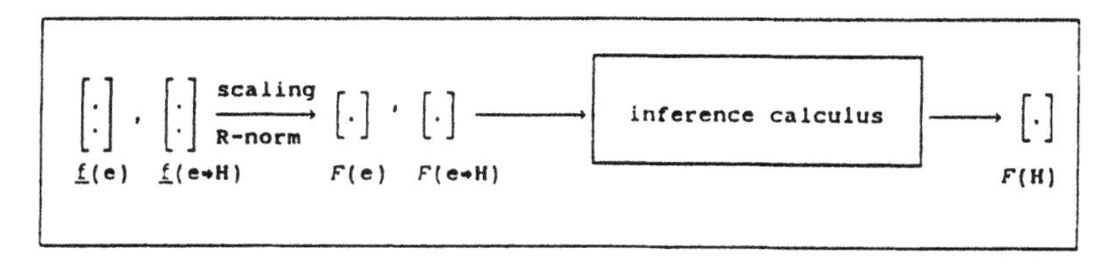

Figure 7. Rule inference on the basis of overall certainties.

Evaluating the four proposed methods it may be concluded that the possibility of inference calculi on the basis of a vectorial mapping (iii) is still an open problem. Method (iv) is less attractive since in fact it does not take into account the presence of various types of uncertainty. Especially the inference calculi (i) and (ii) seem useful. The advantage of method (ii) is that it suffices to use just one inference calculus for the various types of uncertainty.

2.2. CLASSIFICATION INFERENCE

In the method mentioned above it is assumed that the expert knowledge is given in terms of general rules along with some certainty measurement. However, depending on the application domain it is sometimes impossible to summarize and generalize all expert knowledge in general rules. Then the induction step is not performed. In that case the expert knowledge is mostly available in the form of pieces of knowledge (samples), e.g. previously analyzed patient cases in a medical diagnostic expert system like ESATS (see section 1). On the basis of these data classification inference methods can be used.

Let ${}^{r}fi(.)$ be the certainty factor, according to uncertainty calculus i, as related to case r, r = 1,...,n. Defining ${}^{r}\underline{f}(.) = \{{}^{r}f1(.), \ldots, {}^{r}fk(.)\}$ in the case of k types of uncertainty, the certainty vectors of evidence and conclusion can now be denoted by ${}^{r}\underline{f}(e)$ and ${}^{r}\underline{f}(H)$, respectively. Assume one is interested in $\underline{f}(H)$ for a given new case of which the certainty concerning the evidence is given by certainty vector $\underline{f}(e)$. The problem can now be formulated as follows:

Given: ${}^{1}\underline{f}(e), \rightarrow {}^{1}\underline{f}(H)$

$\cdot \qquad \cdot$

${}^{r}\underline{f}(e), \rightarrow {}^{r}\underline{f}(H)$

$\cdot \qquad \cdot$

${}^{n}\underline{f}(e), \rightarrow {}^{n}\underline{f}(H)$

Determine: $\underline{f}(H)$, on the basis $\underline{f}(e)$.

In fact, inference is now reduced to a classification problem. In the case that ${}^{r}\underline{f}(e)$ and ${}^{r}\underline{f}(H)$ for each r are known, the k elements of the certainty vector of the evidence can be considered as the coordinates of a point, say ${}^{r}x$, in a k-dimensional space Ψ. The certainty vector of the conclusion ${}^{r}\underline{f}(H)$ can be considered as a continuous or discrete "label" assigned to point ${}^{r}x$. In the same way a given sample with specific values of $\underline{f}(e)$ for which $\underline{f}(H)$ should be determined, can also be represented as a point, say y, in the k-dimensional space Ψ. The classification problem now is the determination of label $\underline{f}(H)$ of y on the basis of the comparison of y with the labeled points ^{r}x, which form the so-called training set. On the basis of the comparison of this point y with the points ^{r}x of the training set, $\underline{f}(H)$ can be determined. For this purpose methods as used in statistical pattern recognition can be applied e.g. the m-nearest neighbour classifier.

The m-nearest neighbour classifier searches for the m nearest neighbours of the new sample to be classified and determines the label e.g. as a weighted average of the labels of the m nearest neighbours. If m = 1 the the classification label is uniquely determined by the label of the nearest sample.

It is mentioned here that there is a clear difference with the way in which pattern recognition is usually applied. Whereas in most applications the classification space is in fact a feature space, wherein the points represent intrinsic features of e.g. objects to be classified, within the present framework the classification space is related rather to the certainty and interpretation of the values of the intrinsic features than to the features itselves. Nevertheless, both can go hand in hand. Compare the sentence "the temperature of patient is 38^{0}" with the sentence "the temperature of the patient is 40^{0}". Here the increase in the physical feature "temperature" is coupled with an increase in the certainty that the patient has fever.

Above, attention was paid to inference in situations whereby the knowledge is available in terms of (patient) cases. Distinction should be made between two types of information: information on samples related to cases, as mentioned above, and information on samples related to judgements of expert panels.

In the situation of case studies one expert is analyzing a specific phenemenon in several cases, e.g. a clinician analyzing Thallium-201 scintigrams of several patients to assess the malfunctioning of the heart muscle which allows him for each patient to infer the presence of coronary artery disease. In the situation of an expert panel each member of the panel analyzes a phenomenon with respect to one specific case, e.g. a group of clinicians, individualy analyzing one Thallium scintigraphic study, resulting in various estimates for the presence of coronary artery disease for some patient.

More general the differences are summerized in the Figure 8. It is remarked here that only for cases real inference is possible. If inference in the case of expert panels is performed it is nothing more than the assessment of the certainty vector with respect to the conclusion on the basis of the individual judgements of the experts. However, in the case of panels the aim is rather acquisition of more general expert knowledge than judgement of one individual case.

CASES	PANELS
- one expert - knowledge about various cases - main aim: classification of new case	- various experts - knowledge about one case - main aim: acquisition of expert knowledge

Figure 8: Difference between knowledge acquisition on the basis of cases and panels.

However, in practice one has often to do with a combination of both types of information, i.e. samples are related to both cases and judgements of expert panels. Compare medical applications, where patient cases are often judged by several medical experts. In this case the problem of classification inference becomes the following. Let ${}^{rs}\underline{f}(.)$ be the certainty vector related to case r as assessed by expert s, then the problem is:

Given for all s: ${}^{1s}\underline{f}(e), {}^{s}\underline{f}(e\Rightarrow H) \rightarrow {}^{1s}\underline{f}(H)$

$${}^{rs}\underline{f}(e), {}^{s}\underline{f}(e\Rightarrow H) \rightarrow {}^{rs}\underline{f}(H)$$

$${}^{ns}\underline{f}(e), {}^{s}\underline{f}(e\Rightarrow H) \rightarrow {}^{ns}\underline{f}(H)$$

Determine: $\underline{f}(H)$ on the basis of $\underline{f}(e)$,

whereby $\underline{f}(e)$ can be the resultant of individual judgements ${}^{s}\underline{f}(e)$ concerning the certainty of the evidence of the new case by the expert panel. In fact now we have a classification problem in a 2k-dimensional space, which can be solved in a similar way as mentioned above, e.g. by using the m-nearest neighbour classifier.

It is remarked here that the certainty vectors of the rule are independent of the cases. For that reason they have no index r.

The question arises whether it is relevant to take into account expert judgements concerning the rule (e⇒H) in the classification process. The judgements about the cases can vary from expert to expert. However, it can be that the experts are consistent in that sense that they do not differ in opinion about the rule. That means ${}^{s}\underline{f}(e\Rightarrow H)$ is constant for all s. Clearly, in that case the certainty vectors concerning the rule play no role at all in the classification process.

However, in general this will not be the case and there will be a relation between on the one hand the judgements about evidence and conclusion by an expert and on the other hand his judgement about the corresponding rule; In practice pessimism of an expert about the validity of a rule will often then coincide with pessimism about the validity of some conclusion. In this typical situation it is important to include information about expert judgements concerning rules in the classification process.

Methods as mentioned above can also be followed if the set of possible labels ${}^{rs}\underline{f}(H)$ is finite or if one deals with overall certainty numbers F instead of certainty vectors.

In situations where the knowledge about the uncertainties is not complete one can use well-known methods from pattern recognition for missing data in order to obtain estimations of the unknown certainties.

3. COMPARISON OF RULE INFERENCE AND CLASSIFICATION INFERENCE

In this section attention will be paid to the relation between classification and rule inference. Moreover, it is discussed how far information about samples (${}^{rs}\underline{f}(e)$, ${}^{s}\underline{f}(e \Rightarrow H)$, ${}^{rs}\underline{f}(H)$) as used in classification inference can help for rule generation and selection of the appropriate reasoning model for the inference calculus.

Rule inference and classification inference differ in the type of knowledge they use as well as with respect to learning aspects. Clearly, in rule inference explicit knowledge is used, whereas this knowledge is implicit for classification inference. The latter means that the relation between evidences and conclusion is given in terms of samples/ cases without the possibility to make it explicit in terms of rules.

With respect to learning the learning aspect is implicit in rule inference. In fact here we only deal with the result of learning, i.e. the rules. Clearly, learning is inherent to classification inference, where information about new cases can easily be added to the training set. One thing and another are summarized in Figure 9.

	Knowledge	Learning
Rule inference	explicit	implicit
Classification inference	implicit	explicit

Figure 9. Rule inference versus classification inference.

The problem of rule generation is the generation of general rules with their corresponding certainty vectors, as used in rule inference, on the basis of samples related to cases and panel judgements, as used in classification inference. Various methods can be applied. It is assumed thet the general form of a rule is determined by the evidences and conclusions which seem to be relevant considering the samples.

The first way for determination of the rule certainty vector is direct estimation of the rule certainty vector by means of a panel. On the basis of individual assessments ${}^{s}\underline{f}(e\Rightarrow H)$ the final rule certainty vector can be obtained by using e.g. the modified R-norm given above and by computing some average $\underline{f}(e\Rightarrow H)$. This method is direct since information about the cases itselves are not used at all.

The other possibility for the determination of the rule certainty vector is indirect. On the basis of all ${}^{rs}\underline{f}(e)$ and ${}^{rs}\underline{f}(H)$ some estimation of ${}^{s}\underline{f}(e\Rightarrow H)$ can be obtained, on the basis of which the final rule certainty vector can be obtained in the way as described above. This method is appropriate if it is difficult for the expert to express his knowledge in terms of general rules. However, the present method can only be applied if one has some idea about the underlying reasoning model of the expert concerning the relation between the certainty of evidences, conclusions and rule. E.g. if all experts are reasoning according to the Shortliffe-Buchanan model, i.e. the certainty of a conclusion is the product of the certainties of evidences and rule, and if it is assumed that all types of uncertainty can be treated by the same model, then for all s estimations of ${}^{s}\underline{f}(e\Rightarrow H)$ are obtained by computing ${}^{rs}\underline{f}(e\Rightarrow H)$ from

$${}^{rs}\underline{f}(e) \otimes {}^{rs}\underline{f}(e\Rightarrow H) = {}^{rs}\underline{f}(H),$$

and by some averaging over all cases.

Remark that if there are some cases for which the certainty of the evidences tends to 1, the certainty vector of the rule is identical to that of the conclusion. This is due to the definition of the certainty of a rule, which is defined as the certainty of the conclusion if the evidence is absolutely certain; this is rather independent of the underlying reasoning model.

If one has now idea about the underlying reasoning model of the experts, then also the reasoning model should be judged.

Let $M({}^{rs}\underline{f}(e),{}^{s}\underline{f}(e\Rightarrow H))$ some parametrical function which maps $({}^{rs}\underline{f}(e),{}^{s}\underline{f}(e\Rightarrow H))$ in such a way that the outcome may be assumed to be an estimation of ${}^{rs}\underline{f}(H)$, then the optimal reasoning model for the modus ponens is that function $M(.,.)$ for which

$$\sum_r\sum_s\left[M({}^{rs}\underline{f}(e),{}^{s}\underline{f}(e\Rightarrow H)) - {}^{rs}\underline{f}(H))\right]^2$$

is minimal. The rule certainty vector is obtained by computing some average $\underline{f}(e\Rightarrow H)$ on the basis of individual assessments ${}^{s}\underline{f}(e\Rightarrow H)$ by the experts.

In fact, this is the most interesting option, since this leads directly to the uncertainty model control as mentioned in section 1. Reasoning model and rule generation are now directly determined by the specific characteristics of the application domain.

Clearly, on the basis of the generated rule and the reasoning model rule inference can be performed.

From an implementation point of view a combination of rule inference and classification inference can be preferable. In general the computational effort in the case of rule inference is lower than in the case of classification inference, whereby we are dealing with a large amount of samples. However, classification inference has the advantage of learning capability by adding new cases to the feature domain. In operational environments initially cases can be used for rule generation. On the basis of the generated rules rule inference can be performed for new cases. The rules are updated only periodically on the basis of new cases.

4. CONCLUSIONS

In this paper it was stated that in practice one has often to do with more than one type of uncertainty. Corresponding to this, two inference calculi were introduced in order to deal with what is called in this paper multitype uncertainty. These two methods are rule inference and classification inference. The first one is related to the cases where the expert knowledge can be made explicitly in terms of general rules. The latter one is appropriate in situations where the expert knowledge is implicit and only available in terms of singular knowledge facts.
Methods were given by means of wich on the basis of samples rules with their corresponding certainty vectors can be generated as well as methods for the generation of reasoning models, which both can be used for rule inference. At this moment implementation of the various uncertainty vector models within ESATS (Expert System for the Analysis of Thallium-201 Scintigrams) is performed in order to assess the performance of both rule and classification inferences in practice (Backer et al (1988)).

REFERENCES

Backer E., Van der Lubbe J.C.A., Krijgsman W. (1988), On modelling of uncertainty and inexactness in expert systems, Proc. Ninth Symp. on Information Theory, Mierlo, the Netherlands, pp. 105-111

Backer E., Gerbrands J.J., Bloom G., Reiber J.H.C., Reijs A.E.M., Van den Herik H.J. (1988), Developments towards an expert system for the quantitive analysis of thallium-201 scintigrams, In: De Graaf, C.N., Viergever, M.A., Eds., Information Processing in Medical Imaging, New York, pp. 293-306

Backer E., Gerbrands J.J., Reiber J.H.C., Reijs A.E.M., Krijgsman W., Van Den Herik H.J. (1988), Modelling uncertainty in ESATS by classification inference, Pattern Recognition Letters 8, pp. 103-112

Bonissone P.P., Tong R.M. (1985), Reasoning with uncertainty in expert systems, Int. J. Man-Machine Studies, 22, pp. 241-250

Buchanan B.G., Shortliffe E.H. (1984), Rule-based expert systems, Massachusetts, 1984

Buxton R. (1989), Modelling uncertainty in expert systems, Int. J. Man-Machine Studies, 31, pp. 415-476

Dubois D., Prade H. (1987), A tentative comparison of numerical approximate reasoning methodologies, Int. J. Man-Machine Studies, 27, pp. 717-728

Ho T.B., Diday E., Gettler-Summa M. (1988), Generating rules for expert systems from observations, Pattern Recognition Letters, 7, pp. 265-271

Prade H. (1985), A computational approach to approximate and plausible reasoning with applications to expert systems, IEEE Pattern Anal. Mach. Intell., Vol PAMI-7, 3, pp. 284-298

Shafer G. (1975), A mathematical theory of evidence, Princeton Univ. Press

Shortliffe E.H., Buchanan B.G. (1975), A model of inexact reasoning in medicine, Math. Biosciences, Vol. 23, 1975, pp. 351-379

A HYBRID BELIEF SYSTEM FOR DOUBTFUL AGENTS

ALESSANDRO SAFFIOTTI*

IRIDIA - Université Libre de Bruxelles
Av. F. Roosvelt, 50 - CP 194/6
1050 Bruxelles - Belgium
E-mail: r01507@bbrbfu01.bitnet

Abstract This paper aims at bridging together the fields of Uncertain Reasoning and Knowledge Representation. The bridge we propose consists in the definition of a Hybrid Belief System, a general modular system capable of performing uncertain reasoning on structured knowledge. This system comprises two distinct modules, *UR-mod* and *KR-mod*: the *UR-mod* provides the uncertainty calculus used to represent uncertainty about our knowledge; this knowledge itself is in turn represented by the linguistic structures made available by the *KR-mod*. An architecture is drawn for this system grounded on a formal framework, and examples are given using Dempster-Shafer theory or probabilities as *UR-mod*, and first order logic or KRYPTON as *KR-mod*. An ATMS-based algorithm for a Hybrid Belief System is hinted at.

Keywords: Uncertain Reasoning; Knowledge Representation; Dempster-Shafer theory; Probability; ATMS.

1. INTRODUCTION

One can hardly miss noticing that fields which are apparently highly related one another, often follow fairly different courses, and that they suffer, on their way on, from a mutual lack of communication. This seems to be the case, in AI, with the two fields of Knowledge Representation (KR) and Uncertain Reasoning (UR). These fields share the general goal of finding ways to represent (possibly uncertain) knowledge, and to use it in the reasoning process. Yet, the literature in each field seems to have scarcely taken into account the problems and the results emerged in the other one.

Let us consider, for instance, the way in which knowledge is represented by most of uncertainty management techniques: roughly speaking, in a typical case *facts* of the object domain are represented by elements of an algebra of statements, and *relations* among them are expressed by means of mathematical functions (e.g. conditional probabilities, belief functions) defined over this algebra. These "representation tools" may sound familiar to somebody working, say, in statistics, but are not even part of the standard vocabulary in the KR field. Hence, the techniques developed in there are often of scarce practical use in KR systems—and the like—because they are defined in languages which we can hardly map into those idioms (production rules, class inheritance, etc.) that characterize KR. To illustrate the point, we could do worst than comparing the use of Mycin-like Certainty Factors (CF) in KR with that of Dempster-Shafer (hereafter, DS) theory. CFs, scant as they are from the viewpoint of formal soundness, epistemic adequacy and cognitive plausibility, are nonethe-

* This research has been partially supported by the ARCHON project, which is funded by grants from the Commission of the European Communities under the ESPRIT-II Program, P-2256; the partners in this project are: Krupp Atlas Elektronik GmbH, Amber, CERN, CNRG-NTUA Athens, Electricity Council Research Centre, Framentec, FWI Univ. of Amsterdam, Iberduero, Iridia Univ. Libre de Bruxelles, JRC Ispra, Labein, Volmac, Univ. of Porto, Queen Mary and Westfield College.

less defined directly in the context of production rules. This has made them the favourite choice in a number of existing KR tools (often disregarding their applicability conditions). On the other hand, through DS theory is regarded by many as more solidly grounded and more epistemically adequate to AI problems than CFs, it is absent from the universe of these tools. As a matter of fact, DS theory is defined in terms (a set of exhaustive and mutually exclusive hypotheses; functions from subsets of this sets to real numbers; a combination operator between these functions) which are not easily translated into the language of, say, production rules.

But if UR cries, KR does not laugh. Most of the concepts developed in the last decade of KR research do not find a place in the formal machineries developed in the UR field. Consider the statements "Birds are animals", "Typically birds fly", "Most of my friends like music" and "Smoke suggests fire": a person working in KR would probably take them as expressing qualitatively different types of knowledge, and advocate different mechanisms for representing them. Yet, a Bayesian would probably code all of them by the single pattern $P(A|B) = x$.

This paper aims at bridging together UR and KR fields. The bridge we propose consists in the definition of a *Hybrid Belief System* (HBS), a general modular system capable of performing uncertain reasoning on structured knowledge. Two modules, *UR-mod* and *KR-mod*, define system's behaviour with respect to knowledge and uncertainty: the *UR-mod* provides the uncertainty calculus used to represent belief about our knowledge; this knowledge itself is in turn represented by the linguistic structures made available by the *KR-mod*. Two essential requirements will be considered in defining the HBS:

1. it must account for both the distinction and the relationship between "knowledge" and "uncertainty", as the two basic components of uncertain knowledge; and
2. it must be general enough to accommodate a number of KR techniques and of UR techniques: it must not make too strong hypotheses on the form in which knowledge and uncertainty are represented.

Notice that the definition of such a system rests on the assumption that uncertain knowledge may be viewed (and dealt with) as just (categorical) knowledge accompanied by uncertainty about it; we will come back to this point in the Conclusions.

The rest of this paper is organized as follows. Section 2 sketches a first architecture for a HBS, and illustrates its behaviour through a simple scenario. Section 3 defines a formal framework for the HBS. Section 4 gives the full architectural definition of a HBS based on this framework, and hints at a possible ATMS-based algorithm for it. Section 5 concludes.

2. THE ARCHITECTURAL VIEWPOINT

From the viewpoint of the general architecture of an AI agent, we regard a HBS as a subsystem of the whole agent which is responsible for the set of beliefs of the agent[1]. The HBS is supposed to interact with the rest of the AI Agent in a query/answer fashion. In particular, following the suggestion of Levesque (1984), we consider two interaction primitives, **Ask** and **Tell**, which represent the only communication channels between the HBS and the rest of the agent. The intended behaviour of these primitives is as follows:

Tell[α, x] updates the uncertain knowledge contained in the HBS, by affirming formula α with certainty x;

Ask[α] questions HBS about the certainty of formula α, i.e. the certainty degree by which α can be derived from the knowledge present in the HBS.

[1] In this sense, Hybrid Belief Systems are reminiscent of Konolige's "Belief Subsystems" (Konolige, 1984)—hence the name. However, there are major differences between HBS and Belief Subsystems; e.g., updating, an operation fundamental for HBS, is not considered in Belief Subsystems.

Notice that α is a well formed formula (or simply formula) of the particular KR language we are using for representing knowledge, while x is a certainty value represented according to the particular *UR-mod* installed (e.g. a number in [0, 1]). The internal structure of an HBS stresses the dichotomy between *knowledge* and *uncertainty* about it. A HBS will actually comprise two modules, called "KR-module" and "UR-module", intended to deal with categorical knowledge and uncertainty about it, respectively.

The situation is depicted in the picture on the right, where *KR-mod* is a KR system (e.g. we can imagine using KL-ONE, Krypton, etc. as a *KM-mod*); *UR-mod* is a system implementing an uncertainty calculus (e.g. we can imagine using Bayesian network propagation, an implementation of belief functions, etc. as a *UR-mod*); *Tell* and *Ask* are two proper modules of the HBS, which implement the interface to the HBS, and provide the co-operation mechanisms between *KR-mod* and *UR-mod*.

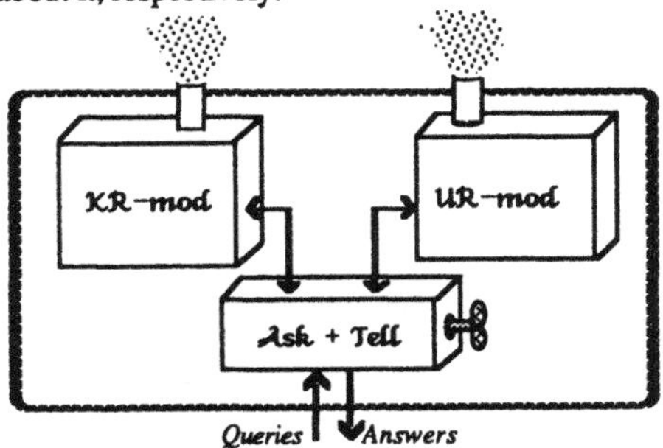

The organization of the cooperation between *KR-mod* and *UR-mod* in the execution of the Ask and Tell primitives constitutes the very core of a HBS. One first, intuitive possibility that shows up is attaching uncertainty information to the formulae used by *KR-mod*, and to use *UR-mod* to manipulate these items of information according to the given UR calculus. In order to do this, any relation that holds between items of uncertainty must be known to *UR-mod*. In particular, we can imagine that the *KR-mod* communicates to *UR-mod* each new formula inferred, and how it has been inferred (i.e. its depending from other formulae). Better that any explanation, a simple example will illustrate this simple idea.

Example 1. We imagine using the M-KRYPTON KR system[2], an extension of the KRYPTON hybrid KR system (Brachman et al., 1985), as our *KR-mod*, and Dempster-Shafer theory of evidence (Shafer, 1976; Smets, 1988) as our *UR-mod*. We want to model the following interesting problem:

Robert is not very learned in Palaeontology: though he is almost sure that brontosauri are animals, he is not quite sure about which are the possible kinds of brontosaurus; for instance, he tends to think that researchers are one of these kinds. Given that Robert knows that Alex is a researcher, how strongly will he believe Alex being an animal?

Here is a list and a graphical representation of the relevant facts to be included in our HBS:

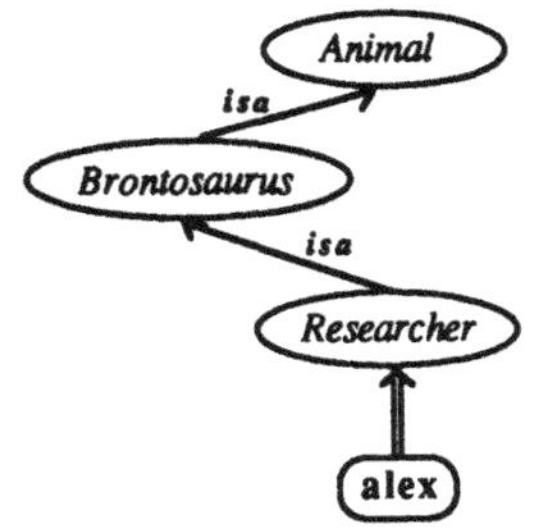

THE FACT "*Brontosauri are animals*"
IS BELIEVED WITH (belief = 0.9 ; plausibility = 1)

THE FACT "*Researchers are brontosauri*"
IS BELIEVED WITH (belief = 0.7; plausibility = 1)

THE FACT "*Alex is a researcher*"
IS BELIEVED WITH (belief = 1 ; plausibility = 1)

These facts are stored by M-KRYPTON, and names are given to them. Binary variables are associated in the *UR-mod* to each name N, with intended semantics "Fact N holds"; information is attached to each variable, giving either its belief values, or a symbolic indication of the dependencies of this variable from other ones.

KR-mod		***UR-mod***	
Name	Formula	Var	Certainty
F1	(IS Brontosaurus Animal)	F1	((yes 0.9) (no 0.0))
F2	(IS Researcher Brontosaurus)	F2	((yes 0.7) (no 0.0))
F3	(Researcher alex)	F3	((yes 1.0) (no 0.0))

---

[2] M-KRYPTON is a KR language developed to model interaction between multiple agents; in particular, it comprises sentences ($\mathbf{B}_i\ \alpha$), read "agent *i* believes that α". Beside expressive power, the interest of M-KRYPTON here lies in its semantics being defined in a possible worlds setting. M-KRYPTON is fully described in (Saffiotti & Sebastiani, 1988).

Let us suppose that we ask now

Ask[`(Animal alex)`].

HBS asks KR-mod to try to deduce `(Animal alex)`; KR-mod generates the following chain of inferences:

```
    (IS Researcher Brontosaurus)     (Researcher alex)
    ---------------------------------------------------
                  (Brontosaurus alex)
(IS Brontosaurus Animal)
-------------------------------------------
             (Animal alex)
```

While doing this, KR-mod communicates each single inference to HBS, which in turn instructs UR-mod to build the corresponding relations between the involved variables. The above table is then updated as follows:

F4	(Brontosaurus alex)	F4	(from (F2 F3))
F5	(Animal alex)	F5	(from (F1 F4))

At this point, KR-mod communicates the name of the asked formula to HBS, which requires a belief value for it to UR-mod. This value is not available, but UR-mod has enough information to compute it.

If we imagine that our UR-mod consists of a network-based implementation of DS theory (Shafer et al, 1987), then the inferences above may be codified by the belief network on the right, where ovals represent variables, and hexagons represent relations among variables (i.e. basic probability assignments defined over the product space of the involved variables).

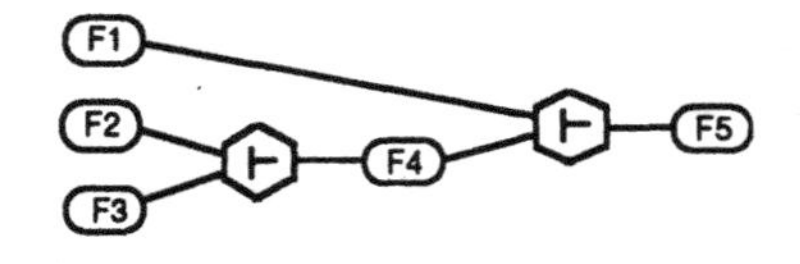

In our case, all the relations are deduction relations $\vdash$; a relation $\beta_1,...,\beta_k \vdash \alpha$ is codified by a bpa over the product space of the truth values for β_1, ..., β_k, α assigning mass 1 to the complement of the subset $\{\beta_1$=true,..., β_k=true, α=false$\}$. The propagation leads to values (0.7, 1) for F4 and (0.63, 1) for F5. Hence

Ask[`(Animal alex)`] = (0.63, 1.0).

Unfortunately, this intuitive type of interaction—to build in *UR-mod* a belief network that "mirrors" the facts and inferences which crowd, in the course of time, *KR-mod*—quickly runs into troubles as we try to push it a little further, or to make it more precise. We mention here three basic difficulties with it. First, translating formulae and inferences in the *KR-mod* to a belief network in the *UR-mod* may sometimes not be an easy matter: some kinds of inference may lead to tangled or ad-hoc translation, and the connections between different fragments of the network become extremely unclear[3]. Second, we do not have a criterion to choose, among different possible translations, the "correct" one. In particular, although we have in general a semantics for the *KR-mod* and for the *UR-mod* separately, we don't have a semantics—neither formal nor intuitive—for the overall integrated system; thus, we do not have any reference with respect to which we can evaluate "correctness". Third, a HBS built this way is not really modular. The translation mechanism is tailored on the specific linguistic structures of a given KR system and their semantics, and the whole translation device would have to be changed, were we to decide to change the *KR-mod* inside a HBS (actually, even the slightest change in the syntax of the KR language will call for a change in some part of our translation procedure). Even worst, the overall translation idea critically relies on the availability of a network-based implementation of the UR calculus we want to use.

3. The Formal Viewpoint

The essential cause of the difficulties above is to be found in the too tight connection between the uncertainty measures and the linguistic structures of the particular KR language we are using: uncertainty measures have been up to now associated to formulae of the KR language. This, upon reflection, is quite unnatural: agents both natural or artificial, entertain belief on the knowledge they talk about, rather than on the linguistic objects used to talk about it. In other words, reasoning and believing are concepts which refer to knowledge; the lan-

[3] The reader wishing to see an example of the kind of acrobatics you may have to perform in order to express certain inferences in a belief network is addressed to (Smets & Hsia, 1990).

guage comes into play only in the moment in which we want to express this knowledge (e.g. in communication). In the logical tradition, a distinction has been made between "formulae" of a language, and "propositions" (i.e. contents) expressed by these formulae. We give a formal account to HBS by isolating the content of our knowledge from the linguistic structures we use to express it along the same lines. The key move is the definition of a space $\mathcal{P}$ of abstract propositions. Given a (formal account of a) UR calculus, and (of) a KR system, we first define the UR calculus in a general way on the space $\mathcal{P}$; then, we use our KR system to map formulae of our KR language to the propositions in $\mathcal{P}$ they connote[4].

We require $\mathcal{P}$ to be a Boolean algebra, with partial order "$\Rightarrow$" (read "entails"). Given a UR calculus, and its set of certainty values $\mathcal{C}$, we consider sets of pairs $<P, m_P>$ with $P \in \mathcal{P}$ and $m_P \in \mathcal{C}$. Intuitively, a pair $<P, m_P>$ represents a partially believed proposition, with certainty degree m_P ; a set of pairs then corresponds to an assignment of certainty values to propositions: we call it a *certainty state*. We also call $\mathcal{B}$ the set of all certainty states: $\mathcal{B} = 2^{(\mathcal{P} \times \mathcal{C})}$. We then require that the given UR calculus can be defined on $\mathcal{B}$ by means of three functions: Ψ^0: $\{\} \rightarrow \mathcal{B}$, which returns an initial certainty state corresponding to total ignorance; Ψ^1: $\mathcal{P} \times \mathcal{B} \rightarrow \mathcal{C}$, which returns the certainty value of a proposition in a given certainty state; and Ψ^2: $\mathcal{P} \times \mathcal{C} \times \mathcal{B} \rightarrow \mathcal{B}$, which returns the new certainty state obtained by acquiring a given proposition with a given certainty value in a (old) certainty state. Intuitively, Ψ^1 defines the *entailment* of our UR calculus, and Ψ^2 its *updating*.

Example 2. We formalize Dempster-Shafer theory in the above framework by defining the Ψ_{DS} triplet on a generic algebra of propositions $<\mathcal{P},\Rightarrow>$[5]. Certainty states will be sets of pairs $<P, m_P>$ with $P \in \mathcal{P}$ and $m_P \in [0,1]$; they correspond to basic probability assignments (bpa) in the standard formulation. We have

$$\Psi^0_{DS}() = \{<\mathcal{U}, 1>\}$$

where $\mathcal{U}$ is the top element of $\mathcal{P}$; this corresponds to the vacuous bpa. $\Psi^1_{DS}(P,\kappa)$ returns a value in [0,1] corresponding to the value of bel(P) for the bpa represented by κ:

$$\Psi^1_{DS}(P, \kappa) = \sum \{m_Q \mid <Q,m_Q> \in \kappa \text{ and } Q \Rightarrow P \}$$

Intuitively, our confidence in the truth of the knowledge represented by P is just the sum of the mass values attributed to any proposition in κ whose truth entails the truth of P. As for the updating part of Ψ_{DS}, we notice that updating is typically performed in DS theory by combining, via to the so-called Dempster's rule of combination, the bpa representing the present state of belief with one representing the new evidence. Accordingly, we combine the old bpa κ with a bpa which allocates the desired amount of belief to the new proposition, and the rest to the universal proposition $\mathcal{U}$.

$$\Psi^2_{DS}(P, x, \kappa) = \kappa \oplus \{<P, x>, <\mathcal{U}, 1-x>\}$$

where $\oplus$ stands for Dempster's rule of combination, recast in terms of pairs $<P, m_P>$.

As a second step, we want to link our UR calculus to a given KR system. To this respect, we regard the roles of a KR system as that of 1) providing a KR language $\mathcal{L}$, and 2) providing a mapping from formulae of $\mathcal{L}$ to the propositions they connote. The most popular way to formally describe a KR system is to define a formal system Σ based on the language $\mathcal{L}$. We first consider having a description of this Σ in terms of a relation $\models$, which defines "truth" of formulae of $\mathcal{L}$ with respect to elements of a given class of mathematical structures used as models of the language. We denote by $\mathcal{M}_\Sigma$ the set of these structures. For instance, if we consider standard first order logic (FOL) as Σ, then $\mathcal{M}_{FOL}$ is the set of standard FOL interpretations $\langle D, V \rangle$. We then let the proposition connoted by α in Σ be the set of all the elements of $\mathcal{M}_\Sigma$ where α holds according to $\models$, i.e. the set $\{\omega \in \mathcal{M}_\Sigma \mid \omega \models \alpha\}$. We then give the operational definition of the Ask, Tell and Empty primitives for a given UR calculus Ψ and a given KR system Σ (with truth relation $\models$)

[4] The full formal framework is detailed in (Saffiotti, 1990a); here, we only sketch it in a form suitable for our goals.
[5] We give only an outline of the full formalization of a HBS using Dempster-Shafer theory (given in Saffiotti, 1990b).

$$\mathbf{Empty}_{\models,\Psi}[] \;\Leftarrow\; \{\, cs \leftarrow \Psi^0() \,\}$$

$$\mathbf{Ask}_{\models,\Psi}[\alpha] \;\Leftarrow\; \{\, \underline{\text{return}}\ \Psi^1(\{\omega \mid \omega \models \alpha\}, cs) \,\}$$

$$\mathbf{Tell}_{\models,\Psi}[\alpha, x] \;\Leftarrow\; \{\, cs \leftarrow \Psi^2(\{\omega \mid \omega \models \alpha\}, x, cs) \,\}$$

where **cs** denotes the current cognitive state of the HBS (the definition implies that cognitive states and certainty states coincide). $\mathbf{Empty}_{\models,\Psi}[]$ resets the current cognitive state to the "empty" one, where nothing is believed (except the valid formulae of $\mathfrak{L}$). $\mathbf{Ask}_{\models,\Psi}[\alpha]$ inspect the current cognitive state to determine to what extent (according to Ψ^1) the proposition connoted by α in entailed (according to $\models$) in it. $\mathbf{Tell}_{\models,\Psi}[\alpha,x]$ builds a new cognitive state obtained by updating (according to Ψ^2) the current one with a new evidence saying: «I believe to a degree x that α is true". Notice that, while we use formulae of $\mathfrak{L}$ to interact with the HBS via the Ask and Tell operations, the UR calculus Ψ works on the propositions connoted by these formulae: we emphasize once again that uncertainty is associated directly to the knowledge it refers to, rather than to the linguistic representations of it.

Example 3. We formalize the primitive operations for a HBS which uses DS theory as its UR component and First Order Logic (FOL) as its KR component. The functions for Ψ_{DS} have been already defined in Ex. 2; the $\models$ relation is the standard truth relation of FOL; the algebra of proposition is given by $<\mathcal{M}_{FOL}, \subseteq>$, where $\mathcal{M}_{FOL}$ is the set of standard FOL interpretations The definitions of Empty, Ask and Tell follow immediately from those of Ψ_{DS}:

$$\mathbf{Empty}_{FOL,DS}[] \;\Leftarrow\; \{\, cs \leftarrow \{<\mathcal{M}_{FOL}, 1>\} \,\}$$

$$\mathbf{Ask}_{FOL,DS}[\alpha] \;\Leftarrow\; \{\, \underline{\text{return}}\ \sum \{m_Q \mid <Q,m_Q> \in cs \text{ and } Q \subseteq \{\omega \mid \omega \models \alpha\}\} \,\}$$

$$\mathbf{Tell}_{FOL,DS}[\alpha, x] \;\Leftarrow\; \{\, cs \leftarrow cs \oplus \{<\{\omega \mid \omega \models \alpha\}, x>, <\mathcal{M}_{FOL}, 1-x>\} \,\}$$

We show the use and appearance of a HBS described by these operations by a simple example.

$\mathbf{Empty}_{FOL,DS}[]$
$\mathbf{Tell}_{FOL,DS}[\, (\forall x.\ (bird(x) \wedge \sim excp(x)) \supset flier(x)\,),\ 1]$
$\mathbf{Tell}_{FOL,DS}[\, (\forall x.\ \sim excp(x)),\ 0.8]$
$\mathbf{Tell}_{FOL,DS}[\, (\forall x.\ penguin(x) \supset (bird(x) \wedge \sim flier(x))\,),\ 1]$
$\mathbf{Tell}_{FOL,DS}[\, bird(Tweety),\ 1]$

Then we have:
$\mathbf{Ask}_{FOL,DS}[\, flier(Tweety)\,] = 0.8$
$\mathbf{Ask}_{FOL,DS}[\, \sim flier(Tweety)\,] = 0$
$\mathbf{Ask}_{FOL,DS}[\, excp(Tweety)\,] = 0$
$\mathbf{Ask}_{FOL,DS}[\, \sim excp(Tweety)\,] = 0.8$

But if: $\mathbf{Tell}_{FOL,DS}[\, penguin(Tweety),\ 1]$
then:
$\mathbf{Ask}_{FOL,DS}[\, flier(Tweety)\,] = 0$
$\mathbf{Ask}_{FOL,DS}[\, \sim flier(Tweety)\,] = 1$
$\mathbf{Ask}_{FOL,DS}[\, excp(Tweety)\,] = 1$

Example 4. Probability values can be expressed in the framework of full DS theory. If we are allowed to use pairs (x_t, x_f) in the Tell operation, with x_t and x_f expressing our belief in the truth of the told propositions and in its falseness, respectively, then a probability value x can be assigned to a formula α (of, say, FOL) by simply performing a $\mathbf{Tell}_{FOL,DS}[\,\alpha, (x, 1-x)]$ operation. Here we just give the definitions of our primitive operations when FOL is used for representing knowledge, and probabilities for representing uncertainty (see Saffiotti, 1990b, for more details). These definitions follow immediately by those in the previous example, by assuming a uniform prior probability distribution over the set of FOL interpretation in the empty cognitive state:

$$\mathbf{Empty}_{FOL,P}[] \;\Leftarrow\; \{\, cs \leftarrow \{<\{\mathfrak{I}\}, 1> \mid \mathfrak{I} \in \mathcal{M}_{FOL}\} \,\}$$

$$\mathbf{Ask}_{FOL,P}[\alpha] \;\Leftarrow\; \{\, \underline{\text{return}}\ \frac{\sum \{x \mid <\{\mathfrak{I}\},x> \in cs \textit{ and } \mathfrak{I} \models \alpha\}}{\sum \{x \mid <\{\mathfrak{I}\},x> \in cs\}} \,\}$$

$$\mathbf{Tell}_{FOL,P}[\alpha, x] \;\Leftarrow\; \{\, cs \leftarrow \{ <\{\mathfrak{I}\},y> \mid <\{\mathfrak{I}\},z> \in cs \textit{ and } y = \begin{cases} zx & \textit{if } \mathfrak{I} \models \alpha \\ z(1-x) & \textit{otherwise} \end{cases} \} \,\}$$

We now switch to considering a proof-theoretic formalization of a HBS, i.e. we consider having a formal account of a KR system Σ in terms of a deduction relation $\vdash$ for it. $\vdash$ specifies which formulae can be deduced by which ones through the deductive apparatus of Σ (normally a set of axioms and inference rules). We define a *possible argument* in Σ to be any consistent set of formulae of $\mathfrak{L}$, and denote by $\mathcal{A}_\Sigma$ the set of all possible arguments in Σ. We then let the proposition connoted by α in Σ will be the collection of all the possible arguments in Σ which prove α. As a consequence, the algebra of propositions $\mathcal{P}$ has now sets of subsets of $\mathcal{L}$ as its elements, and again $\subseteq$ as its partial order. The operations for a HBS with a given UR calculus Ψ and a given KR system Σ (with deduction relation $\vdash$) are then defined by:

$$\mathbf{Empty}_{\models,\Psi}[] \;<=\; \{\ cs \leftarrow \Psi^0()\ \}$$
$$\mathbf{Ask}_{\models,\Psi}[\alpha] \;<=\; \{\underline{\text{return}}\ \Psi^1(\{\omega \mid \omega \vdash \alpha\}, cs)\ \}$$
$$\mathbf{Tell}_{\models,\Psi}[\alpha, x] \;<=\; \{\ cs \leftarrow \Psi^2(\{\omega \mid \omega \vdash \alpha\}, x, cs)\ \}$$

4. A Hybrid Belief System

We can now use the formal account given above to settle down in more precise terms the architecture of a HBS sketched in Section 2. We regard *UR-mod* as a system implementing a given UR calculus on a space $\mathcal{P}$ of uninterpreted propositions; *UR-mod* is interested in (and only in) evaluating and updating the certainty state of these propositions, while it does not care about their meaning. Who cares about meaning is *KR-mod*: it provides the mapping between the formulae of $\mathcal{L}$ that we use for representing knowledge and the knowledge items we refer to (i.e. elements of $\mathcal{P}$), and it guarantees that the semantics of $\mathcal{L}$ is mirrored in the entailment relation $\Rightarrow$ on $\mathcal{P}$. However, *KR-mod* is not concerned with truth (or certainty) at all: it draws valid inferences regardless the context where they are performed. In short, we regard the task of *KR-mod* in a HBS as that of associating each formula α of the KR language $\mathcal{L}$ we use to the proposition of $\mathcal{P}$ it connotes (written $\|\alpha\|$); the task of *UR-mod* is then to implement a UR calculus Ψ on the space $\mathcal{P}$. We draw below the corresponding revised architecture for a HBS.

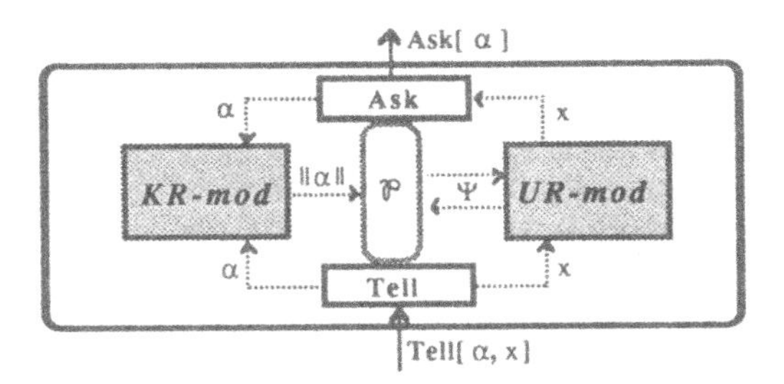

The objects exchanged between *KR-mod* and *UR-mod* are propositions, i.e. either sets of interpretations, or sets of possible arguments. When propositions are sets of possible arguments, the role of *KR-mod* is not so far from a plausible behaviour for a KR system. Though, $\|\alpha\|$ includes <u>all</u> the possible arguments for α according to the given deduction theory, so being in general computationally intractable; in particular, $\|\alpha\|$ is irrespective of what is actually believed in the current cognitive state. Two steps may be undertaken:

1) to only consider possible arguments which are "reasonable" with respect to what is actually believed in the current cognitive state; and
2) to let *KR-mod* provide "fragments" of possible arguments, generated while performing inferences; the reconstruction of full possible arguments from these fragments is then performed outside it.

Step 1 implies that *KR-mod* must access cs to decide which (possible) inferences to draw in order to find possible arguments (alternatively, knowledge may be cached in a Knowledge Base in *KR-mod*). Step 2 greatly weakens the demands on *KR-mod*, switching to a requirement more easily fulfilled by typical KR systems.

If the fragments above correspond to reports of single inference steps, they can be seen as ATMS justifications (deKleer, 1986), and the burden of reconstructing full possible arguments from them may be carried by an ATMS linked to *KR-mod*. Intuitively, and very roughly, the possible arguments for α correspond to the label built by the ATMS for the node[6] γ_α, given a set of justifications communicated by *KR-mod* while trying to deduce α. Telling α means justifying α by an assumption A, with an attached certainty value x. Asking α means computing the certainty for the label of γ_α, according to Ψ. This algorithm is summarized below:

Algorithm 1

Tell[α, x]	1. Ask *KR-mod* to try to deduce both α and $\sim\alpha^{(i)}$; in the deduction, each inference $\beta_1,..,\beta_{k-1} / \beta_k$ performed by *KR-mod* will be communicated to the ATMS as a justification $\gamma_{\beta_1},..,\gamma_{\beta_{k-1}} \Rightarrow \gamma_{\beta_k}$. 2. Create two new ATMS assumptions Γ_A and $\Gamma_{\bar{A}}$, and communicate justifications $\Gamma_A \Rightarrow \gamma_\alpha$; $\Gamma_{\bar{A}} \Rightarrow \gamma_{\sim\alpha}$; and $\Gamma_A, \Gamma_{\bar{A}} \Rightarrow \perp$ to the ATMS. 3. Store x as the certainty value attached to assumptions Γ_A and $\Gamma_{\bar{A}}$ according to Ψ^2.
Ask[α]	1. Ask *KR-mod* to try to deduce both α and $\sim\alpha$ while communicating justifications as above. 2. Ask the ATMS the labels of $\gamma_{\sim\alpha}$ and of γ_α. 3. Compute the value of $\Psi^1(\alpha,cs)$, given that the possible arguments in cs that prove α are the environments in the label of $\gamma_\alpha^{(ii)}$.
Notes:	i) "try to deduce" *may involve finding all the potential deductions, or only those grounded on believed knowledge.* ii) *Some precaution may be needed in order for the environments to be disjoint (e.g. Laskey-Lehner, 1989).*

From the architectural point of view, we can depict the situation corresponding to Algorithm 1 as shown on the right: $\mathbb{A}$ stores the certainty values associated to assumptions in ATMS; ATMS and $\mathbb{A}$ together may then be seen as the place where the current cognitive state is physically stored. In addition, the possibility for the *KR-mod* to cache knowledge (in an adequate format) in an internal knowledge base KB has been indicated.

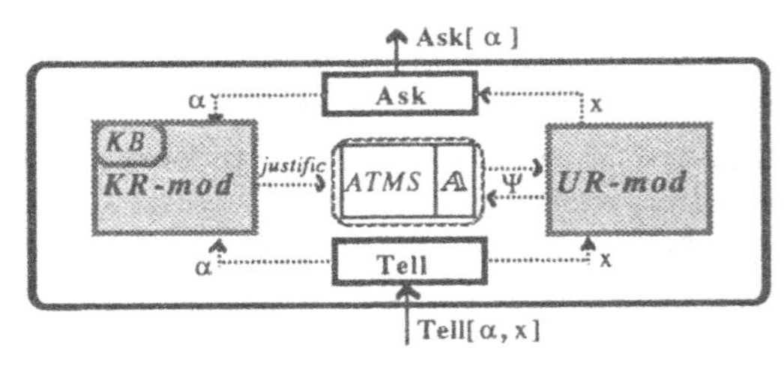

Example 5. We are now in a position to deal with the scenario of Example 1 by using more formal tools. We first consider the case in which we want to use a description of M-KRYPTON in terms of a truth relation $\models$. Thus, propositions are modelled as sets of possible worlds; i.e. $\mathcal{M}_{MK}$ is the set of Kripke worlds <M,s> with M a "hybrid Kripke structure" for M-KRYPTON (see Saffiotti & Sebastiani, 1988). The primitive operations for a HBS using M-KRYPTON and DS theory are then defined by:

$$\mathbf{Empty}_{MK,DS}[] \Leftarrow \{ cs \leftarrow \{\langle \mathcal{M}_{MK}, 1\rangle\} \}$$
$$\mathbf{Ask}_{MK,DS}[\alpha] \Leftarrow \{ \underline{\text{return}}\ \sum \{m_Q \mid \langle Q, m_Q\rangle \in cs \text{ and } Q \subseteq \{\omega \mid \omega \models \alpha\}\} \}$$
$$\mathbf{Tell}_{MK,DS}[\alpha, x] \Leftarrow \{ cs \leftarrow cs \oplus \{\langle\{\omega \mid \omega \models \alpha\}, x\rangle, \langle \mathcal{M}_{MK}, 1-x\rangle\} \}$$

where $\models$ is now the truth relation for M-KRYPTON. In order to be able to follow the evolution of cognitive states, we simplify restrict our attention to worlds where D contains the only individual <u>alex</u>.

The picture on the right illustrates the worlds in $\mathcal{M}_{MK}$: the horizontal lines identify the worlds in which Alex is in the extension of the predicate "Researcher"; the vertical lines, those in which he is in the extension of "Brontosaurus"; and the grey zone comprises the worlds where Alex is in the extension of "Animal".

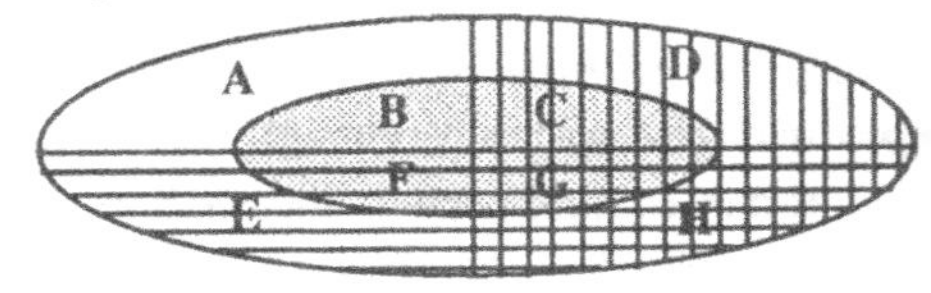

We then perform the following operations (the corresponding current cognitive state is indicated):

$\mathbf{Empty}_{MK,DS}[]$ — cs = $\{\langle \mathcal{M}_{MK}, 1\rangle\}$

$\mathbf{Tell}_{MK,DS}$[(IS Researcher Brontosaurus), 0.7] — cs = $\{\langle \mathcal{M}_{MK}, 0.3\rangle, \langle A+B+C+D+G+H, 0.7\rangle\}$

$\mathbf{Tell}_{MK,DS}$[(IS Brontosaurus Animal), 0.9] — cs = $\{\langle \mathcal{W}_{MK}, 0.03\rangle, \langle A+B+C+D+G+H, 0.07\rangle, \langle A+B+C+E+F+G, 0.27\rangle, \langle A+B+C+G, 0.63\rangle\}$

$\mathbf{Tell}_{MK,DS}$[(Researcher alex), 1] — cs = $\{\langle E+F+G+H, 0.03\rangle, \langle G+H, 0.07\rangle, \langle E+F+G, 0.27\rangle, \langle G, 0.63\rangle\}$

We then have, for instance:

---

6 We use the ATMS notation here, and write γ_α to refer to the ATMS node associated to formula α; also, Γ_A will denote the ATMS node associated with the assumption A.

$\mathbf{Ask}_{MK,DS}$[(Brontosaurus alex)] = 0.07+0.63 = 0.7
$\mathbf{Ask}_{MK,DS}$[(Animal alex)] = 0.63

Example 6. We now turn to re-analysing the problem above in a proof-theoretic setting. This means that we consider having an account of M-Krypton in terms of a deduction relation $\vdash$. Our primitives are then given by

$$\mathbf{Empty}_{MK,DS}[] \Leftarrow \{ cs \leftarrow \{<\mathfrak{A}_{MK}, 1>\} \}$$
$$\mathbf{Ask}_{MK,DS}[\alpha] \Leftarrow \{ \underline{return}\ \Sigma \{m_Q \mid <Q,m_Q> \in cs \text{ and } Q \subseteq \{\omega \mid \omega \vdash \alpha\}\} \}$$
$$\mathbf{Tell}_{MK,DS}[\alpha, x] \Leftarrow \{ cs \leftarrow cs \oplus \{<\{\omega \mid \omega \models \alpha\}, x>, <\mathfrak{A}_{MK}, 1-x>\} \}$$

In order to make the example manageable, we have to restrict our attention to a particular case: so, we consider a $\mathfrak{L}$ comprising the only concept names "Researcher", "Animal" and "Brontosaurus", and the only individual name "alex". We use abbreviations to denote formulae: "R>B" stands for "(IS Researcher Brontosaurus)", and "Aa" for "(Animal alex)". So, $\mathfrak{L}$ = {Ra,Ba,Aa,R>B,B>R,R>H,H>R,B>H,H>B}. Furthermore, we list only "minimal" possible arguments π for each given formula α, i.e. such that for each $\beta \in \pi$ it is not the case that $\pi \vdash \beta$. We then have:

$\mathbf{Empty}_{MK,DS}$[] cs = {<$\mathfrak{A}_{MK}$, 1>}
$\mathbf{Tell}_{MK,DS}$[(IS Researcher Brontosaurus), 0.7] cs = {<$\mathfrak{A}_{MK}$ 0.3>, <{{R>B},{R>A,A>B}}, 0.7>}
$\mathbf{Tell}_{MK,DS}$[(IS Brontosaurus Animal), 0.9] cs = {<$\mathfrak{A}_{MK}$ 0.03>, <{{R>B},{R>A,A>B}}, 0.27>,
<{{B>A},{B>R,R>A}}, 0.07>, <{{R>B,B>A},{R>B,B>R,R>A},
{R>A,A>B,B>A},{R>A,A>B,B>R}}, 0.63>}
$\mathbf{Tell}_{MK,DS}$[(Researcher alex), 1] cs = {<{{Ra},{Aa,A>R},{Aa,A>B,B>R},{Ba,B>R},{Ba,B>A,A>R}}, 0.03>,
<{{Ra,R>B},{Aa,A>R,R>B},{Aa,A>B,B>R,R>B},
{Ba,B>R,R>B},{Ba,B>A,A>R,R>B},{Ra,R>A,A>B},
{Aa,A>R,R>A,A>B},{Aa,A>B,B>R,R>A},{Ba,B>R,R>A,A>B},
{Ba,B>A,A>R,R>A,A>B}}, 0.27>, ... *(other 18 lines)*

As a result, we get again:

$\mathbf{Ask}_{MK,DS}$[(Brontosaurus alex)] = 0.07+0.63 = 0.7
$\mathbf{Ask}_{MK,DS}$[(Animal alex)] = 0.63

The cognitive states above show that, notwithstanding the simplifications and the minimal size of our example, sets of possible arguments tend to have an absolutely unmanageable size. We show how feasibility can be gained by running our example using Algorithm 1 above (and considering the corresponding architecture). In the following table, we summarize the justifications communicated to the ATMS in the Tell and Ask operations[7].

Interaction with the HBS	*Justifications created*
$\mathbf{Tell}_{MK,DS}$[(IS Researcher Brontosaurus), 0.7]	R>A, A>B => R>B $A_{R>B}$ => R>B
$\mathbf{Tell}_{MK,DS}$[(ISBrontosaurus Animal), 0.9]	B>R, R>A => B>A $A_{B>A}$ => B>A
$\mathbf{Tell}_{MK,DS}$[(Researcher alex), 1]	Aa, A>R => Ra Ba, B>R => Ra B>A, A>R => B>R A_{Ra} => Ra
$\mathbf{Ask}_{MK,DS}$[(Brontosaurus alex)] = 0.7	Ra, R>B => Ba Aa, A>B => Ba A>R, R>B => A>B Label of Ba = {{A_{Ra} ,$A_{R>B}$}}

5. CONCLUSIONS

We have defined Hybrid Belief Systems, modular systems which use the linguistic structures of a given KR language for representing knowledge, and a given UR calculus for representing uncertainty about this knowledge. These systems allow us to build intelligent agents in which the choice of a KR technique and of a UR technique may be performed independently; this means that we can choose the two techniques according to the

---

[7] Because in our example all the plausibility values are 1, the ATMS nodes and assumptions relative to negated sentences are not receiving any mass; we therefore omit them in the table.

type of knowledge and to the type of uncertainty that characterize the object domain, and then put them together in an integrated system. This approach is motivated by the conviction that, for both KR and UR, no technique can be absolutely marked as "the best", but each technique is more or less geared to some type of knowledge (resp. uncertainty) (Saffiotti, 1987).

Hybrid belief systems are a good candidate to be an integration paradigm between KR and UR; this paradigm has born out of, and entails, a conception of the role of UR in the whole AI agent: uncertainty is seen as meta-knowledge about the validity of our knowledge (with respect to an intrinsically certain reality), dealt with by an independent cooperating specialized module. Notice that, by assuming that, we are implicitly sticking to an epistemic interpretation of uncertainty: let us consider statement (a) «I am 80% sure that reaserchers are brontosauri» and (b) «80% of reaserchers are brontosauri». (a) illustrates the notion of uncertainty of interest here: we can see it as a categorical implication ("∀x.reasercher(x)→brontosaurus(x)") accompanied by information about its uncertainty. On the contrary, statement (b) expresses, from our point of view, just categorical knowledge which refers to a statistical (but with a firm epistemic status) proposition. An interesting problem is whether other types of uncertainty (vagueness, for instance) are captured by this notion or not. For the case of vagueness, the answer is affirmative, if we accept the interpretation of the vagueness of the sentence "Enzo is rich" as pertaining to the adequacy (in our mind) of the description "rich" to the individual "Enzo", rather than to the ontological fuzziness of the predicate "rich" (cf. Schefe, 1980).

ACKNOWLEDGEMENTS.

This research has benefit from discussions with Robert Kennes, Yen-Teh Hsia, Bruno Marchal, Philippe Smets and Nic Wilson. Fabrizio Sebastiani has an early responsibility in my all-hybridizing impetus.

REFERENCES

Brachman, R.J., Pigman Gilbert, V. and Levesque, H.J. (1985) "An Essential Hybrid Reasoning System: Knowledge and Symbol Level Accounts of Krypton", Proc. of IJCAI-85: 532-539.

de Kleer, J. (1986) "An Assumption-Based Truth Maintenance System" *Artificial Intelligence* **28**: 127-162.

Konolige, K. (1984) "Belief and Incompleteness", *SRI Int. Techn. Note 319* (Stanford, CA).

Laskey, K.B. and Lehner, P.E. (1989) "Assumptions, Beliefs and Probabilities", *Artificial Intelligence* **41**(1).

Levesque, H.J. (1984) "Foundations of a Functional Approach to Knowledge Representation", *Artificial Intell.* **23**.

Saffiotti, A. (1987) "An AI view of the treatment of uncertainty", *The Knowledge Engineering Review* **2**(2): 75-97.

Saffiotti, A. (1990a) "A Hybrid Framework for Representing Uncertain Knowledge", Procs of the Eighth AAAI Conference (Boston, MA).

Saffiotti, A. (1990b) "Using Dempster-Shafer Theory in Knowledge Representation", Procs. of the Sixth Conference on Uncertainty in AI (Cambridge, MA).

Saffiotti, A. and Sebastiani, F. (1988) "Dialogue Modelling in M-KRYPTON, a Hybrid Language for Multiple Believers", Proc. of the 4th IEEE Conf.on AI Applications: 56-60.

Schefe, P. (1980) "On Foundations of Reasoning With Uncertain Facts and Vague Concepts", *Int. Journal of Man-Machine Studies* **12**:35-62.

Shafer G. (1976) *A Mathematical Theory of Evidence* (Princeton University Press, Princeton).

Shafer, G., Shenoy, P.P. and Mellouli, K. (1987) "Propagating Belief Functions in Qualitative Markov Trees", *International Journal of Approximate Reasoning* **1**: 349-400.

Smets, P. (1988) "Belief Functions", in: Smets P., Mamdani E.H., Dubois D. and Prade H. (Eds.) *Non-Standard Logics for Automated Reasoning* (Academic Press, London).

Smets, P. and Hsia, Y. (1990) "Default Reasoning and the Transferable Belief Model", Proc. of the 6th Conf. on Unc. in AI (Cambridge, MA).

HOW TO REASON WITH UNCERTAIN KNOWLEDGE

Nico Roos

Delft University of Technology (TU-Delft)
Department of Computer Science
P.O. Box 356, 2600 AJ Delft, the Netherlands

National Aerospace Laboratory (NLR)
Amsterdam, the Netherlands

Abstract

In this paper a reasoning process is viewed as a process of constructing a partial model of the world we are reasoning about. This partial model is a syntactic representation of an epistemic partial semantic model. In such a partial model different views on the world we are reasoning about can be represented. Multiple views can be the result of updating a partial model with information containing a disjunction or information described by sentences like: 'most humans have brown eyes'. If a proposition does not have to hold in every view, we cannot be certain of it. To express this two certainty measures for conclusions based on a partial model will be defined, a *probability* and a *likelihood* measure. The former will be used for conclusions that express an *expectation* while the latter will be used for conclusions that express an *explanation*. To show the value of the approach various applications of these certainty measures will be descfibed.

Key-word: management of uncertainty.

1 Introduction

In the literature a number of different definitions for the 'certainty' of a proposition can be found. A problem with these definitions is their intended meaning. One of the first attempts to deal with uncertain knowledge is based on the Bayesian probability theory [9]. In this theory probabilities are either viewed as representing relative frequencies or as representing subjective belief values. In the former view one has to be able to assign a correct a priori probability value to every proposition. As J. T. Nutter remarks [15], it is not always possible to know these probabilities. In the latter view a probability describes the belief of a reasoning agent in a proposition [5, 6]. There is, however, no inter-subjective interpretation of such a belief value. Hence, there is no reason why two persons with the same knowledge should agree on a belief value assigned to a statement. The same holds for the belief masses used in the Dempster-Shafer theory [18] and the certainty factors in the certainty factor model of EMYCIN [2]. Another approach, based on the probability theory, is the probabilistic logic of J. Los [13] and of N. J. Nilsson [14]. They define a probability distribution over a set of models. Objections against the use of these logics concerns the intractability of the reasoning process. In [14], Nilsson describes how the probability of a proposition can be determined. Because of the complexity of calculations involved, approximations have to be used.

Until recently the tractability of the computations involved, when reasoning with uncertain knowledge, did not get much attention. To model tractable reasoning and to guarantee the correctness of the conclusions derived, a partial model of the world we are reasoning about is used here.

2 Partial models

To be able to guarantee the correctness of the uncertainty of the conclusions derived, the context the conclusions are based upon, has to be clear. This implies that an agent has to know all logical consequences of his knowledge. To guarantee this, a partial model will be used. The idea behind the use of a partial model is that reasoning should be viewed as a process of extending a partial model and, in this way, approximate the complete model of the real world. Conclusions derived are statements which are satisfied by the resulting partial model. According to P. N. Johnson-Laird [11] humans also use something like a partial model, and use it for reasoning in a similar way as it is described here. Another approach closely related to reasoning by constructing a partial model has been described by D. W. Etherington et al. [8]. They propose the use of a partial model, which they call a 'vivid' knowledge base, to model tractable reasoning.

Like a traditional semantic model a partial model consists of a set of objects and relations over these objects. Since it is a *partial* model neither the objects nor the relations are completely defined. This are, however, not the only differences. Firstly, a partial model must also be able to express our ignorance about the world. For example, it must be able to express that a friend of John goes to Paris by car or by train using only one object to denote this friend. So a partial model must be a epistemic partial model consisting of a set of objects and a set of view on the world, where each view consists of partially defined relations over the objects. Secondly, a partial model must be able to represent the uncertainty expressed by a formula. For example, when no information is available about two relations α and β, the formula $\alpha \vee \beta$ expresses two equally likely views on the world we are reasoning about. This must also be expressed by a partial model after it is being updated using this information. A more complicated situation arises when we want to update a partial model the the formula $\alpha \vee (\beta \vee \gamma)$. This this formula expresses also two equally likely views on the world, α and $\beta \vee \gamma$, where the latter view consists of two other equally likely views. This suggests that a partial model should consists of a set of objects and a tree of views. It can, however, be shown that we cannot correctly update such a partial model with new information. The partial model that will be used here is an epistemic partial model where each view on the world does not only contain partially defined relations but may also contain formulas. These formulas function as constraints on a view. So, updating an empty partial model with $\alpha \vee (\beta \vee \gamma)$ results in a partial model containing two views, one containing α and one containing $\beta \vee \gamma$. If a formula (constraint) occurs in every view of a partial model, then we may update the partial model with this formula.

The formulas that will be used here defer from the formulas used in first order logic. A disjunction in FOL is a binary operator. Therefore, it can only denote two alternative descriptions of the world. To be able to describe three or more alternatives, ***n-place disjunction*** operators are needed.

Sentences containing a disjunction are not the only way to give alternative descriptions of the world. Also sentences like: 'most humans have brown eyes' and 'it often rains in the Netherlands' give alternative descriptions of the world. Here alternative descriptions of the world arise because these sentences do not specify which objects of a class of objects satisfy a property. In the formulas used here, these sentences can be expressed by ***quantified descriptions***. Furthermore, ***indefinite descriptions*** will be used instead of existential quantifiers.

Now after this very brief discussion of a partial model and of the formulas used to update it, their definitions will be given. For a more detailed discussion, confer [17].

Definition 1 Let *Names* be a finite set of names and let $\{o_i \mid i \in \mathbb{N}\}$ be an enumerable set of anonymous objects, $Names \cap \{o_i \mid i \in \mathbb{N}\} = \emptyset$. Then the domain of objects D for a partial model or a formula is defined as:

$$D = Names \cup \{o_i \mid i \in \mathbb{N}\}$$

Definition 2 Since the information described by a formula may refer to the objects in the current partial model, these objects have to be available. Let O_{ex} be these objects to which a formula can refer and which are called the ***external objects.***

The formulas are recursively defined as follows.

- $\varphi = \langle r, o_1, ..., o_n \rangle$ is an n-place relation where $r \in O_{ex}$ is a relation symbol and $o_1, ..., o_n \in O_{ex}$ are the arguments of the relation.
- $\varphi = \langle \neg, \psi \rangle$ is the negation of the formula ψ with external objects, the set O_{ex}.
- $\varphi = \langle \wedge, \psi_1, ..., \psi_n \rangle$ with $n \geqslant 2$ is a conjunction of formulas ψ_i with external objects, the set O_{ex}.
- $\varphi = \langle \vee, \psi_1, ..., \psi_n \rangle$ with $n \geqslant 2$ is a disjunction of formulas ψ_i with external objects, the set O_{ex}.
- $\varphi = \langle \#, n, o, \psi(o) \rangle$ is a description of a group of objects where $n \in \mathbb{N}$ is a natural number, $o \in D$ is an object and $\psi = \psi(o)$ is a formula with external objects, the set $O_{ex} \cup \{o\}$.
- $\varphi = \langle \mathrm{a}, o, \psi_1(o), \psi_2(o) \rangle$ is an indefinite description where $o \in D$ is an object that occurs in both the formulas ψ_1 and ψ_2, and where $\psi_1(o) = \psi_1$ and $\psi_2(o) = \psi_2$. The set of external objects for the formulas ψ_1 and ψ_2 is the set $O_{ex} \cup \{o\}$.
- $\varphi = \langle \%, p, o, \psi_1(o), \psi_2(o) \rangle$ is a quantified description where the percentage $p \in [0,1]$ is a rational number, $o \in D$ is an object that occurs in both the formulas ψ_1 and ψ_2, and where $\psi_1(o) = \psi_1$ and $\psi_2(o) = \psi_2$. The set of external objects for the formulas ψ_1 and ψ_2 is the set $O_{ex} \cup \{o\}$.

Now the formulas have been defined, the definition of the partial model can be given. This partial model consists of a set of objects and a set of views. Each view consists of a set of formulas that have to be satisfied by it.

Definition 3 Let D be the domain of objects for a partial model. Then, a partial model $\mathcal{M}$ is a tuple $\langle O, V \rangle$ where:

- $O \subseteq D$ is the set of known objects,
- $V = \{V_1, ..., V_m\}$ is a set of views V_i where V_i is a set of formulas that have to be satisfied by this view.

To be able to update a partial model with a formula containing new information, an updating function is needed. This function selects a new partial model, which is at least as informative as as the original one. This partial model must also contain the information described by the formula, but no more than that. The partial models that satisfy this property are all equivalent [17]. The partial models selected, may not contain more objects than strictly necessary and may not contain redundant views. If a partial model $\mathcal{M}$ is being updated with a formula φ, this is denoted by $\mathcal{M}[\varphi]$.

For a formula a satisfiability relation with respect to a partial model can be defined. Since we only have a *partial* model, formulas not satisfied by the model, do not need to be *false*. Therefore, it is not sufficient to define a satisfiability relation for the formulas that are *true* ($\models^+$). We also need a satisfiability relation that defines when a formula is *false* ($\models^-$), i.e. a formula that cannot be satisfied by any extension of the partial model.

Definition 4 Let φ be a formula and let $\mathcal{M} = \langle O_{\mathcal{M}}, \{V_1, ..., V_m\} \rangle$ be a partial model.

A formula φ is *true* in $\mathcal{M}$, $\mathcal{M} \models^+ \varphi$, if and only if φ is true in each view of $\mathcal{M}$, i.e. for each V_i: $\langle O_{\mathcal{M}}, V_i \rangle \models^+ \varphi$.

$\langle O_{\mathcal{M}}, V_i \rangle \models^+ \varphi$ if and only if either $\varphi \in V_i$ or one of the following conditions is satisfied.

- $\langle O_{\mathcal{M}}, V_i \rangle \models^+ \langle \neg, \psi \rangle$ if and only if: $\langle O_{\mathcal{M}}, V_i \rangle \models^- \psi$.
- $\langle O_{\mathcal{M}}, V_i \rangle \models^+ \langle \wedge, \psi_1, ..., \psi_n \rangle$ if and only if for each ψ_j: $\langle O_{\mathcal{M}}, V_i \rangle \models^+ \psi_j$.
- $\langle O_{\mathcal{M}}, V_i \rangle \models^+ \langle \vee, \psi_1, ..., \psi_n \rangle$ if and only if for some ψ_j: $\langle O_{\mathcal{M}}, V_i \rangle \models^+ \psi_j$.

- $\langle O_{\mathcal{M}}, V\rangle \models^{+} \langle \mathrm{a}, x, \psi_1(x), \psi_2(x)\rangle$ if and only if for some $o \in O_{\mathcal{M}}$: $\langle O_{\mathcal{M}}, V_i\rangle \models^{+} \psi_1(o)$ and $\langle O_{\mathcal{M}}, V_i\rangle \models^{+} \psi_2(o)$.

A formula φ is *false* in $\mathcal{M}$, $\mathcal{M} \models^{-} \varphi$, if and only if φ is false in each view of $\mathcal{M}$, for each V_i: $\langle O_{\mathcal{M}}, V_i\rangle \models^{-} \varphi$.

$\langle O_{\mathcal{M}}, V_i\rangle \models^{-} \varphi$ if and only if either $\langle \neg, \varphi\rangle \in C$ or one of the following conditions is satisfied.

- $\langle O_{\mathcal{M}}, V_i\rangle \models^{-} \langle \neg, \psi\rangle$ if and only if: $\langle O_{\mathcal{M}}, V_i\rangle \models^{+} \psi$.
- $\langle O_{\mathcal{M}}, V_i\rangle \models^{-} \langle \wedge, \psi_1, ..., \psi_n\rangle$ if and only if for some ψ_j: $\langle O_{\mathcal{M}}, V_i\rangle \models^{-} \psi_j$.
- $\langle O_{\mathcal{M}}, V_i\rangle \models^{-} \langle \vee, \psi_1, ..., \psi_n\rangle$ if and only if for each ψ_j: $\langle O_{\mathcal{M}}, V_i\rangle \models^{-} \psi_j$.
- $\langle O_{\mathcal{M}}, V_i\rangle \models^{-} \langle \mathrm{a}, x, \psi_1(x), \psi_2(x)\rangle$ if and only if $\langle O_{\mathcal{M}}, V_i\rangle \models^{+} \langle \%, 0, x, \psi_1(x), \psi_2(x)\rangle$.

Notice that a quantified description and a description of a group of objects φ can only be true or false in a view if respectively $\varphi \in V_i$ or $\langle \neg, \varphi\rangle \in V_i$. It is not possible to define their meanings in terms of their constituents. The reason for this is illustrated by the following example. Suppose that we know 5 balls to be red. Does this imply that there only exist 5 red balls? Since partial models are incomplete, we cannot answer such a question.

In the next two sections, two certainty measures will be defined for conclusions based on a partial model, a *probability* and a *likelihood* measure. The probability measure will be used for *expectations* and the likelihood measure for *explanations*. An example of both kinds of conclusions is respectively 'John probably loves a woman' and 'it is likely that the patient has a brain tumor'. How such conclusions can be derived is described in the following sections.

3 Expectations

An expectation like: 'John probably loves a woman' can be derived from the quantified description 'Most men love a woman'. The question is how to formalize this deduction. One possibility is to assume that John is randomly chosen from the class of all men and that the quantified description is independent of other known information; e.g. that it is not overruled by some other quantified description. The main objection against this approach is the demand that John has to be randomly chosen. When John is a colleague of yours, he cannot be considered to be randomly chosen from the class of all men.

The approach taken here, is based on the ideas of R. Carnap [3]. A measure for the expectation of a formula will only be derived from the information available. So we have to define how the probability measure is based on the set of views. To motivate the definition below, first consider the following situation. Suppose that some of the views of a partial model satisfy the formula for which we want to determine a probability measure. Then definition of the probability measure should satisfy the following conditions.

- The probability measure should be proportional to the number of views that satisfy the formula.
- The measure should be inversely proportional to the total number of views.
- Since we have no reason to prefer one view to another, the insufficient reason argument of Bernoulli and Laplace [9] can be applied on the views.

The probability measure defined below satisfies these requirements.

Definition 5 Let $\mathcal{M} = \langle O, \{V_1, ..., V_n\}\rangle$ be a partial model and let a formula φ be known in every view V_i, i.e. for every view $\langle O, V_i\rangle$: either $\langle O, V_i\rangle \models^{+} \varphi$ or $\langle O, V_i\rangle \models^{-} \varphi$. Then the probability measure $Pr(\varphi \mid \mathcal{M})$ for a formula φ with respect to a partial model $\mathcal{M}$ is defined as follows:

$$Pr(\varphi \mid \langle O, \{V_1, ..., V_n\}\rangle) = \frac{1}{n} \cdot \sum_i Pr(\varphi \mid V_i)$$

and

$$Pr(\varphi \mid \langle O, V_i \rangle) = \begin{cases} 1 & \text{if } \langle O, V_i \rangle \models^+ \varphi \\ 0 & \text{if } \langle O, V_i \rangle \models^- \varphi \end{cases}$$

Given this definition, the following properties can be proven.

Property 6 Let $\mathcal{M} = \langle O, \{V_1, ..., V_n\} \rangle$ be a partial model and let φ and ψ be formulas that are known in $\mathcal{M}$, i.e. for each V_i:

$$\langle O_{\mathcal{M}}, V_i \rangle \models \varphi \text{ or } \langle O_{\mathcal{M}}, V_i \rangle \models \neg\varphi$$

and

$$\langle O_{\mathcal{M}}, V_i \rangle \models \psi \text{ or } \langle O_{\mathcal{M}}, V_i \rangle \models \neg\psi.$$

Then the axioms of probability are satisfied. These axioms are:

- $0 \leqslant Pr(\varphi \mid \mathcal{M}) \leqslant 1$.
- $Pr(\varphi \vee \neg\varphi) = 1$.
- If $\mathcal{M} \models \neg(\varphi \wedge \psi)$, then: $Pr(\varphi \vee \psi \mid \mathcal{M}) = Pr(\varphi \mid \mathcal{M}) + Pr(\varphi \mid \mathcal{M})$.

Using the definition, it is possible to derive that John probably loves a woman if the partial model contains the fact 'John is a man' and the quantified description 'Most men love a woman' is used.

Property 7 Let $\mathcal{M} = \langle O_{\mathcal{M}}, \{V\} \rangle$ denote a partial model and let φ denote the quantified description: $\langle \%, p, x, \psi_1(x), \psi_2(x) \rangle$. Furthermore, let all the objects described by the class ψ_1 be known in the partial model, let ob be one of these objects and let ψ_2 be unknown for these objects.

Then: $Pr(\psi_2(ob) \mid \mathcal{M}[\langle \%, p, x, \psi_1(x), \psi_2(x) \rangle]) = p$.

In his book *Knowledge in Flux* [10] P. Gärdenfors discusses the dynamic behaviour of probabilistic models. He formulates four postulates that should be satisfied by a probabilistic model when new information is received. These postulates, which imply conditionalization, are:

1. If $\vdash \neg(\alpha \vee \beta)$, then:

$$Pr(\varphi \mid \mathcal{M}[\alpha \vee \beta]) = a \cdot Pr(\varphi \mid \mathcal{M}[\alpha]) + (1 - a) \cdot Pr(\varphi \mid \mathcal{M}[\beta])$$

   where $a = \dfrac{Pr(\alpha \mid \mathcal{M})}{Pr(\alpha \vee \beta \mid \mathcal{M})}$.

2. $Pr(\alpha \mid \mathcal{M}[\alpha]) = 1$.
3. If $\vdash \alpha$, then $Pr(\varphi \mid \mathcal{M}[\alpha]) = Pr(\varphi \mid \mathcal{M})$.
4. $\mathcal{M}[\alpha]$ exists if and only if $Pr(\alpha \mid \mathcal{M}) = 0$.

Property 8 If the probability measure that occurs in Gärdenfors's postulates can be determined, then these postulates are satisfied.

Now I will discuss two examples that have been used to defend respectively the Dempster-Shafer theory [18] and the transferable belief model [20]. In [18] G. Shafer discusses an example illustrating that a Bayesian cannot always assign a consistent probability measure.

Example 9 *Life near Sirius?* Are there or are there not living things in the orbit of the star Sirius? Some scientists may have evidence on this question, but most of us will profess complete ignorance about it. So, let α denote the possibility that there is such life, then we know that $\alpha \vee \neg\alpha$ will hold. When we incorporate this information in the partial model,

$$Pr(\alpha \mid \mathcal{M}_1) = \frac{1}{2}.$$

We can also consider the question in the context of a more refined set of possibilities. We might raise the the question whether there exist planets around Sirius. Let this be denoted by β Shafer considers three possibilities in his example, viz. α, $\neg\alpha \wedge \beta$ and $\neg\alpha \wedge \neg\beta$. If we should update the partial model with $\alpha \vee (\neg\alpha \wedge \beta) \vee (\neg\alpha \wedge \neg\beta)$, we would get the same inconsistent probability measures for α as Shafer does. The formula $\alpha \vee (\neg\alpha \wedge \beta) \vee (\neg\alpha \wedge \neg\beta)$ states that we consider three distinct possibilities. This is not what we actually considered. What we did consider, however, was life or no life and planets or no planets.

When we update $\mathcal{M}_1$ with $\beta \vee \neg\beta$, the probability of α in the resulting partial model $\mathcal{M}_2$ will be:

$$Pr(\alpha \mid \mathcal{M}_2) = \frac{1}{2}.$$

The next example was used by P. Smets to illustrate the difference between his transferable belief model and a Bayesian approach when new information is received [20]. Smets's transferable belief model is based on the Dempster-Shafer theory and is intended to model changes of belief when new information comes available. In the transferable belief model, there is a distinction between a *credal* level and a *pignistic* (betting) level. At the credal level, belief masses are assigned to subsets of the frame of discernment. A belief mass assigned to a set can be transferred to its subsets when new information is received. When one is asked to make a bet, the belief masses assigned to a set, have to be divided over its elements using the insufficient reason argument; i.e. the probabilities at the pignistic level are being determined.

Example 10 *Mr Jone's murdering* Big Boss has decided that Mr Jone has to be murdered by one of the three persons present in his waiting room and whose names are Peter, Paul and Mary. Big Boss has decided that the killer on duty will be selected according to the result of a dice tossing experiment: if the result is even, the killer will be a female; if the result is odd, the killer will be a male. We, the judges, know who were in the waiting room and know about the story of the dice tossing experiment, but ignore what was the result and who was selected. We also ignore how Big Boss would have decided between Peter and Paul if the result given by the dice had been odd.

If we update a partial model using the information available, the probability that the killer is a female and the probability that the killer is a male will both be equal to 0.5.

Then we learn that if Peter was not the killer, he would go to the police station at the time of the killing in order to get a perfect alibi. Peter indeed went to the police station, so he is not the killer. Now the question is what is the probability that the killer is a female and what is the probability that the killer is a male, given this new information.

If we update the partial model using this new information, the probability of Peter being the killer and the probability of Paul being the killer will change. The probability that the killer is a female and the probability that the killer is a male, however, will still both be equal to 0.5.

This example shows that the probability measure defined here, like the transferable belief model but unlike a Bayesian model, results in intuitively sound conclusions. Since the same results follow from the probability measure defined here without the need of using two levels, the probability measure defined here seems to be preferable to the transferable belief model.

The probability measure defined here can be used to model default reasoning. According to L. Shastri [19] and F. Bacchus [1], some default rules are actually quantified descriptions. For thi subclass default reasoning can be realized in a similar way as is described by L. Shastri [19]. The two problems Shastri wants to solve, are: pre-emption and multiple inheritance. In case of pre-emption, the properties of a class can overrule the properties of its superclasses. This is stated in the following result.

Theorem 11 Let $\mathcal{M} = \langle O_{\mathcal{M}}, \{V_1, ..., V_n\}\rangle$ be a partial model and let $\varphi = \langle \%, p, x, \alpha(x), \gamma(x)\rangle$ and $\psi = \langle \%, q, x, \beta(x), \gamma(x)\rangle$ be two quantified descriptions. Furthermore, let all the objects described by the classes α and β be known in the partial model, let the class described by α be a subclass described of the class described by β, let *ob* be an object of the class described by α and let γ be unknown for any object of the class described by β.

Then: $Pr(\gamma(ob) \mid \mathcal{M}[\varphi \wedge \psi]) = Pr(\gamma(ob) \mid \mathcal{M}[\varphi])$.

Example 12 Let Pierre be a Quebecois. Furthermore, let it be known that most Quebecios are not native English speakers and that every Quebecois is a Canadian. Finally, let it be known that most Canadians are native English speakers. Then, according to Theorem 11, Pierre is probably not a native English speaker.

In his article *Objective probabilities* [12], H. E. Kyburg describes a model for assigning probabilities to formulas. To determine these probabilities, he introduces reference classes of objects in his model. For a reference class, one can specify what the percentage of objects is that satisfies some property. To be able to determine the probability that an object possesses this property, Kyburg introduces an axiom that is essentially the same as the theorem described above. Kyburg, however, neither takes into account the number of objects in a reference class nor the number of objects that are known to possess the property. Therefore, he must implicitly assume that a reference class contains infinitely many objects. Another axiom Kyburg introduces is that equivalent formulas must have the same probability. Clearly, this also holds for the model described here.

In case of multiple inheritance, an object inherits conflicting properties of two unrelated classes. The following theorem confirms our intuitions about multiple inheritance. It shows that unlike Shastri's solution, we need not know the distribution of objects in some superclass of the two classes from which the conflicting properties are inherited. We simply can make a decision by comparing the percentages of objects of the classes for which the conflicting properties hold.

Theorem 13 Let $\mathcal{M} = \langle O_{\mathcal{M}}, \{V_1, ..., V_n\}\rangle$ be a partial model and let $\varphi = \langle \%, p, x, \alpha(x), \gamma(x)\rangle$ and $\psi = \langle \%, q, x, \beta(x), \langle \neg, \gamma(x)\rangle\rangle$ be two quantified descriptions. Furthermore, let all the objects described by the classes α and β be known in the partial model, let *ob* be the only object known to belong to both classes and let γ be unknown for any object of these classes.

Then $p > q$ implies: $Pr(\gamma(ob) \mid \mathcal{M}[\varphi \wedge \psi]) > \frac{1}{2}$.

The following example has been described by D. S. Touretzky, J. F. Horty and R. H. Thomason. They gave this example as a counter example for off-path pre-emption. The solution they suggested to derive the correct conclusion, is being described in the example below.

Example 14 Let John be a marine chaplain. Furthermore, let it be known that the percentage of beer drinking marines is p and that the percentage of beer drinking chaplains is q. According to Theorem 13, if $p > (1 - q)$, it is more likely for John to be a beer drinker than not.

4 Explanations

An explanation tries to describe a cause (a disease, a malfunction) for anomalies (symptoms) observed. There may, however, exist more than one cause that can explain the anomalies observed. Since we are interested in the actual cause, we need a method to discriminate between the possible causes. One possibility to discriminate between the possible causes is to determine their probabilities. To be able to do so we have to know their a priori probabilities. As was argued by J. T. Nutter [15], it is often not possible to know these probabilities. Furthermore, in a study carried out by A. Tversky and D. Kahneman [21], it was observed that humans do not use a priori probabilities either. Although this cannot be used as an argument for neglecting a priori probabilities, it is an indication that a priori probabilities may not be necessary for explanations. A stronger argument to neglect a priori probabilities is that there are cases in which the use of a priori probabilities can result in wrong decisions. Consider, for example, the situation in which two diseases can explain the same symptoms. If one of the diseases is common and requires no medication, while the other disease is very rare and will kill a patient when no medication is given, then, by using a priori probabilities the latter disease will never be considered.

Although a priori probabilities are not used here, this does not imply that they cannot be used at all when they are known. When reasoning on a meta level about the likelihood measures, it is still possible to use these a priori probabilities. So, defining a likelihood measure independent of the a priori probabilities, enables us to reason about possible causes with or without using the a priori probabilities.

Instead of using a probability measure, here the compatibility of a possible cause with the current state of knowledge is determined. This means that we have to determine the views in which we cannot believe in the possible cause. This compatibility is expressed by an *unlikelihood* measure. Now a likely cause for the anomalies observed can be determined by showing that all other possible causes are unlikely. This approach can be viewed as a generalization of the *falsification principle*.

Like the probability measure, the unlikelihood measure of a formula is also determined by considering the set of views of a partial model.

- The unlikelihood measure should be proportional to the number of views in which a formula is false.
- The measure should be inversely proportional to the total number of views.
- Since we have no reason to prefer one view to another, the insufficient reason argument of Bernoulli and Laplace [9] should be applied on the set of views.

Definition 15 Let $\mathcal{M} = \langle O, \{V_1, ..., V_n\} \rangle$ be a partial model. The unlikelihood measure $UL(\varphi \mid \mathcal{M})$ for a formula φ with respect to a partial model $\mathcal{M}$ is defined as follows:

$$UL(\varphi \mid \langle O, \{V_1, ..., V_n\} \rangle) = \frac{1}{n} \cdot \sum_i UL(\varphi \mid V_i)$$

and

$$UL(\varphi \mid \langle O, V_i \rangle) = \begin{cases} 1 & \text{if } \langle O, V_i \rangle \models^- \varphi \\ 0 & \text{otherwise.} \end{cases}$$

For this likelihood measure the following property can be proven.

Observation 16 Let $\mathcal{M}$ be a partial model. If $Pr(\varphi \mid \mathcal{M})$ is determined, then

$$Pr(\varphi \mid \mathcal{M}) = 1 - UL(\varphi \mid \mathcal{M}).$$

5 Diagnostic reasoning

Using the unlikelihood measure, an efficient diagnostic reasoning process can be realized when we only try to determine one likely cause. For this diagnostic reasoning process an abstraction hierarchy of possible causes (diseases) is needed. The abstraction hierarchy of possible causes is used as a search tree. In this search tree the unlikelihood measure is used as an evaluation function. In an ideal situation we can find a specific cause in $\mathcal{O}(\log n)$ steps, where n denotes the number of possibl causes. The worst case is, of course, $\mathcal{O}(n)$ steps. The use of an abstraction hierarchy is not only motivated by the wish to realize an efficient diagnostic reasoning process. In a study of existing expert systems, carried out by W. Clancey [7], it was observed that such a hierarchy is implicitly implemented in these systems.

Given an abstraction hierarchy for the possible causes, the following diagnostic reasoning process can be used to determine a specific cause.

1. Start with the most abstract possible cause.
2. Determine the most abstract refinements of the possible cause.
3. Determine for each refinement the unlikelihood measure.
4. For each refinement which is not proven to be unlikely, i.e. its unlikelihood measure is above some threshold value, one can repeat step 2 until one reaches the specific causes, or until there are no possible causes which are not proven to be unlikely.

The correctness of this reasoning process depends on the following theorem.

Theorem 17 The unlikelihood measure for some abstract possible cause $\langle \mathrm{a}, o, \langle cl, o\rangle, \langle d, o\rangle\rangle$ is a lower bound for each of its refinements. Here a refinement is either $\langle \mathrm{a}, o, \langle cl', o\rangle, \langle d, o\rangle\rangle$ or $\langle d, inst\rangle$ where cl' subclass of class cl and $inst \in Names$ is an instance of the class cl for any partial model $\mathcal{M}$.

$$UL(\langle \mathrm{a}, o, \langle cl, o\rangle, \langle d, o\rangle\rangle \mid \mathcal{M}) \leqslant UL(\langle \mathrm{a}, o, \langle cl', o\rangle, \langle d, o\rangle\rangle \mid \mathcal{M})$$

and

$$UL(\langle \mathrm{a}, o, \langle cl, o\rangle, \langle d, o\rangle\rangle \mid \mathcal{M}) \leqslant UL(\langle d, inst\rangle \mid \mathcal{M})$$

This theorem confirms our intuition that if a possible cause turns out to be unlikely, then each of its refinements will also be unlikely. For example, if a lung disease is unlikely, tuberculosis will also be. The strategy behind the diagnostic reasoning process was already suggested by B. Chandrasekaran and M. C. Tanner [4]. Here, this strategy is given a sound foundation. Notice that because only lower bounds are used for the unlikelihood measure, one can use linguistic percentages; e.g. 'most', 'many', etc., and linguistic unlikelihood measures; e.g. 'impossible' and 'unlikely', instead of numbers. These linguistic percentages and unlikelihood measures can be linked using a simple table.

6 Conclusion

In this paper a model for reasoning with uncertain knowledge is presented. The main topics of this model are:

- Reasoning is viewed as a process of constructing a partial model. This makes it possible to assure that all relevant information is taken into consideration when deriving an uncertain conclusion. Uncertain conclusions are, therefore, always correct relative to this partial model.
- Uncertain conclusions are solely derived from the information available. The uncertainty expresses the support for the conclusion given this information.
- Two kinds of uncertain conclusion are distinguished here, *expectations* and *explanations*.
- An efficient diagnostic reasoning process can be realized using the *explanations* and an abstraction hierarchy of possible causes. This reasoning process can also be used when the uncertainty is specified using linguistic values.

An area in which further research is required is the meta reasoning process which controls the diagnostic reasoning process. The diagnostic reasoning process described in this paper does not follow from the treatment of uncertain knowledge described in the paper. Therefore, the diagnostic reasoning process, which is described here, is only one possibility. For example, further research is needed for the diagnosis of multiple possible causes.

References

[1] Bacchus, F. 1989, A modest but well founded inheritance reasoner, IJCAI-89, 1104-1109.

[2] Buchanan, B. G., Shortliffe, E. H. 1984, Rule-based expert systems: the MYCIN experiment of the Stanford Heuristic Programming Project, Addison-Wesley Publishing Company .

[3] Carnap, R. 1950, Logical foundations of probability, The University of Chicago Press, Chicago

[4] Chandrasekaran, B., Tanner, M. C. 1986, Uncertainty handling in expert systems, in: Kanal L. N., Lemmer, J. F. (eds), Uncertainty in Artificial Intelligence, North-Holland, Amsterdam 35-46.

[5] Charniak, E. 1983, The Bayesian basis of common sense medical diagnosis, AAAI-83 70-73.

[6] Cheeseman, P. 1985, In defense of probability, IJCAI-85 1002-1009.

[7] Clancey, W. C. 1984, Classification problem solving, AAAI-84 49-55.

[8] Etherington, D. W., Borgida, A., Brachman, R. J., Kautz, H. 1989, Vivid knowledge and tractable reasoning: preliminary report, IJCAI-89, 1146-1152.

[9] Fine, T. N. 1973, Theories of probability, Academic Press, New York.

[10] Gärdenfors, P., *Knowledge in Flux: Modeling the Dynamics of Epistemic States*, Bradford Books MIT Press, Cambridge MA (1988).

[11] Johnson-Laird, P. N. 1983, Mental models, Toward a cognitive science of language inference and consciousness, Cambridge University Press, Cambridge.

[12] Kyburg, H. E. 1983, Objective probabilities, AAAI-83 902-904.

[13] Los, J. 1963, Semantic representation of the probability of formulas in formal theories, Studia Logica 14 183-196.

[14] Nilsson, N. J. 1986, Probabilistic logic, Artificial Intelligence 28 71-87.

[15] Nutter, J. T. 1987, Uncertainty and probability, IJCAI-87 373-379.

[16] Peng, Y., Reggia, J. A. 1986, Plausibility in diagnostic hypothesis: the nature of simplicity AAAI-86 140-145.

[17] Roos, N., *Models for reasoning with incomplete and uncertain knowledge* PhD thesis, Delft (1991).

[18] Shafer, G. 1976, A mathematical theory of evidence, Princeton University Press, Princeton.

[19] Shastri, L. 1989, Default reasoning in semantic networks: a formalization of recognition and inheritance, Artificial intelligence 39 283-355.

[20] Smets, P., Transferable belief model versus bayesian model, *ECAI-88* (1988) 495-500.

[21] Tversky, A., Kahneman, D. 1982, Judgement under uncertainty: heuristics and biases, In Wendt, D., Vlek, C. (eds), Utility, probability and human decision making, D. Reidel Publishing Company, Dordrecht, Netherlands 141-162.

COMPUTATION AND UNCERTAINTY IN REGULATED SYNERGETIC MACHINES

Panos A. Ligomenides

Cybernetic Research Lab., EE Dept.,
University of Maryland, College Park, MD 20742
and
CAELUM Research Corporation,
Silver Spring, MD 20901

ABSTRACT

The uncertainty which is involved in inductive inference, is often dealt with by simulating probability, possibility, or belief theories with digital computers, and by using heuristic methods of inference based on different kinds of logic (binary, multivalued or fuzzy). We suggest that uncertainty may be managed naturally by synergetic (co-operative), self-organizing, dynamic physical systems, which are trained and are regulated to function as pattern-association "computing machines".

List of key word: Uncertainty, Synergetic computing, Pattern processing.

1. CAUSAL AND PROBABILISTIC MACHINES

We can speak of modeling the functioning of a machine in a discrete space-time view, as shown in figure 1. Looking at the "world" in this fashion, we may view the following operation rule: The machine is at a state s_i, represented by the point i of the space-time view, which is some function $s_i = N_i(s_j, \ldots, s_k, \ldots)$ of the states at the points j,k,.. in some neighborhood of i. Notice that if N_i involves only a few points behind in time, then all we've done is to describe the operation of cellular automata, because it means that we calculate a state from states at earlier times.

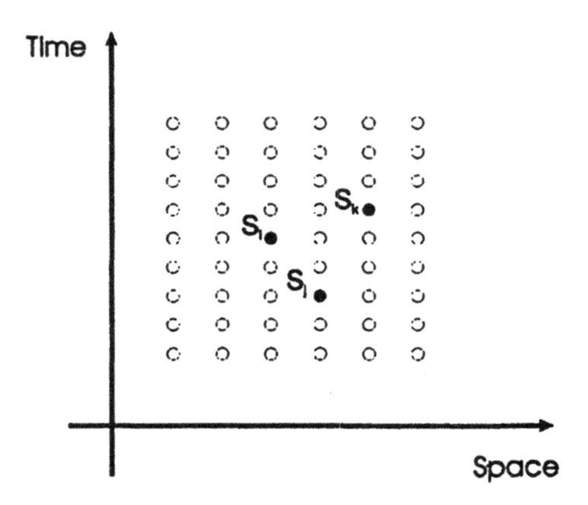

Figure 1. Discrete Space-Time View

Actually, it appears that classical physics is causal. From information of the past, if you include two pieces of data at each point (both momentum and position, or the position at two different times in the past), then you can calculate the future, in principle. So, classical physics is local, causal and reversible. As such, it is also quite adaptable to digital computer simulation, except for - or to the extend of - the discreteness and so on, resulting from going from state-to-state in time and from measuring quantities with limited precision.

Now, let us consider simulating probability with machines, i.e. probabilistic machines that simulate probabilistic behavior of the physical world. Can we have a computer that simulates probability? One way to do that, to have a machine that simulates a probability theory, would be to have the machine *calculate* the probabilities (from measured constraints) and then interpret the numbers as they might represent nature. For example, consider the case of a diffusing particle,

$$P(x,t) = -D \cdot \nabla^2 P(x,t)$$

where x is a general space variable and $P(x,t)$ is the probability that the particle is found at x at a time t. We could discretize the time and probability (with adequate resolution), make an algorithm, and solve this differential equation like we solve any old field equation. Limited by k digits means that when the probability of something happening is less than 2^{-k}, then you say that it does not happen at all. That's OK. With enough digits k we could allow ourselves to do that. After all, if the probability that the coffee will jump out of the cup is something like 10^{-700}, we say it is not going to happen at all, and we are not caught wrong very often. But the difficulty is this: The probability of finding a system of R components in a state $(x_i,...,x_R)$, at time t may be described with R k-digit numbers for each configuration of the system. If the system behaves like a cellular automaton with N cells, a number dictated by the resolution of measurement, then the computer should describe N^R configurations in order to compute probabilities, and no computer today is big enough to do that, even for modest values of R and N. Also, consider that doubling the size of the automaton would require exponentially explosive growth of the simulating computer. Note that the case of recognition in a high-dimensional feature space is like that. Therefore, when it comes to simulating probabilistic physical systems with many components, using a digital computer, we find it impossible [1].

Is there any other way to do the job? The other way to simulate a probabilistic nature might be by a machine which itself is probabilistic. If you use a conventional digital computer, this may mean that you may have to "randomize" somehow the results each time, which is not an easy task. However, a self-organizing, dynamic dissipative structure, whose phase transitions at bifurcation-points are "unpredictable", may be the answer. Such probabilistic (or possibilistic) machines, whose outputs (final states) are not unique functions of the inputs (initial states) are feasible *using nature itself*, i.e. using physical systems that behave appropriately. You have then defined machines that behave the way you want them to behave, regardless of how or whether by imitation, as long as the imitator does it with the same probabilities. How do you know the probabilities? You find out by repeated experiments. This now, leads us to consider using bifurcating, multistable, dissipative physical systems, as computing machines.

2. THE PARADIGM OF COMPUTATION

"Computation" may be viewed as a basic symbolic exercise with dual hypostasis. Whether by man or machine, computation is a coherent, well regulated *physical* activity, which is manifested and contextually interpreted as a metaphorical and denotative *symbolic* activity, as illustrated in figure 2. In essence, we may say that a physical system "computes its next state" if we have a "predictive model" f, which produces the final state y when given the initial state x, $f: x \rightarrow y$ [6,11]. Symbolic models of computation may be formulated at various *abstraction levels* and within different contextual references, as also illustrated in Figure 2 [14,15].

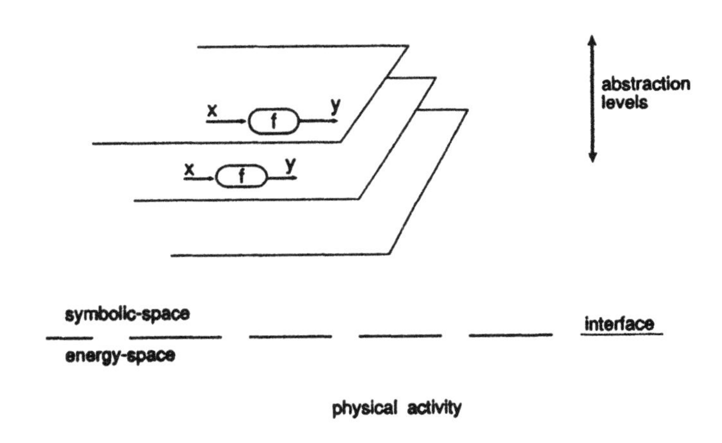

Figure 2. Computation Paradigm

As we shall see later, this generalized concept of computation involves the ability of the "computing machine" to simulate its environment (and parts of itself) by *compressing* sensory information into "internal models". These are *pattern processing* machines [9,12], whose function is holistic, not dominated by the operation of any individual component. They may be realized by certain kinds of dynamical physical systems, whose functional complexity and diversification may increase evolutionarily by self-organization through the exchange of energy *and information* with their environment.

3. COMPUTATION IN SELF-ORGANIZING DISSIPATIVE STRUCTURES

Recent investigations have established that there is a profound unity between the concepts underlying the subjects of irreversible thermodynamics, bifurcation analysis of nonlinearly coupled rate equations, fluctuation theory, quantum optics, oscillating reactions and fluid mechanics. The main theme that has been recurring throughout these diverse investigations is the occurrence of nonequilibrium phase transitions and the concomitant *emergence of cooperativity (synergy) and order* through the formation of dissipative structures [16]. When contextually interpreted, sensory order attains "meaning" and it is assimilated as *knowledge*. Synergetic dissipative structures that self-organize may, then, be used to store, retrieve and process contextually interpreted (i.e. "lawful") order [2-5,14-15].

Many nonlinearly coupled dissipative systems may change their bifurcating and multistability properties far-from-equilibrium, as the distance from equilibrium increases. System dynamics in these cases is described by the nonlinear rate equation,

$$\frac{dq_i}{dt} = Q_i(\{q_j\}, r) + F_i$$

where q represents the state variables, r represents the control parameters, Q represents the non-linear coupled functions, and F represents the delta-correlated and Gaussian distributed fluctuations. The equation generally includes space and time derivatives of the states q_j, as in the case of the reaction-diffusion equation. When r is changed beyond some critical value, a system's stable state-pattern can become unstable and can be replaced by a new stable state-pattern.

A good deal of effort has been invested in the mathematical analysis of such equations, especially in the analysis of the reaction-diffusion equation [2-5,16].

$$\frac{dq_i}{dt} = N_i(\{q_j\}, r) - D \cdot \nabla^2 q_i + F_i$$

General evolution trends of these systems have been analyzed and established. Relations between the distance from equilibrium and the emergence of order have been investigated, and the effects of fluctuations have been elicited [16]. Thermodynamic treatments have dictated the most natural choices of variables, parameters and physical restrictions in the operation and description of these systems. It has been found that the transition dynamics of the system may be determined (at least to the extend needed for qualitative analysis) only by *a few* control parameters; thus we successfully attain an *information compression* in the description of the system's complex dynamic behavior. The evolving values of the control parameters determine the system's spatio-temporal and functional patterns and their trajectories in state space (q), or in control space (r).

The stage is now set for viewing these phenomena under the scope of *computation* and knowledge engineering realizations. An important question is whether we are in a position now to exploit the wealth of "computational machine" possibilities and to construct certain physical-/chemical/biological systems with controlled bifurcation diagrams and dynamics of state-transition, which will provide us with desired computational behaviors.

4. FROM SENSORY PATTERN TO LANGUAGE: MUTUAL SIMULATION IN A FLUCTUATING ENVIRONMENT

Self-organization has to do with the generation of new "forms", with diversification and increase of complexity. The assignment of "meaning" to patterns (symbols, spatio-temporal arrangements, icons, images and combinations of such) is the beginning act on the way to developing a *language*. Notice that the association of meaning to spatio-temporal or relational patterns, discerned in sensory data, is more than a static assignment of meaning to pattern. It may also be functional or it may concern the evolution of dynamic physical, chemical or biological patterns.

Recently, we have come to realize that the contextual interpretation of "order" (i.e. the generation of information) is directly related with self-organization in certain kinds of hierarchical physical (chemical or biological) systems [2-5,16]. The general case of "cognition" (by self-organization and internal modeling) can be stated as follows: Consider two systems, each defined as an ensemble of interacting units, which are "communicating" by energetic interactions and by *informational exchanges* in a fluctuating environment. The important aspect of the symbolic interactions between these two communicating systems is *not copying but simulation*, i.e. the modeling of one system by the other. In this kind of "communication", cognitive processes, functioning within cognitive physical systems, *compress* the sensory complexity generated by the observed system (as it is discerned in the order which is embedded in the sensory data) and reveal *information* by triggering internal representations. Such mutual "recognition" of self-organizing physical systems is illustrated in Figure 3. In this figure we show the cognitive function as an "interface language" between the physical world of energy and the symbolic world of information. Biological organisms develop and use interface languages as tools for probing, modeling and controlling their (generally fluctuating) environment (which may include semi-autonomous parts of themselves). Thus, the essential role of "languages" is to reduce recursively the ever-increasing complexity of the evolving physical environment, by revealing information (forming internal models) which subsequently is used for purposes of goal-seeking control and self-optimization.

The above rather grant view of "language" goes beyond the limited definition of a (syntactic) language as "a set of aperiodic sequences of discrete symbols, which are stochastically interdependent through some given rules of grammar and syntax". In this current view, we see languages as self-organizing hierarchical cognitive devices, which "compute" the order which is discerned in sensory data, by associating it with their own increased organization, diversification and complexity. Their activity and evolution manifest the relationship between the dynamics of the physical and the symbolic worlds.

Although we are very much in the dark when it comes to our understanding the workings of the convolution of the sensory data with the dynamics of the cognitive device, we now know that the generation and interpretation of *form* (i.e. the semantic associations of patterns) are natural properties of synergetic physical systems. Although we can tell very little about the relation of human and biological languages to the neuronal spikes, to the synaptic processes and to the continuous electrochemical interactions in our brain, we are beginning now to understand the languages of "intelligent" computation, more as being physical processes and less (if any) as representing "logical" manipulations of discrete abstract symbols. The concepts of nonlinear dynamics, "fluctuations", "attractors", etc, are beginning to advance our understanding about the cognitive pathways of self-organizing systems into their dynamic sensory world environments. The management of uncertainty in recognition, abstraction, generalization, search and decision making, may be found more in the dynamics of self-organizing synergetic systems, and less in logical processing of abstract symbols.

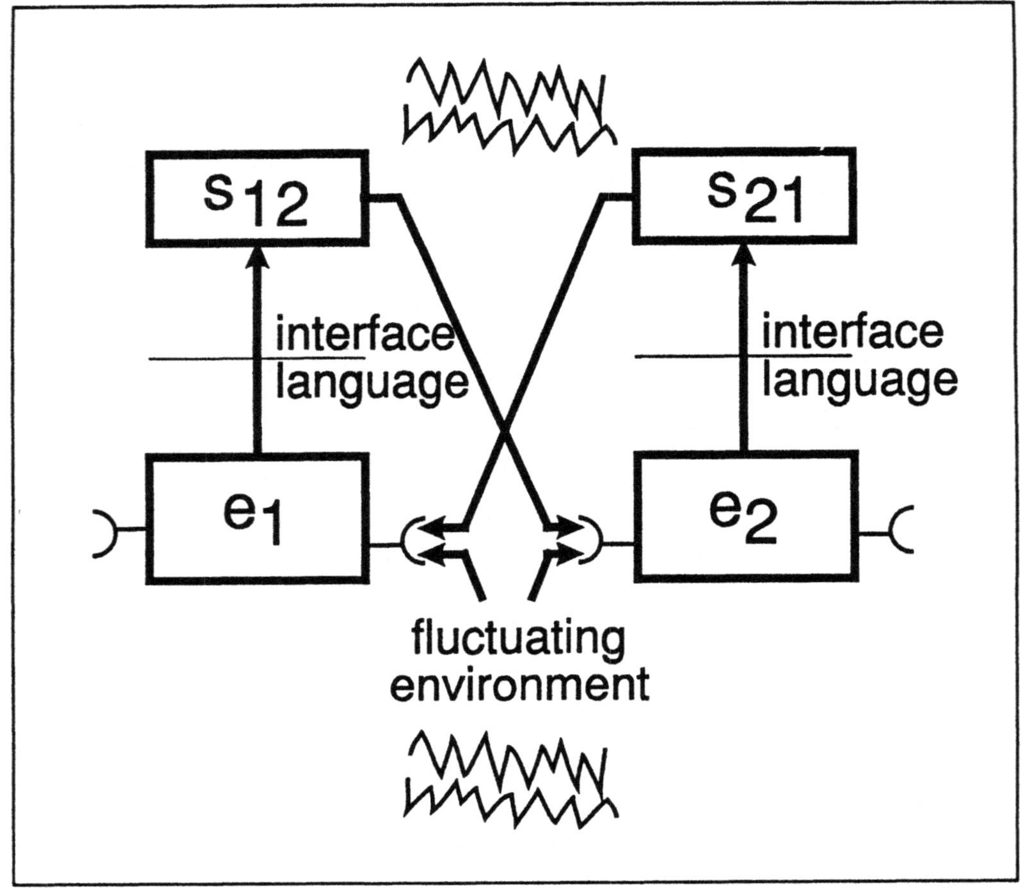

Figure 3. Communication between self-organizing physical systems, e_1 and e_2, in a fluctuating environment. Each system formulates internal representation of the other (s_{12} ans s_{21}) embedded in its organization. Information carrying physical signals communicate states of the physical systems e_1, e_2: $s_{12} = e_1 \otimes s_{21}$, $s_{21} = e_2 \otimes s_{12}$.

5. CLASSIFICATION OF COMPLEX PATTERNS

Processing of extensive data is required for classification of patterns which are characterized by the "values" (in the numerical or the qualitative sense) of a large number of properly selected features. A minimal but adequate set of measurable features must be selected for clear distinguishability of the sensory patterns. Because of measurement noise and pattern variability the features are seen as random variables with a certain probability distribution. For digital computer processing, the probability distribution function of the feature vector is determined from measurements, on the basis of some methodology, as for example the Maximum Information Entropy Principle. Such determinations are normally performed under conditions of uncertainty concerning the completeness or validity of the sensory data. In complex situations the size of the information processing task makes digital computer processing impractical.

One may see the process of *recognition* as a process of communication and internal modeling between the observed and the observer. This approach is intimately related to the theory of evolution, that is the study of successive changes in large scale systems through the exchange of energy *and information* with their environment. Certain kinds of large scale physical systems may evolve from stable subassemblies, as they *change* when destabilized by continuously bifurcating, and by jumping from singularity to singularity. They enter into *internal modeling transactions* effected by communication as they function on at least two hierarchical levels: an energy level that performs a cross-correlation, and a higher "cognitive" level that interprets the result of cross-correlation in terms of internally embedded form (modeling). This process involves the dynamic formation and cognition of *collective* properties, and therefore it underscores the mutual internal modeling and information compression (a prerequisite for simulation) between two dynamical hierarchical systems.

The dynamical co-operative (synergetic) systems which are capable of simulating their environment are characterized by entropy production, progressive differentiation, increasing complexity and self-organization [14-16]. Through their characteristic device of "attractors", synergetic "computing" systems promise to cary out very complex functional repertoires with very simple hardware. If controlled properly, they can become versatile information processors.

Another characteristic of synergetic computers acting as pattern associators is that they may produce information not only by cascaded bifurcation, but also by successive iterations of progressive refinement of attractors, giving rise to ever-increasing resolution of recognition. Attractors are ideal as information storage devices; however, what makes them outstanding for pattern recognition is that they are also ideal as information compression devices. Especially concerning the chaotic attractors functioning within a cognitive system, they promise considerable information dimensionality (for large dynamic storage capacity) and they serve as potent information compressors.

6. PROGRESSIVE REFINEMENT OF ATTRACTORS

It has been suggested [5] that pattern recognition is a progressive multistep process, in which the receiver (regognizer) takes an active part. In the first step (early or preattentive recognition) the pattern (on which attention is focused) is received (perceived) at a most global level [8-14]. At that high level of abstraction, several "categorical" attractors can be reached by the system. These attractors are "broad", in the sense that they discriminate among broadly outlined patterns (configurations), eg "triangle, rectangle, circle. Once the system has responded to the input pattern by reaching one of the top level (maximum abstraction level) broad abstractors, the system is driven progressively into more refined attractors by focusing its attention on successive explorations of more and more features. For example, the global shape of contour lines may lead the system into the base of the attractor "circular", (rather than "triangular", or "rectangular"). Further progressive focusing of attention (reception and processing of additional features), takes the system into progressively refined attractors, such as "apple", "face", "wheel", "tree", etc. Progressively the system "seeks" more details (features), such as color or other identifiable embedded objects, which are also progressively recognized.

Notice that this approach to pattern recognition is different from the "conventional" one, where the patterns are first decomposed into "primitives" or "features", which are subsequently identified and integrated. In the progressive recognition and refinement approach, we begin with recognition of the maximally abstracted pattern (eg the contour), and subsequently proceed to more and more details. In support of the progressive refinement approach, we refer to the ability of humans to supplement interrupted contour lines so that they may be "seen" as continuous ones.

7. MANAGING UNCERTAINTY

The management of uncertainty and ambiguity that enters the "communication" with their environment and the formulation of internal models (Figure 3) is endogenous in synergetic computing physical systems. It is inherently related to their capacity to analogically **associate patterns**, which are embedded in their initial and terminal conditions; also it is related to their concomitant ability to recover parts of spatio - temporal patterns, to abstract and generalize sensory information by feature extraction and classification, and to predict pattern evolution and cause-effect relations by temporal associations; all this done in the presence of uncertainty.

It may be noted also that computation, associated with recognition of objects, is performed along an "is a part of" hierarchy of composition and abstraction. In modeling objects of its environment or of itself, a cognitive device self-organizes and generates a description, which may move up the hierarchy of composition with a reduction in the number of degrees of freedom. The reduced complexity of description is traded with some increase in ambiguity and uncertainty. The interacting variables and parameters of the description at higher hierarchical levels of composition and abstraction represent "collective" properties of the dynamics going on at the level below. In a functional framework, each higher level of the composition hierarchy receives selective information from below, and in turn exercises feedforward (efferent) control on the dynamics below.

Today's computational needs in Artificial Intelligence include such operations as inductive inference, abstraction and generalization, and require means for the **management of uncertainty**. They call for computational systems which are not just complicated, but which are complex in the sense that their functioning is collaborative and non-reducible to simpler systems. **Synergetic** systems, large enough so that the system's behavior is no longer dominated by the behavior of a single component, display "emergent holistic behaviors" properties of the system as a whole [2-5,11]. As we continue building computing machines involving larger numbers of locally interacting components of constantly smaller sizes, requiring lesser waste between function and implementation, being constrained by three-dimensional connectivity and by finite propagation delays, we tend to make our computing systems look more and more like natural physical systems. We see the beginning of convergence between the physical substratum and the computational model. The task of designing computing machines begins now to look more like finding the physical parts of our huge ongoing physical universe, constantly "computing the next stage", which, when properly regulated, would do what we want them to do. From this generalized viewpoint, in order to develop a more fundamental understanding of what may be referred to as "Physics of Computation", we may have to ask new kinds of questions about computation and about nature, that may even extend Physics and Biology into new directions.

8. CONCLUSIONS: SYNERGETIC COMPUTING MACHINES

Complex synergetic systems abound in the real world. A class of bifurcating, multistable, nonlinear physical systems, whose controllable dynamics may be expressed by forms of the reaction-diffusion rate equation, may possess extended computational capabilities. Their suitability for applications in Artificial Intelligence is now being investigated. A better understanding of functions and organizations is also emerging from the investigations, concerning the biological neuronal networks of the cerebral cortex. Engineering specifications of such synergetic systems for inferential computations are currently being developed. The representation of knowledge and the management of uncertainty within the framework of synergetic computers are also currently being researched in our Cybernetic Research Laboratory.

REFERENCES

[1] Feynman, R.P.(1982), "Simulating Physics with Computers", *Intl. J. Theor. Phys.*, 21,6/7:467-489.

[2] Haken, H.(1983), *Synergetics, An Introduction*, Springer Ser. Synergetics, vol.1, 3rd ed., Springer.

[3] Haken, H.(Ed),(1985), *Pattern Formation by Dynamic Systems and Pattern Recognition*, Springer Ser. Synergetics, vol.5, Springer.

[4] Haken, H.(1987), *Advanced Synergetics*, Springer Ser. Synergetics, vol.20, 2nd corr. print., Springer.

[5] Haken, H.(1988), *Information and Self-Organization*, Springer Ser. Synergetics, vol.40, Springer.

[6] Ligomenides, P. A.(1982), "Symbolic Space Determinations on Physical Limitations", *Int'l, J. of Theoretical Phys. 21* (12), 973-978.

[7] Ligomenides, P. A.(1986), "On-Line Recognition and Labeling of Behavioral Modalities", Proc. of the Int'l IEEE Conf. on Systems, Man and Cybernetics, Atlanta, GA, October 14-17.

[8] Ligomenides, P. A(1987)., "Modeling Uncertainty in Human Perception", in *Lecture Notes in Computer Science: Uncertainty in Knowledge-Based Systems*, R.R. Yager and B. Bouchon (Eds), Springer-Verlag.

[9] Ligomenides, P. A.(1987), "Procedural and Progressive Formal Models of Human Perception", Proc. of the IEEE Conf. on Systems, Man and Cybernetics, Alexandria, VA, October 20-23.

[10] Ligomenides, P. A.(1987), "Resemblance, Fuzziness and Uncertainty in Experiential Knowledge Engineering", Proc. of the CAFSFM Int'l Symp. on Fuzzy Sys. and Knowl. Engin., Guangzhou, China, July 10-16.

[11] Ligomenides, P. A.(1988), "Intelligent Robotics: Recognition and Reasoning", (Plenary session paper) Proc. of the IEEE Workshop on Languages for Automation, University of Maryland, pp 44-59, August 29-31.
[12] Ligomenides, P. A.(1990), "On-Line Visual Prosthesis for a Decision Maker", Proc. of SPIE-OPTCON'90 Conf. on Advances in Intelligent Robotic Systems, Boston, Mass., Nov. 2-4.
[13] Ligomenides, P.A.(1988),"Modeling Experiential Knowledge with Procedural Schemata of Holistic Perception", in *Lecture Notes in Computer Science: Uncertainty and Intelligent Systems*, B.Bouchon,L.Saitta and R.R.Yager (Eds), Springer-Verlag.
[14] Ligomenides, P.A.(1990),"Wholistic Processing of Uncertainty in Synergetic Computing Systems", Proc. IEEE Int'l Symp. on Uncertainty Management and Analysis (ISUMA'90), U of Md, Dec.3-5.
[15] Ligomenides, P.A.(1991),"Physical Systems as Computing Machines", *Heuristics*, vol.4, No1.
[16] Nicolis, G. and I. PRIGOGINE, (1977), *Self-Organization in Nonequilibrium Systems*, Wiley, New York.

8. UNCERTAINTY IN INTELLIGENT SYSTEMS

INDUCTIVE LEARNING FROM INCOMPLETE AND IMPRECISE EXAMPLES

Janusz Kacprzyk and Cezary Iwański

Systems Research Institute, Polish Academy of Sciences
ul. Newelska 6, 01-447 Warsaw, Poland

Abstract. We present an application of fuzzy logic with linguistic quantifiers, mainly of its calculus of linguistically quantified propositions due to Zadeh, in inductive learning under imprecision and errors. The classification into the positive and negative examples is allowed to be to a degrees (of positiveness and negativeness), between 0 and 1. The value of an attribute in an object and in a selector need not be the same allowing for an inexact matching. Errors in the data may exist though their number may be not precisely known. A new inductive learning problem is formulated as to find a concept description which best satisfies, say, almost all of the positive examples and almost none of the negative ones.

Keywords: machine learning, inductive learning, learning from examples, imprecision, linguistic quantifier, fuzzy logic.

1. INTRODUCTION

Inductive learning (from examples) is basically a process of inferring a *description* (concept description, classification rule, hypothesis, rule,...) of a class (concept) from individual elements of the class called *positive examples*; some *negative examples* are usually added too.

The *examples* (objects) are commonly described by a set of "*attribute - value*" pairs. For instance (cf. Shaw, 1987), in a banking context, the examples (customers) may be described by the attributes "assets", "total debt" and "annual growth rate", and the concept (class) "good customer" may be described by the following concept description (to be derived via inductive learning), written for convenience in Michalski's (1973, 1980, 1983) variable - valued logic formalism

$$[assets > \$1,000,000]\ [total\_debt < \$250,000]\ [annual\_growth\_rate > 10\%] \rightarrow [class : "Good"] \quad (1)$$

to be read as: IF ("his/her assets exceed $1,000,000" and "his/her debt is less than $250,000" and "his/her annual growth rate exceed 10% ") THEN ("he/she is a good customer").

There are two main criteria for the evaluation of inductive learning procedures:

- *completeness*, i.e. that a concept description must correctly describe *all* the positive examples
- *consistency*, i.e. that a concept description must describe *none* of the negative examples.

Moreover, some additional criteria may be used as, e.g. *convergence*.

This general inductive learning scheme is often inapplicable. First, a *crisp classification* into the positive and negative examples is often artificial, and some *grade of positiveness/negativeness*, say between 0 and 1, may be more adequate. Second, a *misclassification* is always possible (and difficult to detect and correct).

The purpose of this paper is to propose a new approach to inductive learning to try to overcome the above two difficulties. Elements of possibility theory and a fuzzy - logic - based calculus of linguistically quantified propositions, both due to Zadeh, are employed.

Among other related approaches, one can mention Bergadano and Bisio (1988), Gemello and Mana (1988) or Raś and Zemankova (1988) though they do not allow to jointly handle imprecision and errors as in the present approach.

2. LINGUISTICALLY QUANTIFIED PROPOSITIONS

A *fuzzy set* A in $X = \{x\}$ is represented by - and often practically equated with - its *membership function* $\mu_A : X \to [0,1]$; $\mu_A(x) \in [0,1]$ is the *membership grade* of x in A, from *full nonmembership* to *full membership* through all intermediate values.

For our purposes the following operations on fuzzy sets are relevant:

- the *complement*

$$\mu_{\neg A}(x) = 1 - \mu_A(x), \quad \forall x \in X \tag{2}$$

- the *intersection*

$$\mu_{A \cap B}(x) = \mu_A(x) t \mu_B(x), \quad \forall x \in X \tag{3}$$

where $t : [0,1] \times [0,1] \to [0,1]$ is the so - called *t-norm* defined as, $\forall a,b,c \in [0,1]$:

(1) $at1 = a$

(2) $atb = bta$

(3) $atb \geq ctd$ if $a \geq c$, $b \geq d$

(4) $atbtc = at(btc) = (atb)tc$

Some examples of t-norms are: $a \wedge b = \min(a,b)$, which is the most commonly used, ab, and $1 - (1 \wedge ((1-a)^p + (1-b)^p)^{1/p}), p \geq 1$;

- the *union*

$$\mu_{A+B}(x) = \mu_A(x) s \mu_A(x), \quad \forall x \in X \tag{4}$$

where $s : [0,1] \times [0,1] \to [0,1]$ is the so - called *s-norm* (*t-conorm*) defined as, $\forall a,b,c \in [0,1]$:

(1) $as0 = a$

(2) - (4) as for a t-norm.

Some examples of s-norms are: $a \vee b = \max(a,b)$, which is the most commonly used, $a + b - ab$, and $1 - (a^p + b^p)^{1/p}, p \geq 1$.

Now we will sketch Zadeh's (1983) calculus of linguistically quantified propositions.

A *linguistically quantified proposition*, exemplified by most experts are convinced, may be generally written as

$$Qy\text{'s are } F \tag{5}$$

where Q is a *linguistic quantifier* (e.g., most), $Y = \{y\}$ is a *set of objects* (e.g., experts), and F is a *property* (e.g., convinced).

Importance B may also be added to (5) yielding

$$QBy\text{'s are } F \tag{6}$$

that is, say, ***most of the important experts are convinced.***

The problem is to find either truth(Qy's are F) in the case of (5) or truth(QBy's are F) in the case of (6).

In Zadeh's (1983) approach the ***fuzzy linguistic quantifier*** Q is assumed to be a fuzzy set in $[0,1]$. For instance, $Q =$ "almost all" may be given as

$$\mu_{\text{"almost all"}}(x) = \begin{cases} 1 & \text{for } x \geq 0.9 \\ 10x - 8 & \text{for } 0.8 < x < 0.9 \\ 0 & \text{for } x \leq 0.8 \end{cases} \tag{7}$$

Property F is defined as a fuzzy set in Y. If $Y = \{y_1, \ldots, y_p\}$, then it is assumed that truth(y_i is F) $= \mu_F(y_i)$, $i = 1, \ldots, p$.

Truth(Qy's are F) is now calculated using the (***nonfuzzy***) ***cardinalities***, the so-called $\sum$Counts in the following two steps:

$$r = \sum Count(F)/\sum Count(Y) = \frac{1}{p}\sum_{i=1}^{p} \mu_F(y_i) \tag{8}$$

$$\text{truth}(Qy\text{'s are } F) = \mu_Q(r) \tag{9}$$

In the case of importance, $B =$ "important" is a fuzzy set in Y,

and $\mu_B(y_i) \in [0,1]$ is a ***degree of importance*** of y_i, from definitely unimportant (= o) to definitely important (= 1), through all intermediate values.

We rewrite first "QBy's are F" as "$Q(B$ and $F)y$'s are B" and truth(QBy's are F) is calculated as:

$$r' = \sum Count(B \text{ and } F)/\sum Count(B) = \sum_{i=1}^{p}(\mu_B(y_i)\ t\ \mu_F(y_i))/\sum_{i=1}^{P} \mu_B(y_i) \tag{10}$$

$$\text{truth}(QBy\text{'s } are \ F) = \mu_Q(r') \tag{11}$$

For details, see the source Zadeh (1983) or Kacprzyk (1987). Another, more sophisticated approach which is not used here, was proposed by Yager (1983) [see also Kacprzyk (1987) or Kacprzyk and Yager (1985)].

3. INDUCTIVE LEARNING UNDER IMPRECISION AND ERRORS

Inductive learning (***from examples***) is to find a ***concept description*** R which describes (covers) ***all*** the ***positive*** examples and ***none*** of the ***negative*** ones. Thus, if $X = \{x\}$ is the set of positive and negative examples, then we seek an R such that

$$\text{"}All \ Px\text{'s are } R\text{" \& "}None \ Nx\text{'s are } R\text{"} \tag{12}$$

notice that "positive" and "negative" are here crisp (of "yes-no" type).

Problem (12) may be unsolvable as mentioned in Section 1 due to misclassification, errors, difficulty in the crisp determination of positiveness/negativeness, etc.

We "soften" here (12) by: find a concept description R such that

$$Q^+ \ \tilde{P}x\text{'s are } R\text{" \& "}Q^- \ \tilde{N}x\text{'s are } R\text{"} \tag{13}$$

where Q^+ is a linguistic quantifier exemplified by ***almost all***, ***most***, etc., Q^- is a linguistic quantifier exemplified by ***almost none***, (***at most***) ***few***, etc., $\tilde{P}$ denotes a ***soft*** positiveness and $\tilde{N}$ a ***soft*** negativeness, i.e. both to a degree between 0 and 1.

Similarly as in the conventional case, in the new formulation (13) the two main evaluation criteria are now:

- Q^+ - *completeness*, i.e. that a concept description must correctly describe (in fact, as well as possible) Q^+ (e.g., *almost all*) of the positive examples,

- Q^- - *consistency*, i.e. that a concept description must not describe more than Q^- (e.g., *almost none*) of the negative examples.

Notice, first, the inclusion of a *soft* positiveness/negativeness, and, second, a natural accounting for some errors in the classification. Since their number cannot be usually known precisely, evaluation like *almost all* and *almost none* are certainly adequate; evidently if we suspect more errors, we may use some "milder" quantifiers as, e.g., *much more than 75%* and *much less than 25%*.

An example (object) is described by the attribute-value pairs. A single attribute-value pair is called a *selector*, $[A\ r\ 'a']$, where A is an *attribute*, r is a *relation* (e.g., '=', '≥', ...), a is a *value*; [*height* = 'high'], [*color* = 'red'], [*temperature* ≥ '150°C'], etc. may exemplify the selectors.

For instance, an example x may be described as

$$x = [height = \text{'190 cm'}][color = \text{'reddish'}][temperature \gg \text{'100°C'}] \quad (14)$$

The value of A_i in a selector and in a particular example need not be the same as, e.g., when $s_i =$ [*height* = 'high'] and $x =$ [*height* = '190 cm']. We allow therefore for a *degree of identity* of the value of A_i in s and x, $\mu_{s_i}(x) \in [0,1]$, from 0 for definitely different to 1 for definitely identical through all intermediate values.

By a *complex*, C_j, we mean the conjunction of a number of different selectors, $s_{j_1}, \ldots, s_{j_k}$, i.e. $C_j = s_{j_1} \cap \ldots \cap s_{j_k}$.

The *degree of covering* x by complex $C_j = s_{j_1} \cap \ldots \cap s_{j_k}$ is defined as, $\forall x \in X$,

$$\mu_{C_j}(x) = \min(\mu_{s_1}(x), \ldots, \mu_{s_k}(x)) \quad (15)$$

or more generally

$$\mu_{C_j}(x) = \mu_{s_{j1}}(x)\ t\ \ldots\ t\mu_{s_{jk}}(x) \quad (16)$$

where $\mu_{C_j}(x) \in [0,1]$.

The *concept description* R is assumed to be the alternative of the complexes, $R = C_1 \cup \ldots \cup C_m$.

The *degree of covering* x by $R = C_1 \cup \ldots \cup C_m$ is defined as, $\forall x \in X$,

$$\mu_R(x) = \max(\mu_{C_1}(x), \ldots, \mu_{C_m}(x)) \quad (17)$$

where $\mu_{C_j}(x)$ is the degree of covering example x by complex C_j defined by (15) or (16); more generally

$$\mu_R(x) = \mu_{C_1}(x)\ s\ \ldots\ s\ \mu_{C_m}(x) \quad (18)$$

The *imprecision in the classification* into the *positive* and *negative* examples is represented by a *degree of positiveness* of x, $\mu_{\tilde{P}}(x) \in [0,1]$, from 1 for "definitely positive" to 0 for "definitely not positive (definitely negative)" and a *degree of negativeness*, $\mu_{\tilde{N}}(x) \in [0,1]$. For technical reasons, $\mu_{\tilde{P}}(x) = 1 - \mu_{\tilde{N}}(x)$. Notice that if $\mu_{\tilde{P}}(x)$, $\mu_{\tilde{N}}(x) \in \{0,1\}$, then we end up with the conventional case. The $\mu_{\tilde{P}}(x)$'s and $\mu_{\tilde{N}}(x)$'s will be interpreted here in terms of degrees of membership in fuzzy sets theory.

We seek now a concept description R^* such that

$$\text{truth}(\text{"}Q^+\ \tilde{P}x\text{'s are } R\text{"} \ \&\ \text{"}Q^-\ \tilde{N}x\text{'s are } R\text{"}) \to \max_R \quad (19)$$

i.e. to find an "(sub)optimal" R^* in the sense that (19) is satisfied (is true) to the highest possible extent.

In a more extended form, (19) may be written as to find an R^* such that

$$\text{truth}(Q^+\ \tilde{P}x\text{'s are } R) \wedge \text{truth}(Q^-\ \tilde{N}x\text{'s are } R) = \overline{\mu}_{Q^+}(R) \wedge \overline{\mu}_{Q^-}(R) =$$

$$= \mu_{Q^+}(\sum_{x\in X}(\mu_{\tilde{P}}(x) \wedge \mu_R(x))/\sum_{x\in X}\mu_{\tilde{P}}(x)) \wedge \mu_{Q^-}(\sum_{x\in X}(\mu_{\tilde{N}}(x) \wedge \mu_R(x))/\sum_{x\in X}\mu_{\tilde{N}}(x)) \rightarrow \max_R \quad (20)$$

where "$\wedge$" may be replaced by, e.g., a t-norm, and

$$\overline{\mu}_{Q^+}(R) = truth(Q^+\ \tilde{P}x\text{'s are } R) = \mu_{Q^+}(\sum_{x\in X}(\mu_{\tilde{P}}(x) \wedge \mu_R(x))/\sum_{x\in X}\mu_{\tilde{P}}(x)) \quad (21)$$

$$\overline{\mu}_{Q^-}(R) = truth(Q^-\ \tilde{N}x\text{'s are } R) = \mu_{Q^-}(\sum_{x\in X}(\mu_{\tilde{N}}(x) \wedge \mu_R(x))/\sum_{x\in X}\mu_{\tilde{N}}(x)) \quad (22)$$

where $\overline{\mu}_{Q^+}(R)$ is the *degree of completeness* and $\overline{\mu}_{Q^-}(R)$ is the *degree of consistency.*

The concept description R is built up *iteratively*, adding in each iteration a new complex to R, i.e. the number of examples covered by R is not decreasing. Moreover, by adding the complexes in a special way, $\overline{\mu}_{Q^+}(R)$ is increasing as quickly as possible, while $\overline{\mu}_{Q^-}(R)$ is decreasing as slowly as possible. This makes the algorithm more efficient.

To proceed to the description of the algorithm, we introduce now the concept of a *typoid.* Suppose that $x = s_1 \dots s_n = [A_1 = 'a_1'] \dots [A_n = 'a_n']$. Assume that A_i takes on its values in a set $\{a_{i_1}, \dots, a_{i_q}\}$. A *typoid* is defined as an artificial example $\tau = s_1^*, \dots, s_n^* = [A_1 = 'a_1^*'] \dots [A_n = 'a_n^*']$ such that each $s_i^* = [A_i = 'a_i^*']$ is determined by

$$\sum_{x\in X}(\mu_{\tilde{P}}(x) \wedge (1 - \mu_R(x)) \wedge \mu_{s_i=[A_i='a_i']}(x)) \rightarrow \max_{a_i\in\{a_{i_1},\dots,a_{i_k}\}} 22 \quad (23)$$

i.e. into τ we put such consecutive selectors which are *most typical* for the examples that are *not covered* by R and are *most positive;* $\mu_{s_i}(x) \in [0,1]$ is a *degree of identity* of A_i's value in s_i and in x.

The algorithm is now:

Step 1. To initialize, set:

a) $\mu_{\tilde{P}}(x) \in [0,1]$, for each example x (evidently, $\mu_{\tilde{N}}(x) = 1 - \mu_{\tilde{P}}(x)$)

b) $R =$ "$\emptyset$" and $C =$ "$\emptyset$" to be meant that the (initial) R contains no complex, and C contains no selectors.

Step 2. $R := R \cup C$, i.e. "add" to the current R a currently formed C, and assume this as the new R.

Step 3. Form a typoid τ as formerly described.

Step 4. Find an example $x^* \in X$ which is both most positive and most similar to τ formed in Step 3, that is $\mu_{\tilde{P}}(x) \wedge sim(x,\tau) \rightarrow \max_{x\in X}$, where $sim : X \times X \rightarrow [0,1]$ is some function expressing the *similarity* between x and τ, from 0 for *full dissimilarity* to 1 for *full similarity* as, e.g. $sim(x,\tau) = \frac{1}{n}\sum_{i=1}^{n}\mu_{s_{i^*}}(x)$.

Step 5. Form C as follows:

Substep 5a. To initialize, set $C =$ "$\emptyset$", and $h_{max} = 0$.

Substep 5b. For each $s_i^*, i \in I'$, where I' is the set of indexes of the attributes not occuring in C, (s_i^* is the i-th selector of x^* found in **Step 4**), calculate

$$h_i^* = h(\overline{\mu}_{Q^+}(R \cup (C \cap s_i^*)), \overline{\mu}_{Q^-}(R \cup (C \cap s_i^*))) \quad (24)$$

where $h : [0,1] \times [0,1] \rightarrow [0,1]$ is an averaging operator, e.g., $h(u,w) = (u+w)/2$;

Substep 5c. Find $h^* = \max_{i\in I'} h_i^*$, and i^* such that $h^* = h_{i^*}^*$;

Substep 5d. If $h^* > h_{max}$, then:

(1) $h_{max} := h^*$

(2) $C := C \cap s_{i^*}$
(3) $I' := I \cup \{i^*\}$
(4) go to **Substep 5b**;
else go to **Step 6**.

Step 6. If $\min(\overline{\mu}_{Q^+}(R \cup C), \overline{\mu}_{Q^+}(R \cup C)) > \min(\overline{\mu}_{Q^+}(R), \overline{\mu}_{Q^-}(R)))$, then go to **Step 2**.

Step 7. Output the final R, and STOP.

Since a typoid is a good starting point, the algorithm is in general very effective and efficient in practice.
Example. Suppose that we have two attributes, A and B, taking on only two values each, a_1 and a_2, and b_1 and b_2, respectively. Suppose that we have four examples, with their respective $\mu_{\tilde{P}}(\cdot)'s$ $[\mu_{\tilde{N}}(\cdot) = 1 - \mu_{\tilde{P}}(\cdot)]$:

$$x_1 = [A =' a_1'] \, [B =' b_1'] \qquad \mu_{\tilde{P}}(x_1) = 1 \tag{25}$$
$$x_2 = [A =' a_1'] \, [B =' b_2'] \qquad \mu_{\tilde{P}}(x_2) = 0.8 \tag{26}$$
$$x_3 = [A =' a_2'] \, [B =' b_1'] \qquad \mu_{\tilde{P}}(x_3) = 0.5 \tag{27}$$
$$x_4 = [A =' a_1'] \, [B =' b_2'] \qquad \mu_{\tilde{P}}(x_4) = 0 \tag{28}$$

We assume for simplicity that $\mu_{Q^+}(u) = u$ and $\mu_{Q^-}(u) = 1 - u$, $\forall u \in [0,1]$, t is "$\wedge$" and s is "$\vee$".
The consecutive steps of the algorithm are now:

Step 1. To initialize, we take the $\mu_{\tilde{P}}(\cdot)$'s as given above, and $R =$ "$\emptyset$" and $C =$ "$\emptyset$".

Step 2. $R := R \cup C =$ "$\emptyset$".

Step 3. We obtain via (23) $\tau = [A = 'a_1'] \, [B = 'b_1']$.

Step 4. We obtain that x_1 is the closest to τ.

Step 5.

Substep 5a. We assume $C =$ "$\emptyset$" and $h_{max} = 0$.

Substep 5b. We obtain first $\mu_{\tilde{P}}(x_1) + \mu_{\tilde{P}}(x_2) + \mu_{\tilde{P}}(x_3) + \mu_{\tilde{P}}(x_3) = 2.3$, and:
$\overline{\mu}_{Q^+}(R \cup (C \cap [A = 'a_1'])) = \overline{\mu}_{Q^+}([A = 'a_1']) \simeq 0.78$
$\overline{\mu}_{Q^-}(R \cup (C \cap [A = 'a_1'])) = \overline{\mu}_{Q^-}([A = 'a_1']) \simeq 0.88$
$\overline{\mu}_{Q^+}(R \cup (C \cap [B = 'b_1'])) = \overline{\mu}_{Q^+}([B = 'b_1']) \simeq 0.65$
$\overline{\mu}_{Q^-}(R \cup (C \cap [B = 'b_1'])) = \overline{\mu}_{Q^-}([B = 'b_1']) \simeq 0.71$

Substep 5c. We obtain:
$h_1^* = (0.78 + 0.88)/2 = 0.83$
$h_2^* = (0.65 + 0.71)/2 = 0.68$ hence $i^* = 1$, i.e. $s_{i^*} = [A = 'a_1']$;

Substep 5d. Since $0.83 > 0.68$, then:

(1) $h_{max} = 0.83$.
(2) $C := C \cap [A = 'a_1'] = [A = 'a_1']$ and we go to

...

Step 7. The final result is $R = [A = 'a_1']$, where $\overline{\mu}_{Q^+}(R) = 0.78$ and $\overline{\mu}_{Q^-}(R) = 0.88$.

4. CONCLUDING REMARKS

The algorithm proposed may be useful in many practical inductive learning problems in which imprecision, misclassification, inexact matching, etc. preclude the use of conventional techniques Applications, mostly in medicine, are very promising and encouraging.

REFERENCES

Bergadano F. and R. Bisio (1988) *Constructive learning with continous - valued attributes.* In B. Bouchon, L. Saitta and R.R. Yager (Eds.): **Uncertainty and Intelligent Systems**, Springer - Verlag, Berlin - Heidelberg - New York, 154 - 162.

Bouchon B., L. Saitta and R.R. Yager (1988) **Uncertainty and Intelligent Systems** [Proceedings of the Second International Conference on Information Processing and Management of Uncertainty in Knowledge - Based Systems (IPMU' 88); Urbino, Italy, 1988], Springer - Verlag Lecture Notes in Computer Science, Berlin - Heidelberg - New York.

Cohen P.R. and E.A. Feigenbaum (1982) **The Handbook of Artificial Intelligence**. Vol. 3 Kaufmann, Los Altos.

Dietterich T.G. et al. (1981) *Learning and inductive inference.* In P.R. Cohen and E.A. Feigenbaum (Eds.): **The Handbook of Artificial Intelligence**, Kaufmann, Los Altos, 323 - 525.

Gemello R. and F. Mana (1988) *Controlling inductive search in RIGEL learning system.* In B Bouchon, L. Saitta and R.R. Yager (Eds.): **Uncertainty and Intelligent Systems**, Springer - Verlag, Berlin - Heidelberg - New York, 171 - 178.

Kacprzyk J. (1987) *Towards 'human - consistent' decision support systems through commonsense knowledge - based decision making and control models: a fuzzy logic approach.* **Computers and Artificial Intelligence** 6, 97 - 122.

Kacprzyk J. and R.R. Yager (1985) *Emergency - oriented expert systems: a fuzzy approach.* **Information Sciences** 37, 147 - 156.

Michalski R.S. (1973) *Discovering classification rules using variable - valued logic system VL1.* **Proc of the 3rd Int. Joint Conference on Artificial Intelligence (IJCAI)**, 162 - 172.

Michalski R.S. (1980) *Pattern recognition as rule - guided inductive inference.* **IEEE Trans. on Pattern Analysis and Machine Intelligence** PAMI - 2, 249 - 361.

Michalski R.S. (1983) *A theory and methodology of inductive learning.* In R.S. Michalski, J. Carbonell and T. Mitchell (Eds.): **Machine Learning**. Tioga Press, Palo Alto, 83 - 133.

Mitchell T. (1982) *Generalization as search.* **Artificial Intelligence** 18, 203 - 226.

Quinlan J.R. (1983) *Learning efficient classification procedures and their applications to chess and games.* In R.S. Michalski, J. Carbonell and T.M. Mitchell (Eds.): **Machine Learning**. Tioga Press, Palo Alto, 463 - 482.

Raś Z.W. and M. Zemankova (1988) *Learning driven by the concepts structure.* In B. Bouchon, L. Saitta and R.R. Yager (Eds.): **Uncertainty and Intellegent Systems**, Springer - Verlag Berlin - Heidelberg - New York, 193 - 200.

Shaw M.J. (1987) *Applying inductive learning to enhance knowledge - based expert systems.* **Decision Support Systems** 3, 319 - 322

Yager R.R. (1983) *Quantifiers in the formulation of multiple objective decision functions.* **Information Sciences** 31, 107 - 139.

Zadeh L.A. (1983) *A computational approach to fuzzy quantifiers in natural languages.* **Computer and Mathematics with Applications** 9, 149 - 184.

SOME ALGORITHMS FOR EVALUATING FUZZY RELATIONAL QUERIES

P. Bosc & O. Pivert
IRISA/ENSSAT
B.P. 447
22305 Lannion
FRANCE

Abstract

An important issue in extending DBMS functionnalities is to allow the expression and execution of flexible queries in order to make these systems able to satisfy user needs more closely. To deal with this problem it is necessary to define new evaluation methods since standard (crisp) algorithms and access paths have proved to be inappropriate. A challenge in solving this problem is to keep the additional costs entailed by these new querying capabilities at a reasonable level. For this purpose we propose a set of query evaluation algorithms based on techniques expected on the one hand to restrict the amount of tuples to be transferred from disk to main memory and on the other hand to limit the computations which apply to these tuples. This work only concerns the evaluation of basic extended relational operators and may be seen as a first step towards the design of a fuzzy query optimizor. For each of the proposed methods, we suggest some useful measures for determining to what extent the corresponding algorithms are acceptable. Finally, we present an experimental framework for measuring the performance of these algorithms.

Keywords

DBMS, imprecise queries, query processing.

1. INTRODUCTION

A stream of growing interest in database research aims at increasing the usability of DBMS's by "fuzzy" querying systems which provide more flexibility, and therefore also allows more accuracy in characterization of information needs. Here, we view the situation in the framework of relational databases where the data are "crisp" and fully known. One criterion can be expressed as a combination of primary terms such as young, big, more or less equal, whose interpretation is based on the fuzzy sets theory. Thus the challenge is no longer one of selecting tuples which reply in a precise way to a question, but is rather one of determining to what extent each tuple satisfies it, by evaluating their degree of satisfaction [BOSC 88b].

It seems that the conceptual and logical aspect of fuzzy querying have almost exclusively commanded the attention of researchers while the aspects of implementation have been neglected. As regards relational DBMS's, the performance factor, summarized by the response time, remains the determining measure. This factor is especially crucial in flexible querying systems since the performance may become unacceptable due to the highly numerical nature of the methods envisaged.

Indeed, the practice of qualifying data by using fuzzy predicates is based on numerical operators, dedicated to the measurement of a satisfaction degree. These operators lead to a considerable increase in the computation time for examining each tuple. Thus it becomes all the more advantageous to have techniques available which can limit the volume of tuples to be transferred from the disk to the central memory, and in the same way reduce the number of calculations to be carried out.

In answer to this problem, one solution lies in implementing associative rather than sequential access paths. Classical access accelerators (indexing, hash-coding, ...), whose principle is based on access to tuples in a relation according to the values of one or more attributes, cannot be directly adapted to fuzzy querying. Indeed, contrary to boolean criteria, fuzzy queries do not refer directly to entry values of the index. For example, the condition *height* > *1,90m* offers direct access to well identified values, which is not the case for the condition *very tall*.

Thus the problem is on the one hand to define access paths adapted to the evaluation of fuzzy query operators (restriction, join, ...), and on the other hand, to study the impact of fuzzy predicates on standard evaluation techniques, in order to define appropriate algorithms for evaluating imprecise queries. The final objective is to propose several operating mechanics of evaluation for each category of fuzzy queries. In a classic DBMS, the evaluation of relational operators is obtained by multiple algorithms using diverse strategies, depending notably on the type of the considered query and on the existing access paths. Similarly, in the case of a fuzzy querying system, it is desirable to have different algorithms for each class of question in order to be able to choose the most appropriate one at the time of evaluation. This paper provides a first step towards the design of a fuzzy query optimizor.

2. QUERY EVALUATION : A GENERAL PRESENTATION

2.1 Introduction

In classical relational DBMS's, query evaluation is seen to comprise three distinct levels. The first one is related to implementing basic operators. For each of these operators, the system has optimized access algorithms which rely on the use of accelerators such as links, index and hash tables, and on the properties of the manipulated operands. The second level concerns global query evaluation. A large number of possible strategies exist for dealing with a question when it is complex ([JARK 84] for example). One of the roles of the system is to establish a plan for efficient execution through adequate arrangement of the basic operators. The third level corresponds to a semantic analysis of the request. It is a question of transforming the query entered by the user into an equivalent one which can be evaluated more efficiently.

Within the framework of a fuzzy querying system, we are led in the same way to define implementation methods for basic operators, and on a higher level, optimization algorithms for complex queries. However, since classical access paths used as accelerators are inefficient when it is a question of evaluating a fuzzy query, it is is necessary to extend the classical concept of indexing. The idea is no longer to carry out associative access on the attribute values but on the membership function values associated with primary fuzzy terms.

2.2 The derivation principle

In most applications, only a small part of the objects is retained when selection takes place. Therefore we shall try to limit access and computation solely for tuples whose degree of membership is greater than the acceptance threshold required by the user. In order to achieve this, the idea is to set up a boolean condition, according to the fuzzy criterion, which reduces the initial group of tuples by only keeping back those situated above the threshold. Then, there remains the application of the fuzzy selection on this group of qualified objects [BOSC 88a].

Let λ be the acceptance threshold appearing in the request. By applying the derived boolean condition, we are trying to obtain the λ-cut of the fuzzy criterion. In practice however, this constraint turns out to be too rigid and can not always be verified. The objective will thus be to determine the smallest possible group including the λ-cut by means of a boolean query called the *envelope of the initial fuzzy query*. The derivation rules are definable for each of the fuzzy operators (modifiers, connectors,...) [BOSC 91]. The aim is to express the boolean envelope of a complex term as a combination of envelopes related to the simple components of this term.

The derivation rules rely on properties of equality or inclusion between the group of tuples corresponding to the left part and those corresponding to the right part. When equality exists between these two envelope sets, derivation poses no problem. This is the case, notably, for the *and* and *or* operators. However, for certain operators, associated rules are obtained so that the initial group is only included in the final set. This is the case for any mean operator m since m(a,b) ≤ max(a,b) allows a general rule whose left part is m and the right part is a disjunction.

There is a risk of seeing a significant degeneration of the envelope set which, after several applications of the derivation rules, can become equal to the entire relation. This will be the role of the optimizor to measure the implications of this risk.

2.3 Extended indexing

In classical DBMS's, one access method commonly used among the mono-relation accelerators is indexing. Thus, it was natural to try to extend it to the case of a fuzzy querying system. The inherent peculiarity of imprecise questions is to not include a condition of explicit membership of a value to a set. Therefore, a selective access can not be directly used on the attribute values. On the other hand, since fuzzy requests express a condition on the tuples' degree of membership, the idea consists of indexing these degrees. This solution is derived from a technique adopted in certain fuzzy information retrieval systems. Thus certain authors [RADE 82], [TAHA 76] propose to index all key words, by linking a list of document references at each degree of compatibility (measuring the relevance of the document with respect to the key word), each document of the list satisfying the key word to the considered degree. The principle of extensive indexing here is to link to each degree of membership for a primary term, a list of adresses of tuples which satisfies the term to the given degree. In order to limit the volume of secondary data and to facilitate updating management, classical indexes can be drawn upon. For each value of the degree of membership of the considered fuzzy term, is represented a list of the corresponding attribute values. In other words a supplementary level of indexing for the attribute (one index per fuzzy term defined on this atttribute) is built to refine the classiscal index. A main problem is therefore to analyze how the λ-cut of a relation is to be calculated using extended indexing with respect to a criterion.

3. SIMPLE OPERATORS

3.1 The fuzzy restriction

We shall try here to describe the fuzzy restriction of the relation R by a fuzzy criterion denoted fc,with the use of an extended index. The use of extended indexing is well adapted for the calculation of the λ-cut when the criterion is formed of a simple term or a conjunction of terms. In this case, one simple and quite effective method, used in standard relational systems, is to determine the best indexed sub-criterion, i.e the one which offers the best (access cost)/(calculation cost) compromise [SELI 79]. As the interpretation of the fuzzy "and" is minimum, in order to evaluate conjunction,we can simply obtain the elements satisfying the most selective term via an extended index and then filter them by submitting them to the membership functions of the other terms present in the criterion [BOSC 89].

For the evaluation of a disjunction, we apply the fuzzy set union operator *max*. However, in this case, we can not determine from a global λ threshold on a criterion, a threshold on each term of which it is composed in order to envisage nested scans of the parts of the corresponding indexes. When the criterion carries at least one non-indexed term, the only course of action is to sequentially scan the entire relation and calculate for each tuple the degree to which it satisfies the criterion. In an adverse case, a solution may be to construct an <address, degree> list from the indexes for each term, and then to proceed to a sort-merge of these lists [BOSC 89]. A request expressing a fuzzy restriction can also comprise other aggregate functions than conjunction and disjunction, e.g arithmetic mean or weighted mean, thus permitting a compensation effect between the fuzzy concepts given in arguments. One way to evaluate averaging functions is to determine, from the global threshold, a threshold on each of the terms and then perform nested scans of the associated indexes.

3.2 The fuzzy join

A fuzzy join is defined by a fuzzy bi-relational condition (e.g. more or less equal, slightly bigger, a little more recent). In the boolean case, the equi-join could be carried out by a synchronous scan of the indexes, with intersection of the related inverse lists. The problem here is that for a given value u of A, one cannot deduct a unique value v of B such that (u, v) satisfies the fuzzy join i.e. such that $\mu_\theta(u,v) \geq \lambda$ where θ is a fuzzy join predicate. One approach that can be used to avoid the need to elaborate the cartesian product of the implied relations, is to draw on the index in order to reduce the number of accesses that have to be carried out on the relations. Theoretically, the evaluation passes through a nested scan of indexes constructed on the join attributes of the relation. In this way, we may obtain the pairs of attributes values which satisfy the acceptance threshold.

Let us consider the join of the relations R(X, A) and S(Y, B) expressed by the fuzzy predicate as (A θ B). Let us then suppose that the secondary indexes I_A on A, and I_B on B are available. The couples of the retained values $(u, v) \in D(A) \times D(B)$ are such that $\mu_\theta(u,v) \geq \lambda$.With a fixed value of u, $\mu_\theta(u,v)$ can be restated as $\mu_{\theta_u}(v)$ and $\mu_\theta(u,v) \geq \lambda$ becomes $v \in \lambda\text{-cut}(\theta_u)$.
Thus, it is possible to restrain the index of the internal loop, thanks to the λ-cut phenomenon. By reaching uniquely the B-values of S (projection) belonging to the λ-cut of θ_u, a complete scan of the index I_B is avoided. Traditional join algorithms (sort-merge, hash-coding, ...) [STON 76] can equally be adapted to the fuzzy join case, at the expense of a relative small loss of efficiency [BOSC 89]. Classical relational systems also use the semi-join concept, notably with a view to optimizing the evaluation of certain joins. The result of the semi-join of R by S is composed of the tuples x of R so that at least one tuple y of S exists which satisfies x.A θ y.B. This idea can be extended to the case where the joining predicate is fuzzy. Then the only problem is to determine the degree of

membership for the remaining tuples. The fact that several tuples y in S can correspond to an x tuple in R means that an aggregation function must be used to obtain a unique degree of satisfaction for x.

4. MORE COMPLEX OPERATORS

4.1 Fuzzy quantification

In this section we propose to describe the evaluation of requests that imply a fuzzy quantifier Qf. A fuzzy quantifier can be defined as a fuzzy set which induces an imprecise constraint on the fuzzy set's cardinality [ZADE 83]. The idea is to translate the request into a form that allows the classic evaluation techniques to be used. When this is possible, we will rely also on the methods defined in the framework of the fuzzy selection in order to carry out an initial restriction. Here two types of evaluation can be distinguished, according to whether the fuzzy quantifier implies one or two criteria.

a) Proposition quantified on one criterion : Qf **are** (fc_1)

The query "find the departments in which most of the employees are well-paid and young" is an example of such a proposition. Two kinds of request appear : those that are either relatively (a few part, most of, ...) or absolutely (a dozen, about 5, none, ...) quantified. In each case, the request implies that the relation is partitioned, and fixes a condition on the cardinality of the obtained partitions. That can be translated by a qualifying SQL-like block of the type :

select λ A **from** R **group by** A **having** Qf **are** fc

1. If quantification is absolute, Qf concerns the absolute cardinality of fc in each partition E Σ count(fc) = $\Sigma_{x \in E} \mu_{fc}(x)$. Qf (sum ($\mu_{fc}$)) can be substituted for the condition "Qf **are** fc". We can avoid considering those tuples whose degree of membership for the term fc is equal to 0; thus it is possible to initially restrain the relation by such terms, in order to reduce the cardinality of the group to be partitioned. The evaluation process may be decomposed into three parts : i) restriction of R by fc_1 producing the intermediate relation R_1, ii) grouping of the tuples of R_1 according to A, and creation of a binary relation R_2(A,B) where B is the sum of μ_{fc} for the tuples of a partition, iii) restriction of R_2 according to the condition $\mu_{Qf}(B) \geq \lambda$, if λ represents the threshold applying to the query.

2. For the case where quantification is relative, Qf will be applied to the relative cardinality of fc in a partition E : Σ count (fc) = $\Sigma_{x \in E} (\mu_{fc}(x)) / n$ where n = $\| E \|$. In this case, it is no longer possible to initially restrict R by fc in each partition, since all the tuples, including those whose degree is zero must be taken into account at the time of the calculation of n. The criterion Qf becomes Qf (sum(μfc)) / count (*)) where count(*) denotes $\| E \|$ the number of elements in the partition E

Now the evaluation comprises only the preceding stages (ii) and (iii), and in (ii) B takes the value : sum(μ_{fc}) / count (*). In both cases, one can see that Qf can be considered as a fuzzy criterion relying on the result of classical aggregate functions *sum* and *count*. Therefore in this case the query evaluation comes within the framework of the evaluation of a standard *having clause*.

b) Proposition qualified on two criteria : Qf(fc_1) **are** (fc_2)

The query "find the departments in which most of the young employees are well-paid" illustrates this kind of proposition. In this case, we are interested in the proportion of tuples verifying fc_1 which also verify fc_2. In other words, Qf regards the relative cardinality of fc_2 in fc_1 which is expressed by :

$$\Sigma Count(fc_2/fc_1) = \frac{\Sigma Count(fc_1 and\ fc_2)}{\Sigma Count(fc_1)} = \frac{\Sigma_E \min(\mu_{fc_1}(x), \mu_{fc_2}(x))}{\Sigma_E \mu_{fc_1}(x)}$$

As we can see, only the tuples who belong to the support of fc1 (i.e. such as $\mu_{fc1}(x) > 0$) are taken into account in these calculations. It is therefore possible to initially restrict the relation by the fuzzy criterion fc_1 by using the preceding methods of access, which leads to a query of the type :

select λ A **from** R **where** fc_1 **group by** A **having** Qf **are** fc_2

The initial restriction includes calculation of the degree of tuples with respect to the term fc_1 and elimination of tuples whose degree is zero. One evaluation method consists of using a random file hash-coded on the grouping attribute A. This method comprises three stages : i) fuzzy restriction of the relation R by fc_2 , projection on the attributes A and A' (the latter being the attribute which concerns fc_2) and obtaining the fuzzy relation R_1 , ii) sequential scan of R_1 so as to calculate for each value of A, $\Sigma_1 = \Sigma$ count (fc_1) and $\Sigma_2 = \Sigma$ count (fc_1 and fc_2) that are written at the corresponding entry to the file, iii) scan the entries *a* of the random file and select *a* such as $\mu_{Qf}(\Sigma_2 / \Sigma_1) \geq \lambda$.

4.2 Fuzzy sets comparison

In this section we consider the selection of partitions based on the comparison of the contents of each partition with a group of tuples generated by a selection. The comparisons are founded on fuzzy sets comparison indices defined in the unity interval [DUBO 82]. These indices are not unique, and they vary according to the nature of the comparison being desired. The groups compared can themselves be fuzzy. The following query is an example : "find the departments where the group of young employees is included in the overall group of badly paid employees".

We shall also be trying to take advantage of the fuzzy restriction mechanisms. The proposed method of evaluation uses a technique derived from the merge-sort [BLAS 76] implementing the parallel scan of the groups concerned. Let S be the fuzzy reference set, and let R be the fuzzy relation that is partitioned on the attribute A. Let E (val) be a partition of R associated to the value val of A.

Val will be qualified if μ_{op} (E(val), S) $\geq \lambda$, where op is a fuzzy set comparison operator. For each value val, E(val) and S are scanned simultaneously so as to calculate the value of the comparison index. If this is not less than the acceptance threshold, val belongs to the fuzzy relation which constitutes the result. The main cost of the algorithm occurs as a result of the multiple scannings of S. However, as this relation comprises only one attribute, it may be kept in main memory for fast processing.

5. PERFORMANCE EVALUATION

In view of the uncertainties which remain despite the efficiency of the proposed methods, we must consider the performances of a system based on these methods. The safest and simplest approach seems to consist of using a benchmark and measuring the consumption of CPU time and the amount of input/output made by the system. The fact that on the one hand, these performances often depend largely on data present in the base and that, on the other hand, the usable operators are numerous and diverse would have made the design of a realistic analytic model very complex. The assessment is mainly concerned with the validity limits of the proposed methods, and with the determination of their efficiency. A primary objective is to estimate the overhead cost resulting from dealing with fuzziness by the system. The most obvious solution to achieve this consists of comparing the evaluation of fuzzy queries with that of analogous boolean queries when they exist. By an analogous query, we understand a query whose semantics is relatively close and which therefore is likely to give a result of an equivalent size. Specific measures for each evaluation method are necessary, as well as these comparisons.

5.1 The derivation method

The main question to examine concerns the evolution of the envelope set. When the derivation rule which is applied is strong (e.g. conjunction, disjunction), everything runs smoothly one can be sure that the obtained set will not contain more tuples than the initial set. On the other hand, if the rule applied is weak (e.g. a mean), the envelope set may degenerate. After application of the rule, this is likely to contain non-satisfying tuples, due to the non-equivalence of the initial and final boolean criteria. At the very worst, after several applications of such rules, the envelope set can become equal to the entire relation. The aim is to assess this danger for each specific case. One can experimentally quantify the "deviation" using the following ratios :

- (number of tuples effectively accessed) / (number of retained tuples) (a)
- number of accessed pages/ number of useful pages (b)

Thus, it is a question of determining how these quantities vary according to the characteristics of the aggregated fuzzy terms (selectivity,...) and of the data base (distribution of the degrees) on the one hand, and on the other hand from the query structure (conjunction of aggregates, disjunction of aggregates,...) and the type of aggregates implied (arithmetic, geometric, harmonic or weighted mean,...). The existence of access paths does not play any role in (a) but does in (b). An interesting point in the derivation method is that the resulting selection, being boolean, can be executed by a conventional querying system, which in this stage, allows the use of the DBMS's classical access paths and optimizor.

5.2 Extended indexing

One way of judging the efficiency of this supplementary level of indexing, is to compare the evaluation of indexed queries with that of non-indexed queries, the measures bearing on the amount of I/O carried out and the amount of CPU time used. Furthermore, we shall try to determine which evaluation method is the most efficient in a given context for each basic operator.

5.3 The benchmark

5.3.1 Presentation

The measures will be carried out with the help of the Wisconsin benchmark designed by Dewitt [BITT 83]. This benchmark facilitates the control of the different relative parameters for the values present in the base (selectivity factors, proportion of doubles,...) and for the organisation of the relations (index, cluster,...).

5.3.2 Measurement and analysis

There are two ways of proceeding when one wishes to measure the time needed to execute a query : stock the selected tuples in a relation (writing on disk) or displaying them at the terminal. Both methods have their disadvantages : in the first case, elimination of duplicates has an effect on the execution time; in the second case, the time taken for the transfer to the terminal has to be taken into account; The solution chosen in [BITT 83] consists of placing the resulting tuples in a relation without eliminating the duplicates. In our case, in the absence of this option, we will have to evaluate beforehand the overhead cost inherent to the chosen method. Another problem lies in the choice of a representative sample of results for each type of query. The comparisons will be carried out on the basis of the CPU time and of the number of I/O measured for some queries that are representative of each class.

5.3.2.1 Fuzzy selections

Usually, it is considered that the speed of a selecting operation depends on the physical organization of the relation, on the selectivity factor, on the existence of specialized hardware, and on the cost of sending the result onto the screen or stocking it in another relation. In our case, only the first two factors will be considered as variable parameters.

5.3.2.2 Fuzzy joins

According to Dewitt's observations, systems' performances vary in accordance with the hypotheses done (index or not, complex or not,...) more for a join than for any other type of query. According to this author, a set of 10 representative join queries on a relation of 10000 tuples is sufficient to prove the effect of the different parameters. One of our objectives will be to determine how the CPU time consumed evolves for each join algorithm according to the size of the relations implied in the join.

5.3.2.3 Queries with fuzzy quantifiers

The problem is to define "equivalent" boolean queries. One way consists of qualifying the partitions with the help of a boolean criterion which brings into play the aggregates sum and count. For example, the query :

select 0.7 no_dep **from** Emp **group by** no_dep **having** the_most **are** young

can be put in parallel with :

select no_dep **from** Emp **group by** no_dep **having** sum (age)/count(*) < 33

When the quantified proposition brings two fuzzy terms into play, or in case of fuzzy sets comparison, it is possible to determine in a similar way comparable boolean queries. The constants which appear in the examples above are adjusted so as to obtain roughly similar results.

6. CONCLUSION

The main aim of this paper is to show how the proposed methods for evaluating fuzzy questions fit into the framework of classical evaluation, and to describe an operating mode for measuring their performances. The implementation of basic fuzzy operators uses an indexing technique which is an extension of the one employed in traditional SQL-like querying systems. As far as the evaluation of a global query is concerned, the proposed strategy, which relies on the derivation of a boolean expression from a fuzzy criterion, can be compared to the query transformation principle used in a boolean querying context. We have identified a number of query forms for which an efficient processing using these methods is possible. In particular, this is the case for fuzzy conjunctions, for certain fuzzy joins, and to some extent, for disjunctions and means. It turns out that a number of query forms requires an exhaustive scan of the relations involved. This is especially the case for queries where fuzzy quantifiers ór fuzzy set comparisons arise. The performances of all these algorithms will be determined by empirical measures carried out with a benchmark in order to specify the overhead that the fuzziness in queries represents for a DBMS, and to establish, by a series of comparative measures, the specific cases where each of the proposed methods can be effectively applied. We have shown that "simple" fuzzy queries can be resolved with only minor overheads.

REFERENCES

Bitton D., Dewitt D.J., Turbyfill C. (1983) Benchmarking database systems : a systematic approach, Proc. VLDB Conference, pp 8-19.

Blasgen M.W., Eswaran K.P. (1976) On the evaluation of queries in a relational database system, IBM Systems Journal, vol 16, pp 363-377.

Bosc P., Galibourg M. (1988a) Flexible selection among objects : a framework based on fuzzy sets, Proc. SIGIR Conference, pp 433-449.

Bosc P., Galibourg M., Hamon G. (1988b) Fuzzy querying with SQL : extensions and implementation aspects, Fuzzy sets and systems 28, pp 333-349.

Bosc P, Pivert O. (1989) Algorithms for flexible selection in relational databases, Proc 10th ASLIB Conference (G-B), pp 211-226.

Bosc P., Pivert O. (1991) Fuzzy querying in conventional databases, to appear in "Fuzzy logic for the management of uncertainty" (L. Zadeh & J. Kacprzyk eds).

Dubois D., Prade H. (1982) A unifying view of comparison indices in a fuzzy set theoritic framework, in "Fuzzy set and possibility theory" (R. Yager ed.).

Jarke M., Koch J. (1984) Query optimization in database systems, ACM Computing Surveys, vol 16, n° 2, pp 111-152.

Radecki T. (1982) Generalized boolean methods of information retrieval, International Journal of Man-Machine Studies, vol 18.

Selinger P.G. et al. (1979) Access path selection in a relational database management system, Proc. of ACM SIGMOD Conference, pp 23-34.

Stonebraker M. et al. (1976) The design and implementation of INGRES, ACM Transactions on Database Systems, vol 1, n° 3, pp 189-222.

Tahani V. (1976) A fuzzy model of document retrieval systems, Information Processing & Management, vol 12, pp 177-187.

Zadeh L.A. (1983) A computational approach to fuzzy quantifiers in natural languages, Computer and Mathematics, vol 9, pp 149-184.

TIME AND INCOMPLETENESS IN A DEDUCTIVE DATABASE

M. H. Williams
Department of Computer Science, Heriot-Watt University
79 Grassmarket, Edinburgh EH1 2HJ

Q. Kong
CITR, Ritchie Research Laboratories
University of Queensland, Australia

Abstract

Much attention has been given to the problem of time and how it can be handled within a logic based system. Another problem which has attracted considerable attention is that of incompleteness - a concept which is particularly relevant in the temporal domain. This paper considers how these two ideas may be brought together to produce a powerful system for dealing with data with a significant temporal variation.

Keywords

Deductive database, incomplete information, temporal database.

1. INTRODUCTION

The idea of using programming languages based on logic has been gaining increasing support in recent years. However, while there are considerable benefits to be obtained from using it as a programming language, it is not completely ideal for this in its present form. One major weakness from which it suffers is the assumption of monotonicity. This is one of the basic presuppositions of logic, namely that all propositions which are used in an argument are universally true. Consequently, if additional facts do become known, it will never be necessary to amend or withdraw conclusions as one becomes aware of new facts.

In other words, classical logic takes no account of changes in the state of a system. However, in attempting to apply logic to computer-based problems, one is frequently faced with situations where the information changes with time. Database systems are a case in point - many such systems have to cope with constantly changing information (e.g. on the state of a firm's operations). Expert system are another class of problems which have to deal with change from two quarters. On the one hand, while the knowledge rules are being refined and tested, they may be changed to reflect the expert system builder's understanding of the nature of the expert knowledge; on the other hand, when the system is used, it may have to deal with a situation in which the information available changes with time. An extreme example of this type is a blackboard system such as might be used for air traffic control or military applications, which is constantly processing new data and revising its beliefs on the state of the system it is attempting to monitor.

Modern logicians paid little attention to the problems of time until the middle of this century [2] since when there has been growing interest in finding solutions to these problems. Approaches put forward vary from purely qualitative to quantitative ones. The crudest approach (from the point of view of coarseness of representation) is that of **tense logic**, in which four modal tense quantifiers (H - *it will be that*, P - *it has been that*, F - *it has always been that*, and G - *it will always be that*) are added to propositional calculus . [5] A larger set of modal tense quantifiers and temporal operators has been put forward by Seranadas . [10]

At the other extreme one has various approaches based on some form of time stamp as used in databases (e.g. Clifford and Warren). [4] However, even in this case the granularity of temporal values may vary considerably. In a coarse-grained model the precision with which time is represented may be at the level of dates (day, month, year) although for some applications one may only require accuracy at the level of the year. In a fine-grained system one requires a more accurate representation such as second/minute/hour/day/month/year or some equivalent form.

The situation calculus [8] is one such approach in which global states are used as explicit parameters of time-varying relationships. However, it suffers from an excessive computational overhead in carrying over virtually all relationships from one state to the next. To avoid this problem Kowalski and Sergot [7] have proposed a system based on events which they term *event calculus*.

This paper attempts to combine the notion of time as it is handled in the event calculus with that of incompleteness or granularity to produce an effective system for dealing with databases and similar applications.

2. DEDUCTIVE DATABASES

A deductive database can be regarded as a set of clauses of first order logic. In general each clause has the form:

$$p_1(\vec{A_1}) \vee p_2(\vec{A_2}) \vee \cdots \vee p_n(\vec{A_n}) \leftarrow q_1(\vec{B_1}) \wedge q_2(\vec{B_2}) \wedge \cdots \wedge q_m(\vec{B_m}) \quad (1)$$

in which $p_i(\vec{A_i})$ and $q_j(\vec{B_j})$ are predicates, $i \geq 0$ and $j \geq 0$. Such a clause represents the knowledge that if $q_1(\vec{B_1}) \wedge q_2(\vec{B_2}) \wedge \cdots \wedge q_m(\vec{B_m})$ is true then it follows that $p_1(\vec{A_1}) \vee p_2(\vec{A_2}) \vee \cdots \vee p_n(\vec{A_n})$ is also true. Specifically, if in a clause $n > 0$ and $m = 0$, i.e. the clause has the form:

$$p_1(\vec{A_1}) \vee p_2(\vec{A_2}) \vee \cdots \vee p_n(\vec{A_n}) \quad (2)$$

it represents a fact stored in the database. If $n = 0$ and $m > 0$, i.e. the clause has the form:

$$\leftarrow q_1(\vec{B_1}) \wedge q_2(\vec{B_2}) \wedge \cdots \wedge q_m(\vec{B_m}) \quad (3)$$

it represents a query applied to the database.

Due to the computational complexity of satisfiability in such system, the type of clause usually used in a deductive database is a restricted form, known as a Horn clause. A Horn clause is a clause which contains only one positive literal, i.e. it gives one definite conclusion. A database which consists entirely of Horn clauses is referred to as a definite deductive database. The clauses in such system have the form:

$$p(\vec{A}) \quad (4)$$

$$p(\vec{A}) \leftarrow q_1(\vec{B_1}) \wedge q_2(\vec{B_2}) \wedge \cdots \wedge q_n(\vec{B_n}) \qquad (5)$$

A relation in such a database is defined as a group of clauses in which the head of each clause has the same predicate name and arity. For example, in the clauses given in (4) and (5), $p(\vec{A})$ represents a relational scheme, p is the relation name, $\vec{A}=<A_1, A_2, ..., A_n>$ $(n \geq 0)$ is a set of values, each of which comes from an attribute domain. Formula (4) can be regarded as a fact, formula (5) as a rule. $q_i(\vec{B_i})$ are predicates which can be defined either by a fact or by a rule in the database. Such a database can be used not only to search for information explicitly stored in it, but also to deduce the information implied by its rules.

For example, suppose one has a database of family relationships, which contains two relations: *father* (X, Y) which indicates that X is Y's father, and *ancestor* (X, Y) which is interpreted as X is Y's ancestor. The two relations are defined as follows:

father (henry, tom)
father (tom, bob)
father (tom, anna)
father (john, rosa)

ancestor (simon, rob)
ancestor $(X, Y) \leftarrow$ *father* (X, Y)
ancestor $(X, Y) \leftarrow$ *father* $(X, Z) \wedge$ *ancestor* (Z, Y).

In this database the relation *father* contains only facts, while *ancestor* contains both fact and rules. Some information, such as *Simon is Rob's ancestor*, is stored in the database directly, and some is not but can be deduced from this database, e.g. *Henry is Anna and Bob's ancestor.*

3. EVENTS AND STATES

In order to take account of the effect of time, a state-transition approach is assumed in which time is represented as a continuous domain which is sampled at discrete points.

In the simplest interpretation, at any given point in time the set of properties describing the system of interest, constitute a *state*. An *event* is a phenomenon which takes place in the universe of discourse at some point in time and which causes a change from one state (corresponding to one set of properties) to the next (corresponding to a slightly different set of properties). In this simple interpretation the transition is assumed to take place instantaneously so that at any moment in time a unique state applies.

Suppose that initially the system is in state S_0 and that with each transition the state number is increased by 1. Then the complete history of the system during the period of interest may be characterised by the sequence of states.

$$(S_0, S_1, \ldots, S_n).$$

Corresponding to each state S_i is a unique time t_i when that state is first entered. Thus the period during which the system is in state S_i corresponds to the half-open interval of time $[t_i, t_{i+1})$. For consistency the total period of interest will be assumed to be the half-open interval $[t_0, t_{n+1})$.

Relating this to logic, the set of properties which constitute a state may be represented as a set of Horn clauses. As a consequence of the above assumptions, each simple fact describing a property of the system will have associated with it a half-open interval of time, $[t_i, t_j)$, during which it is true. One simple way of representing this information is by linking the clause to two separate clauses recording the beginning and end of this interval.

For example, consider a system consisting of a set of employees over a period $[t_0, t_4)$. Suppose that employee number 1, Smith, is present for the entire period, Jones (number 2) was appointed at time t_1 but left at time t_3 whereas Evans (number 3) was appointed at time t_1 and was still employed at time t_4. Green (number 4) was present from time t_2 to t_4. The set of clauses representing this system may be written as follows:

$employee\,(t_a, 1, smith)$
$employee\,(t_b, 2, jones)$
$employee\,(t_c, 3, evans)$
$employee\,(t_d, 4, green)$
$start\,(t_a, t_0)$
$start\,(t_b, t_1)$
$start\,(t_c, t_1)$
$start\,(t_d, t_2)$
$end\,(t_b, t_3)$.

From this it is clear that at time t_0 there will be one employee (*Smith*) in the system, at time t_1, there will be three (*Smith*, *Jones* and *Evans*), etc. Where the end dates are missing, the end of the interval in which that property holds, has not yet been encountered.

Alternatively the transitions affecting a system may be considered from the point of view of events. In such a case there will generally be a set of rules R which determine the consequences of these events on the properties of the system. Thus if during some period $[t_0, t_n)$ one has a sequence of events E where E_i gives rise to the transition from state S_i to state S_{i+1}, the complete history of the system may be characterised by the initial state, the sequence of events which have caused transitions to occur and the set of rules R which translate these into changes in the state of the system, viz.

$$S_0, (E_0, \cdots, E_n), R.$$

Since in this simple interpretation events are expected to occur instantaneously, it may be assumed that if event E_i occurs at time t_j, the consequence of this event will take effect at time t_j. With this in mind each simple fact which contains information on an event which takes place at time t must be qualified by a parameter which indicates the time of the event.

To illustrate this, consider the example used by Kowalski [6] of engaging and losing academic staff. The basic set of events (E_i) are

hire (*e* 1, *mary*, *lecturer*)
time (*e* 1, 10/5/1970)
leave (*e* 2, *john*, *lecturer*)
time (*e* 2, 1/6/1975)
leave (*e* 3, *mary*, *professor*)

$time(e3,\ 1/10/1980)$
$promote(e4, mary, lecturer, professor)$
$time(e4,\ 1/6/1975).$

This tells us that the event in which Mary was hired as a lecturer took place on 10th of May 1970, the event in which John gave up his lecturer position took place on 1st June 1975, etc. In order to relate this to the state of the system, the following set of rules (R) is introduced:

$rank(after(E), X, Y) \leftarrow hire(E, X, Y)$
$rank(before(E), X, Y) \leftarrow leave(E, X, Y)$
$rank(after(E), X, Y) \leftarrow promote(E, X, Y, Z)$
$rank(before(E), X, Z) \leftarrow promote(E, X, Y, Z).$

This tells us that for some period *after(e)* the rank of x is y if x is hired as y at time e, and so on. To define the beginning and end times of these periods the following additional clauses are required:

$start(after(E), E)$
$end(before(E), E).$

For example, *start(after(e1), e1)* states that the period *after(e1)* begins at the time of event $e1$. Finally, we also require

$end(after(E), E') \leftarrow after(E) = before(E')$
$start(before(E'), E) \leftarrow after(E) = before(E')$
$after(E) = before(E') \leftarrow rank(after(E), X, Y) \wedge rank(before(E'), X, Y) \wedge E < E'$
$\qquad \wedge \neg((rank(after(E^*), X, Y')) \wedge E < E^* \wedge E^* < E')$
$\qquad \wedge \neg((rank(before(E^*), X, Y')) \wedge E < E^* \wedge E^* < E').$

Information about an occurrence may usually be represented either as an event or as a property of a set of states. For example, the information about the hiring, promotion and departure of lecturers in the above example could be reexpressed as

$rank(t_a, mary, lecturer)$
$rank(t_b, mary, professor)$
$rank(t_c, john, lecturer)$
$start(t_a,\ 10/5/1970)$
$end(t_a,\ 1/6/1975)$
$start(t_b,\ 1/6/1975)$
$end(t_b,\ 1/10/1980)$
$end(t_c,\ 1/6/1975).$

The treatment of time as an entity rather than as a property enables missing values to be handled in a natural way. For example, one might rewrite the latter representation to express time as a property as follows:

$rank(10/5/1970,\ 1/6/1975, mary, lecturer)$
$rank(1/6/1975,\ 1/10/1980, mary, professor)$
$rank(?,\ 1/6/1975, john, lecturer).$

However, in this case one is faced with the problem of dealing with the missing value for John.

4. PARTIAL INFORMATION

The general problem of incompleteness of information is well-known. Two particular categories of incomplete information are particularly important, viz.

(a) Missing values. In this case the value of an attribute is completely unknown. Kowalski and Sergot have studied this case in detail for the event calculus.

(b) Incomplete values. In this case the value of an attribute is not known exactly but it is known that it ranges over a specified subset of the attribute domain.

An incomplete value can be regarded as a special kind of value which lies part way between an exact value and a missing value - i.e. a value which ranges over a subset of the attribute domain. Conversely, a missing value can be regarded as a special kind of incomplete value in that it represents a value which ranges over the whole attribute domain. The notion of incompleteness corresponds to the notion of granularity of information.

To illustrate this, consider the problem of accuracy of values in any continuous domain. For example, suppose that an attribute has the value 50. Does this mean a value which is somewhere between 45 and 55, or a value between 49.5 and 50.5 or even a value between 49.999995 and 50.000005? Clearly the accuracy of a number is very important in knowing how to interpret the number.

This is certainly true of time. For some events one might record their occurrence to within a small fraction of a second, for others the date (day/month/year) is sufficient. Yet others may only require the month and year or even simply the year. This concept of incompleteness is not restricted to temporal values but could apply to other domains as well, for example that of a student having an age between 18 and 24.

Thus for many applications one may wish to cater for varying degrees of granularity within the system. For example, when an employee is entered into the system, an employee record may be created with a time stamp which is accurate to within a fraction of a second. However, the record may also contain an attribute recording the date of birth of the employee (correct to the nearest day) or his or her age (correct to the nearest year). When querying such a system one may wish to find all employees whose records had been inserted in some particular 1 second interval during a day, or all those whose age is over 60. To deal with such variations in accuracy, the idea of a range of values is introduced.

5. REPRESENTATION OF INCOMPLETE DATA

A **range** is a new data type representing incomplete information, which can be defined as

Range ::= LowerBound .. UpperBound
LowerBound ::= Numerical value
UpperBound ::= Numerical value.

A **range** represents a particular numerical value which is not known exactly but is known to be not less than the lower bound of the range and not greater than the upper bound of the range. For a discrete ordered domain a range represents some point from a set of points taken from the domain. For example, given the domain $\{t_0, t_1, \ldots, t_n\}$, the range $t_2..t_5$ represents some value from the set $\{t_2, t_3, t_4, t_5\}$. Similarly the fact that *John is in his late thirties* can be described as:

$$age\,(john\,,35..39)$$

in which "35..39" stands for a value in the range 35 to 39. It has the same logical meaning as:

$$\exists_{Age\,(35 \leq Age \leq 39)}\; age\,(john\,, Age\,).$$

The addition of incomplete data necessitates a slight change to the unification mechanism which is used in . [9, 3] Two new definitions of unifiable [12, 13] can be given as follows:

Definition 1 *Unifiable*: If there is a substitution $\theta = \{t_1/v_1, t_2/v_2, \ldots, t_n/v_n\}$, such that an empty clause can be deduced from two literals, $\neg p\,(X_1, \ldots, X_n)$ and $p\,(Y_1, \ldots, Y_n)$, i.e.

$$\{\neg(p\,(X_1, \ldots, X_n)\theta), p\,(Y_1, \ldots, Y_n)\theta\} \vdash \Box.$$

then these two literals are said to be unifiable, and the substitution θ is called a unifier.

A weaker condition is:

Definition 2 *Possibly unifiable*: Two literals $\neg p\,(X_1, \ldots, X_n)$ and $p\,(Y_1, \ldots, Y_n)$ are said to be *possibly unifiable* if there exists a substitution θ, a subclause of $p\,(Y_1, ..., Y_n)\theta$ denoted as $p\,(Y'_1, ..., Y'_n)$ and a subclause of $p\,(X_1, ..., X_n)\theta$ denoted as $p\,(X'_1, ..., X'_n)$ such that

$$\{\, \neg p\,(X'_1, ..., X'_n), p\,(Y'_1, ..., Y'_n)\} \vdash \Box$$

In general there two kinds of queries which one might want to put to a database. The first is a *definite query* which seeks to know whether something is definitely true (this corresponds to standard inference or deduction in logic, and the definition of *unifiable* is relevant). The second is a *possible query* which seeks to know whether something could possibly be true (here the definition of *possibly unifiable* is relevant).

To enable users to indicate out what kind of answers they want, two functions *def* and *pos* are provided. If a query is preceded by the function *def* , then the notion of *definitely unifiable* will be applied during query evaluation, which will follow the normal process of logical deduction and generate a *definite answer*. If a query is preceded by the function *pos* , then the definition of *possibly unifiable* will be applied during query evaluation. The answer generated is a *possible answer*.

To illustrate this, consider an example database of information about the engagement and loss of academic staff. Each event in this database has associated with it a unique event identifier and this is used to relate it to the time of the event. For example, the fact that *Tom was hired as a lecturer on the 1st of October 1975* can be expressed in the database as:

hire (*e* 1, *tom* , *lecturer*).
time (*e* 1, 1/10/1975).

the fact that *Tom ceased to be a lecturer in January, 1980 (exact day unknown)* can be expressed as:

$leave(e2, tom, lecturer).$
$time(e2, 1/1/1980..31/1/1980).$

Suppose that the database has a relation *employee_record* (*Start_time*, *End_time*, *Name*, *Rank*) which states that a person *Name* had a *Rank* from *Start_time* to *End_time*. It is defined as:

$$employee\_record(T1, T2, Name, Rank) \leftarrow hire(E1, Name, Rank) \wedge time(E1, T1) \wedge leave(E2, Name, Rank) \wedge time(E2, T2).$$

If a query asks whether the date when Tom ceased to be a lecturer was definitely after 15th January 1980, i.e.

$$?-def(employee\_record(T1, T2, Name, lecturer) \wedge later(T2, 15/1/1980))$$

the answer would be "no". But if the query asks whether the date when Tom ceased to be a lecturer was possibly after 15th January 1980, i.e.

$$?-pos(employee\_record(T1, T2, Name, lecturer) \wedge later(T2, 15/1/1980))$$

the answer would be "yes", because it is possible that he ceased being a lecturer after the middle of January.

In order to handle queries such as this, one needs to define comparison operators which operate on temporal data (including incomplete temporal data). In the case of the above example these are needed in the definition of the predicate *later*.

6. COMPARISON OF INCOMPLETE VALUES

If the domain of an attribute is a discrete set of values, and the value of an attribute is a single element from this set, then comparison of one attribute value against another is straightforward. However, if the domain of an attribute is ordered, and each attribute value represents a range of domain values, then comparison of two attribute values is more complex. For this purpose a set of operators is introduced.

Let $A \otimes B$ stands for a comparison expression in which A and B are two ranges and $\otimes$ is one of the comparison operators ($<$, $\leq$, $>$, $\geq$, $=$ and $\neq$). Two functions *def* and *pos* can be applied to the comparison expression to specify which kind of inference is required. For example $def(A > B)$ stands for the fact that whatever the value that A may take within the range specified, it is definitely greater than any value which B may take within the range specified and $pos(A > B)$ stands for the fact that the value taken by A within the range specified may possibly be greater than the value taken by B within the range specified. The whole set of operators can be given in logic notation as follows:

1) $def(A \otimes B)$ means that

$$\forall_{a \in A} \forall_{b \in B} (a \otimes b)$$

2) $pos(A \oplus B)$ means that

$$\exists_{a \in A} \exists_{b \in B} (a \oplus b)$$

These 12 operators are described in table 1. One thing which should be pointed out is that the *range* here is an incomplete data item which represent a particular unknown value ranging over the *range*. Two ranges having the same lower bound and the same upper bound might represent different values. In other words they are not definitely equivalent to each other.

Other operators including the set of temporal operators given by Allen [1] can be expressed in the same way as above and are shown in table 2.

7. EXAMPLES OF TIME-DEPENDENT INCOMPLETENESS

The combination of *time* and *incomplete information* together can enhance the power of the database. Consider the following examples of engaging and losing academic staff. The database might have such information as:

hire (*e* 1, *mary*, *lecturer*)
time (*e* 1, 10/5/1970)
leave (*e* 2, *john*, *lecturer*)
time (*e* 2, 1/6/1975)
leave (*e* 3, *mary*, *professor*)
time (*e* 3, 1/10/1980)

and the following rules:

$employee(Event, Name, Rank) \leftarrow hire(Event, Name, Rank)$

$employee(Event, Name, Rank) \leftarrow promote(Event, Name, OldRank, Rank)$

$employee(Event, Name, Rank) \leftarrow leave(Event_1, Name, Rank) \wedge$

$(\neg hire(Event_2, Name, Rank)) \wedge$

$(\neg promote(Event_3, Name, OldRank, Rank)) \wedge new\_event(Event)$

$employee(Event, Name, Rank) \leftarrow promote(Event_1, Name, Rank, NewRank) \wedge$

$(\neg hire(Event_2, Name, Rank)) \wedge$

$(\neg promote(Event_3, Name, OldRank, Rank)) \wedge new\_event(Event)$

$start(Event, Time) \leftarrow time(Event, Time)$

$start(Event, Start..End) \leftarrow employee(Event, Name, Rank) \wedge end(Event, Time_1) \wedge$

$employee(Event_1, Name, Rank_1) \wedge$

$start(Event_1, Time_2) \wedge earlier(Time_1, Time_2) \wedge$

$(\neg(employee(Event_2, Name, Rank_2) \wedge start(Event_2, Time) \wedge$

$earlier(Time_1, Time) \wedge earlier(Time, Time_2))) \wedge$

$make\_range(Time_1, Time_2, Start, End)$

$end(Event, Time) \leftarrow employee(Event, Name, Rank) \wedge$

$promote(Event_1, Name, Rank, NewRank) \wedge time(Event_1, Time)$

$end(Event, Time) \leftarrow employee(Event, Name, Rank) \wedge$

$leave(Event_1, Name, Rank) \wedge time(Event_1, Time)$

$end(Event, Start..End) \leftarrow employee(Event, Name, Rank) \wedge start(Event, Time_2) \wedge$

$employee(Event_1, Name, Rank_1) \wedge$

$end(Event_1, Time_1) \wedge earlier(Time_1, Time_2) \wedge$

$(\neg(employee(Event_2, Name, Rank_2) \wedge end(Event_2, Time) \wedge$

$earlier(Time_1, Time) \wedge earlier(Time, Time_2))) \wedge$

$make\_range(Time_1, Time_2, Start, End)$

$employee\_record(Time_1, Time_2, Name, Rank) \leftarrow employee(Event, Name, Rank) \wedge$

$start(Event, Time_1) \wedge end(Event, Time_2)$

$employee\_record(Time_1, after, Name, Rank) \leftarrow employee(Event, Name, Rank) \wedge$

$start(Event, Time_1) \wedge (\neg end(Event, Time_2))$

$employee\_record(before, Time_2, Name, Rank) \leftarrow employee(Event, Name, Rank) \wedge$

$(\neg start(Event, Time_1)) \wedge end(Event, Time_2)$

$earlier(Time_1..Time_2, T3) \leftarrow earlier(Time_2, Time_3)$

$earlier(Time_1, Time_2..Time_3) \leftarrow earlier(Time_1, Time_2)$

$earlier(Day_1/Month_1/Year_1, Day_2/Month_2/Year_2) \leftarrow Year_1 < Year_2$

$earlier(Day_1/Month_1/Year, Day_2/Month_2/Year) \leftarrow Month_1 < Month_2$

$earlier(Day_1/Month/Year, Day_2/Month/Year) \leftarrow Day_1 \leq Day_2.$

Suppose that one then wishes to add the fact that *Mary was promoted from lecturer to professor in June 1975*. Because the system is able to deal with incompleteness of this nature, the above information can be stored into the database as:

$$promote\,(e\,4,\, mary\,,\, lecturer\,,\, professor\,)$$
$$time\,(e\,4,\, 1/6/1975..30/6/1975).$$

If the user asks whether Mary was definitely a lecturer on 31st May 1975, i.e.

$$?-def\,(employee\_record\,(T\,1, T\,2, mary\,,\, lecturer\,) \wedge earlier\,(T\,1,\, 31/5/1975) \wedge earlier\,(31/5/1975,\, T\,2)).$$

the answer will be *yes*. If the user asks whether she was definitely a lecturer on 3rd June 1975, i.e.

$$?-def\,(employee\_record\,(T\,1, T\,2, mary\,,\, lecturer\,) \wedge earlier\,(T\,1,\, 3/6/1975) \wedge earlier\,(3/6/1975,\, T\,2))$$

the answer will be *no* because from the data available one cannot be certain that Mary was definitely a lecturer on that date. However, if one asks whether she might possibly have been a lecturer on 3rd June 1975, i.e.

$$?-pos\,(employee\_record\,(T\,1, T\,2, mary\,,\, lecturer\,) \wedge earlier\,(T\,1,\, 3/6/1975) \wedge earlier\,(3/6/1975,\, T\,2))$$

the answer will be *yes*.

8. CONCLUSION

If time is to be handled adequately in a deductive database, it must be coupled with a mechanism for dealing with incompleteness or uncertainty. The reason for this is partly the fact that the accuracy with which a temporal value is specified, either in a database clause or in a query, defines a level of granularity which is important in finding a solution. This is compounded by the inevitable human difficulties in recording exactly the times of past and future events.

A model is proposed in this paper which combines the notion of incompleteness with that of time, to provide a powerful and consistent approach to solving the problem. The event calculus provides a natural representation of temporal information but on its own does not deal with incomplete information, apart from simple null values. On the other hand, the approach to handling incomplete information described in [12] provides a useful extension to definite clauses in a deductive database. The combination of the two approaches into a single uniform model provides a powerful mechanism for dealing with time.

In order to handle incompleteness, two different kinds of query, definite and possible, have been introduced to enable users to specify their different requirements. This also necessitates two slightly different inference mechanisms, each with its own unification process. These are described in the paper.

The two different forms of query have been realized in a query language, called Squirrel, which is an extension of SQL which caters for rules [11] A database system with Squirrel as its query language has been implemented using the two different inference mechanisms referred to. An overview of the complete system is given in [14]

ACKNOWLEDGEMENTS

This work arises out of work funded under the Alvey programme, and the authors wish to acknowledge the support of the Science and Engineering Research Council and of ICL. Particular thanks are due to Dr. J.M.P.Quinn of ICL and Mr.J.R.Lucking of STC. This work is being taken forward as part of an Esprit project, European Declarative Systems, and once again thanks are due to the EC for their support.

References

1. Allen J.F. (1983), "Maintaining knowledge about temporal intervals," *Communications of the ACM*, **Vol. 26**, (11), pp. 832-843, (1983).

2. Burgess J.P. (1979), "Logic and Time," *Journal of Symbolic Logic*, **Vol. 44**, (4), pp. 566-581, (December 1979).

3. Chang C.L. and Lee R.C.T. (1973), *Symbolic logic and mechanical theorem proving*, Academic Press, New York, N.Y., (1973).

4. Clifford J. and Warren D.S. (1983), "Formal semantics for time in databases," *ACM Transactions on Database Systems*, **Vol. 8**, (2), pp. 214-254, (June 1983).

5. Fiadeiro J. and Sernadas A. (1986), "The infolog linear tense propositional logic of events and transactions," *Information Systems*, **Vol. 11**, (1), pp. 61-85, (1986).

6. Kowalski R.A. (1986), "Database updates in the event calculus," Imperial College, pp. 1-29, (1986).

7. Kowalski R.A. and Sergot M. (1986), "A logic-based calculus of events," *New Generation Computing*, **Vol. 4**, (1), pp. 67-95, OHMSHA, LTD. and Springer-Verlag, (1986).

8. McCarthy J. and Hayes P.J. (1969), "Some philosophical problems from the standpoint of artificial intelligence," in *Machine Intelligence*, ed. B. Meltzer and D. Michie, Edinburgh University Press, Edinburgh, (1969).

9. Robinson J.A. (1965), "A machine-oriented logic based on the resolution principle," *Journal of ACM*, **Vol. 12**, (1), pp. 23-41, (1965).

10. Sernadas A. (1980), "Temporal aspects of logical procedure definition," *Information Systems*, **Vol. 5**, pp. 167-187, (1980).

11. Waugh K.G., Williams M.H., Kong Q., Salvini S., and Chen G. (1990), "Designing SQUIRREL: an extended SQL for a deductive database systems," *The Computer Journal*, **Vol. 33 (6)**, pp. 535-546, (1990).

12. Williams M.H. and Kong Q. (1988), "Incomplete information in a deductive database," *Data and Knowledge Engineering*, **Vol. 3**, pp. 197-220, (1988).

13. Williams M.H., Kong Q., and Chen G. (1988), "Handling incomplete information in a logic database," in *UK IT 88 Conference Publication*, pp. 224-227, (1988).

14. Williams M.H., Chen G., Ferbrache D., Massey P., Salvini S., Taylor H., and Wong K.F. (1988), "Prolog and deductive databases," *Knowledge-Based Systems*, Vol. 1 (3), pp. 188-192, (1988).

Table 1.
Operators for comparing ranges of possible values

operator $\otimes$	full name	explanation $A \otimes B$
pos_l	possibly_less_than	$lb[A] < ub[B]$
pos_g	possibly_great_than	$ub[A] > lb[B]$
pos_e	possibly_equal_to	$ub[A] \geq lb[B] \vee lb[A] \leq ub[B]$
pos_le	possibly_less_or_equal_to	$lb[A] \leq ub[B]$
pos_ge	possibly_great_or_equal_to	$ub[A] \geq lb[B]$
pos_ne	possibly_not_equal_to	$lb[A] \neq lb[B] \vee ub[A] \neq ub[B]$
def_l	definitely_less_than	$ub[A] < lb[B]$
def_g	definitely_great_than	$lb[A] > ub[B]$
def_e	definitely_equal_to	$lb[A] = lb[B] = ub[A] = ub[B]$
def_le	definitely_less_or_equal_to	$ub[A] \leq lb[B]$
def_ge	definitely_great_or_equal_to	$lb[A] \geq ub[B]$
def_ne	definitely_not_equal_to	$ub[A] < lb[B] \vee ub[B] < lb[A]$

Table 2. Allen's operators

operator $\otimes$	explanation $A \otimes B$
before	A def_l B
after	A def_g B
equal	A def_e B
meets	$ub[A] = lb[B]$
overlaps	$lb[A] < lb[B] < ub[A] < ub[B]$
during	$lb[B] < lb[A] < ub[A] < ub[B]$
starts	$(lb[A] = lb[B]) \wedge (ub[A] < ub[B])$
ends	$(lb[B] < lb[A]) \wedge (ub[A] = ub[B])$

AN APPROACH TO THE LINGUISTIC SUMMARIZATION OF DATA

Ronald R. Yager
Machine Intelligence Institute
Iona College
New Rochelle, NY 10801

Kenneth M. Ford
Institute for Human & Machine Cognition
The University of West Florida
Pensacola, FL 32514

Alberto J. Cañas
Institute for Human & Machine Cognition
The University of West Florida
Pensacola, FL 32514

Abstract

We introduce a new approach to the summarization of data based upon the theory of fuzzy subsets. This new summarization allows for a linguistic summary of the data and is useful for both numeric and non-numeric data items. It summarizes the data in terms of three values: a summarizer, a quantity in agreement, and a truth value. We also discuss a procedure for investigating the informativeness of a summary. Finally, we describe *Summarizer*, an implementation of this new approach to the summarization of data.

Keywords

Summarization, fuzzy sets, database, linguistic summaries.

1. INTRODUCTION

Our ability to summarize provides us with an important capacity for obtaining a grasp on the meaning of large collections of data. Moreover, it provides a starting point for our capacity to make useful inferences on the basis of large aggregations of data or experience. This capability to summarize enables humans to construct functional representations of our complex environment in a manner amenable to the anticipation of future events. The statement that "many Chinese like rice," which is a summarization of some observations, allows us to make inferences about the viability of opening a rice shop in China.

In addition, the facility to summarize data and/or observations has much to do with the human ability to *communicate* our observations of the world in a useful and comprehensible manner. We would like to endow our mechanical reasoning systems with some facsimile of the human's capacity for abstraction and generalization.

On a continuum, at one extreme lies a large mass of undigested data while at the other extreme lies the usual summarization in terms of the mean or average value of the data. The mean is helpful in understanding the content of data, but, in some respects, it may provide too terse a summarization. The variance provides a method of judging the validity of the mean as the summary. However, in many instances, especially in situations involving presentations to non-quantitatively oriented people (e.g., corporate management, military leaders), an alternative form of summarization may be useful. Ideally, this new method of summarization would be especially practicable if it could provide us with summaries that are not as terse as the mean, as well as treating the summarization of non-numeric data.

In this paper we shall present a new perspective on the summarization of data based upon the theory of fuzzy sets [1-5]. This approach will provide us with a linguistic summary of the data. These personalized linguistic summaries will be less terse than the mean, and in addition, will allow us a myriad of possible ways to summarize data. Finally, the *Summarizer,* an implementation of the proposed method of summarization is discussed.

2. SUMMARIZING DATA

Assume V is some observable quality or attribute that can take values in the set $X = \{x_1, x_2, \ldots\}$. We allow V to admit numeric or non-numeric values. (For example, V could be age, salary, hair color, years of education or any other conceivable quality.) Let $Y = \{y_1, \ldots, y_n\}$ be a set of objects that manifests the quality V. We shall use $V(y_i)$ to indicate the value of the quality V for object y_i thus $V\colon Y \rightarrow X$.

The data we wish to summarize consists of a collection $D = \{V(y_1), V(y_2), \ldots, V(y_n)\}$ (i.e., the observations of property V for elements in the set Y). If Y is a group of n people and V is the quality age, then the data would be the ages of the n people.

Def. A *summary* of a data set consists of three items:

(1) A summarizer, S.
(2) A quantity in agreement, Q.
(3) A measure of the validity of the summary, T.

Given a data set D, we can hypothesize any appropriate summarizer S and any quantity in agreement Q; the measure T will indicate the truth of the statement that Q objects satisfy the statement S. As we shall see, the summarizer, the quantity in agreement, and T will play roles analogous to the mean, the variance, and the confidence respectively.

Example: If D is a set of data representing the ages of people, we can hypothesize the summary:
S = "about 25," Q = "most."

Then T, obtained by a procedure described below, will indicate the truth of the statement, "Most people in D are about 25." It is important to note that we are not restricted in our selection of S to the mean but can select any value of S useful for the purpose for which we are summarizing the data. A similar statement holds for the value Q. It is with the evaluation of T that we determine the validity of the summary.

Example: If $D = \{25, 13, 12, 19, 37, 25, 56, 45, 73\}$ is a set of ages of a group of people, we can discuss summaries in the form:

(1) (S = "about 15," Q = "some") ... Hence, some people are about 15,
(2) (S = "middle age," Q = "most") ... Hence, most people are at middle age.

Again, the evaluation of T will determine the validity of the summary. T will be a number in the unit interval such that the closer T is to 1, the more truthful or valid the proposed summary.

3. ON THE FORM OF SUMMARIZER S

As noted earlier, we shall summarize our data by means of a summarizer S. These summarizers may take linguistic values (e.g., some useful linguistic summaries could be "old," "young," "about 30," "middle age," "over 40," "exactly 15") [3]. The capability to employ linguistic values for summarizers is based upon the faculty to quantitatively represent these linguistic values as fuzzy subsets of the base set X, a set containing all the possible observed values of our data.

Assume X is a set of elements. A fuzzy subset H of X is a generalization of the idea of a subset of X. The fuzzy subset H associates with each element $x \in X$ a value $H(x) \in [0, 1]$, called the membership grade of x in H. $H(x)$ indicates the degree to which x satisfies the concept signified by H. If H were an ordinary subset, then $H(x)$ would be restricted to the set $\{0, 1\}$.

Consider the set: $X = \{1, 2, 3, 4, 5, 6, 7, 8, 9, 10\}$.

We could represent the value "near 5" as $\{0/1, .5/2, .7/3, 1/4, 1/5, 1/6, .7/7, .3/8, 0/9, 0/10\}$.

We could represent the value "large" as $\{0/1, 0/2, 0/3, 0/4, 0/5, 0/6, .3/7, .5/8, .8/9, 1/10\}$.

We could represent the value "exactly 3" as $\{0/1, 0/2, 1/3, 0/4, 0/5, 0/8, 0/9, 0/10\}$.

Thus, we observe that linguistic values may be associated with fuzzy subsets expressing the meaning of the words in terms of membership grades. In the context of our methodology for summarizing data, a summarizer S is associated with a fuzzy subset over X. The meaning of the summarizer is equivalent to the membership function of the fuzzy subset. In applications (as discussed later), a user suggesting a new (i.e., previously undefined) summarizer is asked to provide a personal meaning for the term S. A variety of knowledge acquisition techniques are employed to elicit a fuzzy subset representing S.

4. ON THE FORM OF THE QUANTITY IN AGREEMENT Q

The second component of the data summary is the quantity in agreement, Q. Q is a proposed indication of the extent to which the data satisfy the summary S. Q can be either an absolute or a proportional type of quantity. Examples of the absolute form of 'quantity in agreement' are: "about five," "at least 30," "exactly 3," "less than 50," "several," etc. With the absolute type of quantity, we are specifying how many pieces of data satisfy S. The relative form of 'quantity in agreement' is exemplified by such terms as "more than half," "most," "at least 25%," and are characterized by indicating what proportion of the data satisfy S.

In either of the two cases, relative or absolute quantities of agreement, we are formulating Q in terms of a linguistic value, just as we have done to express S. Thus Q can also be represented as a fuzzy subset. However, in the case of a relative quantity, Q is specified as a fuzzy subset of the unit interval $[0, 1]$, and for an absolute quantity, Q is specified as a fuzzy subset of the set R^+ of real numbers.

It is important to note that the meaning of the quantity Q is its fuzzy subset representation. As is the case for the summarizer S, since a user is providing the quantity Q, he/she must also provide its meaning (in terms of any fuzzy subset desired).

5. CALCULATION OF THE TRUTH OF A SUMMARY

Assume we have a set of data $D = \{d_1, d_2, \ldots, d_n\}$ corresponding to the readings on some variable V for all the elements in a population of n people where each d_i is drawn from some set X. For example, if the V is a subject's weight, then X could be a set of real numbers. If V is hair color, then X could be a set of hair colors. Let S be a proposed summarizer of the elements in D, expressed as a user-defined fuzzy subset of X. Finally, let Q be a quantity in agreement expressed as a fuzzy subset of either $[0, 1]$ or R^+. We will show a method for determining the validity T of a proposed summary (S, Q) in light of the data D.

We shall first consider the case where Q is a relative quantifier, $-Q$: $[0, 1] \rightarrow [0, 1]$. The procedure for obtaining T in this case is as follows:

(1) For each $d_i \in D$, calculate $S(d_i)$, the degree to which d_i satisfies the summarizer S.

(2) Let $r = \frac{1}{n} * \sum_{i=1}^{n} S(d_i)$, the proportion of D that satisfies S.

(3) Then $T = Q(r)$, the grade of membership of r in the proposed quantity in agreement.

Example: Assume we have a population of six people, and we have measured the ages of these people:

$D = \{25, 37, 22, 36, 31, 30\}$.

Assume a proposed summary of this data is:

Summarizer = "about 30," Quantity in agreement = "most."

The value "about 30" is defined by the user as the fuzzy subset S, where the membership function is:

$S(x) = e^{-(\frac{x-30}{6.6})^2}$.

Thus, we get:

$d_1 = 25$ $S(d_1) = .56$
$d_2 = 37$ $S(d_2) = .32$
$d_3 = 22$ $S(d_3) = .23$
$d_4 = 36$ $S(d_4) = .44$
$d_5 = 31$ $S(d_5) = .98$
$d_6 = 30$ $S(d_6) = 1$

and therefore $r = 3.53/6 = .588$.

If the user's definition for "most" is defined by the fuzzy set where $Q(r) = r^2$, then

$T = Q(.58) = (.58)^2 = .3364$.

Thus, the validity of the summary "about 30" for "most" elements in D, given the definition of "most" and "about 30," is .3364.

If we choose a different summarizer for our data, then we may obtain a different value for T (i.e., truthfulness or validation of the summarizer). For example, if our summarizer S is "at least 25 years old" and our quantity in agreement is still "most"; then the following results are obtained:

Since S is such that $S(d) = 1$ for $d \geq 25$ and $S(d) = 0$ for $d < 25$
then $S(d_1) = S(d_2) = S(d_4) = S(d_5) = S(d_6) = 1, S(d_3) = 0$
then $r = 5/6 = .833$
and then $Q(.833) = .69$.

Thus given our data, and our definition for "most," the validity of the summary "most people in this sample are over 25" is .69. In the case where our amount in agreement Q is an absolute quantity, then our procedure remains the same, except that in Step 2, $r = \sum_{i=1}^{n} S(d_i)$, the total amount of satisfaction to S.

6. SOME CONCEPTS FROM FUZZY SET THEORY

Since in our model we exploit fuzzy subsets to represent the linguistic terms used to specify a summarization S and a quantity in agreement Q, we must first present some concepts from fuzzy set theory.

Def: If F is a fuzzy set of X, then F is said to be normal if there exists some X such that $F(x) = 1$.

Def: Assume F and G are two fuzzy subsets of X. Then F is said to be contained in G, denoted $F \subset G$, if $F(x) \leq G(x)$ for all $x \in X$.

Def: If F is a fuzzy subset of X, the α level set of X, denoted F_α, is the crisp subset of X defined by $F_\alpha = \{x / F(x) \geq \alpha, x \in X\}$.

Def: If F is a fuzzy subset of X, then the negation of F, denoted $\overline{F}$, is also a fuzzy subset of X in which $\overline{F}(x) = 1 - F(x)$. For example, if the fuzzy set T represents "tall," then $\overline{T}$ represents "not tall."

Def: If F is a fuzzy subset of X, then
(1) There exists some $n > 1$, such that F^n, where $F^n(x) = (F(x))^n$, is representative of "very F."
(2) There exists some $0 < n < 1$, such that F^n, where $F^n(x) = (F(x))^n$ is representative of "sort of F." Note that if $n_1 > n_2$, then $F^{n_1} \subset F^{n_2}$.

Def: Assume X is a set on which there exists a negation operation, N, such that $N(x) = \overline{x}$.
If F is a fuzzy subset of X, then the antonym of F, denoted $\widehat{F}$, is the fuzzy subset of X, such that $\widehat{F}(x) = F(\overline{x})$. In particular, if X is the unit interval, then $\widehat{F}(x) = F(1 - x)$. Examples of antonym pairs would be true-false, tall-short, big-small.

Def: Assume F is a normal fuzzy subset of the set of real numbers. Then a representative value for F, denoted $R(F)$, is defined as:

$$R(F) = \int_0^1 M(F_\alpha)\, d\alpha,$$

where $M(F_\alpha)$ is the mean value of α level sets F_α [6].

Def: Assume F is a fuzzy subset of the finite set X. Then the specificity of F is defined as:

$$S(F, X) = \int_0^1 \frac{1}{\text{Card } F_\alpha}\, d\alpha,$$

where card F_α is the number of elements in the α level set F_α.

The specificity measures the degree to which F suggests one and only one element of X as its manifestation.

We note the following properties of specificity:
(1) $0 \leq S(F, X) \leq 1$,
(2) $S(F, X) = 1$, if $F(x) = 1$ for one element in X and $F(x) = 0$ for all the rest,
(3) If A and B are two normal fuzzy subsets of X, such that $A \subset B$, then $S(A) \geq S(B)$,
(4) If F and $\widehat{F}$ are antonyms, then $S(\widehat{F}, X) = S(F, X)$ [7].

7. PROPERTIES OF THE SUMMARIZATION

Most naturally used quantities in agreement fall into one of three classes. Assume Q is a fuzzy subset of X (X is either R^+ or the unit interval I).
(1) Q is said to be a monotonically non-decreasing quantifier if $r_1 > r_2 \Rightarrow Q(r_1) \geq Q(r_2)$. Examples of this type of Q are "at least 30%," "almost all," "most."
(2) Q is said to be a monotonically non-increasing quantifier if $r_1 > r_2 \Rightarrow Q(r_1) \leq Q(r_2)$. Examples of this type of Q are "at most 30%," "few," "almost none."

(3) Q is said to be a unimodal type quantifier if there exist two values $a \leq b \in I$, such that for all $r < a$, Q is monotonically non-decreasing, for all $r > b$, Q is monotonically non-increasing and for all $r \in [a, b]$, $Q(r) = 1$. Examples of this type of Q are "close to 30%," "exactly 5," "some."

Theorem 1: Assume we have a set of data D. Let S and Q be a proposed summary and quantity in agreement having an associated truth value T. Consider another proposed summary $\overline{S}$ and $\hat{Q}$, where $\overline{S}$ is the negation of S and $\hat{Q}$ is the antonym of Q. If T' is the associated truth value of this proposed summary, then $T = T'$.

Proof:

$$T = Q\left(\frac{1}{n} \sum_{i=1}^{n} S(d_i)\right)$$

$$T' = \hat{Q}\left(\frac{1}{n} \sum_{i=1}^{n} \overline{S}(d_i)\right) = \hat{Q}\left(\frac{1}{n} \sum_{i=1}^{n} (1 - S(d_i))\right)$$

$$T' = \hat{Q}\left(1 - \frac{1}{n} \sum_{i=1}^{n} S(d_i)\right) = Q\left(\frac{1}{n} \sum_{i=1}^{n} S(d_i)\right)$$

This result implies, for example, that if "many" and "few" are antonyms, then given a data set D about heights, the summarization "tall" for "many" will have the same truth value as the summarization "not tall" for "few."

Theorem 2: Assume we have a data set D. Let S and Q be a proposed summarization having the truth value T. The summarization S and $\overline{Q}$, where $\overline{Q}$ is the negation of Q, has the truth value T' where $T' = 1 - T$.

Proof: $T' = \overline{Q}(r) = 1 - Q(r) = 1 - T$.

These two theorems provide an equivalence rule for summaries, where two summaries are said to be equivalent (or more precisely semantically equivalent) when they are supplying the same information about the data set D.

Equivalence Rule:

The summary (S, Q, T) is equivalent to the summary $(\overline{S}, \hat{Q}, T)$, and both are equivalent to the summary $(S, \overline{Q}, 1 - T)$, thus $(S, Q, T) \Leftrightarrow (\overline{S}, \hat{Q}, T) \Leftrightarrow (S, \overline{Q}, 1 - T)$.

Theorem 3: Assume D is a collection of data drawn from a set X. Let S be a proposed summary described as a fuzzy subset of X. Let Q_1 and Q_2 be two quantities in agreement, such that $Q_1 \subset Q_2$. Then if T_1 is the truth of the summary S with quantity of agreement Q_1, and if T_2 is the truth associated with S and Q_2, then $T_2 \geq T_1$.

Proof: In both cases, since the S value is the same for each d_i, then they have the same value for r. However, since $Q_1 \subset Q_2$, then $Q_1(r) \leq Q_2(r)$ and hence $T_1 \leq T_2$.

This theorem implies that there exists a tradeoff between the specificity with which we state our value Q and the degree of truth associated with our summarization. Thus, we run the risk of forcing our summarization to be false, if we attempt to be too specific in the formulation of Q.

Theorem 4: Assume D is a collection of data drawn from a set X:

(1) Let Q be a monotonically non-decreasing quantity in agreement, and let S_1 and S_2 be two summaries, fuzzy subsets of X, such that $S_2 \subset S_1$. Then if T_1 is the truth of summary S_1, Q, and if T_2 is the truth of summary S_2, Q, then $T_1 \geq T_2$.

(2) If Q is monotonically non-increasing and $S_2 \subset S_1$, then $T_2 \geq T_1$.

Proof: Since $S_2 \subset S_1$, then for any observation d_i, $S_1(d_i) \geq S_2(d_i)$, and hence:

$$\sum_{i=1}^{n} S_1(d_i) \geq \sum_{i=1}^{n} S_2(d_i).$$

Therefore, $r_1 \geq r_2$.
If Q is monotonically non-decreasing, $r_1 \geq r_2 \Rightarrow Q(r_1) \geq Q(r_2)$, hence $T_1 \geq T_2$.
If Q is monotonically non-increasing, $r_1 \geq r_2 \Rightarrow Q(r_1) \leq Q(r_2)$, hence $T_1 \leq T_2$.

Thus, this theorem states that if Q is a monotonically increasing type quantity, such as, "at least 30%" or "almost all," and if we are too specific in the statement of our summarizer S, we run the risk of obtaining a low validation of our summary.

8. INFORMATIVENESS OF A SUMMARY

Assume a data set D drawn from a measurement space X. In providing a summary of the data, we are attempting to present the information contained in the data set in a more concise manner, one that is easier for the human mind to comprehend. However, with regard to a specific data set, the act of summarization by its very nature looses some information. In a sense, the data set itself is its own most informative description. A question that naturally arises is whether a summarization about the data set D is informative. A first inclination might be to assume that the truth associated with a summarization is the indication of the informativeness of a summary. The following example is intended to illustrate the inappropriateness of this.

Example: Assume $X = \{1, 2, 3, 4, 5\}$. Then consider the summary:
S = "greater than or equal to 2"
Q = "few," T = False.

Even though we have obtained a false result, we have learned a lot about the data. In the following, we shall present an outline of an approach to indicating the informativeness of a summarization with respect to the capturing of the original data set D.

Assume we have a summary (S, Q, T) of some unknown data set D having elements drawn from some measurement space X. Let the elements of this summary be S, a fuzzy subset of X, Q a fuzzy subset of I, and T a truth value from the set I.

What are the possible sets that could have been our original data set? Let $\mathcal{D}$ be a set whose members are all possible collections of elements drawn from X^*. The elements of $\mathcal{D}$ are multisets consisting of n elements. The first observation we can make is that $D \in \mathcal{D}$. That is, D is an element of this set. It is a collection of n values drawn from X. Our summary (S, Q, T) provides us with more information about our unknown data set D. If $d_1, \ldots, d_n$ are the elements of our unknown collection, then

$$T = Q\left(\frac{1}{n} \sum_{i=1}^{n} S(d_i)\right).$$

This implies the following criteria: Let $Q^{-1}(T)$ be the set of elements in the unit interval such that $r \in Q^{-1}(T)$ implies $Q(r) = T$, then only those elements in $\mathcal{D}$ satisfying $\frac{1}{n} \sum_{i=1}^{n} S(d_i) \in Q^{-1}(r)$ are possible manifestations of our unknown data set. Let us denote this subset of $\mathcal{D}$ by $\mathcal{D}_s$, $\mathcal{D}_s \subset \mathcal{D}$. We shall call $\mathcal{D}_s$ the set of possible data sets.

We note that if $\mathcal{D}_s$ has just one element, that is, one possible data set, then our summarization can be used to exactly obtain the unknown data set. As the number of elements in $\mathcal{D}_s$ increases, we are less certain as to which is the unknown data set D. We should note that if $\mathcal{D}_s$ is empty, then our summary is incompatible with $\mathcal{D}$.

In situations when X is finite, then $\mathcal{D}$ is finite, and we can use our measure of specificity to indicate the informativeness of the data summary with regard to the set D. Let us denote $I(S, Q, T)$ as the measure of information from the summary about D. Since $(S, Q, T) \rightarrow \mathcal{D}_s$, we can say that:

(1) $I(S, Q, T) = 0$ if card $\mathcal{D}_s = 0$.

(2) $I(S, Q, T) = 1/K$ if card $\mathcal{D}_s = K$.

We note that this is very much related to Shannon's entropy, which would be measured for this as $-ln\ K$.

Assume we have two summaries associated with this unknown data set: (S_1, Q_1, T_1) and (S_2, Q_2, T_2). Each summary suggests a subset of $\mathcal{D}$ of possible values of D; these may be denoted as $\mathcal{D}_{S_1}$ and $\mathcal{D}_{S_2}$. The possible values for the data set D suggested by this pair of summaries is $\mathcal{D}_s = \mathcal{D}_{S_1} \cap \mathcal{D}_{S_2}$.

Since $\mathcal{D}_s \subset \mathcal{D}_{S_1}$ and $\mathcal{D}_s \subset \mathcal{D}_{S_2}$, then if card $\mathcal{D}_{S_1} = K_1$ and card $\mathcal{D}_{S_2} = K_2$ and card $\mathcal{D}_s = K_3$ and hence $\mathcal{D}_s$ is telling us more about the potential value for D. Thus, generally, as we add more and more summaries, we gain more information about the possible value for the original data set D.

9. PROPERTIES OF INFORMATIVE SUMMARIES

Let us recapitulate. We have a data set D. In providing a summary of this set, we are reducing the amount of information we know about this set for the convenience of being able to see what the data is saying in a more compact form. However, each particular summary reduces the information we know about the data set in a different way. In some instances, our goal is to provide informative summaries. That is, summaries that give as much information about the data set as possible while still providing a summary. Let us continue our investigation of the possibility of measuring the informativeness of summary triplets (S, Q, T).

First, consider the case where Q is "all," has membership one for $r = 1$ and zero elsewhere. Assume that S has membership of one for $x^* \in X$ and zero for all others. In this case, Q and S are highly specific, and Q is increasing. If $T = 1$, we can use this summary to easily deduce that all elements in D are s^*. Thus, this summarization is highly informative about the data set D.

Reconsider Q as defined above, but let us now assume S to be less specific. In particular, there exists some subset A of X for which S has membership one. If the truth of this summary is one, we can only say that D is made up of elements from the set A. In particular, the larger A, that is, the less specific S, the larger the possible set of potential values for D. Thus, the more specific S in this case, the more informative the summarization.

Reflect again on the case where S is highly specific, that is, there exists only one value having non-zero membership, x^*. Let Q be a monotonically-increasing quantifier. Consider two possible associated truth values $T_1 > T_2$. Since Q is increasing, then r_1, associated with T_1, must be such that $r_1 \geq r_2$, where r_2 is associated with T_2. Since $r_1 \geq r_2$, this in turn fixes more of the elements in the unknown D as having membership grade 1 and less having membership grade 0 for the first case than for the second. Since, for the situation when membership grade is 1, we know with certainty the observed value was x^*, while when membership grade is 0 there exist a multitude of possible values for our observation, we can conclude then that the situation with T_1 is more informative about D.

Consider a situation where we have two increasing quantities Q_1 and Q_2, such that $Q_1 \subset Q_2$, that is, Q_1 is more specific than Q_2. Let S again be the highly specific summary, and let our truth be T. Since $Q_1 \subset Q_2 \Rightarrow Q_1(r) \leq Q_2(r)$, then if $Q_1(r_1) = Q_2(r_2) = T$, then $r_1 > r_2$. This implies that for Q_1, more of the elements of D are fixed at x^*.

From this discussion, we can conclude that one way to achieve highly informative summaries is to obtain summarizations in which Q is monotonically non-decreasing, S and Q are highly specific and T is high, close to 1. However, our quest for these types of summarizations are moderated by the theorems, in which we have shown that as S and Q become more specific, T becomes smaller. Thus, there appears some optimal summary, not easily apparent, that provides the most information. Further, we note that the elements in the data set D itself impose some limit on the degree to which any summary can be informative.

Let us again assume Q to be increasing, but this time very unspecific. That is, there exists only one value of r having membership grade 0. Because of the monotonicity, this must occur at $r = 0$. Let S be very unspecific: for example, $S(\hat{x}) = 0$, while for all other $x \in X$, $S(x) = 1$. If for this situation, T is low (for example, $T = 0$), then we can see that since $Q(r) = T = 0$, then $r = 0$. For $r = 0$, this implies that $S(d_i) = 0$ for all i, and hence, from our characterizations of S, under this condition all the d_i's must be equal to $\hat{x}$, and hence, D is completely known. By continuing in this fashion, we can conclude that if Q is monotonically non-decreasing and if S and Q are highly non-specific, then if T is false, we obtain a highly informative summarization.

However, we note that these two situations are very closely related. We make the following observations:

(1) If Q is monotonically non-decreasing, then $\hat{Q}$ (the antonym of Q), where $\hat{Q}(r) = Q(1 - r)$ is such that $\hat{Q}$ is monotonically non-increasing and the specificity of $\hat{Q}$ is the same as Q.

(2) If Q is monotonically non-increasing, then $\bar{Q}$ (not Q) is monotonically non-decreasing. Furthermore, if Q is non-specific but with at least one element having 0 membership grade, then $\bar{Q}$ becomes specific.

(3) Similarly, if S is non-specific but with at least one element having 0 membership, then $\bar{S}$ becomes highly specific.

Consider a summary S, Q and T in which S is non-specific, Q is monotonically non-increasing and non-specific and T is low. As we showed in Theorem 1, the summary $\bar{S}$, $\hat{Q}$ will have the same truth T as S, Q. However, $\bar{S}$ becomes specific, and $\hat{Q}$ while remaining non-specific becomes monotonically non-increasing.

Finally, consider the summary $\bar{S}$, $(\bar{\hat{Q}})$ from our previous Theorem 2 with truth $T' = 1 - T$. In this final case, we have $\bar{S}$, $(\bar{\hat{Q}})$ and T', where $\bar{S}$ is highly specific, $\bar{\hat{Q}}$ monotonically increasing and highly specific and T' high. Thus, we see for summaries of monotonically increasing quantities Q, that if T is high, then we have a highly informative summary when both S and Q are highly specific, while in cases where T is low, we have highly informative summaries if $\bar{S}$ and $\bar{\hat{Q}}$ are highly specific.

As an attempt to get a measure of the informativeness of a summary (S, Q, T), we can suggest the following: $I = [T * Sp(Q) * Sp(S) \vee (1 - T) * Sp(\bar{\hat{Q}}) * Sp(\bar{S})]$, where T is the truth and Sp indicates the specificity of the associated fuzzy set. The larger I, the more informative.

Let us now turn our attention to monotonically non-increasing quantifiers such as "few" or "less than 20%." Assume Q is monotonically non-increasing and specific. For example, $Q(r) = 1$ if $r = 0$ elsewhere $Q(r) = 0$; S is non-specific (for example, $S(x^*) = 0$, while for all other $x \in X$, $S(x) = 1$). If our truth value is $T = 1$, we can see that $Q(r) = 1$, which implies that $r = 0$, in turn implying that x^* is the only possible value for elements in D. This situation may be illuminated by an argument analogous to the earlier case where Q is monotonically non-decreasing. A summary T, Q, S in which Q is highly specific and monotonically non-increasing, in which S is highly specific, and in which T is very true gives a lot of information concerning the members of D. For example, if we have $X = \{0, 1, 2, 4, 5\}$, Q = "very few," S = "at least one," $T = 1$, then we know that the elements in D are zero.

This observation could have been obtained from our equivalency rule for summarizers. Assume we have a summary (S_1, Q_1, T_1) where Q_1 is monotonically non-decreasing and highly specific and S_1 highly specific and T_1 is highly true. We have previously shown that this is a very informative type of summary.

Consider the summary $(\bar{S}, \bar{\hat{Q}}_1, T_1)$. This summary is equivalent to the summary (S_1, Q_1, T_1) and therefore must be equally informative about the elements in D. However, this is a new summary in which quantity in agreement $\bar{\hat{Q}}_1$ is still highly specific but monotonically non-increasing, and the truth T_1 is high, but the summarizer S_1 non-specific.

We have also shown that if (S_2, Q_2, T_2) is a summary in which S_2 is non-specific, Q_2 is monotonically non-decreasing and non-specific, and T_2 is false, then this is a highly informative summary about the elements in D. If we apply our equivalency rule, we get $(\bar{S}_2, \hat{Q}, T_2)$, in which $\hat{Q}_2$ is now monotonically non-increasing and non-specific, while $\bar{S}_2$ is specific and T_2 is false. This new summary must also be highly informative about the elements in D.

Thus if (S, Q, T) is a summary of some data set D, and Q is a monotonically non-increasing quantity, then a possible indicator of the informativeness of this summary is

$$I(S, Q, T) = [T * Sp(Q) * Sp(\bar{S}) \vee (1 - T) * Sp(S) * Sp(\bar{Q})].$$

Let us now consider the information content in a unimodal type summary such as "about 40%" or "near 5." Consider the case where S is highly specific (for example, $S(x^*) = 1$ while all other x's have membership grades 0). Q is unimodal and highly specific (for example, $Q(a/n) = 1$, $Q(r) = 0$ for all other $r \in [0, 1]$) and $T = 1$. Analyzing this case, we see that since $T = 1$, then $Q(r) = 1$, thus $r = a/n$. Since S is non-zero only for x^*, this implies that a of the readings in D must be x^* while the remaining $n - a$ readings can be anything but x^*.

From this, we can conclude that for summaries with unimodal quantities Q that are highly specific, summarizers S that are highly specific and truth that is high, the information content about D is high. Furthermore, the closer the location of the peak of Q is to 1 along the r axis, the more information. For example, "very close to 60% of the sample is almost 30 years old is highly true" tells us more about the elements in D than the statement, "close to 40% of the sample is almost 30 years is true."

We have shown then that a summary (S_1, Q_1, T_1) will be informative about the data set D if S_1 is highly specific, Q_1 is unimodal, highly specific and centered about a and in which T_1 is high. We also note that the higher a is, the more informative, thus the limit (when $a = 1$), is the most informative.

Applying our equivalence rule to this situation, we get a new summary $(\bar{S}_1, \hat{Q}_1, T_1)$, which is equivalent to (S_1, Q_1, T_1). However, we note that $\bar{S}_1$ is broad, and Q_1 is still unimodal and specific but its center is about $1 - a = b$. From this, we can conclude that for unimodal summaries, good information is obtained when the summarizer is broad (unspecific), Q is specific but centered about a low value and T is high.

From this, we can hypothesize a measure of informativeness for unimodal summaries. Assume (S, Q, T) is a unimodal summary. Let a be the center of the range where $Q(r) = 1$. Then a measure of the informativeness can be expressed as $I = \text{Max}\,[a * T * Sp(Q) * Sp(S), (1 - a) * T * Sp(Q) * Sp(\bar{S})]$.

10. SUMMARIZER

Summarizer is an implementation of the method of summarization described above. *Summarizer* seamlessly integrates the ability to perform summarizations with the typical query environment found in relational databases. It is intended to be usable in 'real world' environments, by technologically naive personnel. Figure 1 shows the system architecture of *Summarizer*. A prototype version has been implemented.

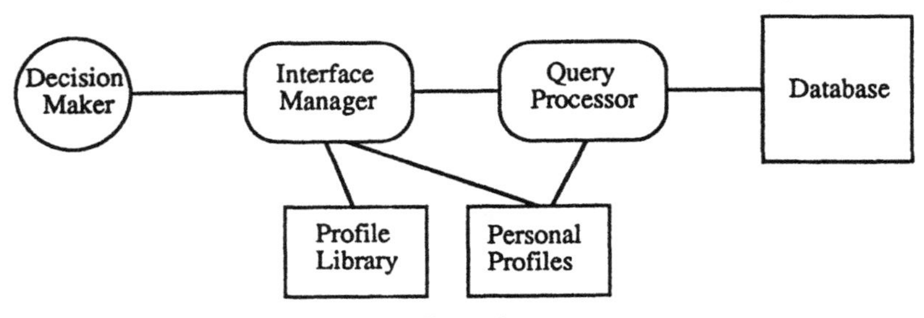

Figure 1.

Assuming that users may have somewhat differing interpretations of a given summarizer and quantity in agreement, a record is kept of every user's interpretations in a Personal Profile library. For each specific summarizer and quantity in agreement, the Personal Profile serves as a repository for the information necessary to carry out the corresponding operations. The Query Processor exploits this information to perform the summarization and quantity in agreement operations according to the particular user's interpretation.

When a user specifies a standard database query, the Query Processor will execute it as usual. If the query includes a summarizer or a quantity in agreement, the Query Processor will retrieve the definition of that summarizer and quantity in agreement from the user's Personal Profile and process the query accordingly.

However, when the query includes an undefined summarizer or quantity in agreement (i.e., not already found in the user's Personal Profile), the Interface Manager will interactively obtain (from the user) the necessary information and update the Personal Profile. The nature of the interaction necessary to elicit this information will depend on the particular summarizer or quantity in agreement, and is specified in the Profile Library. The Profile Library is needed because a different knowledge acquisition approach may be appropriate for each summarizer and quantity in agreement.

For example, partial information on the quantity in agreement *most* may be obtained by asking the user to select *most* of the points from the set in Figure 2.

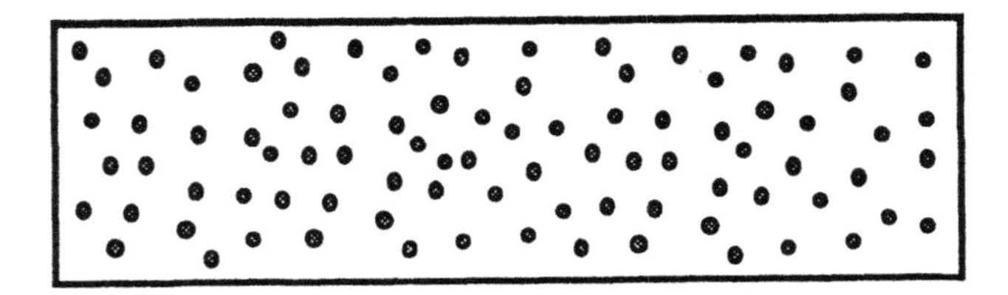

Figure 2.

This interaction is followed by a session in which a sequence of screens (see Figure 3) containing a collection of points are presented to the user. For each screen, *Summarizer* has selected (on the basis of the user's response to the interaction illustrated in Figure 2) a different number of points to propose to the user as potentially representing the quantity in agreement *most*. In each case the user is asked, with respect to the current screen, whether he/she considers that *most* of the points have been selected. It is important to determine not only what *most* means to the user, but also what is *not most*. Thus, *Summarizer* attempts to elicit the upper and lower boundaries of a quantity in agreement (in this case *most*).

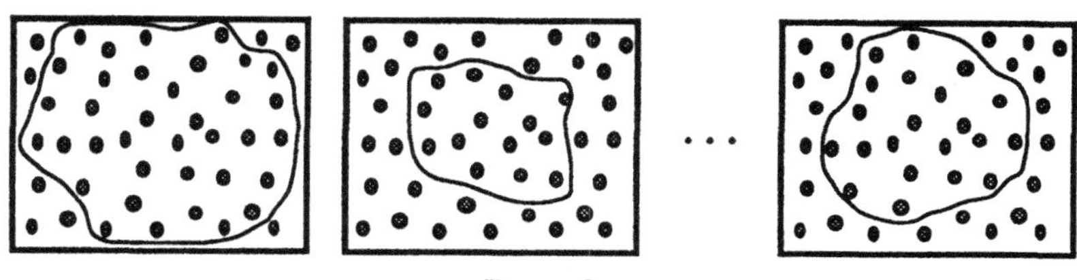

Figure 3.

Figure 4 shows another interaction by which partial information on the summarizer *about* might be obtained by asking the user to indicate which points are *about* 5 in the rulers. This interaction should be fast and easy – if it is slow or cumbersome, the user may become discouraged in his/her efforts to use the system and refrain from trying different summarizers and quantities in agreement. After the Personal Profile entry is completed, the query will be resolved and the user may apply the newly defined summarizer or quantity in agreement in future queries.

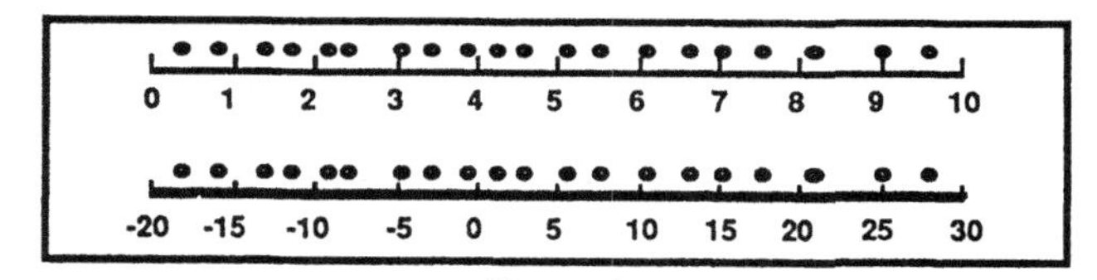

Figure 4.

The availability of personal definitions for the quantity in agreement poses some interesting alternatives. For example, for a query such as "Are most employees about 60 years old?" *Summarizer* not only presents the results, but can also search the Personal Profile for other quantities in agreement having a higher *T* value (i.e., the validity or truth of the statement) than the quantity in agreement used in the original query. Likewise, *Summarizer* can offer alternative summarizations on the basis of *I*, the informativeness of the summarization. Thus, upon consideration of the original summarization (*S, Q, T*) *Summarizer* can suggest alternatives that are either more valid or more informative. In Figure 5 the user has specified the query, "Is the age of most employees about 60?". In this case the system has indicated a very low validity of only .05 for the original query, however, *Summarizer* has proposed a new quantity in agreement (i.e., *a few*) having a substantially higher validity (0.82).

In the prototype of *Summarizer*, the user profile is based on the assumption that each user is consistent in applying a summarizer and/or a quantity in agreement under different contexts (i.e., for different attributes of the database). For example, if the user specifies "Are most fossils about 10,000 years old?," "Are most people in this room about 30 years old?" and "Are most salaries in the Research Department about $50,000?," we currently assume that *most* and *about* have the same meaning under the same context (age), as well as under a different context (salary). (Since this assumption is quite possibly false, we are empirically investigating this issue.) As a result, each summarizer or quantity in agreement need be defined only once in the personal profile for that particular user. That definition will be applied to all objects or attributes.

Query:

Quantity in Agreement	Attribute		Summarizer	Value
Most	age-of-employee	**entries are**	about	60

Response:

Query: Most age-of-employee about 60
Validity: .05

Other relevant information:
Few age-of-employee about 60
Validity: .82

Figure 5.

11. CONCLUSION

The ability to summarize is an important mechanism in the analysis of information. It provides a means of usefully answering particular questions about collections of data, and additionally, a way of formatting the information so as to enable an analyst to comprehend the content of the data in a holistic manner. The procedure presented in this paper has a number of advantages over the classic (i.e., mean) way of summarizing data. Among these are the ability to summarize non-number as well as numeric data, the ability to provide numerous different summaries for special purposes, and the ability to provide linguistic summaries. The work described in this paper is part of a continuing research effort aimed at the development and unification of the prerequisite underlying theoretical foundations for an adequate approach to the summarization of data. In addition, we have designed and constructed *Summarizer* – a working testbed for the ideas in this paper.

ACKNOWLEDGMENTS

Many people have influenced the development of this work. In particular we want to thank John Coffey and Jeff Yerkes.

REFERENCES

1. L.A. Zadeh, "Fuzzy Sets," *Information and Control*, Vol. 8, 1965, pp. 328-353.
2. L.A. Zadeh, "Fuzzy Sets as a Basis for a Theory of Possibility," *Fuzzy Sets and Systems*, Vol. 1., 1978, pp. 3-28.
3. L.A. Zadeh, "PRUF–A Meaning Representation Language for Natural Languages," *Int. J. of Man-Machine Studies*, Vol. 10, 1978, pp. 395-460.
4. R.E. Bellman, and L.A. Zadeh, "Local and Fuzzy Logics," in Dunn and Epstein, eds., *Modern Uses of Multi-Valued Logics*, Reidel, Dordect, Holland, 1977.
5. R.R. Yager, "Quantified Propositions in a Linguistic Logic," in *Proc. 2nd Int. Seminar on Fuzzy Set Theory*, Linz, 1981, pp. 69-124.
6. R.R. Yager, "A Procedure for Ordering Fuzzy Subsets of the Unit Interval," *Information Science*, Vol. 24, 1981, pp. 143-161.
7. R.R. Yager, "Measuring Tranquility and Anxiety in Decision Making," *Int. J. of General Systems*, Vol. 8, 1982, pp. 139-146.

MANAGEMENT OF UNCERTAINTY IN THE ATTACHMENT PROBLEM IN NATURAL LANGUAGE PROCESSING

Paolo TERENZIANI, Leonardo LESMO, Elisabetta GERBINO
Dipartimento di Informatica - Universita' di Torino
Cso. Svizzera 185 - 10149 Torino - Italy -

ABSTRACT

The problem of choosing the most plausible attachment for a prepositional phrase (PP) or a relative clause (RC) is one of the most significant challenges to a NL Processing system, since it presupposes an in-depth understanding of the interaction between the different sources of knowledge involved in the task of processing a sentence (in particular, we consider syntactic and semantic knowledge). In the paper, we argue that traditional categorical approaches do not adequately face the intrinsic complexity of the attachment problem and we describe a more flexible approach, based on management of uncertainty: both syntax and semantics provide a Confidence Degree (CD) for the linguistic connections, and a weighed combination of these CDs allow the system to select the most plausible connection when different alternatives are possible.

1 INTRODUCTION

As Winograd states in [24], the research in Natural Language Understanding (NLU) is being carried on today within a new paradigm: the computational paradigm. Its main differences with respect to the previous (generative) one, stands in the "attention to process organization" and the "relevance of non-linguistic knowledge".

The aim of this paper is not to take into account all the problems that non-linguistic knowledge conveys into NL analysis, but to make clear that the in-depth understanding of the respective roles of syntactic and semantic knowledge sources and the clarification of the way they interact to construct the interpretation of NL sentences is fundamental to building NL processing systems.

If we consider the evolution of NL processing systems [3], it becomes apparent that the role of syntactic knowledge can vary from being completely neglected to being the basis of the process.

In the former case we have mainly semantic approaches [19]. These approaches present some problems with respect to the perspicuity of the model, since the structural information has to be duplicated for the different entities or represented in a procedural form.

In the latter case (see for example [27]) the semantic interpretation is "appended" to the syntactic analysis in order to build the semantic representation of the input sentence. It is widely accepted that this way of using semantics is highly inefficient, since the number of alternative parses is often so high (especially in case of sentences with prepositional specifications) that it is not cost effective to delay the intervention of semantics [18]; on the contrary, it is preferable to use semantics as soon as possible in order to provide constraints for the syntactic analysis[1].

[1]Many psycholinguistic researches support an highly integrated approach to NL processing. Tyler and Marslen-Wilson [21], for example, suggest the reality of separate lexical, syntactic, semantic and pragmatic knowledge and propose a model where their interaction is bottom-up (i.e. the lower levels of knowledge have priority over the higher levels) and follows the principle of "do it as early as possible".

But, if one wants to achieve a full integration between syntactic and semantic analysis another basic problem arises: how can this integration be achieved?
The analysis of the attachment problem provides basic insights to this problem and constitutes an important test-bed for verifying and comparing integrated NL processing systems.
The following examples show some structurally ambiguous sentences:
(1a) The boy eats the ice-cream **with the spoon**
(1b) The boy eats the ice-cream **with strawberries**
(2a) The dog of the blonde girl **that is studying in the park** is very nice
(2b) The dog of the blonde girl **that is barking in the park** is very nice
(3) The friend of the boy **who ate the ice-cream** is very nice
Sentences 1a and 1b, as well as 2a and 2b, have the same surface form, and the attachment of the bold parts of the sentence (the PPs and the RCs respectively) can only be determined by considering semantic information. On the contrary, in sentence (3) both the attachments of the RC (to "friend" or to "boy") are semantically correct. In this case, the syntactic principle of Right Association[2] leads to prefer the attachment to the nearest noun in the sentence (e.g. "boy"). The examples show that the choice of the most plausible attachment for PPs and RCs can only be performed in a satisfactory way by considering both syntactic and semantic information.

A first integrated approach to the attachment problem (and, more generally, to the problem of structural disambiguation) using Augmented Transition Networks (ATN) parsers is given by the LUNAR system [26]. When traditional ATN parsers have to make a choice between alternative paths, they choose one at random, and if that path succeeds they will never go back to try the others. In LUNAR, it is proposed to eliminate semantically anomalous parses by the addition of a semantic checker: the checker's finding the parse to be semantically ill-formed is treated exactly as a syntactic failure (the parser backtracks and tries another path).
In [17], it is pointed out that the semantic checker as conceived in these approaches is not adequate because it can not make comparative judgements: it accepts the first result that is acceptable, without considering that some better choice might exist. This limitation is overcome in PARSIFAL [17]. PARSIFAL cannot backtrack, and must therefore detect structural ambiguity whenever it arises and decide irrevocably which alternative is better. Therefore, all the alternatives are considered at the same time, and thus a comparative judgment can be made.
A different approach to the problem is given in [9], where structural ambiguity is considered as a "closure problem"[3]. Ford, Bresnan and Kaplan propose a closure theory based on lexical preference: each verbal lexical entry contains markers for the cases for which the fillers are normally provided (based, presumably, on frequency) and (in neutral contexts) these verb expectations allow to discriminate which constituents must be closed. For example, *keep* would be marked as [AGENT keep PATIENT STATE]. This would explain why the preferred reading of "The women kept the dogs on the beach" is that with "on the beach" as the STATE case of "keep".
One of the limitations of this approach is that it only considers the expectations of the verb, and therefore it does not give a way of choosing between NP heads if there is more than one of them. This limitation is overcome in the approach proposed in [22], which is developed in the general framework of Preference Semantics [23]. In [22], preferences are associated not only to the semantic concepts corresponding to the verbs, but also to nouns and prepositions. For example, the expectations of "to buy" and of "ticket" are as follows (the notation is not the original one; we indicate with "A(B)" that the selectional restriction for the case A is B):
[buy RECIPIENT(HUMAN)]
[ticket DIRECTION(PLACE)]

[2]It is also called principle of Low Right Attachment: the new constituent is attached as low and as far to the right in the parse tree as possible [10].

[3]In parsing, a constituent of the parse tree is said to be "open" if it has not been declared complete, and so other constituents may still be attached to it. When a constituent is complete, it is "closed". In most natural languages (included English and Italian) it is almost always true that the attachment of a constituent (say C1) to another one (say C2) causes the closure of all the sub-constituents of C2 that are not closed yet. Therefore, the attachment disambiguation can be seen as the process of deciding which open constituents should be closed.

and allow to select the correct attachments for sentences like "I bought a ticket to John" and "I bought a ticket to New York". Expectations for prepositions are more complex, and are associated to each possible meaning of the considered prepositions.
In all these approaches the integration between different knowledge sources for the task of selecting one of the alternative attachments is categorical, in the sense that (in the most complex cases) the choice is made through the introduction of an algorithm which evaluates the considered (syntactic, semantic, pragmatic) preference criteria in a fixed order, until one of them is satisfied and allows to choose one of the alternative attachments. In such a case, the preference criteria of the lower levels are not tryed any more (see for example the algorithm proposed in [11], page 174).

On the other hand, we claim that, in some cases, greater flexibility is required.
Let us consider, for example, sentence (4).
(4) A boy with a horse **that is dancing in the park** is very kind
In (4) the RC can, in principle, refer to "horse" and to "boy". If we suppose that both the boy and the horse, being living entities, can dance, then the categorical approaches considered before would choose the reading that "the horse is dancing in the park". In fact, since both the attachments would be semantically correct, syntactic preferences should be considered, and the rightmost attachment would be chosen (because of the principle of Right Association). This is, in principle, a possible reading of (4), but certainly it is not the most plausible one. The problem is that, in analysing sentence (4), a human hearer applies his knowledge that a boy is a preferable agent of the action of dancing than a horse and, since this preference is stronger than the criteria given by the Right Association principle, he chooses to attach the RC to "boy". In other words, both semantics and syntax give indications about the plausibility of the attachment, and the choice is made after having weighed both of them.
In order to obtain greater flexibility we propose an approach based on management of uncertainty: in our system both syntax and semantics provide a Confidence Degree (CD) for the linguistic connections. A weighed combination of these CDs allows the system to choose the most plausible path when different alternatives are possible.

The paper is organised as follows. In the second section a brief overview of the GULL system is presented. In the third section, we describe what kind of syntactic and semantic information contribute to the weighting. Finally, in section four, we analyse how structurally ambiguous sentences are processed in the extended system: the syntactic generation of alternative attachments, the parallel activation of the semantic module on them, the computation of the final weights and the decision process.

2 *OVERVIEW OF THE SYSTEM*

In this section the basic architecture of the system is described, without considering the extensions we introduced in order to deal with the attachment problem.
GULL (General Understander of Likely Languages) [6, 12, 20] is a Natural Language Understanding System for the Italian language which has been developed since 1986 at the Dipartimento of Informatica of the University of Torino as a successor of the FIDO system [14-16].
The basic principle of GULL's architecture is the coincidence of the two concepts of "structured representation of a sentence" and "status of analysis"[4] (see [14]). Of course, this principle implies that all information present in the input sentence must also be present in its structured representation. Actually, what happens is that new pieces of information, which are implicit in the "linear" input form, are made explicit in the result of the analysis. These pieces of information are extracted using the syntactic knowledge (how the constituents are structured), the lexical knowledge (number and gender of nouns, tense and person for verbs ...) and the semantic knowledge.
The main advantage of such an approach is that the whole interpretation process is centered around a

[4]These two concepts have usually been considered as distinct. For example in ATNs the parse tree is held in a register, but the global status of a partial interpretation also includes the contents of the other registers, the stack containing the return states of the pushed constituents, the contents of the HOLD list.

single structure: the dependency structure of the constituents composing the input sentence. This enhances the modularity of the system: the various knowledge sources are mutually independent while the control flow can be designed in such a way that all knowledge sources contribute, by cooperating in a more or less synchronized way, to the overall goal of comprehension.

In GULL the syntactic structures are represented as dependency trees (or "head-and-modifier" tree, see [24]), where all nodes, and not only the leaves, are associated with surface words. This representation has been chosen because, in comparison with ordinary parse trees, it simplifies the interaction between syntax and semantics. In fact it gives the semantic interpretation component a structured representation which is easier to analyse because it is closer to the direct representation of the relationships between the different constituents of the input sentence.
The dependency tree gives a "case representation" of the input sentence (see [8]): in verbal nodes (REL nodes) each pointer corresponds to a case of the verb; in REF nodes (corresponding to nouns and pronouns) the dependent structures represent the specifications of the node. Several other types of nodes are defined: ADJ (for ADJectives), DET (DETerminers), MOD (MODifiers; for adverbs), CONN (CONNections; mainly prepositions) and TOP (which specifies the top-level clausal structure of the sentence).
The syntactic knowledge source of the GULL system is composed of a set of condition-action rules. where the condition examines the current status of the analysis (i.e. the parse tree that has already been built and the input words) whereas the action extends the dependency tree representing the syntactic structure of the input sentence.

The semantic component of the system is activated as soon as a semantically interpretable constituent is analysed by syntax. This fact allows, among other things, to choose between alternative provisional attachments as soon as possible.
The semantic component consists in a set of condition-action rules. The condition part of the rules considers the constituent just analysed and its surrounding context (e.g. its mother node and its descendants) and match them against the domain knowledge. This match binds some variables in the rules which are then used in the action part in order to build up the semantic representation [20].
The domain knowledge consists in a KL-ONE like semantic net that represents the selectional restrictions (in the paper, we adopt the notation [CONCEPT ROLE1(FILLER1) ... ROLEn(FILLERn)] for indicating that the concept CONCEPT has roles ROLE1, ROLEn whose selectional restrictions are FILLER1, FILLERn respectively).
Moreover, a Preplate Table [23] is added in order to encode semantic information about prepositional specifications. A list of preplates is associated with each preposition, and each preplate indicates

- the relational concept of the semantic net representing a meaning of the preposition
- the cases of the relational node involved in the prepositional specification, and their respective selectional restrictions

For example, a preplate entry for the preposition "of" is:
of: [BELONG PAT(THING) AGT(PERSON)]
It indicates that a possible meaning of the prepositional specification "A of B" (where A and B are two nouns) is BELONG, where A is the PAT (patient) and must be a THING and B is the AGT (agent) and must be a PERSON.

3 WEIGHED SYNTACTIC AND SEMANTIC INFORMATION

Basically, all knowledge sources must be extended in order to provide also Confidence Degrees. A crucial point of the approach is the determination of such CDs. These data (especially verb preferences) could come from textual analysis or from formal experiments on people's preferences such as the one Ford, Bresnan and Kaplan [9] ran (Connine, Ferreira, Jones, Clifton and Frazier [4] have collected a large set of suitable data). However, for our prototype system, we think that our intuition should suffice.

As regards the syntactic information, the system uses a weighed form of the Right Association principle: to each provisional attachment a SyCD (Syntactic Confidence Degree) is associated, whose

value is the inverse of the distance of the attachment, starting from the current node and climbing up the tree.
Actually, some other pieces of syntactic information are taken into account; for example agreement is very useful in determining the attachment of a RC. Moreover, we are currently investigating the possibility of assigning different syntactic weights to the prepositions, since it seems that, for example, the links established between two nouns via an "of" preposition are stronger than the one provided by an "in" preposition.

In order to provide the desired semantic information, Domain Knowledge has been enriched by associating a weight (whose value ranges in the interval (0..1]) to its semantic links. The CDs are associated both with the semantic net and with the Preplate Table. For example
[0.3 DANCE AGT(HORSE)]
states that a HORSE is the AGENT of DANCE with CD equal to 0.3 and
of: [1 BELONG PAT(THING) AGT(PERSON)]
states that "THING of PERSON" is a valid semantic path, with CD equal to 1.
As a consequence, the operation of matching a constituent against the Domain Knowledge is no longer a categorical (yes/no) process: if the constituent is semantically correct, a SeCD (Semantic CD) representing its "plausibility" is returned.

It is not possible to describe the actual organization of the semantic net here, but at least two basic considerations must be provided.
First of all, we consider the specifications of a noun independently of each other (e.g. in the sentence "the boy with blue eyes in the classroom" the specifications "boy with blue eyes" and "boy in the classroom" are analysed independently of each other, and they correspond to different paths in the semantic net). On the contrary, verbal case frames are considered globally[5]. This means, among other things, that a SeCD is associated to a whole case frame, an not to its single cases (e.g. [1 TEACH AGT(TEACHER) PAT(COURSE)] specifies that 1 is the CD of the whole case frame).
Secondly, we distinguish (according to the proposal in [9]) between "closed" case frames (e.g. case frame to which no more cases can be added) and "open" case frames (the converse). This distinction is basic in order to capture the fact that [0.7 GIVE AGT(PERSON)] is a correct "open" case frame, but it is an incorrect "closed" one, since necessary cases (e.g. the patient) are missing.

4 PROCESSING STRUCTURALLY AMBIGUOUS SENTENCES IN THE EXTENDED SYSTEM

This is the central section of the paper, and illustrates the overall organization of the extended system: it describes the generation of alternative attachments when analysing ambiguous sentences, the evaluation of the syntactic and semantic CDs for each attachment (starting from the weighed knowledge sources described in the previous section), their combination in order to determine the plausibility of the global configurations of dependency tree and, finally, the choice of the most plausible structural hypothesis.
The system has been extended on the basis of the following principles:

- all the possible alternative attachments must be generated
- the choice between different attachment must be made as soon as possible
- the choice must be performed considering all the available syntactic and semantic pieces of information, and on the basis of a "global" analysis (in the sense that the alternative attachments are not considered in isolation, but taking into account the global configuration of dependency tree generated by selecting each one of the attachments).

While the first two principles are strongly supported by cognitive considerations [1], we think that the last point requires some words of explanation. In section 3, we showed that the cases of a verb must be analysed globally and that it is important to distinguish between open and closed case frames. Moreover, it must be considered that the choice of a provisional attachment affects the global

[5]This fact enables the system to detect, for example, the uncorrectness of sentences like "the boy the dog plays" where two different agents appear simultaneously.

configuration of the dependency tree, since all the nodes below the chosen attachment (that are not yet closed) must be closed. All these considerations point out that the choice of the most plausible attachment for a PP or a RC can not be performed only on the basis of local (semantic and syntactic) plausibility of the alternative attachments. On the contrary, the plausibility of the whole configurations of dependency tree generated by selecting each one of the alternatives must be compared.

Whenever more than one attachment of a PP or a RC are possible during the analysis of an input sentence, the extended system operates as follows:

A) all the possible attachments are generated (these attachments are provisional, since only one of them will remain in the final dependency tree)

B) for each attachment, the local syntactic and semantic CDs are evaluated together with the syntactic and semantic CDs of the attachments of the configuration generated by selecting that provisional attachment

C) for each attachment, the global syntactic and semantic CDs of the corresponding configuration is evaluated by combining the CDs of all the attachments in the configuration

D) the global syntactic and semantic CDs are weighed in order to evaluate the global plausibility of each configuration, and the attachment which corresponds to the configuration with highest plausibility is chosen.

(A) ATTACHMENT GENERATION

The basic process for determining the attachment of a newly entered node of the dependency tree is still used in the extended system: in this case, it is applied in order to determine all possible attachments and to generate them. This consists in climbing up the rightmost path of the dependency tree and considering all the REF and REL nodes for PPs and only REF nodes for RCs (until an empty REL node or the TOP node is encountered). The alternative attachments are marked as "provisional" in order to distinguish them from the final definitive attachments. See for example, in fig.1, the "provisional" dependency tree for the sentence:
(6) "Mary loves a boy with blue eyes that is playing in the garden with a ball",
when the PP "with a ball" is being analysed (note that, at this point of the analysis, the choices between alternative attachments of the RC "that play" and of the PP "in the garden" have been already performed).
Provisional dependency trees appears as graphs, since nodes can be attached, via provisional links, to more than one node of the tree. Actually, the graph is a compact and implicit way for representing a set of trees, each of which is obtained by selecting one attachment for each set of alternative provisional links.

(B) EVALUATION OF SYNTACTIC AND SEMANTIC CDs

For each provisional attachment, the local syntactic CD (SyCD) and semantic CD (SeCD) must be evaluated, and the resulting value stored in the attachment.
As described in the previous section, SyCDs are a weighed form of the Right Association Priciple: to each provisional attachment is associated a value which is the inverse of the distance of the attachment, starting from the current node and climbing up the tree (for example, in fig.1 SyCD(#1)=1 SyCD(#2)=0.5, SyCD(#3)=0.33 and SyCD(#4)=0.25).

The SeCDs are obtained by activating the semantic module of the system in parallel on each provisional alternative attachment. The semantic rules have been modified in order to access, in the condition part, the weighed Domain Knowledge (semantic net in case of attachments to verbs, preplate table in case of prepositional specification) and in order to return the CD obtained by the match. The action part of the rules which builds up the semantic representation is delayed: only when the final attachment has been chosen the action part of the rule correspondent to that attachment is executed.
For each provisional attachment, the system computes the semantic and syntactic "plausibility" of all the attachments of the configuration of dependency tree generated by selecting that attachment and not considering its alternatives. Note that only the "variant" part of the tree are considered: all the

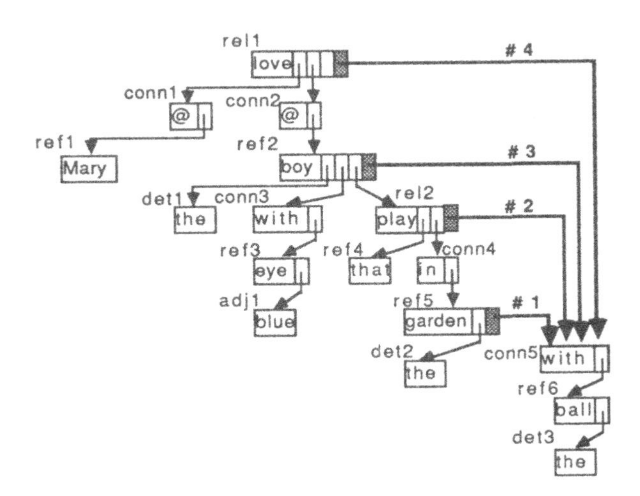

fig.1 Provisional dependency tree of the sentence: "Mary loves a boy with blue eyes that is playing in the garden with a ball". The bold arcs represent the provisional attachments.

invariant parts (such as, for example, the path ref2-conn3-ref3-adj1 in fig.1) are neglected, since they do not contribute to the goal of discriminating between the alternative provisional attachments.
The algorithm which computes the set of SeCDs corresponding to each configuration of dependency tree when a PP (or RC) is being analysed and alternative provisional attachments have been generated is the following (henceforth, L(N1,N2) denotes the link -attachment- from node N1 to node N2, and S(Ni) denotes the set of all the alternative links to the node Ni):

FORALL L(N,N1) $\in$ S(N1)
CASE OF
- the type of N is REF
  SETOF-SeCDs(L(N,N1)) <--
    {SeCD of the attachment L(N,N1)} $\cup$
    {SeCDs of the "open" case frames of nodes Mi, without the addition of case N1 |
    L(Mi,N1) $\in$ S(N1) AND the type of Mi is REL AND Mi is higher than N in the tree} $\cup$
    {SeCDs of the "closed" case frames of nodes Mi, without the addition of case N1 |
    L(Mi,N1) $\in$ S(N1) AND the type of Mi is REL AND Mi is lower than N in the tree}
- the type of N is REL:
  SETOF-SeCDs(L(N,N1)) <--
    {SeCD of the "open" case frame of N, with the addition of case N1} $\cup$
    {SeCDs of the "open" case frames of nodes Mi, without the addition of case N1 |
    L(Mi,N1) $\in$ S(N1) AND the type of Mi is REL AND Mi is higher than N in the tree} $\cup$
    {SeCDs of the "closed" case frames of nodes Mi, without the addition of case N1 |
    L(Mi,N1) $\in$ S(N1) AND the type of Mi is REL AND Mi is lower than N in the tree}

The previous algorithm requires some explanations. The semantic evaluation of an attachment or a case frame corresponds to an activation of the semantic module of the system in order to check the semantic correctness of the attachment (case frame) in the domain knowledge, and the resulting CD is returned. Since the cases of a verb are not independent of each other, the whole case frame of a REL node which is the starting node of a provisional attachment must be considered as "variant". Therefore, its SeCD must contribute in the evaluation of the global plausibility of the configurations generated by choosing each one of its alternative attachments. Moreover, this SeCD must be

evaluated by considering the case frame "open" or "closed" depending on whether the considered alternative attachment starts from an upper or lower node than the REL node in the dependency tree. Actually, the algorithm is optimised, since the semantic evaluation of an attachment (or of an open or closed case frame) is performed only once.
As the result of the application of such an algorithm, all the SeCDs of the "variant" attachments of the configuration generated by choosing a particular provisional attachment are stored in a special slot (SETOF-SeCDs) of such an attachment. Let us consider an example.
In the following, we will adopt the shorthands:

- SeCD(REFx, REFy) stands for the SeCD of the prepositional specification represented by the path connecting node REFx to node REFy of the dependency tree;
- ClosedSeCD(REL, REF1, ... REFn) and OpenSeCD(REL, REF1, ... REFn) stands for the CD associated to the closed and the open case frame of the verb REL with cases REF1 ... REFn.

In the example shown in fig.1, after the execution of the previous algorithm, the SeCDs corresponding to the configurations generated by the attachment #1, #2, #3 and #4 (stored in the SETOF-SeCDs of these attachments) are respectively:
#1: {SeCD(ref5,ref6), OpenSeCD(rel2,ref4,ref5), OpenSeCD(rel1,ref1,ref2)}
#2: {OpenSeCD(rel2,ref4,ref5,ref6), OpenSeCD(rel1,ref1,ref2)}
#3: {ClosedSeCD(rel2,ref4,ref5), SeCD(ref2,ref6), OpenSeCD(rel1,ref1,ref2)}
#4: {ClosedSeCD(rel2,ref4,ref5), OpenSeCD(rel1,ref1,ref2,ref6)}

(C) GLOBAL CDs OF THE CONFIGURATIONS

For each configuration, the global SeCD (henceforth GSeCD) is computed by suitably combining the SeCDs of the "variant" attachments in the configuration, stored (by the execution of the previous algorithm) in the corresponding provisional attachment. Since all the attachments in a configuration are in AND, the function which combines their SeCDs must be a t-norm [7].
An immediate solution should be to combine the SeCDs with a product (probability theory) or a minimum operator (possibility theory). In a previous paper [13], it has been claimed that, in some applications, none of them gives the desired results. In particular, it was argued that, when using product, the final result decreases when a high number of values are considered, even if these values are high (say 0.9), while, when using the minimum, the result is not sensitive to values higher than the minimum (intuitively, the case of n low values should differ from the case of n-1 high values and 1 low value; for further discussions about these topics see also [2]).
Therefore we adopted the formula for combining ANDed values proposed in [13], which has as a lower bound the product and, as an upper bound, the minimum:

$$e(\text{AND } (T_1\ T_2\ \dots\ T_n)) = \alpha+\beta * (\beta-\alpha)$$
$$\text{where } \alpha = \prod_{j=1}^{n} e(T_j) \quad \text{and} \quad \beta = \min_{j=1}^{n} e(T_j)$$

The resulting evidence degree consists of the product of the evidence degrees of the n elementary terms increased by a quantity which is a part of the difference between the minimum (β: upper bound) and the product (α: lower bound) proportional to the minimum.
This formula presents some advantages: the result is sensitive also to values different from the minimum and it does not tend to zero as the number of terms different from 1 increases.

As regards the global SyCD (henceforth GSyCD) of a configuration of dependency tree, it can be equalized to the SyCD of the corresponding provisional attachment, and no more computation is needed.

(D) WEIGHT OF SeCD AND SyCD AND SELECTION OF THE MOST PLAUSIBLE ATTACHMENT

At the end, for each configuration a weighed combination of its final syntactic and semantic weights is computed following the general rule:

Total weight := W1 * GSeCD + W2 * GSyCD
where W1 + W2 = 1

We have not determined the most suitable values for W1 and W2 yet. Actually, we are analysing the impact of changing their values on the overall performance of the system.
The total weights of the configurations are compared in order to choose the most plausible attachment.

Finally, it must be noted that, in the system, the choices between different attachments are made as soon as possible. This implies that, for particular kinds of sentences (typical of the Italian language), some provisional (and possibly wrong) choices are made. In fact, in Italian, a PP can occur between the subject noun phrase (NP) of the sentence and the verb. For example, let us consider the sentence
(7) Il ragazzo che studia **in giardino** gioca - The boy who studies **in the garden** plays -
The PP "in giardino" (in the garden) can, in principle, refer to "studia" (studies) or to "ragazzo" (boy) or also to "gioca" (plays). When the PP "in giardino" is being analysed, the evaluation of the (syntactic and semantic) CDs of the attachment to "gioca" must be delayed until the verb of the sentence will be analysed. Nevertheless, the partial choice between the attachments to "studia" and to "ragazzo" is not delayed, and it is made considering the plausibility of the corresponding configurations as previously shown in the paper (let #i the chosen attachment). Only when the main verb of the main sentence (i.e. "gioca") will be analysed, the plausibility of the configuration generated by the attachment of "in giardino" to "gioca" will be evaluated, and the choice between this attachment and #i will be made[6].

5 CONCLUSIONS

In the paper we pointed out (considering the attachment problem) the necessity of a flexible interaction between syntax and semantics in NL processing, and we showed how such flexibility has been achieved by weighting the contributions of these knowledge sources.
Other knowledge sources may affect the selection of the correct attachment. For instance, there are approaches that consider as basic the information provided by the context in which sentences are uttered (e.g. the previously uttered sentences) and propose to choose between alternative attachments by considering the "Principle of Parsimony" (i.e. choose the attachment that leaves the minimum number of presuppositions unsatisfied [5]) or the more strict "Principle of Referential Success" (i.e. choose the attach that minimizes the number of entities referred to and not already introduced in the context [25]). Absity [11] is an outstanding example of system considering the Principle of Referential Success. It adopts a complex algorithm for selecting the most plausible attachment that considers different sources of information (the Priciple of Referential Success, syntactic and semantic preferences).
Although we recognize the importance of contextual information for discriminating between alternative attachments [1], we argue that the syntactic and semantic criteria described in the paper play a basic role [11], especially in cases in which the principle of Referential Success does not apply (e.g. with indefinite descriptions, or with definite attributive or generic descriptions). Furthermore, we claim that the modularity and flexibility of our approach should make easier the task of considering also contextual information, and of integrating it in the general interpretation process of the system.

The GULL system is implemented in FRANZ LISP and runs on SUN workstations, under the Unix operating system.

REFERENCES

[1] Altmann, G. and Steedman, M., "Interaction with context during human sentence processing," *Cognition* **30** pp. 191-238 (1988).

[2] Bhatnagar, R.K. and Kanal, L.N., "Handling Uncertain Information: A Review of Numeric and Non-numeric Methods," pp. 3-26 in *Uncertainty in Artificial Intelligence*, ed. L.N. Kanal, J.F. Lemmer,North-Holland (1986).

[6] The algorithm described in the previous section is not applicable for computing the CDs of the configurations when a REL (verbal) node is being filled. Another (more complex) algorithm is applied, but it will not be shown here for space constraints.

[3] Charniak, E., "Six topics in Search of a Parser ," pp. 1079-1087 in *Proc.7th IJCAI*, Vancouver (1981).

[4] Connine, C., Ferreira, F., Jones, C., Clifton, C., and Frazier, L., "Verb frame preferences: Descriptive norms," *Journal of psycholinguistic research* **13(4)** pp. 407-428 (1975).

[5] Crain, S. and Steedman, M., "On Not Being Led Up The Garden Path: The Use of Context by the Psychological Parser," in *Syntactic Theory and How People Parse Sentences*, ed. D.R. Dowty, L.J. Karttunen, A.M. Zwicky,Cambridge Univ. Press (1984).

[6] DiEugenio, B. and Lesmo, L., "Representation and interpretation of determiners in natural language," pp. 648-654 in *Proc. IJCAI 87*, Milano (1987).

[7] Dubois, D. and Prade, H., *Fuzzy sets and systems: theory and applications*, Academic Press (1980).

[8] Fillmore, C., "The case for case," in *Universals in linguistic theory*, ed. Bach and Harms,Holt, Rinehart and Winston, New York (1968).

[9] Ford, M., Bresnan, J.V., and Kaplan, R.M., "A competence based theory of syntactic closure," in *The mental representation of grammatical relations*, ed. Bresnan ,MIT Press (1982).

[10] Frazier, L. and Fodor, J.D., "The sausage machine: A new two-stage parsing model," *Cognition 6 (4)*, pp. 291-235 (December 1978).

[11] Hirst, G., *Semantic interpretation and the resolution of ambiguity*, Cambridge University Press (1987).

[12] Lesmo, L., Berti, M., and Terenziani, P., "A Network Formalism for Representing Natural Language Quantifiers," pp. 473-478 in *Proc. European Conference on Artificial Intelligence 88*, Munich (1988).

[13] Lesmo, L., Saitta, L., and Torasso, P., "Evidence Combination in Expert Systems," *Man-Machine Studies* **22** pp. 307-326 (1985).

[14] Lesmo, L. and Torasso, P., "Analysis of Conjunctions in a Rule Based Parser," pp. 180-187 in *Proc. 23rd Annual Meeting of Association Computational Linguistics*, (1985).

[15] Lesmo, L. and Torasso, P., "A Flexible Natural Language Parser based on a two-level Representation of Syntax," pp. 114-121 in *Proc. 1st Conf. ACL Europe*, Pisa (1983).

[16] Lesmo, L. and Torasso, P., "Weighted Interaction of Syntax and Semantics in Natural Language Analysis," pp. 772-778 in *Proc. 9th Int. Joint Conf. on Artificial Intelligence*, Los Angeles (1985).

[17] Marcus, M., *A Theory of Syntactic Recognition for Natural Language*, MIT Press (1980).

[18] Sagalowicz, D., "Mechanical Intelligence: Research and Applications," Tech. Report, SRI Int., Menlo Park (1980).

[19] Schank, R., *Conceptual information processing*, North Holland, Amsterdam (1975).

[20] Terenziani, P., "A Rule Based Approach to the Semantic Interpretation of Natural Language," *to appear in Computers and Artificial Intelligence*, (1991).

[21] Tyler, L.K. and Marslen-Wilson, W.D., "Speech Comprehension Process," pp. 169-184 in *Perspectives on mental representation: Experimental and Theoretical studies of cognitive processes*, ed. Mehler Walker, Garret,Lawrence Erlbaum Associates (1982).

[22] Wilks, Y.A., Huang, X., and Fass, D., "Syntax, preference and right attachment," pp. 779-784 in *Proc.9th IJCAI*, Los Angeles (1985).

[23] Wilks, Y.A., "A preferential Pattern-Seeking Semantics for Natural Language Inference," *Artificial Intelligence* **6** pp. 53-74 (1975).

[24] Winograd, T., *Language as a Cognitive Process - Syntax -*, Addison-Wesley (1983).

[25] Winograd, T., *Understanding Natural Language*, Academic Press, New York (1972).

[26] Woods, W.A., Kaplan, R.M., and Nash-Webber, B.L., "The Lunar Sciences Natural Language Information System: Final Report," Tech. Report 2378, BBN Inc., Cambridge (1972).

[27] Woods, W.A., "Semantics for a question answering system," *Outstanding dissertations in Computer Sciences*, (1979).

A TOPOLOGICAL APPROACH TO SOME CLUSTER METHODS*

J. Jacas J. Recasens
Sec. Matemàtiques i Informàtica
E.T.S. d'Arquitectura de Barcelona
Univ. Politècnica de Catalunya
Diagonal 649. 08028 Barcelona. Spain

Abstract.

One of the most usual ways of classifying the elements of a set is to cluster them according to some kind of "proximity measure". Proximity is a topological concept and therefore it is natural to ask for topological structures that lead to cluster methods.

Using this idea, we construct some families of cluster methods starting on from a kind of V_D–spaces.

In order to relate the elements of these families, morphisms between cluster methods are defined.

Keywords: Cluster Analysis, V_D–space, single linkage, Numerical Stratified Clustering, Indistinguishability operators, triangular norm, triangular conorm.

INTRODUCTION.

Most cluster methods start from a dissimilarity matrix. The entries of a dissimilarity matrix $\mathfrak{m}$ are commonly treated as distances between OTU's (Operational Taxonomical Unities) [10] and consequently metrical ideas are applied to define cluster methods.

But the entries of $\mathfrak{m}$ can also be considered as degrees of proximity and topological techniques can be used to construct and to give theoretical substract to some cluster methods. This point of view is exploded in this paper in order to study and construct some families of these methods.

In a set X of objects to be classified according to a dissimilarity matrix $\mathfrak{m}$, it is natural to identify at each level $k \in [0,1]$, the points p, q of X such that $\mathfrak{m}(p,q) \leq k$. This leads to consider in section 1 a special kind of neighbourhoods $N_p(k,h)$ that endow X with the structure of a V_D-space, i.e. given two neighbourhoods U, V of $p \in X$, there exists a neighbourhood W of p such that $W \subset U \cap V$. In sections 2 and 3, making use of the V_D-spaces techniques, some families of cluster methods related to Max-Min and Max-T closure (T being a t-norm) are defined [4,6].

In order to compare these methods, morphisms are introduced in section 4.

Research partially supported by the DGICYT, project n. PS.87-0108

0. PRELIMINARIES.

In this section, we recall some well known concepts. For a more detailed exposition on these topics, readers are referred to [8] and to [5], [11] for parts a) and b) respectively.

a) Given a set X, $C(X)$ will denote the set of dissimilarity matrices on X (i.e., the set of symmetric fuzzy relations $\mathfrak{m}$ on X such that $\mathfrak{m}(p,p) = 0 \; \forall p \in X$) and $\Sigma(X)$ the set of symmetric and reflexive binary relations on X. A numerically stratified clustering on X is a map

$$c : [0,1] \to \Sigma(X)$$

that satisfies:

(i) $\lambda \leq \lambda' \Rightarrow c(\lambda) \subset c(\lambda')$
(ii) $c(\lambda) = X \times X$ for large enough λ.
(iii) Given $\lambda \in [0,1]$ there exists $\delta > 0$ such that $c(\lambda + \delta) = c(\lambda)$.

$NSC(X)$ will denote the set of all numerically stratified clusterings on X.

A cluster method is a map

$$\mathfrak{C} : C(X) \to NSC(X).$$

b) A De Morgan triple is a triple (T, S, φ) where T is a t-norm, S a t-conorm and φ a strong negation related by

$$T \circ (\varphi \times \varphi) = \varphi \circ S.$$

If $\mathfrak{m}$ is a dissimilarity matrix on X, then $M = \varphi \circ \mathfrak{m}$ is a reflexive and symmetric fuzzy relation on X.

For a given t-norm T, the Max-T product $M \circ N$ of two fuzzy relations M, N on X is defined in the same way as the Max-Min product; namely by

$$(M \circ N)(x,y) = \sup_{z \in X}(T(M(x,z), N(z,y)).$$

Since Max-T is associative, we can define

$$M^n = \underbrace{M \circ M \circ \ldots \circ M}_{n \text{ times}}$$

The T-closure of a reflexive and symmetric fuzzy relation M is

$$M^c = \sup_{n \in N} M^n.$$

It can be shown that if X is finite of cardinal s, then

$$M^c = \sup_{n \in \{1,\ldots,s\}} M^n$$

1. DISSIMILARITY TOPOLOGY.

Let X be a finite set of objects to be classified and $\mathfrak{m}$ a dissimilarity matrix on X, at each level $k \in [0,1]$ we identify the points $p, q \in X$ such that $\mathfrak{m}(p,q) \leq k$. Therefore, given a dissimilarity matrix $\mathfrak{m}$, we define a neighbourhood of a point p of X in the following way:

Given a fixed $k \in [0,1]$, for each $h \in (0,1]$ $N_p(k,h)$ will denote the neighbourhood of p given by

$$N_p(k,h) = \{q \in X \mid \mathfrak{m}(p,q) < k+h\}.$$

It is trivial to show that if U, V are neighbourhoods of p, then there exists a neighbourhood W of p with $W \subset U \cap V$. Therefore, X is endowed with the structure of a V_D–space [9].

Proposition 1.1. This structure defines a topology if and only if $\mathfrak{m}$ is an ultrametric [3].

Proof: We have to show that if $q \in N_p(k,h)$, then there exists $h' \in (0,1]$ such that $N_q(k,h') \subset N_p(k,h)$. Taking $h' = h$, if $x \in N_q(k,h)$ then $\mathfrak{m}(x,p) \leq Max(\mathfrak{m}(p,q), \mathfrak{m}(q,x)) < Max(k+h, k+h) = k+h$ and therefore $x \in N_p(k,h)$. □

In the setting of the theory of V_D–spaces we define the concept of contiguity:

Definition 1.1. Given $A \subset X$, $p \in X$ is *contiguous* to A if and only if there exists $q \in A$ with $\mathfrak{m}(p,q) \leq k$.

The concept of contiguity allows us to define a Čech closure $\mathcal{C}^k$ [9] in the power set of X, denoted $\mathcal{P}(X)$, as follows:

$$\begin{aligned} \mathcal{C}^k : \mathcal{P}(X) &\rightarrow \mathcal{P}(X) \\ A &\rightarrow \mathcal{C}^k(A) \end{aligned}$$

where $\mathcal{C}^k(A)$ is the set of points of X contiguous to A.

1.1 Properties of $\mathcal{C}^k$:

Given $A, B \in \mathcal{P}(X)$ we have

1.1.1. $\mathcal{C}^k(\emptyset) = \emptyset$

1.1.2. $\mathcal{C}^k(A \cup B) = \mathcal{C}^k(A) \cup \mathcal{C}^k(B)$

1.1.3. $A \subset \mathcal{C}^k(A)$

1.1.4. $A \subset B \rightarrow \mathcal{C}^k(A) \subset \mathcal{C}^k(B)$

1.1.5. $\mathcal{C}^k(A) = \bigcup_{p \in A} \mathcal{C}^k(\{p\})$

1.1.6. $k \leq k' \Rightarrow \mathcal{C}^k(A) \subset \mathcal{C}^{k'}(A)$.

Proof: Straightforward. □

The subsets $A \subset X$ such that $\mathcal{C}^k(A) = A$ are called $\mathcal{C}^k$–closed. In order to built a topology in X, we define a Kuratowski closure map $\mathcal{C}^k_{\mathfrak{K}}$ as follows:

$$\begin{aligned} \mathcal{C}^k_{\mathfrak{K}} : \mathcal{P}(X) &\rightarrow \mathcal{P}(X) \\ A &\rightarrow \mathcal{C}^k_{\mathfrak{K}}(A) \end{aligned}$$

where $\mathcal{C}^k_{\mathfrak{K}}(A)$ is the intersection of all $\mathcal{C}^k$–closed sets that contain A.

1.2 Properties of $\mathcal{C}^k_{\mathfrak{R}}$:

Given $A \in \mathcal{P}(X)$ we have:

1.2.1. $\mathfrak{m}$ is an ultrametric if and only if $\mathcal{C}^k = \mathcal{C}^k_{\mathfrak{R}}$ $\forall k \in [0,1]$.

1.2.2. $A \subset \mathcal{C}^k(A) \subset \mathcal{C}^k_{\mathfrak{R}}(A)$

1.2.3. $(\mathcal{C}^k)^n(A) \subset \mathcal{C}^k_{\mathfrak{R}}(A)$, where $(\mathcal{C}^k)^n(A) = \underbrace{\mathcal{C}^k(\mathcal{C}^k(\ldots\mathcal{C}^k(A)))}_{n \text{ times}}$

1.2.4. $\mathcal{C}^k_{\mathfrak{R}}(A) = \{x \in X \mid \exists n \in \mathbb{N} \text{ such that } x \in (\mathcal{C}^k)^n(A)\}$.

Proof:
1.2.1. is a consequence of Proposition 1.1.
1.2.2 and 1.2.3 are straightforward.
1.2.4: Let $M = \{x \in X \mid \exists n \in \mathbb{N} \text{ such that } x \in (\mathcal{C}^k)^n(A)\}$.

i) $M \subset \mathcal{C}^k_{\mathfrak{R}}(A)$: If $x \in M$, then there exists $n \in \mathbb{N}$ such that $x \in (\mathcal{C}^k)^n(A)$ and from 1.2.3 follows that $x \in \mathcal{C}^k_{\mathfrak{R}}(A)$.

ii) $\mathcal{C}^k_{\mathfrak{R}}(A) \subset M$: If sufices to prove that M is $\mathcal{C}^k$-closed, which follows from properties 1.1.1. 1.1.6. □

2. CLUSTER METHODS RELATED TO SINGLE LINKAGE.

The closure operator $\mathcal{C}^k$ allows us to define a family $\{\mathfrak{C}_n\}_{n\in\mathbb{N}}$ of cluster methods that has the *single linkage* method as a particular case.

From property 1.2.4 the three following propositions can be deduced:

Proposition 2.1. Let s be the cardinal of X. Then $(\mathcal{C}^k)^n = \mathcal{C}^k_{\mathfrak{R}}$ for each $n \geq s-1$.

Proposition 2.2. Given $p, q \in X$, either $\mathcal{C}^k_{\mathfrak{R}}(\{p\}) = \mathcal{C}^k_{\mathfrak{R}}(\{q\})$ or $\mathcal{C}^k_{\mathfrak{R}}(\{p\}) \cap \mathcal{C}^k_{\mathfrak{R}}(\{q\}) = \emptyset$.

Therefore $\mathcal{C}^k_{\mathfrak{R}}$ defines a partition of X. If $\mathfrak{R}$ is its corresponding equivalence relation, we have

Proposition 2.3. $\mathfrak{R}$ is the equivalence relation associated to the *single linkage* at level k of the dissimilarity matrix $\mathfrak{m}$. In other words:

Given $p, q \in X$, $p\mathfrak{R}q$ if and only if there exists a chain $x_0, x_1, \ldots x_r \in X$ with $x_0 = p$, $x_r =$ and $\mathfrak{m}(x_{i-1}, x_i) \leq k \quad \forall i = 1, \ldots, r$.

On the other hand, for each $n \in \mathbb{N}$ we can define a cluster method as follows:

Definition 2.1 (of the nth cluster method). For each $k \in [0,1]$, let

$$\mathbf{C}^k_n = \bigcup_{p \in X} \{(p, x) \mid x \in (\mathcal{C}^k)^n(\{p\})\}.$$

Property 1.1.6 assures us that we have a numerical stratified clustering

$$c_n : [0,1] \to \Sigma(X)$$
$$k \to \mathbf{C}_n^k$$

and therefore a Cluster method $\mathfrak{C}_n$.

It is worth noting that given a dissimilarity matrix $\mathfrak{m}$ the numerical stratified clustering c_n coincides with the α-cuts of the Max-Min product M^n of $M = \varphi \circ \mathfrak{m}$, φ being a strong negation.

From Properties 1.1.3, 1.2.3 and Proposition 2.1 follows that the family $\{\mathfrak{C}_n\}_{n\in\mathbb{N}}$ of cluster methods is in fact finite and gives a gradation in the sense that when n increases, $\mathfrak{C}_n$ becomes closer to *single linkage* method.

3. CLUSTER METHODS RELATED TO A t-CONORM.

Given a De Morgan triple (T, S, φ) and a dissimilarity matrix $\mathfrak{m}$, the Max-T closure of the matrix $\varphi \circ \mathfrak{m}$ can be constructed in the same way as the usual transitive closure (Max-Min) [11].

This procedure generates a cluster method that will be called the *Max- T cluster method.*

Given a t-conorm S, we define $\mathfrak{S}(a) = a\ \forall a \in [0,1]$ and for $n \geq 2$ recursively

$$\mathfrak{S}(a_1, a_2, \ldots, a_n) = S(a_1, \mathfrak{S}(a_2, \ldots, a_n)) \qquad \forall a_1, a_2, \ldots a_n \in [0,1].$$

Using this notation, the cluster methods defined in the last section can be generalized in the following way:

Given a dissimilarity matrix $\mathfrak{m}$, and a t–conorm S, for any $n \in \mathbb{N}$, $k \in [0,1]$, $C_S^{k,n}$ is a map defined by:

$$C_S^{k,n} : X \longrightarrow \mathcal{P}(X)$$
$$p \longrightarrow C_S^{k,n}(p)$$

being $C_S^{k,n}(p) = \{q \in X | \exists x_0, x_1, \ldots x_n \in X$ with $x_0 = p, x_n = q$ and $\mathfrak{S}(\mathfrak{m}(x_0, x_1), \mathfrak{m}(x_1, x_2), \ldots,$ $\mathfrak{m}(x_{n-1}, x_n)) \leq k\}$.

In particular

$$C_S^{k,0}(p) = \{p\}$$
$$C_S^{k,1}(p) = \{q \in x | \mathfrak{m}(p,q) \leq k\}, \quad \forall k \in [0,1]$$

$C_S^{k,n}$ defined in X can be extended to a closure map in the power set of X also denoted by $C_S^{k,n}$:

$$C_S^{k,n} : \mathcal{P}(X) \longrightarrow \mathcal{P}(X)$$
$$A \longrightarrow C_S^{k,n}(A) = \bigcup_{p\in A} C_S^{k,n}(p).$$

It is easy to prove that the closure maps $C_S^{k,n}$ satisfy properties 1.1.1.–1.1.6. of section 1, and if $S(x,y) =$Max(x,y) then $C_S^{k,n} = (C^k)^n$.

In a similar way, it can be shown that, for each $n \geq s-1$, $C_S^{k,n} = C_S^{k,n+1}$ where s is the cardinal of X.

For each $n \in \mathbb{N}$ we can define a cluster method as follows:

Definition 3.1. (of the nth cluster method). Given a t–conorm S, for each $k \in [0,1]$ let

$$\mathbf{C}_S^{k,n} = \bigcup_{p \in X} \{(p,x) \mid x \in C_S^{k,n}(\{p\})\}.$$

The map

$$c_{n,S} : [0,1] \longrightarrow \Sigma(X)$$
$$k \longrightarrow \mathbf{C}_S^{k,n}$$

is a numerical stratified clustering and therefore, we have a cluster method $\mathfrak{C}_{n,S}$.

For each t–conorm S, the family $\{\mathfrak{C}_{n,S}\}_{n \in \mathbb{N}}$ is finite and gives a gradation of subdominant methods [8] such that when n increases, $\mathfrak{C}_{n,S}$ tends to the *Max–T cluster method*, being $(T, S, \varphi$ a De Morgan triple [11].

If $S(x,y) = \text{Max}(x,y)$, then $\mathfrak{C}_{n,S}$ coincides with the Cluster method $\mathfrak{C}_n$ defined in section 2.

4. MORPHISMS OF CLUSTER METHODS.

Morphisms are introduced in order to compare and relate the cluster methods constructed in the preceeding sections.

Given two sets X, Y and a map $f : X \to Y$, a pair of maps (both denoted by f^*) can be defined as follows:

$$f^* : C(Y) \to C(X)$$
$$\mathfrak{m} \to f^*\mathfrak{m} = \mathfrak{m} \circ (f \times f)$$

$$f^* : NSC(Y) \to NSC(X)$$
$$c \to f^*c : [0,1] \to \Sigma(X)$$

where $f^*c(t) = \{(a,b) \in X \times X | (f(a), f(b)) \in c(t)\}, \ \forall t \in [0,1]$.

Definition 4.1. Given two cluster methods $\mathfrak{C}, \mathfrak{C}'$ in the sets X, Y respectively, a homometry between $(X, \mathfrak{C})$ and $(Y, \mathfrak{C}')$ is a map $f : X \to Y$ such that

$$f^*\mathfrak{C}'\mathfrak{m}(t) \subset \mathfrak{C} f^*\mathfrak{m}(t) \quad \forall \mathfrak{m} \in C(Y) \quad \forall t \in [0,1] \tag{1}$$

If $\forall t \in [0,1]$, equality holds then f will be called a strict-homometry. Therefore f is a strict homometry if and only if the following diagram commutes:

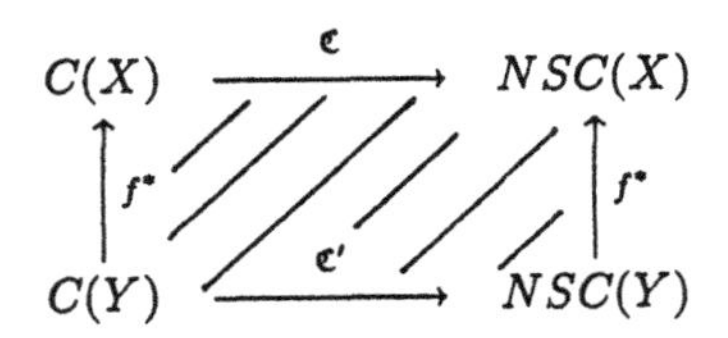

An <u>isometry</u> is a bijective strict-homometry.

<u>Definition 4.2.</u> Given two cluster methods $\mathfrak{C}, \mathfrak{C}'$ in a set X, we say that $(X, \mathfrak{C})$ is <u>similar</u> to $(X, \mathfrak{C}')$ if and only if there exists a monotonic map $F\colon [0,1] \to [0,1]$ such that

$$\mathfrak{C}\mathrm{m}F(t) \subset \mathfrak{C}'\mathrm{m}(t) \qquad \forall \mathrm{m} \in C(X) \quad \forall t \in [0,1].$$

If $\forall \mathrm{m} \in C(X)$ and $\forall t \in [0,1]$ the equality holds, we say that $(X, \mathfrak{C})$ is <u>strictly-similar</u> to $(X, \mathfrak{C}')$. In this case, the following diagram commutes $\forall \mathrm{m} \in C(X)$:

$$\begin{array}{ccc} [0,1] & \xrightarrow{\mathfrak{C}'\mathrm{m}} & \Sigma(X) \\ & {}_{F}\searrow \quad \nearrow_{\mathfrak{C}\mathrm{m}} & \\ & [0,1] & \end{array}$$

Summing up this two definitions we have

<u>Definition 4.3.</u> Given two cluster methods $\mathfrak{C}, \mathfrak{C}'$ defined in the sets X, Y respectively, a <u>morphism</u> $\gamma : (X, \mathfrak{C}) \to (Y, \mathfrak{C}')$ is a pair $\gamma = (f, F)$ where f is a map of X into Y and F a monotonic map from [0,1] into itself such that

$$\mathfrak{C}f^*\mathrm{m}F(t) \subset f^*\mathfrak{C}'\mathrm{m}(t) \quad \forall \mathrm{m} \in C(Y) \quad \forall t \in [0,1].$$

If $\forall \mathrm{m} \in C(Y)$ and $\forall t \in [0,1]$, the equality holds then γ is called a <u>strict-morphism</u>. Therefore γ is a strict–morphism if and only if the following diagram commutes:

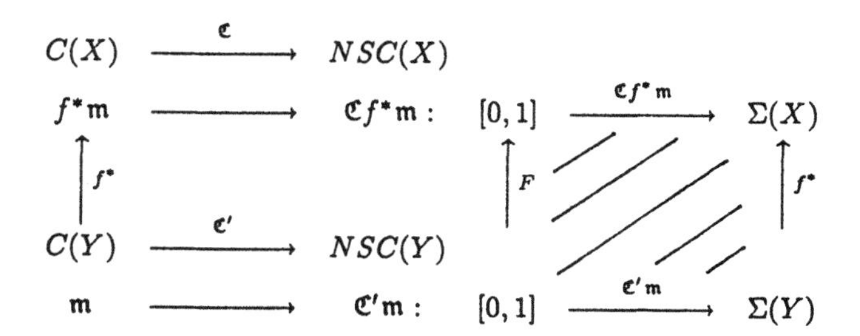

A bijective strict morphism is called an <u>isomorphism</u>. If in definition 4.3, we take $X = Y$ and $f = j$, then $(X, \mathfrak{C})$ is similar to $(X, \mathfrak{C}')$. Finally, if $F = j$, then γ is a homometry.

It is easy to prove that the maps

$$C_s^{k,n} : \mathcal{P}(X) \to \mathcal{P}(X)$$

defined in the last section satisfy the relation

$$C_s^{k,n}(A) \subset C_s^{k,n}(A) \quad \forall A \in \mathcal{P}(X) \quad \text{if } n \leq m.$$

Therefore it follows

Proposition 4.1. If $n \leq m$ then $\gamma = (j,j) : (X, \mathfrak{C}_{n,s}) \rightarrow (X, \mathfrak{C}_{m,s})$ is a morphism of cluster methods.

If $\mathfrak{SL}$ denotes the *single linkage* method, the following two properties can also be proved

Proposition 4.2. For each t-conorm S and each $n \in \mathbb{N}$, there exists a monotonic decreasing map $F : [0,1] \rightarrow [0,1]$ such that $\gamma = (j,F) : (X, \mathfrak{C}_{n,s}) \rightarrow (X, \mathfrak{SL})$ is a morphism of cluster methods.

Proposition 4.3. Let S, S' be two t-conorms such that Max$\leq S \leq S'$, and n, m two integers such that $n \leq m$. Then the following diagram of cluster morphisms commutes:

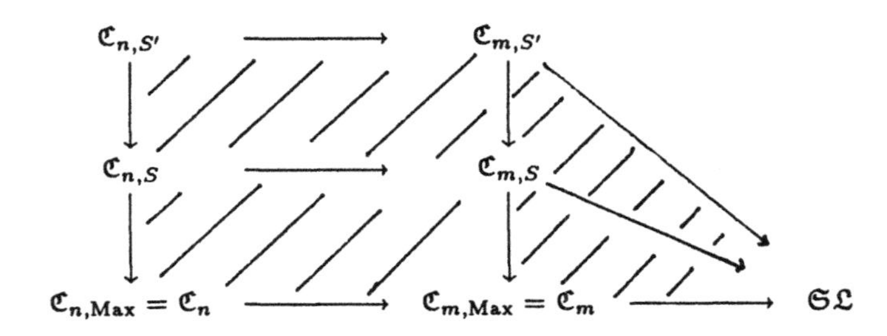

SUMMARY.

We have defined some families of cluster methods using topological ideas and showed that the use of these techniques associated to fuzzy relations is a powerful tool in order to built a comprehensive theory of Cluster Analysis methods.

This is the first time, as far as the authors know, that morphisms are used in Cluster Analysis and in a forthcoming paper they will be studied in a more general frame.

REFERENCES.

[1] Alsina, C., Trillas, E. (1978) Introducción a los Espacios Métricos Generalizados. Fund. J. March. Serie Universitaria 49.

[2] Bouchon, B., Cohen, G., Frankl, P. (1982) Metrical properties of fuzzy relations. Problems of Control and Information Theory 11, 389-396.

[3] Defays, D. (1975) Ultramétriques et relations floues. Bull. Societé Royale de Sciences de Liège, 1-2, 104-118.

[4] Jacas, J. (1990) Similarity Relations - The Calculation of Minimal Generating Families. Fuzzy Sets and Systems 35, 151-162.

[5] Jacas, J., Valverde, L. (1987) A Metric Characterization of T-transitive Relations, Proceedings of FISAL-86, Palma de Mallorca 81-89.

[6] Jacas, J., Valverde, L. (1990). On Fuzzy Relations, Metrics and Cluster Analysis in: J.L. Verdegay & M. Delgado Eds., Approximate Reasoning. Tools for Artificial Intelligence (Verlag TÜV, Rheinland) I5R, 96.

[7] Janowitz, H.F., Schweizer, B. (1988) Ordinal amb Percentile Clustering. Technical Report. Dept. of Mathematics and Statistics. Univ. Massachusetts.

[8] Jardine, N., Sibson, R. (1971) Mathematical Taxonomy. Wiley, New York.

[9] Schweizer, B., Sklar, A. (1983) Probabilistic Metric Spaces. North-Holland, New York.

[10] Sneath, P.H.A., Sokal, R.P. (1973) Numerical Taxonomy. W.H. Freeman & Co., San Francisco.

[11] Valverde, L. (1985) On the structure of F-indistinguishability operators. Fuzzy Sets and Systems, 17, 313-328.

A FUZZY KNOWLEDGE-BASED SYSTEM FOR BIOMEDICAL IMAGE INTERPRETATION

E. Binaghi, A. Della Ventura, A. Rampini, R. Schettini
Istituto di Fisica Cosmica e Tecnologie Relative - CNR - Milano
Via Ampere 56, 20133 Milano

Abstract

A general purpose knowledge-based system for biomedical image interpretation is presented. The system acquires knowledge directly from the experts by means of a user friendly dialogue. The knowledge introduced tailors the system to a particular biomedical application. Frame representation technique is used for the representation of descriptive knowledge and a fuzzy reasoning strategy, based on fuzzy production rules, is adopted to manipulate the certain and uncertain knowledge contained into Frame Slots and to deduce interpretations. A detailed description of the application of the system to the analysis of CT images of vertebrae for the quantity evaluation of the bone mineral content is provided.

Keywords:

biomedical image interpretation, knowledge-based systems, frames, fuzzy reasoning.

1. INTRODUCTION

The recent developments of medical technology have made it possible to investigate physiological systems and organs by means of a variety of digital imaging devices. Several experiments have addressed the problem of developing software systems which can interpret biomedical images and, in some cases, support medical diagnosis. These systems, which incorporate specia knowledge and simulate the decision making of human experts have been amply experimented with AI techniques for knowledge representation and logic inference [1,3,5].

A general purpose knowledge-based system for biomedical image interpretation is here proposed. The system acquires knowledge directly from the experts by means of a user friendly dialogue. The knowledge introduced tailors the system to a particular biomedical application. Frame representation technique is used for the representation of descriptive knowledge and a fuzzy reasoning strategy, based on fuzzy production rules is adopted to manipulate the certain and uncertain knowledge contained into frame slots and to deduce interpretations.

In the following sections, after a presentation of the general concepts of frame representation and approximate reasoning techniques adopted, a detailed description of our system is provided. The system is applied to a problem of analysis of CT images of vertebrae, to provide a working example.

2. KNOWLEDGE REPRESENTATION AND UNCERTAINTY MANAGEMENT

In dealing with biomedical image interpretation, domain knowledge contains: descriptive knowledge regarding medical descriptions of the anatomical parts concerned, feature descriptions and static relations among objects, behavioural knowledge regarding the criteria on which data manipulation and comparison with the model are based.

The choice of the most suitable representation is based on these considerations:

- descriptive knowledge is strongly structured, presenting a large number of items and complex relations between them; attributes characterizing descriptive knowledge may involve procedures for the assignment of values to objects which are instances of models;
- behavioural knowledge regarding criteria for "manipulation" and "comparison" actions, may be naturally embedded in rules; rules are fired to decide the next action to be performed, on the basis of the current problem status.

Both types of knowledge may be affected by an intrinsic uncertainty due to the presence of medical concepts that are intrinsically qualitative, the presence of objects with fuzzy features, and the presence of imprecision introduced during the imaging process. An approach that provides for all these aspects consists in combining the use of frames [6] for descriptive knowledge representation with the use of production rules [7] for the representation of behavioural knowledge. With regard to the uncertainty, the problem can be adressed by using Fuzzy Set Theory and Fuzzy Logic formulated by Zadeh [8]. Production systems involved in biomedical image interpretation may be represented as fuzzy production systems so that each rule is represented as a fuzzy conditional statement and the reasoning employed is fuzzy reasoning capable of making deductions with certain and uncertain knowledge.

3. DESCRIPTIVE KNOWLEDGE: FRAME-BASED REPRESENTATION

We use frames for the explicit representation of descriptive knowledge about objects included in biomedical images. This knowledge concerns a-priori anatomical knowledge about intrinsic properties of human organs, and image processing knowledge about properties of the images involved in the application. It concerns multi-typed objects characterized by a set of attributes. We may distinguish value attributes, specifying what value or range of values the attribute of a given object should have, and relational attributes denoting taxonomical links among objects. An explicit representation of both relational and value attributes is obtained by structuring the frames with - a Link Part, describing relations with other frames and - a set of Slots, indicating the attributes defining the concepts represented by the frames [6] . Our knowledge base contains two kinds of frames, Model Frames representing models of anatomical parts (Model Frames), Object Frames representing the description of specific objects in the current image (Object Frames) and created when an object is recognized as an istance of a given anatomical model. To assist the presentation of the methodology with an explanatory example, we refer to the problem of the analysis of CT images of the vertebra. The expected result of the analysis is a quantitative evaluation of the bone mineral content related to metabolic pathologies of the skeleton structures, such as osteoporosis. The analyst's procedure consists in identifying the spongy bone, outlining the Region of Interest (ROI), excluding the high vascularity area, and computing the bone mineral content in the ROI by the conversion factors obtained from analysis of calibration phantom regions [9]. Figure 1 shows a sketch of the vertebra.

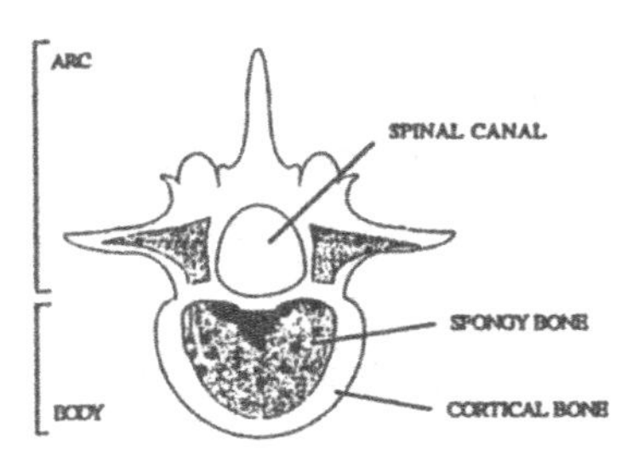

Fig.1 Vertebra sketch

In this application context, the knowledge base contains Model Frames representing anatomical parts such as lombar vertebra, cortical bone, spongy region. Object frames are related to those objects into the current images that are recognized or created in accordance with the goals of the process. In particular, Objects Frames for cortical bone and spongy region are created as a result of the recognition procedure. Frames representing models are predefined on the basis of the

knowledge provided by the experts and describe classes of objects; in each Model-Frame, Own-Slots list characterizing attributes of the class, and Member Slots describe those attributes that assume a significant value for interpretation of the fina image. Object Frames are instances of Model Frames and contain a Member-of link to the class to which the object represented by the frames belongs. Slots of the Object Frames are Own Slots and represent the attributes of the object; these slots are inherited from the Member-Slots of the corresponding Model Frame.
In order to deal with situations in which knowledge related to a given object is qualitative or intrinsically imprecise, we model uncertain attributes as fuzzy variables [10] and introduce Fuzzy Slots in Model and Object Frames. Fuzzy Own-Slots of Model Frames contain the specification of a fuzzy variable in terms of: $< label\ range\ \mu >$, where *label* is the name of the variable, *range* is the set of values the variable may take, and μ is the membership function of the fuzzy set associated to that variable. The membership function expresses the distribution of the possibility associated with the variable values in the range: for each value in the range the membership function gives the degree of acceptance of the value for the given feature. The contents of Fuzzy Slots of Object Frames are: $< label, v, \mu(v) >$, where *label* is the name of the attribute, v is the value measured and $\mu(v)$ is the degree of possibility that the attribute may take the value v. This membership function is inherited from the corresponding Model Frame. An example of the Model Frames and Object Frames involved in the analysis of CT images of vertebrae is given in Table 1. Own Slots in the Spongy Region Model Frame represent features useful for the recognition task: they contain the procedure computing values that correspond to each feature of the objects. The procedures are executed with appropriate parameters during the interpretation of recognition rules. The Object Frame SP-RE inherits the Member Slots of the Spongy Region Model Frame. The inheritance mechanism activates procedures for the computation of values in the Own-Slots of SP-RE.

4. BEHAVIOURAL KNOWLEDGE: PRODUCTION RULES AND FUZZY REASONING

Rules are formalized and stored in the Knowledge Base in terms of fuzzy conditional statements, which may also be nested, having the following general form [8] :

$IF\ C_1\ AND/OR\ C_2 ... AND/OR\ C_n\ THEN\ D_1\ AND\ D_2 ... AND\ D_n$

where C_i is a fuzzy or a non fuzzy proposition [9] denoting conditions that must be satisfied and D_i is a fuzzy or a non-fuzzy proposition denoting a sentence or an action to be performed. A significant part of the rules is dedicated to recognition tasks. Antecedents C_i denote conditions under which the attributes in Own-Slots of a given Model-Frame match the attributes of the object to be recognized; the consequent part contains only one proposition D denoting the creation of an object frame that is an instance of the given Model Frame. The rule for recognizing the candidate object Obj-i as Spongy Region of the vertebra is as follows:

Rule: Spongy Region Recognition Rule;
IF OBJ-i is Below Object Frame.Spinal Canal
THEN
IF OBJ-i is Surrounded by Object Frame.Cortical Bone AND OBJ-i is Circular AND Ray(OBJ-i) is a Model Frame.Spongy-Region.Ray
THEN
Create Object-Frame(Link part: Member of Spongy Regions, Slot-Part:(Inherit(Own Slot, Member Slot of Spongy Region), FOR ALL).

Candidate objects Obj-i are represented in the Knowledge Base as Low-Level Frames and are created at the end of a primary description phase applied to regions resulting from segmentation procedure.

The evaluation of the antecedent of the above rule implies the execution of procedures contained in the Own Slots of Spongy Region Model Frames. Procedures Below and Surrounded require that Object frames for Spinal Canal and Cortical Bone respectively be already recognized. A metarule in the Knowledge-Base subordinate the activation of the recognition rule to this condition. Spinal Canal is recognized by a thresholding procedure and Cortical Bone is recognized as the region containing pixels with maximum density values. The fuzzy antecedent condition $Obj - i$ *is circular* expresses the discriminant condition that the shape of the spongy region is circular, conceiving circular a fuzzy concept with an intrinsic variability in its definition.

MODEL FRAME: SPONGY REGION;

PART OF: LOMBAR VERTEBRA;
PARTS: ROI; VESSEL REGION;
OWN-SLOT: Position-Below
Type: Boolean; Value: Below;
OWN-SLOT: Position-Surrounded
Type: Boolean; Value: Surrounded-by
FUZZY OWN-SLOT: Shape-Circular
Label: Circular Range:(Real; (0,1))- $\mu_{Circular}$
FUZZY OWN-SLOT: Shape-Ray
label: Ray Range:(Integer; (0,30)) - μ_{Ray}
MEMBER-SLOT: Density
Type: real; Value:
MEMBER-SLOT: Position-Centroid
Type: Array (1..2) of Integer; Value:
FUZZY MEMBER-SLOT: Homogeneity
Type: Real; Value:

OBJECT FRAME:SP-RE;

MEMBER OF: SPONGY REGION;
OWN-SLOT: Density
Type: Real Value:
OWN-SLOT: Position of Centroid
Type: Array (1..2) of Integer; Value:
FUZZY OWN-SLOT: Homogeneity
Type: real; Value:

Table 1

The evaluation of the degree of circularity of a given object is made by applying a measure of fuzzy entropy [11] to the accumulation matrices produced by Hough transform procedure. Entropy provides a measure of the degree of "spreading" in accumulation matrices and then a measure of the degree of circularity. Entropy values are the universe of discourse of a fuzzy set whose membership function provides the degree of circularity of a given object in correspondence with its entropy value.

The rule interpretation process involves a reasoning activity organized in the following steps:

- Computation of the global degree of certainty of the antecedent of the rule, concerning the evaluation of fuzzy and non-fuzzy condition C_i in the antecedent of the rule; non-fuzzy conditions are translated into fuzzy conditions by introducing fuzzy sets with descrete membership functions [12] ; rules pertaining to fuzzy AND/OR composition are applied to combine and propagate contributions to the certainty derived from a single condition.
- Fuzzy Inference; for each of the candidate objects, the degree of satisfaction (DoS) in performing the action specified in the consequent part of the rules is computed by applying the Compositional Rule of Inference [8] . A decision-making activity is performed on the DoSs; in the case of recognition rule an Object Frame is created for a candidate object Obj-i if:

$$DoS_{obj-i} = \max_k DoS_{obj--k} AND\ DoS_{obj-i} \geq t$$

where t is a threshold value on the degree of satisfaction. heuristically determined by a calibration phase involving reasoning on a training set of images.

5. ORGANIZATION OF THE SYSTEM

A software system incorporating the described methodology has been implemented. The following requirements have been taken into account in the design of the system: easy definition and implementation of the knowledge base; adaptability to different classes of biomedical images; easy management of image processing functions. Figure 2 shows the architecture of the system highlighting the structural parts and the interactions between them.

The Interface Management System supports a multi-window user-program interface. The fixed parts of the messages are stored in the Message Base and read at appropriate moments. The user interacts directly with this sub-system, providing data for the knowledge representation. The Control Module supervises activations of the sub-parts of the system and implements overall control strategy. The Data Management System manipulates data in the Working Memory, and, reading information from the Knowledge Base, performs recognition tasks and provides interpretation results. The image processing library module contains image processing routines which are shared by all the biomedical applications to which the system may be tailored.

The system has two functioning modes: knowledge acquisition mode and reasoning mode. In the Knowledge acquisition mode the system acquires descriptive and behavioural knowledge from the expert. The knowledge introduced tailors this system to particular applications. The elicitation of data from experts is organized in the form of a structured interview. A detailed in-depth first sequencing of questions is proposed so that image processing knowledge and biomedical knowledge may be acquired. Questions are grouped by specific topics. The first topic concerns the acquisition of information about numerical preprocessing of the input image up to the segmentation phase. Basing on the the acquired information, the system select appropriate procedures from the Image Processing Library Module and prepares preprocessing and segmentation procedures. The second topic in Knowledge Acquisition is the acquisition of information regarding "description strategy". The system implements multilevel description strategies in which different sets of features are organized at different levels of abstraction.

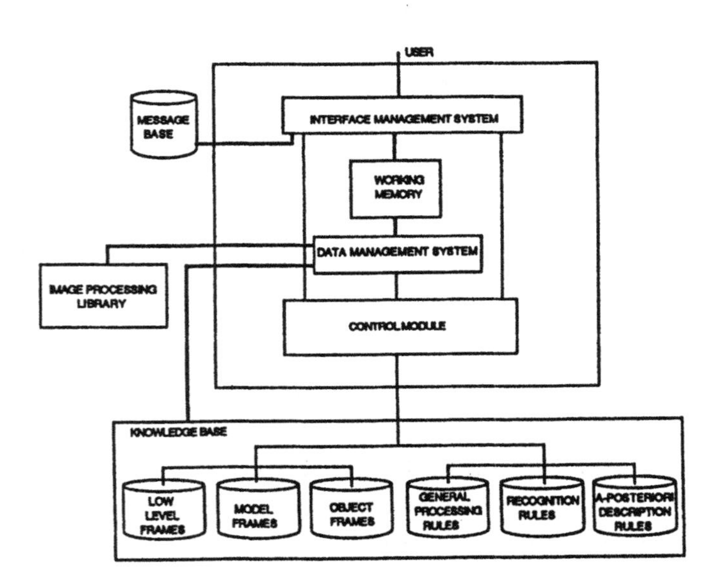

Fig. 2 System Architecture

In the case of the analysis of CT images of vertebrae, two description levels are needed: - low level description, or the encoding of binary digital images resulting from the segmentation phase - high level dscription, or the characterization of structures in the image.

The result of knowledge acquisition concerning low level description is the creation of Low Level Frame schema containing a slot to which a routine producing a description useful in the next phases is referred.

In regards high level description, the system acquires descriptive domain knowledge including anatomical properties of the parts of the body of interest and organizes it in Frame structures.

The third topic concerns the acquisition of Behavioral Knowledge in the form of rules.

The system organizes the acquisition of rules in three sub-steps dedicated to the acquisition of general processing rules, recognition rules and a-posteriori description rules respectively.

The acquired rules are written in the Knowledge Base and included in the Processing List, Recognition List and A-posteriori Description List respectively.

In the Reasoning Mode, the system receives the input image and performs the interpretation task in a sequence of phases: preprocessing, segmentation, description, recognition and final interpretation.

The criteria for deciding the next phase to be performed and, during a phase, the next action to be fired, are represented by the system by means of meta-rules evaluating current "problem status".

The rules of the rule-list in the Knowledge Base appropriate to the active phase are loaded into the working memory and included into the Current List.

A run-down of the interpretation of the Spongy-Region Recognition Rule mentioned in section 4 is here given to exemplify the reasoning activity of the system in the recognition phase.

Figure 3 shows the input image to which the example refers. Figure 4 shows the result of the segmentation phase performed with a Laplacian zero-crossing function.

The image contains the candidate objects Obj-1, Obj-2 and Obj-3 to which the rule is applied.

The degree of certainty of the antecedent of rule is computed:

- for Obj-1: $\delta_1 = \min(\mu_{Below}(1), \mu_{surrounded}(1), \mu_{circular}(0.9), \mu_{Ray}(15)) = = \min(1,1,0.3,0.7) = 0.3$
- for Obj-2: $\delta_2 = \min(\mu_{Below}(1), \mu_{surrounded}(1), \mu_{circular}(0.6), \mu_{Ray}(11)) = = \min(1,1,0.8,0.7) = 0.7$
- for Obj-3: $\delta_2 = \min(\mu_{Below}(1), \mu_{surrounded}(1), \mu_{circular}(0.5), \mu_{Ray}(20)) = = \min(1,1,0.7,0.5) = 0.5$

The DoS for Obj-1, Obj-2 and Obj-3 is computed: $DoS_1 = 0.3$ DoS sub 2 = 0.7 DoS sub 3 = 0.5

The decision-making rule mentioned in section 4 is applied with threshold value t=0.5. Object Obj-2 is recognized as Spongy Region.

Figure 5 shows the final interpreted result of the analysis.

This result is obtained by applying a rule that modifies the recognized spongy region to better approximate the region of interest.

The system has been implemented in a standard Prolog environment.

Time consuming and procedure oriented functions have been implemented in C-language and linked directly to Prolog modules.

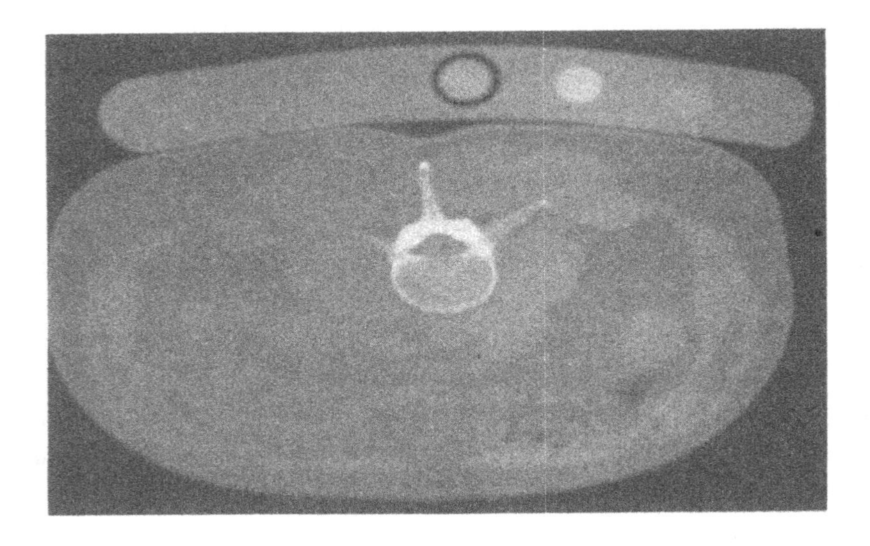

Fig. 3 Input CT image: lombar section of the human body

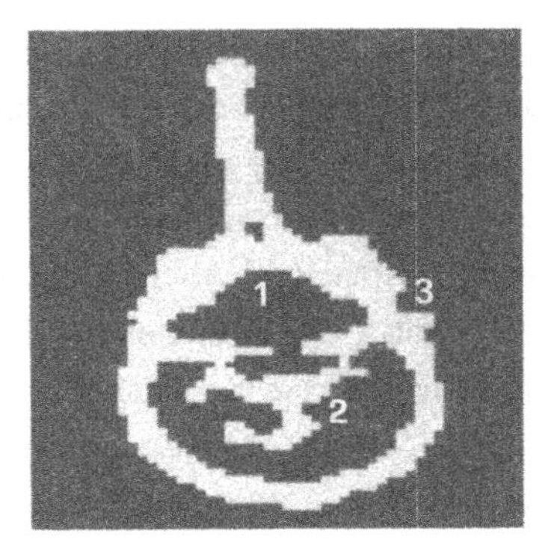

Fig.4 Result of the segmentation

6. CONCLUSIONS AND FUTURE PROSPECTS

The system presented above has been applied to the analysis of CT images of vertebrae. This application successfully adressed the problem of accelerating the quantitative analysis of the vertebral bone mineral content and of providing more objective procedures building up a consensus among experts. The system is currently being applied to the classification of colposcopic images for the early detection of preneoplastic and neoplastic lesions. The development of this application is based on results obtained in a preliminary work in which the possibility of automatizing medical procedures concerned has been assessed [4]. Results to date are promising but exhaustive data are not yet available.

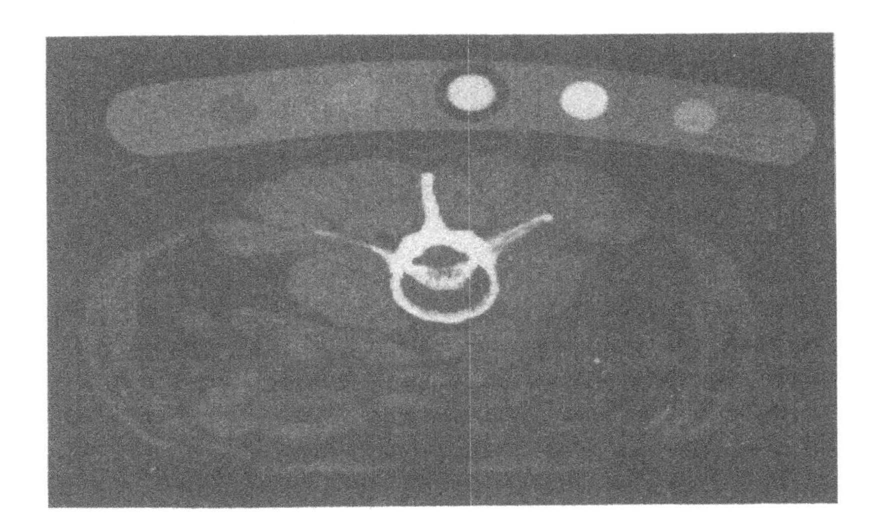

Fig.5 Result of the Spongy Region refinement

REFERENCES

[1] Adlassnig K., Kolarz G., Scheithauer W., 1985, Present State of the Medical Expert System Cadiag-2, Methods of Information in Medicine, F.K. Schattauer Verlag GmbH.

[2] E. Binaghi, 1990, A fuzzy Logic Inference Model for a Rule-Based System in Medical Diagnosis, Expert Systems, Vol.7,No.3, 134-141.

[3] Davis R., Buchanan B., Shortliffe E., 1977, Production Rules as a Representation for a Knowledge-Based Consultation Program, Artificial Intelligence, 8, 15-45.

[4] A.Della Ventura, G.Pennati, M.Sideri, 1989, Computer Aided Screening of Subjects at Risk for Cervical Neoplasia, in Recent Issues in Pattern Analysis and Recognition, Lecture Notes on Computer Science, Vol.399, 338-350, Springer-Verlag.

[5] R. Fikes R. and T. Kehler T., 1985, The Role of Frame-Based Representation in Reasoning", Comm. of ACM, vol.28.

[6] Kalender W., Klotz E., Suess C., 1987, Vertebral Bone Mineral Analysis: An Integrated Approach with CT, Radiology, Vol.164, 419-423.

[7] H. J. Levesque, R. J. Brachman, "A fundamental tradeoff in knowledge representation and reasoning", in R.J. Brachman, H.J. Levesque, Readings in knowledge representation, Morgan Kaufmann Publishers, pp. 42-70, 1985.

[8] S. Pal, D. Dutta Majumder, 1986, Fuzzy Mathematical Approach to Pattern Recognition, Wiley Eastern Limited.

[9] A. Rosenfeld, 1986, Dialog: Expert Vision Systems: Some Issues, Computer Vision, Graphics and Image Processing, Vol. 34, 99-117.

[10] Zadeh L., 1965, Fuzzy Sets, Information and control, 8, 1965, 338-353.

[11] Zadeh L., 1981, PRUF - a meaning representation language for natural languages", in E. H. Mamdani and B.R. Gaines, Fuzzy Reasoning and its Applications, Academic Press.

MANAGEMENT OF CAOTIC SYSTEMS WITH THE MODEL FOR THE REGULATION OF AGONISTIC ANTAGONISTIC COUPLES

E. BERNARD-WEIL
(Clinique Neuro-Chirurgicale de l'Hôpital de la Pitié,
83, Bd. de l'Hôpital, 75013, Paris)

Abstract
The elementary model for the regulation of agonistic antagonistic couples (MRAAC), formed by two state and two control non-linear differential equations, has been built to simulate the behavior of agents with opposite and cooperative actions in order to be able to check an imbalance, defined in relation to reference values, if it occured. The agonistic antagonistic (AA) networks associates several MRAAC in an AA fashion. In this paper, we mainly consider the possibility of quasiperiodic and strange attractors (SA) (obtained by some cyclic inputs), and we propose the notion of "balanced" and "imbalanced" SA. In case of "imbalance", the whole of the network may be theoretically balanced again by the use of only a couple of control variables, with or without turning the SA's into periodic attractors.

Key words
Control, strange attractors, chaos, networks, agonistic antagonistic models.

The most important part of our researches about the model for the regulation of agonistic antagonistic couples concerns the behavior of one couple of agents, but, already in (2), we had built a "super model" associating several elementary couples in such manner that their association happened under the rules of the agonism antagonism (for instance, two elementary couples instead of the two agents of the elementary couple) (fig. 1 and 2).

This model was made to answer a theoretical question related to our medical praxis: knowing the fact that very numerous biological couples act on the same receptor (here the cell with regards to its mitogenic or multiplication activity) - for instance hormono-mediators, likely neuro-mediators, or immuno-mediators, as well as coupled intracellular activities between stimulatory and inhibitory agents, or eliciting expression-repression of a gene -, was it possible, in the case of a global imbalance concerning more or less each subsystem, to reach a global equilibration by acting at the level of one only of these subsystems?

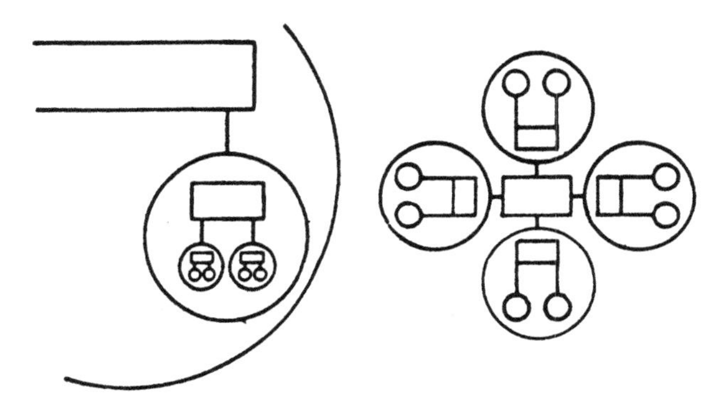

Figure 1 Figure 2

Then such an orientation of our modeling revealed the possibility to observe with the "supermodel", now called "agonistic antagonistic network", the appearance of strange attractors, with the idea that it was allowed to identify balanced and imbalanced strange attractors and to propose a control in the second case.

II-THE ELEMENTARY MRCAA (brief recall)

II.1. The MRAAC is a general or function model, here the balance function, common on principle to various concrete systems. The balance function concerns an antagonistic balance (of the type $x = y$) and an agonistic function (of the type $x + y = m$, m being a constant parameter or variable in relation to time): therefore the MRAAC is a model of equilibration and of growth (or of decrease). It does not include physical parameters, but phenomenological. It was directly built as associating a describing model and a control model (in case of imbalance).

II.2. After 15 years of using the empirical model, a formal modeling was undertaken, by the way of non-linear differential equations:

$$\dot{x} = \sum_i k_i (u + r)^i + \sum_i c_i (v + s)^i$$
$$\dot{y} = \sum_i k'_i (u + r)^i + \sum_i c'_i (v + s)^i \qquad (\underline{1})$$

$$\dot{X} = \sum_i \bar{k}_i (u + r)^i + \sum_i \bar{c}_i (v + s)^i + \sum \lambda_i (X - \bar{X})^i$$
$$\dot{Y} = \sum_i \bar{k}'_i (u + r)^i + \sum_i \bar{c}'_i (v + s)^i + \sum \lambda'_i (Y - \bar{Y})^i \qquad (\underline{2})$$

$u(t)=x(t)-y(t)+n$; $r(t)=X(t)-Y(t)$; $v(t)=x(t)+y(t)-m$; $s(t)=X(t)+Y(t)$; x = endogenous cortisol for instance; y = endogenous vasopressine for instance; X = exogenous cortisol; Y = exogenous vasopressin; x and X, y and Y are respectively of the same "nature"); k_i, c_i, n, m, λ_i are constant (or not) parameters (in general, $n = 0$);

($\underline{1}$) = state equations; ($\underline{2}$) = control equations; other inputs are added: p(t), antagonistic stimulus (for instance osmotic stimulus) and q(t), agonistic stimulus (for instance volemic stimulus or stress) (so we have u(t)=x(t)-y(t)+p(t); v(t)= x(t)+y(t)-m+q(t)). The λ_i's are = 0 only in case of cyclic control (cf. infra).

The model may allow to restore the balance after a perturbation of the system, seing that one of the singular points of the system (or physiological point) (u,v) = (0,0). According to the diversity of the parametric fields, multiple phase-portraits are possible: stability or not of the physiological point (under an asymptotical form or under the form of a limit-cycle), existence, unicity or multiplicity, stability or not of the pathological points (u = v = 0) (cf. bi- and tri-dimensional phase-portraits in (5)).

If a change in the parametric field leads to the appearance of some pathological singular point, then the control by the variable X, Y must intervene. Consequently, the parametric field of equations ($\underline{2}$) has to be chosen in order that the singular point (u+r = v+s = 0) would be in turn stable.

II.3.1. After simulation of a pathological state (asymptotical imbalance y(t) > x(t)) in relation to a change in the parameter values (that may be due to iterated simulated stress acting on k_i, c_i, now variables parameters and no longer constant) one might verify that adding to the system X(t) alone is unable to restore the balance (x(t) + X(t) remains lower than y(t)), while the simultaneous adding of X(t) and Y(t) (the hormone already in excess in the system !) allows to restore an asymptotical balance (x(t)+X(t)=y(t)+Y(t)=m/2): the values of X(t) and of Y(t) were obtained by the simulation of the whole of the mathematical model ($\underline{1}$) and ($\underline{2}$) (cf. curves in (5)).

II.3.2. More realistic is the cyclic control, reestablishing for instance the circadian rythm of the hormones (cf. the comparison betwwen the stability of asymptotical and of cyclic control in (6)). More realistic too, as hormonal dosages have showed the fact (4), is the existence of an imbalance in relation to a distortion of the physiological limit-cycle, and not to the appearance of a pathological attractor. In these conditions, the simulation with the MRAAC reveals that one may reestablish the physiological limit-cycle with some rythmic impulses of X(t) and Y(t) (fig. 3).

Figure 3

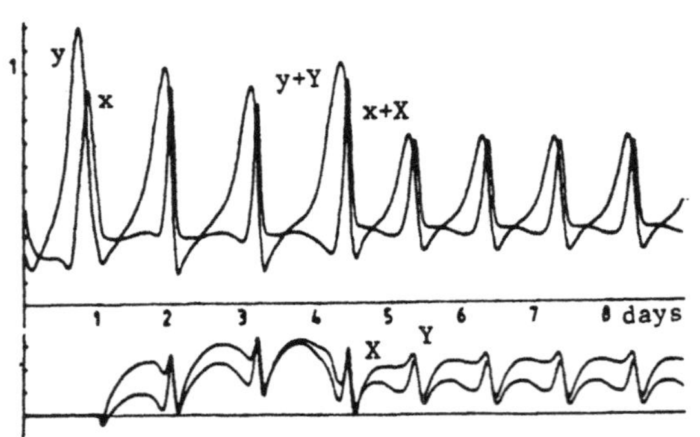

The choice of the control parameters was made in general by minimization of the following objective function:

$$J_{k_i, c_i, \lambda_i, \bar{X}, \bar{Y}} = \Sigma\,(\bar{x}_j - (x_j + X_j))^2 = \Sigma\,(\bar{y}_j - (y_j + Y_j))^2 \quad \bar{x}_j, \bar{y}_j = \text{experimental values} \qquad (3)$$

The λ_i's avoid a "drift" of the order 4 limit-cycle (x,y,X,Y). The principles of this type of control may be summarized by saying that, by "surimposing" the therapeutical phase-portrait to the pathological phase-portrait (both imbalanced), it is possible to form again the physiological phase-portrait (balanced). Or we may state that this type of control consists in a kind of shaping the epigenetic landscape.

III-THE AGONISTIC ANTAGONISTIC NETWORK (AAN)

III.1. The scope of this chapter was not to directly solve a concrete problem about the control of an imbalance in the case of the elementary MRAAC, as in chapter II, but to undertake for the moment only a theoretical research: a) if a network is formed by several elementary MRAAC, we have firstly to precise what type of behavior of the variables may occur: asymptotical, periodic, quasi-periodic or chaotic (seing that we have more than 2 variables); b) the simulation of the model may show some states of balance or imbalance, in relation to the agonistic and antagonistic reference values; in case of chaotic behavior, there is apparently no singular point that complies, or not, with these norms, but one will find some strange attractors that may be considered as balanced or imbalanced (by the calculus of the mean values of the variables); c) the control of a state of imbalance could be considered as it was done in chapter II, by means of control equations similar to the state equations; a particular case will concern the reestablishment of the balance in a wholly imbalanced AAN by means of only a set of control equations (corresponding to a couple of variables alone); we will try also to check the behavior of an "imbalanced" strange attractor, a problem different of the transformation of this one in a quasiperiodic attractor (even both results can be associated: cf. infra).

The domain of chaotic phenomena is now spreading over, from the climatic attractor (13) up to physico-chemical and biological processes (1) (8) (9) (10) (12) (14) (16). Chaos in not considered as the proof of an abnormal state, in the contrary it seems for many authors: well-spring of information, reserve of periods. Therefore, the following simulations could help to better understand these behaviors and better act, it necessary, in front of them.

In order to formalize the diagram of the figure 1, one may use the following equations, with $u_i = x_i - y_i$; $r_i = X_i - Y_i$; $v_i = x_i + y_i - m_i$ + synchronizer (synchronizer = $A_i \sin(2\pi/T_i + \Phi_i)$; $s_i = X_i + Y_i$; k_i, c_i, $\bar{k}_i$, $\bar{c}_i$, $\hat{k}_i$, $\hat{c}_i$, Φ_i, λ_i, $\bar{X}_i$, $\bar{Y}_i$ = constant parameters.

$$\dot{x}_i=\sum_j k_{ij}(u_i+r_i)^j + \sum_j c_{ij}(v_i+s_i)^j + \sum_k\left[\bar{k}_{ijk}\sum_j(\sum_{n=1}^{n=k}(u_n+r_n))^j + \bar{c}_{ijk}\sum_j(\sum_{n=1}^{n=k}(v_n+s_n))^j\right]$$

$$\dot{y}_i=\sum_j k'_{ij}(u_i+r_i)^j + \sum_j c'_{ij}(v_i+s_i)^j + \sum_k\left[\bar{k}'_{ijk}\sum_j(\sum_{n=1}^{n=k}(u_n+r_n))^j + \bar{c}'_{ijk}\sum_j(\sum_{n=1}^{n=k}(v_n+s_n))^j\right] \qquad (\underline{4})$$

$$\dot{X}_i=\sum_j \hat{k}_{ij}(u_i+r_i)^j + \sum_j \hat{c}_{ij}(v_i+s_i)^j + \sum_j\lambda_{ij}(X_i-\bar{X}_i)^j \qquad i=1,2,3..t$$

$$\dot{Y}_i=\sum_j \hat{k}'_{ij}(u_i+r_i)^j + \sum_j \hat{c}'_{ij}(v_i+s_i)^j + \sum_j\lambda'_{ij}(Y_i-\bar{Y}_i)^j \qquad j=1,2,3..q \quad k=2,4,8..r$$

But we have rather used some close equations that correspond as a matter of fact to only a AA combination of several AA elementary models (fig. 2):

$$\dot{x} = \sum_j k_{ij}(u_i+r_i)^j + \sum_j c_{ij}(v_i+s_i)^j + \bar{k}_{ij}\sum_i(\sum_j(u_i+r_i)^j + \bar{c}_{ij}\sum_i(\sum_j(v_i+r_i)^j$$

$$\dot{y} = \sum_j k'_{ij}(u_i+r_i)^j + \sum_j c'_{ij}(v_i+s_i)^j + \bar{k}'_{ij}\sum_i(\sum_j(u_i+r_i)^j + \bar{c}'_{ij}\sum_i(\sum_j(v_i+r_i)^j$$

$$\dot{X}_p = \sum_j \hat{k}_{pj}(u_p+r_p)^j + \sum_j \hat{c}_{pj}(v_p+s_p)^j + \sum_j\lambda_{pj}(X_p-\bar{X}_p)^j \qquad (\underline{5})$$

$$\dot{Y}_p = \sum_j \hat{k}'_{pj}(u_p+r_p)^j + \sum_j \hat{c}'_{pj}(v_p+s_p)^j + \sum_j\lambda'_{pj}(Y_p-\bar{Y}_p)^j$$

One notices in both cases that each element of a couple varies on one hand according to a proper parametric field, of the same type as in (1) and (2), and on the other hand according to a parametric field determining a "interconnecting structure", itself with an AA pattern. One can find in (9) some examples of interconnection results: according to the parametric values, we might observe for instance a checking of separate imbalanced elementary models due to a good interconnection, or the inverse phenomenon.

III.2. Firstly, a AAN without synchronizer was used (preceding simulations) to simplify the problem. In fact, the presence of a synchronizer in the agonistic expressions v(t), nearly always added in the simulations with the elementary MRAAC is going to bring a new dimension, i.e. a caotic dimension.

On the figure 4 is represented a tri-dimensional phase-portrait (x_1, y_1, y_2) of SMRAAC (close to a balanced functioning) with limit-cycles (x_1,y_1) and (x_2,y_2).

Without synchronizer (fig. 4a), there are quasi-periodic attractors, as a Poincaré's section (with a given value of y_2) reveals the fact (at the bottom of the figure); on the figure 4b, one finds an attractor that seems very close to this of the figure 11a in despite of the addition of a synchronizer, but, in these conditions, the attractor exhibits its strange nature, as the Poincaré's section now very different of the preceding section, proves the fact. These remarks agree with other researches concerning the appearance of caotic behaviors in non-linear différentiel equations, particularly when they are forced (11) (15).

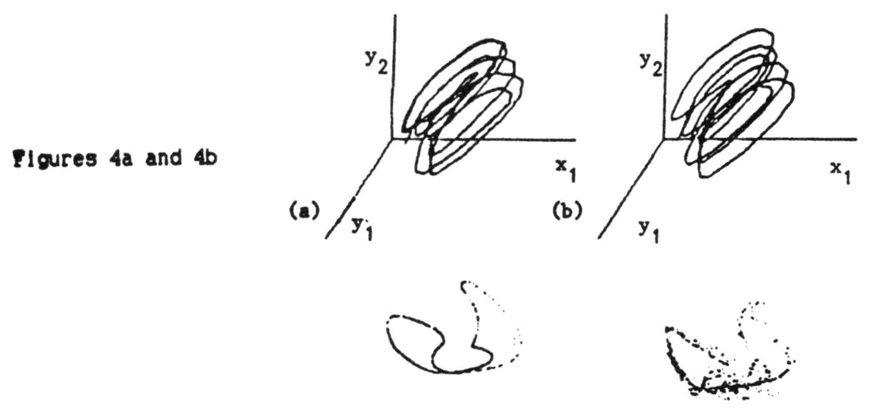

Figures 4a and 4b

III.3. Now we have to consider the occurence of imbalanced AA SA that have been also simulated with the AAN, due to a choice of a new parametric field (for the elementary AA models and for the interconnecting expression). It could seem astonishing to propose the idea of a balanced or imbalanced SA, seing that the successive values of the variables x_i, y_i in a SA cannot be predictible in principle. Nevertheless, the simulations proved that such an "imbalanced" model showed, for a lasting period of observation or even with different initial conditions, some comparable average values of the variables. Consequently, the idea to control these pathological SA by the use of control variables in the same way as in the case of imbalanced limit-cycles (fig. 3) should represent a new stage in the field of the AA control methods.

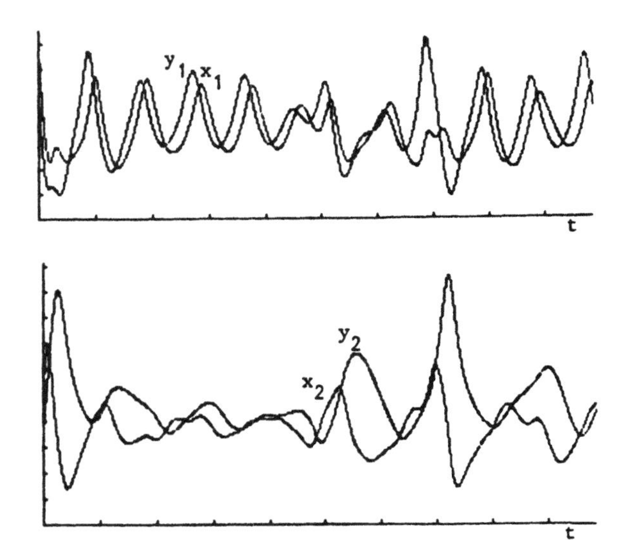

Figure 5

Figure 5 shows two imbalanced SA in a AAN (cf. the parameter values in the Addendum). The proof of their chaotic nature was done: a) by the result of a Poincaré's section; b) by the power spectrum; c) by the fact that the greater exponent of Lyapounov was positive (figure 6).

Figure 6

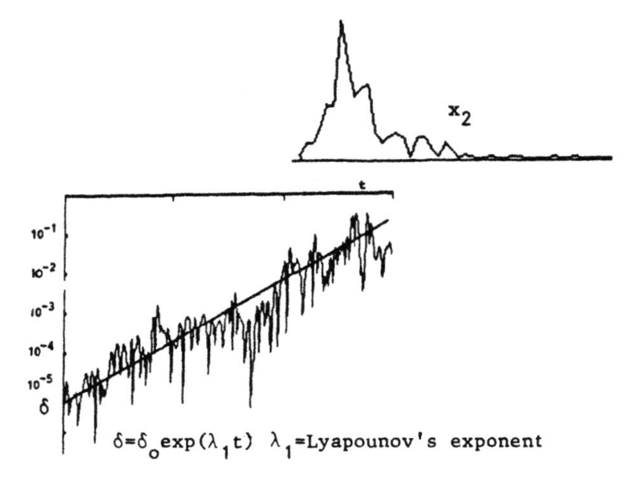

Control upon only a couple could be made either with control variables X_1, Y_1, or with control variables X_2, Y_2.

Figure 7 shows the effect of the first type of control upon the behaviour of both couples (it was noticeably that the action at the level of only the first couple allowed to check the more pronounced imbalance of the second couple).

Figure 7

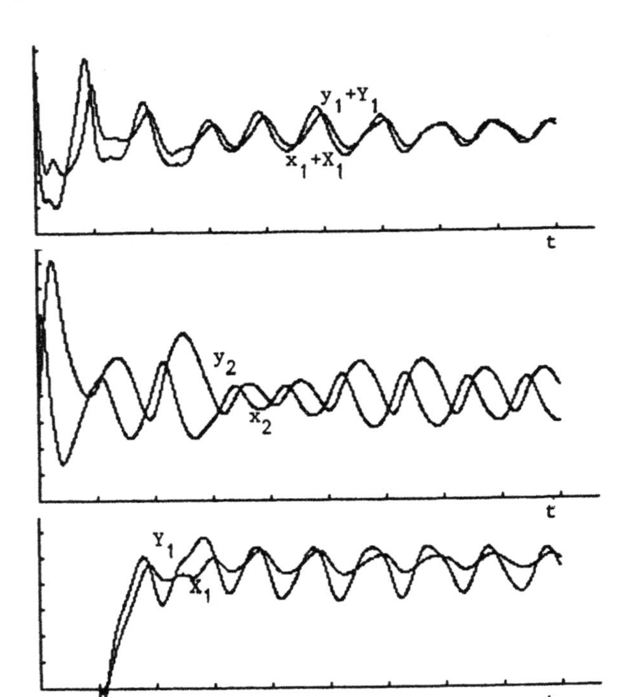

Meanwhile, one may wonder what is the meaning of this kind of simulation: if in the case of limit-cycle - or even of quasi-periodic cycles - it was understandable that the solving proposed by the simulation could be carried back to the real simulated system (by taking into account the convenient time of a period to begin the control), the problem seems very different in the case of imbalanced SA.

However we remark that the solving proposed by the simulation (curves of X, Y) could evoke a quasiperiodic control. From which, the effects of periodic inputs directly added in (5), without using the control equations, has been tested (fig. 8): some improvement of the imbalances was noticed, in the same time as the disappearance of the caotic behavior (the identification of the parameters of 2 Fourier's series of dimension 4 was performed, with initial conditions suggested by the figure 6).

But we have to emphasize the fact that the end of the control of a caotic system is not to necessarily turn a SA into a quasiperiodic attractor (cf. supra). How to create a convenient control SA that, added to an "imbalanced" SA at any time, could reestablish a global "balance" of the system without destroying its caotic behavior? An orientation of the researches in order to give an answer to this crucial question seems to us useful and perhaps urgent.

Figure 8

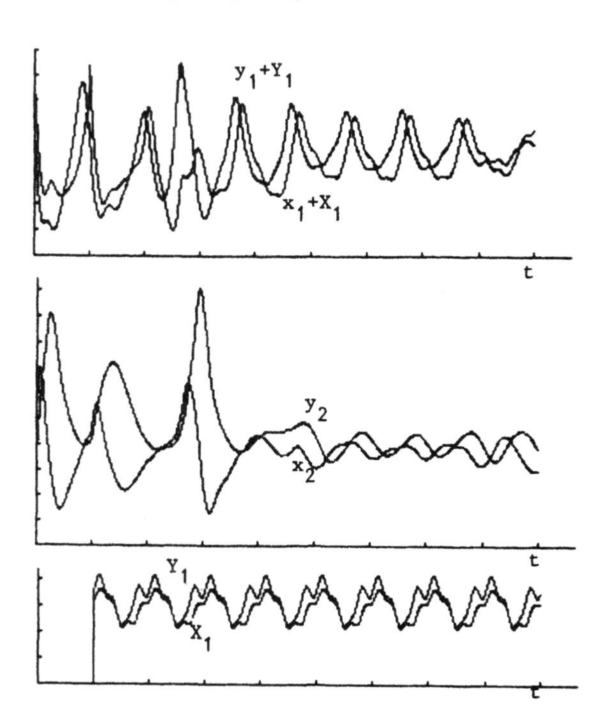

III.4. Finally, one may examine the parametric conditions that lead to a chaotic state from periodic and quasi-periodic oscillations in the AAN's.

We mainly studied the influence of the frequency of the synchronizers. In the proposed simulations, the natural frequency of the couples x_1, y_1, x_2, y_2 (not forced) corresponded nearly to a period of 96 (with a lenght step of .1 in the

Runge-Kutta method). By measuring the value of the greater exponent of Lyapounov for diverse frequencies of the synchronizers (the same for both synchronizers), we remarked that a periodic attractor was noted until nearly twice the natural frequency, then making way for chaotic attractors. If the values of the T_is were close to the natural frequency, it was still possible to observe some SA if the frequencies of the two synchronizers were different (cf. parameter values in the Addendum).

Figure 9

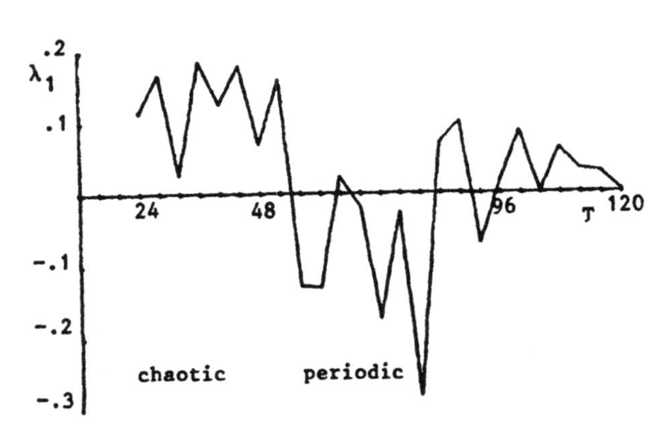

REFERENCES

(1) A. BABLOYANTZ and A. DESTEXHE, Proc.Natl.Acad.Sc.USA, 83 (1986): 3513 - 3517.

(2) E. BERNARD-WEIL, in Petits Groupes et Grands Systèmes, Hommes et Techniques, Paris, pp. 143 - 153.

(3) E. BERNARD-WEIL, in Régulations physiologiques: Modèles Récents, G. CHAUVET, éd., Masson, Paris, 1986, pp. 133 - 155.

(4) E. BERNARD-WEIL, P. VAYRE and J.L. JOST, Ann.Endocrinol., 47(1986), 201- 203.

(5) E. BERNARD-WEIL, in Mathematical Models in Medicine, M. WITTEN, ed., Pergamon Press, New York, t.I, pp. 1587 - 1600.

(6) E.BERNARD-WEIL, Précis de Systémique Ago-Antagoniste. Introduction aux Stratégies Bilatérales, L'Interdisciplinaire, Limonest, 1988.

(7) E. BERNARD-WEIL, Rev.Intern.Systém., 2 (1988), 399 - 416.

(8) M. DUBOIS, P. AATEN et P. BERGE, La Recherche, 18 (1987), 190 - 201.

(9) A.L. GOLDBERGER and B.J. WEST, in Chaos in Biological Systems, H. DEGN, A.V. HOLDEN and L.F. OLSEN, eds, Plenum Press, 1987, pp. 1 - 4.

(10) A. GOLDBETER and O. DECROLY, Am.J.Physiol., 245 (1983), R478-R483.

(11) H. KAWAKAMI, Proc. IEEE Int.Symposium on Circuits, vol. 1, pp.378 -381.

(12) J.G. MILTON, A. LONGTIN, A. BEUTER et al., J.Theor.Biol., 138 (1980), 129 - 147.

(13) C. NICOLIS and G. NICOLIS, Nature, 311 (1984), 529 - 532.

(14) I. PROCACCIA, Nature, 333 (1988), 618 - 623.

(15) K. TOMITA, Phys.Rep., 86 (1982), 1 - 25.

(16) P. TRACQUI, Des concepts de la dynamique non-linéaire à l'auto-organisation des systèmes biologiques, Thèse de Doctorat ès-Sciences, Université Paris-VI, 1990.

ADDENDUM-Parameter values for the simulations of figures 6 and 7

$k_{11} = -.475$; $c_{11} = .2$; $k_{12} = -2.5$; $c_{12} = .5$; $k_{13} = -2.8$; $c_{13} = .36$; $k'_{11} = -.6$;
$c'_{11} = -.01$; $k'_{12} = -.01$; $c'_{12} = -.5345$; $k'_{13} = -.2$; $c'_{13} = -2$.

$k_{21} = .29$; $c_{21} = -.29$; $k_{22} = .399636$; $c_{22} = .453472$; $k_{23} = .5$; $c_{23} = -1$; $k_{21} = .310$
$k'_{21} = .3105$; $c'_{21} = .2705$; $k'_{22} = .797456$; $c'_{22} = -.125794$; $k'_{23} = 1$; $c'_{23} = .6$.

$\bar{k}_{11} = -.01$; $\bar{c}_{11} = .15$; $\bar{k}_{12} = 0$; $\bar{c}_{12} = -.5$; $\bar{k}_{13} = -1$; $\bar{c}_{13} = 0$; $\bar{k}'_{11} = -.1$; $\bar{c}'_{11} = -.1$;
$\bar{k}'_{12} = 0$; $\bar{c}'_{12} = 0$; $\bar{k}'_{13} = 0$; $\bar{c}'_{13} = -1$.

$\bar{k}_{21} = .05$; $\bar{c}_{21} = -.04$; $\bar{k}_{22} = .8$; $\bar{c}_{22} = .3$; $\bar{k}_{23} = .25$; $\bar{c}_{23} = 0$; $\bar{k}'_{21} = .01$; $\bar{c}'_{21} =$
$.05$; $\bar{k}'_{22} = .2$; $c'_{22} = -.1$; $\bar{k}'_{23} = .5$; $\bar{c}'_{23} = 0$.

$\hat{k}_{11} = -.3331424$; $\hat{c}_{11} = -.1313686$; $\hat{k}_{12} = -.2823283$; $\hat{c}_{12} = -.380071$; $\hat{k}_{13} = .2589844$;
$\hat{c}_{13} = .3489473$; $\hat{k}'_{11} = -.6808183$; $\hat{c}'_{11} = -.5121212$; $\hat{k}'_{12} = .0438177$; $\hat{c}'_{12} = -.1758889$;
$\hat{k}'_{13} = .30193229$; $\hat{c}'_{13} = -.0107469$.

$\bar{X} = -.5382119$; $\bar{Y} = .2798437$; $\lambda_1 = .3965617$; $\lambda_2 = -.3626675$; $\lambda_3 = -.1215371$;
$\lambda'_1 = -.7610956$; $\lambda'_2 = -.9205176$; $\lambda'_3 = -.3710565$.

$A_1 = .1$; $A_2 = .05$; $\phi_1 = -.5235$; $\phi_2 = -.8$; $T_1 = 96$; $T_2 = 99$.

Initial conditions: $x_{1o} = .8637$; $y_{1o} = .7512$; $x_{2o} = .7031$; $y_{2o} = .4823$; 0 for the control variables.

m_1, $m_2 = 1$.

9. DECISION-MAKING UNCER UNCERTAINTY

FUZZY MULTICRITERIA TECHNIQUES: AN APPLICATION TO TRANSPORT PLANNING

ALAN D. PEARMAN*, JAVIER MONTERO** & JUAN TEJADA**

* School of Business and Economic Studies
The University, Leeds LS2 9JT (U.K.)

** Dept. Statistics & O.R., Faculty of Mathematics
Complutense University, Madrid 28040 (Spain)

Abstract: This paper deals with the problem of allocating a fixed budget between a set of competing investment proposals. Based on previous practical experience with problems of this type, an analysis using the fuzzy outranking approach promised to be potentially effective. A decision aid computer package has been developed and applied to a case study. It proved to be flexible and to give additional insights into the structure of the set of alternatives.

Keywords: Multicriteria Decision Making, Fuzzy Outranking Relation, Transport Planning.

1 - INTRODUCTION

In recent years, a range of mathematically based techniques, termed multicriteria methods, has been derived with the aim of helping decision makers make more effective choices from among the options available to them. In particular, a number of such techniques stress the importance of achieving a realistic balance between the power which a rigorous mathematical foundation can give and the reality that most public sector decisions involve factors which are very difficult to quantify in a numerically precise form. Fuzzy multicriteria techniques, because of their potential for representing imprecise preferences in a formal yet flexible way, represent a potentially valuable tool for the analysis of such decisions.

Within transport planning, a common administrative arrangement in many countries is for local highway planning agencies to have devolved to them responsibility for smaller-scale investments, subject to a centrally determined budget. Individually, such projects are typically quite small; the work reported here, for example, is primarily concerned with schemes costing no more than one or two million pounds and the normal range of costs is often much lower even than this. Failure to allocate the available resources in an appropriate fashion is potentially a substantia misallocation of public funds. In normal circumstances, there are many more candidate schemes than available funds will allow to be undertaken. A means of selecting preferred schemes is thus required.

Small, local highway schemes can be of many different types -junction improvements; road straightening; setting up traffic signals, etc-. They are not all directly concerned with achieving faster, cheaper traffic flows through the road network. There are frequently proposals to be considered whose principal purpose is to alleviate some environmental problem or to reduce accidents or to improve access to some facility. However, all are competing against each other under the same budget constraint and for this reason must (implicitly or explicitly) be compared and ranked. Any evaluation framework must therefore incorporate all the potentially decision-relevant criteria likely to occur for any of the types of scheme which might be proposed. Both to ensure consistency of policy and to facilitate use by non-specialists, the criteria and their weights need to be established in advance, before knowing the specific set of candidate projects which are to be ranked. In this respect, the analysis reported here is more difficult than many multicriteria choice problems, both because of the difficulty of specifying measurement scales suitable to such a wide range of possible schemes and because of the size of the resulting evaluation model.

2 - PREVIOUS EXPERIENCE

During the late 1970's and early 1980's, many local authorities in Great Britain developed what came to be termed Priority Assessments Techniques (PAT's) to provide guidelines for selecting between competing highway improvement projects. Earlier work (Simon *et al.*, 1988) examined the range of techniques that had been developed and, based on that work, formulated a PAT (Pearman *et al.* ,1989) with an explicit multicriteria formulation, using a straightforward additive value function. This model (COMPASS) was implemented through a linked series of spreadsheet programs.

Building on experience of PAT's already in operation, the structure chosen for COMPASS was a hierarchical one, with a value tree having four aggregate levels of impact (safety; traffic; environment; local planning and development) which in turn were sub-divided into 11 and then 32 lower-level impacts. The 32 are all measurable, some objectively and some judgementally. Examples include numbers of accidents saved and travel cost and time saving (each disaggregated by vehicle class), all of which are normally assessed on an objective scale, but also impacts such as visual intrusion and disruption to existing activity patterns, which are judgementally assessed in a 0-10 scale.

The need to admit judgemental assessment is occasioned in part by the nature of the criteria, but also by the need to keep strict control of the cost of the appraisal process itself. For relatively low cost schemes, complex and hence expensive traffic forecasts, pollution estimates, etc. cannot be justified, as their cost could easily represent a substantial proportion of any potential scheme benefits. Apart from its contribution to clarifying the set of criteria being used and convincing decision makers that the full range of relevant criteria has been taken into account, a major reason for adopting a hierarchical structure is to facilitate the evaluation of smaller schemes. Where a scheme is too small to justify the expense of evaluation using the full set of (potentially) 32 lowest-level criteria, some or all aspects of the project may be evaluated at a higher (more aggregate) level in the tree. This may be done either by replacing a group of lowest-level criteria by a single judgemental assessment or by choosing one of the relevant lowest-level criteria to stand as a proxy for all the rest. In this way, with minor local adjustment to the weighting of criteria, small projects can be evaluated directly alongside larger ones using the same basic evaluation framework and broadly equivalent scoring and criteria weights. This allows a wide range of disparate projects to be ranked against each other, a major objective for COMPASS, since concern had been expressed that existing procedures were leading to an under-evaluation of small, unspectacular but cost-effective schemes, allowing higher-profile alternatives undue precedence in the ranking process.

With the 32 criteria established and appropriate measurement scales developed, the next important step in the modeling process was to determine the relative weights to be given to unit changes in the different scores. Even exploiting the hierarchical structure of the impacts, this was not a simple task. Both the size of the problem and the difficulty in clarifying what levels of real changes correspond to what scores in the judgementally assessed impacts contribute to the problem. For the initial applications of COMPASS, the weights were established in one of two ways. One was essentially a "pricing out" procedure (Keeney & Raiffa, 1976); the other used Saaty's analytic hierarchy method, asking users to make pairwise comparisons of criteria first within each of seven related groups of lowest-level criteria and then successively up the tree between representative criteria selected from lower-level groupings.

As well as being linear in the criteria weights, the model used, in its basic form, uses linear scales for each of the individual criteria, i.e., a given level of improvement is scored equally, independently of what was the initial score on the criterion concerned. For some clearly non-linear phenomena like noise levels, this was a major simplification. It can only be justified as an approximation against the setting of an evaluation exercise where many of the scores are likely to be correct only to a broad level of magnitude and by the existence of sensitivity testing modules within the COMPASS structure that permit the consequences of doubts about the accuracy of any of the input data to be explored.

Overall project ranking within COMPASS is based on a calculation of project effectiveness: capital cost ratio. This choice is made because in all probable applications the capital cost constraint is likely to be binding. Hence, drawing a direct comparison with cost-benefit analysis and capital budgeting in general (see, e.g., Pearce & Nash, 1981), ranking by ratio provides an effective heuristic for establishing the best overall set of projects to implement. Especially in the presence of some schemes whose capital cost may be high relative to the budget available, ranking by effectiveness with capital cost simply weighted and netted out as a negative contribution to effectiveness will not in general lead to an optimal selection of schemes.

The original COMPASS program has been tested on data provided by a number of local authorities. The feedback has been positive, while at the same time identifying areas of potential improvement. In particular, one matter of concern derives from the difficulty in establishing accurate measures of project effectiveness on each of the different criteria. A second concern is the extent to which decision makers view projects in their own right as individual schemes, rather than as members of a package of schemes contributing to an overall improvement in traffic conditions in the town for which they are responsible.

This paper responds to these two issues by exploring the use of a fuzzy multicriteria model for prioritizing highway schemes (although the previous pure additive model had the advantage of being relatively straightforward to implement and to explain to non-specialist decision makers, trial applications to priority assessment exercises have suggested attempting an alternative procedure based on a fuzzy multicriteria approach). As will be explained more fully in subsequent sections, this approach firstly allows for a more flexible assessment of whether one project's criteria scores really do establish that it is preferable to another. Additionally, a method is derived for testing sets of projects against each other, where the assessment of the sets looks at their aggregate achievement in each of the 32 dimensions of impact, rather than at what is contributed by any one component project from the set in isolation. The functioning of the method is illustrated using data on a group of twelve projects supplied by a British local authority (see the APPENDIX with the original data). The projects represented are broadly typical, with the one exception that they are clearly divided into low-cost and high-cost subsets, with relatively little representation of the more common medium-scale schemes. It is clear that high-cost and low-cost schemes are particularly difficult to compare, and that the way of including capital cost in the model will be important.

3 - THE BASIC MODEL

This alternative PAT model has been derived from the method proposed by Siskos *et al.*, 1984, but the output -final information to be analyzed by the decision maker-is expanded according proposals given in Montero & Tejada, 1986, in order to get a better knowledge of projects. The initial set of data -expected scores G(i,j) under the ith criteria if project j is developed- has been given by a group of specialists, including the relative weight of each criteria and the threshold of significant difference and the veto threshold for the scores under each criterion. Based on this set of parameters, the following fuzzy relations are then defined:

a) *Partial fuzzy outranking relations:* they provide the strength of relationship between any two projects, in outranking terms, regarding only one criterion. In this way, a value PO(i,j,k) is assigned to represent the degree to which project k is outranked by the project j, taking into account only the ith criterion. It depends on the significance threshold, and as a first step it was defined as PO(i,j,k)=1 if G(i,k)≤G(i,j), PO(i,j,k)=0 if G(i,k)≥G(i,j)+S(i) and linear interpolation for intermediate values.

b) *Partial fuzzy discordance relations:* they provide information about incomparability phenomena between two projects, due to a given criterion. In relation to the outranking of a project j by the project k, a value PD(i,j,k) will show how discordant is an unfavorable difference in performance under the ith criterion, depending on the veto threshold and the significance threshold (as a first step it was taken as PD(i,j,k)=1 if G(i,k)≥G(i,j)+V(i), PD(i,j,k)=0 if G(i,k)≤G(i,j)+S(i) and also linear interpolation otherwise).

c) *Fuzzy concordance relation:* it provides the weighted aggregated outranking, obtained from all the partial outranking relations and taking into account the criteria weights W(i) ∀i, previously standardized.

$$C(j,k) = \sum_i W(i)*PO(i,j,k) \qquad \forall j,k$$

d) *Fuzzy outranking relation:* it aggregates the fuzzy concordance relation with the partial discordance relations. Final analysis will be based on its associated domination structure (see Orlovsky ,1978). Initially the following expression was proposed:

$$D(j,k) = C(j,k) * \mathit{min}_i\{1-PD(i,j,k)\}$$

4 - THE OUTPUT

The initial idea was to adapt an *outranking method* to our particular transport planning problem. Two important difficulties were found in such an implementation: the specific formulas to be used and the way of including cost (it should not be considered just as another 33rd criterion, and if included in the basic formulas, it becomes too easily either decisive or non-relevant). At its final stage, the complete treatment is provided by a set of programs; in order to get more insight, complementary fuzzy relations are also provided (*similarity, strict preference and incomparability*) at each level, together with other useful indices (*ratio effectiveness*, for example).

Based on the fuzzy set of nondominated alternatives

$$ND(j) = 1 - max_{i}\{D(k,j)-D(j,k)\}$$

(Orlovsky, 1978), two outputs are included in a first step (see Montero & Tejada, 1986):

a) *Hierarchical representation with δ-level choice sets*, which allows us to consider slight errors in the evaluation of such a fuzzy set of nondominated projects. Orlovsky's choice set will appear as the 1-level set, when it is non-empty.

b) *Successive discarding analysis*. Worst projects -those with the lowest degree of nondomination- are successively discarded, and then a new nondomination structure is re-evaluated without taking into account those projects.

Cost is included, as a second step, in two alternative ways, and then both previous treatments a) and b) are developed:

c) *Effectiveness as a fuzzy preference relation*, to be aggregated at the fuzzy outranking relation level, through an interactive aggregation rule.

$$D_{pe}(j,k) = D(j,k) * min\ [1, Cost(k)/Cost(j)]$$

d) *Effectiveness as a fuzzy set*, to be aggregated also through an interactive aggregation rule, but at the fuzzy set of nondominated alternatives level.

$$ND_{fe}(j) = ND(j) * CK/Cost(j)$$

where CK is defined in order to get an appropriate scale (for example, minimum cost of the projects under consideration).

TABLE 1

Project	Cost	RE X1000	Rank	ND	Rank	FE X100	Rank	PE X100	Rank
1	2870	0.1500	7	0.4464	6	0.3889	8	57.345	4
2	3710	0.0813	10	0.4473	5	0.3014	9	44.632	6
3	6030	0.0238	11	0.4630	4	0.1920	12	13.435	10
4	850	0.1136	9	0.0908	12	0.2673	11	9.089	12
5	3300	-0.037	12	0.3884	7	0.2943	10	46.638	5
6	3600	0.1473	8	1.0000	1=	0.6944	7	100.00	1=
7	201	0.7625	6	0.2471	9	3.0736	6	26.812	8
8	50	3.5650	2	1.0000	1=	50.000	2	100.00	1=
9	30	2.4697	3	0.1851	11	15.428	3	39.138	7
10	105	0.9041	5	0.2689	8	6.4251	4	10.948	11
11	25	7.1837	1	0.5449	3	54.496	1	86.355	3
12	87	1.5457	4	0.2124	10	6.1000	5	15.044	9

In the above TABLE 1 it is shown a small portion of the output for the complete set of our twelve *single* projects: in the RE column appear the ratio effectiveness indices, ND are the nondomination degrees without costs, FE are the nondomination degrees based on (c), and PE are the nondomination degrees (d).

Since we are not just looking for a *single* project, but for a satisfactory set of projects that can be developed with a certain budget, each group of projects should be considered as one project.

e) *Aggregated projects:* families of different projects are considered if the total cost is within an appropriate range -in this way the influence of cost is minimized-. The complete set of outputs should be then obtained for such families of aggregated projects, with appropriate scores (it was previously checked that the associated scores of any group of out single projects could be evaluated according to the additive assumption).

Obviously, the number of aggregated groups can become too large too easily. The following procedure was then applied, but it is clear that the model is open to *ad hoc* changes by the decision maker -even a tentative process with different cost ranges-, taking into account all the information available at each moment:

1.- Compare, by applying the PAT1 program, all *maximal* sets of aggregated projects within a fixed range in cost -that is, sets of projects that can be jointly developed within the budget and such that no other project can be added while remaining in the fixed cost range-. There were found 19 maximal groups in the *a priori* fixed range (7.5 to 8.0 million pounds). This PAT1 program is able to find all maximal sets in a fixed range of cost within any family of projects, developing the complete analysis for them.

2.- Consider just a few of those maximal aggregated projects, and apply the PAT2 program to each one in order to study if some projects inside them should be rejected (this PAT2 program develops the complete analysis for all the subsets within any family of projects). After a deep analysis of the PAT1 output -containing the summarized structure of all maximal groups- three maximal groups were chosen:

$$A = \{2,6,7,8,9,10,11,12\}$$

$$B = \{1,2,4,7,8,9,11\}$$

$$C = \{1,4,6,7,8,9,10,11,12\}$$

and no *single* project could be clearly rejected within them -remaining in the required range in cost-. As an example, see TABLE 2 for group A, where five subsets were found in the cost range:

$$A1 = \{2,6,7,8,9,10,11,12\}$$

$$A2 = \{2,6,7,8,9,10,12\}$$

$$A3 = \{2,6,7,8,10,11,12\}$$

$$A4 = \{2,6,7,8,10,12\}$$

$$A5 = \{2,6,7,9,10,11,12\}$$

TABLE 2

Project	Cost	RE X1000	Rank	ND	Rank	FE	Rank	PE	Rank
A1	7808	0.2109	1	1.0000	1	0.9929	1	1.0000	1
A2	7783	0.1885	4	0.2475	3	0.2465	3	0.2505	3
A3	7778	0.2022	2	0.5659	2	0.5641	2	0.5697	2
A4	7753	0.1797	5	0.0618	4	0.0618	4	0.0684	4
A5	7758	0.1893	3	0.0010	5	0.0000	5	0.0072	5

3.- Apply the PAT3 program in order to compare maximal aggregated projects. The decision maker should take the decision based on the PAT3 output (TABLE 3 shows a small portion of this output), where the complete analysis for any selection of aggregated projects is made (an initial proposal was to develop {2,6,7,8,9,10,11,12}, and it was considered an acceptable choice -an *a priori* idea was to develop principally cheap projects but only one expensive project-).

TABLE 3

Project	Cost	RE X1000	Rank	ND	Rank	FE	Rank	PE	Rank
A	7808	0.2109	2	1.0000	1	1.0000	1	1.0000	1
B	7928	0.2073	3	0.4473	3	0.4406	3	0.4469	3
C	7818	0.2395	1	0.6897	2	0.6889	2	0.6897	2

In this paper only a part of the outputs are included, always assuming linear interpolation in the basic formulas; other expressions -e.g., logistic curves- and fuzzy numbers are also being investigated in order to improve the appropriateness model to our particular problem and in order to increase its sensitivity. In any case, robustness can be studied through simulation, by including random errors in the initial data -though it can be complex since the number of parameters required in the model is usually large, even in a small case like the one analyzed here-. The complete programs can be obtained from the authors on request.

Finally, once again the difficulty of how capital cost should be included when analyzing problems of this kind should be pointed out. Aggregating projects allows a reasonable way of avoiding the difficulty in our case.

5 - GENERAL COMMENTS

This package of fuzzy multicriteria techniques has been applied to a to a particular transport planning problem in the U.K.; it is basically, an adaptation from the outranking method proposed by Siskos *et al.*, 1984, initially developed for a complex decision making problem involving radioactive protection measures for a nuclear power plant, but with *ad hoc* formulas and an expanded output.

Some *a priori* advantages of building a fuzzy outranking relation should be noted:

1) On the one hand, any multicriteria technique should always be understood just as an aid to reaching a final decision; such a decision will be usually taken directly by the decision-maker, who only in special cases would leave it to be taken *automatically*. A classical value function approach gets in fact a solution through a representation on the real line; but it seems to be rather fictitious for any complex multicriteria decision making problem and, finally, such a classical additive model has a clear tendency to supply the decision-maker with extra work, providing not much insight about the structure of the set of alternatives (it does not take into account any kind of conflict, similarities or incomparabilities). In a way, our final set of fuzzy outranking relations can be considered as an intermediate level between the initial mess of data and the classical value function aggregation. Moreover, the whole package of programs can be structured by the decision-maker in many different ways, and such a flexibility should always be required to any decision making aid method.

2) On the other hand, Siskos's proposal takes out the main drawback in constructing a fuzzy preference relation: the decision maker's difficulties in establishing direct pairwise comparison values. Fuzzy outranking relations are defined from the scores, weights and the veto and significance thresholds; since scores will be evaluated by specialists, and weights can be known by applying some standard technique, then the decision-maker should define directly only the values of both thresholds. Moreover, this approach allows us to model incomparability phenomena, a relevant characteristic when dealing with practical multicriteria problems (a single criterion can inhibit direct outranking between alternatives).

The particular package of programs applied to our data provides a more complete knowledge about the structure of single and aggregated sets of projects through a flexible process.

Acknowledgement

This paper has been supported by *Dirección General de Investigación Científica Técnica* (national grant PB88-0137) and under a British-Spanish *Acción Integrada* (project 180/07).

REFERENCES

Keeney, R.L. & Raiffa, H. (1976): *Decisions with Multiple Objectives, Preferences and Value Tradeoffs.* John Wiley, New York.

Montero, J. & Tejada, J. (1986): *Fuzzy preferences in decision making*; in B. Bouchon & R.R. Yager (eds.): *Uncertainty in Knowledge-Based Systems*. Springer-Verlag, Berlin.

Orlovsky, S.A. (1978): *Decision making with a fuzzy preference relation*. Fuzzy Sets and Systems 1, 155-167.

Pearce, D.W. & Nash, C.A. (1981): *The Social Appraisal of Projects*. MacMillan, Basingstoke.

Pearman, A.D.; Mackie, P.J.; May, A.D. & Simon, D. (1989): *The use of multicriteria techniques to rank highway investment proposals*; in A.G. Lockett & G. Islei (eds.): *Improving Decision Making in Organizations*. Springer-Verlag, Berlin.

Simon, D.; Mackie, P.J.; May, A.D. & Pearman, A.D. (1988): *Priority assessment technique for British local authority highway schemes*. Transportation Research Record 1156, 10-17.

Siskos, J.; Lochard, J. & Lombard, J. (1984): *A multicriteria decision making methodology under fuzziness*; in H.J. Zimmermann, L.A. Zadeh & B.R. Gaines (eds.): *Fuzzy Sets and Decision Analysis*. North-Holland, Amsterdam.

APPENDIX: DATA SET

PROJECTS (One criteria per row)

1	2	3	4	5	6	7	8	9	10	11	12	W(i)	S(i)	V(i)
0.400	0.433	0.367	0.000	0.133	0.833	0.000	0.100	0.067	0.000	0.167	0.033	0.004	0.200	1.000
0.133	0.133	0.067	0.067	0.067	0.200	0.067	0.067	0.200	0.000	0.067	0.000	0.024	0.200	1.000
0.167	0.000	0.000	0.000	0.000	0.000	0.000	0.000	0.000	0.000	0.000	0.000	0.162	0.200	1.000
0.000	0.067	0.033	0.000	0.233	0.167	0.000	0.000	0.000	0.000	0.000	0.000	0.004	0.200	1.000
0.000	0.000	0.000	0.000	0.200	0.200	0.000	0.000	0.000	0.000	0.000	0.000	0.024	0.200	1.000
0.000	0.000	0.000	0.000	0.000	0.000	0.167	0.000	0.000	0.000	0.000	0.000	0.162	0.200	1.000
0.889	2.662	4.637	0.000	0.449	1.428	0.000	0.150	0.000	0.000	3.219	3.131	0.016	1.000	5.000
3.557	2.219	1.546	0.000	0.300	1.428	2.859	0.400	0.999	1.306	1.609	2.088	0.020	1.000	5.000
1.334	2.219	0.000	0.963	0.599	1.428	0.000	0.000	0.000	0.000	0.000	0.000	0.016	1.000	5.000
3.557	1.775	2.319	0.000	0.449	0.952	0.953	0.200	2.498	2.938	4.024	1.044	0.020	1.000	5.000
24.50	7.800	7.500	10.50	-21.0	47.30	5.300	3.300	0.000	0.000	2.200	2.300	0.002	10.00	50.00
3.100	1.200	1.100	1.100	-2.90	5.400	0.700	0.500	0.000	0.000	0.300	0.300	0.004	1.000	5.000
3.000	1.200	0.400	0.900	-1.60	2.000	0.700	0.300	0.000	0.000	0.200	0.200	0.005	1.000	5.000
0.100	0.100	0.100	0.100	-0.20	0.700	0.000	0.000	0.000	0.000	0.000	0.000	0.002	0.200	1.000
0.400	0.100	0.200	0.000	-0.30	0.800	0.000	0.100	0.000	0.000	0.100	0.100	0.002	0.600	3.000
0.000	0.000	0.000	0.000	0.000	0.000	0.000	0.000	0.000	0.000	0.000	0.000	0.002	0.200	1.000
52.90	16.80	16.20	22.60	-67.8	102.2	11.40	7.000	0.000	0.000	4.700	5.000	0.002	20.00	100.0
10.40	4.100	3.700	3.700	-14.0	17.80	2.300	1.600	0.000	0.000	1.100	1.000	0.002	4.000	20.00
10.80	4.200	1.500	3.300	-8.30	7.200	2.500	1.100	0.000	0.000	0.700	0.700	0.002	3.000	15.00
1.600	1.600	1.600	-4.90	1.600	11.30	0.000	0.000	0.000	0.000	0.000	0.000	0.002	2.000	10.00
-6.20	-11.9	-47.8	-19.0	-0.50	-2.80	-8.80	-4.50	-0.70	-1.30	-11.7	-7.80	0.002	10.00	50.00
0.176	0.436	0.000	0.000	0.798	0.046	-.001	0.000	0.000	0.000	0.000	0.000	0.071	0.200	1.000
0.023	0.033	0.000	0.000	0.000	0.030	0.002	0.000	0.000	0.000	0.000	0.000	0.040	0.100	0.500
-.017	-.033	0.000	0.000	0.000	0.000	0.000	0.000	0.000	0.000	0.000	0.000	0.009	0.100	0.500
-.001	0.004	0.000	0.000	0.001	0.060	0.000	0.000	0.005	0.001	0.000	0.000	0.048	0.100	0.500
0.002	0.001	0.002	0.000	0.005	0.001	0.001	0.000	0.014	0.002	0.004	0.003	0.028	0.010	0.050
-1.00	-.008	0.000	0.000	0.000	-.008	0.000	0.000	0.000	0.000	0.000	0.000	0.029	0.200	1.000
-.006	-.002	0.000	0.000	-.002	-.002	-.002	0.000	0.000	0.000	0.000	0.000	0.034	0.010	0.050
-.003	-.003	-0.02	0.000	-0.12	-.001	0.000	0.000	0.000	-.002	0.000	0.000	0.006	0.100	0.500
0.000	0.000	0.000	0.000	0.000	0.000	0.000	0.000	0.000	0.000	0.000	0.000	0.057	0.010	0.050
0.000	0.000	0.000	0.000	0.000	0.000	0.000	1.000	0.000	0.000	0.000	0.000	0.116	0.200	1.000
0.250	0.600	0.200	0.450	0.250	0.200	0.200	0.300	0.000	0.150	0.200	0.200	0.085	0.200	1.000

COSTS

1	2	3	4	5	6	7	8	9	10	11	12
2870.	3710.	6030.	850.	3300.	3600.	201.	50.	30.	105.	25.	87.

FUZZY LOGIC APPROACH TO MODELLING IN ECOSYSTEM RESEARCH

A. Salski, C. Sperlbaum
Ecosystem Research Centre, University of
Kiel, Schauenburgerstr. 112, 2300 KIEL 1, Germany

Abstract
The problem of uncertainty often appear in modelling of ecosystems, in particular uncertainty of data and even uncertainty in relations between system components. To solve these problems and the problems of subjectivity in evalation of ecological paramaters a fuzzy logic approach has been proposed. The paper shows some first results obtained by investigation into this problems.
Keywords
Ecosystem, modelling, fuzzy sets.

1. Introduction. Uncertainty in ecosystems

In ecosystem research scientists study the complex structure and dynamic in representative ecosystems. From the geographical point of view ecosystems are locally restricted, but in reality they are open to external effects and therefore are mutually connected by various interactions. The centre of research is investigating the structures of the community of life and the composition of the material in ecosystems.

Since april 1988 these problems have been investigated in Ecosystem Research Centre at the University of Kiel (Windhorst et al.,1989). As a representative place of the specific regional geographical and agricultural conditions the region of the Bornhöveder Seenkette in Schleswig-Holstein has been chosen. The main part of the incurred data are data either registrated automatically or analysed in a laboratory. These data are transmitted into the central database (relational database system ORACLE) which represents the basis of statistic analysis and of simulation models.

Not all ecosystem parameters are measurable. The values of such parameters can be obtained by special estimation methods. Problems of uncertainty often appear in these situations, in particular uncertainty of data and even uncertainty in relations between system components. In this paper a fuzzy logic approach to solve these problems has been proposed.

2. Knowledge Representation

The knowledge-based system is divided into a body of knowledge and an inference mechanism. The body of knowledge consists of two parts: facts about the problem and rules to represent expert knowledge for solving problems (Leung et al., 1988).

The facts can be generated at the beginning of the session by interactive asking or reading from a database or they can be created during the process as a result of the system inference.

As ecosystem scientists prefer to use vague, ill-defined natural language to describe their knowledge and to estimate nonmeasurable parameters, a set of linguistic rules as knowledge representation has been chosen. The fuzzy expressions in these rules are defined by fuzzy sets, which are formulated in collaboration with the specific experts.

3. Design of a fuzzy knowledge-based model

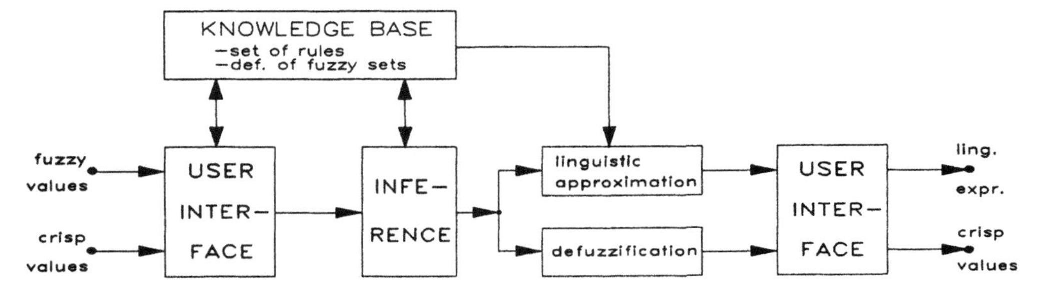

Fig.1: Simulation on the basis of a fuzzy knowledge-based model.

Figure 1 shows the information flow in the simulation process based on the fuzzy model. Non-fuzzy, numerical values (results of measurements) or fuzzy expressions (results of subjective evaluation or estimation) are allowed for the process input. The main part of the fuzzy model is a knowledge base with a set of linguistic rules and definitions of fuzzy sets describing the system to be modelled. A fuzzy output value is computed on the basis of these rules using a certain inference method. Then this fuzzy output value can be transformed into a numerical value (so-called defuzzification) or approximated to one of the fuzzy, linguistic values we have defined for the output variable. This approximation can be accomplished by means of distance calculation between fuzzy sets.

Below a step-by-step description follows of how to obtain a fuzzy model described above in a schematic and systematic way. Some explanations for the scheme to be used are given further (see also Salski, 1987).

Choosing preliminary assumptions

a) determine the shape of the membership function (linear or nonlinear)
b) choose the inference method
c) choose the defuzzification method
d) choose the approximation method

Obtaining the fuzzy model

a) define the input and output variables
b) define the space of values of input and output variables
c) choose the quantification levels (quantization of the space of input and output values)
d) define the fuzzy sets which are used to describe the terms of the linguistic values
(definition of the shapes of the fuzzy sets)
e) obtain linguistic rules
- check completeness of the set of rules

- check degree of 'covering' of the input value's space
- check degree of interactivity of the rules

f) adjust the model's parameters (improvement of the model's performance).

Some explanation for the scheme above

It is necessary to note that the preliminary choice of the quantification levels and fuzzy set parameters has a subjective character. The number of quantification levels (point c) has a great influence on the accuracy of a fuzzy model. Beyond a certain number of levels this influence is small. The chosen number may be checked during the stage of parametric optimization of the model (see point f).

One of the parameters which has an essential influence on the model's accuracy is the so-called degree of fuzziness of fuzzy sets. This value can be corrected during the stage of optimization of the model (see point f).

The major problem of fuzzy modelling is finding an appropriate set of linguistic rules describing the system to be modelled (point e). The formulation of these linguistic rules has a subjective character. There are two main methods to obtain these rules in ecosystem research:

1) They may be taken directly from the expert's experience. Nevertheless it is a very difficult process to define these rules on this basis, because the knowledge of the expert might be too complex to be written as a limited set of rules.
2) A basic set of rules is formulated and changed by the expert based on experiments of the process.

The set of linguistic rules should be complete and provide a correct answer for every possible input value. Thus the obvious condition should be satisfied that the sum of all input values (union of fuzzy sets) should 'cover' the value space of the input variable. If this condition is not satisfied the number of linguistic rules should be increased. If this is impossible one can try to increase the degree of fuzziness of the fuzzy sets (see point d). The other conditions in point e can be checked using the method presented in (Czogala et al.,1981).

To improve the model's performance one can adjust the parameters of the model, such as:

- number of quantization levels (see p. c)
- parameters of the fuzzy sets (see p. d)
- degree of 'covering' of the value's space (see point e).

The relation between these parameters and the performance of the fuzzy simulation is not yet completely known theoretically. In a concrete application one can check it by way of experiments (as in Salski et al.,1987).

4. Applications in ecosystem research

Two applications in zoology and soil science described below are just developed in the ecosystem research centre at the University of Kiel.

4.1 Preliminary assumptions

The preliminary assumptions described below are chosen for both the models described in point 4.2 and 4.3.

a) The shape of the membership functions of the fuzzy sets representing the values of the input and output variables is chosen to be the linear, trapezoidal form.

b) The chosen inference method:

The value of the output variable Y is calculated based on max-min composition:

$Y = X \circ R$

with membership function μ_Y

$$\mu_Y = \max_{(x \in X)} (\min(\mu_X, \mu_R(x, y)))$$

where X is fuzzy set in space X, the value of input variable (fact), $x \in X$, $y \in Y$,

Y is fuzzy set in space Y, the value of the output variable (conclusion),

R is fuzzy matrix calculated by the method showed below,

o is max-min operator,

μ_X is membership function of X,

μ_R is membership function of R .

For calculation of the fuzzy matrix R of any rule in the form:

"IF X is F1 THEN Y is F2"

the method proposed by Mizumoto, Fukami and Tamaka (Leung et al.,1988) is used. This method has been found to be closer to human intuition and reasoning than other methods.

$$R = (F_1 \times Y \rightarrow_S X \times F_2) \wedge (\overline{F_1} \times Y \rightarrow_G X \times \overline{F_2})$$

$$\mu_{F_1}(x) \rightarrow_S \mu_{F_2}(y) = \begin{cases} 1 & if \;\; \mu_{F_1}(x) \leq \mu_{F_2}(y) \\ 0 & if \;\; \mu_{F_1}(x) > \mu_{F_2}(y) \end{cases}$$

$$\mu_{F_1}(x) \rightarrow_G \mu_{F_2}(y) = \begin{cases} 1 & if \;\; \mu_{F_1}(x) \leq \mu_{F_2}(y) \\ \mu_{F_2}(y) & if \;\; \mu_{F_1}(x) > \mu_{F_2}(y) \end{cases}$$

where X, Y are values' spaces of F_1 and F_2, $x, y \in X, Y$,

F_1 is a fuzzy set in the antecedent part of the rule (in X),

F_2 is a fuzzy set in the consequent part of the rule (in Y),

$\overline{F}$ is the complement of the fuzzy set F

$\times$ is Cartesian product,

$\wedge$ is intersection of two fuzzy sets.

In this paper the problem of multiple input variables, associated with "and" in the antecedent part and with a single output variable in the consequent part is considered:

"IF X_1 is A_1 AND X_2 is A_2 THEN Y is A_3".

In this case the rule above is broken up into two rules:

"If X_1 is A_1 THEN Y is A_3" and

"IF X_2 is A_2 THEN Y is A_3".

Then the value of the output variable is calculated by taking the fuzzy union of the output values obtained from the rules with single input variable.

c) The "centre of gravity" method is used here as defuzzification method. Then the crisp value C (as result of the defuzzification operation) is calculated as follows:

$$C = \frac{\int_Y y \mu_Y(y) d_y}{\int_Y \mu_Y(y) d_y}$$

where Y is a fuzzy set in Y, $y \in Y$,

μ_Y is membership function of Y.

d) The linguistic approximation of the values of the output variables is evaluated by the normalized Euklidean distance measure d:

$$d(F_1, F_2) = \frac{1}{n} \sqrt{\sum_{i=1}^{n} (\mu_{F1}(z_i) - \mu_{F2}(z_i))^2}$$

where F_1, F_2 are fuzzy sets in the space Z, $z_i \in Z$ for i = 1,..,n.

μ_{F1}, μ_{F2} are membership functions of F_1, F_2.

4.2 Fuzzy model of the breeding success of larks

The purpose of this application is the examination of the following questions: Is the existence of the population of larks guaranteed? How does the population of larks change under certain conditions like vegetation structure and actual stock of larks?

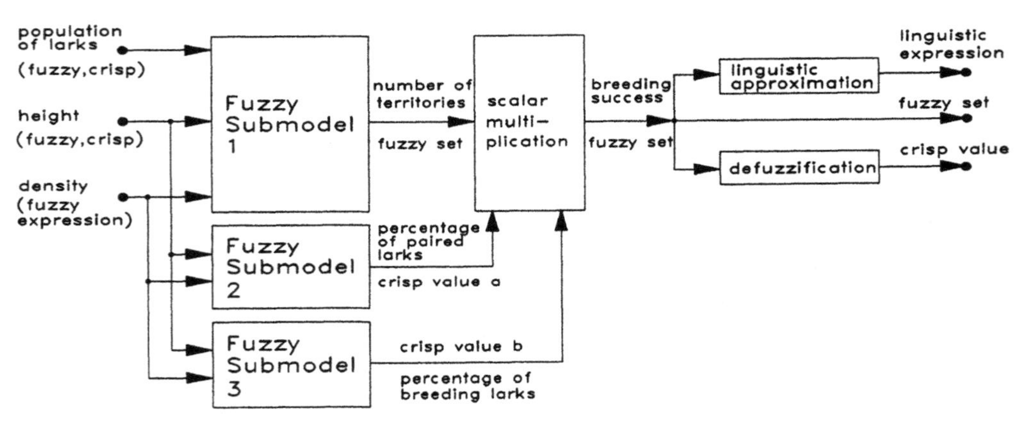

Fig.2: Fuzzy model of the relation between vegetation levels, population of larks and breeding success.

The input variables of the model at Fig.2 are the current population of larks (that is the number of the remaining larks of the preceding year), current height of vegetation and density of one planttype. The values of the variables "population of larks" and "height" can be crisp or fuzzy. The values chosen for the input variable "density" are linguistic expressions, like "standard" or "high".

The knowledge base of all three fuzzy submodels at Fig.2 contains the sets of fuzzy rules. The fuzzy submodel 1 at Fig.2 illustrates the valuations of the environment conditions by the male lark, which is reflected in the number of territories. The output variable of the fuzzy submodel 1, number of territories is an important process variable and the model showed at Fig. 2 allows to check it. For the submodel 1 125 fuzzy rules are defined in the form:

IF "height" is "low" AND "population of larks" is "very high" AND "density" is "minimal smaller than standard"

THEN "number of territories" is "high".

The fuzzy submodel 2 represents the way how the female larks valuate the environment conditions for the offered territories and how these conditions result in the breeding success. The output variable 'a' of this model is the percentage of the paired larks defined as part of the population of larks. For the submodel 2 25 rules are defined.

The fuzzy submodel 3 illustrates how many female larks actually want to breed. The output variable 'b' is the percentage of breeding larks defined as part of the population of paired larks. There are also 25 rules defined.

Scalar multiplication of a fuzzy set T and scalar v is defined here as the fuzzy set B with membership function

$$\mu_B(x) = \mu_T(x/v)$$

where T is number of territories (fuzzy set in X),

B is number of breeding female larks (fuzzy set in X),

X is the space of values of number of territories, $x \in X$,

v is crisp value, here v=ab.

There are provided three forms of output values: fuzzy set, linguistic expression and crisp value calculated based on the definitions showed in point 4.1.

The comparison between results of the simulation and first results of field research are showed in the tables 1 and 2 below.

If the planttype is "pea" and the input variables have the values:

height of vegetation: "average"

population of larks : 7

density : "standard"

the results are as follows:

Tab.1

output variable	type of value of the output variable	Simulation results	field research results
number of territories	crisp	6.54	6
	linguistic approximated	"medium"	"about 6"
number of paired larks	crisp	3.17	3
breeding success	crisp	1.44	1
(number of breeding larks)	linguistic approximated	"very low"	"very low"

For "wayside" and for the values of the input variables:
height of vegetation: "about 30"
population of larks : 12
density : "standard"
the output values are as follows:

Tab.2

output variable	type of value of the output variable	Simulation results	field research results
number of territories	crisp	11.62	12
	linguistic approximated	"large"	"large"
number of paired larks	crisp	11.04	11
breeding success (number of breeding larks)	crisp	10.89	11
	linguistic approximated	"high"	"high"

4.3 Fuzzy model of soil qualifiers

In soil science particular soil qualifiers are measured, estimated and evaluated in order to estimate the soil water regime.

In our project in Kiel for that purpose about 100 characteristic drilling profiles were determined for which all soil qualifiers are precisely measured. Besides about 1600 drilling profiles were digged in order to investigate the estimation of the soil properties of the whole area. For these 1600 profiles the soil qualifiers are not measured due to temporal and financial reasons, but approximately estimated and classified. The conventional classification, especially the division into classes is a fundamental part of the data reduction process. During this process a loss of information arises, because the classifications have been done by identifying the central concept of a class with the whole class and the boundary values marking the class are sharply defined.

Therefore it is sensible to design another model that allows to enter of not only exact data from measurement but also imprecise data (for ex. "very small" or "about 3.0") and any combination of them. An approach to solve this problem is to use the methodology provided by the fuzzy set theory.

In this application exemplarily the investigation and estimation of the soil characteristic, namely available water capacity of the root zone "nFKWe" is considered.

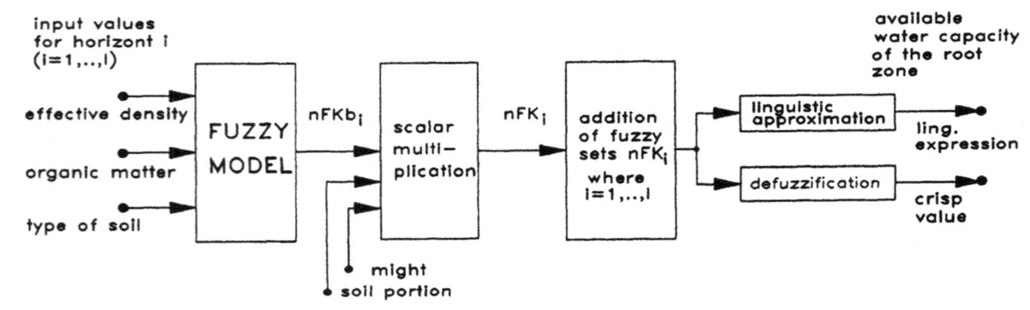

Fig.3 Fuzzy model of the relation between mineral soil qualifiers and the available water capacity of the root zone.

Each profile is subdivided into various horizons. Per profile the value of the effective rooting depth specifies the number of horizons, for which the available water capacity "nFK" is evaluated.

There are developed two models of the evaluating process, one for the mineral soil and one for peat soil. For a mineral soil the value of the basic available water capacity "nFKb" is evaluated from the input variables : "type of soil", "effective density" (which estimates the soil structure) and "organic matter". For a peat soil the input variable of the model of the "volume of solid matter" and the output variable is the basic value of the available water capacity.

The value of the "nFK" of each horizon is the scalar multiplication (see point 4.2) of the fuzzy value of the variable "nFKb" with the crisp values "might" and "soil portion".

For each profile the values of "nFK" of all horizons are summed up in addition. Addition of two fuzzy sets F_1 and F_2 is defined as the fuzzy set F with membership function

$$\mu_F(z) = \max_{z=x+y}(\min(\mu_{F_1}(x), \mu_{F_2}(y)))$$

where x,y,z are elements of the space of the values of the fuzzy sets F, F_1 and F_2.

After summing up all values of the output variable "nFK" the evaluation process ends with the result of the available water capacity of the root zone which is alternatively linguistic approximated or defuzzificated.

5. Final remarks

The problem of uncertainty in ecological modelling seems to be an essential problem. The paper shows some first results obtained by investigation into this problem based on fuzzy approach. Because of promising results the authors will continue this research and wish to design the next modells and also to develop a modelling support system based on fuzzy logic.

Acknowledgements

The authors wish to thank W. Daunicht (subject zoology) and U. Schleuß (subject soil science) for their cooperation in formulation of the knowledge base.

References:

Windhorst,W., Schaefer,W., Salski,A., Meyer,M.(1989) Erfassung, Verwaltung und Auswertung von Daten im Projektzentrum Ökosystemforschung der Christian-Albrechts-Universität zu Kiel. In Proc. of4-th Sym. Informatik im Umweltschutz, Karlsruhe'89, Springer Verlag, 251-262.

Leung,K.S.,Lam,W.(1988) Fuzzy concepts in expert systems, IEEE Computer (v. 21), 43-56.

Salski,A.(1987) The design of fuzzy logic control. Delft University of Technology, Rep.N-273, 1-12.

Czogala,E.,Pedrycz,W.(1981) Some problems concerning the construction of algorithms of decision-making in fuzzy systems, Int. J. Man-Machine Studies (v.15), 201-211.

Salski,A.,Noback,H.,Stassen,H.G.(1987) A model of the navigator's behaviour based on fuzzy set theory. In Patrick,J., Duncan,K.D. (eds): Training, Human Decision Making and Control, North-Holland, 205-222.

Burrough,P.A.(1989) Fuzzy mathematical methods for soil survey and land evaluation. Journal of Soil Science (v.40), 477-492.

AN APPLICATION OF FUZZY MULTI-CRITERIA DECISION MAKING FOR TOPOLOGICAL DESIGN OF LARGE NETWORKS

Aaron Kershenbaum and Teresa Rubinson
Polytechnic University
333 Jay Street
Brooklyn, New York

Abstract: In this paper, techniques of fuzzy multiple attribute decision making are applied to the design and analysis of large network topologies. A theoretical approach is presented that is computationally efficient, and broadly applicable to a wide class of large network design problems. This approach was developed to allow rapid analysis of performance trade-offs inherent in a broad range of topological network structures, and to assess and identify the sensitivity of design selection to changes in the decision criteria.
Keywords: Large network analysis, topological network design, fuzzy logic, fuzzy multiple attribute decision making, Ordered Weighted Aggregation (OWA) Operators, fuzzy membership evaluation, uncertainty

1. INTRODUCTION

In this paper, we examine an approach to topological design and evaluation of large mesh networks. We shall present methods to rapidly produce high quality topological designs, and techniques to compare and rank these designs relative to multiple decision criteria.

Our focus is on the design of basic topological network structures, and hence, we do not consider details of any particular implementation of technology. We present a theoretical approach that is intended to be computationally efficient, and broadly applicable to a wide class of large network design problems.

The techniques presented in this paper can be used to develop network designs that are useful in their own right, or as starting solutions for more complex, and detailed design procedures. Furthermore, our approach provides a tool to allow rapid analysis of performance trade-offs inherent in a particular design, and to assess and identify sensitivity in design selection due to change in the decision criteria.

2. NETWORK DESIGN

2.1. Statement of Design Problem

The topological design problem addressed here can be characterized as:

Given:
- link costs between all possible network nodes
- specification of inter-node traffic requirements
- constraints on link capacities and utilization rates

Find:
- good, feasible topological designs
- associated rankings of these designs, relative to a multiple criteria objective function encompassing several network performance factors (e.g., cost, delay, reliability)

It should be noted that in classical approaches to this problem, a "good" design is one that meets stated requirements and constraints at minimum cost [5]. Here, a "good" design is one that meets as many performance criteria as possible, while accommodating performance requirements and constraints.

2.2. Current Approaches

In general, the network design problem as described above is very difficult, even with many simplifying assumptions. Optimal solutions for problems of this type typically employ exhaustive search or branch-and-bound procedures [1]. Given N potential node locations and K possible line capacities, there are $K*2^{N(N-1)/2}$ possible network topologies. Thus, optimal solution techniques for this problem are unsuitable for the large networks with which we are concerned. Many heuristic techniques have been developed to solve specific topological design problems; however, most of these techniques are incremental, and find only a local minimum [2]. Thus, if the initial starting solution is poor, the final solution may be far from optimal.

Compounding the difficulty of network design is the uncertainty and imprecision inherent in the decision criteria and problem data. For example, in designing a new network, traffic requirements may be unknown and actual performance requirements may be difficult to estimate. It is desirable to provide a design that is robust and capable of supporting actual requirements that are, within reasonable limits, different from requirements used in planning the network. In providing a solution, most existing techniques "optimize" with respect to a single decision variable (which is usually cost) [2,5]. In reality, network designers seek to optimize and balance trade-offs between multiple performance criteria. For example, it is desirable to reduce delay and increase reliability as much as possible, if the resulting cost impact is not too great. Techniques that attempt to address this issue by utilizing a weighted or multiple linear combination objective function are problematic in that they force a precise numeric specification of decision trade-offs that is at best unnatural, and at worst, contrary to actual design practice. Most commonly used design techniques do not explicitly evaluate the impacts of multiple, and possibly conflicting, objectives driving actual design processes.

Worse yet, results provided by existing approaches do not easily admit to scrutiny. Given the imprecision and uncertainty that often accompanies initial specification of design requirements, and the nature of existing solution procedures, it is quite possible that mistakes will be made in the network design. If mistakes are made, then a good design may be rejected or a poor, infeasible one accepted. Current techniques do not tell us how far from the optimal solution we are, nor do they characterize the region of feasibility or the stability of the proposed solution.

3. APPROACH

3.1. <u>Overview of Approach</u>

In this paper, we address the fact that, far from knowing an optimal design solution, "no one even knows what a good network looks like in many realistic situations" [2]. We suggest a procedure that will efficiently provide: 1) a variety of design solutions, 2) an evaluation of how "good" the proposed designs are, and 3) a mechanism to examine the feasibility of initial design constraints and the resulting impacts on the final network configuration.

Our procedure consists of three major steps. The first step involves application of a powerful heuristic design procedure - MENTOR - developed by Kershenbaum, Kermani, and Grover. This step is used to quickly generate a variety of network alternatives for a specific design problem. The second step involves calculation of fuzzy performance measures for the resulting designs. The final step evaluates the proposed designs and their performance using a fuzzy multiple attribute decision tool.

3.2. Step One: Production of Network Designs

Our procedure utilizes the MENTOR algorithm to produce topological network designs. Features of the MENTOR algorithm that are particularly noteworthy for the purposes of this study are [2]:

- Computational efficiency. Mentor's computational complexity is of order (N^2), which is significantly faster than other currently used algorithms.
- High solution quality. Mentor has been demonstrated to produce high quality solutions that are competitive with other commonly used heuristic procedures.

Details of the Mentor algorithm and its implementation can be found in [2]. To briefly summarize, MENTOR achieves its efficiency, in large measure, by application of a simple and highly effective rule: send traffic over direct routes between nodes when the traffic exceeds a specified threshold; otherwise, send the traffic via a tree. MENTOR builds the tree to route all traffic not handled by direct routes, by computing a network center and application of a heuristic that uses a parameter - set to a value between 0 and 1 - to adjust computed link lengths.

MENTOR can rapidly produce design recommendations satisfying performance requirements and constraints. The results of the MENTOR algorithm can be varied by adjusting the parameter used in computing the tree described above, to provide alternative designs for the same network requirements. Therefore, in our procedure, we use MENTOR in an iterative fashion, making incremental adjustments in the design parameters, to produce a number of alternative designs to satisfy the stated requirements.

3.3. Step Two: Calculation of Network Performance Measures

The second part of our procedure involves calculation of performance measures for the networks generated by MENTOR. To implement this step, we need to identify measures that will allow comparison of one network design to another. Secondly, because of our requirement of computational efficiency, we need measures that can be easily computed. Finally, we require that each measure of performance have a corresponding fuzzy membership function.

For the sake of exposition, we propose three performance measures to evaluate proposed designs: cost, reliability, and delay. These measures were chosen because they are readily calculated from information collected during execution of the MENTOR algorithm. Other measures of network performance are possible and may be desirable for the design process. This is an area requiring further study, and is one in which we are actively engaged. Our approach does not preclude changes to the performance criteria - and in fact, many commonly used performance criteria can be formulated quite naturally in the required membership form - and is extensible to the extent that the new performance criteria are measurable without major computational difficulty.

After identifying relevant performance criteria, we construct a fuzzy membership function mapping all possible values of the performance criteria in the unit interval [0,1]. There are a number of approaches described in the literature for the practical estimation of membership functions; however, a definitive approach has yet to be developed [2,5]. We are currently investigating the empirical validity of the techniques suggested below for determination of the membership functions required for our analysis of network performance.

Norwich and Turksen have presented in [4] a method that they have used to construct membership functions based on interviews with subjects. Norwich and Turksen suggest a "direct rating method", and a "reverse rating" method to periodically verify results obtained by the direct rating method. The subject defining a membership function is asked to adjust the position of a pointer on a horizontal line segment, to indicate how strongly s/he agrees that the object or concept under evaluation corresponds to a descriptive term. The subject is presented with random combinations of, for example, cost values and "low cost", and is then asked to rate each combination by sliding the pointer to the appropriate position on the segment. "For the reverse rating procedure, the subject is told that we will randomly set the pointer to some position in the segment and that the rating thus created will have the same meaning as in the direct rating procedure." The details of this method are described fully in [4].

From the discrete data points collected in this manner, a generalized, continuous membership function must be specified. This generalization incorporates the idea that the membership function represents an implicit analytical definition. Zimmermann [9], and Kochen and Badre [3] have suggested a general model of membership where the membership function, $u(x)$, is defined:

$$u(x) = 1/(1 + d(x))$$

Conceptually, this model measures the deviation from a perceived ideal. In this model, the amount of deviation, $d(x)$, from the standard is measured by:

$$d(x) = e^{(a-bx)}$$

As $d(x)$ approaches infinity, $u(x)$ approaches zero, indicating complete deviation from the standard. As $d(x)$ approaches zero, $u(x)$ approaches one, indicating complete compliance with the standard. To tailor this model to a particular context, parameters a and b must be estimated from experimental data using a least squares analysis.

The membership model defined by Zimmermann above has many required theoretical properties, and is a useful starting point for defining a membership function. However, an analysis of the curves provided with this model revealed, that in fact, they do not fit well with the membership functions we derived for our network performance criteria using experimental data collected in interviews with an experienced network designer. This lack of fit with observed data is in large measure due to the insensitivity of the model formulation, owing to the fact that it is specified by only two parameters (a and b). For the network performance measures required in our study, we have found that it is often possible to define a polynomial function that agrees more closely with observed data. We are currently investigating the merits of improved sensitivity in the membership function and the resulting impacts on the final decision outcome.

We now review how the membership function for each performance criterion is determined. First, the membership functions are estimated from data collected in interviews with experienced network designers, using the direct and reverse rating methods suggested by Norwich and Turksen. In our example, three context specific variables are defined over the universe of discourse: "low cost", "good reliability", "low delay." Assume that the following data was obtained from interviews with an experienced network designer:

Variable: Good Delay

u(x)	x (Delay in Seconds)
1.0	.5
.9	1.0
0.0	4.0

Through mathematical analysis, it was determined that this data fits well with a polynomial of the form:

$u(x) = ax^2 + bx + c$ (where a = .0286, b = .157, and c = 1.086)

It should be noted that this polynomial has a much better fit with observed data than does the corresponding membership model provided by Zimmermann. The final form of our membership function has the general appearance indicated in the figure below.

In summary, in this phase of our procedure, membership functions are constructed for each performance criterion, through empirical investigation and use of an implicit analytical model of the membership function. These functions are then used to evaluate proposed designs from Mentor by mapping ratings of each performance criterion into the corresponding membership function. The major effort of this phase is in the determination of the performance criteria and the associated membership functions. Once the membership functions have been defined and implemented, corresponding membership values for observed performance data can be easily computed.

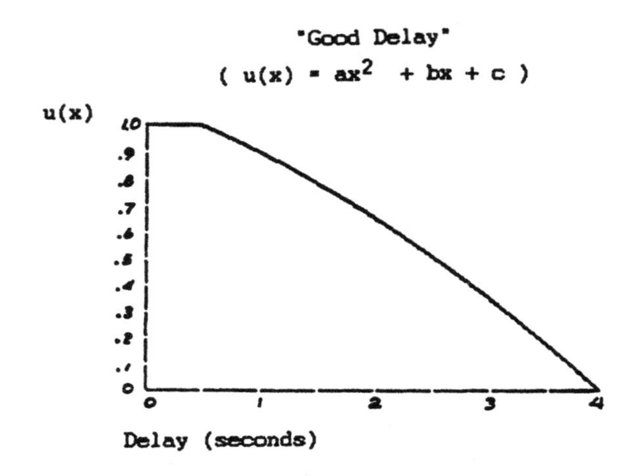

3.4. Step Three: Ranking of Proposed Designs

The final phase of our methodology involves ranking each network design proposal according to its relative merits as expressed in a multi-criteria objective function. This is a Multi-Attribute Decision Making (MADM) problem, as defined by Zimmermann, that is solved by [9]:

1) Aggregation of judgments with respect to all goals and decision alternatives, according to some decision function F(x):

$F(x) = u^{"}(x) * u^{'}(x) * u^{""}(x)$

where:

$u^{"}(x)$ = "low cost network" membership value for alternative x

$u'(x)$ = "reliable network" membership value for alternative x

$u^{""}(x)$ = "low delay" membership value for alternative x

$*$ = generic operator, and does not denote multiplication

2) Rank ordering of the decision alternatives according to the aggregated judgments, such that the optimal $F_j^*(x) = \max (F(x_j)$ (where j corresponds to the u(x) for a particular network design).

One of the most critical aspects of judgement aggregation is the selection of the aggregation operator. We propose use of Yager's Ordered Weighted Averaging Aggregation (OWA) operators for our analysis. Yager defines an OWA operator as [8]:

A mapping F from $I^n \longrightarrow I$ (where I =[0,1]) is called an OWA operator of dimension n if associated with F, is a weighting vector $W = [W_1, W_2, .W_3]$ such that: 1) $W_i \in (0,1)$

2) $\sum_{i=0}^{n} w_i = 1$

3) $F(a_1, a_2, \ldots, a_n) = W_1 b_1 + \ldots + W_n b_n$ (where b_i is the ith largest element in the collection $a_1, \ldots a_n$).

Yager's OWA operators have a number of interesting properties that make them particularly well suited to the problem at hand. One feature is that they allow "combining of criteria under the guidance of a quantifier...That is, the requirement that 'most' of the criteria be satisfied corresponds to one of these OWA operators" [8]. For example, one may wish that all criteria be satisfied in the evaluation procedure, or that at least one criterion is satisfied. These examples correspond, respectively to "AND" and "OR" fuzzy operators. To mitigate these extremes, Yager has suggested OWA operators as fuzzy quantifiers (e.g. "most" or "some" criteria must be satisfied). A second important feature of Yager's OWA operators is that they are relatively easy to implement. The most difficult aspect in defining an OWA operator is the determination of the weighting vector W. Once W has been empirically determined (See [8] for Yager's suggestions on the determination of W), the decision function F is computed as the inner product of W and the ordered vector B. Besides numerical efficiency, these operators also exhibit necessary theoretical properties, including monotonicity, commutivity, symmetry, and idempotency.

One of the major objectives of our work is to examine the impacts on the resulting design of decision trade-offs. In order to address this issue, we must incorporate representation of the relative importance of each decision criterion in the overall decision evaluation F'(x). Yager has suggested an approach in [8], which we adopt, i.e.:

$F'(x) = u(a_1, a_2, a_n)$

where the satisfaction of criterion A is:

$a = H(A_n(x), I_n) = (I_n \vee p) * (A_n(x))^{(I_n \vee q)}$

and:

q = degree of "or-ness" (see [8] for formal definition)

p = 1 − q

I_n = degree of importance of A_n (See [6,7] for calculation of I_n based on the maximum eigenvalue of pair-wise criteria comparisons)

We now demonstrate with an example. Three networks are to be evaluated. Performance data for each network is provided in the table below, with the corresponding membership ranking in parentheses.

Design	Cost ($)	Delay (Seconds)	Reliability (%)
1	200 (.7)	2 (.8)	.98 (.9)
2	150 (.8)	3 (.5)	.92 (.3)
3	600 (.1)	1 (1)	.99 (1)

We will examine the results when two different OWA operators are used in the decision function: a pure "and" operator and a simple mean operator. This implies that the respective weighting vectors for these operators are W_1 = [0,0,1], and W_2 = [1/3, 1/3, 1/3]. It also implies that (p = 1, q = 0) for the "and" operator and that (p = 1/2, q = 1/2) for the mean operator [8].

Assume that the importance of each criteria with respect to each other has been determined:

- o Low cost is more important than delay.
- o Low cost is more important than reliability.
- o Low delay is somewhat more important than reliability.

Using Yager's example in [7], the following pair-wise criteria importance matrix is constructed:

	Reliability	Cost	Delay
Reliability	1	1/3	1/2
Cost	3	1	3
Delay	2	1/3	1

The maximum eigenvalues for this matrix are computed, associating importance to reliability, cost, and delay, respectively:(I1 = .48, I2 =1.77, I3 =.75).

The resulting decisions for the "and" and mean operator are summarized below. B corresponds to the ordered argument vector including the criteria importance, W is the OWA weighting vector, and F indicates the evaluation of the decision function.

"And" Operator

Design	B	W	F
1	[.95,.85,.53]	[0,0,1]	.53
2	[.67,.59,.56]	[0,0,1]	.56 *
3	[1,1,.02]	[0,0,1]	.02

Note: Design 2 is the preferred solution

"Mean" Operator

Design	B	W	F
1	[.94,.63,.47]	[.3,.3,.3]	.68 *
2	[1.2,.45,.27]	[.3,.3,.3]	.64
3	[.75,.5,.03]	[.3,.3,.3]	.43

Note: Design 1 is the preferred solution, and is, in fact, the design that is the most balanced in satisfying the criteria.

To summarize, a decision function is formed that aggregates decision criteria using an OWA operator. The importance of each criteria is included in the decision function according to [6,7]. The decision function is then evaluated with respect to each performance criteria for each proposed design. The alternative with the highest overall rating is selected as the best design. Selection of this alternative completes our procedure.

4. CONCLUSION

In conclusion, we have presented a procedure which can be automated for rapid production and evaluation of alternative topological network designs to meet stated requirements. By incorporating a fuzzy multiple criteria decision model, our approach encompasses a more natural and multi-dimensional view of decision trade-offs in the network design process, extending the notion of optimality to include several decision variables expressed in both numeric and linguistically qualified (e.g., low cost", "low delay", "good reliability", etc.) terms. By allowing rapid iteration, our procedure provides a powerful adaptive tool to elucidate truly important underlying design considerations. For example, if our procedure identifies networks that meet all the design criteria, our selection of a design is easy. If, however, our procedure can not produce designs satisfying the design constraints fully, then two possibilities exist. There may exist true design compromises between potential configurations, reflecting differences to which we may or may not be indifferent. Alternatively, we may not be able to identify any networks to satisfy the design constraints. This, in turn, implies that the design constraints are incompatible and require reformulation. Our procedure allows easy adjustment of the design criteria so that this kind of evaluation can be made. In summary, we have outlined a procedure that accounts for imprecision and uncertainty in the design requirements, thereby providing significant insights into the design sensitivity and overall quality.

REFERENCES

[1] Boorstyn, R., and Frank, H., "Large-Scale Topological Optimization," Transactions on Communications, Vol. COM-25, No. 1, January 1977, pp. 29-47.
[2] Kershenbaum, A., Kermani, P. and Grover, G.,"MENTOR: An Algorithm for Mesh Network Topological Optimization and Routing," IBM Research Report, PC 14764 (#66171), July 14, 89, 28 pages.
[3] Kochen, M. and Badre, A., "On the precision of adjectives which denote fuzzy sets," J. Cybern, 4, No. 1, 1976, pp. 49-59.
[4] Norwich, A., and Turksen, I., "A Model for the Measurement of Membership and the Consequences of its Empirical Implementation," Fuzzy Sets and Systems 12, North-Holland, 1984, pp. 1-25.
[5] Tannenbaum, A., Computer Networks, 1st Edition, Prentice-Hall, pp. 32-90.
[6] Yager, R., "Fuzzy decision making including unequal objectives," Fuzzy Sets and Systems, Vol. 1, 1978, pp. 87-89.
[7] Yager, R., "Multiple Objective Decision Making Using Fuzzy Sets," Int. Journal of Man-Machine Studies, 1977,9, pp. 375-382.
[8] Yager, R., "On Ordered Weighted Averaging Aggregation Operators in Multicriteria Decisionmaking," IEEE Transactions on Systems, Man, and Cybernetics, Vol. 18, No. 1. January/February 1988, pp. 183-190.
[9] Zimmermann, H.J., Fuzzy Sets, Decision Making, and Expert Systems, Kluwer-Nijhoff Publishing, copyright 1987.

A DISTANCE MEASURE FOR DECISION MAKING IN UNCERTAIN DOMAINS

F. Esposito[1], D. Malerba[1], G. Semeraro[2]

[1] Istituto di Scienze dell'Informazione, Università di Bari
via G. Amendola n.173, 70126 Bari, Italy

[2] CSATA - Tecnopolis Novus Ortus
70010 Valenzano (BA), Italy

Abstract

A novel definition of syntactic distance between structural symbolic descriptions is proposed. It is based on a probabilistic interpretation of the canonical matching predicate. By means of this distance measure it is possible to cope with the problem of matching noise affected descriptions or imprecise rules. Furthermore, an extension of the syntactic distance which manages incomplete descriptions is presented. Finally, the application of the syntactic distance to the problem of classifying digitized office documents by using their page layout description is shown.

Keywords: pattern matching, syntactic distance, concept recognition, incomplete descriptions.

1. INTRODUCTION

A fundamental task performed by most knowledge-based systems is pattern matching. It consists of searching for a correspondence between two descriptions, generally the description of a new event and the premise of a rule in a knowledge base, in such a way that, when they have some common properties, the proper action, indicated in the consequence of the rule, can be taken.

Unfortunately, when dealing with complex problems from the real world, the process of pattern matching becomes more difficult for a variety of reasons. First of all, noise can affect the description of the new event modifying the values of some attributes or deforming the relationships among the subparts of which the event consists. In addition, the rules in the knowledge base can be imprecise due to the difficulties an expert faces in encoding into a formal representation language concepts that are inherently imprecise and whose interpretation is often context-dependent. Lastly, the available information may be incomplete due to either insufficient human knowledge or malfunction of the equipment during the measurement of a feature.

In such situations a *canonical (strict) matching predicate* is inadequate; a *flexible matching* which takes on values in a continuous range and represents a degree of similarity (*measure of fitness*) between the event and the premise of the rule appears to be more appropriate. The degree of similarity is strictly connected to the concept of *distance*, as the more distant two objects are, the less similar they can be considered.

Several distance measures were proposed in the fields of pattern recognition [1], [2], image analysis

[3] and machine learning [4], [5]. They can be roughly classified according to the following criteria:

- *representation language* (used for expressing the premises and consequences of rules): propositional logic, first-order predicate logic, feature vectors, attributed graphs;
- *kind of problem*: pattern matching in knowledge-based systems, concept acquisition, pattern classification, discriminant analysis, conceptual clustering, numerical taxonomy;
- *theoretical approach*: geometrical, syntactical, probabilistic, entropical, hybrid;
- *kind of recovered deformation*: local or structural.

This last criterion requires further explanation. Generally speaking, an object can be recursively decomposed into subparts, until atomic parts, called *primitives*, are detected. As a result, the complete description of the object is given by the attribute values of each composing part and the relationships among the subparts. Distance measures coping with *local deformations* can consider only differences between the attributes of the two descriptions to be matched, while those dealing with *structural deformations* can also manage differences at the level of mutual relationships among subparts.

In this paper a definition of *syntactic distance* (*SD*) suitable for structural deformations is proposed. It is defined, according to a top-down evaluation scheme, on the space of *well formed formulas* (*wff's*) of a variable-valued logic system, the VL_{21}, which is a multi-valued version of a first-order predicate logic with typed variables [6]. The approach followed consists of computing the probability of whether two wff's may match perfectly, assuming that the probability distribution of the domains of descriptors is known.

An extension of the syntactic distance for coping with the problem of incomplete descriptions in decision making is also introduced. Finally, we describe an application of the proposed distance to the classification of digitized office documents through the use of automatically generated physical page layout descriptions.

2. THE DEFINITION OF A SYNTACTIC DISTANCE

The basic component of the VL_{21} representation language is the *selector* or *relational statement*, written as:

$$[L = R]$$

where:

- L, called *referee*, is a function symbol with its arguments;
- R, called *reference*, is a disjunction of values of the referee's domain.

Function symbols of referees are called *descriptors* and they are n-adic functions ($n \geq 1$) which map onto one of three different type of domains: *nominal*, *linear* and *tree-structured*. In a nominal domain no relation is imposed on its values, while a total order relation must be satisfied by the values of a linear domain. In a tree-structured domain, values can be considered as nodes of a tree since a hierarchical (partial order) relation is defined on them. In literature, 1-adic functions are often named *attributes* while n-adic functions, with $n>1$, are commonly called *relations*, because they establish correspondences among the subparts of an object. Selectors can be combined by applying different operators, some of which are: AND, OR, and *decision operator* (::>). Such combinations are used to define the premises and consequences of decision rules.

Below we give the definition of syntactic distance according to our notation.

Let F1 and F2 be two VL_{21} wff's, then the syntactic distance between F1 and F2, SD(F1,F2), is:

$$SD(F1,F2) = 1 - Flex\_Match(F1,F2) \qquad (1)$$

where Flex_Match(F1,F2) denotes the probability that F1 perfectly matches F2, that is:

$$Flex\_Match(F1,F2)=P(Match(F1,F2)) \qquad (2)$$

Here Match represents the canonical matching predicate defined on the space $\mathfrak{S}$ of VL_{21} wff's:

$$Match: \mathfrak{S} \times \mathfrak{S} \rightarrow \{false, true\}$$

while:

Flex_Match: $\mathcal{S} \times \mathcal{S} \rightarrow [0,1]$.

The definition (2) marks the transition from deterministic matching to probabilistic.

Since the main application of the syntactic distance is noise-affected concept recognition, from now on F1 will denote the description of a concept and F2 the observation to be classified. Moreover, the kind of isomorphism used in concept recognition is the *specializing* one (*s-isomorphism*) [7], therefore the match of F1 and F2 consists in searching for a substitution such that:

$$F2 \Rightarrow \sigma(F1) \tag{3}$$

Flex_Match is computed according to the following top-down evaluation scheme:

1) F1 is a disjunction of conjuncts: $F1 = Or\_atom_1 \vee \ldots \vee Or\_atom_n$. Then

$$\text{Flex\_Match}(F1,F2) = P(\text{Match}(Or\_atom_1 \vee \ldots \vee Or\_atom_n), F2) =$$

$$= P(\text{Match}(Or\_atom_1, F2) \vee \text{Match}(Or\_atom_2, F2) \vee \ldots \vee \text{Match}(Or\_atom_n, F2)) =$$

where Or_atom_i is the generic conjunction of selectors in F1. By replacing the symbol Match(Or_atom_i,F2) with E_i we have:

$$= \sum_{i=1}^{n} P(E_i) - \sum_{i=1}^{n-1}\sum_{j=i+1}^{n} P(E_i \wedge E_j) + \sum_{i=1}^{n-2}\sum_{j=i+1}^{n-1}\sum_{k=j+1}^{n} P(E_i \wedge E_j \wedge E_k) + \ldots + (-1)^{n-1} \cdot P(E_1 \wedge E_2 \wedge \ldots \wedge E_n) =$$

Here we assume that all the elementary events E_i are mutually independent, that is:

$$P(E_i \mid E_j) = P(E_i) \qquad \forall\, i \neq j$$

$$P(E_i \mid E_j \wedge E_k) = P(E_i) \qquad \forall\, i \neq j \neq k$$

$$\ldots$$

since the consideration of high-order joint probabilities is generally impractical. In such a case, it follows that

$$P(E_i \wedge E_j) = P(E_i)P(E_j) \qquad \forall\, i \neq j$$

$$P(E_i \wedge E_j \wedge E_k) = P(E_i)P(E_j)P(E_k) \qquad \forall\, i \neq j \neq k$$

$$\ldots$$

There could be two clear exceptions to this assumption:

- *Or_atom_i is a specialization of another Or_atom_j*, then: P(Match(Or_atom_i,F2)|~Match(Or_atom_j,F2))=0. However, we never expect such a specialization to occur within the concept description F1, otherwise one of the or_atoms would be redundant.
- *E_i and E_j are mutually exclusive: $E_i \wedge E_j = \varnothing$*, then: $P(E_i \mid E_j) = P(E_j \mid E_i) = 0$. Mutual exclusion of the or-atoms of the same generalization is very unusual when dealing with structural descriptions. For instance, given the following concept description:

  F1: [on_top(x1,x2)=true][touch(x1,x2)=false]

  ∨

  [on_top(x1,x2)=false][touch(x1,x2)=true]

  a possible observation might be:

  F2: [on_top(x1,x2)=true][touch(x1,x2)=false]

  [on_top(x2,x3)=true]

  [on_top(x3,x4)=false][touch(x4,x5)=true]

  which is covered by both or-atoms. These considerations could be easily extended to the case of three or more elementary events.

As a result we have:

$$= \sum_{i=1}^{n} P(E_i) - \sum_{i=1}^{n-1}\sum_{j=i+1}^{n} P(E_i)\cdot P(E_j) + \sum_{i=1}^{n-2}\sum_{j=i+1}^{n-1}\sum_{k=j+1}^{n} P(E_i)\cdot P(E_j)\cdot P(E_k) + \ldots + \prod_{i=1}^{n} (-1)^{n-1}\cdot P(E_i) \tag{4}$$

where $P(E_i) = P(\text{Match}(\text{Or\_atom}_i,F2)) = \text{Flex\_Match}(\text{Or\_atom}_i,F2)$. In some situations, when the computation of formula (4) is computationally expensive, Flex_Match(F1,F2) may be approximated by its lower limit, that is the maximum value of $\text{Flex\_Match}(\text{Or\_atom}_i,F2)$, $i \in \{1,2, \ldots, n\}$.

2) F1 is a conjunction of selectors: $F1 = Sel_1 \wedge Sel_2 \wedge \ldots \wedge Sel_k$ $(k>0)$. Then by definition of matching, we have:

$$\text{Match}(F1,F2) = \text{Match}(Sel_1,F2) \wedge \text{Match}(Sel_2,F2) \wedge \ldots \wedge \text{Match}(Sel_k,F2)$$

where, now, theelementary events $\text{Match}(Sel_i, F2)$, $i=1,2,\ldots,k$, are not independent because of the presence of the variables as arguments of a descriptor. In order to optimize the matching algorithm it is possible to decompose F1 into two parts:

$$F1 = F1' \wedge F1''$$

so that:

- $F1' = Sel_1 \wedge Sel_2 \wedge \ldots \wedge Sel_i \wedge \ldots \wedge Sel_m$ is a conjunction of selectors such that the referee of Sel_i contains the maximum non-null number of variables not appearing in the referees of Sel_1, Sel_2, ..., Sel_{i-1}
- F1" is the conjunction of the remaining selectors of F1.

Consequently, we have:

$$\text{Flex\_Match}(F1,F2)=P(\text{Match}(F1',F2) \wedge \text{Match}(Sel_{m+1},F2) \wedge \ldots \wedge \text{Match}(Sel_k,F2)).$$

Instead of searching for a substitution such that the logical implication (3) is true, we weaken the condition of strict matching that:

$$F2 \Rightarrow \sigma(F1') \tag{5}$$

Under such a hypothesis, the elementary events $\text{Match}(Sel_{m+1},F2)$, ..., $\text{Match}(Sel_k,F2)$ become independent since the substitution has already bounded the variables in F1', therefore:

$$\text{Flex\_Match}(F1,F2) = \max_{\sigma_j} \prod_{i=m+1}^{k} \text{Flex\_Match}_j(Sel_i,F2) \quad \text{for each } \sigma_j \text{ such that } F2 \Rightarrow \sigma_j(F1') \tag{6}$$

where Flex_Match_j is the approximate matching computed according to the substitution σ_j.

This formula must be interpreted as follows:

while varying the considered substitution σ_j, which is responsible for the consistent binding of the variables in F1', the flexible matching between F1 and F2 is computed as the highest value given by the multiplication of the degree of similarity between each selector of F1" and F2.

When it is not possible to find a substitution satisfying (5) then we can set Flex_Match(F1,F2)=0, since F1 and F2 have no similarities, not even at the level of the subcomponents.

The constraint that F1' matches F2 by a canonical matching procedure is necessary in order to avoid an exponential growth of computational time, since the problem of matching two VL_{21} wff's is NP-complete. This heuristic method could be interpreted as follows: there must be at least some correspondences between F1 and F2, that is F1' is a conjunction of *Must-relationships* [8].

An extension of formula (5) could take into account the *weights* w_i associated with all the descriptors:

$$\text{Flex\_Match}(F1,F2) = \begin{cases} \max_{\sigma_j} \prod_{i=m+1}^{k} w_i \cdot \text{Flex\_Match}_j(Sel_i,F2) & \text{for each } \sigma_j \text{ such that } F2 \Rightarrow \sigma_j(F1') \\ 0 & \text{otherwise} \end{cases} \tag{7}$$

where $w_i \in [0,1]$ represent the user's preferences or descriptor relevances.

Henceforth, F2 will be supposed to be a conjunctive VL_{21} wff since the matching procedure between F1 and any disjunctive wff F2 can be reduced to several matching procedures between F1 and each or-atom of F2.

3) F1 is a selector: $F1 = [f(x_1, ..., x_t)=g_1, g_2, ..., g_y]$, where f is a t-adic function symbol and $g_1, g_2, ..., g_y$ are the domain values that f can take on when applied to $x_1, ..., x_t$.

Flex_Match$_j$(F1,F2) is determined by considering the degree of similarity between the selector $\sigma_j(F1)=Sel_{F1}$ and the corresponding selector of F2, $Sel_{F2}=[f(z_1, ..., z_t)=e]$, which has the same referee as Sel_{F1}. It should be pointed out that the reference of Sel_{F1} can contain more than one element since F1 is the description of a concept, whereas the reference of Sel_{F2} usually contains only one element since F2 describes an observation. Indeed, a multiple-value reference for Sel_{F2} denotes uncertainty in the measurement process, and it should be dealt in a different way. Consequently:

$$Flex\_Match_j(F1,F2)=Flex\_Match(Sel_{F1},Sel_{F2}) \quad (8)$$

and Flex_Match(Sel_{F1},Sel_{F2}) computes the degree of similarity between the references of Sel_{F1} and Sel_{F2}. As we search for an s-isomorphism, the similarity between the references of Sel_{F1} and Sel_{F2} is equal to 1 if and only if the reference of Sel_{F2} is more specific than that of Sel_{F1}. The notion of *specialization* is intended as *set inclusion*, if the descriptor f is a nominal or linear one. This interpretation can be easily extended to tree-structured descriptors: each single element in the reference of F1 is replaced by all the values representing the leaves of the subtree having just that element as its root. When the set inclusion does not hold, the definition of Flex_Match(Sel_{F1},Sel_{F2}) takes into account the probability that the value in the reference of Sel_{F2} becomes equal to one of the y values in the reference of Sel_{F1}. Therefore we give the following definition:

$$Flex\_Match(Sel_{F1},Sel_{F2}) = \max_{p\in[1,r]} P(EQUAL(g_i, e)) \quad (9)$$

where EQUAL(x,y) denotes the matching predicate defined on any two values x and y of the same domain The probability of the event P(EQUAL(g_i, e)) can be defined as the probability that an observation *e* may be considered a distortion of g_i, that is :

$$P(EQUAL(g_i,e)) = P(\delta(g_i,X) \geq \delta(g_i,e)) \quad (10)$$

where:

- X is a random variable assuming values in the domain D_f of f ;
- δ is a distance defined on the domain itself.

When no information is available on the probability distribution of X, we can assume that each value from the domain of f has the same probability, that is 1/C, where C is the number of elements of D_f.

The definition of δ has to be specialized according to the type of VL_{21} descriptor. In particular we propose for *nominal descriptors*:

$$\delta(x,y) = \begin{cases} 0 & \text{if } x=y \\ 1 & \text{otherwise} \end{cases} \quad (11)$$

and for *linear descriptors* with a finite domain:

$$\delta(x,y) = | ord(x) - ord(y) | \quad (12)$$

where *ord(x)* denotes the ordinal number given to the value x of the domain of the descriptor f.

In fact, if the referee of F1 is a linear descriptor with a totally ordered domain $D_f = \{y_0, y_1, ..., y_{C-1}\}$, it is always possible that consecutive elements of D_f are assigned consecutive integers (starting from 0 or from any other integer); for instance: $ord(y_0)=0$, $ord(y_1)=1$, ..., $ord(y_{C-1})=C-1$.

The specialization of δ has repercussions on the definition of P(EQUAL(g_i,e)), which is adapted to nominal, linear and tree-structured descriptors.

For *nominal descriptors* it is defined by:

$$P(\,EQUAL(g_i, e)\,) = \begin{cases} 1 & \text{if } g_i = e \\ (C-1)/C & \text{otherwise} \end{cases} \quad (13)$$

For *linear descriptors* it becomes:

$$P(\,EQUAL(g_i, e)\,) = \begin{cases} [1+ord(e)+(C-2ord(g_i)+ord(e))\cdot step(C-1-2ord(g_i)+ord(e))]/C & \text{if } g_i>e \\ 1 & \text{if } g_i=e \\ [C-ord(e)+(2ord(g_i)-ord(e)+1)\cdot step(2ord(g_i)-ord(e))]/C & \text{if } g_i<e \end{cases} \quad (14)$$

where:

$$step(x) = \begin{cases} 0 & \text{if } x < 0 \\ 1 & \text{otherwise} \end{cases} \quad (15)$$

Derivations of formulas (13) and (14) are omitted.

For *tree-structured descriptors*:
each element in the references of Sel_{F1} and Sel_{F2} is replaced by the values representing the leaves of the subtree which has that element as its root. Then formulas (13) or (14) are adopted, depending on whether the generalization hierarchy for the descriptor is unordered or ordered, respectively. The only changes to be made both in (13) and in (14) consist in replacing C with the cardinality of $LEAVES(D_f)$, where $LEAVES(D_f)$ represents the set composed of all the leaves of the tree representing the domain of the tree-structured descriptor f.

As pointed out by Shapiro and Haralick, [9], the definition of a distance measure between structural descriptions allows exploitation of the Bayesian decision framework. This is even more so that is possible when the definition of a distance is based on a probabilistic interpretation of the matching predicate. The measure of fitness, Flex_Match(F1,F2), actually computes the probability that a new event F2 may come from the class described by F1, P(F2|F1), or in other words, it computes the probability that any observation of the concept described by F1 would be farther from the centroid F1 than the F2 being considered. If Flex_Match(F1,F2) is too small, it signals the possibility that F2 is not an instance of the class described by F1, even though that is the closest.

Decision-making based on a syntactic distance is more expensive than a true/false matching procedure, therefore a *multilayered framework* is more beneficial. At first, a canonical matching procedure is applied in order to classify an observation F2. Three outcomes are possible:

a) *single match*: no further processing is required; the observation is assigned to the class represented by the matched recognition rule;
b) *no match*: a flexible matching is realized through the syntactic distance; provided that its value is not greater than a fixed threshold, the observation is classified as belonging to the class with the lowest syntactic distance;
c) *multiple match*: the roles of F1 and F2 are swapped for each matched rule in order to evaluate the differences between the non-matched parts of F2 and F1 themselves; the observation is classified into the category with the lowest value of SD(F2,F1).

3. THE PROBLEM OF INCOMPLETE DESCRIPTIONS

Generally, in real-world problems, descriptions may be incomplete for a number of reasons, which can be shortly summarized in the following three points:

- real ignorance of a feature value (*unknown value* or "?");
- a feature value is irrelevant for the problem at hand ("*don't care*" value);
- the value of a feature does not make sense at all (*meaningless value*).

The definition of the proposed similarity measure can be extended in order to deal with the first situation, i.e. *unknown* values.

Let $D_f = \{y_0, y_1, \dots, y_{C-1}\}$ be the domain of a descriptor f, then:

$$P(EQUAL(g,?)) = P(\,\delta(g,X) \geq \delta(g,Y) \cap Y \in D_f\,) =$$

$$= \sum_{i=0}^{C-1} P(\,\delta(g,X) \geq \delta(g,Y) \mid Y = y_i\,)\, P(Y = y_i) =$$

$$= \sum_{i=0}^{C-1} P(\,\delta(g,X) \geq \delta(g,y_i))\, P(y_i) \qquad (16)$$

As $P(EQUAL(g,y_i))$ specializes according to the type of descriptor, we can conclude that for *nominal descriptors*:

$$P(EQUAL(g,?)) = 1/C \sum_{i=0}^{C-1} P(\,\delta(g,X) \geq \delta(g,y_i)) =$$

$$= [(C-1)(C-1)/C + 1]/C = (C^2 - C + 1)/C^2 \qquad (17)$$

according to formula (13).

For *linear descriptors* with a finite domain:

$$P(EQUAL(g,?)) = 1/C \sum_{i=0}^{C-1} P(\delta(g,X) \geq \delta(g,y_i)) =$$

assuming that $ord(y_i)=i$ and $ord(g)=s$ and so by applying formula (14) we have:

$$= 1/C \sum_{i=0}^{C-1} \{step(s-i) \cdot step(i-s) +$$

$$+ (1-step(i-s)) \cdot [(C-2s+i) \cdot step(C-1-2s+i) + i+1]/C +$$

$$+ (1-step(s-i)) \cdot [(2s-i+1) \cdot step(2s-i) + C - i]/C \} \qquad (18)$$

Some interesting properties may be easily drawn out of this:

Property 1. For each nominal domain D_f whose values have the same probability, P(EQUAL(g,?)) does not depend on g, and for each $y \in D_f - \{g\}$:

$$P(EQUAL(g,y)) < P(EQUAL(g,?)) < P(EQUAL(g,g)).$$

Property 2. For each finite linear domain D_f whose values have the same probability, there are two values y_s and y_t, with $s < t$, so that for each $y \in \{y_s, y_{s+1}, y_{s+2}, \dots, y_t\}$:

$$P(EQUAL(g,?)) \leq P(EQUAL(g,y)).$$

In short, these properties state that some values of the domain D_f are less similar to g than the *unknown* value.

Formulas given above are valid for both *attributes* and *relations*, with the difference that *unknown* attribute values are omitted in the description of an observation, whereas *unknown* relation values are explicitly specified in order to distinguish them from *default* values.

4. APPLICATION DOMAIN

The syntactic distance has been implemented and tested on digitized office document classification [10], [11]. The main idea consists of the recognition of the layout structure of a generic printed page dealing only with geometrical characteristics, that are automatically detected through a segmentation and layout analysis process. The page layout recognition is carried out by using a process of inductive generalization, in which some meaningful examples of document classes, relevant for a specific office, are used to train the system. An empirical learning algorithm, inspired by the STAR methodology based on Michalski's approach to hypotheses generation, is used [6].

The system supplies VL_{21} decision rules which characterize the classes of documents in terms of numerical and symbolic features (descriptors), so allowing the automatic classification of new documents.

Since noisy documents are handled also, it is not possible to use a strict matching procedure for classifying the test documents, so a syntactic distance defined for VL_{21} descriptions was introduced.

In Fig. 1 an example of document page layout is reported. It was produced by segmenting the document through RLSA techniques and grouping together some segments (or blocks) which satisfy some predefinite requirements, such as closeness, equal width, same type, etc. The document is described in the VL_{21} language by using the following descriptors:

CONTAIN_IN_POS(Doc,Block), WIDTH(Block), HEIGHT(Block),
TO_RIGHT(Block1,Block2), ON_TOP(Block1,Block2),ALIGN(Block1,Block2).

The next processing step concerns the classification of the document by matching its description with the hypotheses generated by the STAR methodology. In Table I the recognition rules for each class and the corresponding flexible matching value (FM) with the document in Fig. 1 are reported.

As an example of the computation of the measure of fitness, let us consider the recognition rule for the third class. It is a disjunction of two or-atoms, therefore formula (4) must be applied. When F1 equals to the first or-atom, then F1' becomes the following conjunction: F1' = [to_right(s1,s2)][to_right(s1,s3)][to_right(s4,s2)] which does not strictly match the description of the testing document, i.e. F2. Consequently, we have:

$\text{Flex\_Match}(\text{Or\_atom}_1,F2)=0$

On the contrary, when F1 equals to the second or-atom, then we find two substitutions such that:

$F2 \Rightarrow \sigma_j(F1')$ (F1' = [on_top(s1,s2)][to_right(s2,s3)])

namely, $\sigma_1 = \{s1 \leftarrow x2, s2 \leftarrow x3, s3 \leftarrow x4\}$ and $\sigma_2 = \{s1 \leftarrow x3, s2 \leftarrow x5, s3 \leftarrow x4\}$. Now, formula (6) is applied and the flexible matching at the level of selectors is computed. According to the substitution σ_1 only the fifth selector of the or-atom does not strictly match the description F2, therefore we have:

$$\prod_{i=3}^{7} \text{Flex\_Match}_1(\text{Sel}_i,F2) = 1 \cdot 1 \cdot 10/11 \cdot 1 \cdot 1 = 0.909$$

since the linear domain of the descriptor *height* has eleven values and we are assuming that each value of the domain has the same probability (this enable us to the use of formula (14)). It is easy to verify that the flexible matching computed according to the substitution σ_2 gives a lower value, so we can conclude that:

$\text{Flex\_Match}(\text{Or\_atom}_2,F2) = 0.909$

according to formula (6) and that the flexible matching between the description of class three and the description of the document is 0.909 according to formula (4).

The highest FM value in the Table I is that associated with class 3, therefore the document is correctly classified. The document is not rejected because the FM values are all higher than the fixed threshold (0.85).

Results of various experimentations with different approaches to empirical learning are extensively discussed in [10]. Here we report only some hints on the system performance:

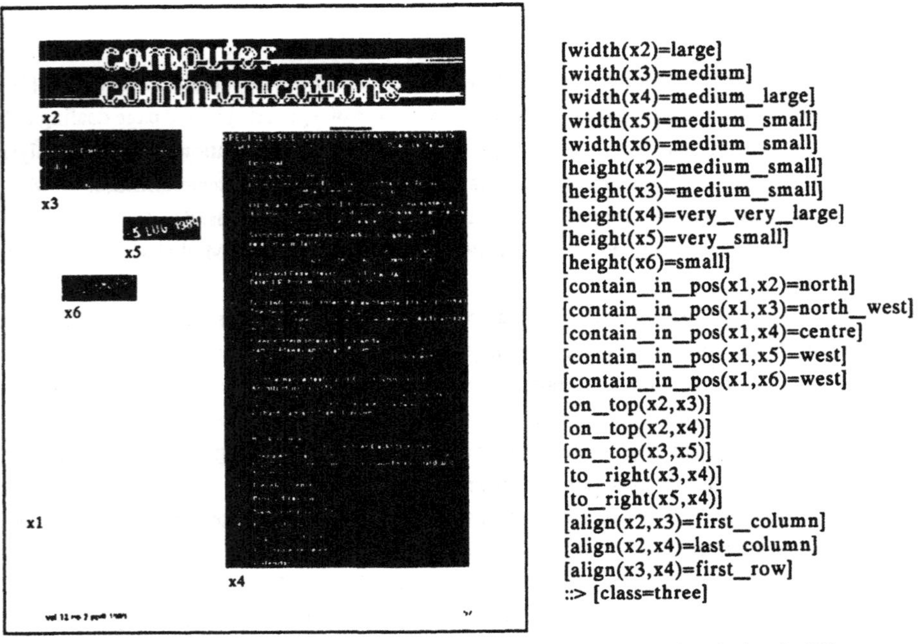

Fig. 1 The page layout of a testing document and its description in VL_{21}.

TABLE I

MEASURE OF FITNESS OF THE TEST DOCUMENT FOR EACH CLASS

CLASS	VL_{21} DESCRIPTION	FM
1	[to_right(s1,s2)][align(s1,s2)=first_column][height(s1)=medium_small][width(s2)=medium]	0.0
2	[to_right(s1,s2)][align(s1,s2)=last_row][height(s1)=very_small][width(s1)=medium_small] [width(s2)=very_small,small]	0.40
3	[to_right(s1,s2)][to_right(s1,s3)][to_right(s4,s2)][to_right(s3,s2)] [height(s2)=very_very_large] OR [on_top(s1,s2)][to_right(s2,s3)][on_top(s1,s3)][align(s1,s2)=first_column] [height(s1)=medium][height(s2)=medium_small][height(s3)=medium .. very_very_large]	0.909
4	[align(s1,s2)=last_column][width(s1)=medium_small][height(s1)=medium_small]	0.444
5	[to_right(s1,s2)][height(s1)=very_large .. greatest]	0.454
6	[to_right(s1,s2)][to_right(s3,s1)][on_top(s1,s4)][on_top(s5,s1)][width(s2)=medium_small] [height(s2)=very_very_small]	0.0
7	[contain_in_pos(s1,s2)=north_west][contain_in_pos(s1,s3)=centre] [align(s2,s3)=first_column][height(s3)=very_very_large,greatest]	0.90
8	[align(s1,s2)=mid_row][to_right(s3,s4)][on_top(s5,s1)][on_top(s1,s3)] [align(s5,s1)=first_column][align(s3,s4)=last_row][width(s2)=small]	0.0

- about 6 secs for generating the symbolic description of a document;
- about 3 secs for matching and classifying a document.

These results refer to an implementation in C on an OLIVETTI PC M280. The total processing time for a single document was less than one minute, including the scanning process.

5. CONCLUSIONS

We have presented a straightforward generalization of the canonical matching predicate to a syntactic distance that represents the probability of a strict matching between two VL_{21} wff's. This syntactic distance is able to cope with structural deformations, that is modifications both in the primitive attributes and in the relationships between component parts. Since the main application of the syntactic distance is concept recognition, a specializing isomorphism between a recognition rule and the observed event is sought. However, the presence of noise or the variability of the phenomenon itself requires that the similarity between two descriptions is evaluated according to a best match procedure. As such a procedure is computationally impractical, we have also described a simple approximate matching between a subpart of the recognition rule and the event to be classified. Moreover, we have proposed a multilayered decision framework based upon the consideration that strict matching is less computationally expensive. In fact, the syntactic distance should be computed when no rule strictly matches the observation, or even when there are multiple matches, in that case the differences between the non-matching parts of the observation and the matching rules are evaluated.

In addition to noise, another source of uncertainty is incompleteness: we have tried to cope with this problem by extending our definition of distance between selectors.

Finally, we have briefly described the application domain that stimulated the definition of the syntactic distance, the page layout document recognition, and some already obtained results.

REFERENCES

[1] SANFELIU, A., & FU, K. S. (1983). A distance measure between attributed relational graphs for Pattern recognition. *IEEE Trans. Syst., Man, and Cybern.*, vol. SMC-13, 353-362.

[2] WONG, A. K. C., & YOU, M. (1985). Entropy and distance of random graphs with application to structural pattern recognition. *IEEE Trans. Pattern Anal. Machine Intell.*, vol. PAMI-7, 599-609.

[3] ESHERA, M.A., & FU, K.S. (1984). A graph distance measure for image analysis. *IEEE Trans. Syst., Man, and Cybern.*, vol. SMC-14, 398-408.

[4] MICHALSKI, R. S., MOZETIC, I., HONG, J., & LAVRAC, N. (1986). The AQ15 Inductive Learning System: An Overview and Experiments, Intelligent Systems Group, Department of Computer Science, University of Illinois, Urbana, IL.

[5] KODRATOFF, Y., & TECUCI, G. (1988). Learning Based on Conceptual Distance. *IEEE Trans. Pattern Anal. Machine Intell.*, vol. PAMI-10, 897-909.

[6] MICHALSKI, R. S. (1980). Pattern Recognition as Rule-Guided Inductive Inference. *IEEE Trans. on Pattern Anal. Machine Intell.*, vol. PAMI-2, 349-361.

[7] LARSON, J. B. (1977). Inductive Inference in the Variable Valued Predicate Logic System VL21: Methodology and Computer implementation. Doctoral dissertation, Dept. of Computer Science, University of Illinois, Urbana, Illinois.

[8] WINSTON, P.H. (1984). *Artificial Intelligence* (2nd ed.). Addison-Wesley, Reading, Mass., 391-414.

[9] SHAPIRO, L.G., & HARALICK, R.M. (1985). A Metric for Comparing Relational Descriptions. *IEEE Trans. on Pattern Anal. Machine Intell.*, vol. PAMI-2, 90-94.

[10] ESPOSITO, F., MALERBA, D., SEMERARO G., ANNESE, E., & SCAFURO, G. (1990). Empirical learning methods for digitized document recognition: an integrated approach to inductive generalization. *Proceedings of the sixth IEEE Conference on Artificial Intell. Applications*, Santa Barbara, CA, 37-45.

[11] ESPOSITO, F., MALERBA, D., SEMERARO, G., ANNESE, E., SCAFURO, G. (1990). An experimental page layout recognition system for office document automatic classification: an integrated approach for inductive generalization. *Proc. of the 10th IEEE Int. Conf. on Pattern Recognition*, Atlantic City, NJ, 557-562.

DECISIONS AND LACK OF PRECISION IN CROP MANAGEMENT: THE ROLE OF PROCESSING BOTH OBJECTS AND PROCEDURES THROUGH SEMANTIC NETWORKS

Marianne CERF[1]; Sébastien POITRENAUD[2]; Jean-François RICHARD[2]; Michel SEBILLOTTE[1], Charles Albert TIJUS[2]

(1) - I.N.R.A. Département Systèmes Agraires et Développement Paris Grignon
16, rue Claude Bernard - 75 231 PARIS CEDEX 05
(2) - Laboratoire de Psychologie Cognitive du Traitement de l'Information Symbolique Université de Paris 8 - 93326 SAINT DENIS CEDEX 08

Abstract

This paper adresses the central issue of technical agricultural decision in crop management intercourse which is a complex task where decision of action is the most crucial topic in regard to environment characteristics, in regard to the objects of the task and in regard to the procedures that have to be applied. Moreover, many object properties are variables which take imprecise values due to: the dynamic character of the cultivation system, more or less accessibility to these values, the presence of intermediary variables and, finally, imprecise instruments.

Starting with a functional diagram of [plant population*soil*weather*techniques] system and a set of observations for sugar beet settling, we describe agronomic and ergonomic knowledge integration in task decomposition. We then show that relevant knowledge can be represented into semantic networks of the objects of the task (where classes factorize common structure of objects through associated procedures with simple and multiple inheritance principles). In such semantic networks, objects are classes and subclasses instances by an inclusion relation which is defined by the procedures (considered as properties of classes of objects) which they may be applied to these objects. Then, we suggest how such a representation involves some farmers' decision making mecanisms in spite of uncertainty.

Keywords

Farmers' decision making, procedural knowledge based-system, semantic network

1.INTRODUCTION

Technical agricultural decision support is acquiring a great importance due to the decrease of stable routines that comes from acceleration of technical and economic evolutions. However, for few cases, in spite of the effort in agronomic aid conception based on the knowledge of biophysical processes, the lack of efficiency of the agronomist advice that farmers were supposed to apply to farm management (Darré, 1985) implies the emergence of new technical agricultural decision support conceptions.

This causes agronomic researchers to focus on analyzing farm-functioning (Capillon & al, 1975; Sebillotte, 1979), and more precisely, farmers' decision processes (Sebillotte & Soler, 1988). The results of these studies have led to recognize that knowledge representation was not consistent between agronomists and farmers: The former are mostly concerned by the understanding of biophysical processes while the latter have to decide and carry out the actions that are necessary during crop management intercourse.

Nevertheless, present technical agricultural decision support do not fit in farmers' decisions processes (Cerf & al, 1990). It can be seen as a lack of research in the analyzing and modelling of farmers know-how. To overcome this problem, we studied agricultural work with an ergonomic and cognitive point of view. Our objectives were to point out what decisions are to be supported, what knowledge is to be acquired by the farmers in order to allow them to adapt their know-how to the numerous and important changes of the technical and economic environment. We also tried to make compatible farmers and agronomists knowledge representation.

As a first step, we represent agronomist knowledge in the way of a functional diagram of the [plant population*soil*weather*techniques] system. On the basis of the diagram and the analysis of agronomic diagnosis methods, we identified imprecision sources and decisions that an agronomist would list.

As a second step, with a procedural knowledge-based language, we analyzed and modelled agricultural work as it is carried out by the farmers for the sugar beet settling. This formalism (Poitrenaud & al, 1990) allows us to define a hierarchy for the different task levels (general goals of agricultural production, biophysical functions to be filled, cultivation operations to be carried out).

As a third step, using this formalism, we point out which decisions farmers really take, and discuss the efficiency of the procedures (as well as the precision of the thresholds which they are associated to) in regard to the specificity of the referent when compared to the concerned object.

Finally, we argue the psychological relevance of such an hypothesis (procedures access through objects) and discuss its interest to understand how farmers face the complexity of the system they manage. We also point out some modelling problems due to the characteristics of the situation and give some tracks in order to make consistent agronomists and farmers knowledge representation.

2. FUNCTIONAL DIAGRAM OF THE [SOIL*PLANT*WEATHER*TECHNIQUES] SYSTEM

The diagram (fig. 1) is based on agronomic knowledge about the sugar beet yield elaboration. The finality of the system (to reach the potential production, meaning that the yield was only limited by climatic factors, and to insure the necessary quality for the sugar beet processing) allows us to define the links between the different parts of the system and to list the variables to be used for crop management. The functionnal diagram shows the high number of variables interacting into causal chains at each moment but

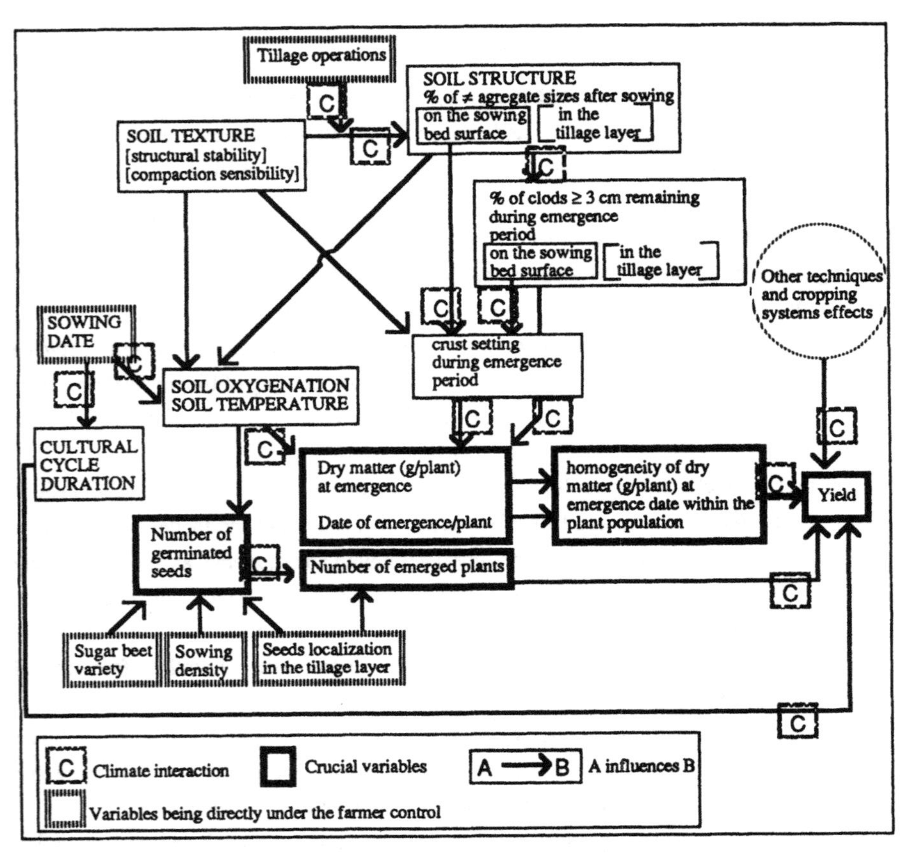

Figure 1. Part of the functional diagram of the [soil*sugar beet*weather* techniques] system

also during the cultural cycle (about 200 days) as well as it points out their simultaneous evolution under climatic events.

2.1. imprecision sources in crop management

The identification of the connections within the [soil*plant*weather*techniques] system allows us to point out the sources of imprecision concerning the different components of the problem space (initial state, means of action, target state, evaluation).

2.1.1. Initial state and imprecision on crucial variables

As shown by the functional diagram, the variables determining the yield (which we call crucial variables) are not always those on which actions can be undertaken. For example, although the established plant number is a major yield component, the number of seeds can not just be equivalent to the emerged plants number target. The farmers have to integrate all connections which might result in a loss of plants after sowing. The length of causal chains and the uncertainty of climatic events which determine the relative weight of the different links are imprecision sources on the population to be sown.

Other sources of imprecision on crucial variables cannot be directly inferred from the functional diagram. The rather poor existing instrumentation for collecting information allows a rough estimation of the values of the crucial variables. Also, the sampling that would be necessary to characterize the high level of heterogeneity of a cultivated field is rarely compatible with the time farmers accept to spend for collecting information.

2.1.2. Imprecision of the operating means

The functional diagram shows the importance of obtaining a sowing bed allowing a homogeneous emergence, the heterogeneity of the plant population being a prejudice for harvest quality (size of the roots, sugar concentration). Nevertheless, the farmers rarely reach this target. This is mostly due to the initial plot heterogeneity as a result of its cultivation history and its soil characteristics as well as to the lack of adaptability of the tools which cannot be adjusted to each local heterogeneity. Another factor is the duration of field work operations (for example, a sowing operation is realized at a 2.5 ha/h speed on rectangular plot of 10 ha) compared to spontaneous evolution of soil processes (the soil moisture might vary significantly in 4 hours) which increases heterogeneity of the obtained sowing bed.

2.1.3. Target state and imprecision due to climatic uncertainty

As drawn, the functional diagram does not allow us to specify the targets to be reached for a specific task as crop settling. To identify and to establish a hierarchy among them, the agronomist has to be able to infer soil reactions under some climatic events. Doing so, the agronomist reduces the weather uncertainty by setting a target state that, in frequency, will optimize the system functioning. However, such a reasoning does not satisfy the farmers as climatic events frequency does not give any information about a specific year climate. The necessity of performing a specific task on several plots, as soils might react differently (or similarly) under climatic actions, increases the imprecision of the target state to be reached on a given plot. The target state may have to be defined in regard to a general production objective for all plots and in regard to the possibility (or impossibility) of optimizing the [soil*plant population*weather] system functioning for each plot.

2.1.4. Evaluation and imprecision due to response latencies

Response latencies do not directly appear on the functional diagram. Nevertheless, they determine the possibility farmers have to evaluate the results of their actions. For example, the effects of the sowing bed characteristics (size of aggregates, soil moisture and temperature) on the percentage of emerged plants, will be perceptible only a few weeks after sowing. The interactions between the obtained sowing bed, the seeds and the climatic or parasitical events appearing after sowing will make the interpretation of plant numbers difficult.

2.2. Functional diagram and decisions

As we already noticed, the task decomposition into goals and subgoals does not directly appear on the functional diagram. The decisions to be taken (as for example, for soil tillage before sowing) have not been listed. As a matter of fact, due to the diversity of soils and of agricultural machinery, the agronomist needs more information about the pedo-climatic context and the available operating means (tillage machines, insecticides and so on) before defining the decisions to be taken. In a first step (Sebillotte, 1978), he identifies couples for each available machine (operating condition*obtained results), in regard to his knowledge about a given machine (for example: a harrow) and soil properties (cohesion, splitting properties). He can then tell, which technique (operating condition*machine) should be used in order to reach a specific target. In fact, decisions, from his point of view, will be a choice of operating conditions and machines. Nevertheless, the agronomist does not often go to this procedural level. More so, such a description is not drawn at the carrying out level with the temporal order of actions. This results in the fact that solving contradictory goals is not even considered: the agronomist does not usually take action planning into account.

This ergonomic approach of farm management points out the extreme complexity farmers have to face. This complexity is partly due to the numerous variables and connections interferring with crop management intercourse. It is also a result of the importance of action planning because different tasks have to be performed almost simultaneously. Such a planning has to take into account available procedures and evolutions of the [plant*soil*weather] system. The difficulty to evaluate the changes due to climatic uncertainty enhances the task complexity. In such a situation, evaluation of the obtained results is delayed, error recoveries and detections are hard. Moreover, a given technical task is infrequently performed (for example, one sows sugar beet once a year) and conditions of performance are rarely similar from one year to another. In such a context, what are farmers' real decisions ? How do they manage to reduce the high number of decisions to be taken ? How do they turn round imprecision ?

3. A SEMANTIC NETWORK ARCHITECTURE BASED ON AN OBJECT-ORIENTED DESCRIPTION

3.1. Semantic of action: objects, procedures and relations

Studies of internal knowledge representation and procedural knowledge are central issues in cognitive psychology. The general cognitive hypothesis underlying our approach is that procedures are tied to objects, so that they are accessed through as properties of classes of objects. In other words, procedures are stored in memory in the same way as structural properties of objects in a semantic network. We consider that the availability of a procedure for a given type of object is recognized by a very early decision: it precedes the processing of how the procedure will be worked out. This hypothesis has some cognitive relevance (Richard, 1983). Following theoretical foundations for human semantic of action (Richard, 1990) and for object-oriented languages (Lecluse & Richard, 1988; Goldberg & Robson, 1983), PROCOPE was designed as a computational architecture to unify declarative and procedural knowledge in semantic networks in order to provide a framework for the development of procedural knowledge-based systems.

There are two basic steps in domain description. The language of description is first based on task decomposition that provides elementary procedures. Then, procedures are used to construct the semantic network of the task microworld (Objects, Procedures, Relations). The first step method is quite similar to some other action-grammar formalisms that have been developed (Frederiksen, 1988; Tauber, 1988; Payne & Green, 1986; Kieras & Polson, 1985; Reisner, 1981). The second step corresponds to a new approach of knowledge representation. The central core of PROCOPE is a knowledge-representation user-interface enabling the novice to evaluate procedures and to solve task-problem by learning. Furthermore it allows (i) to describe into a single point of view agronomic and agricultural knowledge and to simulate expert performance, (ii) to demonstrate how agronomic decisions have corresponding translations into the ergonomic point of view of farmers, and (iii) to point out what really are farmers' decision processes in the planning of field work operations carrying out.

3.2. Elementary procedures from task decomposition

As illustrated in fig. 2, the format of a procedure is composed of three parts (the operating conditions, the procedure itself and the exit conditions). The hierarchy of goal decomposition into subgoals from the level of farmers' tasks to primitive actions is a directed graph where the vertices are elementary procedures labels and the edges are subgoal decompositions with {AND} operators found while computing a given procedure and with {OR} operators that select alternative subgoals. A procedure being one of the goal decomposition into subgoals, this allows to represent alternative sets of subgoals. As primitive agricultural actions satisfy biophysical constraints, the graph integrates agronomist knowledge at intermediary levels.

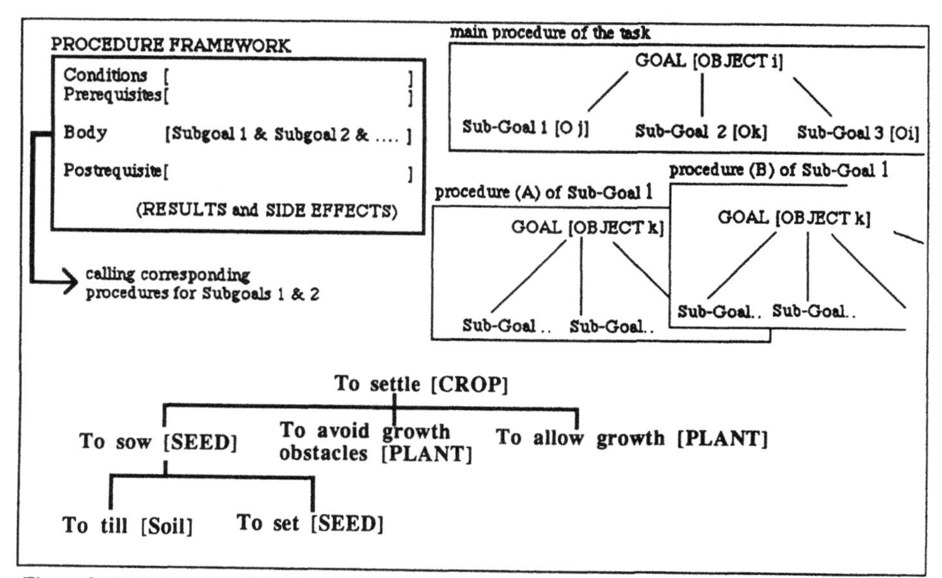

Figure 2- Task decomposition framework of the main goal of the task into subgoals connected by temporal constraints and framework of a procedure : each subgoal has at least a corresponding procedure that is similarly described into subgoals. In bold, a short example is given for settle [crop]. Up to that point, procedural semantic of objects is unended because objects are disseminated through the decomposition tree with redundance.

Thus, as shown in fig.2, the same general procedure for "settle [crop]" is involved whether the crop is [spring wheat] or [sugar beet]. As a matter of fact, the distinction between the two different procedures appears when more specific procedures are evoked in order to carry out these two kinds of tasks. Soil types, seed varieties and/or climatic events specifications contribute to distinguish between various procedures.

3.3. Construction of a semantic network with objects and associated procedures

A PROCOPE microworld description is a semantic network of task objects, in regard to the procedures applied to them. This network is structured by the identity relation ("X" and "Y" belong to the same class if all that can be done on "X" can be done on "Y", the reverse being true as well) and by the inclusion relation (" X" is a kind of "Y", if it is possible to do on "X" what is possible to do on "Y"). Classes factorize common object structures through associated procedures with simple and multiple inheritance principles. Given the objects of a domain with associated procedures, PROCOPE produces a hierarchy of abstract and concrete classes represented as a directed graph [G: $< N, P, R>$] which is defined to consist of a set of nodes "N" (representing classes of objects), a set of procedures "P" and a set of inclusion relations "R" between nodes. Node n_i corresponds to a subclass label assignment of node n_j if and only if $p(n_i)$, procedures of node n_i, are included in the set of $p(n_j)$, procedures of node n_j. In such a semantic network, procedures are domain-dependant and object-specific.

As it appears from the data we collected on the field (1988-1990) during sugar beet sowing period, representing farmers' declarative and procedural knowledge, using PROCOPE allows to point out the fact that farmers do not only categorize the objects of their work (agricultural machinery, soils, plants) on the basis of the biophysical characteristics of the processus (fig.3). They also categorize them in reference to the work these objects are associated to. For instance, while an agronomist distinguishes heavy and loamy soils owing to their behaviour under climatic actions, farmers will also give different procedures to till the plots presenting such soils. But while the former will advise to use different techniques, the later might perform the same actions on each of these plots as they belong to the same farm block, meaning that he treats them as elements of the same given class "plots located in the block far from the farm" to which he associates a specific procedure.

3.4. Semantic network knowledge representation consequences on cognitive mechanisms

Some cognitive mechanisms emerge as consequences of processing the PROCOPE knowledge based system. This involves procedural knowledge access (to select a procedure), categorization (to select an object), classification (to add a class or review the structure of the semantic network) and planning (to roughly organize the different procedures in order to anticipate the carrying out of the task).

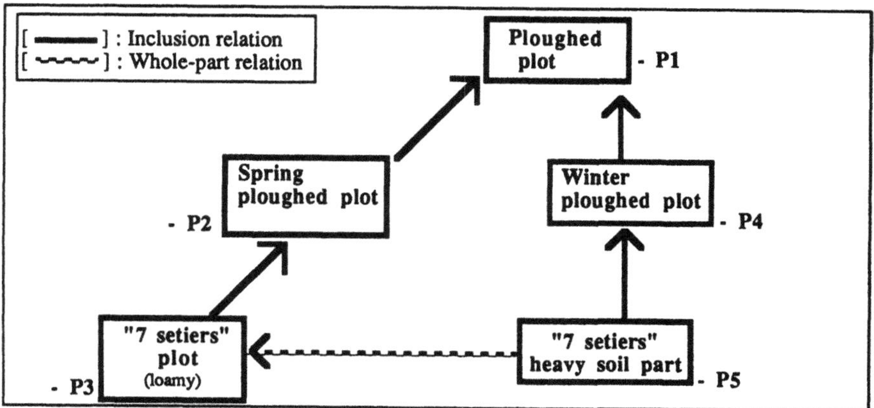

Figure 3 - Procope semantic network representation with inclusion and whole-part relations. At this point, procedural semantic of objects merges in regard of applied procedures. For instance, in bold is given a part of the [plot] network where :
P1 - {To sow : to till [object] & to sow [plant : sugar beet]}, {To till: to use [equipment : tillage machine] & to work [object]}, {To work : to make [operation : tillage operation]}.
P2 - {To till : to use [equipment : harrows & croskill roller] & to work [object]}, {To work : to make [operation : 1st tillage] using [equipment : harrow] & (climatic conditions before starting) to make [operation : 2nd tillage] using [equipment : harrow/croskill-roller combination]}.
P3 - {To work : ...(...) to make [operation : crossing 2nd tillage]}.
P4 - {To till : to use [equipment : drag & harrow & clod-crusher] & to work [object]}. {To work : to make[operation : 1st tillage] using [equipment : drag] & (climatic conditions before starting) to make [operation : 2nd tillage] using [equipment : harrow/clod-crusher combination]}.
P5 - {To repare : to till after sowing[object]}, {to till after sowing : (climatic conditions before starting) & to use [equipment : croskill roller].
Procedures are linked to objects. For intance, searching for "to sow [plant : sugar beet]" that is : to use [equipment : sowing machine] while adjusting [operation : seed depth & distance between

3.4.1. Procedure access

To select a procedure, the referent and the goal to be pursued have to be known. Then a bottom-up process includes selecting among candidate procedures that satisfy the goal, using evaluation of prerequisites and side effects, in two-steps (i) from the given class of the object (ii) to the highest level of the graph: if the procedure linked to a class of object cannot be applied (due to the cost or to the present year environment) then the procedure of the superordinate class is tested and so on. The semantic network operates as a search tree in case of multiple inheritance.

3.4.2. Categorization and classification

To select the appropriate object, the semantic network operates as a decision tree in a top-down process with a categorization strategy guided by the presence of properties that classify objects. If the target object is not recognized as an element of a class of the semantic network, a procedure is then chosen by comparing the properties of a possible source object and those of the target one. If necessary, some adjustments are made to the procedure which is associated to the source object. The latter might be the object whose properties are present in the target object (superordinate class) or the one whose properties least differ from those of the target object.

Given a default set of candidate objects, meaning that no classes are relevant either to correctly categorize an object or to apply a new procedure, a classification process will occur in order to add a new class.

3.4.3. Planning

Task decomposition that produces successive levels of abstraction where procedures can be seen as macro-operators, allows hierarchical planning. A given task can be solved at successive levels: each one being under the control of the higher level of abstraction. PROCOPE first maps the task into the highest level, then maps it into successively more detailed levels until the performing of primitive actions. At any level, if prerequisites cannot be satisfied or if side effects give contradictory results, PROCOPE backtracks to select another procedure among other candidates through the semantic network.

3.5. Semantic network knowledge representation consequences on crop management

Investigating decision making for sugar beet settling, our hypothesis is that semantic networks allow farmers to decrease uncertainty relevant to farm management as well as to reduce decision making, because most of the decisions that have to be taken are already connected to the network nodes. The capability to support uncertainty is critically important in knowledge-based representation. We have seen that the functional diagram of the sugar beet sowing task allows to point out some characteristics such as (i) interaction of dynamic biophysical processes within a plot, (ii) asynchronous evolution of these biophysical processes between the plots, (iii) uncertainty of the biophysical processes evolution due to climatic variability and uncertainty of climatic event effects on : 1) these processes and 2) the consequences of the actions on yield elaboration and (iiii) the existence of long causal chains between the variables on which actions can be carried out and variables crucial to plant population growth and development. These characteristics provide an ergonomic uncertainty at many levels: in processing the current state, on the effects of operators applied to the current state, on the moment that the target state should be achieved (reducing or increasing criteria of the goal expression), and on causes evaluation of obtained results.

3.5.1. Uncertainty in processing the current state

In PROCOPE expert semantic network, a crucial variable is a variable justifying a procedure. Most of the objects are already defined by such variables because these typify object classes. Furthermore, a new captured information that changes the value of an object crucial variable (due to climatic effect, for instance) is indicative of changes for the subgraph classes structured by this variable. This inference rule avoids adding unnecessary measures by propagation of variable value changes into the given subgraph.

3.5.2. Uncertainty on the effects of operators applied to the current state

Due to the preexistent organization of states as classes in the semantic network and/or as preconditions of operators, the process of selecting a procedure mainly consists of selecting the object class. Classification reduces diversity and heterogeneity of states by compromising in order to define prototypes that summarizes instances of classes. The cognitive point of view is that human operator's semantic network has to be limited in class numbers.

3.5.3. Uncertainty on the moment that the target state should be achieved and uncertainty on causes evaluation of obtained results

One major line of agronomic observations is that farm management requires very sophisticated reasoning about [plant*soil*weather*operators] interactions. Farmers' expectancies are restricted unless they handle a more sophisticated description of these interactions, particularly with additional reflections on climatic events frequency and on causal phenomena. In the PROCOPE expert semantic network, reasoning about complex interactions is replaced by additional knowledge including weather classification. Similarly, uncertainty remains about the cause evaluation of obtained results because, in PROCOPE learning principles, assimilation is not primarily based on reflections about possible causes, but on new classifications guided by obtained results (objects are structured in classes by procedures coupled with their main result and their side-effects). Lately, causal knowledge merges from differences in procedures application and is expressed in the semantic network.

4. DISCUSSION

Crop system management can be seen as a typical situation of some professional activities. These are characterized by the deferment of the given task performance evaluation in such a way that improving the efficiency through expertise acquiring is not easy (frequent lack of retroaction).Thus the evaluation and the improvement of performance are quite difficult (such as: how to interpret the effects of actions), and might be erroneous (such as: to make a mistake about the cause) and/or hazardous (such as: what needs to be questionned). In such professional situations, capability is not only founded on large declarative knowledge but also on procedural one; it also depends on the availibility of a rough representation of work organization.

Thus primary design objectives of knowledge representation should be to enable (i) the learning of more relevant procedures (in regard to the objects which they are applied to) (ii) the evaluation of performed procedures and (iii) the improvement of planning activities. With PROCOPE formalism, description for carrying out of procedures allows: to track learning with assimilation of new information that might change the network structure, to compare and evaluate graphs or subgraphs isomorphism (for instance between expert and novice) and to evaluate what has to be learned in order to apply correct procedures to objects, given that procedure applicability is justified by the properties of objects. As well, PROCOPE formalism allows: to simulate the carrying out of tasks issued of described procedures and their interactions with the external world and to compare the results of automatic performance issued from various semantic networks. Thus, knowledge representation underlying PROCOPE seems suitable to be used in the conception of "on-line aid" for effective planning and technical decision support.

Model consistency, for decision making and planning in crop management intercourse, has still to be evaluated through the ability (i) to describe personal agricultural knowledge and to simulate individual performances of the sowing-task (ii) and to provide both agronomic and ergonomic help in an interactive simulation. Nevertheless, we already started to design a procedural knowledge-based system where the agronomist's and the farmers' knowledge have been integrated in a common description which will enable us to provide technical and agricultural learning support. In addition, we conducted interviews with farmers during sugar beet settling on their plots. Thinking-aloud protocol experiments have also been conducted. The data analysis shows that farmers describe and apply procedures on objects specifically defined as instances of classes. Instances of classes coupled with associated procedures have been drawn from the interview records. We have been able to construct classifications of agricultural objects on the basis of procedures which can be applied to them. For instance, sugar beet is like rape seed, in the way of sowing; "cranettes" are like limestone heavy soils, in the way of workability during tillage and so on. This allows to construct and to implement semantic networks of some of the subjects (eight farmers and an expert with both agronomic and agricultural knowledge).

5. Conclusion

Complex tasks are under the scope of cognitive psychologists (Hoc, 1989; Samurçay and Rogalski, 1988; Tijus, 1988). The work presented here is one of the first attempts to construct a procedural knowledge-based system of a complex agricultural task that allows to simulate the farmers' activity and to produce aid to novices that integrate agronomic and ergonomic knowledge. The principles underlying PROCOPE are issued from a theoretical explanation of learning by doing (Richard, 1990; George 1983; Anzaï & Simon, 1979) with the competence of self-modification (Anzaï, 1984). However, if our model is a clear proof of the interest of a complex task semantic network approach, it actually does not model the dynamic character of the [plant population*soil*weather*techniques] system. While farmers operate in a dynamic environment, our model assumes that dynamic processes are treated by the operators as discrete processes. In planning, and oftentimes in order to act, farmers have to decide "when" a variable will take a given value. He must then anticipate dynamic evolution and eventually has to take into account the speed of variation. We still don't know how farmers elaborate such a planning and if they do. We need more research to construct adaptive models to simulate tasks conducted in dynamic environment. Similarly, if we actually model reactive planning, PROCOPE does not fit the case where no procedure is available which means that farmers have to construct a new procedure. By working on farmers' constrainsts at many levels of the task decomposition and given a hierarchical list of constrainsts, we may open up new research area for very complex task conducting including both reactive planning and pure problem solving.

REFERENCES

ANZAI, Y. (1984). Cognitive control of real-time event-driven systems. Cognitive Science (8), 221-254.

ANZAI, Y. & SIMON, H.A. (1979). The theory of learning by doing. Psychological Review, 80, 124-140.

CAPILLON, A., SEBILLOTTE, M. & THIERRY, J. (1975). Evolution des exploitations agricoles d'une petite région. Elaboration d'une méthode d'étude. Chaire d'agronomie I.N.A. P.G., C.N.A.S.E.A.

CERF, M., POITRENAUD, S., RICHARD, J.F., SEBILLOTTE, M. & TIJUS, C.A. (1990). Comment modéliser la conduite des cultures. Rapport intermédiaire de recherche M.R.T.

FREDERIKSEN, C. H. (1988). The representation of procedures: Acquisition and Application of Procedural Knowledge. Actes du Colloque International "Informatique Cognitive des Organisations" ICO 89. Québec 12-15 juin 1989.

GEORGE, C. (1983). Apprendre par l'action, Paris, P.U.F.

GOLDBERG A., ROBSON D. (1983). Smalltalk-80, the language and its implementation, Reading Mass. Addison-Wesley.

HOC, J.M. (1989). Strategies in controlling a continuous process with long response latencies: needs for computer support to diagnosis. International Journal of Man-Machine Studies, N° 30, 47-67.

KIERAS, D.E. & POLSON, P. (1985) An approach to the formal analysis of user complexity. International Journal of Man-Machine Studies, 22 (4) 365-394

LECLUSE, C. & RICHARD, P. (1988). Modeling inheritance and genericity in Object Oriented Databases. Rapport technique Altaïr 18-88.

MULLER, P. (1984). Le technocrate et le paysan. Les éditions ouvrières (eds).

PAYNE, S.J. & GREEN, T.R.G (1986). Task action grammars: a model of the mental representation of task languages. Human Computer Interaction, 2, 93-133.

POITRENAUD, S., RICHARD, J.F., TIJUS, C.A., TAGREJ, M. & PICHANCOURT, I. (1990). La description des procédures: leur décomposition hiérarchique et leur rôle dans la catégorisation des objets. 4ème colloque de l'ARC "Progrès de la recherche cognitive", 28-30 mars 1990.

REISNER, P. (1981). Formal grammar and human factors design of an interactive graphics system. IEEE Trans. Software Engineering, 7 (2), 229-240.

RICHARD, J.F. (1983). Logique du fonctionnement et logique de l'utilisation. Rapport de recherche INRIA N° 202.

RICHARD, J.F. (1990). Les activités mentales: comprendre, raisonner, trouver des solutions. Paris, Colin (ed).

SAMURCAY, R. & ROGALSKY, J. (1988). Designing systems for training and decision aids: cognitive task analysis as a prerequisite. In "Human-Computer Interaction" - Interact' 87, Bullinger, H.J. & Shackel, B. (eds). Elsevier Science Publishers, B.V., North-Holland, 113-152.

SEBILLOTTE, M. (1978). La collecte des références et les progrès de la connaissance agronomique. In "Exigences nouvelles pour l'agriculture: les systèmes de culture pourront-ils s'adapter", sous la direction de Boiffin, J., Huet, P. & Sebillotte, M., Dec. 78, 466-493. INAPG-ADEPRINA.

SEBILLOTTE, M. (1979). Analyse du fonctionnement des exploitations agricoles, trajectoires et typologie. In "Eléments pour une problématique de recherche sur les systèmes agraires et le développement". Assemblée constitutive du département S.A.D., Toulouse, 20-30.

SEBILLOTTE, M. & SOLER, L.G. (1988). Le concept de modèle général et la compréhension du comportement de l'agriculteur. C.R. Ac. Agri. Fr., 74, 4, 59-70.

TIJUS, C.A. (1988). Cognitive processes in artistic creation: toward the realization of a creative machine. Leonardo, 21, 167-172.

TAUBER, M. (1988). On mental models and the user interface in human-computer interaction in Working with Computers: Theory versus Outcome. Van Der Veer G.C., Green T.R.G., Hoc J.M. & Murray D.M. (eds). Londres Academic Press.

10. NEURAL NETWORKS

A NEURAL NETWORK EXPERT SYSTEM WITH CONFIDENCE MEASUREMENTS

Stephen I. Gallant

ICANN
51 Fenno St.
Cambridge, MA 02138
USA

Yoichi Hayashi

Department of Computer & Information Science
Ibaraki University
Nakanarusawa, Hitachi-shi
Ibaraki 316
Japan

Abstract

We show two ways to attach confidence measurements to MACIE, a connectionist expert system model. One technique involves multiple generation of neural network knowledge bases and parallel execution of expert system inference engines. Required modifications for forward and backward chaining and explanations by If-Then rules with confidence measurements are given.

In the second approach we show that inferences made by the MACIE connectionist expert system with *partial information* have the same confidence as inferences made with *full information* under certain standard probability models, including distribution-free Valiant-style learning.

Keywords: confidence, MACIE, neural network, expert system, Valiant model

1 Introduction

We have previously described how to generate a neural network knowledge base from training examples and how a neural network inference engine, MACIE, can use this knowledge base for making inferences from partial information, for backward chaining to obtain additional information, and for explaining inferences by If-Then rules [2]. One drawback of this approach is that all decisions are made crisply, without giving the user any measure of how much confidence to attach to such inferences.

In this paper we show two ways to extend MACIE to give confidence measurements for all inferences and explanations. In contrast with some conventional expert systems, inferences and explanations for MACIE will have clearly defined interpretations using standard probability theory. It is hoped that adding such confidence information will be helpful for a human who is required to make a final decision based (in part) upon MACIE's recommendations.

This paper also extends work reported in [1]. There it was shown how to produce confidences for inferences by using a set of training examples that explicitly contained confidence information. Here we do not require that training examples contain such confidence information. See also previous work on constructing a fuzzy neural expert system for medical diagnosis [8] and on extracting fuzzy If-Then rules from networks using the notion of 'linguistic relative importance' [9,10,11].

2 The Parallel Execution Method

Generating a neural network from training examples is a probabilistic process [1]. For most approaches the choice of a different random number seed can produce a different network.

How different?

For some sets of inputs ("clear cases") almost all networks will make the same inferences. For other sets of inputs ("doubtful cases") there will be some disagreement among networks as to whether an inference is justified (based upon the current partial information) or disagreement as to what that inference should be.

Our basic approach is to take advantage of this disparity by generating a *collection* of neural network models (from the same set of training examples). We then run MACIE on all of the networks in parallel. This gives additional information about any output variable at any stage in the consultation, namely

- the number of models that classify this output as true (N1)
- the number of models that classify this output as false (N2)
- the number of models that require more information to reach a conclusion (N3)

For example if $N1/(N1 + N2 + N3) = .85$ we can now give the user an inference such as:

"Administer drug Quackomycin is true (C=.85)"

The confidence is an estimate as to the fraction of neural network models (generated from a given set of training examples using a particular learning algorithm) that will produce the inference in question. (It may be helpful to inform the user only of inferences with confidence greater than some threshold, say .7.)

We must also specify how backward chaining is performed using such a parallel approach, i.e. how to determine the next unknown variable that the system asks the user in order to obtain additional information. Here a simple solution is to perform backward chaining using each of the individual networks, and then to ask for the variable most requested by the individual systems (with ties broken arbitrarily).

The last part of MACIE that needs modification is the production of `If-Then` rules as justification for its inferences. To do this the user must first specify a 'confidence threshold' that gives the minimum confidence for an acceptable rule. This value must be no greater than the confidence for the variable currently being explained. For example we could ask for a rule justifying our previous inference of "Administer Quackomycin" that has a confidence of at least .65.

To produce such a rule we can use a greedy algorithm, resetting known input variables to values of 'unknown' until no additional input variables can be eliminated without dropping the confidence level below the given threshold. (Eliminating clauses makes the rule more generally applicable.) We then form an `If-Then` rule from the remaining variables as in [2]. Such a rule gives an explanation in the following sense:

1. Whenever its conditions are fulfilled then its conclusion will be inferred with at least the confidence specified.

2. This rule's conditions apply to the inference in question.

2.1 Example

To demonstrate this approach we generated 10 knowledge bases (using different random number seeds) for a noisy fault detection problem described in [5,6]. We then simulated the parallel execution method using multiple MACIE runs. A value of .7 was used for the cutoff for producing inferences and .4 for rule explanations. Figure 1 gives a short sample run of the system, and figure 2 shows explanations.

```
MACIE/Rulex Version 6.2
Copyright (c) 1990 S. I. Gallant, US pat. 4,730,259

Enter initial values for input, intermediate or output vars..

Format:  Variable number, value, ...

Numbers and names of variables:
         LIKELIHOOD
        1:  Hydrogen Gas reading #1 low
        2:  Hydrogen Gas reading #2 low
        3:  Oxygen Gas reading low
        4:  No power to lemon squeezer
        5:  No juice flowing
        6:  No power to refrigeration unit
        7:  Temperature not low enough
        8:  Taste bad (ignoring temperature)
        9:  Problem with Hydrogen supply
       10:  Problem with Oxygen supply
       11:  Fuel cell malfunction:  not working
       12:  Fuel cell malfunction:  water clogged
       13:  Short in power line to lemon squeezer
       14:  Short in power line to refrigeration unit
       15:  Lemon squeezer malfunction
       16:  Refrigeration unit malfunction
       17:  System functions correctly

 1t  2t  8f  5t  { these variables are initialized }

 Has cooling stopped?
        {requested by 5 of 10 models}
 --> y)es, n)o, u)nknown, ?)explain, i)nformation on vars.

y


 Has power to the lemon squeezer been interrupted?
        {requested by 3 of 10 models}
 --> y)es, n)o, u)nknown, ?)explain, i)nformation on vars.

y

In 8 of 10 cases the following inference holds.
In 2 of 10 cases more information was required to reach
an inference.
In 0 of 10 cases a different inference was made.

CONCLUDE:  (9) Problem with Hydrogen supply TRUE.
```

Figure 1: Sample run of the parallel execution method for obtaining confidence measurements (edited).

2.2 Discussion

In essence this method uses a consensus approach where several expert systems run simultaneously and vote for inferences. This helps the user in two ways. First he or she has a measure of whether any inference is a "clear case" or a "doubtful case" as measured with respect to the neural network

```
Would you like to try a new case?
 --> y)es, n)o, ?)explain, i)nformation on vars.

i

          LIKELIHOOD
     T  1:  Hydrogen Gas reading #1 low
     T  2:  Hydrogen Gas reading #2 low
        3:  Oxygen Gas reading low
     T  4:  No power to lemon squeezer
     T  5:  No juice flowing
        6:  No power to refrigeration unit
     T  7:  Temperature not low enough
     F  8:  Taste bad (ignoring temperature)
     T  9:  Problem with Hydrogen supply
     F 10:  Problem with Oxygen supply
     F 11:  Fuel cell malfunction:  not working
     F 12:  Fuel cell malfunction:  water clogged
     F 13:  Short in power line to lemon squeezer
     F 14:  Short in power line to refrigeration unit
     F 15:  Lemon squeezer malfunction
     F 16:  Refrigeration unit malfunction
     F 17:  System functions correctly

Would you like to try a new case?
 --> y)es, n)o, ?)explain, i)nformation on vars.

?

       '#' for explanation of variable number # (E.g. '3')

9

In 4 of 10 cases the following rule holds.
In 6 of 10 cases more information was required to reach
an inference.
In 0 of 10 cases a different inference was made.

If Hydrogen Gas reading #1 low is TRUE
   and Hydrogen Gas reading #2 low is TRUE
   and No juice flowing is TRUE
   and Temperature not low enough is TRUE
   and Taste bad (ignoring temperature) is FALSE

Then Conclude:  Problem with Hydrogen supply is TRUE.
```

Figure 2: Explanation of inference (edited).

---

generation process. Second, the risk that using only one neural network will produce an incorrect answer due to bad luck in the generation process is reduced.

We can also use this approach for producing a set or rules with confidence measures from a set of training examples. We generate a collection of networks and use a search procedure that employs the mechanism for giving `If-Then` rule justifications discussed above. The addition of confidence measures to rules extends previous work with the RULEX program described in [3].

3 Probability Bounds for Inferences

There is a second and perhaps more useful way to get confidence information. The key observation is that MACIE's inferencing method for partial information is conservative, in that an inference based upon partial information is never violated by additional information (more variables) becoming known. More specifically, MACIE makes an inference only when the currently known partial information is sufficient to guarantee the ultimate positivity or negativity of a cell's weighted sum, and hence determine the output (± 1) for that cell.

Now suppose we have tested a network with data drawn from some population and have determined probability estimates of how well the network performs with *full* samples, i.e. samples with all inputs specified. Say we have 99% confidence that the network is correct at least 85% of the time. Then this estimate will also apply to inferences based upon *partial* information (where some inputs are unknown), provided the partial information comes from full samples drawn from the same population used in testing. Thus if we make an inference having seen only 5 out of the 25 inputs, we could still have 99% confidence that the inference would be correct with probability $\geq .85$. The reason is that this is our confidence in the full 25-input sample and MACIE is guaranteed to make the same inference with the full sample.

Similarly any statistical test of the performance of the network using full samples applies to partial inferences. In particular we have shown elsewhere how to derive Valiant-style distribution-free generalization bounds for various connectionist models [7,4]. These same bounds also apply to inferences.

Thus if we train a single-cell model having 10 inputs on 3626 training examples and the resulting network correctly classifies 95% of the *training* data, then we have at least 90% confidence that the network will correctly classify at least 90% of new test data drawn from the same population ([7], Table 4). The data may contain noise, provided that training examples are independently noisy.

4 Concluding Remarks

We have shown two methods for adding confidence measurements to connectionist expert systems. The first method is based upon parallel execution of neural network models and bears resemblance to Hisdal's model [12,13] of how humans come up with similar confidence estimates. The second method provides standard probabilistic estimates of correctness for inferences based upon partial information.

References

[1] Frydenberg, M. & Gallant, S. I. Fuzziness and Expert System Generation. International Conference on Information Processing and Management of Uncertainty in Knowledge-Based Systems, Paris, France, June 30 - July 4, 1986. Extended paper reprinted in *Uncertainty in Knowledge-Based Systems*, B. Bouchon & R. Yager Editors, Springer-Verlag: Berlin 1987.

[2] Gallant, S. I. Connectionist Expert Systems. *Communications of the ACM*, Vol. 31, Number 2, Feb. 1988, 152-169. (Japanese translation in *Neurocomputer*, published by Nikkei Artificial Intelligence, pg. 114-136, 1988.)

[3] Gallant, S. I. Example-Based Knowledge Engineering with Connectionist Expert Systems. Proc. IEEE MIDCON, Dallas, Texas, August 30-September 1, 1988, 32-37.

[4] Gallant, S. I. *A Connectionist Learning Algorithm With Provable Generalization and Scaling Bounds*. Neural Networks, Vol.3, pp. 191-201, 1990.

[5] Gallant, S. I. Automated Generation of Expert Systems for Problems Involving Noise and Redundancy. AAAI Workshop on Uncertainty in Artificial Intelligence, Seattle, Washington, July 10-12, 1987. Pg. 212-221.

[6] Gallant, S. I. Bayesian Assessment of a Connectionist Model For Fault Detection. AAAI Fourth Workshop on Uncertainty in Artificial Intelligence, St. Paul, MN, August 19-21, 1988. Pg. 127-135.

[7] Gallant, S. I. Perceptron-Based Learning Algorithms. *IEEE Transactions on Neural Networks*, Vol. 1, No. 2, June 1990, 179-192.

[8] Hayashi, Y. A Neural Expert System with Automated Extraction of Fuzzy If-Then Rules and Its Application to Medical Diagnosis. In Touretzky, D.S. and Lippman, R. (eds.) *Advances in Neural Information Processing Systems*, Vol. 3, San Mateo, CA: Morgan Kaufmann, 1991 (in press).

[9] Hayashi, Y. & Nakai, M. Reasoning Methods Using a Fuzzy Production Rule with Linguistic Relative Importance in an Antecedent. (in Japanese) Trans. IEE of Japan, Vol. 109-C, No. 9, Sept. 1989, 661-668.

[10] Hayashi, Y. & Nakai, M. Automated Extraction of Fuzzy IF-THEN Rules Using Neural Networks. (in Japanese) Trans. IEE of Japan, Vol. 110-C, No. 3, March 1990, 198-206.

[11] Hayashi, Y. & Imura, A. Fuzzy Neural Expert System with Automated Extraction of Fuzzy If-Then Rules from a Trained Neural Network. Proceedings of the First Int. Symposium on Uncertainty Modeling and Analysis, pp. 489-494, Maryland, Dec. 3-5, 1990.

[12] Hisdal, E. Are Grades of Membership Probabilities? First World Congress on Fuzzy Sets and Systems (IFSA '85), Palma de Mallorca, July 1-6, 1985.

[13] Hisdal, E. Infinite-Valued Logic Based on Two-Valued Logic and Probability. International Journal of Man-Machine Studies. Part 1.1: Vol. 25 (1986) 89-111, Part 1.2: Vol. 25 (1986) 113-138, Part 1.3-Part 1.6: to appear.

DEALING WITH UNCERTAINTY IN A DISTRIBUTED EXPERT SYSTEM ARCHITECTURE

Luca Console*, Claudio Borlo**, Alberto Casale***, Pietro Torasso*

Dipartimento di Informatica - Universita' di Torino
Corso Svizzera 185 -10149 Torino (Italy)

ABSTRACT. In this paper we describe the main characteristic of EXODUS: a distributed diagnostic expert system architecture. Knowledge in EXODUS is represented by means of a connectionist-like network and no centralized inference engine has been defined. Control is distributed among the nodes and evidence combination and propagation play a fundamental role in the reasoning process. Such an architecture is proposed as a mechanism for reaching very fastly a set of hypothesis accounting for the data characterizing a diagnostic problem (mimicing the ability of human experts who can provide initial solutions to a problem very easily). In the paper we describe in detail the architecture of the system and the mechanisms we introduced for dealing with uncertainty. In the final part of the paper we suggest that such a form of associational reasoning should be integrated with some form of deep reasoning (able to provide detailed explanations and to solve complex cases).

1. INTRODUCTION

In many types of problem solving, human experts exhibit an astonishing ability to provide (at least) a (partial) solution to a particular problem without any apparent cognitive effort and by using just a very limited amount of relevant data extracted from a large variety of data possibly related with the particular problem under examination. In some cases these experts have significant difficulties in articulating the reasoning process they have followed to reach the conclusion. At some extent, it seems that problems are solved in an almost automatic way and without a real investigation of all the possible alternative solutions that in principle could have been applied. Just in few cases the human expert has to switch to another form of reasoning in which different alternatives have to be investigated in depth in order to interpret data in a reasonable way[1].

The ability of human expert in solving cases stresses the need of investigating not only aspects related to "Competence" (that is the amount of knowledge of the expert) but also aspects related to "Performance" (how this knowledge is used and what kind of "compilation" has to be applied to large bodies of knowledge in order to make them easily usable). In fact, when one compares the

---

* Current affiliation: Dipartimento di Matematica e Informatica, Universita' di Udine.

** Current affiliation: Andersen Consulting - Torino.

*** Current affiliation: Cap Sesa Innovation, Paris, France.

The research described in this paper has been partially supported by CNR and MPI. The authors are grateful to A.F. Rocha (Univ. Campinas - Brasil) for many helpful discussions and suggestions.

[1] Apart from the case of diagnostic systems (which will be discussed in the following), similar results have been obtained in analyzing natural language: apart from a few kind of sentences (the so-called "garden-paths") humans are able to understand sentences in real time with no apparent efforts and without recognizing the huge amounts of potential alternative interpretations. This capability has suggested some researchers to develop new approaches to syntactic analysis so that a NL interpreter can parse most sentences in a deterministic way [9, 10].

reasoning process of real human expert with the one of non experts in the field, it is apparent that non experts evaluate a larger number of alternatives and to use a greater amount of data in order to reach a conclusion[2].

In the field of diagnostic expert system this problem has been investigated over the last years and now many researchers agree in recognizing the opportunity of adopting expert system architectures where there is room for both "heuristic" and "deep" knowledge. Heuristic knowledge is intended to provide the system with the capability of jumping from data to conclusions in a few steps and exploring only limited parts of the knowledge base, while "deep" knowledge is intended to provide the system with a behavioral or causal model of the domain (for a discussion of the role of different types of knowledge in expert systems see [16, 17]).

However the solutions currently adopted for representing heuristic knowledge are only partially adequate since in most cases heuristic knowledge is represented by means of production rules or prototypical descriptions based on frame-like structures. These formalisms are quite well understood and a large variety of inference engines working on them is easily available. While these formalisms are able to model heuristic knowledge, they have some difficulties in capturing all the characteristics of heuristic reasoning and, in particular, the capability of easily jumping to conclusions without requesting additional data or exploring too many alternatives.

A reason for that seems to be the existence of a centralized control which is in charge of deciding what part of knowledge has to be used at each cycle (this is apparent in production system where the inference engine has to solve conflicts among competing rules). This seems to prevent the possibility of immediately reaching a conclusion even if the data sufficient for supporting the conclusion itself are available.

In frame systems the control is potentially more distributed, but usually frame-systems are more adequate to produce expectations on data given a particular hypothesis than to reach a conclusion given a set of data. In order to provide a partial solution to these problems, most frame systems have been supplemented with a trigger mechanism which allows to evoke the relevant diagnostic hypotheses in a data-driven way.

These remarks can be applied to many systems which combine different forms of knowledge and, in particular, to CHECK: an architecture for diagnostic problem solving combining heuristic and causal knowledge that we developed over the last few years and in which heuristic knowledge is represented via a taxonomy of diagnostic hypotheses modeled by means of frame-like structures [17].

In recent years many researches have investigated the connectionist approach which advocates an architecture where a large number of simple highly connected processors work in parallel in order to carry on particular tasks (in most cases perceptual tasks) [4, 11].

Some characteristics of the connectionist approach seem to be suitable to model some properties of the reasoning process followed by the human expert in solving "simple" cases: in particular the ability of jumping to conclusions as soon as some significant datum is provided as well as the ability of changing mind (switching to another conclusion) when other data which better support another hypothesis are provided.

Another attractive characteristic of connectionist model regards the ability of adapting the models itself to the environment due to their learning capabilities. This seems to correspond to the need of adapting a-priori knowledge to the particular environment (many studies have shown that a decision support system exhibiting a high accurateness in a given situation may have poor performance when transferred to another location; see, for example, the discussion in [7] regarding the use of connectionist models in medical domains).

Connectionist models have been mainly used in the past to model perceptual tasks, while only recently some attempts have been made to move them to cognitive tasks in which reasoning plays a major role [1, 5, 12, 15, 18].

Shastri's work [15], for example, has shown how the knowledge in a semantic net can be represented with a connectionist network and how one can perform some inferences (such as inheritance and recognition) on this knowledge still adopting a connectionist approach. Moreover, some

---

[2] Some studies provide empirical evidence of this observation (see, for example, the discussion in [8]).

interesting works about connectionist expert systems have been recently proposed, see, e.g., Gallant [5] Peng & Reggia [12] and Ahuja, Soh & Schwartz[3] [1].

These recent works, however, disregarded some typical features of the connectionist approach: mainly, the adoption of nodes with a very limited computational capability, the fact that all the knowledge is represented on the weights on the arcs, the absence of a-priori knowledge, the strong impact of learning algorithms. The main efforts consisted in representing knowledge not only on the arcs but also in the nodes and in defining some a-priori shape of the network elicited from the interaction with a human expert.

The aim of the paper is to describe the main characteristics of a distributed system in which knowledge is represented in an associative way and some basic properties of a connectionist approach are maintained. In particular, we present EXODUS (EXpert Object oriented DistribUted System), a distributed (and potentially parallel) expert system architecture suitable for heuristic classification and diagnostic problem solving. Knowledge is represented in EXODUS by means of a connectionist-like network and no centralized inference engine has been defined. Control is distributed among the nodes and the inference process is carried on by means of message propagation inside the network.

Approximate reasoning and evidence combination (or, more generally, the mechanisms for dealing with uncertainty) play a fundamental role in the reasoning process (in some sense they are the "core" of the reasoning process). As we shall see in the following, the behavior of a node depends only on its evidence value and thus all the inferential process can be defined as a form of evidence combination and propagation.

EXODUS presents some analogies with classical connectionist networks: in particular, it uses a net of simple units connected by means of weighed arcs, the nodes have a potential and the propagation mechanisms are based on threshold functions. However, as well as Shastri's and Gallant's works, EXODUS exhibits some significant differences with respect to traditional connectionist models:

- The nodes are more complex from a computational point of view.
- Since the structure of the net and the weights on the arcs may have a meaning for the human expert, some form of a-priori knowledge (elicited from the human expert) can be put into the network.
- There is not an unique type of arc.
- Approximate reasoning mechanisms based on possibility theory [19] have been extended to include co-operation and competition mechanisms; positive and negative evidence measures are taken into account separately in a bidimensional evidence space.

2. *KNOWLEDGE REPRESENTATION*

Two different types of knowledge are represented in EXODUS: knowledge about data and knowledge about diseases.

Knowledge about data is represented in a quite traditional way. In particular, data are organized into a taxonomy: high level nodes in the taxonomy represent classes of findings (such as, for example, in a medical application the classes "clinical data", "physical examinations", ...) while the leaves in the taxonomy represent specific findings (such as, for example, "fever", "liver pain", ...).

A set of attributes can be associated with each class; such attributes provide a prototypical description of the class itself. In our language, we imposed that the set of admissible values of each attribute must be a finite set of linguistic values. However, it is possible to define such linguistic values by means of fuzzy sets. Moreover, one can also specify that two linguistic values are in antinomy (i.e. mutually exclusive). Let us consider, for example, the finding "fever"; such a finding can be characterized using the two attributes "intensity (which can assume the linguistic values "absent", "low", "medium", "high" and "very_high") and "duration" (which can assume the values "limited"

[3] The last two works, in particular, reinterpreted in terms of connectionist models Reggia's "covering set" theory [13], which was originally proposed as a "logic-based" approach to diagnosis.

and "protracted"). The linguistic values for the attribute "intensity" can be defined as fuzzy sets over the domain defined by the real numbers in the interval [35,44]. Moreover one can specify that the value "absent" is in antinomy with all the other values.

Attributes are inherited through the class-subclass relationships defining the hierarchy of data. This means that each leaf in the hierarchy is characterized by a set of attributes: different instances of a datum can be created by associating specific values to (some of) the attributes of the datum.

Let us consider again the finding "fever"; we can define the instances "fever low and protracted", "fever high", "fever protracted" (notice that we can define also partial instances in which not all the attribute assume a specific value).

Knowledge about diseases is represented by means of an inferential network (relating findings to diseases). Such a network can be characterized as a multi-layered network in which three different types of nodes can be recognized:

- "*finding*" nodes, constituting the lowest layer of the network;
- "intermediate" nodes, constituting the intermediate layers of the network (many intermediate layers can be present in our inferential network);
- "*hypothesis*" nodes constituting the highest layer of the network.

Each "finding" node is in a one-to-one correspondence with one instance of a datum in the hierarchy of data (so that "finding" nodes represent the link between the two knowledge bases of EXODUS).

Each "hypothesis" node represents a diagnostic hypothesis (fault) in the diagnostic system.

"intermediate" nodes have no specific meaning (even if in some cases they may correspond to syndromic states).

Every node in the network has a "*potential*" given by a point in the space that defines the evidence of the node. In particular, we chose to represent the evidence of a node using a four-valued logic (characterized by the four values "Unknown", "True", "False", "Contradiction"). The evidence has two components: the positive evidence (e+) and the negative evidence (e-); both of them are real values restricted to the interval [0,1]. At any time, therefore, the evidence of a node is a point in the square $[0,1]\times[0,1]$ (see [14] for a discussion on bidimensional evidence values). In figure 1 we have reported such an evidence square and the meaning of the extreme points in the square.

The evidence's square is subdivided into regions by linear thresholds, see figure 2 (as we shall see, a node has a different behavior during the inferential process according to the region into which its evidence falls)[4]. In particular, we recognize:

- two activation regions, the "True" and "False" regions. The fact that the evidence of a node "N" falls into the "True" region corresponds to the truth of the concept represented by the node (i.e. the presence of a finding or of a hypothesis). The fact that the evidence of a node "N" falls into the "False" region corresponds to the falsity of the concept represented by the node (i.e. the absence of a finding or the exclusion of a hypothesis).

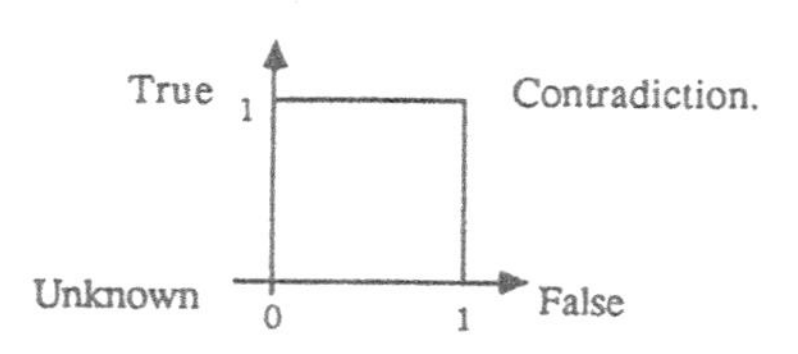

Figure 1 - The evidence square

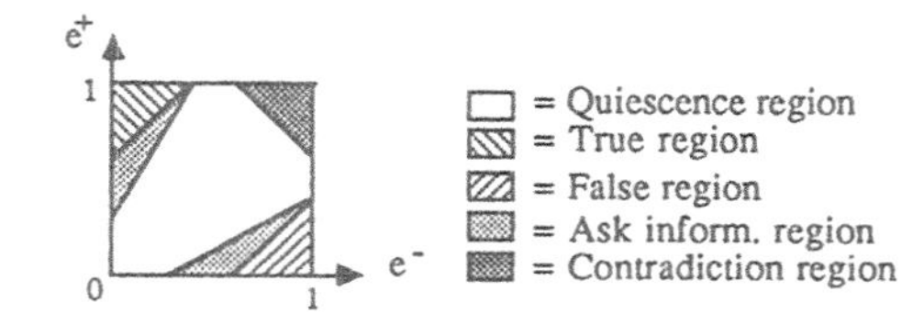

Figure 2 - Regions in the evidence square

[4] The angular coefficient of the threshold may vary depending on the type of the node. In particular, the activation region of an intermediate node is smaller than the activation region of a finding node (moreover the dimension decreases moving from lower to higher layers). This is in accordance with the fact that as far as we move towards diagnoses, we must have a greater support in order to activate a node.

- two "ask-information" regions;
- a large "quiescence" region.

We shall discuss more precisely in the following sections the meaning of the different regions.

The nodes in the network can be connected by means of weighed arcs. Each arc connects a finding or intermediate node to an intermediate or hypothesis node (so that no arc can enter a finding node or exit from a hypothesis node). The weight associated with an arc connecting N_1 to N_2 is a real number in the interval [-1,1] and is interpreted as a strength of sufficiency (respectively to confirm N_2 if the weight is positive or to exclude N_2 if the weight is negative).

The arcs exiting from a node are subdivided into two classes: the "True" arcs and the "False" arcs which are associated respectively with the "True" and "False" regions of the evidence square of a node. Given an arc "a" connecting N_1 to N_2, the meaning of "a" is the following:

- if "a" is a "True" arc, then the truth of N_1 has an influence on the evidence of N_2 (more specifically, a positive influence if the weight of "a" is positive, a negative influence if the weight of "a" is negative).
- if "a" is a "False" arc, then the falsity of N_1 has an influence on the evidence of N_2 (more specifically, a positive influence if the weight of "a" is positive, a negative influence if the weight of "a" is negative).

It is interesting to notice that our inferential networks strongly resemble connectionist networks ("True" arcs correspond to excitatory links while "false" arcs correspond to inhibitory links), even if the nodes of our networks correspond to structured concepts rather than to unstructured one.

Since we are interested in performing multiple faults diagnosis (through a differential process), we have introduced two further types of arcs: competition and co-operation links between hypotheses.

Competition is a symmetric relationship between hypotheses. The fact that two hypotheses H and H' are in competition means that they tend to exclude each other, i.e. that they cannot be considered together as part of a multiple fault. As a consequence the confirmation of one of the hypothesis must exclude the other. A weight S (real number belonging to [0,1]) can be associated with a competition link to represent the strength of the competition.

Co-operation is a relation (not necessarily symmetric) between two hypotheses H and H'. The fact that a co-operation link connects H to H' (in such a case we say that H co-operate with H') means that each time H can be confirmed then also H' must be taken into account. A weight S (real number belonging to [0,1]) can be associated with a co-operation link to represent the strength of the co-operation. We can have H co-operating with H' at strength S and H' co-operating with H at strength $S' \neq S$. If $S >> S'$ (and S close to 1) then there is a cause-effect relation between H and H'.

As we shall see in the following sections, competition and co-operation links are very important to perform a differentiation between the hypotheses activated by the data characterizing a diagnostic problem and to get some suggestions about which hypotheses can be grouped together to form a multiple-fault solution to the diagnostic problem itself (see also the comments in section 5).

This concludes a brief review on the architecture of EXODUS. A graphical description of such an architecture has been reported in figure 3.

2. REASONING IN EXODUS

In this section we describe the inferential process in EXODUS networks. Basically, reasoning in EXODUS corresponds to the propagation of evidence values in the inferential network. In such a process we can recognize three different phases:

- the data acquisition phase, in which initial data are volunteered by the user and, as a consequence, initial evidence values are associated with some of the "finding" nodes in the inferential network (notice that the initial evidence value of each node is (0,0) so that all the nodes are initially quiescent);
- the inferential phase, in which the evidence values are propagated forward in the network (i.e. from lower layers to upper ones) and further data may be requested to the user;

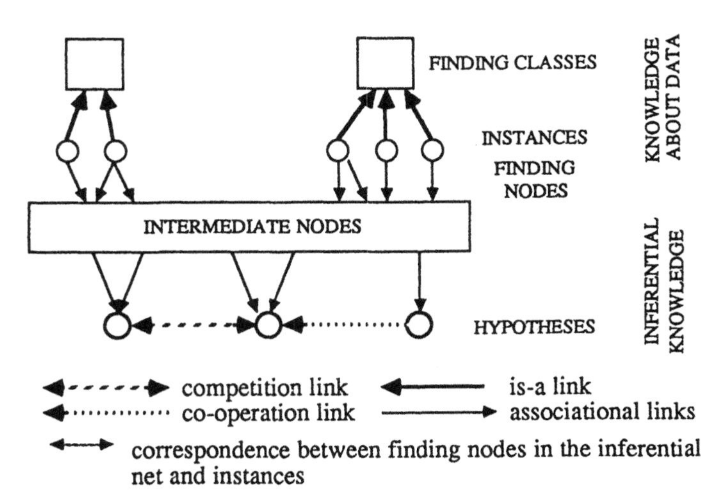

Figure 3 - Architecture of EXODUS

- the differential phase, in which the hypotheses are differentiated and a multiple-fault solution to the diagnostic problem is suggested by taking into account the competition and co-operation links between the hypothesis nodes activated in the inferential phase.

The three phases will be analysed in more details in the following subsections.

3.1. DATA ACQUISITION PHASE

Initial data are volunteered by the user. In particular, each datum provided by the user is an instance of a class F in the hierarchy of data. The system matches the information provided by the user and the instances of F in the hierarchy so that an evidence value can be associated with each instance of F (and, as a consequence, with each "finding" node in the inferential network). The match can be a fuzzy match in case the linguistic values associated with the attribute of a finding are defined by means of fuzzy sets. It is interesting to notice that the explicit representation of antinomy (mutual exclusion) relationships between linguistic values allows us to associate both positive and negative evidence values to the instances of a finding (i.e. to the "finding" nodes). For example, suppose the user specifies that a patient has a "high" fever; we can associate a negative evidence values to the instances of "fever" in which the attribute "intensity" assumes the value "absent", since we know that the values "absent" and 'high" of the attribute "intensity" of fever are in antinomy (i.e. mutually exclusive).

3.2 INFERENTIAL PHASE

Let us analyse now how inference is performed in EXODUS networks, i.e. how evidence values are propagated in the network.

As we noticed in section 2, the evidence square of each node is subdivided into regions and the behavior of the node depends on its evidence value (i.e. the behavior of a node is determined by the region in which its evidence falls). Let us analyse the behavior of the different types of nodes.

Behavior of a "finding" node

Consider a finding node N in the inferential network. The node is initially quiescent; when an evidence value is assigned to the node, it behaves as follows:

- if the evidence value falls into the "quiescence" region, the node remains quiescent;

- if the evidence value falls into one of the "True" or "False" regions, the node activates itself and:

  (1) if the evidence value of N falls into the "True" region, N must send its evidence to all the nodes N_i such that a "True" arc connects N to N_i. Since a unique value has to be sent on the connections, N must compute its combined evidence value e_N (i.e. a unique value combining e_N^+ and e_N^-). The function we use in such a combination is the following:

  $$e_N = f_+(e_N^+, e_N^-) = e_N^+ - k e_N^-$$

  where "k" is the angular coefficient of the threshold of the "True" region. In other words, the idea is to project (e_N^+, e_N^-) onto the e^+ axis using a line parallel to the threshold.
  Notice that since such a combination is made only in the "True" region, many of the problems deriving from the combination of positive and negative evidence values do not arise here (see [14] for a discussion on such problems).
  The combined evidence e_N of N is propagated on all the "True" arcs exiting from N. In particular, given the "True" arc a_1 connecting N to N_i, with weight R_i, the value which is propagated from N to N_i is the value

  $$X = e_N * R_i.$$

  (2) if the evidence value of N falls into the "False" region, the combined evidence e_N of N is propagated on all the "False" arcs exiting from N. In such a case the combined evidence value is computed using a function function f_- similar to f_+ (in particular, f_- can be obtained from f_+ by changing e_N^+ and e_N^- and using as the parameter "k" the angular coefficient of the threshold of the "False" region). For each "False" arc a_i and connecting N to N_i with weight R_i, the value which is propagated from N to N_i is the value

  $$X = e_N * R_i.$$

Behavior of an "intermediate" or "hypothesis" node

Each intermediate and hypothesis node N in the inferential network is initially quiescent (i.e. its initial evidence degree is $(e_N^+, e_N^-) = (0,0)$). When N receives a value X from one of the nodes of a lower layer, it behaves as follows:

- If $X>0$, then e_N^+ is incremented as follows (we use the Bernoulli additive function which is consistent with the interpretation of the weights on the arcs entering N as strengths of sufficiency):

  $$e_N^+ = e_N^+ + X*(1-e_N^+)$$

- If $X<0$ then e_N^- is incremented (using the same formula)
- The new evidence value falls into one of the regions of the node's evidence (e_N^+, e_N^-); in particular:

  - If the evidence value falls into the "Contradiction" region then the whole inferential process is stopped and the contradiction is reported to the user (this may correspond either to a contradiction in the knowledge base or to a contradiction in the data characterizing the specific case under examination).

  - If the evidence value falls into the "True" or "False" region, the node propagates its evidence value towards higher level nodes in the same way as discussed before in the case of finding nodes.

  - If the evidence value falls into the "Ask Information" region, the node N tries to enter an "ask information" mode. We imposed that such a mode can be entered only once during the evolution of a network. If the mode can be entered, N propagates an "ask information" messages to every node N_i such that an arc connects N_i to N (i.e. "ask information" messages are propagated downward on every arc entering N). The node N does not wait for the answers; such answers (i.e. the evidence values of the nodes connected to N) will be sent to N when available. When such evidence value are provided, the evidence of N will be recomputed.

Asking Information

As we noticed before, when the evidence value of a node N falls into the "ask information" region for the first time, N sends an "ask information" message to every node N_i such that there is an arc connecting N_i to N. When a node N_i receives an "ask information" message, it tries to enter the "ask information" mode. If the mode can be entered, N_i behaves as follows:

- If N_i is an intermediate node, it propagates the "ask information" message downward.
- If N_i is a finding node, information about the finding is requested to the user. In particular, the system ask information about all the attributes of the finding. In other words, since N_i is an instance of a finding F in the hierarchy of data, information about F is asked to the user (so that information about a finding is asked to the user only once).

Each time new information is gathered from the user, the evidence values of the "finding" nodes in the inferential network may be updated (this is the case, for example, of the instances of a finding F, when information about F is asked to the user). Such new evidence values are immediately propagated into the inferential network. Notice that this leads to recomputing the evidence values of some of the nodes in the network ("intermediate" nodes and/or "hypothesis" nodes). As a result of the recomputation of the evidence value of a node N, it may be the case that the evidence of N moves from one region into another. If the evidence value moves from the "quiescence" region into an activation region, then N becomes active and in turn it propagates its evidence value in the network. A more subtle case is the one where the evidence of N moves from an activation region into the "quiescence" region; in such a case, in fact, we have to consider that the evidence value previously propagated by N has not to be taken into account any more. In such a case, the value 0 is propagated on all the arcs exiting from N, so that the nodes receiving such a value will recompute their evidence value excluding the contribution from N.

It is interesting to notice that, as a consequence of the recomputations of the evidence values of the nodes when new information is provided by the user, the evolution of the inferential network is nonmonotonic (diagnosis is in general a nonmonotonic process, see the comments in [2, 3]).

3.3. DIFFERENTIAL PHASE

When the evolution of the network stops, the hypotheses have got a certain potential: before choosing the final set of hypotheses representing the solution to the diagnostic problem under examination, the competition and co-operation links between the hypotheses are taken into account.

Competition is a symmetric relation of strength S between two hypotheses H and H'. The idea is that when two hypotheses compete, they tend to exclude each other. From the point of view of evidence evaluation, this means that the effect of competition is that of increasing the negative evidence pushing the hypotheses towards a larger degree of uncertainty. If the positive evidence values are similar, both H and H' move towards uncertainty; if H dominates, H' will be discarded. A strength S=1 corresponds to mutual exclusion; in such a case if H is true (e(H)=(1,0)) and H' unknown (e(H')=(0,0)) then H' becomes false (e(H')=(0,1)).

This means that the increase of the negative evidence of H' must be proportional to the strength of competition and to the evidence of H (if H has not been established, i.e. if $e_H^+ < e_H^-$ then the increase is null). One (simple) way to compute such an increase is the following:

$$e_{H'}^- = e_{H'}^- + (1-e_{H'}^-)*S*(e_H^+ - k e_H^-) \quad \text{(if the evidence of H is in the "True" region)}$$
$$= e_{H'}^- \quad \text{(otherwise)}$$

where again k is the angular coefficient of the threshold of the "True" region of H.

A similar increase is performed on the negative evidence of H (competition is symmetric).

Co-operation is a relation (not necessarily symmetric) between two hypotheses H and H'. The idea is that if H co-operate with H', then the establishment of H should contribute to the establishment of H' (but the vice-versa is not necessarily true). From the point of view of evidence combination, the effect of the co-operation is that of increasing the positive evidence $e_{H'}^+$ of H' in relation to truth of H. In the extreme case where H co-operate with H' with strength 1, if H is true (e(H)=(1,0)) and H' unknown (e(H')=(0,0)), H' becomes true (e(H')=(1,0)).

This means that the increase of the positive evidence of H' must be proportional to the strength of co-operation and to the evidence of H (the function we use is similar to the one used for competition).

4. IMPLEMENTATION ISSUES

EXODUS has been implemented using SMALLTALK; we chose an object-oriented language because the philosophy of such a language seems to be closer to the idea of having a set of objects (the nodes of the network) communicating via message passing and to the idea of having a distributed architecture.

In particular, we defined a SMALLTALK class for each type of node in the inferential network (so that the nodes in the network are instances of such classes). The behavior of a node is defined by a set of methods (for example, there is a method which is activated when a message with the evidence of a lower level node is received, a method for recomputing the evidence value of the node, a method for sending such an evidence value to upper level nodes). Evidence propagation in the inferential network has been thus obtained via message passing between objects.

As we noticed, the architecture of EXODUS is intrinsically a distributed one and all the propagations could be performed in a parallel way. In the SMALLTALK implementation such an intrinsic parallelism of the model has been simulated by means of a scheduler which has the role of sequentializing the messages to be sent.

5. DISCUSSION

In this paper we have presented a distributed architecture for diagnostic problem solving. In particular, we have shown how heuristic knowledge can be integrated within a connectionist approach. According to this paradigm there is a shift in attention from structural and logical criteria for organizing knowledge (for example, "triggers", "necessary findings", "supplementary findings", "validation rules" adopted in CHECK) to mechanisms for evidence combination. These mechanisms play a fundamental role in EXODUS inferential process: the behavior of a node, in fact, depends only on its evidence value.

We are currently testing the system on a medical knowledge base (in the domain of leprosis). As regards the future developments of the system, we plan to move along two different directions:

- on the one hand, we plan to study and analyse the learning algorithms developed for connectionist networks and thus to extend EXODUS with learning capabilities;
- on the other hand, we are evaluating the possibility of replacing the current heuristic level of CHECK with EXODUS.

The second point deserves some further comments. We strongly believe that complex problems can be solved only with the co-operation of different knowledge sources and reasoning mechanisms. The adoption of suitable mechanisms for hypothesis generation is of fundamental importance for diagnostic problem solving, but other forms of knowledge (causal or functional models) have to be used for explanation purposes and for solving complex (unusual) cases. In particular, we agree with Gutknecht and Pfeiffer [6] and we advocate the integration between connectionist models and deep models as a powerful approach to the development of diagnostic expert systems. In EXODUS we are working to the integration between the system described in this paper and the causal level of CHECK [2, 17]. The former should be used to generate diagnostic hypotheses in a very fast way; the latter should be used for solving complex cases and for generating precise explanations (see [17] for a discussion on the role of associational and deep causal knowledge in diagnostic problem solving). One of the problem in the integration is the design of a clear interface between the two system. In our project knowledge about data (i.e. the taxonomy defining data) should be used by both the causal and the connectionist system thus providing a common base for the two systems. Moreover, we defined a one to one relationship between the hypothesis nodes in EXODUS and the nodes defining the diagnostic hypotheses in the causal models (in CHECK causal models hypotheses are concept which are defined in terms of states of the modeled system via a DEFINED_AS arc). In such a way the activation of a hypothesis node in EXODUS network directly corresponds to the activation of a group of states in the causal model: confirming the hypothesis corresponds to confirming such states, i.e. to determining a causal path starting from a set of initial causes which contains such states and account for all the observed findings (see the precise definitions in [2, 3]).

REFERENCES

[1] Ahuja, S.B., Soh, W-Y, and Schwartz, A., "A Connectionist Processing Metaphor for Diagnostic Reasoning," *International Journal of Intelligent Systems* 4(2) pp. 155-180 (1989).

[2] Console, L., Theseider Dupre', D., and Torasso, P., "A Theory of Diagnosis for Incomplete Causal Models," pp. 1311-1317 in *Proc. 11th IJCAI*, Detroit (1989).

[3] Console, L. and Torasso, P., "Hypothetical Reasoning in Causal Models," *International Journal of Intelligent Systems* **5**(1) pp. 83-124 (1990).

[4] Feldman, J. and Ballard, D., "Connectionist Models and their properties," *Cognitive Science* **6** pp. 205-254 (1982).

[5] Gallant, S., "Connectionist expert systems," *Communications of the ACM* **31**(2) pp. 152-169 (1988).

[6] Gutknecht, M. and Pfeiffer, R., "Experiments with a Hybrid Architecture: Integrating Expert Systems with Connectionist Networks," pp. 287-299 in *Proc. 10th Int. Work. on Expert Systems and Their Applications (Conf. on 2nd Generation Expert Systems)*, Avignon (1990).

[7] Hart, E. and Wyatt, J., "Connectionist Models in Medicine: An Investigation of their Potential," pp. 115-124 in *Lecture Notes in Medical Informatics 38*, Springer Verlag (1989).

[8] Leao, B. and Rocha, A.F., "A Methodology proposed for Knowlede Acquisition," pp. 1042-1049 in *Proc. Int. Congress Medical Informatics Europe 87*, Rome (1987).

[9] Lesmo, L., Magnani, D., and Torasso, P., "A Deterministic Analyser for the Interpretation of Natural Language Commands," pp. 440-442 in *Proc. 7th IJCAI*, Vancouver (1981).

[10] Marcus, M., *A Theory of Syntactic Recognition for Natural Language*, MIT Press (1980).

[11] McClelland, J. and Rumelhart, D., *Parallel Distributed Processing, Vol. 2*, MIT Press (1986).

[12] Peng, Y. and Reggia, J., "A connectionist model for diagnostic problem solving," *IEEE Trans. on Systems, Man and Cybernetics* **SMC-19**(2) pp. 285-298 (1989).

[13] Reggia, J.A., Nau, D.S., and Wang, P.Y., "Diagnostic expert systems based on a set covering model," *Int. J. of Man-Machine Studies* **19** pp. 437-460 (1983).

[14] Rollinger, C.H., "How to represent evidence - Aspects of Uncertain Reasoning," pp. 358-361 in *Proc. IJCAI 83*,, Karlsruhe (1983).

[15] Shastri, L., "A connectionist approach to knowledge representation and limited inference," *Cognitive Science* **12** pp. 331-392 (1988).

[16] Steels, L., "Components of Expertise," *AI Magazine* **11**(2) pp. 30-49 (1990).

[17] Torasso, P. and Console, L., *Diagnostic Problem Solving: Combining Heuristic, Approximate and Causal Reasoning*, Van Nostrand Reinhold (1989).

[18] Touretzky, D.S. and Hinton, G., "Symbols among the neurons," pp. 238-243 in *Proc. 9th IJCAI*, Los Angeles (1985).

[19] Zadeh, L.A., "Fuzzy Sets as a Basis for a Theory of Possibility," *Fuzzy Sets and Systems* **1** pp. 3-28 (1978).

THE COMBINATORIAL NEURAL NETWORK: A CONNECTIONIST MODEL FOR KNOWLEDGE BASED SYSTEMS

RICARDO JOSÉ MACHADO and ARMANDO FREITAS DA ROCHA

Rio Scientific Center - IBM Brasil
Estr. da Canoa, 3520, Rio de Janeiro
22610 - Brasil

Biology Institute - Unicamp
Campinas
13100 - Brasil

Abstract

This paper describes the Combinatorial Neural Model, a high order neural network suitable for classification tasks. The model is based on the fuzzy sets theory, neural sciences and expert knowledge analysis results. The model presents interesting properties such as: modularity, explanation capacity, concomitant knowledge and data representation, high speed of training, incremental learning, generalization capacity, feature selection, processing of uncertain and incomplete data, fault tolerance.

Keywords

Neural networks, Learning, Connectionist expert systems

1. INTRODUCTION

In this paper we intend to explore the application of the connectionist approach to the field of knowledge processing, focusing the relevant task of classification - that underlies many problem domains such as diagnosis, prediction, debugging, monitoring, interpretation, selection, etc. We introduce the *Combinatorial Neural Model* (CNM) [5], a connectionist model able to perform heuristic learning. Heuristic learning is a keystone for human acquisition of knowledge and is based on the recognition of environment regularities through the repeated observations of the external world. The model gets inspiration from the neural sciences, fuzzy logic[10], and from recent research on expert knowledge analysis [2 3 4 8 9].

2. THE BASIC MODEL

As usual in connectionist systems, our model is formed by a set of parallel processing units (*cells* or *"neurons"*) that are connected in a network according to a defined topology. These units execute a very simple processing, that in our model can be of two types: *fuzzy-AND*, *fuzzy-OR*. Each cell represents a symbolic concept of the problem domain, such as evidences, hypotheses, problem states, etc. Each cell computes a numerical value belonging to the interval [0,1], called its *state of activation* (or cell output), based on the cell input information. The state of activation can be interpreted as the degree of belief (or acceptance) the system places in the concept represented by the specific cell. The processing units are connected through arcs (*"synapses"*) forming *"neural networks"*. These synapses are characterized by a strength value (*weight*) varying from 0 (not connected) to 1 (fully connected). Inhibitory synapses may be included in the network. They would perform the fuzzy negation on the arriving signal X, transforming it in 1-X, before attenuating it by the synaptic weight.

In the proposed model the network is a feedforward network (no cycles allowed), with three or more layers. The first is the *Input Layer* receiving symbolic data from the environment or from the user. This input information is presented to the system with a degree of belief or matching, varying between 0 (null belief) and 1 (full belief). The last is the *Output Layer*, formed by fuzzy-OR cells representing the different *hypotheses* (*classes*) existing in the problem domain. The fuzzy-OR cells used in this layer implement the mechanism of competition between the different pathways coming from the lower layers.

The intermediate layers called *Combinatorial Layers* are formed by hidden fuzzy-AND cells that associate different combinations or patterns of input data. Cells belonging to the combinatorial layers represent intermediate abstractions, able to reduce the computational complexity of performing the classification task. This network architecture is quite similar to the knowledge graphs elicited from experts by the application of the heuristic knowledge acquisition technique of Rocha et al [2 3 4 8 9]. In this technique, experts express their knowledge about each hypothesis of the problem domain, by selecting a set of appropriate evidences and building an acyclic weighted AND-OR graph from these evidences to the specific class (called *knowledge graph*). The similitude between these structures makes easy to engrave expert knowledge into the neural network by the conversion of knowledge graphs to a CNM network.

The proposed model can be characterized by its: Processing units properties, Network properties, Dynamic network properties.

3. PROCESSING UNITS PROPERTIES

Figure 1 presents the two basic types of processing units: fuzzy-AND and fuzzy-OR cells, where

X_j j=1,...n represent the inputs for the cell.

Y is the computed state of activation and also the output of the cell.
W_i represents the strength (weight) of the connection (synapsis) from input i to the unit. The weights can be interpreted as the degree of membership of the concept represented in the input to the concept represented in the output of the cell.

Figure 1 shows also how to compute the state of activation for fuzzy-AND and fuzzy-OR neurons using the classical rules of fuzzy logic adapted for incorporating the influence of the synaptic weights.

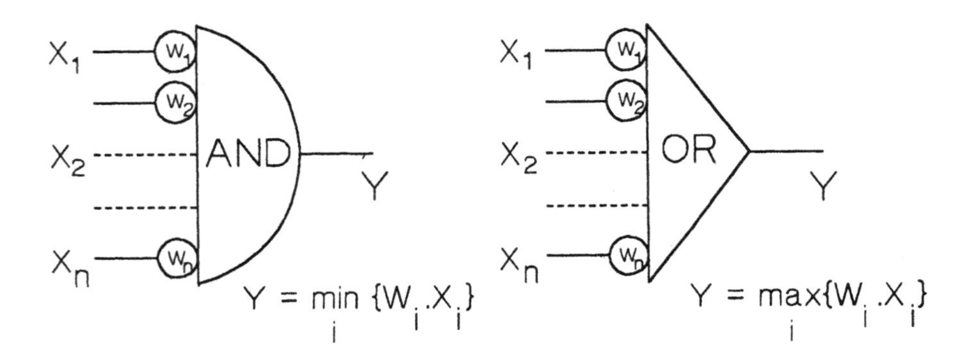

Figure 1 - Fuzzy AND and OR cells

4. NETWORK PROPERTIES

We will focus in this paper on networks composed of only three layers, since the inclusion of additional layers does not increase the generality of the model. It is always possible to produce a three layers model equivalent to models with a larger number of layers. This property comes from the use of the *minimum* operator for modeling the fuzzy-AND cells.

4.1. Entering Symbolic Data - The Input Layer

Each neuron of the input layer represents the value of an attribute of an object, that is, an evidence or finding, for instance "the temperature of John is normal". In this situation each element of the set F_k of the possible findings of an attribute k is represented by a different neuron. The degree of confidence on each of these propositions is represented by the state of activation of the neuron. This approach presents the following interesting abilities:

- To represent the situation of ignorance (unknown or unavailable evidence).

For that it is sufficient to set equal to zero all the inputs corresponding to the findings of the set F_k of an attribute k.

- To codify complex attributes (eg. colour) through the representation of its elementary components (eg. primitive colors) in different neurons.
- To reason non-monotonically when representing only positive evidences (as medical experts usually do).

4.2. Associating Symbolic Data - The Combinatorial Layer

A fundamental characteristic in the reasoning of human experts is to chunk input evidences in clusters of information using the logical connective AND for representing regular patterns of the environment. This is an important reason for including in the network model a hidden layer (the *Combinatorial Layer*) representing different combinations of input data. The *Primitive* or *Complete Version* of the model includes all combinations of input evidences from C(n,1) to C(n,n), where n is the number of cells of the input layer. The complete version of the model is clearly infeasible because of the combinatorial explosion, here expressed in a spatial form, that requires 2^n cells in the hidden layer for each hypothesis.

Our proposal to contain the combinatorial explosion in the intermediate layer is to limit the size of the clusters. We define the *Model of Order m*, that includes in the hidden layer only the combinations with length equal or smaller than a threshold number m. We suggest to make m up to the magical number of Miller [7] (seven plus or minus two). This proposal seems to be a natural mechanism for containing the combinatorial explosion, and there is considerable support for this proposal from the Psychology and from the analysis of the experts behavior (human experts never aggregate more than 9 evidences when providing knowledge graphs or production rules).

5. DYNAMIC NETWORK PROPERTIES

The network has two important operation modes: the *Consultation Mode* and the *Learning Mode*.

In the *Consultation Mode* all cells operate in parallel. The available evidences are presented to the neurons of the input layer with their corresponding belief degrees. This information is propagated forward through the network until the hypotheses nodes, according to the inference rules presented in figure 1. The resulting possibility degrees of the hypotheses nodes are the response to the consultation. The hypotheses whose possibility degrees surpass a predefined *acceptance threshold* are presented to the user as the problem solution. Note that the system may express its indecision declaring that one object is similar to several classes, as humans frequently do in ambiguous situations. The total time of computation is directly proportional to the number of layers, since we deal with feedforward networks. The computational complexity of the consultation is O(n+s) on a serial computer, where s is the number of synapses and n is the number of neurons.

In the *Learning Mode* the network constructs inductively internal complex representations of its environment using a punishment and reward mechanism inspired in the operant conditioning. This corresponds to a major goal of research on neural networks that is to discover efficient learning procedures. The algorithm finds the weights that make the model to present the desired behavior based on a set of examples. The learning algorithm must be capable of modifying connection strengths in such way that internal units which are not part of the input or output come to represent important features of the task domain. We present next an inductive learning algorithm proposed for the combinatorial neural network model.

5.1. Discovering the Expert Knowledge - the Learning Rule

We depart from an untrained network with the topology described before and from a training data base containing examples (cases with its correct classification). The system works according a supervised learning mechanism and is able to proceed in only one *iteration* (training data base sweep). The learning process is performed in two steps executed by the following specific algorithms: *Punishment and Reward Algorithm*, and *Pruning and Normalization Algorithm*

Punishment and Reward Algorithm

- Set to each synapsis of the network an accumulator for rewards and an accumulator for punishments, both with initial values equal to zero.
- Set all network arc weights equal to 1.
- For each example case from the training data base do:
  - Propagate the evidence beliefs from the input nodes until the hypotheses layer.
  - For each synapsis reaching an hypothesis node do:
    - If the reached hypothesis node corresponds to the correct class of the case then backpropagate from this node until the input nodes, increasing the rewards accumulator of each traversed arc by the product:
      current evidential flow . destiny neuron activation
      else backpropagate from the hypothesis node until the input nodes, increasing the punishments accumulator of each traversed arc by the product:
      current evidential flow . destiny neuron activation
  - Endfor
- Endfor

The training algorithm uses a simple version of the backpropagation mechanism and is able to work with boolean or fuzzy training data. The arc rewards and punishments are calculated according to the Hebb's law of Neurophysiology. The algorithm produces, after processing all cases of the training data base, a trained network that corresponds to the initial network with numbers in the range [-T, +T] attached to its arcs. (T is the number of cases of the training data base). This network, to become operant, must pass by a pruning and a normalization process, presented next:

Network Pruning and Normalization Algorithm

- For each synapsis in the network do:
  - Compute the net accumulator value
    NETACC = rewards accumulator value - punishments accumulator value
  - If NETACC $\leq$ 0 then - Remove the synapsis from the network
    - Endfor
  - If punishment accumulator value of the arc < 0
    then compute the arc weight as
    NETACC / maximum NETACC of the class subnetwork
    else compute the arc weigth as
    $\sqrt{T_{acc}}$ + (1-$\sqrt{T_{acc}}$).NETACC / maximum NETACC of the class subnetwork
  - If arc weight < pruning threshold then delete the arc
- Endfor

The goal of the pruning process is to remove all the weak and negative synapses (those ones with low or negative *net accumulator values*). This is done by defining a *Pruning Threshold* for synaptic weights (that should not exceed $\sqrt{T_{acc}}$). The goal of the normalization process is to normalize the net accumulator values for the interval [0, 1], producing an operant network. The algorithm identifies all pathognomonic pathways and preserves them in the network assigning weights for its synapses with value larger than $\sqrt{T_{acc}}$ (acceptance threshold), what guarantees automatically the acceptance of the corresponding class. The choice of the pruning and acceptance thresholds are important for the performance of the network. They can be determined empirically testing the model systematically for several values of threshold and selecting the network that delivers the best performance. However the system is very robust concerning these choices.

We will present next one simple example to demonstrate the operation of the training and pruning algorithms.

John, Diana, Mary and Peter as an example

This example shows how to train a network to perform differential medical diagnosis. We will deal only with positive evidences, reasoning by default as usually physicians do. The training data base is formed by the following four cases of patients having the disease d_1 or d_2, and presenting or not the symptoms s_1, s_2, s_3, s_4:

John(d_1, s_1, s_2, s_3) Diana(d_1, s_1, s_2, s_4)
Mary(d_2, s_1, s_3, s_4) Peter(d_2, s_2, s_3, s_4)

We depart from an untrained network of order 3 with 4 input cells for the symptoms and 2 output cells for the diseases. Figure 2 shows the pruned network using the pruning threshold 0.4. We can observe the appearance of two different structures in this network: Node A represents a strong knowledge structure, called *knowledge germ*. Nodes B and C represent the factual information about the patients John and Diana. Hence we have living together in the same network knowledge and factual data. This is a remarkable property of this structure that can be simultaneously a knowledge base and a data base. If we prune the network using threshold 0.6 only the knowledge germs will remain. Note that it is possible to extract easily rules from this network, for instance:

If the patient presents symptoms s_1 and s_2 ***then*** *the patient has the disease d_1 with confidence degree 1.*
If the patient presents symptoms s_3 and s_4 ***then*** *the patient has the disease d_2 with confidence degree 1.*

These two rules are able to discriminate all cases of the diseases d_1 and d_2.

Additional testing of the model was done with several toy problems. For instance a combinatorial network of order 2 learned to solve the XOR problem in one only iteration, while an order 3 network learned to solve the T-C problem, also in one iteration.

The model was also applied to the medical problem of diagnosing the renal syndromes: Uremia, Nephritis, Calculosis and Hypertension, based on 58 evidences taken from the patient history and physical examination. Expert knowledge was used to indicate which evidences should be used for the construction of each class untrained sub-network. The model was evaluated using the holdout technique: from a sample of 378 patients, 250 randomly selected cases were used to train an order 2 network, and the other 128 cases were used to test the system. Assuming the hypothesis with highest activation as the response of the system, it was obtained a misclassification error rate equal to 0.1171 (a result statistically significant even at significance levels smaller than 0.000001). All errors were nephrite cases misclassified as uremia. These errors are justifiable since the discrimination between these diseases requires laboratory tests data - that were not available in this experiment.

This problem domain allows also to demonstrate an important capability of the Combinatorial Neural Model: to reduce the input dimensionality by selecting the best features. If we increase the pruning threshold in a certain range, the system discards irrelevant features while maintaining the same performance. It was possible to cut the input dimensionality to less than a half (58 to 27 evidences), without increasing the misclassification rate.

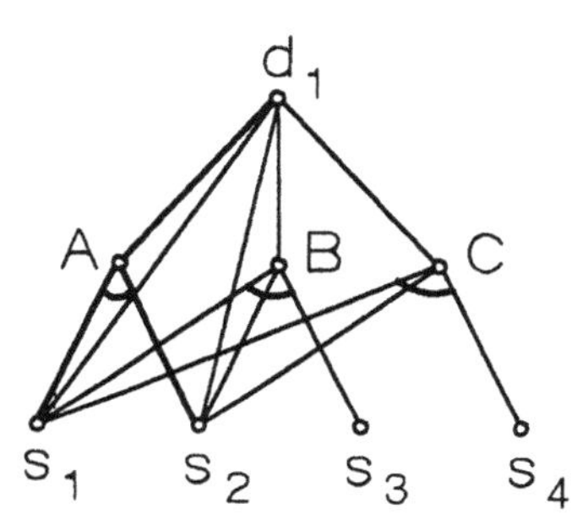

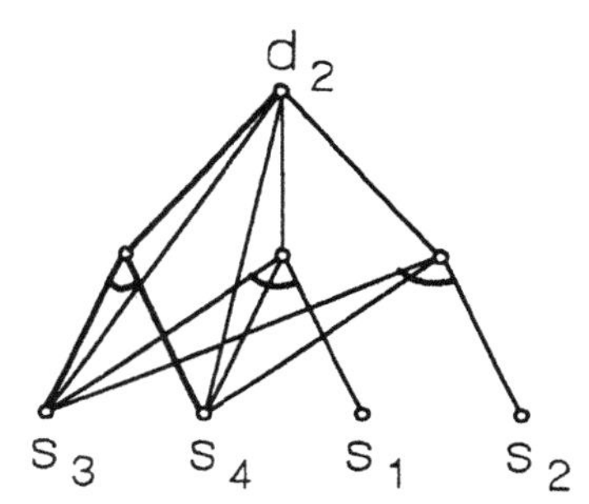

Figure 2 - Pruned network using threshold 1 in the medical diagnosis example. (Thick arcs have weight 1 and thin arcs have weight 0.5).

6. CONCLUSION

Connectionist models are drawing increasing interest also in the field of Expert Systems. The powerful learning techniques available in neural networks offer an attractive alternative for the automatic construction of knowledge bases, reducing in this way the effort required for expert systems development. The Combinatorial Neural Model introduced in this paper represents an important bridge between the fields of artificial neural networks and of knowledge engineering, allowing the construction of connectionist expert systems able to inherit desirable capabilities from both fields. The Combinatorial Neural Model is based on the fuzzy sets theory instead the usual threshold logic units, and used a network architecture heavily inspired in the heuristic knowledge graphs commonly provided by human experts. This neural model presents interesting properties such as: parallel distributed processing, modularity, explanation capacity, knowledge and data representation, reasoning by examples in the limit situation, high speed of training, incremental learning, forgetting capacity, generalization capacity, feature selection, processing of uncertain and incomplete data, ability to reason non-monotonically and fault tolerance.

The system is highly modular. Each cluster (fuzzy-AND cell) can be seen as an independent individual module that competes with other modules for establishing the classification of the case. Since each cluster can be seen as a heuristic fuzzy rule, it is quite simple to provide an explanation of the response of the system, just showing the chain of rules derived from the clusters responsible for the activation of the winning hypothesis.

The model presents the remarkable capacity of accommodating in the same network knowledge about the problem domain as well factual data. The system can reason non-monotonically representing and processing only positive evidences, as frequently human experts do.

The clusters combining information (fuzzy-AND cells in the untrained network) can be seen as tools or building blocks available to be used by the system. The system through the learning process needs only to select which of the available tools better capture the regularities of the problem domain. In this way, we liberate the network from the hard task of deciding which tools are necessary and of creating them. According to Giles [1], this process constitutes a major part in the learning procedure for networks utilizing backpropagation. This characteristic of our model allows the realization of the training process in one iteration only. The system is not disturbed with problems of local minima and convergence speed that may plague other models [1].

The model allows also to perform incremental learning. Due to the use of synaptic accumulators for rewards and punishments, at any moment we can interrupt the training and put the network into operation, continuing the learning later with additional cases. It is also possible to engrave into the network an initial knowledge taken from human experts, and improve this knowledge with additional experience acquired during the operation of the system.

During the training phase, we observe the appearance of clusters of information with very high strength, called *germs of knowledge* [3 4], that better capture the regularities of the environment. Each of these germs is accompanied by a set, called *halo*, of functionally similar but weaker clusters. The clusters of the halo are formed by the different subsets of the elements of the germ or by a mixture of elements of the germ with other evidences [3 4]. The notion of halo gives the network the capacity of generalization and of graceful decay. Cases not previously seen by the network are processed adequately by the knowledge germs. Even cases that do not present evidences matching exactly a knowledge germ can be reasonably processed by the clusters of the halo with some loss of confidence. In the limit situation the system would reason by examples using the factual data existing in the network. In ambiguous situations the model emulates humans accepting more than one hypothesis.

The model is able to reduce the input dimensionality by pruning irrelevant features during the learning process. This corresponds to one of the fundamental problems in statistical pattern recognition, that is to determine which attributes should be employed for the best classification result.

The network is robust concerning uncertain and incomplete input information, as well as errors in the synaptic weights. It gracefully decays when these faults increase.

The model order limitation for containing the network spatial combinatorial explosion is generally an adequate mechanism for high level cognitive tasks. However the tractability of very large domains requires the reduction of the cluster maximum order at the untrained network. This may result into an initial structure insufficient to represent the necessary heuristic knowledge for accurate performance. For solving this problem an attractive alternative is the application of a genetic algorithm that allows a controlled generation of high order clusters required to accommodate the heuristic knowledge, whenever necessary [6].

REFERENCES

[1] Giles C.L. and Maxwell T. (1987) Learning, invariance, and generalization in high-order neural networks, Applied Optics, 26 (1987), 4972-4978

[2] Greco G. and Rocha A.F. (1987) The fuzzy logic of text understanding, Fuzzy Sets and Systems 23 (1987), 347-360

[3] Leão B.L. and Rocha A.F. (1990) Proposed methodology for knowledge acquisition: a study on congenital heart diseases diagnosis, Methods of Inf. in Medicine 29 (1990), 30-40

[4] Machado R.J., Rocha A.F. and Leão B.F. (1991) Calculating the mean knowledge representation from multiple experts. To appear in Multiperson Decision Making Models Using Fuzzy Sets and Possibility Theory, (Fedrizzi M., Kacprzyk J., eds), Kluwer Academic Publishers

[5] Machado R.J. and Rocha A.F. (1989) Handling knowledge in high order neural networks: the combinatorial neural model, IBM Rio Scientific Center Technical Report CCR076

[6] Machado R.J., Rocha A.F. et al (1990) Heuristic learning expert system, Proceedings of 7º Simpósio Brasileiro de Inteligência Artificial, Campina Grande, (1990), 1-7

[7] Miller G.A. (1956) The magical number seven, plus or minus two: some limits on our capacity for processing information, Psychological Review 63 (1956), 81-97

[8] Rocha A.F. et al (1989) Handling uncertainty in medical reasoning, III International Fuzzy Sets Association Conference, USA

[9] Rocha A.F., Laginha M.P.R., Machado R.J., Sigulem D., Anção M. (1990) Declarative and procedural knowledge - two complementary tools for expertise. In Approximate Reasoning Tools for Artificial Intelligence (Verdegay J.L., Delgado M., eds), Verlag Tuv

[10] Zadeh L.A. (1965) Fuzzy sets, Information and Control 8 (1965), 338-353

A MEDICAL DECISION AID BASED ON A NEURAL NETWORK MODEL

M. E. COHEN
Department of Mathematics
California State University
Fresno, California 93740 USA

D. L. HUDSON
Section on Medical Information Science
University of California, San Francisco
Fresno, California 93703 USA

Abstract

Development of neural networks was one of the early topics of investigation in artificial intelligence in the 1950's. Research in this area lay dormant for a number of reasons, including the discovery that the single-layer perceptron-type networks could not represent some basic logical operations. Also, the state of computer hardware at that time could not accommodate neural networks of useful dimensions. In the 1980's interest in neural networks has increased enormously, due to advance in computer hardware and in theoretical developments which overcame shortcomings of the original perceptron model. In this paper, a neural network learning algorithm is described in the context of the development of a decision making aid. The model can accommodate a variety of data types. The approach is illustrated in a specific example in medical decision making.

Key Words

Neural networks, computer learning, decision making, connectionist expert systems

1. INTRODUCTION

Investigation into techniques of automated decision making has occupied researchers almost since the invention of the modern electronic computer (Feigenbaum et al. 1971). This research found numerous applications in medicine, where more traditional computational techniques had not met with great success (Shortliffe 1976). In all medical applications, the problem of dealing with uncertain information becomes evident from the outset (Zadeh 1983). Early approaches included rule-based expert systems, often with the use of certainty factors in an attempt to accommodate uncertainty (Shortliffe 1975) and pattern classification techniques which derived weighting factors directly from data (Ben-Bassat et al. 1980). Later approaches have employed a number of sound theoretical methodologies for handling uncertainty (Sanchez et al. 1989, Adlassnig 1980, Kacprzyk and Fedrizzi 1988, Yager). While rule-based techniques produced expert systems with "human-like" reasoning abilities as well as explanation capabilities, the difficult problem remained of obtaining an accurate and reliable knowledge base for each application. Pattern classification, on the other hand, did not have the advantages of providing a seemingly logical progression of thought, but had the advantage of extracting knowledge directly from data, without the necessity of expert intervention. Thus each approach has its advantages and disadvantages.

A very early approach, neural networks, centered on developing models patterned after biological nervous systems, with emphasis on the parallel nature of natural information processing (Hubel and Wiesel 1962). At the heart of these methods is a learning algorithm which permits extraction of knowledge from examples (Rosenblatt 1961, Nilsson 1965). Recently, there has been renewed interest in these techniques (Rummelhart and McClelland 1986, Kohenen 1987, Grossberg 1987). In this paper, an automated medical decision making aid is described which was developed using a neural network learning algorithm developed by the authors (Cohen et al. 1989). It is illustrated in an application for determination of prognostic factors in malignant melanoma of the skin, the most lethal form of skin cancer, which is unfortunately increasing, particularly in the sun belt regions (Blois et al. 1983, Blois et al 1982).

In the next section, a short background on the history and basic theory of neural networks is provided, followed by a description of the learning algorithm developed by the authors. These techniques are then illustrated in a medical decision making system which results in an expert system which combines the neural network decision function with expert-supplied rules.

2. NEURAL NETWORKS

Neural network research is divided into two general categories of researchers: those who wish to accurately model nervous system structures of humans and animals, and those who utilize known information about these structures to build computerized systems, although interaction between these groups is resulting in better models of both types. Only the second category will be discussed here.

The basic constructs of the neuron which are useful for neural network research are its input and output structures. Basically, a neuron is made up of the body or soma, dendrites, through which input is received, and one single axon, through which output is produced. Neurons transmit messages to one another through biochemical messengers which exhibit either positive (excitatory) or negative (inhibitory) properties. If the net result exceeds a certain threshold, the axon fires an electrical impulse, which in turn causes a release of biochemical transmitters at the next connection (Butter 1968). Thus messages are transmitted through the system.

The second aspect of the nervous system which is useful in implementing artificial neural networks is the parallel structure of natural information processing systems. These structures were initially applied in the implementation of visual information processing systems, building upon the structures elucidated by Hubel and Wiesel (1962).

A basic neural network has a structure similar to figure 1, with each node (representing a neuron) receiving many inputs. The nodes at the bottom are input nodes. Subsequently, all nodes assume a value calculated as the sum of their inputs according to the equation:

$$v_i = \sum_{j=0}^{m} w_{i,j} \, n_j \tag{1}$$

The $w_{i,j}$'s determine the weight which should be given to each incoming node. The top layer represents the output nodes, which provides the final result.

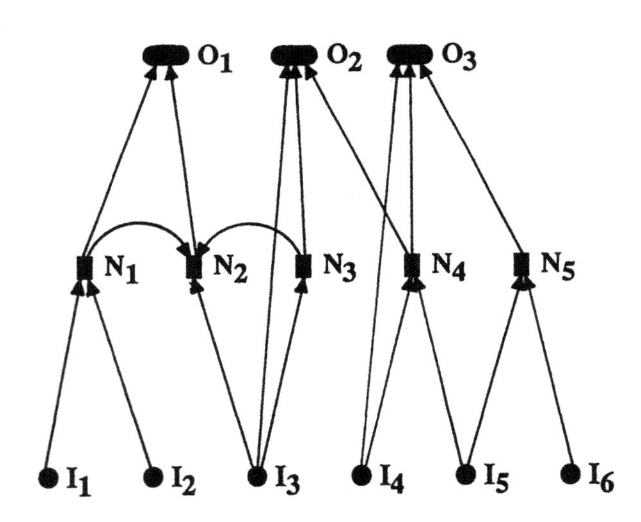

Figure 1: Feed Forward Network with Interaction Nodes (Hidden Layer)

A major aspect of neural network research is determination of the weighting factors in equation (1). These are ascertained through some type of learning algorithm. These learning algorithms fall into two general categories: supervised learning in which data of known classification are used to adjust the weights iteratively until correct classification results, and unsupervised learning which uses data of unknown classification and searches for natural clusters within the data. Within each of the categories, a number of approaches have been developed, including Hebbian learning (Kosko 1986), statistical approaches, generalized perceptron models (Smith et al. 1988), supervised (Cohen et al. 1989) and unsupervised learning techniques (Parker

1986). A number of books and articles are available which provide good overviews of the field (Rummelhart and McClelland 1986, Grossberg 1987, Kohenen 1987).

3. LEARNING ALGORITHM

The learning algorithm described here uses a non-statistical approach consisting of an iterative supervised learning algorithm. Weights are determined through a supervised learning approach based on previous work of the authors (Cohen and Hudson 1987). In this approach, data of known classification are used to determine weighting factors. The method is an adaptation of the potential function approach to pattern classification. The potential functions are the general class of Cohen multidimensional orthogonal polynomials defined by the equation

$$C_n(x_1,\ldots,x_m) = \frac{m!}{n!(m-n)!}\sum_{k=1}^{n}\frac{(-1)^k(m-k)!}{(n-k)!(m-n)!}\sum_{i_k=k}^{m}\ \sum_{i_{k-1}=k-1}^{i_k-1}\cdots\sum_{i_2=2}^{i_3-1}\ \sum_{i_1=1}^{i_2-1}$$

$$\sum_{p=1}^{k}\frac{x_{i_p}^{a(n,i_p)}\,[a(n,i_p)+v_{i_p}]}{v_{i_p}} \tag{2}$$

$$\int_0^1\cdots\int_0^1 C_n\, C_r\, x_1^{v_1-1}\cdots x_m^{v_m-1}\, dx_1\ldots dx_m = \begin{cases}0 & n=r\\ A & n\neq r\end{cases} \tag{3}$$

where m is the dimensionality of the data, a_i, $i=1,\ldots,k$ are parameters which may be arbitrarily selected to yield desired characteristics, A is the normalization constant, and v_i, $i=1,\ldots,m$ are assigned values corresponding to the components of the first feature vector. The potential function is then defined by

$$p(x,x_k) = \sum_{i=1}^{\infty} Ø(x)\, Ø_i(x_k) \tag{4}$$

where $Ø_i(x)$ are the components of the Cohen polynomials, which themselves form an orthogonal set

$$\int_{-1}^{1}\cdots\int_{-1}^{1} Ø_n(x_1,\ldots,x_r)\, Ø_m(x_1,\ldots,x_r)\, dx_1\ldots dx_r = \begin{cases}0 & n=m\\ B & n\neq m\end{cases} \tag{5}$$

where B is the normalization constant and x_k is the kth feature vector. A feature vector is made up of r component values, where r is the number of nodes (for example, test results) which are included. A number of special cases can be derived from this orthogonal series. In order to obtain orthogonal polynomials, the series a_k must be chosen to assume integral values. However, it should be noted that this function is capable of generating non-integral series, which can be used to develop networks in which nodal values contribute to fractional powers. This important property is illustrated as it applies to the neural network model.

The supervised learning proceeds iteratively, until a separation of the data into correct categories is accomplished. In the process, weights are adjusted for each of the features. The decision hypersurface takes the form

$$D(\mathbf{x}) = \sum_{i=1}^{m} w_i \, x_i^{a} + \sum_{i=1}^{m} \sum_{\substack{j=i \\ i \neq j}}^{m} w_{i,j} \, x_i^{b} \, x_j^{c} \tag{6}$$

In network form, the w_i's would provide weighting factors for connections of nodes x_1 and x_2 and the $w_{i,j}$'s the weighting factors for the interaction for these nodes. All weights in equation (6) are with respect to the same output node, thus the weighting factors indicate the originating node only, for simplicity of notation. Once weighting factors have been determined, in the decision making model, equation (6) produces a numerical value. The larger the absolute value, the more certain one can be that the vector belongs to that category (Hudson et al. 1989).

The maximum and minimum values for the decision surface D(x) must be determined. Let $A_i = \{m_1,...,m_k\}$, the set of all values which x_i can assume, where $m_i > 0$ for all i. Then to obtain the maximum value $D_{max}(x)$:

$$\begin{aligned} &\text{If } w_i > 0, \text{ let } x_i' = \max\,[A_i] \\ &\text{If } w_i < 0, \text{ let } x_i' = 0 \end{aligned} \tag{7}$$

for all i=1,...,n. Then

$$D_{max}(\mathbf{x}) = \sum_{i=1}^{m} w_i \, x_i' + \sum_{i=1}^{m} \sum_{\substack{j=1 \\ i \neq j}}^{m} w_{i,j} \, x_i' x_j' \tag{8}$$

Similarly, $D_{min}(x)$ is obtained by the following:

$$\begin{aligned} &\text{If } w_i > 0, \text{ let } x_i' = 0 \\ &\text{If } w_i < 0, \text{ let } x_i' = \max[A_i] \end{aligned} \tag{9}$$

and apply equation (6).

All decisions are then normalized by

$$D_n(\mathbf{x}) = \begin{cases} D(\mathbf{x})/D_{max}(\mathbf{x}) & \text{if } D(\mathbf{x}) > 0 \quad \text{(class 1)} \\ D(\mathbf{x})/|D_{min}(\mathbf{x})| & \text{if } D(\mathbf{x}) < 0 \quad \text{(class 2)} \\ 0 & \text{if } D(\mathbf{x}) = 0 \quad \text{(indeterminate)} \end{cases} \tag{10}$$

The result is a value between -1 and 1, inclusive, which gives a degree of membership in that category. The values are then shifted to give an answer between 0 and 1, inclusive by

$$V(\mathbf{x}) = [1 + D_n(\mathbf{x})]/2 \tag{11}$$

4. DATA ANALYSIS

In attempt to refine the parameters which are important in determining prognosis, 1756 cases of melanoma were examined, from the melanoma clinic at University of California, San Francisco, established by the late M. S. Blois, M.D. Each case contained several hundred parameters, but for the purposes of this analysis, 109 were considered relevant. The cases were subdivided as follows:

Total number of cases:	1756
Stage 1 cases:	1567
Cases eliminated:	
Acral-lentiginous/mucosal	53
Missing values	143
Remaining cases	1371

The remaining 1371 cases were then grouped:

Group 1 (surviving)	1177
Group 2 (died from melanoma)	167
Group 3 (died from other causes)	27

Only stage 1 cases were considered, since stage 2 and 3 cases were too advanced to collect valid data. In order to make a valid comparison, group 1a was formed, which consisted of those surviving after 5 years, with no stage 3 symptoms. This group contained 304 patients. The initial comparison was made between group 1a and group 2.

The variables selected by the model as contributing to the difference in prognosis were the following:

x_1: Thickness of the tumor
x_2: Clark's Level
x_3: Gender of the patient
x_4: Skin thickness
x_5: Location on body
x_6: Lymph node involvement
x_7: Mitotic rate

Use of the above seven variables resulted in correct classification into group 1a or group 2 of 79% of all cases, which compared favorably with previous statistical methods which yielded approximately 65% accuracy. The resulting network is shown in figure 2, with the connection weights determined by the learning algorithm. The .3 and the .5 indicate the power of each x_i for the nodes at that level.

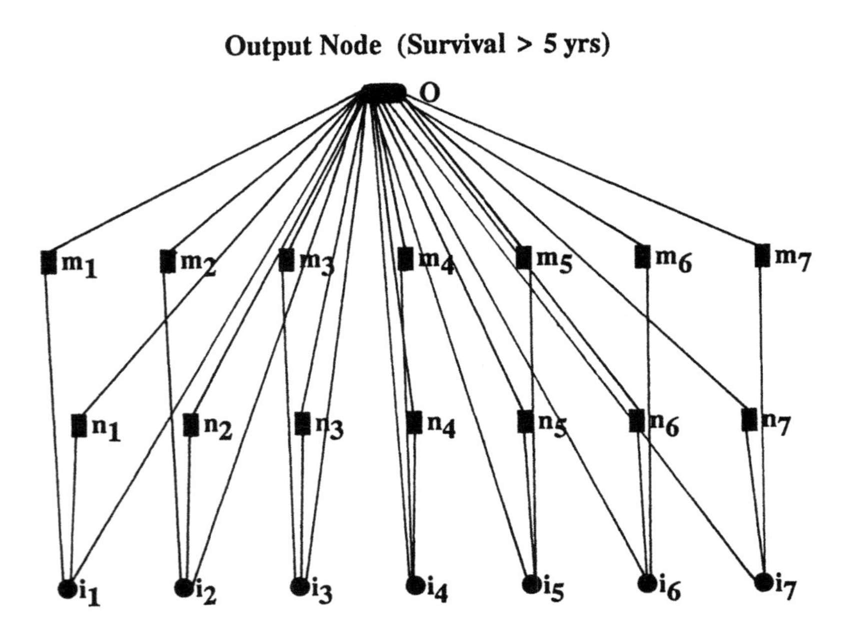

Levels 1: Input Node
2: Interaction Nodes (factor .5)
3: Interaction Nodes (factor .3)
4: Output Node (Prognosis)

Figure 2: Neural Network

The resulting equation is

$$D(\mathbf{x}) = -15.8 + 37.1x_1^{.3} + 1.8x_2^{.3} + \ldots - 2.0x_7^{.3} - 10.1x_1^{.5} + 1.7x_2^{.5} + \ldots + 9.9x_7^{.5} + 7.3x_1^{.3}x_2^{.3} + \ldots + 4.5x_6^{.3}x_7^{.3} \quad (12)$$

5. EXPERT SYSTEM

One advantage of the above procedure is that it allows the reduction of variables from 109 to 7. In a prospective decision aid, the values of these 7 variables can then be elicited directly in question format, producing an interaction in the form of an expert system, as shown in figure 3. Note that the data entered by the user can be either binary, categoric, or continuous. The algorithm will function properly as long as there is an ordering on each variable.

At this stage, expert-supplied rules can be combined with the neural network algorithm to take advantage of all possible sources of information, whether it be inherent in the data or ascertained from the years of experience of experts in the field.

In the following consultation, if a yes/no response is indicated, a number between 0 and 10, inclusive, may be entered instead to indicate a degree of presence of the symptom. A value of 0 corresponds to no; a value of 10 corresponds to yes.

Name of patient: **RW**
Age (Years): **47**

Thickness of tumor (mm): **0.85**
Level (1,2,3,4,5): **2**
Gender (M/F): **F**
Skin thickness (mm): **1.8**

Using the code:
1: Upper extremities
2: Lower extremities
3: Lower back or chest
4: Abdomen
5: Head or neck

Location on body: **3**
Lymph node involvement (y/n): **n**
Mitotic rate (1,2,3): **1**

This patient has a favorable 5-year prognosis.

Figure 3: Sample Run for Elicitation of Pertinent Variables (User responses are in bold face.)

6. CONCLUSION

The method described here incorporates a combination of techniques for establishment of a knowledge base. The learning algorithm can be applied to an accumulated data base of information to extract parameters which are important for decision making, along with appropriate weighting factors. The method is applicable for any decision making process, whether two-category or multi-category. In the final expert system, the neural network approach can be expanded laterally, or to additional levels, to accommodate decision making strategies proposed by experts in the form of rules.

The above method was initially implemented for the analysis of testing modalities in carcinoma of the lung. A prospective study is underway to increase the sample size for this project. The melanoma application has the advantage of a very large data base which has been collected over a period of fifteen years, and thus the reliability of the network is significantly increased. The method lends itself directly to any application for which suitable data are available.

7. REFERENCES

K.P. Adlassnig, A fuzzy logical model of computer-assisted medical diagnosis, Math. Inform. Med., 19, 3 1980, 141-148.

M. Ben-Bassat, R. Carlson, et al., Pattern-based interactive diagnosis of multiple disorders: The MEDAS system, IEEE Transactions on Pattern Analysis and Machine Intelligence, PAMI-2, 1980, 149-160.

M. S. Blois, R. W. Sagebiel, M. S. Tuttle, T.M. Caldwell, H.W. Tayler, Judging prognosis in malignant melanoma of the skin, Annals of Surgery, 198, 2, 1983, 200-206.

M.S. Blois, R.W. Sagebiel, R.M. Abarbanel, T.M. Caldwell, M.S. Tuttle, Malignant melanoma of the skin, The association of tumor depth and type, and patient sex, age, and site with survival, Cancer, 52, 7, 1982, 1330-1341.

C.M. Butter, Neuropsychology: The Study of Brain and Behavior, Brooks/Cole, Belmont, CA, 1968.

M.E. Cohen, D.L. Hudson, M.F. Anderson, A neural network learning algorithm with medical applications, in Computer Applications in Medical Care, 13, IEEE Computer Society Press, 1989.

M.E. Cohen, D.L. Hudson, M.F. Anderson, A neural network learning algorithm with medical applications, in Computer Applications in Medical Care, L.C. Kingsland, Ed., 13 IEEE Computer Society Press, 1989, 307-311.

M.E. Cohen, D.L. Hudson, Use of pattern classification in medical decision making, Lecture Notes in Computer Science, 286, B. Bouchon, R. R. Yager, Eds., Springer-Verlag, 1987, 245-254.

E.A. Feigenbaum, B.G. Buchanan, J. Lederberg, Generality and problem solving: A case study using the DENDRAL program, Machine Intelligence, 6, 1971, 165-190.

S. Grossberg, The Adaptive Brain, vols. 1 and 2, North Holland, 1987.

D.H. Hubel, T.N. Wiesel, Receptive fields, binocular interaction, and functional architecture of the cat visual cortex, J. Physiology, 160, 1, 1962, 106-154.

D. L. Hudson, M. E. Cohen, M. F. Anderson, Neural network techniques in a medical expert system, International Fuzzy Set Association, 1989, 476-479.

J. Kacprzyk, M. Fedrizzi, Combining Fuzzy Imprecision with Probabilistic Uncertainty in Decision Making, Springer-Verlag, 1988.

T Kohenen, Self-organization and associative memory, Springer-Verlag, New York, 2nd Ed., 1984.

B Kosko, Differential Hebbian learning, Proc. American Institute of Physics: Neural Networks for Computing, 1986, 277-282.

N.J. Nilsson, Learning Machines, McGraw Hill New York, 1965.

D.B. Parker, G-maximization: An unsupervised learning procedure for discovering regularities, in Proc. American Institute of Physics, Neural Networks for Computer, 1986.

F. Rosenblatt, Principles of Neurodynamics, Perceptrons, and the Theory of Brain Mechanisms, Spartan, Washington, 1961.

D.E Rummelhart, J.L. McClelland, and the PDP Research Group, Parallel Distributed Processing, vols. 1 and 2, MIT Press, Cambridge, 1986.

E. Sanchez, R. Bartolin, Fuzzy Inference and Medical Diagnosis, A case study, First Annual Meeting, Biomedical Fuzzy Systems Association, 1989, 1-8.

E.H. Shortliffe, Computer-Based Medical Consultations, MYCIN, Elsevier /North Holland, New York, 1976.

J.W. Smith, et al., Using the ADAP learning algorithm to forecast the onset of diabetes mellitus, in Computer Applications in Medical Care, 12, IEEE Computer Society Pres, 1988, 261-265.

Shortliffe, E.H., A model of inexact reasoning in medicine, Math. Bioscience, 23, 1975, 251-279.

R.R. Yager, On treatment selection in the face of uncertainty in expert systems, Iona College, Machine Intelligence Institute, Technical Report #MII-704, 1-24.

L.A. Zadeh, The role of fuzzy logic in the management of uncertainty in expert systems, Fuzzy Sets and Systems, 11, 1983, 199-227.

A FUZZY NEURON BASED UPON MAXIMUM ENTROPY ORDERED WEIGHTED AVERAGING

by Michael O'Hagan
ORINCON Corporation
9363 Towne Centre Dr.
San Diego, CA 92121
(619) 455-5530 x535
mikeoh@nosc.mil

Abstract

There have been several interesting attempts to blend fuzzy set logic and neural networks. These include Shiue and Grondin [3], Yager [4], Taber and Deich [7], Kosko [10], Oden [8], and more recently Yamakawa and Tomoda [1] and Langheld and Goser [9]. This paper extends the work of Yamakawa and Tomoda by employing Maximum Entropy Ordered Weighted Averaging (MEOWA) operations at the soma of a fuzzy neuron. The MEOWA is an extension by O'Hagan [6, 2] of Yager's [4, 5] Ordered Weighted Averaging (OWA) operators. The advantage of using the MEOWA operator in a fuzzy neuron is that a single type IC appears to be sufficient to build a neural pattern recognition system. Neural inputs and weights are fuzzy numbers rather than the usual scalar values of "standard' neural networks. Fuzzy logic operations replace the usual summing, dot product, and nonlinear "squashing" operations in the soma. Fuzzy neurons appear to be able to perform pattern recognition with reduced resolution inputs relative to standard neural networks: O(m) + O(n) vs. O(mn).

Keywords

Fuzzy neural networks, neural integrated circuits, squashing function, fuzzy logic operations, aggregation theory, ordered weighted averaging, maximum entropy ordered weighted averaging, membership functions, character recognition.

1. Introduction

Recently , there have been several interesting descriptions in recent technical literature blending both fuzzy set logic and neural networks. These are delineated in the abstract above. NASA Houston have sponsored two symposia on both fuzzy logic and neural networks in 1989 and 1990. There seems to be a natural synergism between the two areas.

The MEOWA aggregation operators, whose use is proposed here, are an extension by O'Hagan [6,2] of Yager's [4,5] original Ordered Weighted Averaging (OWA) operators. The MEOWAs represent an easily computed set of weights that can represent an interesting class of English language quantifiers ranging from "there exists" for ORness values near 1.0, to "on the average" for values of ORness near 0.5, to "for every" for values of ORness near 0.0. The advantage of using the MEOWA operator in a fuzzy neuron is that a single type IC appears to be sufficient to build a fuzzy neural pattern recognition system. Also, within the proposed pattern recognition tasks, it may be quite useful to perform MEOWA fuzzy compositions at a 0.90 and 0.10 ORness level rather than the more conventional MAX-MIN fuzzy compositions used in [1].

2. Fuzzy Neurons

The fuzzy neuron is similar in structure to the conventional scalar-valued ANN model and is shown in Figure 1 (adapted from [1]). In contrast to the "standard" model artificial neural networks (ANNs), these fuzzy extentions use fuzzy numbers as inputs and weights rather than scalar values, and in addition, use

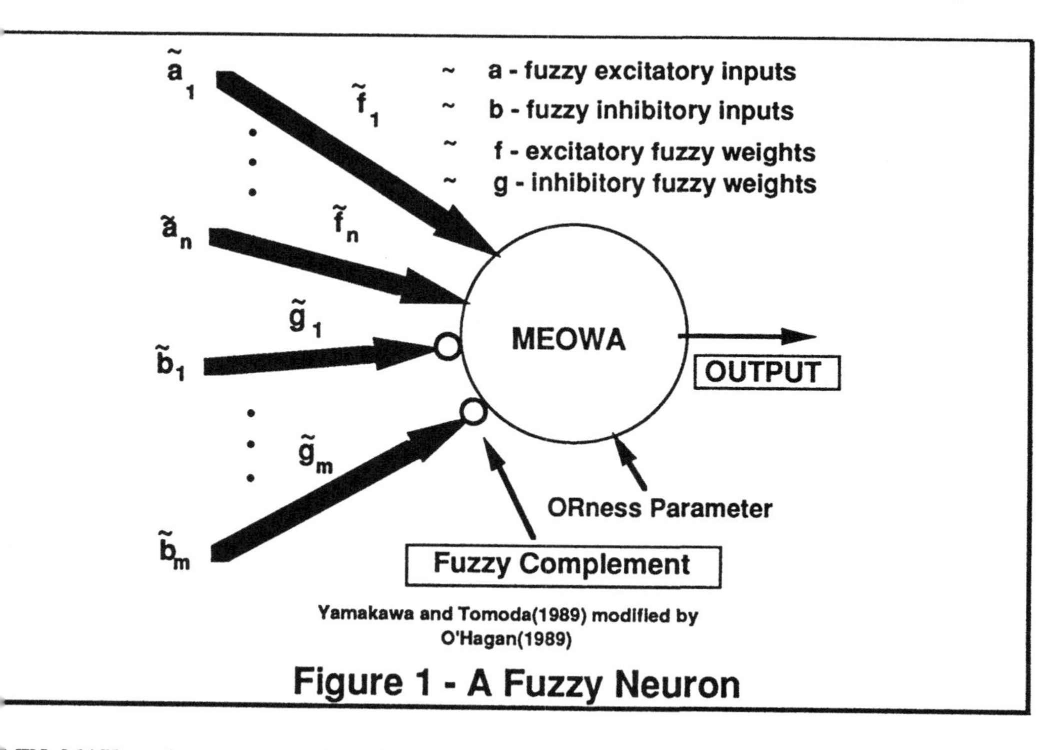

Figure 1 - A Fuzzy Neuron

MIN, MAX, or the more general MEOWA fuzzy set logic operators in the soma rather than the usual summing, thresholding and nonlinear "squashing" of the "conventional" ANN model. In [1] Yamakawa and Tomoda do not use a threshold value. By extending the model to include a more general MEOWA aggregation at the soma that can range smoothly between Max (fuzzy OR) and Min (fuzzy AND) as a function of a scalar parameter corresponding to ORness (optimism or certainty factor (CF)), we gain some flexibility in the operation of the fuzzy neuron by aggregating inputs with varying degrees of ORness or optimism. The addition of a threshold may also be explored for this type fuzzy neuron.

The original application of the fuzzy neuron proposed by Yamakawa and Tomoda [1] using the Min/Max operators was directed to character recognition. Seven linear observations, 3 vertical and 4 horizontal, were used as input (see Figs. 2 thru 6 below [adapted from [1]]). These observations were called cross-detecting lines. An illustration of such a detecting line in the character frame is given in Figure 2 where various written forms of the Number "3" are shown.

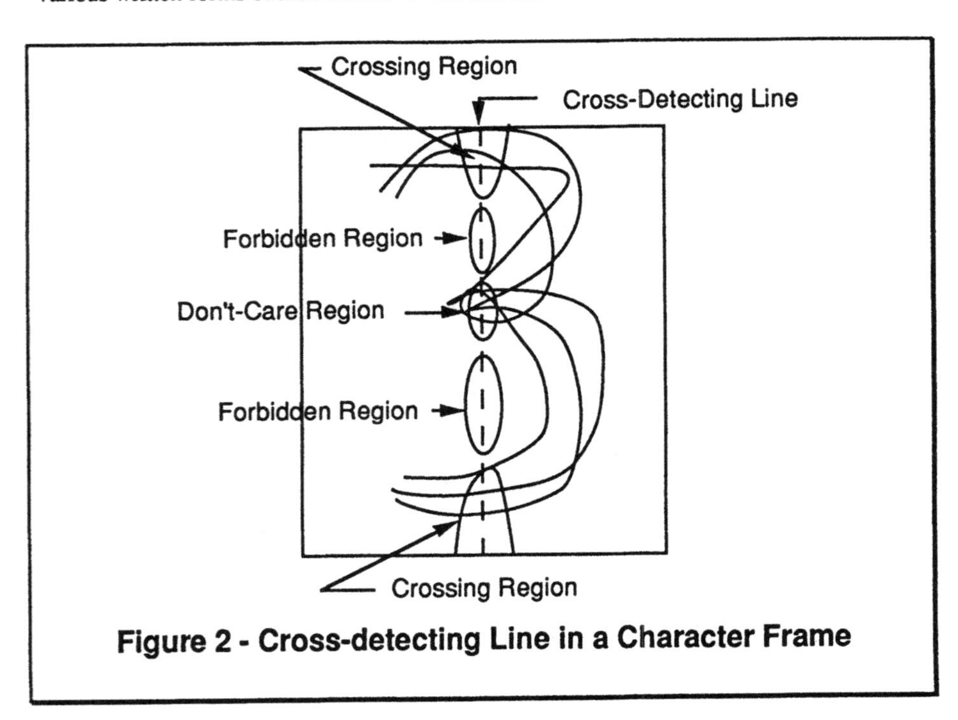

Figure 2 - Cross-detecting Line in a Character Frame

3. Development of Membership Functions within Regions

There are three types of regions on the cross-detecting line: a **crossing region**, a **forbidden region** and a **don't-care region**. The crossing regions denote line segments where a given character should cross; forbidden regions are line segment areas where a given character should not cross; and the don't care regions denote a line segment where a given character might only uncertainly cross depending on the penmanship of the writer and might in general cross in an acute rather than right angle. The number and positions of the cross-detecting lines may vary with the characters to be recognized.

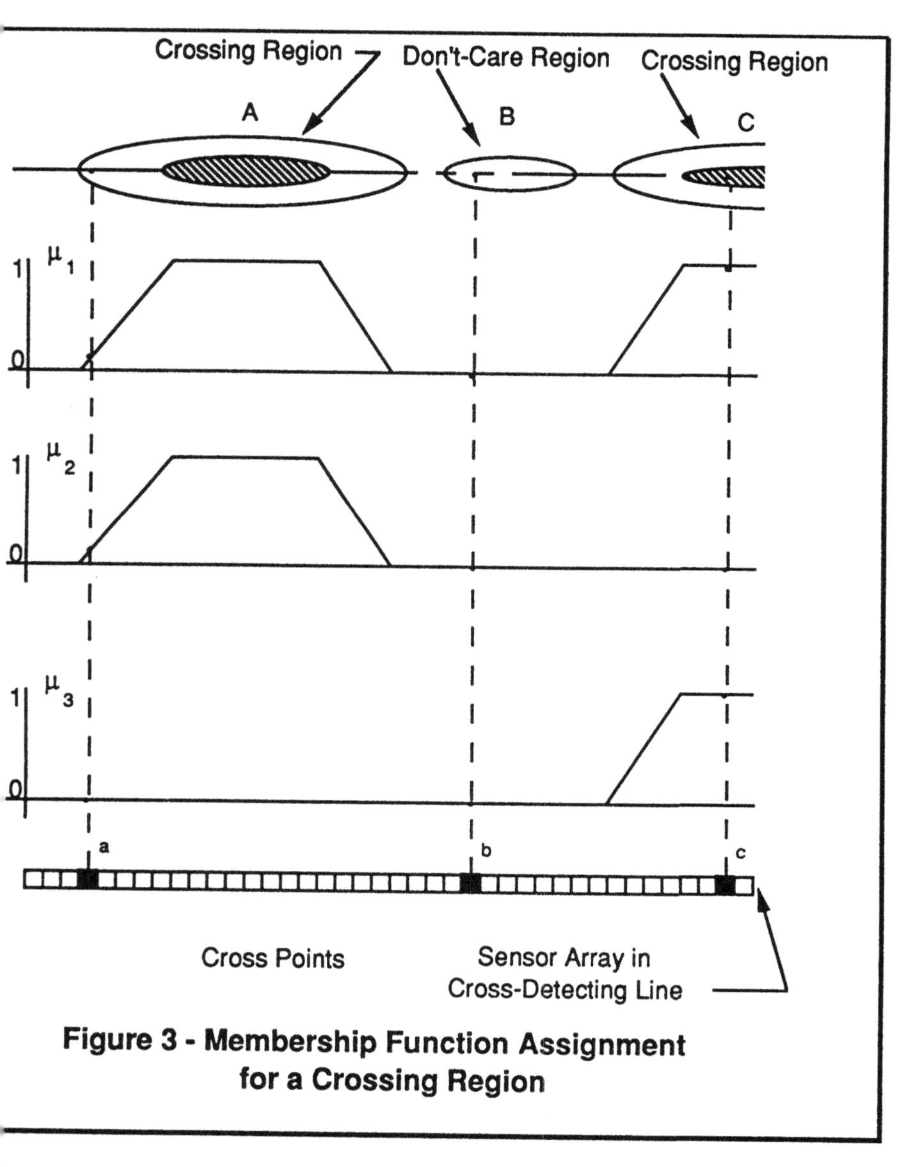

Figure 3 - Membership Function Assignment for a Crossing Region

The crossing membership function shown in Figure 3 for $\mu_1()$ has three components: $\mu_1(a)$, $\mu_1(b)$, and $\mu_1(c)$. The minimum value (fuzzy AND) of these membership grades does not properly evaluate the membership of these three points satisfying the features of a particular character, because the cross point in the "Don't-Care" region (B) has a zero value (also the point "b" is not significant). However, the points "a" and "c" are significant within the crossing regions A and C. We are thus led to decompose the $\mu_1()$ fuzzy membership function into two separate and distinct membership functions $\mu_2()$ and $\mu_3()$, which give the grades for "a" and "c" satisfying crossing regions A and C, respectively. In other words,

$$\mu_1() = \{\mu_2(a) \wedge \mu_2(b) \wedge \mu_2(c)\} \vee \{\mu_3(a) \wedge \mu_3(b) \wedge \mu_3(c)\}$$

gives a proper evaluation (fuzzy composition) of the three points crossing and satisfying some particular character's features for this position in the retina. Figure 4 shows a membership function for two forbidden regions separated by a don't-care region. This would be an example of the middle of the crossing line shown for the character "3" in Figure 2. Figure 4 shows a membership function for two forbidden regions. "d", "b", and "e" are cross points between a cross-detecting line and a character crossing obtained by a sensor array. In this case, the membership grade of rejection is obtained by:

$$\mu_4() = \{\mu_5(d) \vee \mu_5(b) \vee \mu_5(e)\} \vee \{\mu_6(d) \vee \mu_6(b) \vee \mu_6(e)\}.$$

Thus the membership grade for satisfaction is given by the complement of the grade for rejection and can be reduced to:

$$\overline{\{\mu_5(d) \vee \mu_5(b) \vee \mu_5(e)\} \vee \{\mu_6(d) \vee \mu_6(b) \vee \mu_6(e)\}} =$$

$$\overline{\{\mu_5(d) \vee \mu_5(b) \vee \mu_5(e)\}} \wedge \overline{\{\mu_6(d) \vee \mu_6(b) \vee \mu_6(e)\}} =$$

$$\{\overline{\mu_5(d)} \wedge \overline{\mu_5(b)} \wedge \overline{\mu_5(e)}\} \wedge \{\overline{\mu_6(d)} \wedge \overline{\mu_6(b)} \wedge \overline{\mu_6(e)}\} =$$

$$\{\overline{\mu_5(d) \wedge \mu_6(d)}\} \wedge \{\overline{\mu_5(b) \wedge \mu_6(b)}\} \wedge \{\overline{\mu_5(e) \wedge \mu_6(e)}\} =$$

$$\{\overline{\mu_4(d)} \wedge \overline{\mu_4(b)} \wedge \overline{\mu_4(e)}\} = \{\mu_7(d) \wedge \mu_7(b) \wedge \mu_7(e)\}.$$

In other words, the satisfaction grade of membership concerning a "Forbidden" region equals the minimum value of the complement grade of membership of the "Forbidden" regions.

4. Integrated Circuit Building Blocks

Yamakawa and Tomoda [1] proposed building a character recognition system from three basic building blocks: an ensemble minimum, an ensemble maximum, and a two-input minimum. Using a more general MEOWA block, the individual parts count can be *reduced to just one* IC type. However, the OWA box requires input of a scalar parameter ("ORness" or CF) in order to perform its ensemble aggregation

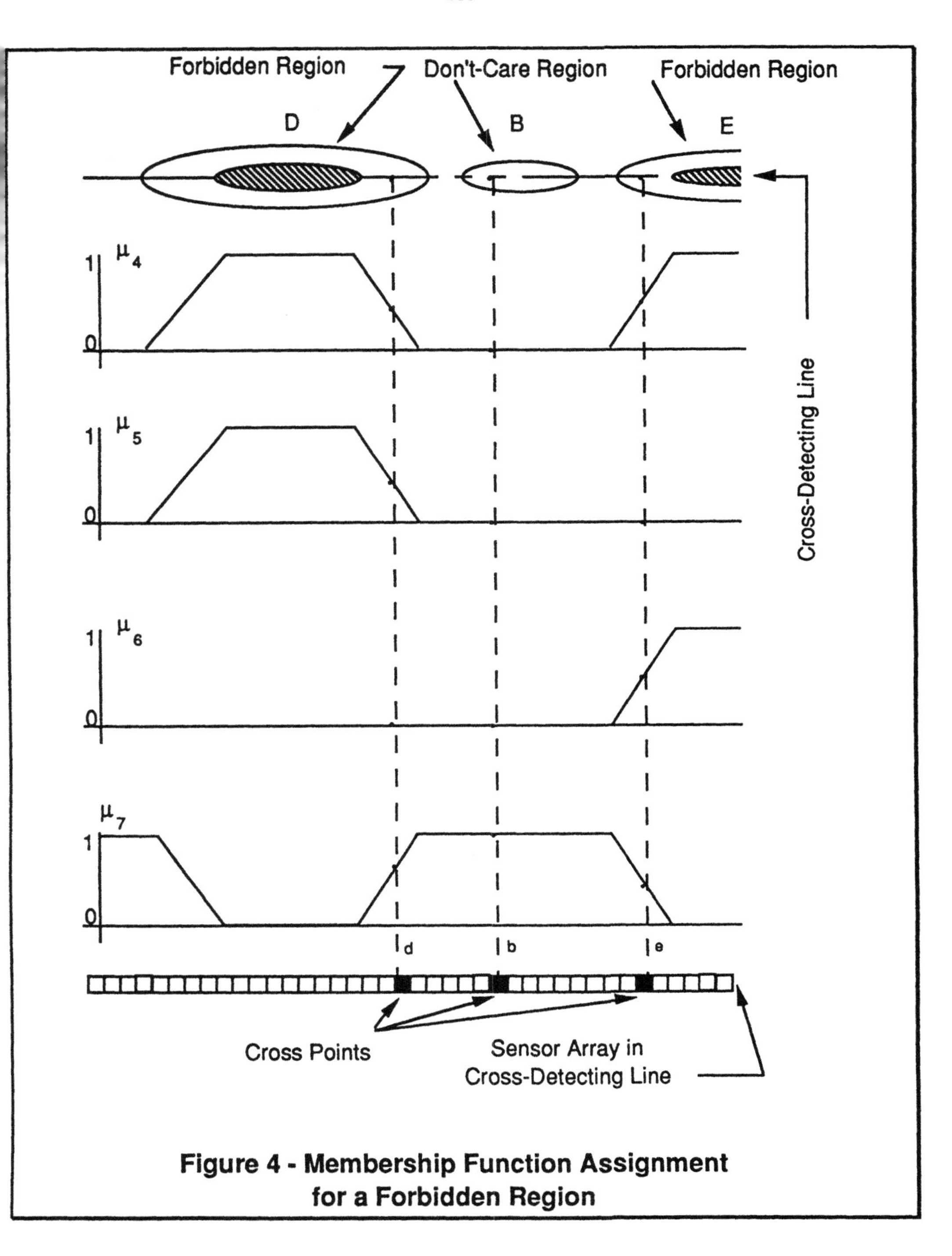

Figure 4 - Membership Function Assignment for a Forbidden Region

function. With "ORness" set to 1.0, the MEOWA performs a Max; with "ORness" set to 0.0, it performs a Min operation; with the "ORness" set between 0.0 and 1.0 it performs an ensemble aggregation between the two extremes. It should be noted that the order of the ME-OWA operation can be quite small, typically no larger than 5 at the first level, due to the fact that zero inputs can be eleminated before application of the sorting operation and subsequent dot product with the weights. The building blocks are shown in Figure 5. An ME-OWA maximum order of eleven is needed at the second stage of the recognition task for '2' as shown below in Figure 8.

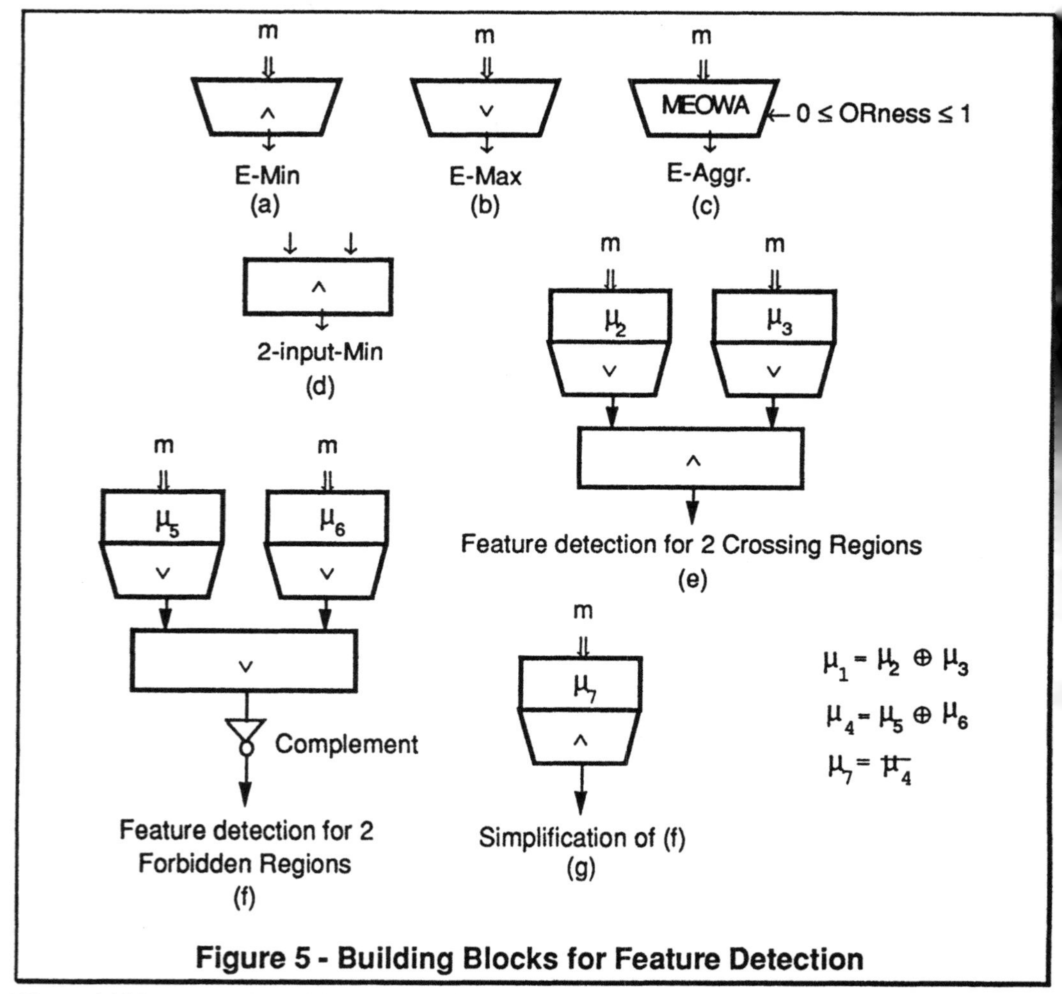

Figure 5 - Building Blocks for Feature Detection

Figure 5 (a) is an ensemble minimum, which takes the minimum value of an m element input. Figure 5 (b) shows the ensemble maximum element, which produces the maximum value of m inputs. Figure 5 (c) shows the proposed MEOWA aggregation block, which produces an ensemble aggregation output corresponding to the ensemble minimum block with ORness parameter input of 0.0; produces output corresponding to the ensemble maximum with ORness parameter input of 1.0; and produces a general ensemble aggregation between these two extremes for 0 < ORness < 1. Figure 5 (d) shows a simple two input minimum block (or corresponding maximum block with a simple change of operator). Figure 5 (e) shows a feature detecting block for a crossing region with two excitatory connections corresponding to Figure 3 above. Figure 5 (f) shows the feature detecting block with two inhibitory connections corresponding to Figure 4 above. Figure 5 (g) shows the simplification possible for a forbidden region by proper complementation of the fuzzy measures. The formulae at the right bottom of the Figure 5 are the membership definitions shown in Figs. 3 and 4.

Figure 6 shows membership function assignment for crossing and forbidden regions for each feature-detecting line in the case of the number "2". Seven cross-detecting lines (four horizontal and three vertical) are assigned on the recognition frame. The written figure "2" should cross with cross-detecting lines 1 and 2 somewhere in the right-hand portions of the lines. For cross-detecting line 3, the "2" should cross in the left-hand portion of the line. Simultaneously, "2" should not cross line 3 at its extreme left or in the right-hand regions (this is shown by use of a dotted membership line in Figure 6). The upper region of vertical cross-detecting line 5 is a don't-care region, however, line 5's lower region is a required crossing region.

Figure 7 shows a fuzzy neuron for recognizing a written number "2" as proposed by Yamakawa and Tomoda [1]. The fuzzy neuron accepts seven buses from sensor arrays in cross-detecting lines, four horizontal lines of m-bits and three vertical lines of n-bits, giving a total of 4m+3n total input lines. This is a large saving in input retina size versus the standard ANN models which would normally use O(m*n) inputs.

A proposed circuit as shown in Figure 8 can be built with MEOWA elements in which all "$\wedge$" are replaced by an ensemble aggregation element with ORness = 0 and all "$\vee$" are replaced by an ensemble aggregation element with ORness = 1. The same number of inputs would be used (4m + 3n) in the first layer; eleven inputs would be used for the second layer ensemble aggregation for recognition of "2". Only a *single type* MEOWA element is necessary for this and all similar character recognition tasks. By *varying the ORness parameters* within the circuit, superior recognition performance may be obtained. This possibility is being investigated with computer simulations of the proposed fuzzy neuron model. The major difference in this aggregation, and other proposed methods, is the need for a sort (and possible zero elemination) before application of a standard dot product using the OWA or MEOWA weights. Atkins [11] has proposed using a Hopfield network implimenting the required sorting operation, hence we have yet another example of the synergy between fuzzy logics (in this case aggregation) and neural networks.

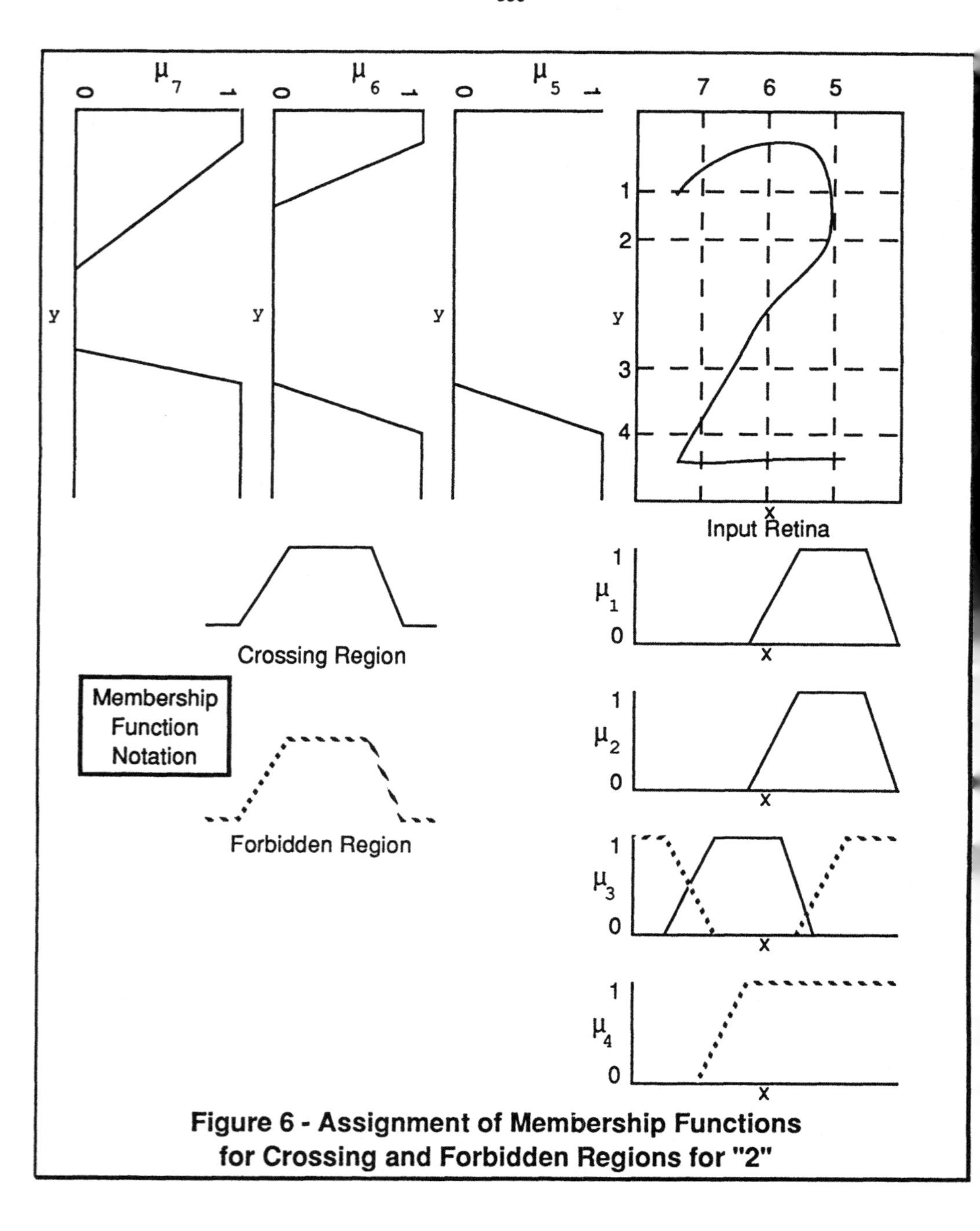

Figure 6 - Assignment of Membership Functions for Crossing and Forbidden Regions for "2"

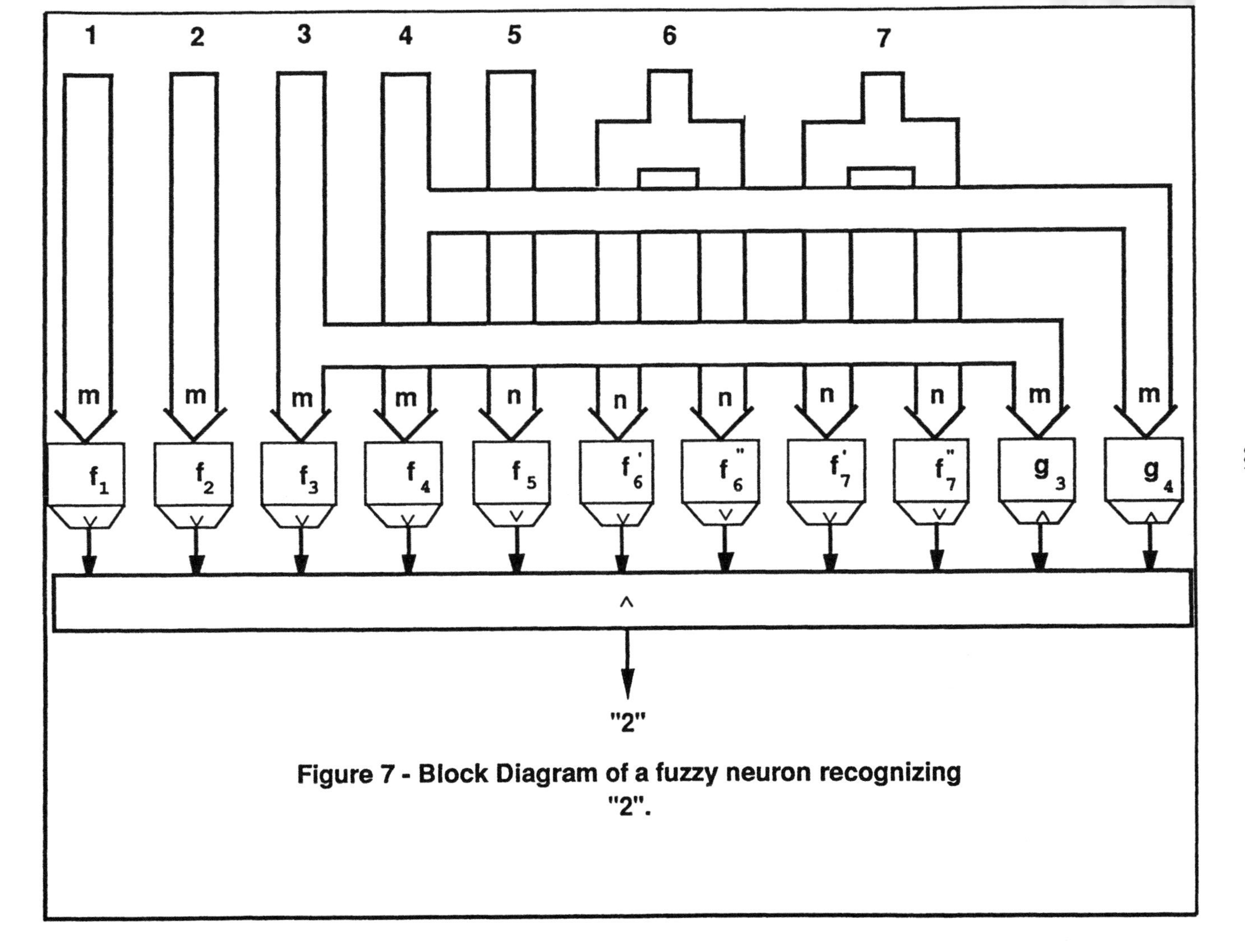

Figure 7 - Block Diagram of a fuzzy neuron recognizing "2".

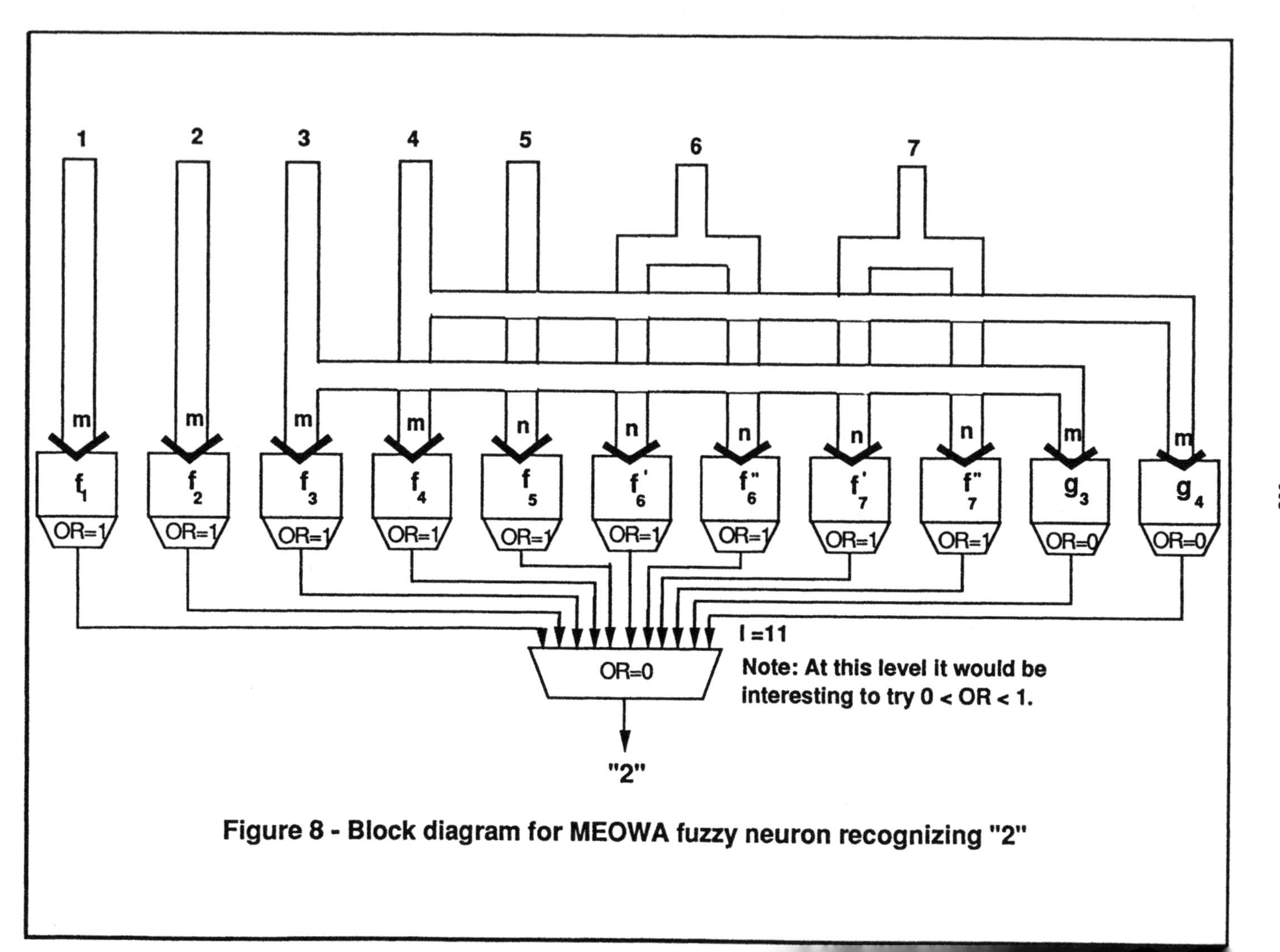

Figure 8 - Block diagram for MEOWA fuzzy neuron recognizing "2"

References

[1] T. Yamakawa and S. Tomoda. (1989) A Fuzzy Neuron and Its Application to Pattern Recognition, In Proc. 3rd Ann. IFSA, 30-38, Univ. of Washington, Seattle, WA, August 1989.

[2] M. O'Hagan. (1988) Aggregating Template or Rule Antecedents in Real-Time Expert Systems with Fuzzy Set Logic. In Proc. 22nd Ann. Asilomar Conf. on Signals, Systems, and Computers, IEEE & Maple Press, 681-689, Pacific Grove, CA, October 31-November 2, 1988. Accepted for publication in IEEE Trans. on Syst., Man, and Cybernetics.

[3] L.C. Shiue and R.O. Grondin. (1987) On Designing Fuzzy Learning Neural-Automata. In Proc. ICNN, 2, 299-307, San Diego, CA, June 1987.

[4] R.R. Yager (1987) On the Aggregation of Processing Units in Neural Networks. In Proc. ICNN, 2, 327-333, San Diego, CA, June 1987.

[5] R.R. Yager. (1988) On Ordered Weighted Averaging Aggregation Operators in Multi-Criteria Decisions, Technical Report MII-705, Machine Intelligence Institute, Iona College, New Rochelle, NY, 1987 and in IEEE Trans. on Systems Man, and Cybernetics, 8,1, 183-190, Jan/Feb 1988.

[6] M. O'Hagan. (1987) Fuzzy Decision Aids. In Proc. 21st Ann. Asilomar Conf. on Signals, Systems, and Computers, IEEE and Maple Press, 2, 624-628, Pacific Grove, CA, Nov 1987.

[7] W.R. Taber and R. O. Deich. (1988) Fuzzy Sets and Neural Networks. In Proc. NASA 1988 First Joint Technology Workshop on Neural Networks and Fuzzy Logic, Houston, TX, May 1988.

[8] G. C. Oden. (1988) FuzzyProp: A Symbolic Superstrate for Connectionist Models. In Proc. ICNN, 1, 293-300, San Deigo, CA, July 1988.

[9] E. Langheld and K. Goser. (1990) Generalized Boolean Operations for Neural Networks. In Proc. IJNCC-90, 2, 159-162, Washington, DC, Jan. 15-19, 1990.

[10] B. Kosko. (1988) Fuzziness and Neural Networks. In Proc. 2nd Ann. IFSA, 5-7, Iizuka, Japan, August 20-24, 1988.

[11] M. Atkins. (1990) Sorting by Hopfield NET. In Proc. IJCNN-90, 2, 65-68, Washington, D.C., Jan. 15-19, 1990.

Lecture Notes in Computer Science

For information about Vols. 1–454
please contact your bookseller or Springer-Verlag

Vol. 455: C.A. Floudas, P.M. Pardalos, A Collection of Test Problems for Constrained Global Optimization Algorithms. XIV, 180 pages. 1990.

Vol. 456: P. Deransart, J. Maluszyński (Eds.), Programming Language Implementation and Logic Programming. Proceedings, 1990. VIII, 401 pages. 1990.

Vol. 457: H. Burkhart (Ed.), CONPAR '90 – VAPP IV. Proceedings, 1990. XIV, 900 pages. 1990.

Vol. 458: J.C.M. Baeten, J.W. Klop (Eds.), CONCUR '90. Proceedings, 1990. VII, 537 pages. 1990.

Vol. 459: R. Studer (Ed.), Natural Language and Logic. Proceedings, 1989. VII, 252 pages. 1990. (Subseries LNAI).

Vol. 460: J. Uhl, H.A. Schmid, A Systematic Catalogue of Reusable Abstract Data Types. XII, 344 pages. 1990.

Vol. 461: P. Deransart, M. Jourdan (Eds.), Attribute Grammars and their Applications. Proceedings, 1990. VIII, 358 pages. 1990.

Vol. 462: G. Gottlob, W. Nejdl (Eds.), Expert Systems in Engineering. Proceedings, 1990. IX, 260 pages. 1990. (Subseries LNAI).

Vol. 463: H. Kirchner, W. Wechler (Eds.), Algebraic and Logic Programming. Proceedings, 1990. VII, 386 pages. 1990.

Vol. 464: J. Dassow, J. Kelemen (Eds.), Aspects and Prospects of Theoretical Computer Science. Proceedings, 1990. VI, 298 pages. 1990.

Vol. 465: A. Fuhrmann, M. Morreau (Eds.), The Logic of Theory Change. Proceedings, 1989. X, 334 pages. 1991. (Subseries LNAI).

Vol. 466: A. Blaser (Ed.), Database Systems of the 90s. Proceedings, 1990. VIII, 334 pages. 1990.

Vol. 467: F. Long (Ed.), Software Engineering Environments. Proceedings, 1969. VI, 313 pages. 1990.

Vol. 468: S.G. Akl, F. Fiala, W.W. Koczkodaj (Eds.), Advances in Computing and Information – ICCI '90. Proceedings, 1990. VII, 529 pages. 1990.

Vol. 469: I. Guessarian (Ed.), Semantics of Systeme of Concurrent Processes. Proceedings, 1990. V, 456 pages. 1990.

Vol. 470: S. Abiteboul, P.C. Kanellakis (Eds.), ICDT '90. Proceedings, 1990. VII, 528 pages. 1990.

Vol. 471: B.C. Ooi, Efficient Query Processing in Geographic Information Systems. VIII, 208 pages. 1990.

Vol. 472: K.V. Nori, C.E. Veni Madhavan (Eds.), Foundations of Software Technology and Theoretical Computer Science. Proceedings, 1990. X, 420 pages. 1990.

Vol. 473: I.B. Damgård (Ed.), Advances in Cryptology – EUROCRYPT '90. Proceedings, 1990. VIII, 500 pages. 1991.

Vol. 474: D. Karagiannis (Ed.), Information Syetems and Artificial Intelligence: Integration Aspects. Proceedings, 1990. X, 293 pages. 1991. (Subseries LNAI).

Vol. 475: P. Schroeder-Heister (Ed.), Extensions of Logic Programming. Proceedings, 1989. VIII, 364 pages. 1991. (Subseries LNAI).

Vol. 476: M. Filgueiras, L. Damas, N. Moreira, A.P. Tomás (Eds.), Natural Language Processing. Proceedings, 1990. VII, 253 pages. 1991. (Subseries LNAI).

Vol. 477: D. Hammer (Ed.), Compiler Compilers. Proceedings, 1990. VI, 227 pages. 1991.

Vol. 478: J. van Eijck (Ed.), Logics in AI. Proceedings, 1990. IX, 562 pages. 1991. (Subseries in LNAI).

Vol. 480: C. Choffrut, M. Jantzen (Eds.), STACS 91. Proceedings, 1991. X, 549 pages. 1991.

Vol. 481: E. Lang, K.-U. Carstensen, G. Simmons, Modelling Spatial Knowledge on a Linguistic Basis. IX, 138 pages. 1991. (Subseries LNAI).

Vol. 482: Y. Kodratoff (Ed.), Machine Learning – EWSL-91. Proceedings, 1991. XI, 537 pages. 1991. (Subseries LNAI).

Vol. 483: G. Rozenberg (Ed.), Advances In Petri Nets 1990. VI, 515 pages. 1991.

Vol. 484: R. H. Möhring (Ed.), Graph-Theoretic Concepts In Computer Science. Proceedings, 1990. IX, 360 pages. 1991.

Vol. 485: K. Furukawa, H. Tanaka, T. Fullsaki (Eds.), Logic Programming '89. Proceedings, 1989. IX, 183 pages. 1991. (Subseries LNAI).

Vol. 486: J. van Leeuwen, N. Santoro (Eds.), Distributed Algorithms. Proceedings, 1990. VI, 433 pages. 1991.

Vol. 487: A. Bode (Ed.), Distributed Memory Computing. Proceedings, 1991. XI, 506 pages. 1991.

Vol. 488: R. V. Book (Ed.), Rewriting Techniques and Applications. Proceedings, 1991. VII, 458 pages. 1991.

Vol. 489: J. W. de Bakker, W. P. de Roever, G. Rozenberg (Eds.), Foundations of Object-Oriented Languages. Proceedings, 1990. VIII, 442 pages. 1991.

Vol. 490: J. A. Bergstra, L. M. G. Feljs (Eds.), Algebraic Methods 11: Theory, Tools and Applicatlons. VI, 434 pages. 1991.

Vol. 491: A. Yonezawa, T. Ito (Eds.), Concurrency: Theory, Language, and Architecture. Proceedings, 1989. VIII, 339 pages. 1991.

Vol. 492: D. Sriram, R. Logcher, S. Fukuda (Eds.), Computer-Aided Cooperative Product Development. Proceedings, 1989 VII, 630 pages. 1991.

Vol. 493: S. Abramsky, T. S. E. Maibaum (Eds.), TAPSOFT '91. Volume 1. Proceedings, 1991. VIII, 455 pages. 1991.

Vol. 494: S. Abramsky, T. S. E. Maibaum (Eds.), TAPSOFT '91. Volume 2. Proceedings, 1991. VIII, 482 pages. 1991.

Vol. 495: 9. Thalheim, J. Demetrovics, H.-D. Gerhardt (Eds.), MFDBS '91. Proceedings, 1991. VI, 395 pages. 1991.

Vol. 496: H.-P. Schwefel, R. Männer (Eds.), Parallel Problem Solving from Nature. Proceedings, 1991. XI, 485 pages. 1991.

Vol. 497: F. Dehne, F. Fiala. W.W. Koczkodaj (Eds.), Advances in Computing and Intormation - ICCI '91 Proceedings, 1991. VIII, 745 pages. 1991.

Vol. 498: R. Andersen, J. A. Bubenko jr., A. Sølvberg (Eds.), Advanced Information Systems Engineering. Proceedings, 1991. VI, 579 pages. 1991.